Linear Optimization and Duality

Linear Optimization and Duality

A Modern Exposition

Craig A. Tovey

CRC Press is an imprint of the
Taylor & Francis Group, an **informa** business
A CHAPMAN & HALL BOOK

First edition published 2021
by CRC Press

6000 Broken Sound Parkway NW, Suite 300, Boca Raton, FL 33487-2742
and by CRC Press

2 Park Square, Milton Park, Abingdon, Oxon, OX14 4RN

CRC Press is an imprint of Taylor & Francis Group, LLC

ISBN: 978-1-4398-8746-2 (hbk)
ISBN: 978-1-315-11721-8 (ebk)

Typeset in Computer Modern font
by KnowledgeWorks Global Ltd.

Contents

Preface xi

About the Author xiii

To the Teacher xv

To the Reader xvii

List of Symbols xxi

Glossary xxv

1 A Gentle Introduction to Optimization and Duality 1
- 1.1 Optimization Problems . . . 1
- 1.2 Duality in Multivariate Calculus, Sudoku, and Matrices . . . 8
  - 1.2.1 An Example of Duality in Optimization . . . 8
  - 1.2.2 A General Definition of Duality . . . 9
  - 1.2.3 Partial Derivatives . . . 10
  - 1.2.4 Duality in Sudoku . . . 13
  - 1.2.5 Duality in Matrices . . . 15
- 1.3 General Approaches to Solving Optimization Problems . . . 16
  - 1.3.1 Saddlepoints and Lagrange's Method . . . 20
- 1.4 Notes and References . . . 26
- 1.5 Problems . . . 27

2 Introduction to Linear Programming 33
- 2.1 What is Linear Programming? . . . 33
  - 2.1.1 Symmetry and Alternative Forms . . . 38
- 2.2 Visualizing LP . . . 41
- 2.3 Presolving LP . . . 52
- 2.4 Notes and References . . . 55
- 2.5 Problems . . . 56

3 Formulating and Solving Linear Programs 61
- 3.1 Three Classic Primal/Dual Formulation Pairs . . . 62
- 3.2 Visualizing Problems, Defining Variables, and Formulating Constraints . 66
- 3.3 LP Modeling Methods . . . 72
- 3.4 More Examples of LP Formulation . . . 81
- 3.5 Using Software to Solve LPs . . . 90
- 3.6 Shadow Prices . . . 95
- 3.7 Notes and References . . . 102
- 3.8 Problems . . . 102

4 Polyhedra **111**
4.1 Separation . . . 114
4.2 Fourier-Motzkin Elimination . . . 116
4.3 Theorems of the Alternative . . . 121
4.4 Extreme Points and Optimal Solutions . . . 123
4.4.1 Degeneracy . . . 130
4.5 Two Ways to Represent Polyhedra . . . 131
4.6 Notes and References . . . 135
4.7 Problems . . . 135

5 The Simplex Method **147**
5.1 The Simplex Method . . . 147
5.2 Degeneracy . . . 154
5.2.1 The Perturbation/Lexicographic Method . . . 155
5.2.2 Bland's Rule . . . 157
5.3 The Full Tableau Method . . . 159
5.4 Phase I: Getting an Initial Basic Feasible Solution . . . 162
5.4.1 Degeneracy at the End of Phase I . . . 165
5.4.2 Phase II: Bringing in the Original Objective . . . 166
5.5 Notes and References . . . 167
5.6 Problems . . . 168

6 Variants of the Simplex Method **175**
6.1 Lower and Upper Bounds . . . 175
6.2 Dual Simplex Method . . . 181
6.2.1 An Example of Both the Primal and Dual Simplex Algorithms, with Geometry . . . 183
6.3 Primal-Dual Algorithm . . . 191
6.3.1 The Primal-Dual Algorithm Applied to Assignment Problems . . . 193
6.4 Self-Dual Parametric Algorithm . . . 195
6.5 Dantzig-Wolfe Decomposition . . . 199
6.6 High Level Computational Issues . . . 200
6.7 Notes and References . . . 202
6.8 Problems . . . 203

7 Shadow Prices, Sensitivity Analysis, and Column Generation **211**
7.1 The Mathematical Justification of Shadow Prices . . . 212
7.2 Sensitivity Analysis . . . 215
7.3 Degeneracy and Shadow Prices . . . 216
7.4 Parametric Programming . . . 218
7.5 Column Generation . . . 221
7.5.1 The Cutting Stock Problem . . . 223
7.6 Problems . . . 226

8 Advanced Topics on Polyhedra **231**
8.1 The Geography of Mount Duality . . . 231
8.1.1 Variants of Theorems of the Alternative and Strong Duality . . . 231
8.1.2 Degeneracy and Complementary Slackness . . . 234
8.1.3 From a Theorem of the Alternative to Strong Duality and Back Again . . . 238
8.1.4 Helly's Theorem – A Hiking Trail on Mount Duality . . . 238

8.1.5 Separation Characterization of Convexity 241
8.1.6 Projection and Representation 242
8.2 Facets of Polyhedra 243
8.2.1 Affine Independence and Dimension 244
8.2.2 Faces and Facets 246
8.3 Column Geometry 250
8.4 Polarity 253
8.5 Problems 256

9 Polynomial Time Algorithms **261**
9.1 Running Time of Algorithms 262
9.2 Some Core Polynomial Time Algorithms 267
9.2.1 The Euclidean Algorithm for Greatest Common Divisor 267
9.2.2 Sorting 267
9.2.3 Searching 268
9.2.4 Linear Equations and Matrices 270
9.2.5 Square Roots 273
9.3 Heaps 276
9.4 Notes and References 281
9.5 Problems 282

10 Speed of the Simplex Method and Complexity of Linear Programming **287**
10.1 Worst-Case Behavior of the Simplex Method 287
10.1.1 Derivation of the Klee-Minty Example 288
10.2 Complexity of Linear Programming: The Ellipsoid Algorithm 293
10.2.1 Polynomial Equivalence of Separation and Optimization 299
10.3 Problems 300

11 Network Models and the Network Simplex Algorithm **305**
11.1 Network Models 309
11.1.1 Examples of Minimum Cost Flow Models 310
11.2 Integrality and Duality 315
11.3 Network Simplex Algorithm 318
11.3.1 Preliminary Graph Concepts 318
11.3.2 Every Basis is a Spanning Tree 321
11.3.3 Computing Basic Solutions 323
11.3.4 Computing Reduced Costs 326
11.3.5 Putting the Pieces Together — The Network Simplex Algorithm . 330
11.4 Problems 332

12 Shortest Path Models and Algorithms **339**
12.1 Modeling Shortest Path Problems 339
12.2 Acyclic Graphs 343
12.3 The Principle of Optimality 345
12.4 Primal and Dual LPs for the Shortest Path Problem 347
12.5 Shortest Paths in Graphs That Might Have Negative Cycles 350
12.6 Dijkstra's Algorithm for Graphs with Nonnegative Costs 356
12.6.1 Running Time of Dijkstra's Algorithm 358
12.7 Problems 360

13 Specialized Algorithms for Maximum Flow and Minimum Cut **365**
13.1 Ford and Fulkerson's Augmenting Path Algorithm . . . 366
13.2 The Edmonds-Karp Modifications to the Ford-Fulkerson Algorithm . . . 369
13.3 Further Improvements of Maximum Flow Algorithms . . . 372
13.4 Maximum Flow as a Linear Program . . . 374
13.4.1 Max Flow Min Cut . . . 375
13.4.2 Bipartite Matching . . . 377
13.4.3 Köenig-Egerváry Theorem . . . 379
13.4.4 Dilworth's Theorem . . . 379
13.5 Problems . . . 383

14 Computational Complexity **387**
14.1 Introduction . . . 387
14.1.1 An Historical View of NP-Hardness . . . 388
14.1.2 Fast, Slow, Easy, and Hard . . . 388
14.2 Problems, Instances, and Real Cases . . . 390
14.3 YES/NO Form . . . 391
14.4 The Nuts and Bolts of NP-Hardness Proofs . . . 393
14.4.1 An Introductory Example . . . 394
14.4.2 The General Form of an NP-Hardness Proof . . . 396
14.4.3 Why this Proof Structure? . . . 397
14.5 Spotting Complexity . . . 397
14.6 Illustrations of Common Pitfalls . . . 402
14.7 Examples of NP-Hardness Proofs . . . 408
14.7.1 The Easiest and Most Common Kind of NP-hardness Proof . . . 408
14.7.2 Specialization . . . 409
14.7.3 Padding and Forcing . . . 410
14.7.4 Gadgets: Transforming 3-SAT to 3-coloring . . . 411
14.8 Complexity of Finding Approximate Solutions . . . 416
14.9 Dealing with NP-Hard Problems . . . 417
14.9.1 Pseudo-Polynomial Algorithms, Dynamic Programming, and Unary NP-Hardness . . . 418
14.9.2 Special Structure . . . 421
14.10 Formal Definition of NP and Other Complexity Classifications . . . 424
14.10.1 Unsolvable Problems . . . 424
14.10.2 co-NP Completeness and Lack of Succinct Characterizations . . . 425
14.10.3 Definitions of NP, co-NP, and NP-Complete . . . 426
14.10.4 Classes of Search Problems . . . 428
14.10.5 P-space Completeness . . . 428
14.10.6 Other Kinds of Difficulty . . . 429
14.11 References and Conclusions . . . 429
14.12 Problems . . . 430

15 Formulating and Solving Integer Programs **435**
15.1 Using Binary Variables to Model Problems Beyond the Scope of LP . . . 436
15.2 Solving IPs . . . 448
15.3 Tighter Formulations . . . 450
15.3.1 Valid Inequalities . . . 453
15.3.2 Facets . . . 453
15.4 Tips . . . 454
15.5 Notes and References . . . 457

15.6 Problems 457

16 Elementary Nonlinear Programming Theory 463

16.1 Multivariate Calculus Prerequisites 464
16.2 First-Order Karush-Kuhn-Tucker Conditions for NLP 469
16.2.1 Global Optimization: A Cursory Look 475
16.3 Problems 477

17 Introduction to Nonlinear Programming Algorithms 481

17.1 Bisection Search 481
17.2 Newton's Method 482
17.2.1 Newton's Method in Higher Dimensions 486
17.2.2 Convergence Rate of Newton's Algorithm 488
17.3 Unconstrained and Linearly Constrained Gradient Search 488
17.3.1 Linear Equality Constraints 490
17.4 Notes and References 491
17.5 Problems 492

18 Affine Scaling and Logarithmic Barrier Interior-Point Methods 495

18.1 Affine Scaling 496
18.2 Logarithmic Barrier and the Central Path 498
18.3 Notes and References 512
18.4 Problems 513

19 Appendices 515

19.1 Linear Algebra and Vector Spaces 515
19.1.1 Linear Combinations, Dimension 515
19.1.2 The Dot Product, Cosines, Projections, Cauchy-Schwartz Inequality 516
19.1.3 Determinants, Inverses 516
19.1.4 Gaussian Elimination 517
19.1.5 Vector Spaces 517
19.1.6 Dimension, Basis 518
19.1.7 Inner Products and Norms 519
19.1.8 Linear Functions and Matrices 519
19.2 Point Set Topology 520
19.3 Summary 523

Bibliography 527

Index 545

Preface

During the more than 35 years I've taught linear optimization, I've found many ways to explain various parts of the material. Eventually, my lectures and in-class activities differed substantially from any of the textbooks available. My students told me that I should write my own book. This text is the result.

It is a classic beginner's writing mistake to write a mathematical proof in the order in which one discovered it. (Euler is an exception – we learn even from his false starts.) I think it is a similar mistake to write a book in the order in which one learned. All of our standard linear programming textbooks present the material in the order in which it was discovered. First, linear programming formulation. Second, the simplex method. Third, duality and sensitivity analysis. Duality is treated as a difficult add-on to the basic theory. Students end up without knowing duality in their bones. This text brings in duality in Chapter 1 and carries duality all the way through the exposition. Chapter 1 gives a general definition of duality that shows the dual aspects of a matrix as a column of rows and a row of columns. The proof of weak duality in Chapter 2 is via the Lagrangian, which relies on matrix duality. The first three LP formulation examples in Chapter 3 are classic dual pairs including the diet problem and 2-person zero sum games. I hope my students will come to know duality in a more profound and instinctive way than I do.

This book departs from convention in another significant way. It is customary when writing a mathematical textbook for all of the results to follow a rigorously proved order. With all due respect to Cauchy *et al.*, however, mathematics is largely empirical. When my students finish the course, I want them to know what is true. At some point they should see rigorous proofs of all results, but the intuitive or empirical knowledge need not come in lockstep with the rigorous derivations. Students see the strong duality theorem in Chapter 2 and again in Chapter 3 on formulation. They don't see the rigorous proof until chapter 4. I'll state mathematical facts with examples and intuitive explanations pages or chapters before their proofs appear.

When I was in graduate school, I used to argue with my fellow student Richard Stone as to whether algebra or geometry better explained mathematics. According to popular learning mode theories, we were both partly correct: "analytic visual" learners learn better from algebra; "global visual" learners learn better from geometry. Thus, a hypothetical student in the room with Richard and me would learn from only one of us. However, the preponderance of independent empirical research contradicts learning mode theory. Instead, people learn the most when they use multiple modes. The hypothetical student would learn more from the two of us than they could from either one. Hence, throughout this book, I explain both visually and algebraically whenever I can. Moreover, I emphasize the relationships between the two, because I believe connecting the two yields an even deeper understanding.

Nowadays it is not enough to know the theory of optimization. Let's be honest – more of our students will use optimization software than will need to know what is "under the hood". The state of the art of optimization software is a glorious thing, but to use the software effectively one must learn about modeling and computational issues. Such knowledge is generally more empirical than mathematical. It is neither acquired nor communicated in the form of definitions and theorems. Accordingly, in parts of the book I will drop my mathematician's voice and instead adopt the voice of the practitioner.

About the Author

Craig A. Tovey is a Professor in the H. Milton Stewart School of Industrial and Systems Engineering at Georgia Institute of Technology. Dr. Tovey received an A.B. from Harvard College, an M.S. in computer science and a Ph.D. in operations research from Stanford University. His principal activities are in operations research and its interdisciplinary applications. He received a Presidential Young Investigator Award and the Jacob Wolfowitz Prize for research in heuristics. He was named an Institute Fellow at Georgia Tech, and was recognized by the ACM Special Interest Group on Electronic Commerce with the Test of Time Award. Dr. Tovey received the 2016 Golden Goose Award for his research on bee foraging behavior leading to the development of the Honey Bee Algorithm.

To the Teacher

I've written this book for a first-year graduate or advanced undergraduate course on linear optimization. It assumes knowledge of elementary linear algebra and single-variable calculus. Part of Chapter 1 teaches the small amount of differential multivariate calculus needed in the rest of the book. The prerequisite linear algebra is covered in an appendix.

The principal novelties of exposition herein are:

- Emphasis on duality throughout.
- Linkage between algebraic and geometric meanings.
- Practical tips for modeling and computation.
- Coverage of computational complexity and data structures.
- Exercises and problems based on the learning theory concept of the zone of proximal development.
- Both rigor and intuition.
- Guidance for the mathematically unsophisticated reader.

To elaborate bluntly on the last point, we always have the same problem when we teach engineering or first-year graduate students. We must deliver a rigorous exposition of the material, but many or most of our students cannot recognize, much less write, a correct formal proof. Everyone tries hard; everyone is frustrated. Some students, though very bright, drop the course or the entire program. You gripe about their lack of mathematical sophistication, and your colleagues cluck sympathetically.

But the next year brings the same problem.

And most textbooks worsen it. They presume the reader has mastered mathematical rigor. Does that presumption help your students attain mastery? No. Instead, it renders them anxious and fearful; it saps both their confidence and enthusiasm.

This book will help them, and it will help you. I've inset questions, special exercises, and actionable reading tips throughout the text, all to guide them to read, think, and comprehend at a higher mathematical level. These interpolations lead them to work small numerical examples of a statement, question the logic of a proof, and test the necessity of a theorem's hypotheses. All questions are answered in footnotes on the same page. All reading tips are set apart in a different font. The exercises are separated from the other problems. A solutions manual will be available to instructors.

As the teacher, you are the key to the solution, and you will have your own methods. To allow this book to help, I ask you do two things. First, to paraphrase a maxim of social work, *meet your students where they are*. Set the atmosphere of the class to be accepting of those who are not mathematically sophisticated. For example, tell them it is OK not to comprehend mathematical notation instantly; it is OK not to grasp the meaning of a definition or theorem immediately. Show them how to build their understanding from 1 by 1 examples to the general. Challenge them to find the flaw in a faulty proof. Second, *insist that learning is ultimately the student's responsibility.* We can't open a student's skull, pour in knowledge, and seal it up. There is no learning without neural activity, and only the student can choose to activate his or her brain. This may be obvious to you, but in my experience many American students schooled under the 2001 NCLB Act don't know it.

All errors are my responsibility. I welcome any corrections and suggestions. Please send them to `cat@gatech.edu` with subject heading LODE.

To the Reader

Linear programming is ubiquitous but nearly invisible. The dog food and the gasoline you buy are blended with linear programming. The airplanes you fly are scheduled by linear programming and its big brother, integer programming. The cows that produce your milk and butter are fed with diets devised daily by linear programming. The paper this book is made from and the cardboard boxes used to ship it are cut based on plans from linear programming. This book aims to make the invisible visible.

In imitation of [222], I've incorporated questions into the text. Questions are to engage you, make your reading a more active process, and help you check that you understand what you are reading. Questions don't require pencil and paper. Their answers are short and all are given in the text. You should answer all the questions on a first reading. The exercises and problems at the end of each chapter are grouped by difficulty. The groupings are based roughly on hierarchies of knowledge in learning theory. The two main ideas I've employed are, first, that learning takes place in the "zone of proximal development" or "ZPD" for short, and second, that there are different levels of mastery of a subject.

The ZPD concept is due to Vygotsky [291]. It is the set of things the learner can do with assistance. Easier things, which the learner can do unaided, don't contribute much to learning, since they are already learned. More difficult things, which the learner cannot do even with aid, have been shown empirically not to be very useful, either. If you have ever felt completely lost even after spending several hours trying to understand something, or you have stared at a problem for two hours and still had a blank sheet of paper in front of you, you may have experienced this phenomenon. It isn't enough to work hard; you have to work smart. And smart means figuring out where your ZPD is and working within that zone.

To help you find your ZPD, I've classed the problems into 4 categories, E, M, D, and I. "E"asy problems are elementary exercises that ask you to apply concepts, theorems, and algorithms directly. For example, if the text defines "singular matrix," an E problem might ask you to determine whether or not a particular small matrix is singular, or to find a matrix that is not singular. If the text gives an algorithm for matrix multiplication, an E problem might ask you to use that algorithm to multiply two specified matrices. "M"edium problems ask you to apply material in non-obvious ways, or require you to fully understand the rigorous mathematical reasoning in the proofs of theorems and the derivations of algorithms. "D"ifficult problems typically ask you to extend or modify results and methods presented in the text, or to determine their range of applicability. "I"ntegrative problems are distinguished from the others by requiring you to combine knowledge from more than one part of the book, or from other subject areas. Aside from that, they may be at "E," "M" or "D" levels.

Mastery of the E level is in general necessary to solve problems on the M level, and likewise mastery at the M level is a good preliminary to solving problems at the D level. Indeed, the E,M,D classification is a rough rendering of several hierarchies of mastery in the education literature. Being able to use existing tools with help, and being able to use existing tools without supervision, are low in these hierarchies. Recognizing errors and knowing which tool to use in different situations are higher in the hierarchy, as are being able to supervise and being able to teach. Design and creativity, such as the ability to invent and build new tools, are very high or at the top in these hierarchies. *Question: Should this be a question or an E, M, D, or I problem?*[1]

Many students, such as those whose background is in engineering, may not yet know how to read a mathematical textbook. Here are some tips:

- Expect to read each page at least twice.
- Anticipate. Whenever you read a definition, stop. Without looking at what follows, think of an instance that satisfies the definition and one that doesn't. Whenever you come to an example, read its setup and try to work it out yourself. Whenever you come to a theorem, think of a simple or an extreme example to see why it makes sense in that case. Whenever you come to a proof, stop reading and try to work out your own proof. If you can't, read the first paragraph, and try to figure out what is going to come next, and so on.
- When you don't understand an abstraction or a proof, play with specific examples. Most people learn by going from the specific to the general. Do not be embarrassed to work with specific numerical examples of tiny matrices, starting with 1×1 matrices, with single-variable linear and other optimization problems, and other simple small cases. There is a good chance that your teachers, who now have such facility with abstractions, acquired their facility by doing just that.
- Draw pictures. Cognitive scientists tell us that humans process visual information 100 to 200 times as efficiently as they do abstract symbolic information. Harness that processing power whenever possible.
- Once you can replicate the statement and proof of a theorem without much peeking at the textbook, challenge the theorem. Examine each assumption and figure out where it is used in the proof. If you don't see any place where it is used, either a more general theorem is true, or you don't fully understand the proof. For each assumption, seek a counterexample to the statement of the theorem if that one assumption were removed. Think of a stronger conclusion than the theorem's, and try to find a counterexample or proof.
- When you get stuck trying to prove something that you are confident is true, do each of the following: check to make sure that you are using all of the assumptions; work out some examples to develop an intuition as to why it is true; and try the main proof techniques that you have not tried yet, such as proof by the contrapositive, proof by contradiction, induction, and proof by an algorithm.
- If you feel totally stuck solving a problem or proving a statement you may be working on something outside your ZPD. You might work for days, learning nothing but fear and dislike of math. Instead, invent an easier version of the problem more likely to be within your ZPD. For example, try the two- or three-dimensional version of a problem in n dimensions. You can get an easier version of an if-then statement either by adding an assumption or by replacing its conclusion with one that is obviously implied by the original conclusion.

References

George Pólya's "How to Solve It" [246] is the classic book on mathematical problem solving.

---

[1] Question.

It teaches how to guess what is true, to concoct easier versions and extreme cases of problems, and many other methods. To learn rigorous mathematical reasoning, the earliest book I know of is by Daniel Solow [268]. It has been followed by several others including [61], [288],[45][153]. I particularly recommend the second edition of John D'Angelo and Doug West's book [63] for both discrete math and real analysis, and Michael Schramm's book [260] for real analysis. The former is simultaneously playful and rigorous; the latter's first chapter is a superb introduction to mathematical reasoning.

Don Knuth rates the exercises on a fine scale of difficulty in his classic series on the art of computer programming [186, 187, 188, 189]. László Lovász's innovative problems and exercises book [215] comprises three sections: problems, hints, and full solutions. It is broad and deep but is written for those who already have mastered the basics of mathematical notation and rigorous reasoning.

Problems

E Exercises

1. Think of an easier version of each of the following problems:
   - A 100-meter rope is suspended between the tops of two poles 80 meters apart, one 20 meters high and the other 25 meters high. Find the height of the lowest point on the rope.
   - Find $\int x^{2k+1}e^x\,dx$ for integer $k \geq 1$.
   - Prove that there exist arbitrarily long sequences of consecutive composite numbers.
   - How many ways can 100 boys and 100 girls be placed in a line so that no two girls are adjacent?
   - Prove that the perigrumble of a zlotsky is more than twice the perigrumble of a mroggle.
   - You roll 24 6-sided dice on the floor. However, each die rolls under the couch where you cannot see it, with probability 1/5 independent of all other dice. What is the probability you will see exactly the same number of 1's, 2's, and 3's?
2. Think of an extreme example to invalidate each of the following formulas:
   - If you travel from A to B at speed r mph and return to B at speed r' mph, your average speed is $\frac{r+r'}{2}$.
   - The sum of the first n terms of an arithmetic sequence with first term a and common difference d is $(a + nd - d)(n/2)$.
   - $x^3 + 100x^2 + 10000 \geq 0$ for all x.
   - The probability n random people all have different birthdays is $(\frac{364}{365})^n$.
   - $n^2 + n + 41$ is prime for all nonnegative integers n.

M Problems

3. Think of a harder version of the question, "Should this be a question or an E, M, D, or I problem?"
4. Suppose you are given a set of 20 problems. Your assignment is to separate them into the E, M, D, and I classes. How difficult is your assignment?

List of Symbols

Logic

$\forall$ for all.

$\exists$ there exists.

$\nexists$ there does not exist.

$\Rightarrow$ implies; only if.

$\Leftarrow$ is implied by; if.

$\Leftrightarrow$ is equivalent to; implies and is implied by; if and only if.

$\neg X$ Not X, the negation of X. X is true $\Leftrightarrow \neg X$ is false.

$\equiv$ is defined as equal to.

Sets

$\in$ is an element of.

$\cap$ set intersection: $S \cap T$ is the intersection of sets S and T.

$\cup$ set union: $S \cup T$ is the union of sets S and T.

$\bigcup_{S \in \mathcal{S}} S$ The union of all sets S in the collection of sets $\mathcal{S}$.

$\bigcap_{S \in \mathcal{S}} S$ The intersection of all sets S in the collection of sets $\mathcal{S}$.

2^S The set of all subsets of the set S, including ϕ. If S is finite $|2^S| = 2^{|S|}$. For all S, 2^S has strictly larger cardinality than S.

$\overline{S}$ The complement of the set S.

$\{x, y, z\}$ The set comprising x, y, and z.

(x, y, z) The ordered set of which x is the first, y the second, and z the third element.

$\{\mathbf{y} : \text{conditions on } \mathbf{y}\}$ The set of all $\mathbf{y}$ satisfying the stated conditions.

$\{\mathbf{y} \in S : \text{conditions on } \mathbf{y}\}$ The set of all $\mathbf{y} \in S$ that also satisfy the stated conditions.

$|S|$ The cardinality of the set S.

$\subset$ is a subset of. By definition $S \subset S$ for all sets S.

$S \subseteq T$ S is a subset of T, and we want to emphasize that S might equal T.

$\subsetneq$ is a subset of but is not equal to.

$\supset$ is a superset of.

$\supseteq, \supsetneq$ Figure these out after reading the preceding four definitions.

ϕ The empty set. By definition, $|\phi| = 0$.

$\blacksquare$ *Q.E.D.*, or *quod erat demonstrandum*, meaning "what was to be demonstrated." Denotes the end of a mathematical proof.

Vectors and Matrices

$\mathbf{x}, \mathbf{y}, \mathbf{b}$ vectors, as distinguished from real numbers x, y, b.

$\mathbf{0}$ A vector of 0's, as distinguished from the integer 0.

$\mathbf{1}$ A vector of 1's, as distinguished from the integer 1.

x_i, y_j The ith element of vector $\mathbf{x}$, the jth element of vector $\mathbf{y}$.

$\boldsymbol{\pi}$ A vector of variables, $(\boldsymbol{\pi}_1, \boldsymbol{\pi}_2, \cdots, \boldsymbol{\pi}_n)$, as distinguished from the mathematical constant π.

$\Re^n$, $\mathcal{Z}^n$ The set of n-vectors of reals and integers, respectively.

$\Re^{m \times n}$ The set of real-valued matrices with m rows and n columns.

$\mathbf{x} \cdot \mathbf{y}$ The usual dot product (inner product) operator on pairs of vectors, defined as $\mathbf{x} \cdot \mathbf{y} \equiv \sum_i \mathbf{x}_i \mathbf{y}_i = \mathbf{x}^T \mathbf{y}$.

$||\mathbf{x}||$ The usual Euclidean norm of a vector $\mathbf{x}$, defined as $||\mathbf{x}|| \equiv \sqrt{\sum_i \mathbf{x}_i^2} = \sqrt{\mathbf{x} \cdot \mathbf{x}}$.

A_{ij} The element in row i and column j of the matrix A.

$A_{i,j}$ Equivalent to A_{ij} and used where "ij" may be ambiguous.

A^T The transpose of the matrix A, defined by $(A^T)_{ij} = A_{ji}$. The transpose of a column vector is a row vector.

$A_{:,j}$ The jth column of A.

$A_{i,:}$ The ith row of A. (Question: what is the difference between $(A_{:j})^T$ and $(A^T)_{:,j}$?[2])

A^{-1} The inverse of the square matrix A.

$\mathbf{0}$ A vector of 0's, as distinguished from the integer 0.

$\mathbf{1}$ A vector of 1's, as distinguished from the integer 1.

$diag(\mathbf{x})$ The diagonal matrix X defined as $X_{i,i} = \mathbf{x}_i \forall i$ and $X_{i,j} = 0\ \forall i \neq j$.

e_i The ith unit vector. Its ith entry is 1; the other entries are 0.

$(\mathbf{x}_1, \ldots, \mathbf{x}_n)$ When inline in the text, the column n-vector whose ith entry is x_i. This avoids the visual awkwardness of displaying a column inline.

$\mathbf{x} > \mathbf{y}$ When $\mathbf{x}$ and $\mathbf{y}$ are vectors in $\Re^n$, $\mathbf{x}_i > \mathbf{y}_i$ for all $i = 1 \ldots n$.

$\mathbf{x} \geq \mathbf{y}$ When $\mathbf{x}$ and $\mathbf{y}$ are vectors in $\Re^n$, $\mathbf{x}_i \geq \mathbf{y}_i$ for all $i = 1 \ldots n$.

[2] The jth column of A turned into a row vector versus the jth row of A turned into a column vector.

$\mathbf{x} \gneq \mathbf{y}$ When $\mathbf{x}$ and $\mathbf{y}$ are vectors in $\Re^n$, $\mathbf{x}_i \geq \mathbf{y}_i$ for all $i = 1 \ldots n$ and $\mathbf{x}_i > \mathbf{y}_i$ for at least one $i : 1 \leq i \leq n$.

$\leqq, <=$ Symbols equivalent to "$\leq$".

$\geqq, >=$ Symbols equivalent to "$\geq$".

Other Mathematics

$B(\mathbf{x}, \alpha)$ The (open) ball around $\mathbf{x}$ of radius $\alpha > 0$; the set $\{\mathbf{y} : \; ||\mathbf{x} - \mathbf{y}|| < \alpha\}$.

$\bar{B}(\mathbf{x}, \alpha)$ The closed ball around $\mathbf{x}$ of radius $\alpha > 0$; the set $\{\mathbf{y} : \; ||\mathbf{x} - \mathbf{y}|| \leq \alpha\}$.

$\Re$ The real numbers.

$\mathcal{Z}$ The integers (both positive and negative).

$f : S \mapsto T$ The function f is a mapping from S to T. For all $s \in S$, $f(s) \in T$.

$\rightarrow$ Approaches or converges to.

$\lim_{x \rightarrow z} f(x)$ The limit of $f(x)$ as x converges to z.

$f^{'}(x)$ The derivative of the function $f : \Re \mapsto \Re$ at x.

$f"(x)$ The second derivative of the function $f : \Re \mapsto \Re$ at x.

$\frac{\partial f}{\partial \mathbf{x}_k}$ The partial derivative with respect to $\mathbf{x}_k$ of the function $f : \Re^n \mapsto \Re$ at $\mathbf{x}$.

$\nabla f(\mathbf{x})$ The gradient $(\frac{\partial f}{\partial \mathbf{x}_1}, \frac{\partial f}{\partial \mathbf{x}_2}, \ldots, \frac{\partial f}{\partial \mathbf{x}_n})$ of the function $f : \Re^n \mapsto \Re$ at $\mathbf{x}$.

f^{-1} The inverse of the function f. If $f : S \mapsto T$ and $t \in T$ then $f^{-1}(t) = \{s : \; f(s) = t\}$. f^{-1} is a function iff f is a bijection; otherwise $f^{-1}(t)$ could be the empty set ϕ or could be a set containing more than one member of S.

$epi(f)$ The epigraph of the function f.

$d(x, y)$ In a metric space, the distance between x and y.

$\sum_{k=\alpha}^{\beta} F(k)$ The sum, as integer k ranges from α to β, of the value $F(k)$.

$\prod_{k=\alpha}^{\beta} F(k)$ The product, as integer k ranges from α to β, of the value $F(k)$.

$G = (V, E)$ G is a graph with vertex set V and edge set E.

$f(n) = O(g(n))$ An asymmetric relation between nonnegative functions f and g. Read "f is big O of g", crudely meaning that $f(n)$ is less than or equal to a constant multiple of $g(n)$. The precise meaning is, there exist $c > 0$ and N such that $f(n) \leq cg(n)$ for all $n \geq N$.

$f(n) = O(1)$ For function f defined on the positive integers, f is bounded by a constant. Read "f is big O of 1".

$f(n) = o(g(n))$ Read "f is little o of g". An asymmetric relation between nonnegative functions f and g, meaning that $\frac{f(n)}{g(n)} \rightarrow 0$ as $n \rightarrow \infty$.

$f(n) = o(1)$ Read "f is little o of 1", meaning $\lim_{n \rightarrow \infty} f(n) = 0$.

$f(n) = \Omega(g(n))$ The $\geq$ version of $O()$. Read "f is Omega of g" or "f is big Omega of g", meaning $g(n) = O(f(n))$.

$f(n) = \theta(g(n))$ Read "f is theta of g", meaning $f(n) = O(g(n))$ and $g(n) = O(f(n))$.

Glossary

affine combination An affine combination of K n-vectors $\mathbf{v}^1 \ldots \mathbf{v}^K$ is $\sum_{k=1}^K \alpha_k \mathbf{v}^k$ where $\alpha_k \in \Re\ \forall k$ and $\sum_{k=1}^K \alpha_k = 1$. An affine combination is a special kind of linear combination for which the sum of the weights α_k equals 1. *Geometrically,* the set of affine combinations of two distinct points is the line that is incident on them; the set of affine combinations of three non-collinear points is the plane that contains them.

affine dimension $S \subset \Re^n$ has affine dimension k if S contains $k+1$ but not $k+2$ affinely independent points.

affine function Informally, a linear function plus a constant. Real-valued function $f : \Re^n \mapsto \Re$ is affine if $f(\mathbf{x}) = \mathbf{a} \cdot \mathbf{x} + a_0$ for some vector $\mathbf{a}$ and real a_0. More generally, function $f : \Re^n \mapsto \Re^m$ is affine if $f(\mathbf{x}) = M\mathbf{x} + \mathbf{b}$ for some matrix M and vector $\mathbf{b}$.

affine independence A set of n-vectors $\mathbf{y}^0, \mathbf{y}^1, \ldots, \mathbf{y}^m$ is affinely independent iff the vectors $\mathbf{y}^1 - \mathbf{y}^0, \mathbf{y}^2 - \mathbf{y}^0, \ldots, \mathbf{y}^m - \mathbf{y}^0$ are linearly independent.

arc an edge in a graph

ball A subset of $\Re^n$ of form $\{\mathbf{y} : ||\mathbf{x} - \mathbf{y}| < \epsilon\}$, denoted $B(\mathbf{x}, \epsilon)$ and called the (open) ball around x of radius ϵ. The *closed ball* $\bar{B}(\mathbf{x}, \epsilon)$ is the closed set $\{\mathbf{y} : ||\mathbf{x} - \mathbf{y}| \leq \epsilon\}$.

basic feasible solution In brief, a basic solution that is feasible. For a polyhedron $P = \{\mathbf{x} | A\mathbf{x} \leq \mathbf{b}\} \subset \Re^n$, $\mathbf{y} \in \Re^n$ is a basic feasible solution of P if $\mathbf{y} \in P$ (i.e. $A\mathbf{y} \leq \mathbf{b}$) and $\mathbf{y}$ is a basic solution of P. It is true that a point is a basic feasible solution of P iff it is an extreme point of P.

basic solution For a polyhedron $P = \{\mathbf{x} : A\mathbf{x} \leq \mathbf{b}\} \subset \Re^n$, $\mathbf{y} \in \Re^n$ is a basic solution of P if there exist n linearly independent constraints from $A\mathbf{x} \leq \mathbf{b}$ that are binding at $\mathbf{y}$. Equivalently, there are n linearly independent rows of A indexed by I such that $A_{i,:}\mathbf{y} = \mathbf{b}_i\ \forall i \in I$.

basis A basis of a linear subspace is any set of linearly independent vectors in the subspace such that every member of the subspace equals some linear combination of those vectors. In $\Re^n$ every set of n linearly independent vectors is a basis, and vice-versa. A basis can also be defined as any maximal linearly independent subset of elements of a space.

bijection A function $f : S \mapsto T$ is a bijection if it is
surjective or *onto*: Informally, it hits all the elements of T. Formally, $\forall t \in T\ \exists s \in S$ such that $f(s) = t$
and
injective or *one-to-one* or $1-1$: Informally, different elements of S map to different elements of T. Formally,$f(s) = f(s') \Rightarrow s = s'$.
If f is a bijection, its inverse f^{-1} is also a bijection.

binding A constraint $f(\mathbf{x}) \geq \alpha$, $f(\mathbf{x}) \leq \alpha$ or $f(\mathbf{x}) = \alpha$ is binding at $\hat{\mathbf{x}}$ if $f(\hat{\mathbf{x}}) = \alpha$.

bounded A set $S \subset R^n$ is bounded iff $\exists\ M$ such that $||s|| \leq M\ \forall s \in S$.

cardinality Informally, the number of elements in a set S, denoted $|S|$. Formally, two sets S, T have equal cardinality $|S| = |T|$ if and only if their elements can be placed in one-to-one correspondence, i.e., there exists a bijection from S to T. The empty set ϕ has cardinality $|\phi| = 0$; a nonempty finite set S has cardinality $|S|$ equal to number of elements it contains. For infinite sets, cardinality is the order of infinity its elements can be put into 1-1 correspondence with. If there exists a bijection between the integers and S we say S is countably infinite or $|S| = \aleph^0 \equiv |\mathcal{Z}|$. The real numbers have cardinality $|\Re| \equiv \aleph^1$. Inductively, if $|S| = \aleph^n$ then the cardinality of the set of its subsets is $|2^S| = \aleph^{n+1}$.

Cauchy sequence A sequence of elements in a metric space $x_1, x_2, \ldots$ is a Cauchy sequence iff for all $\epsilon > 0$ there exists integer N such that $d(x_i, x_j) < \epsilon$ for all $i \geq N, j \geq N$. That is, eventually all terms in the sequence are within ϵ of each other.

closed A set S is closed iff its complement $\overline{S}$ is open. It is true that S is closed iff it contains all its cluster points. In $\Re^n$ (or any complete metric space) S is closed iff every Cauchy sequence of points in S converges to a point in S.

cluster point x is a cluster point of a set S in a metric space iff for all $\epsilon > 0$ there exists a point $y \in S$ whose distance to x is less than ϵ. That is, $\forall \epsilon > 0\ \exists y \in S$ such that $d(x, y) < \epsilon$. By definition every point in S is a cluster point of S.

compact A set E is compact iff every covering of E by open sets possesses a finite subcovering. That is, if $\mathcal{S}$ is a collection of open sets such that $E \subseteq \bigcup_{S \in \mathcal{S}} S$ then there exists a finite subset $\mathcal{T} \subset \mathcal{S}$ such that $E \subseteq \bigcup_{S \in \mathcal{T}} S$. In $\Re^n$ sets are compact if and only if they are closed and bounded.

complement The complement of a set S is the set of all elements not in S. You have to know what set S is a subset of to know what the complement $\overline{S}$ is. The complement of $\{0\}$ with respect to the nonnegative integers is the set $\{1, 2, 3, \ldots\}$. The complement of $\{0\}$ with respect to $\Re$ is the union of two intervals, $(-\infty, 0) \bigcup (0, \infty)$.

complete A metric space E is complete iff every Cauchy sequence of elements in E converges to an element in E. The integers are obviously complete; the rationals are not complete; the reals are complete.

concave function A real-valued function $f : \Re^n \mapsto \Re$ is concave iff $-f$ is convex. Only affine functions are both concave and convex.

cone $C \subset \Re^n$ is a cone iff for all $\mathbf{x} \in C$ and all real $\alpha \geq 0$, $\alpha\mathbf{x} \in C$. Informally, a cone is a set closed under nonnegative scaling.

continuous Informally, a function $g : E \mapsto F$ is continuous if it maps points close in E to points close in F; when $E = F = \Re$, you can draw the function without lifting your stylus from the wax tablet, or your finger from the tablet screen. In nonstandard analysis, the formal definition is intuitive: if x and $x^{'}$ are infinitesimally close then $g(x)$ and $g(x^{'})$ are infinitesimally close. In standard analysis the definition is less intuitive. Formally, for metric spaces E, F equipped with distance functions d_E, d_F, g is continuous if for all open subsets $S \subset F$, the inverse image $g^{-1}(S)$ is open in E. Equivalently, S closed in E implies $g(S)$ closed in F; or, for all $x \in E$ and all $\epsilon > 0$, there exists $\delta > 0$ such that $d_E(x, y) < \delta \Rightarrow d_F(g(x), g(y)) < \epsilon$.

contradiction, proof by A method to prove "A implies B". Assume A is true and B is false; deduce a logical contradiction or impossibility such as "$1 = 2$" or "S is empty and non-empty." This method is logically valid because it proves that the negation of "A implies B" is impossible.

contrapositive A method to prove "A implies B". Assume B is false; deduce that A must be false. This method is logically valid because $\{A \Rightarrow B\} \Leftrightarrow \{\neg\{A \text{ and } \neg B\}\} \Leftrightarrow \{\neg\{\neg B \text{ and } A\}\} \Leftrightarrow \{\neg B \Rightarrow \neg A\}$.

convergence of a sequence The sequence $x_1, x_2, \ldots$ of points in a metric space E converges to $y \in E$ iff for all $\epsilon > 0$ there exists an integer N such that $d(x_i, y) < \epsilon$ for all $i \geq N$. That is, for all $\epsilon > 0$, eventually all points in the sequence are within ϵ of y.

convex combination A convex combination of K n-vectors $\mathbf{v}^1 \ldots \mathbf{v}^K$ is a linear combination $\sum_{k=1}^{K} \alpha_k \mathbf{v}^k$ where $\alpha_k \geq 0\ \forall k$ and $\sum_{k=1}^{K} \alpha_k = 1$. A convex combination is a special kind of affine combination in which all weights α_k are nonnegative.

convex function A real-valued function $f : \Re^n \mapsto \Re$ is convex iff $f(\alpha\mathbf{x} + (1-\alpha)\mathbf{y}) \geq \alpha f(\mathbf{x}) + (1-\alpha) f(\mathbf{y})$ for all $\mathbf{x} \in \Re^n, \mathbf{y} \in \Re^n$, and $0 \leq \alpha \leq 1$. Equivalently, f is a convex function if its epigraph $\{(\mathbf{x}, x_0) : f(\mathbf{x}) \geq x_0\}$ is a convex set in $\Re^{n+1}$. *Geometrically,* f is convex if for all x, y the line segment between $(x, f(x))$ and $(y, f(y))$ lies on or above the graph of the function f. For $n = 1$, if f is twice differentiable, f is convex iff its second derivative is nonnegative everywhere.

convex set A set $S \subset \Re^n$ is convex iff $\mathbf{x} \in S, \mathbf{y} \in S$, and $0 \leq \alpha \leq 1$ imply $\alpha\mathbf{x} + (1-\alpha)\mathbf{y} \in S$. Geometrically, the line segment between any two points in S is contained in S.

countable A set S is countable if it is finite or countably infinite.

countably infinite Has the same cardinality as the integers.

diagonal matrix A square matrix M such that $M_{i,j} = 0\ \forall i \neq j$.

dimension 1. The dimension of a subspace is the least number of vectors needed to span the subspace, equal to the maximum cardinality of a set of linearly independent vectors in the subspace. More succinctly, the cardinality of a minimal spanning vector set, equal to the cardinality of a maximal linearly independent vector set. 2. Dimension of a polyhedron, see *affine dimension.*

direction Vector $\mathbf{d} \in \Re^n$ is a direction of a set $S \subset \Re^n$ if for all points $\mathbf{s} \in S$ and all real $\theta \geq 0$, the point $\mathbf{s} + \theta\mathbf{d}$ is in S. In words, from any point in S, moving any (positive) amount in the direction $\mathbf{d}$ keeps you within the set S. Some authors require S to be nonempty. The set of directions of S is called the *recession cone* of S.

dot product The dot product of two vectors $\mathbf{x}, \mathbf{y}$ both in $\Re^n$ is $\mathbf{x} \cdot \mathbf{y} = \mathbf{y} \cdot \mathbf{x} \equiv \sum_{i=1}^{n} \mathbf{x}_i \mathbf{y}_i$ $= ||\mathbf{x}|| ||\mathbf{y}|| \cos\theta$, where $0 \leq \theta \leq \pi$ is the angle between $\mathbf{x}$ and $\mathbf{y}$. Other inner products exist but the dot product is the only inner product used in this text.

epigraph The epigraph of a function $f : \Re^n \mapsto \Re$ is the set $\{(\mathbf{x}, x_0) : f(\mathbf{x}) \geq x_0\}$. Geometrically, the epigraph is the set of points on or above the function.

extreme point A point $\mathbf{x} \in S \subset \Re^n$ is an extreme point of S if it is not a convex combination of other points in S. For sets S that are convex, $\mathbf{x}$ is an extreme point of S if it is not a convex combination of two other points in S, that is, there do not exist $\mathbf{u} \in S, \mathbf{v} \in S, \mathbf{u} \neq \mathbf{x}, \mathbf{v} \neq \mathbf{x}$ and $\alpha : 0 \leq \alpha \leq 1$ such that $x = \alpha\mathbf{u} + (1-\alpha)\mathbf{v}$.

feasible region The set of all feasible solutions.

feasible solution A solution that satisfies all constraints.

graph, directed graph A set V of elements called *vertices* or *nodes*, together with a set E of vertex pairs, called *edges* or *arcs*, denoted $G = (V, E)$. If the members of E are unordered pairs, i.e. $(u, v) = (v, u)$, G and its edges are called *undirected.* Graphs are undirected unless specified otherwise. If $E \subset V \times V$ comprises ordered vertex pairs, G is a *directed graph* and an edge $(u, v) \in E$ is said to be *directed* from u to v.

halfspace Geometrically, one of the two "halves" of space separated by a hyperplane. In two dimensions, a line separates the plane into two half-planes; in three dimensions, a plane separates space into two half-spaces. Formally, for any vector $\mathbf{c} \in \Re^n, \mathbf{c} \neq \mathbf{0}$, and real $\alpha \in \Re$, the sets $\{\mathbf{x} : \mathbf{c} \cdot \mathbf{x} \geq \alpha\}$ and $\{\mathbf{x} : \mathbf{c} \cdot \mathbf{x} \leq \alpha\}$ are the two *closed halfspaces* defined by the hyperplane $\mathbf{x} : \mathbf{c} \cdot \mathbf{x} = \alpha$. The sets $\{\mathbf{x} : \mathbf{c} \cdot \mathbf{x} > \alpha\}$ and $\{\mathbf{x} : \mathbf{c} \cdot \mathbf{x} < \alpha\}$ are the two *open halfspaces* defined by that hyperplane.

hyperplane Geometrically, the generalization of a line in 2D and a plane in 3D to n dimensions. Formally, for any nonzero vector $\mathbf{c} \in \Re^n$ and any real number α, the set of points $\mathbf{x}$ satisfying the equation $\mathbf{c} \cdot \mathbf{x} = \alpha$ is a hyperplane.

iff if and only if, words equivalent to the symbol $\Leftrightarrow$.

induction A method of proof to prove a proposition $P(n)$ true for all integers $n \geq n_0$. Step 1 (base case): prove $P(n_0)$. Step 2 (inductive step): prove that for every $n \geq n_0$, if $P(n)$ is true, then $P(n+1)$ is true. *Strong induction* is the variant in which Step 2 proves that if $P(m)$ is true for all $n_0 \leq m \leq n$, then $P(n+1)$ is true.

inner product A mapping $\langle\rangle : \Re^n \times \Re^n \mapsto \Re$ such that for all $\mathbf{x}, \mathbf{y} \in \Re^n$: $\langle\rangle(\mathbf{x}, \mathbf{x}) \geq 0$; $\langle\rangle(\mathbf{x}, \mathbf{x}) = 0 \Leftrightarrow \mathbf{x} = \mathbf{0}$; $\langle\rangle(\mathbf{x}, \mathbf{y}) = \langle\rangle(\mathbf{y}, \mathbf{x})$; $\langle\rangle(\alpha\mathbf{x}, \mathbf{y}) = \alpha\langle\rangle(\mathbf{x}, \mathbf{y})$. See dot product.

LHS Left-hand-side. The LHS of the inequality "$A\mathbf{x} + \mathbf{c} \geq \theta\mathbf{b}$" is "$A\mathbf{x} + \mathbf{c}$".

line A set of form $\{\mathbf{x} + \theta\mathbf{y} : \theta \in \Re\}$ where $\mathbf{x}$ and $\mathbf{y}$ are vectors in $\Re^n$.

linear combination A linear combination of K n-vectors $\mathbf{v}^1 \ldots \mathbf{v}^K$ is any weighted sum of the vectors, $\sum_{k=1}^{K} \alpha_k \mathbf{v}^k$ where $\alpha_k \in \Re\ \forall k$. Define the $n \times K$ matrix V by setting its kth column $V_{:,k} = \mathbf{v}^k$. Then every product of form $V\alpha$ where $\alpha \in \Re^K$ is a linear combination of $\mathbf{v}^1 \ldots \mathbf{v}^K$.

linear dependence Any set of n-vectors that is not linearly independent is linearly dependent.

linear function $f : \Re^n \mapsto \Re^m$ is linear iff $f(\alpha x + \beta y) = \alpha f(x) + \beta f(y)$ for all $x \in \Re^n, y \in \Re^n, \alpha \in \Re, \beta \in \Re$. The definition is equivalent if β is replaced by 1. A function f is linear iff there exists a matrix M such that $f(x) = Mx\ \forall x$.

linear independence A set of n-vectors $v^1 \ldots v^K$ is linearly independent iff the sole linear combination of those vectors that equals $\mathbf{0}$ is $\sum_{k=1}^{K} 0v^k$. Equivalently, no member of the set can be expressed as a linear combination of the others.

maximal Set S is *maximal* with respect to property $\mathfrak{P}$ if (i) S has property $\mathfrak{P}$; (ii) every set $S^{'}$ that strictly contains S, $S \subsetneq S^{'}$, does not have property $\mathfrak{P}$. Informally, you can't add any elements to S and retain property $\mathfrak{P}$.

metric space A set E is a metric space with respect to a distance function $d : E \times E \mapsto \Re$ iff $d(x,y) = d(y,x) \geq 0 \; \forall x,y \in E$, $d(x,y) = 0 \Leftrightarrow x = y$, and $d(x,y) + d(y,z) \geq d(x,z) \; \forall x,y,z \in E$. The last condition is called the triangle inequality.

minimal Set S is *minimal* with respect to property $\mathfrak{P}$ if (i) S has property $\mathfrak{P}$; (ii) every strict subset $\check{S}$ of S, $\check{S} \subsetneq S$, does not have property $\mathfrak{P}$. Informally, you can't remove any elements from S and retain property $\mathfrak{P}$.

mixed graph A graph containing both directed and undirected edges.

n**-vector** A vector (ordered list) containing n entries. A vector written $\mathbf{x}$ is always a column vector. Its transpose $\mathbf{x}^T$ is a row vector.

node A vertex in a graph.

open A set S is open iff for all $s \in S$ there exists $\epsilon > 0$ such that $t \in S$ for all t within distance less than ϵ of s. *Geometrically,* for every $s \in S$ there is an open ball around s that is contained in S. *Algebraically,* S is open iff $\forall s \in S \exists \epsilon > 0$ such that $d(t,s) < \epsilon \Rightarrow t \in S$.

orthogonal Vectors $\mathbf{x}, \mathbf{y}$ are orthogonal if their dot product is zero, $\mathbf{x} \cdot \mathbf{y} = 0$. Geometrically, they are mutually perpendicular. Two sets S, T in $\Re^n$ are orthogonal iff every member of S is orthogonal to every member of T, $\mathbf{s} \cdot \mathbf{t} = 0 \; \forall \mathbf{s} \in S, \; \mathbf{s} \in T$.

orthogonal complement The orthogonal complement of $S \subset \Re^n$, denoted $S^\perp$, is the set of vectors orthogonal to every element of S, $\{t : t \cdot s = 0 \; \forall s \in S\}$ Equivalently, $S^\perp$ is the maximal set orthogonal to S. The sum of the dimensions of S and $S^\perp$ equals n.

polyhedron A set of form $\{\mathbf{x} : A\mathbf{x} \leq \mathbf{b}\}$. Equivalently, a finite intersection of closed halfspaces.

polytope A bounded polyhedron.

ray A set of form $\{\mathbf{x} + \theta\mathbf{y} : \theta \geq 0\}$ where θ is real and $\mathbf{x}$ and $\mathbf{y}$ are vectors in $\Re^n$.

recession cone See *direction.*

RHS Right-hand-side. The RHS of the inequality "$A\mathbf{x} + \mathbf{c} \geq \theta\mathbf{b}$" is "$\theta\mathbf{b}$".

slack A constraint is slack at solution $\mathbf{x}$ if it is satisfied but not binding at $\mathbf{x}$.

solution A set of values assigned to the variables, not necessarily feasible.

span The span of a set of vectors in $\Re^n$ is the set of all linear combinations of those vectors.

s.t. "subject to" or "such that."

subset A set S is a subset of a set T iff for every $s \in S$ it is true that $s \in T$. More concisely, $S \subset T \Leftrightarrow s \in T \; \forall s \in S$.

tight binding

tree A graph $G = (V, E)$ with the properties:

- $|E| = |V| - 1$;
- G is connected;
- G has no cycles.

Any two of these properties imply the third.

uncountable Not countable; cannot be placed in 1-1 correspondence with the integers or any subset of the integers.

uniformly continuous A function $g : E \mapsto F$ is uniformly continuous if for all $\epsilon > 0$ there exists $\delta > 0$ such that for all $x \in E$, $d(x, y) < \delta \Rightarrow d(g(x), g(y)) < \epsilon$. In the definition of the weaker continuity property, the choice of δ can depend on x.

vertex 1. See *graph*. 2. An extreme point of a polyhedron.

zlotsky An animal native to the country Zembla.

Chapter 1

A Gentle Introduction to Optimization and Duality

1.1 Optimization Problems

Preview: This section tells you what an optimization problem is. It defines key terms that will be used throughout the book, including **solution, constraint, feasible, objective function** and **optimal solution.**

A typical first-year calculus problem asks for the dimensions of a rectangle with a perimeter of 100 feet that has maximum area. The standard solution is to let y denote the width of the rectangle in feet, calculate the area of the rectangle to be $y(50-y)$ square feet, set the derivative of the function $f(y) = 50y - y^2$ to 0, and get $50 - 2y = 0 \Rightarrow y = 25$. The answer is therefore a width of 25 feet, a length of $50 - y = 25$ feet, and an area of 625 square feet. If you are conscientious, you will check the second derivative $f''(25) = -2 < 0$ to be sure your answer is a maximum rather than a minimum. You will also subconsciously notice that the width and length are both positive numbers, which makes sense.

This calculus problem is an example of an optimization problem. To write it in a standard form, let x_1 denote the width of the rectangle in feet, and let x_2 denote the length of the rectangle in feet. The variables x_1 and x_2 are called the *variables* of the problem. Define the function $f(x_1, x_2) = x_1 x_2$, which equals the area of the rectangle. The function f is called the *objective function.* Write the optimization problem

$$\begin{aligned} &\text{maximize} \quad x_1x_2 \quad \text{subject to} && (1.1)\\ &2x_1 + 2x_2 = 100 && (1.2)\\ &x_1 \geq 0 \quad , \quad x_2 \geq 0 && (1.3)\end{aligned}$$

Lines (1.2) and (1.3) are called the *constraints* of the problem. The first is an *equality constraint*; the second is a pair of *inequality constraints.* Figure 1.1 depicts the set of values of the variables that satisfy constraints (1.2) and (1.3). This set is called the *feasible region.* In a first-year calculus course, constraints (1.3) are usually ignored. This book will show you how to handle them properly. In general, an optimization problem has a vector of variables $\mathbf{x} = \{x_1, x_2, \ldots, x_n\}$, a real-valued objective function $f(\mathbf{x})$ (that is, for every $\mathbf{x}$, $f(\mathbf{x})$ is a particular real number), and a set of constraints. It should always be possible for a constraint to be expressed using constraints of the form $g(\mathbf{x}) \geq 0$ where $g(\mathbf{x})$ is a real-valued function. Sometimes it is more convenient to use a form other than $g(\mathbf{x}) \geq 0$. We usually

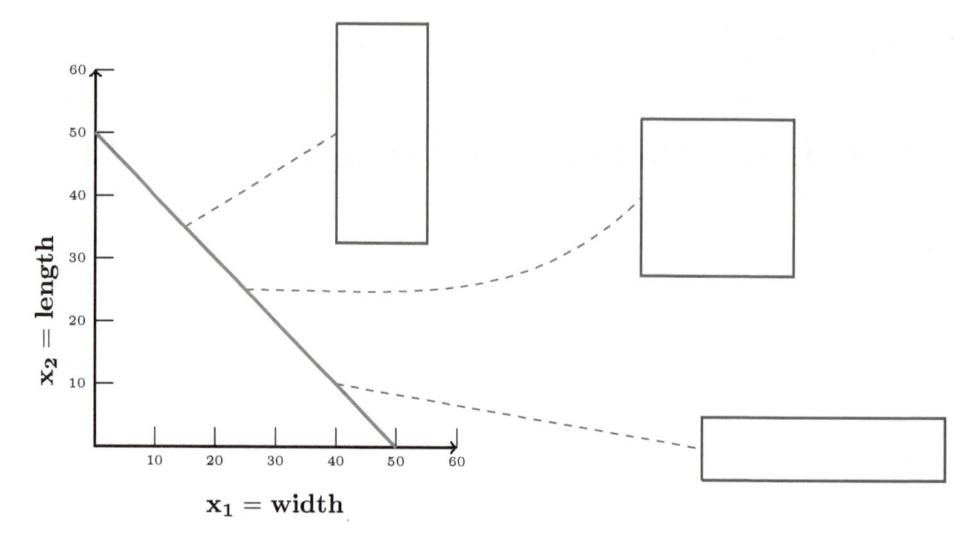

FIGURE 1.1: The feasible region defined by constraints (1.2) and (1.3) is the line segment connecting (0,50) with (50,0). The optimal solution $(25, 25)$ corresponds to the square; the suboptimal solutions $(40, 10)$ and $(15, 35)$ correspond to the wide and long rectangles, respectively.

prefer the more concise form $g(\mathbf{x}) = 0$ to the equivalent pair $g(\mathbf{x}) \geq 0$ and $-g(\mathbf{x}) \geq 0$. If a variable x_i must have an integer value, we will write "x_i integer" rather than "$x_i - \lfloor x_i \rfloor = 0$." In this book, optimization problems will have a finite number of constraints. The general form of an optimization problem employs these three forms of constraints, and permits the goal to be either minimization or maximization. *Question: Read Definition 1.1. What is* $h_1(\mathbf{x})$ *in the calculus problem* (1.1)–(1.3)*?*[1]

Definition 1.1 (Optimization Problem) *Let* $\mathbf{x} \in \Re^n$ *denote a vector of* n *real variables. Let* $f : \Re^n \mapsto \Re$, $g_j : \Re^n \mapsto \Re, j = 1 \ldots m$ *and* $h_j : \Re^n \mapsto \Re, j = 1 \ldots m'$ *be real-valued functions on the domain* $\Re^n$. *Let* $I \subseteq \{1, 2, \ldots, n\}$ *be a subset of the variable indices. The general optimization problem is*

$$\text{minimize or maximize} \quad f(\mathbf{x}) \quad \text{subject to} \tag{1.4}$$

$$g_i(\mathbf{x}) \geq 0 : \quad 1 \leq i \leq m \tag{1.5}$$

$$h_i(\mathbf{x}) = 0 : \quad 1 \leq i \leq m' \tag{1.6}$$

$$x_i \text{ integer:} \quad i \in I. \tag{1.7}$$

Functions g_j, h_j *need not be defined on the entire domain* $\Re^n$, *as long as they are defined to take real values for all* $\mathbf{x} \in \Re^n$ *that satisfy* (1.7). *Function* f *does not need to be defined on the entire domain* $\Re^n$, *as long as it is defined to take real values for all* $\mathbf{x} \in \Re^n$ *that satisfy* (1.5),(1.6) *and* (1.7).

The word "solution" has a peculiar meaning in optimization. Any set of values assigned to the variables is called a *solution*. A solution is *feasible* if it satisfies all the constraints. Otherwise it is *infeasible*. In Figure 1.1, the points $(60, 70)$, $(-1, -3)$ and $(60, -10)$ are all

[1] $2x_1 + 2x_2 - 100$.

infeasible solutions. The feasible solutions compose the *feasible region.* If the feasible region is not empty, the problem is called *feasible*; otherwise it is called *infeasible.* The value of the objective function at a solution is called the *objective value* or simply the *value* of the solution. If the value of a particular feasible solution $\mathbf{x}^*$ is at least as good as the value of any other feasible solution, then $\mathbf{x}^*$ is called an *optimal solution*, and its value is called the *optimal value* or simply the *value* of the problem. A solution that has worse value than some other feasible solution's value is *suboptimal*; a solution that has better value than any feasible solution's value is called *superoptimal.* A superoptimal solution must be infeasible, but a suboptimal solution may be either infeasible or feasible. *Question: why isn't "suboptimal" (respectively "superoptimal") defined more concisely as having worse (respectively better) value than the optimum solution's?*[2] Definition 1.2 states these and related definitions in terms of the optimization problem (Definition 1.1). All of this terminology is standard except my terms "better" and "worse". I use these terms throughout the book to avoid separate statements about minimization and maximization.

[2]There might not be an optimal solution.

Definition 1.2 (Optimization Terminology) *Let $n, m, m', f, g_j : j = 1 \ldots m, h_j : j = 1 \ldots m', I$ be the components of an optimization problem as in Definition 1.1. Then the following terminology applies.*

Variable *Each* $\mathbf{x}_i : i = 1, \ldots, n$ *is a variable .*

Solution *Any* $\mathbf{x} \in \Re^n$*, that is, any set of values assigned to the variables.*

Constraint *Any of the statements from* (1.5),(1.6),(1.7)*. For example, "*$g_6(\mathbf{x}) \geq 0$*," "*$h_5(\mathbf{x}) = 0$*," and "*x_2 *integer" could all be constraints.*

Satisfy, Violate *A solution* $\mathbf{x}$ *satisfies a constraint if the constraint is true for* $\mathbf{x}$*; otherwise* $\mathbf{x}$ *violates the constraint.*

Feasible Solution *A solution that satisfies all of the constraints.*

Infeasible Solution *A solution that does not satisfy one or more of the constraints.*

Feasible Region *The set of all feasible solutions. It may be empty.*

Feasible Problem, Infeasible Problem *The optimization problem is feasible if its feasible region is not empty. Otherwise, it is infeasible.*

Better *If minimizing, "better" means "*$<$*", i.e., "strictly less than." If maximizing, "better" means "*$>$*", i.e., "strictly greater than."*

Worse *If minimizing, "*$>$*." If maximizing, "*$<$*."*

At least as good *Not worse.*

Optimal Solution *A solution* $\mathbf{x}^*$ *is optimal if it is feasible and if* $f(\mathbf{x}^*)$ *is at least as good as* $f(\mathbf{x})$ *for all feasible solutions* $\mathbf{x}$*.*

Suboptimal Solution *Solution* $\mathbf{x}$ *is suboptimal iff there exists a feasible solution* $\mathbf{y}$ *that has better objective value than* $f(\mathbf{x})$*.*

Objective Function *The function* f*. For every feasible solution* $\mathbf{x}$*,* $f(\mathbf{x})$ *must be a real number.*

Problem Value *If an optimal solution* $\mathbf{x}^*$ *exists, the value of the problem is* $f(\mathbf{x}^*)$*.*

Unbounded Optimum, Unbounded Problem *A problem has unbounded optimum if for all* $M \in \Re$ *there exists a feasible solution* $\mathbf{y}$ *such that* $f(\mathbf{y})$ *is better than* M*. If this is the case we say the problem is unbounded. It would be more accurate but less concise to say that the problem value is unbounded.*

Bounded *The problem is bounded iff it is not unbounded. Infeasible problems are by definition bounded, but are rarely called so because the term "infeasible" conveys more information.*

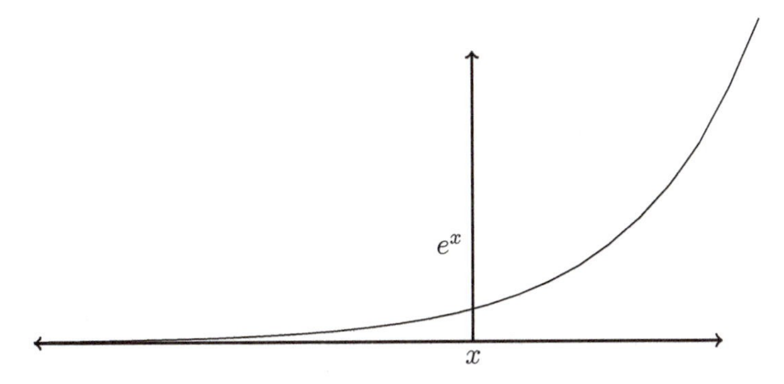

FIGURE 1.2: The function e^x increases without bound as x increases, and decreases without reaching a minimum as x decreases.

Example 1.1 The problem $\max f(x) = e^x$ is unbounded because $e^t > t$ for all t (see Figure 1.2).

Example 1.2 The problem $\max f(x) = \sin x$ subject to $x \geq 0$ has value 1, which is attained at the infinitely many optimal solutions $x = (2k+1)\pi : k = 0, 1, 2 \ldots$. The problem $\max f(x) = \sin x$ subject to x integer has no optimal solution because no integer multiple of π is an integer, yet $\sin k$ can be arbitrarily close to 1 for integer k (see Figure 1.3).

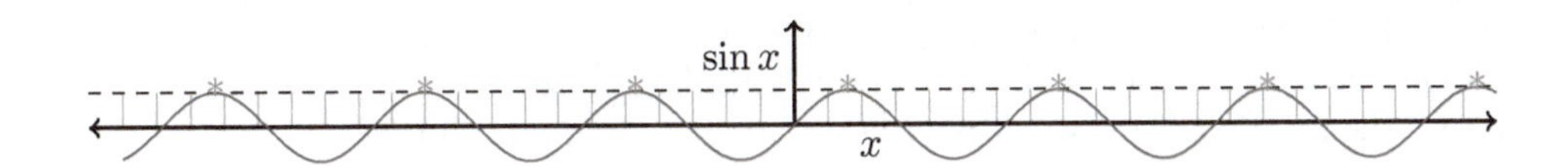

FIGURE 1.3: Function $\sin x$ has maximum value 1 but never attains that value for any integer x. (In this figure, it appears that $\sin 11 = 1$. However, $\sin 11 < 0.99999020$.)

Reading Tip: *If you are learning how to do mathematical proofs, examples like this are good opportunities to practice. Write a rigorous proof that the problem is unbounded. Don't skip this because the example seems obviously correct to you. The obviousness is what makes this a good opportunity. Your entire focus will be on your proof technique.*
Set up the proof by plugging in the definitions. By Definition 1.2 the problem is unbounded if for all $M \in \Re$ *there exists a feasible solution* y *such that* $f(y)$ *is better than* M. *Let* $M \in \Re$. *We must find a feasible* y *such that* $f(y) = e^y$ *is better than* M. *The problem is a maximization, so by Definition 1.2,* $e^y > M$ *is required.*
Now that you've plugged in the definitions, set $y = M$. *This value of* y *is feasible because the problem puts no constraints on* y. *Set* $t = M$ *in the inequality* $e^t > t$ *to get* $f(y) = e^y = e^M > M$, *as required.*
To prove the inequality $e^t > t$, *examine Figure 1.2. Geometrically the inequality means that the function* $f(t) = e^t$ *is strictly above the function* $g(t) = t$. *At* $t = 0$, $e^t = 1 > 0 = t$. *At all* $t \geq 0$ *the slope of* $f(t)$ *is* $e^t \geq 1$, *while the slope of* $g(t)$ *is 1. At all* $t \leq 0$ *the slope of* $f(t)$ *is* $e^t \leq 1$ *while the slope of* $g(t)$ *is 1. Hence* $f(t) \geq t + 1 > t$ *for all* t.

Example 1.3 The optimization problem

$$\begin{aligned} \max x_1^2 + x_2^2 \quad & \text{subject to} \\ |x_1| \leq & 3 \\ |x_2| \leq & 4 \end{aligned}$$

can be cast in the form of Definition 1.1 as

$$\begin{aligned} \max x_1^2 + x_2^2 \quad & \text{subject to} \\ g_1(x_1, x_2) = 3 - |x_1| \geq & 0 \\ g_2(x_1, x_2) = 4 - |x_2| \geq & 0 \end{aligned}$$

where $m' = 0$ and $|I| = 0$. This is a valid formulation. However, its constraint functions g_1 and g_2 are not differentiable. We might prefer to reformulate the problem as

$$\begin{aligned} \max x_1^2 + x_2^2 \quad & \text{subject to} \\ g_1(x_1, x_2) = x_1 + 3 \geq & 0 \\ g_2(x_1, x_2) = x_2 + 4 \geq & 0 \\ g_3(x_1, x_2) = 3 - x_1 \geq & 0 \\ g_4(x_1, x_2) = 4 - x_2 \geq & 0 \end{aligned}$$

The solution $(0.5, -2.1)$ is feasible; the solution $(-3.3, 2.8)$ is infeasible. The feasible region is a rectangle centered at $(0, 0)$ with sides of length 6 parallel to the x_1 axis and length 8 parallel to the x_2 axis, as in Figure 1.4. The problem is feasible because this rectangle is not empty. The solutions $(3, 4), (3, -4), (-3, 4), (-3, -4)$ are all optimal with objective function value 25. The problem value is therefore 25. If the problem were to minimize $f(x_1, x_2)$ subject to the same constraints the solution $(0, 0)$ would be the unique optimal solution. If the problem were to minimize $f(x_1, x_2)$ subject to no constraints the solution $(0, 0)$ would still be the unique optimal solution.

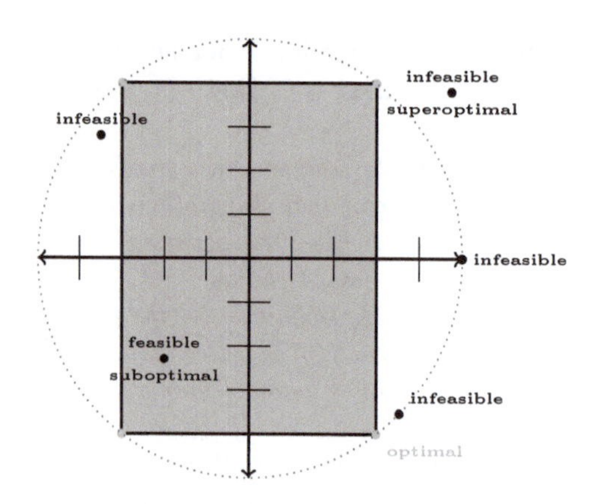

FIGURE 1.4: The rectangular feasible region of Example 1.3 has four optimal solutions.

Not all optimization problems have optimal solutions. This can happen in several ways. First, some problems are infeasible. Second, some problems are unbounded. You may be tempted to say that the optimal value of an unbounded problem is infinity, but don't do it. It is mathematically incorrect. Third, a feasible problem that is not unbounded can still fail to have an optimal solution, because feasible solutions might have values arbitrarily close, but never equal, to some value. For example, there is no optimal solution to the problem of minimizing $1/x$ subject to $x \geq 1$. As x increases, the objective value gets arbitrarily close to 0 but is always strictly positive.

Example 1.4 The problem $\min f(x) = e^x$ has no optimal solution because $e^x > 0$ for all $x \in \Re$ but as $x \to -\infty$, e^x gets arbitrarily close to 0. See Figure 1.2. To be precise, for any $x \in \Re$, let $y = x - 1$. Then $e^y < e^x$. Hence x is not optimal.

▶ *The preceding examples depend on $|x|$ being permitted to get arbitrarily large. If constraints of form $g(x) > 0$ were permitted, a simple example with a very limited feasible region such as minimizing x subject to $0 < x < 1$ would suffice to show that no optimal solution might exist, even if the problem is feasible and not unbounded. This is the main reason we don't ordinarily permit constraints of that form. From a mathematical point of view, this simple example works because the feasible region is open rather than closed. If the feasible region is bounded and closed, and if the objective function is well-behaved in the sense that it is continuous, then by a well-known theorem in real analysis the optimization problem must have an optimal solution. The terms bounded, open, closed, and continuous are all defined in the Glossary.* ◀

Example 1.5 For variables x_1, x_2, suppose the constraints are x_1 and x_2 integer ($I = \{1, 2\}$), and $x_1^2 + x_2^2 - 10 \geq 0$. Figure 1.5 illustrates some of the feasible solutions.

- For the objective $\min f(x_1, x_2) = 5|x_1| + 6|x_2|$ the solutions $(4, 0)$ and $(-4, 0)$ are optimal, the solution $(3, 1)$ is suboptimal, and the solution $(1, -2)$ is infeasible.
- For the objectives $\max f(x_1, x_2) = 3x_1 + 4x_2$ and $\min f(x_1, x_2) = 3x_1 + 4x_2$ the problem is unbounded.
- The objective function $f(x_1, x_2) = \frac{x_1}{x_2+0.5}$ is legal because it has a real value at all feasible solutions, even though it has no real value at $(5.0, -0.5)$ (and at infinitely many other infeasible solutions).
- The objective function $f(x_1, x_2) = \frac{x_1}{x_2}$ is illegal because it does not have a real value at $(4, 0)$ (and at infinitely many other feasible solutions).
- For the objective
$$\min f(x_1, x_2) = \begin{cases} 1 : x_2 = 0; \\ |\frac{x_1}{x_2} - 3| : x_2 \neq 0; \end{cases}$$
the optimization problem has value 0, with infinitely many optimal solutions $\pm(3k, k) : k \geq 1, k$ integer.
- In contrast, for the objective
$$\min f(x_1, x_2) = \begin{cases} 1 : x_2 = 0; \\ |\frac{x_1}{x_2} - \sqrt{3}| : x_2 \neq 0; \end{cases}$$
the optimization problem has no optimal solution because there exist rational numbers arbitrarily close to $\sqrt{3}$, but $\sqrt{3}$ is irrational, not equal to the ratio of any two integers.

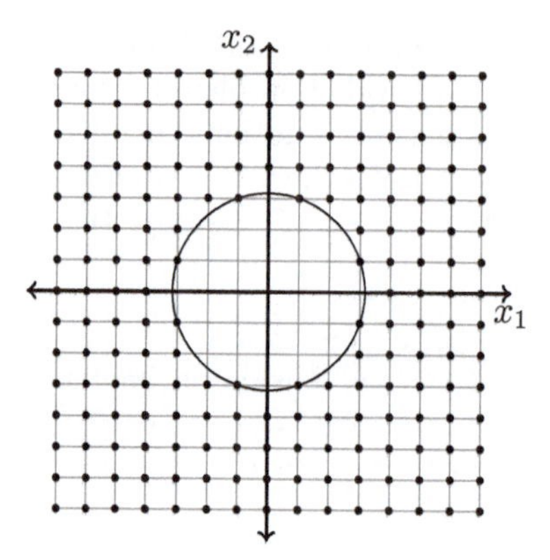

FIGURE 1.5: The feasible solutions of Example 1.5 are integer pairs on or outside the circle.

Summary: Memorize the terms defined in this section. Most make intuitive sense. However, the term "solution" means any set of numbers assigned to the variables. A solution can be suboptimal or even infeasible. Also, remember the different ways that an optimization problem may fail to have an optimal solution: infeasible, unbounded, or feasible and bounded but not reaching the bound.

1.2 Duality in Multivariate Calculus, Sudoku, and Matrices

Preview: Duality is a general method to see things from two points of view. This section defines duality and applies it to multivariate differential calculus, Sudoku puzzles, and matrices.

1.2.1 An Example of Duality in Optimization

For your first example of duality, revisit the area-maximization problem at the start of this chapter. How do you know that the square has maximum area over all rectangles with perimeter 100? Let x_1 and x_2 be the width and length of the rectangle, respectively. The standard explanation, as stated earlier, is to substitute $50 - x_1$ for x_2, giving the single-variable function $50x_1 - x_1^2$ depicted in Figure 1.6. By analytic geometry or calculus the area function is maximized at $x_1 = 25$ which implies $x_2 = 25 = x_1$.

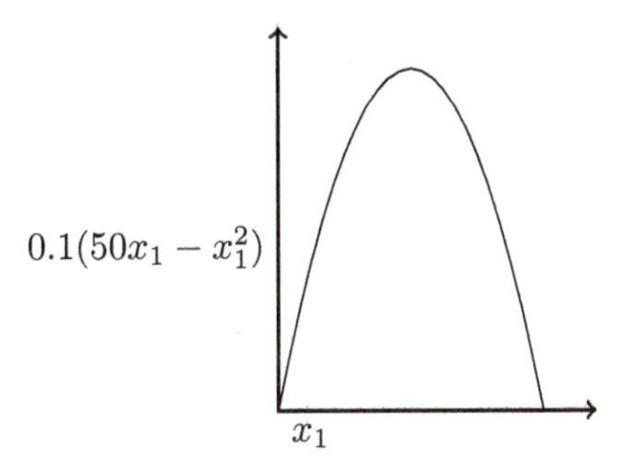

FIGURE 1.6: The area of the rectangle as a function of one variable, the width x_1.

Figure 1.7 depicts a different way to explain why the square maximizes area. Instead of simplifying the problem by reducing the number of variables from two to one, make the problem more complicated by perturbing the constraint (1.2) on the perimeter.

Figure 1.7 shows possible impacts of increasing the perimeter, the right-hand-side (RHS) of (1.2) from 100 to 120. At the solution $x_1 = 40, x_2 = 10$ we could increase the width x_1 by 10, which would increase the area by $10x_2 = 100$. Increasing the length x_2 by 10 instead would increase the area by $10x_1 = 400$. Figure 1.7 depicts the potential additional areas as rectangles with horizontal and vertical dotted lines, respectively. Therefore, the impact of changing the right-hand-side (RHS) from its current value 100 is ambiguous. This ambiguity is why $x_1 = 40, x_2 = 10$ is not optimal. For if we decrease the width x_1 by 1 and simultaneously increase the length x_2 by 1, the net effect on area will be an increase of approximately $40 - 10 = 30$.

As you will see in Section 1.3.1, duality requires that at optimality, the impact on the objective function of changing a constraint's RHS should have an unambiguous numerical value. *Question: What is the numerical value of the impact on the objective function of changing the RHS of the constraint $x_1 \geq 0$?*[3]

At the solution $x_1 = 15, x_2 = 35$, the impact on the objective of changing the perimeter is again ambiguous. Increasing the width gains more area than increasing the length. That's why that solution is suboptimal. The solution $x_1 = 15 + \theta, x_2 = 35 - \theta$ has better objective value (for any small $\theta > 0$). Only at the square solution $x_1 = x_2 = 25$ do changes in width and length have equal effect. Only that solution is optimal.

1.2.2 A General Definition of Duality

A function of two variables $f : A \times B \mapsto C$ takes two input values, $a \in A$ and $b \in B$, and produces an output value $f(a, b) = c \in C$. The function f can be viewed as a collection of functions, one for each $a \in A$, that map from B to C. That is, think of f as the set of functions $\{g_a : B \mapsto C | a \in A\}$ where $g_a(b) \equiv f(a, b) \forall a \in A, \forall b \in B$. Exchange the roles of A and B to view f as the the set of functions $\{h_b : A \mapsto C | b \in B\}$ where $h_b(a) \equiv f(a, b)\ \forall b \in B, \forall a \in A$. These two ways of viewing f are duals of each other. (This definition is taken from the famous Chapter 0 of [214].) See Figure 1.8.

For example, $f(x, y) = 5x + x^2y - 10y^3$ can be viewed as a set of functions of y, one for each real value of x. For $x = 1$, $g_1(y) = 5 + y - 10y^3$. For $x = 20$, $g_{20}(y) = 100 + 400y - 10y^3$. It is equally valid to view f as a set of functions of x, one for each real value of y. For $y = 0$, $h_0(x) = 5x$. For $y = -1$, $h_{-1}(x) = 5x - x^2 + 10$. *Question: what is $h_3(x)$?*[4]

[3] Zero, with no ambiguity.
[4] $5x + 3x^2 - 270$.

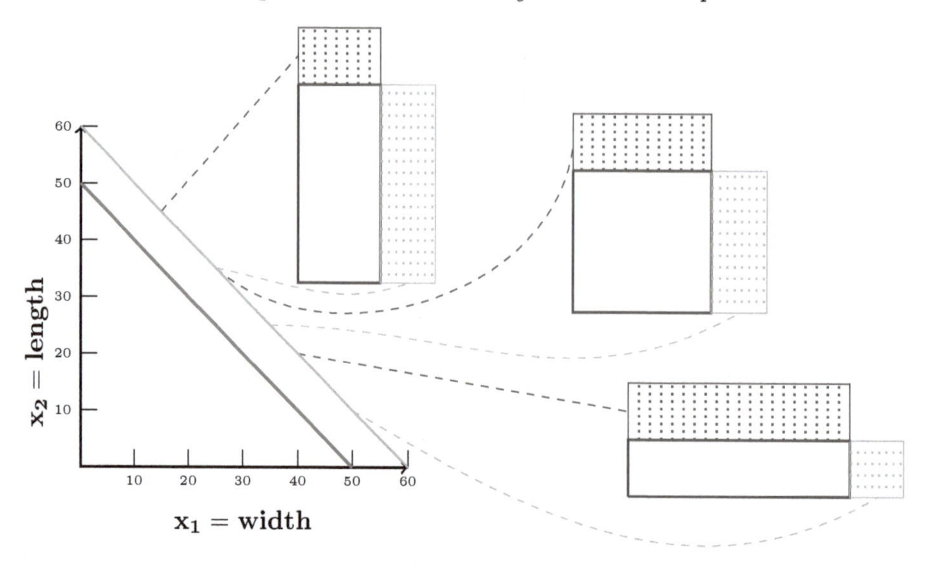

FIGURE 1.7: The increase in area when increasing the width by 10 versus increasing the length by 10. Only the square experiences the same amount of increase. Why does that show that the non-square rectangles do not maximize area?

1.2.3 Partial Derivatives

You can figure out how to take derivatives in multivariate calculus by combining single-variable differentiation with duality. Consider the function from the previous example, $f(x, y) = 5x + x^2y - 10y^3$. For each value of x, $g_x(y)$ is a single-variable function, a function of y. For $x = 1$, $g_1(y) = f(1, y) = 5 + y - 10y^3$. Therefore, $g_1'(y) = 1 - 30y^2$. For $x = 20$, $g_{20}(y) = 100 + 400y - 10y^3$, and therefore $g_{20}'(y) = 400 - 30y^2$. For each specific x one could find the value $g_x'(y)$ similarly. Let's find all these values in one fell swoop. Treat x as a constant in the function $f(x, y) = 5x + x^2y - 10y^3$. Then $g_x'(y) = x^2 - 30y^2$. This is called the *partial derivative* of f with respect to y. The standard notation for this partial derivative is

$$\frac{\partial f}{\partial y} = x^2 - 30y^2.$$

Viewing f as a set of functions of x, one for each value of y, you can, similarly, calculate $h_0'(x) = 5$ and $h_{-1}'(x) = 5 - 2x$. In general, for any value of y, the partial derivative of f with respect to x equals

$$\frac{\partial f}{\partial x} = 5 + 2xy.$$

Question: If $f(x, y) = 2 + 5xe^y - x^3y^3$, what is $\frac{\partial f}{\partial y}$? What is $\frac{\partial f}{\partial x}$?[5]

Reading Tip: *If you are not already familiar with partial derivatives, invent a few examples to make sure you understand them. Start with a very simple example like $f(x, y) = 3x$. At $y = 1$, $f(x, y) = f(x, 1) = 3x$. At $y = 1.1$, $f(x, y) = f(x, 1.1) = 3x$. Since $f(x, y)$ does not change when y changes, $\frac{\partial f}{\partial y}$ should equal 0. If you treat x as a constant, $3x$ is a constant. Hence its derivative with respect to y is 0. That makes sense. At $x = 1$, $f(x, y) = f(1, y) = 3$. At $x = 1.1$, $f(x, y) = f(1.1, y) = 3.3$. This is consistent with $\frac{\partial f}{\partial x} = 3$. Next, try the examples $f(x, y) = 3x + 4y$, $f(x, y) =$*

[5] $5xe^y - 3x^3y^2$; $5e^y - 3y^3x^2$.

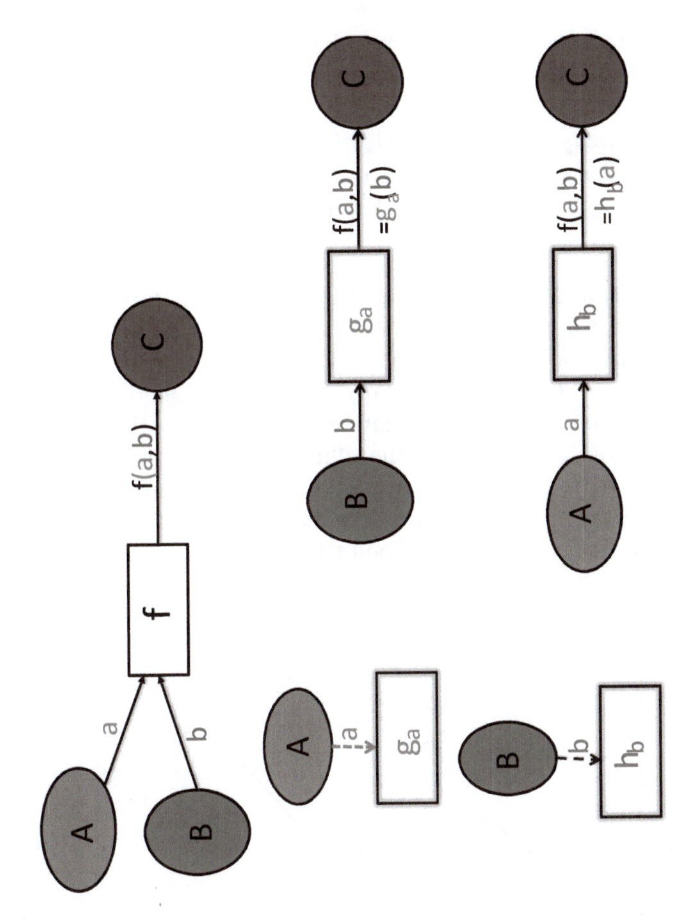

FIGURE 1.8: The primal views function $f : A \times B \mapsto C$ as a set of functions from B to C, one for each $a \in A$. The dual views f as a set of functions from A to C, one for each $b \in B$.

xy, $f(x, y) = e^{2x+3y}$, and $f(x, y) = e^{xy}$. Check yourself by computing the partial derivatives of $f(x, y) = 5x + x^2y - 10y^3$ from the example in the text.

The pair of partial derivatives, written as a vector, is called the *gradient* of f and is denoted ∇f. That is, for a function $f(x, y)$, the gradient is

$$\nabla f(x, y) \equiv \left(\frac{\partial f}{\partial x}, \frac{\partial f}{\partial x} \right).$$

In the example above, $\nabla f(x, y) = (5 + 2xy, x^2 - 30y^2)$.

Generalize the idea of partial derivatives to real-valued functions of n variables $x_1, x_2, \ldots, x_n$ by defining, for all i, the partial derivative

$$\frac{\partial f}{\partial x_i}$$

as the derivative of f with respect to variable x_i where the other variables $x_j : j \neq i$ are treated as constants. Let $\mathbf{x}$ denote $(x_1, x_2, \ldots, x_n)$. The gradient of f generalizes to

$$\nabla f(\mathbf{x}) \equiv \left(\frac{\partial f}{\partial x_1}, \frac{\partial f}{\partial x_2}, \ldots, \frac{\partial f}{\partial x_n} \right).$$

The gradient of a function has an important meaning in optimization. The gradient at $\mathbf{x}$, $\nabla f(\mathbf{x})$, is the direction in which the function increases the most, in the immediate vicinity of the point $\mathbf{x}$. Why is this so?

Reading Tip: *If you don't already know this property of gradients, try a few numerical examples. Starting at* $(0,0,0)$ *what direction increases* $f(x_1,x_2,x_3) = x_1 + 2x_2 + 3x_3$ *the most? Starting at* $(1,100)$ *what direction increases* $f(x_1,x_2) = x_1^2 + x_2^2$ *the most? How about starting at* $(100,1)$*? How about the function* $x_1^2 - x_2^2$*?*

A real-valued *affine* function of vector variable $\mathbf{x}$ is a function of form $f(\mathbf{x}) = \mathbf{a}\cdot\mathbf{x} + \mathbf{b}$. Differential calculus is concerned with approximating functions by affine functions. In one-dimensional calculus, the function $f(x)$ of a single variable x is approximated at $x = t$ by the affine function $f(t) + (x-t)f'(t)$. Geometrically, (see Figure 1.9) the affine function is the line tangent to f at the point t. In multivariate calculus the function $f(\mathbf{x})$ of a vector $\mathbf{x}$ of variables is approximated at $\mathbf{x} = \mathbf{t}$ by the affine function $f(\mathbf{t}) + (\mathbf{x}-\mathbf{t})\cdot\nabla f(\mathbf{t})$. Geometrically, (see Figure 1.10) if $\mathbf{x} = \{x_1, x_2\}$ we visualize the graph of the function $\mathbf{x}, f(\mathbf{x})$ in three dimensions; the affine function is the plane tangent to f at the point $\mathbf{t}$. In general it is the hyperplane tangent to f at the point $\mathbf{t}$.

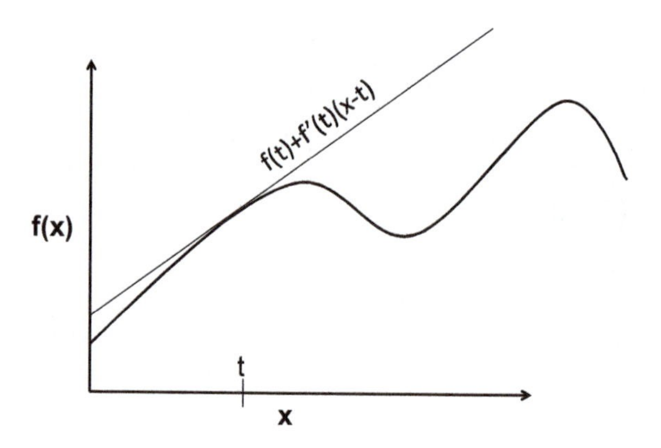

FIGURE 1.9: The derivative $f'(t)$ of a single-variable function f at t provides a line tangent to f at t which approximates f near t.

It is no coincidence that $f(t)+(x-t)f'(t)$ is the sum of the first two terms of the Taylor series of f at t, which you know from one-variable calculus, and that $f(\mathbf{t}) + (\mathbf{x}-\mathbf{t})\cdot\nabla f(\mathbf{t})$ is the sum of the first two terms of the multivariate Taylor series. The first $k+1$ terms of the Taylor series for a function f sum to a kth order polynomial that approximates f. If f is a kth or lower order polynomial, the approximation is exact.

To apply this affine approximation to optimization, suppose we are at the solution $\mathbf{t} \in \Re^n$ to the optimization problem $\max_{\mathbf{x}\in\Re^n} f(\mathbf{x})$. If we take a step from $\mathbf{t}$ to $\mathbf{t}+\mathbf{d}$, the objective function will change by $f(\mathbf{t}+\mathbf{d}) - f(\mathbf{t}) \approx \mathbf{d}\cdot\nabla f(\mathbf{t})$. (The smaller the step size $||d||$ the more accurate this approximation is apt to be.) For a given step length $\epsilon > 0$, what choice of $\mathbf{d}$ will increase the approximation of f the most? By the basic property of the dot product, $\mathbf{d}\cdot\nabla f(\mathbf{t}) = ||\mathbf{d}||||\nabla f(\mathbf{t})||\cos\theta$, where θ is the angle between $\mathbf{d}$ and $\nabla f(\mathbf{t})$. Therefore, the step $\mathbf{d}$ in the direction $\nabla f(\mathbf{t})$, namely $\epsilon\frac{\nabla f(\mathbf{t})}{||\nabla f(\mathbf{t})||}$ is the step of length ϵ that maximizes the increase in the approximation of f at $\mathbf{t}$. *Question: what step decreases the approximation of f the most?*[6]

If $\mathbf{t}$ is an optimum solution to $\max_{\mathbf{x}\in\Re^n} f(\mathbf{x})$ and f is differentiable, then $\nabla f(\mathbf{t}) = 0$. This fact is the generalization to n dimensions of the condition $f'(t) = 0$ which must hold at the maximum of a single variable differentiable function f.

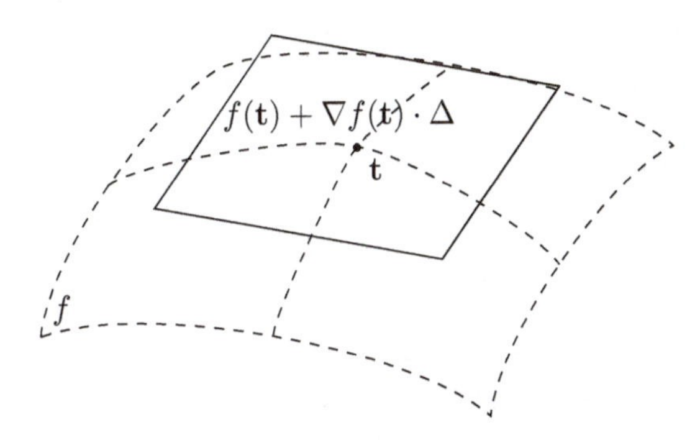

FIGURE 1.10: The gradient $\nabla f(\mathbf{t})$ of a two-variable function f at $\mathbf{t}$ provides a plane tangent to f at $\mathbf{t}$ which approximates f near $\mathbf{t}$ by the affine function $f(\mathbf{t}+\Delta) \approx f(\mathbf{t}) + \nabla f(\mathbf{t}) \cdot \Delta$.

1.2.4 Duality in Sudoku

To define Sudoku games, we first must define submatrices.

Definition 1.3 (Submatrix) *Let A be a matrix with row indices $j \in \{1 \ldots m\} \equiv J$ and column indices $k \in \{1 \ldots n\} \equiv K$. Then the matrix A' is a submatrix of A if and only if there exist $\phi \neq J' \subseteq J$ and $\phi \neq K' \subseteq K$, such that $A' = A_{J',K'}$. That is, the rows of A' are the rows of A that have indices in J', restricted to elements in columns that have indices in K'. Equivalently, the columns of A' are the columns of A that have indices in K', restricted to elements in rows that have indices in J'.*

Example 1.6 Let

$$A = \begin{bmatrix} 11 & 12 & 13 & 14 \\ 15 & 16 & 17 & 18 \\ 19 & 20 & 21 & 22 \\ 23 & 24 & 25 & 26 \end{bmatrix}.$$

Then the following arrays are all submatrices of A, with index sets J', K' equal to $\{1,2,3,4\}, \{1,2,3,4\}$; $\{2\}, \{3\}$; $\{1,3\}, \{3,4\}$; $\{2,4\}, \{1,2,4\}$, respectively.

$$A; \begin{bmatrix} 17 \end{bmatrix}; \begin{bmatrix} 13 & 14 \\ 21 & 22 \end{bmatrix}; \begin{bmatrix} 15 & 16 & 18 \\ 23 & 24 & 26 \end{bmatrix};$$

[6]The opposite step $\mathbf{d} = -\frac{\epsilon \nabla f(\mathbf{t})}{||\nabla f(\mathbf{t})||}$.

Question: how many submatrices does an $m \times n$ matrix have?[7]

Sudoku is a puzzle on a 9×9 matrix which is subdivided into nine 3×3 submatrices by lines between rows 3 and 4, rows 6 and 7, columns 3 and 4, and columns 6 and 7. Each cell of the matrix must contain an integer between 1 and 9. Initially a few of the cells contain integers and the rest are empty. The goal of the puzzle is to fill in the empty cells so that each integer occurs once in each of the following submatrices:

- $M_{i,:}$ $i = 1, 2, \ldots, 9$ (each column)
- $M_{:,j}$ $j = 1, 2, \ldots, 9$ (each row)
- $M\{i, i+1, i+2\}, \{j, j+1, j+2\}$ $i = 1, 4, 7; j = 1, 4, 7$ (each 3×3 subdivision)

Figure 1.11 shows an example of a Sudoku puzzle.

1	2		3		4			5
		9						
				8				
	1					2	9	
		2					7	
			1	7				
				9		4		
								6
								2

FIGURE 1.11: A Sudoku puzzle

Here is a useful duality in Sudoku. Define the solution to a Sudoku puzzle as a function $f : A \times B \mapsto C$ where

$A = \{1, 2, 3, 4, 5, 6, 7, 8, 9\}$, the set of possible values assigned to cells.

$B = \{(i, j) : 1 \leq i \leq 9; 1 \leq j \leq 9\}$, the set of cells.

$C = \{T, F\}$, representing "True" and "False."

$f(a, b) = T$ if value a is assigned to cell b; $f(a, b) = F$ if not.

For each integer $a \in A$, the function $g_a : B \mapsto C$ tells which cells a is assigned to. For each cell $b \in B$, the function $h_b : A \mapsto C$ tells what integer is assigned to b. These two sets of functions are duals of each other. Either set of functions describes the solution. Sometimes the $g_a()$ functions are more helpful, and sometimes the $h_b()$ are. In the example in Figure 1.11, think in terms of the $g_a()$ functions to find out where the 2 is in the center submatrix $M_{\{4,5,6\},\{4,5,6\}}$. One of its nine cells must contain a 2, but two are already occupied, and six of the remaining seven are ruled out by the 2's elsewhere in M. Hence $g_2((6, 6)) = f(2, (6, 6)) = T$. In contrast, think in terms of the dual $h_b()$ functions to find out what value must be assigned to cell $(1, 5)$, the middle cell in the top row. The entries in row 1 rule out $1, 2, 3, 4, 5$. In terms of the function h_b, $h_{(1,5)}(1) = h_{(1,5)}(2) = h_{(1,5)}(3) = h_{(1,5)}(4) = h_{(1,5)}(5) = F$. The entries in column 5 rule out $7, 8, 9$. In terms of the function h_b, $h_{(1,5)}(k) = F$ for $k = 7, 8, 9$. Therefore, $h_{(1,5)}(6) = f(6, (1, 5)) = T$. If you get stuck solving a Sudoku puzzle, observe your reasoning to see if you have been thinking only in terms of the $g_a()$ or only in terms of

[7] $(2^m - 1)(2^n - 1)$.

the $h_b()$. If so, try the dual reasoning. Students in one of my courses tried out this duality on several beginner-level and several expert-level puzzles. They were able to solve all of the easy puzzles using just the $g_a()$ point of view, but each difficult puzzle forced them to switch between the dual points of view.

1.2.5 Duality in Matrices

Describe an $m \times n$ matrix M as a function $f : A \times B \mapsto C$ where

$A = \{1, 2, \ldots m\}$, the row index set.

$B = \{1, 2, \ldots n\}$, the column index set.

$C = \Re$, the set of possible cell values.

$f(a, b) = M[a, b]$ for all $a \in A, b \in B$.

For each row index a, $g_a : B \mapsto \Re$ is the n-vector of values in row a. For each column index b, h_b describes the m-vector of values in column b. These two sets of functions, depicted in Figure 1.12, are duals of each other. Each fully describes the matrix M, one as a column of rows, the other as a row of columns. Neither description is better than the other. It is best to keep both descriptions in mind. In this book, M_{ij} denotes the element in row i and column j of the matrix M; $M_{i,:}$ denotes the ith row of M; $M_{:,j}$ denotes the jth column of M.

$$A \left\{ \begin{matrix} 1 \\ 2 \\ \vdots \\ m \end{matrix} \right. \overbrace{\begin{matrix} 1 \; 2 \; \cdots \; n \end{matrix}}^{B} \left[\quad \right] = \begin{bmatrix} g_1(B) \\ g_2(B) \\ \vdots \\ g_m(B) \end{bmatrix} = \begin{bmatrix} h_1(A) & h_2(A) & \ldots & h_n(A) \end{bmatrix}$$

FIGURE 1.12: A matrix is a set of rows, one for each column index value. Dually, it is a set of columns, one for each row index value.

Linear algebra provides relationships between the two descriptions of a matrix. (If you do not know the terms in the following theorem, consult the Glossary or Appendices.) One surprising but fundamental result is the following.

> **Theorem 1.2.1** *For any matrix M, the dimension of the space spanned by the rows of M equals the dimension of the space spanned by the columns of M.*

Note on terminology: the space spanned by the rows (columns) of M is often called the row space (column space). The dimension of the row (column) space is called the row (column) rank of M. Since the two are equal, they are also called the rank of M.

In the succeeding chapters we will often encounter terms of form $\mathbf{y}^T M \mathbf{x}$, where M is an $m \times n$ matrix, $\mathbf{y} \in \Re^m$, and $\mathbf{x} \in \Re^n$. Depending on whether we first multiply $\mathbf{y}^T$ by M or M by $\mathbf{x}$, we view this term as $(\sum_{j=1}^{n} \mathbf{y} \cdot M_{:,j}) \cdot \mathbf{x}$ or as $\mathbf{y} \cdot (\sum_{i=1}^{m} M_{i,:} \cdot \mathbf{x})$. The former is the dot product of a linear function of $\mathbf{y}$ with the vector $\mathbf{x}$; the latter is the dot product of a linear function of $\mathbf{x}$ with the vector $\mathbf{y}$. These two views of the same term will be key to duality in linear programming.

Reading Tip: *If you aren't sure you fully understand this duality, do one or more examples until you see the pattern. Start with a completely numerical example like*

$$M = \begin{bmatrix} 2 & 3 & 4 \\ 5 & 6 & 7 \end{bmatrix},$$

$\mathbf{y} = (10, 20)$, *and* $\mathbf{x} = (\sqrt{2}, \sqrt{5}, \pi)$. *(Notice I chose all different numbers.) Be sure you understand why the formula calls for* $\mathbf{y}^T$ *rather than* $\mathbf{y}$. *If this example is confusing, switch to an even simpler example where* M *is a* 1×3 *matrix.*

Summary: Duality gives you two different perspectives on the same thing. Applying duality to functions of two variables yields the concept of a partial derivative. Compute the partial derivative $\frac{\partial f}{\partial dx_i}$ by treating x_i as the only variable and $x_j : j \neq i$ as constants. The gradient $\nabla f \equiv \{\frac{\partial f}{\partial dx_1}, \ldots, \frac{\partial f}{\partial dx_n}\}$ of a function f points in the direction of maximum increase of f, and the dot product $\nabla f(\mathbf{t}) \cdot \mathbf{d}$ approximates the change in f when moving from $\mathbf{t}$ to $\mathbf{t} + \mathbf{d}$. The key duality of a matrix is that it is both a row of columns and a column of rows: $\mathbf{y}^T M$ is a linear combination of the rows of M; $M\mathbf{x}$ is a linear combination of the columns of M.

1.3 General Approaches to Solving Optimization Problems

Preview: How do we solve optimization problems?

The set of possible solutions to an optimization problem is usually infinite, or at least very large. It is typically either impossible or impractical to try out all possible solutions. In general, we solve optimization problems by finding mathematical conditions that must be satisfied by the optimal solutions, or by at least one optimal solution if there are multiple optima. These conditions should be such that they are satisfied by only a finite and hopefully small subset of the possible solutions. We restrict our search to this small subset. If we search this subset thoroughly, we are guaranteed to find an optimal solution if one exists.

For example, in the calculus problem from the beginning of the chapter, we wanted to maximize the function $f(y) = 50y - y^2$. The mathematical condition we imposed was $f'(y) = 0$. The reasoning behind this condition applies to the maximization of any real-valued differentiable function of a single variable. First, if f attains its maximum at y, then y must be a local maximum of f. In mathematical terms, if $f(y) \geq f(z)\ \forall z$, then there exists $\epsilon > 0$ such that $f(y) \geq f(z)\ \forall z$ in the range $y - \epsilon \leq z \leq y + \epsilon$. Second, since f is differentiable, if y is a local maximum of f, then $f'(y) = 0$. The condition $f'(y) = 0$ is called a *necessary* condition because every optimal solution must satisfy it. The condition is not *sufficient* because a solution might satisfy the condition yet not be optimal.

The reverse directions of this reasoning would not be correct. First, many differentiable functions have local maxima that are not maxima. *Question: give an example.*[8] Second, for many functions there is a point at which the derivative is zero, but the point is not a local maximum. *Question: give an example.*[9] These examples show that the condition is, in general, not sufficient to guarantee optimality. That is OK, because the purpose of the condition is to restrict the search for an optimal solution to a manageably small set of possibilities. The specific calculus problem we solved was so simple that only one solution satisfied the condition. Usually there are several such solutions and each must be checked.

As derived in Section 1.2, in multivariate calculus the variable x generalizes to a vector $\mathbf{x}$ of variables. The condition $f'(y) = 0$ for optimization without constraints generalizes to the condition $\nabla f(\mathbf{x}) = 0$. The most commonly used conditions to restrict the search for an optimal solution to an unconstrained optimization problem are $\nabla f = 0$ and being a local optimum. The latter condition is stronger than the former if f is continuously differentiable. On the other hand the latter condition applies to nondifferentiable functions such as $f(x) = |x|$. The precise definition of a local optimum is as follows:

Definition 1.4 *The point $\mathbf{x}^*$ is a local minimum (respectively maximum) of $f : \Re^n \mapsto \Re$ if there exists $\epsilon > 0$ such that $||\mathbf{x}^* - \mathbf{x}|| < \epsilon \Rightarrow f(\mathbf{x}^*) \leq f(\mathbf{x})$ (respectively $f(\mathbf{x}^*) \geq f(\mathbf{x})$). A local minimum (maximum) in a minimization (maximization) problem is called a local optimum.*

Local Optima and Convex Functions

Convex functions compose a profoundly important class which includes linear functions.

Definition 1.5 *The function $f : \Re^n \mapsto \Re$ is convex if for all $\mathbf{x} \neq \mathbf{y}$ and $0 < \lambda < 1$, $f(\lambda\mathbf{x} + (1-\lambda)\mathbf{y}) \leq \lambda f(\mathbf{x}) + (1-\lambda)\mathbf{y}$. If the inequality always holds strictly f is strictly convex. The function g is (strictly) concave iff the function $-g$ is (strictly) convex.*

Geometrically, a function is convex if the line segment between any two points on the graph of the function lies on or above the function (see Figure 1.13). All linear functions are convex but not strictly convex since the inequality in the definition holds with equality everywhere. The function in Figure 1.13 is not strictly convex because it behaves linearly within some ranges. For example, for small $\epsilon > 0$, the line segment between $(x_1^{'}, f(x_1^{'}))$ and $(x_1^{'} + \epsilon, f(x_1^{'} + \epsilon))$ lies on the graph of f.

Example 1.7 The functions $f(x) = \frac{1}{2}x^2 + x - 3$ (Figure 1.14) and $f(x_1, x_2) = (x_1 - .2)^2 + (x_2 + .5)^2$, (see Figure 1.15) are strictly convex. The function $f(x_1, x_2) = 11x_1 + 3x_2^2$ (Figure 1.16) is convex but not strictly convex because f behaves linearly if x_2 is fixed and x_1 changes.

► *Here is a definition equivalent to 1.5. See Figure 1.13*

[8] $f(y) = y + \sin y - \cos y$ at $y = \pi/2$.
[9] 0 is a local minimum of $f(y) = y^2$ and neither a local maximum nor minimum of $f(y) = y^3$.

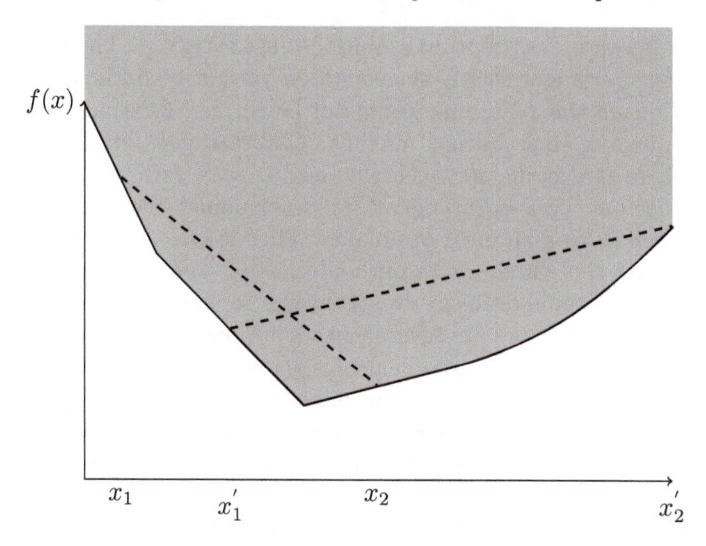

FIGURE 1.13: For all x_1, x_2 the (dashed) line segments between $(x_1, f(x_1))$ and $(x_2, f(x_2))$ lie on or above the graph of the convex function $f(x)$. Equivalently, the shaded region above the graph of f is a convex set.

Definition 1.6 *The epigraph of a function $f : \Re^n \mapsto \Re$ is the $n+1$-dimensional set epi$(f) \equiv \{(\mathbf{x}, y) : y \geq f(\mathbf{x})\}$. Geometrically the epigraph is the set of points on or above the graph of the function f. Then f is a convex function iff epi(f) is a convex set.*

Reading Tip: *I've asserted the equivalence of Definitions 1.5 and 1.6, without proof. In general, an assertion unaccompanied by a proof in a mathematical book should be provable by plugging in definitions, with no additional mathematical reasoning. In math jargon, Definition 1.6 is a "mere restatement" of Definition 1.5. Do not skip over restatements. Prove their correctness. Doing so will help you master concepts and definitions. Writing your proof in full is also an excellent exercise in writing formal proofs because all of your focus is on the formalism.*

◀

Convex functions have the very useful property that a local minimum is a global minimum. Definition 1.5 implies that maxima of concave functions behave like minima of convex functions.

Theorem 1.3.1 *If $\mathbf{x}^*$ is a local minimum of a convex function $f : \Re^n \mapsto \Re$ and $\mathbf{x}^* \in \mathcal{S} \subseteq \Re^n$ then $\mathbf{x}^*$ is an optimal solution to the problem*

$$\min_{\mathbf{x} \in \mathcal{S}} f(\mathbf{x}).$$

Proof: Since $\mathbf{x}^* \in \mathcal{S}$, the point $\mathbf{x}^*$ is feasible. Suppose to the contrary there exists $\mathbf{y}^* \in \mathcal{S}$ with $f(\mathbf{y}^*) < f(\mathbf{x}^*)$. We will obtain a contradiction to the local minimality of $\mathbf{x}^*$. The geometric idea is to examine the values of f on the segment between $(\mathbf{x}^*, f(\mathbf{x}^*))$ and $(\mathbf{y}^*, f(\mathbf{y}^*))$. The segment must slope downwards from $(\mathbf{x}^*, f(\mathbf{x}^*))$ and must lie on or above the function. Hence the function must decrease from $\mathbf{x}^*$, contradicting local minimality. To translate the geometric idea into precise algebra, let $f(\mathbf{y}^*) < f(\mathbf{x}^*)$ and let $\epsilon > 0$ as guaranteed by Definition 1.4. If $||\mathbf{x}^* - \mathbf{y}^*|| < \epsilon$ then by definition $f(\mathbf{y}^*) \geq f(\mathbf{x}^*)$, a contradiction. Otherwise let $\lambda = \frac{\epsilon}{2||\mathbf{x}^* - \mathbf{y}^*||}$. Consider the point $\mathbf{z} = \lambda \mathbf{y}^* + (1 - \lambda)\mathbf{x}^*$ which by construction is on the

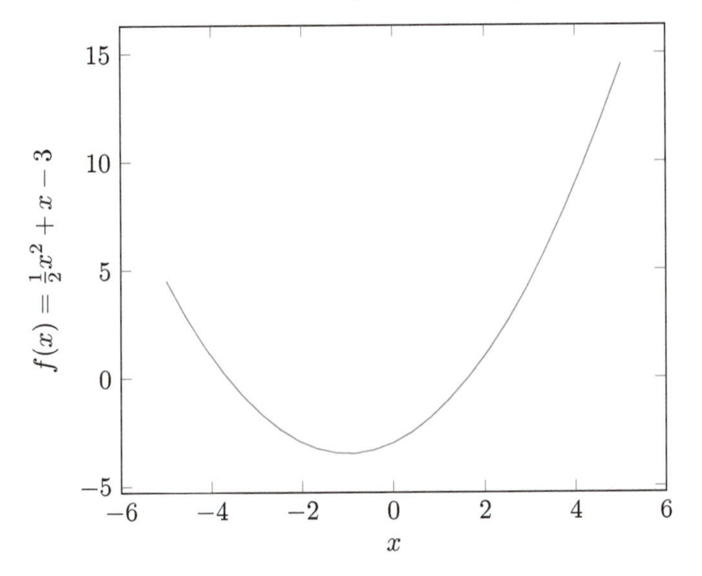

FIGURE 1.14: Strictly convex function from $\Re \mapsto \Re$.

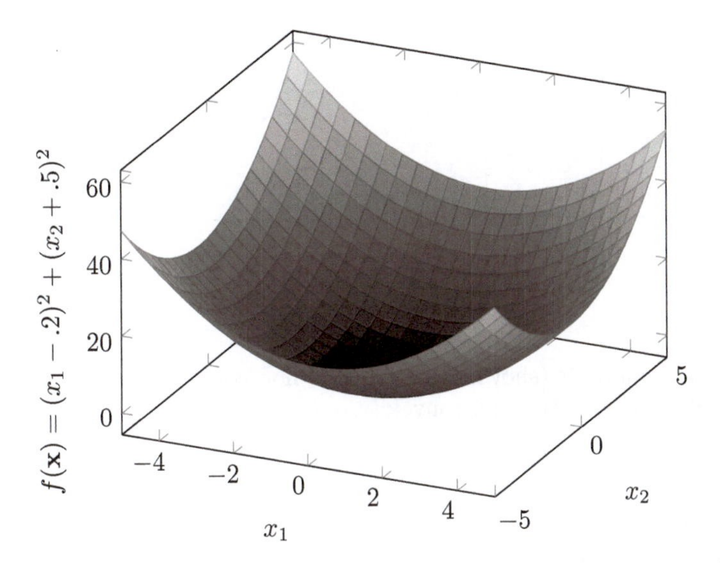

FIGURE 1.15: Strictly convex function from $\Re^2 \mapsto \Re$.

line segment between $\mathbf{x}^*$ and $\mathbf{y}^*$, within distance $\epsilon/2$ of $\mathbf{x}^*$. On the one hand, by convexity $f(\mathbf{z}) \leq \lambda f(\mathbf{y}^*) + (1-\lambda)f(\mathbf{x}^*) = f(\mathbf{x}^*) + -\lambda(f(\mathbf{x}^*) - f(\mathbf{y}^*)) < f(\mathbf{x}^*)$. On the other hand, $||\mathbf{x}^* - \mathbf{z}|| < \epsilon \Rightarrow f(\mathbf{z}) \geq f(\mathbf{x}^*)$, a contradiction. ■

Corollary 1.3.2 *If the function f in Theorem 1.3.1 is strictly convex, then $\mathbf{x}^*$ is the unique optimal solution.*

Proof: If $\mathbf{x}^* \neq \mathbf{y}^*$ and $f(\mathbf{x}^*) = f(\mathbf{y}^*)$, then by strict convexity $f(\frac{\mathbf{x}^*+\mathbf{y}^*}{2}) < f(\mathbf{x}^*)$. Then $\mathbf{x}^*$ would not be an optimal solution to the problem

$$\min_{\mathbf{x} \in \Re^n} f(\mathbf{x}),$$

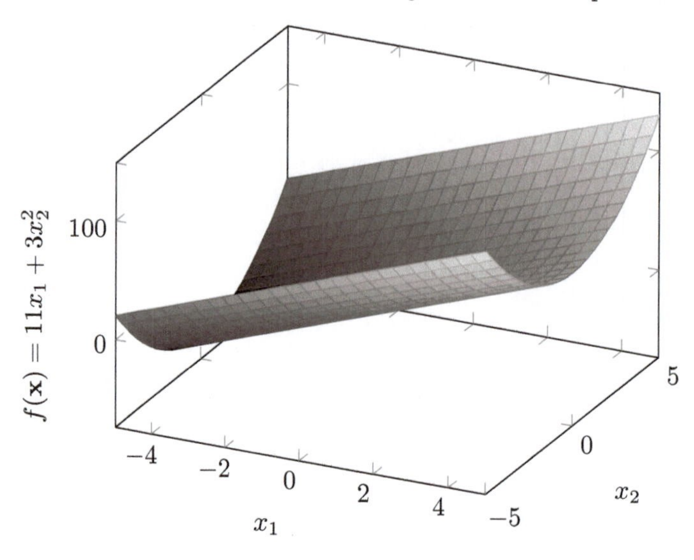

FIGURE 1.16: Non-strictly convex function from $\Re^2 \mapsto \Re$.

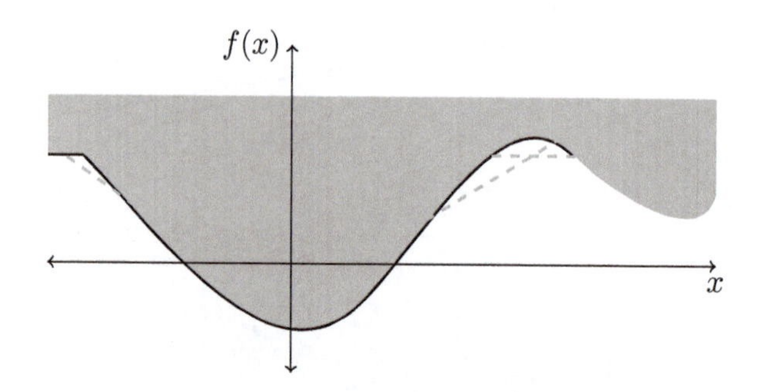

FIGURE 1.17: The epigraph (shown shaded) of a non-convex function is not convex. Dashed line segments illustrate the non-convexity.

a contradiction to Theorem 1.3.1. ■

Reading Tip: *Look at the conditions of the theorem. Why is* $\mathbf{x}^* \in \mathcal{S}$ *a condition? Why is* f *convex a condition? Draw an example where* f *is not convex and* $\mathbf{x}^*$ *is not an optimal solution. Why is* $\mathcal{S}$ *not required to be a convex set? In particular, why isn't* $\mathbf{z} \in \mathcal{S}$ *required?*

Theorem 1.3.1 is the principal reason convex function minimization has a special place in the field of optimization. Indeed, there is a powerful duality theory for convex functions, developed mainly by Fenchel [93, 94] and Rockafellar [251, 252, 253]. But there is no corresponding strong theory for general functions.

1.3.1 Saddlepoints and Lagrange's Method

Constraints often make an optimization problem very tough to solve. One general strategy for handling constraints is to convert them into costs or penalties in the objective function. The classical example of this strategy is Lagrange's multiplier method for optimization subject to equality constraints. Lagrange studied the optimization problem

$$\text{minimize} \quad f(\mathbf{x}) \quad \text{subject to} \tag{1.8}$$

$$h_i(\mathbf{x}) = 0 : \quad 1 \le i \le m. \tag{1.9}$$

His genius was to construct a new function, called the *Lagrangian* in his honor, out of f and the h_i. The Lagrangian employs a vector $\boldsymbol{\pi}$ of m additional variables (sometimes called *Lagrange multipliers*) and is defined as

$$L(\mathbf{x}, \boldsymbol{\pi}) \equiv f(\mathbf{x}) - \sum_{i=1}^{m'} \pi_i h_i(\mathbf{x}). \tag{1.10}$$

The first part of the intuitive idea of the Lagrangian is to add a cost for violation of the constraints (1.9) instead of directly enforcing them. The optimizer must minimize the value of $L(\mathbf{x}, \boldsymbol{\pi})$ instead of $f(\mathbf{x})$, but he is no longer required to satisfy the constraints (1.9). Let $\mathbf{x}^*$ be an optimal solution to (1.8),(1.9). Since by definition $\mathbf{x}^*$ is feasible, the optimizer can set $\mathbf{x} = \mathbf{x}^*$ to get $L(\mathbf{x}, \boldsymbol{\pi}) = f(\mathbf{x}^*)$ regardless of the value of $\boldsymbol{\pi}$. That is good, but not enough. We want minimizing the Lagrangian to be equivalent to solving (1.8),(1.9). How do we prevent the optimizer from getting a value less than $f(\mathbf{x}^*)$?

The second part of the intuitive idea is to let an adversary control $\boldsymbol{\pi}$. The adversary wants the Lagrangian to be large rather than small. She can't prevent $L(\mathbf{x}, \boldsymbol{\pi}) \le f(\mathbf{x}^*)$ but she tries to prevent $L(\mathbf{x}, \boldsymbol{\pi}) < f(\mathbf{x}^*)$. If she is smart and the optimizer is smart, they will choose values $\bar{\boldsymbol{\pi}}$ and $\bar{\mathbf{x}}$ respectively so that changing $\bar{\boldsymbol{\pi}}$ won't increase the Lagrangian (she can't do better versus $\bar{\mathbf{x}}$), and changing $\bar{\mathbf{x}}$ will not decrease the Lagrangian (he can't do better versus $\bar{\boldsymbol{\pi}}$). Such a solution is called a *saddlepoint*, which term deserves a formal definition.

Definition 1.7 (Saddlepoint) *Let $L(\mathbf{x}, \boldsymbol{\pi})$ be a real-valued function of vector variables $\mathbf{x}$ and $\boldsymbol{\pi}$. The point $(\bar{\mathbf{x}}, \bar{\boldsymbol{\pi}})$ is a saddlepoint of L if*

$$\forall \mathbf{x}, \ \forall \boldsymbol{\pi} \quad L(\bar{\mathbf{x}}, \boldsymbol{\pi}) \le L(\bar{\mathbf{x}}, \bar{\boldsymbol{\pi}}) \le L(\mathbf{x}, \bar{\boldsymbol{\pi}}). \tag{1.11}$$

Example 1.8 Return again to the area maximization problem of Sections 1.1 and 1.2.1. As before, let x_1 and x_2 denote the width and length of the rectangle, respectively. We ignore the inequality constraints $x_1 \ge 0, x_2 \ge 0$ until a later chapter. The perimeter $2x_1 + 2x_2$ is constrained to equal 100. The Lagrangian is

$$L(x_1, x_2, \pi_1) = x_1 x_2 - \pi_1(2x_1 + 2x_2 - 100)$$

and its gradient is

$$\nabla L(x_1, x_2, \pi_1) = \begin{pmatrix} x_2 - 2\pi_1 \\ x_1 - 2\pi_1 \\ 2x_1 + 2x_2 - 100 \end{pmatrix}$$

The solution to $\nabla L = \mathbf{0}$ is $x_1 = x_2 = 25, \pi_1 = 12.5$. We've derived the optimal x_i values twice already. The Lagrangian gives us more information: *The cost of enforcing the constraint* $2x_1 + 2x_2 = 100$ *is* $\pi_1 = 12.5$. If the perimeter 100 increased by an infinitesimal amount d, the optimal area would increase almost exactly by $12.5d$. *Question: verify that the area would increase by the amount stated.*[10] Figure 1.7 shows that the effects

[10] x_1, x_2 each increase by $\frac{d}{4}$. The area increases to $(25 + \frac{d}{4})^2 = 25^2 + 12.5d + \frac{d^2}{16}$.

of increasing the width x_1 and the length x_2 are equal only when $x_1 = x_2$. Setting $\nabla L = \mathbf{0}$ forces $x_2 = 2\pi_1 = x_1$, giving a square shape at the optimum. In essence, the Lagrangian requires π_1 to have a consistent meaning, a specific numerical value, which can only be when the shape is square, $x_1 = x_2$.

Example 1.9 Let $f(\mathbf{x}) = (x_1 - x_2)^2$, $h_1(\mathbf{x}) = x_1 - 5$, and $h_2(\mathbf{x}) = x_2 - 2$. The only feasible solution is $\mathbf{x}^* = (5, 2)$ with $f(\mathbf{x}^*) = 9$. If $\boldsymbol{\pi} = (0, 0)$ then $L(1, 1, \boldsymbol{\pi}) = 0 < 9$. If $\boldsymbol{\pi} = (4, 3)$ then $L(2, 2, \boldsymbol{\pi}) = 0 - 4(2 - 5) - 3(2 - 1) = 6 < 9$. Many other values of $\boldsymbol{\pi}$ also allow the optimizer to do better than $f(\mathbf{x}^*)$. But it is possible to prevent a value less than 9. Set $\boldsymbol{\pi} = (6, -6)$. Then $L(x_1, x_2, \boldsymbol{\pi}) = (x_1 - x_2)(x_1 - x_2 - 6) + 18$. Substitute $y = x_1 - x_2$ and solve the resulting single-variable optimization problem to get $x_1 - x_2 = y = 3 \Rightarrow L(x_1, x_2, \boldsymbol{\pi}) = 9$. This example illustrates that $\boldsymbol{\pi}$ must be selected carefully to foil the optimizer. The saddlepoint $(\bar{\mathbf{x}}, \bar{\boldsymbol{\pi}}) = (5, 2,\ 6, -6)$.

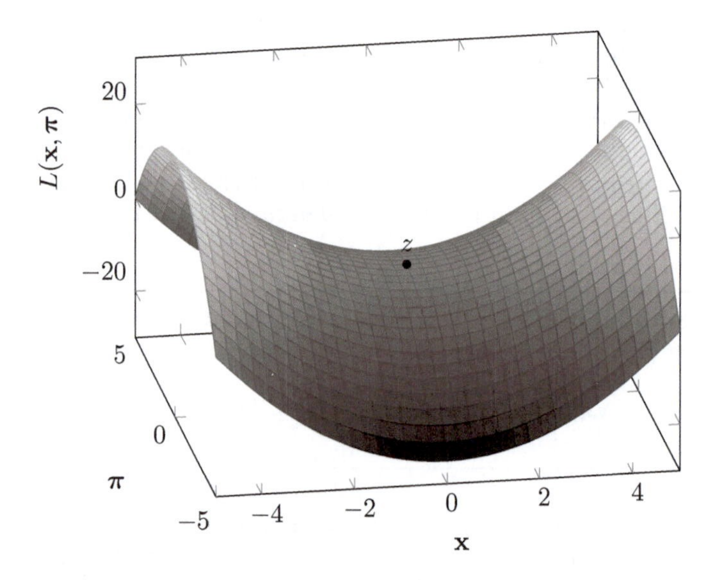

FIGURE 1.18: At the saddlepoint $z = (\mathbf{x}, \boldsymbol{\pi})$, neither player can improve his or her situation. The function $L(\mathbf{x}, \boldsymbol{\pi})$ increases if the minimizer moves $\mathbf{x}$ left or right; the function decreases if the maximizer moves $\boldsymbol{\pi}$ forward or back.

Figure 1.18 illustrates the geometric meaning of a saddlepoint, marked as z in the figure, for the function $x^2 - \pi^2$. Think of the figure as the side view of a saddle as on a horse facing to your right. The $\mathbf{x}$ dimension runs from the back to the front of the saddle; the $\boldsymbol{\pi}$ dimension runs from side to side; the vertical dimension (the height) is the value of the function L. At the saddlepoint z, the adversary has chosen $\bar{\boldsymbol{\pi}} = 0$ to bisect the saddle, separating its left half from its right half. At that value of $\boldsymbol{\pi}$, the optimizer (who minimizes) faces a convex parabolic curve defined by the bisecting plane's intersection with the saddle. The best he can do is to choose the $\mathbf{x}$ value at the saddlepoint. Conversely, at the saddlepoint the optimizer has chosen $\bar{\mathbf{x}} = 0$ to cut the saddle in half, separating the front from the back. The adversary (who maximizes) faces a concave parabolic curve. The best she can do is to choose the $\bar{\boldsymbol{\pi}} = 0$ value at the saddlepoint.

The first-order conditions consistent with the inequalities (1.11) are that the partial derivatives of $L(\mathbf{x}, \boldsymbol{\pi})$ with respect to each $\mathbf{x}_i$ and each $\boldsymbol{\pi}_j$ are all zero. This is succinctly expressed as $\nabla L(\mathbf{x}, \boldsymbol{\pi}) = 0$. The portion of these equations involving the derivatives with respect to the $\boldsymbol{\pi}$ variables are precisely the feasibility constraints $h_j(\mathbf{x}) = 0$. The novelty of these conditions lies in the other equations.

Algorithm 1.3.3 Lagrange's Method for Equality Constrained Optimization
To solve

$$\min f(\mathbf{x}) \quad \textit{subject to} \quad h_j(\mathbf{x}) = 0 : \; j = 1, \ldots, m$$

where $\mathbf{x} \in \Re^n$ *and* $f : \Re^n \mapsto \Re$ *and* $\forall i \;\; h_i : \Re^n \mapsto \Re$ *are real valued functions, define*

$$L(\mathbf{x}, \boldsymbol{\pi}) \equiv f(\mathbf{x}) - \sum_{j=1}^{m} \pi_j h_j(\mathbf{x})$$

where $\boldsymbol{\pi} \in \Re^m$ *and solve*

$$\nabla L(\mathbf{x}, \boldsymbol{\pi}) = 0$$

for $\mathbf{x}$ *and* $\boldsymbol{\pi}$.

Example 1.10 Minimize $x_1^2 + 4x_2^2 - 40x_2$ subject to $x_1 = 12 + 2x_2$.
Solution: Define $h_1(\mathbf{x}) = x_1 - 2x_2 - 12$.
$L(\mathbf{x}, \boldsymbol{\pi}_1) = x_1^2 + 4x_2^2 - 40x_2 - \boldsymbol{\pi}_1(x_1 - 2x_2 - 12)$. $\nabla L(\mathbf{x}, \boldsymbol{\pi}_1) = 0$ is equivalent to

$$\begin{aligned} 2x_1 - \pi_1 &= 0; \\ 8x_2 - 40 + 2\pi_1 &= 0; \\ x_1 - 2x_2 - 12 &= 0; \end{aligned}$$

Combine the 1st and 2nd equations to get $2x_1 = \pi_1 = -4x_2 + 20 \Rightarrow x_1 = -2x_2 + 10$. Plug into the 3rd equation: $-2x_2 + 10 - 2x_2 - 12 = 0 \Rightarrow x_2 = -0.5 \Rightarrow x_1 = 11 \Rightarrow \pi_1 = 22$. The objective value is $121 + 1 + 20 = 142$.

It is evident what the values of $x_1, \ldots, x_n$ found by Lagrange's method mean. They are the solution to the optimization problem (1.8),(1.9). The values of the additional variables $\pi_1, \ldots, \pi_m$ have an important but less obvious meaning. *They are the costs of enforcing the equality constraints* (1.9). To see why this must be so, we first need a precise definition.

Definition 1.8 (Cost of Enforcing a Constraint) *Let* v^* *be the optimal value of an optimization problem of form* (1.4),(1.5),(1.6). *Select a constraint* $g_j(\mathbf{x}) \geq 0$ *(or* $h_j(\mathbf{x}) = 0$*). For* $\epsilon < 0$ *let* $v^*(\epsilon)$ *be the optimal value of the same optimization problem except that the selected constraint is changed to* $g_j(\mathbf{x}) \geq \epsilon$ *(or* $h_j(\mathbf{x}) = \epsilon$*). (The value of an infeasible problem is treated as* ∞*.) The cost of enforcing the selected constraint is*

$$\lim_{\epsilon \to 0} \frac{v^*(\epsilon) - v^*}{\epsilon}$$

if the limit exists.

Example 1.11 For the problem $\min 300x_1$ subject to $4x_1 \geq 31$ the cost of enforcing the constraint is $300/4 = 75$. For the problem $\min x_1^2$ subject to $x_1 \geq 5$, the cost of enforcing the constraint is 10 because $\frac{df}{dx_1} = 10$ at $x_1 = 5$. If the constraints were changed from inequality to equality constraints the costs would be the same.

Question: If $g_j(\mathbf{x}^*) > 0$ for an optimal solution $\mathbf{x}^*$, what is the cost of enforcing constraint $g_j(\mathbf{x}) \geq 0$?[11]

Theorem 1.3.4 *If Algorithm 1.3.3 finds a saddlepoint $(\bar{\mathbf{x}}, \bar{\boldsymbol{\pi}})$ of L, then $\bar{\mathbf{x}}$ is an optimum solution and the values $\bar{\boldsymbol{\pi}}$ equal the costs of enforcing the constraints $h_j(\mathbf{x}) = 0$ at $\bar{\mathbf{x}}$.*

Proof: For the Lagrangian of the optimization problem with equality constraints (1.4),(1.6) suppose the maximizing player gives the variable π_j a value α strictly more than the cost of enforcing $h_j(\mathbf{x}) = 0$ at an optimal solution $\mathbf{x}^*$. By definition there exists $\epsilon > 0$ and a solution $\mathbf{x}^*(\epsilon)$ with $\frac{f(\mathbf{x}^*(\epsilon)) - f(\mathbf{x}^*)}{\epsilon} < \alpha$ and $\mathbf{x}^*(\epsilon)$ feasible with respect to (1.6) except that $h_j(\mathbf{x}^*(\epsilon)) = \epsilon$. Then

$$L(\mathbf{x}^*(\epsilon), \boldsymbol{\pi}) = f(\mathbf{x}^*(\epsilon)) - \alpha\epsilon < f(\mathbf{x}^*) = L(\mathbf{x}^*, \boldsymbol{\pi}).$$

Therefore, if the maximizing player who assigns values to $\boldsymbol{\pi}$ makes π_j greater than the cost of enforcing constraint $h_j(\mathbf{x}) = 0$, the minimizing player can make the Lagrangian less than $f(\mathbf{x}^*)$.

Similarly, if the maximizing player sets π_j to a value β strictly less than the cost of enforcing $h_j(\mathbf{x}) = 0$ at $\mathbf{x}^*$, then there exists $\epsilon < 0$ and a solution $\mathbf{x}^*(\epsilon)$ with $\frac{f(\mathbf{x}^*) - f(\mathbf{x}^*(\epsilon))}{|\epsilon|} > \beta$ and $\mathbf{x}^*(\epsilon)$ feasible with respect to (1.6) except that $h_j(\mathbf{x}^*(\epsilon)) = \epsilon$. Then

$$L(\mathbf{x}^*(\epsilon), \boldsymbol{\pi}) = f(\mathbf{x}^*(\epsilon)) - \beta\epsilon = f(\mathbf{x}^*(\epsilon)) + \beta|\epsilon| < f(\mathbf{x}^*) = L(\mathbf{x}^*, \boldsymbol{\pi}).$$

Therefore, if the maximizing player sets π_j less than the cost of enforcing $h_j(\mathbf{x}) = 0$, the minimizing player can make the Lagrangian less than $f(\mathbf{x}^*)$.

The minimizing player can always achieve $L(\mathbf{x}, \boldsymbol{\pi}) = f(\mathbf{x}^*)$, regardless of the maximizer's choice of $\boldsymbol{\pi}$, by choosing $\mathbf{x} = x^*$. On the other hand, the algorithm cannot select $\mathbf{x}$ such that $f(\mathbf{x}) < f(\mathbf{x}^*)$ since the $\frac{\partial f}{\partial \pi_j} = 0$ constraints enforce feasibility of $\mathbf{x}$. ∎

Example 1.12 The optimal solution to $\min 50x_1$ subject to $x_1 = 10$ is $x_1^* = 10$ with value 500. The Lagrangian function is $L(\mathbf{x}_1, \pi_1) = 50x_1 - \pi_1(x_1 - 10)$. At $\pi_1 = 51$, set $x_1 = 11$ to get $L(11, 51) = 550 - 51 = 499 < 500$. At $\pi_1 = 49$, set $x_1 = 9$ to get $L(9, 49) = 450 - 49(-1) = 499 < 500$.

Example 1.13 *Continuation of Example 1.10.*
The value $\pi_1 = 22$ tells us that the cost of enforcing the constraint $h_1(\mathbf{x}) = 0$ is 22. If $h_1(\mathbf{x}) = 0$ is changed to $h_1(\mathbf{x}) = 1 \Leftrightarrow x_1 = 2x_2 + 13$, the estimated impact on the objective value is an increase of 22.
To calculate the exact impact, re-solve the problem with $h_1(\mathbf{x}) = x_1 - 2x_2 - 13$. As

[11] Zero.

before, $x_1 = -2x_2 + 10$. Plug into $h_1(\mathbf{x}) = 0$ to get $-2x_2 + 10 - 2x_2 - 13 = 0 \Rightarrow x_2 = -0.75 \Rightarrow x_1 = 11.5$. The objective value is $132.25+2.25+30 = 164.5$, compared with the estimated value $142 + 22 = 164$. The estimate is too large by a mere $\frac{164.5-164}{164} \approx 0.3\%$. The value $\boldsymbol{\pi}_1 = 22$ is the instantaneous rate of change in the objective if $h_1(\mathbf{x})$ changes. Therefore, it should tend to give very accurate estimates for very small changes in $h_1(\mathbf{x})$, but it is apt to be less accurate for large changes in $h_1(\mathbf{x})$. To illustrate, calculate the objective value subject to the constraint $x_1 - 2x_2 - 20 = 0$. The predicted increase in objective value is $(20 - 12)(22) = 176$. To get the exact value, again $x_1 = -2x_2 + 10$. Substitute into $x_1-2x_2-20 = 0$ to get $-2x_2+10-2x_2-20 = 0 \Rightarrow x_2 = -2.5 \Rightarrow x_1 = 15$. The objective is $225 + 18.75 + 100 = 343.75$, an increase of 201.75. The estimate is too small by $\frac{343.75-318}{343.75} \approx 7.5\%$.

Classically, Lagrange's method is justified by the implicit function theorem of calculus. In Chapter 4 we will use the matrix duality of Section 1.2.5 to justify Lagrange's method.

Lagrange's method turns a constrained optimization problem into the problem of solving a system of simultaneous equations. This calls for an additional variable for each constraint. A related idea that does not appear to require additional variables is to replace constraints with penalty terms in the objective. A penalty term should be zero if the corresponding constraint is satisfied. For example, one could add the term $\sum_{j=1}^{m} K(g_j(\mathbf{x}))^2$ to a minimization objective, where K is a large positive number. In practice, implementations usually solve the problem multiple times, letting K increase, having proved that the solution to the penalty problem converges to a solution of the original constrained problem as $K \to \infty$. *Question: Is K a variable or a constant?*[12] We call K a *parameter*, something that is a constant from one perspective and a variable from another perspective. Each time we solve the penalty problem for a specific value of K, we think of K as a constant. But from the perspective of the overall solution procedure, K is a variable. The penalty terms must be "nice" enough that it is computationally feasible to minimize the original objective function plus the penalty functions. Thus, penalty terms are ordinarily twice-differentiable, and often are convex. A similar idea for inequality constraints $g_j(\mathbf{x}) \geq 0$ is to make a barrier at $g_j(\mathbf{x}) = 0$ with a cost function that increases without bound, e.g., $-K \log g_j(\mathbf{x})$, as $g_j(\mathbf{x})$ approaches 0.

The next few chapters of this book are about linear programming problems, which have infinitely many possible solutions. Chapter 4 will define a finite subset of these solutions, called *basic* solutions, and prove that if there exists an optimal solution, then there exists an optimal solution that is basic. Being basic is a mathematical condition that helps make it manageable to solve linear programming problems. The subsequent Chapter 5 derives the simplex method, a procedure that traverses a sequence of basic solutions *en route* to an optimal basic solution.

When it isn't obvious how to find solutions that meet the mathematical conditions that must be satisfied by some or all of the optimal solutions, we usually search for them numerically. Many search methods are iterative improvement algorithms. Start at iteration $i = 1$ with some solution $\mathbf{x}^1$. Perform a computation that finds a new solution $\mathbf{x}^{i+1}$ that is better than $\mathbf{x}^i$, that comes closer to satisfying the mathematical conditions. Increment i and repeat until $\mathbf{x}^i$ satisfies the conditions exactly or to the desired degree of accuracy. When there are multiple mathematical conditions, a search method might require all $\mathbf{x}^i$ to satisfy a few of the conditions and work towards satisfying the others. For example, in linear programming we will see that if an optimal solution exists, there must be an optimal solution that satisfies three conditions, called primal feasibility, dual feasibility, and

[12] Both. Keep reading.

complementary slackness. One algorithm, the simplex algorithm, always enforces the first and third conditions, and works towards achieving the second. Another algorithm, the dual simplex algorithm, always enforces the second and third conditions, and works towards achieving the first. Some other algorithms always enforce the first and second conditions, and work towards achieving the third. If you want to invent a novel algorithm for a problem, analyze the known algorithms from this point of view to see if there is a combination that has not been tried.

Another idea that is sometimes used instead of iterative improvement is to start with a large region to search and repeatedly eliminate portions of the search region from consideration, until what remains is small enough to be dealt with directly. The tricky part of implementing this idea can be to maintain a succinct description of the remaining search region. The ellipsoid algorithm is a good example in which this difficulty is overcome. The branch-and-bound trees of integer programming, which can become unmanageably large, are a good example in which this difficulty is not always overcome.

Yet another idea is to solve a simple problem related to the one at hand, and gradually modify the simple problem to become the true one, tracking the change in the solution as the modifications are made. Homotopy methods and cutting plane methods in integer programming can be described in this way.

Summary: Optimization problems are usually solved in two steps. First, determine mathematical conditions that must be satisfied by (all or some) optimal solutions. Second, search the solutions that satisfy these conditions. Another key idea, exemplified by Lagrange's method for equality constraints, is to move constraints into the objective function. The additional variables in Lagrange's method take values equal to the costs of enforcing the constraints. Other key ideas include iterative improvement to search for solutions, penalty and barrier functions, iterative reduction in the search space, and tracking the solution as a simple problem is modified to become the actual problem.

1.4 Notes and References

Loomis and Sternberg's definition of duality [214] encompasses many dualities. Generally speaking, it applies when there are two types of objects, such as points and lines, individuals and social groups, and there is some notion of intersection between an element of the first type and an element of the second type. In the cases just mentioned, a point can be on a line, or equivalently a line can pass through a point; a group has various individuals as members, but equivalently an individual belongs to various groups. A point may be characterized by the lines that pass through it, while a line may be characterized by the points it passes through; as Breiger [33] observes, people may be characterized by the set of social groups they belong to, while a social group may be characterized by the set of people who belong to it. In these dualities, the essence is the intersections themselves. See the I Problems for more examples.

On the other hand, this does not encompass all dualities. As best I understand, it does not encompass the wave/particle duality of light in quantum mechanics. Also, it is not equivalent to the mathematical duality between a set S and its complement $\overline{S}$. According to that duality, the essence is neither S nor $\overline{S}$. Rather, it is their distinction, the separation of the universal set into S and $\overline{S}$, that is essential. This is the duality between a graph and its complement, and it is Yin-Yang duality, according to which the essences of light, good, alive, and I are the distinctions between light and dark, good and bad, alive and dead, and I and not I, respectively.

I've chosen to emphasize the economic meaning of the dual variables by defining the cost of enforcing a constraint. The usual way to define duals in nonlinear programming is in terms of perturbations to the problem. This is essentially the same definition with a difference in emphasis. Either way, whether or not the dual behaves well is a matter of stability of the problem with respect to perturbations. If a small perturbation changes the problem from having a finite optimum to having an unbounded optimum, the problem is not stable.

1.5 Problems

E Exercises

1. For the problem Maximize $x + y$ subject to $0 \leq x \leq 1; 0 \leq y \leq 1$, find a feasible solution, an infeasible solution, a superoptimal solution, a feasible suboptimal solution, an infeasible suboptimal solution, and the unique optimal solution.

2. For each of the following problems, identify: the variables, the objective function, the constraints, a feasible solution, an infeasible solution, and the feasible region.

   (a) Minimize $2x_1 - x_2$ subject to $x_1 \geq 0; x_1 + x_2 \leq 10; x_2 \geq 3$.

   (b) Maximize x^3 subject to $-1 \leq x \leq 1$.

   (c) Minimize $\log z_1 + z_2^2$ subject to $z_1 \geq 1$ and $z_1 + z_2 \leq 4$.

   (d) Maximize $u_1 - u_2$ subject to $|u_2| \geq 0$ and $u_1^2 + u_2 \leq 0$.

   (e) Minimize $7x_1+4x_2+5x_3$ subject to $0 \leq x_i \leq 1\ \forall i$; $x_i(1-x_i) = 0\ \forall i$; $2x_1+x_2+x_3 \leq 15$.

3. Compute the following product two different ways and verify that the result is the same.

$$(x_1, x_2)\begin{bmatrix} 4 & -1 \\ 2 & 5 \end{bmatrix}\begin{pmatrix} x_1 \\ x_2 \end{pmatrix}$$

4. Compute the following product two different ways and verify that the result is the same.

$$(x_1, x_2)\begin{bmatrix} a & b \\ c & d \end{bmatrix}\begin{pmatrix} x_1 \\ x_2 \end{pmatrix}$$

5. Compute $\frac{\partial f}{\partial x}$, the partial derivative of function f with respect to x, and $\frac{\partial f}{\partial y}$, the partial derivative of function f with respect to y, for the following functions:

   (a) $f(x, y) = 5xy + x^2 + y^3$

(b) $f(x, y) = 4x^2y - 7xy^2$

(c) $f(x, y) = (x - y)^3$

(d) $f(x, y) = x^y$

(e) $f(x, y) = (3x^2 - xy + 2y^4)^4$

(f) $f(x, y) = e^x e^{-2y} e^{3xy}$

6. Reformulate the area maximization problem of Example 1.8 to have constraint $x_1 + x_2 = 50$. Write the Lagrangian $L(x_1, x_2, \pi_1)$ and solve $\nabla L = \mathbf{0}$. Why is your value for π_1 different from its value in the example? Explain the apparent inconsistency.

7. You wish to enclose the maximum possible rectangular area with 120 feet of fencing. One side of the rectangle will be formed by a sheer rock face and hence will need no fencing.

   - Model this scenario as an optimization problem.
   - Solve the problem with single-variable calculus.
   - Solve the problem by a geometric argument as in Figure 1.7.
   - Solve the problem with Lagrange's method. Compute from your value of the Lagrange multiplier the approximate optimal area if the amount of fencing increased to 124. How close is the approximation to the exact value? Repeat for fence lengths 117 and 80.

8. (a) Write the Lagrangian $L(\mathbf{x}, \boldsymbol{\pi})$ for the problem Minimize $x_1 + x_2^2 + x_3^3$ subject to $x_1 + x_2 + x_3 = 3$ and $x_1^2 + x_2^2 + x_3^2 = 3$.

   (b) Find ∇L, the gradient of the Lagrangian.

   (c) Write the equations $\nabla L = 0$.

9. (a) Write the Lagrangian for the problem Minimize $\sum_{i=1}^n ix_i^i$ subject to $\sum_{i=1}^n x_i = 0$, $\sum_{i=1}^n x_i^2 = n$, and $\prod_{i=1}^n x_i = 1$.

   (b) Find ∇L, the gradient of the Lagrangian.

   (c) Write the equations $\nabla L = 0$.

M Problems

10. Find a high school geometry textbook, and carefully read the definitions of points and lines. Verify that, despite the suggested visual intuitive descriptions, neither has a precise definition except in relation to the other. Which axiom destroys the otherwise symmetric relationship between points and lines?

11. Many mathematical theorems are proved by interchanging summations, that is, by the equation

$$\sum_{i=m}^{n} \sum_{k=s}^{t} x_{i,k} = \sum_{k=s}^{t} \sum_{i=m}^{n} x_{i,k}.$$

Explain why this is a duality as defined in this chapter.

12. Explain why the following equation is a duality:

$$\sum_{i=1}^{n} a_i \sum_{k=1}^{t} b_k = \sum_{k=1}^{t} b_k \sum_{i=1}^{n} a_i$$

13. Prove that matrix multiplication is associative. Hint: The interchange of summations is the crux of the duality.

14. Use Lagrange's method to find the isosceles triangle with maximum area subject to perimeter 18. Verify that the value of your Lagrange multiplier gives a good estimate if the perimeter were decreased to 15.

15. In this problem you will see why the Optimization Problem cannot in general be solved by simply setting $\nabla f(\mathbf{x}) = 0$.

    (a) Solve the problem $\min x_1 x_2$ subject to constraint (1.2), $2x_1 + 2x_2 = 100$ graphically or with other elementary reasoning.

    (b) What would you find if you reduced the problem to one variable and used calculus?

    (c) Why does the problem have no optimal solution?

    (d) Solve the problem $\min x_1 x_2$ subject to constraints (1.2),(1.3) graphically or with other elementary reasoning.

    (e) Treat the problem you solved in part (15d) as a single-variable problem with objective function $f(\mathbf{x}) = x_1(50 - x_1)$. Why are the two optimal solutions you found optimal, even though both fail to satisfy $f'(\mathbf{x}) = 0$? Conclude that even for single-variable problems, the condition $f'(\mathbf{x}) = 0$ is not adequate to solve the Optimization Problem.

16. For each of the following problems, determine whether the problem is infeasible, is unbounded, is feasible but has no optimal solution, or has an optimal solution.

    (a) $\min x^2$ subject to $x - 10 \geq 0$.

    (b) $\max x^2$ subject to $x + 10 \geq 0$.

    (c) $\min x^2$ subject to $x^2 - 5 \geq 0; 8 - x^2 \geq 0; x$ integer.

    (d) $\min \sin x$ subject to $x \geq 0, x$ integer.

    (e) $\min |\sin x|$ subject to $x \geq 0, x$ integer.

    (f) $\min |\sin x|$ subject to $x - 1 \geq 0, x$ integer.

    (g) $\max x^4$ subject to $|x| - 2 \geq 0$.

    (h) $\min x^3$ subject to $|x| - 1 \geq 0$.

17. An upper bound of a set $S \subset \Re$ is a real value α such that $\alpha \geq s \ \forall s \in S$. A "least upper bound," or "l.u.b." of a set $S \subset \Re$ is an upper bound of S such that for all $\beta < \alpha$, β is not an upper bound of S. An l.u.b. is also called a "supremum" or "sup". Prove that no set $S \subset \Re$ can have two distinct l.u.b.s.

18. (Continuation) Assume that if S is not empty and S has an upper bound, then S has an l.u.b. Prove that if an optimization problem is feasible, has no optimum solution and is not unbounded then there exists $\alpha \in \Re$ such that every feasible solution has objective value worse than α, and for all $\epsilon > 0$ there exists a feasible solution y such that $|f(y) - \alpha| < \epsilon$.

19. Employ the assumption that $\sqrt{2}$ is irrational to prove that the objective

$$\min f(\mathbf{x}) = \left| \sqrt{2} - \frac{x_1}{x_2 + 0.25} \right|$$

has no optimum value subject to the constraints x_1, x_2 integer.

20. Show that the definition of $\frac{\partial f}{\partial x_1}$ given in this chapter for a real-valued function $f(x_1, x_2)$ is consistent with

$$\lim_{\epsilon \to 0} \frac{f(x_1 + \epsilon, x_2) - f(x_1, x_2)}{\epsilon}.$$

21. Define the second order partial derivative

$$\frac{\partial^2 f}{\partial x_1 \partial x_2} \equiv \frac{\partial(\frac{\partial f}{\partial x_1})}{x_2}.$$

For each function in Problem 5, compute the second order partial derivatives $\frac{\partial f}{\partial x \partial y}$ and $\frac{\partial f}{\partial y \partial x}$, and determine whether or not they are equal.

22. As in Problem 21 define

$$\frac{\partial^2 f}{\partial x_1 \partial x_2} \equiv \frac{\partial(\frac{\partial f}{\partial x_1})}{x_2}.$$

Prove that for any polynomial $p(x_1, x_2)$, $\frac{\partial^2 p}{\partial x_1 \partial x_2} = \frac{\partial^2 p}{\partial x_1 \partial x_2}$. Define a class of functions in terms of convergence of Taylor series for which the same equality holds.

23. Let $A = \Re^n$, $B = \Re^n$, and $C = \{1, -1\}$. Define the function $f : A \times B \mapsto C$ to be $f(\mathbf{a}, \mathbf{b}) = 1$ iff $\mathbf{a} \cdot \mathbf{b} > 0$ and $f(\mathbf{a}, \mathbf{b}) = -1$ iff $\mathbf{a} \cdot \mathbf{b} \leq 0$. For each $\mathbf{a} \in A$ let the function $g_{\mathbf{a}} : B \mapsto C$ be defined as $g_{\mathbf{a}}(\mathbf{b}) = f(\mathbf{a}, \mathbf{b})$. Similarly let $h_{\mathbf{b}} : A \mapsto C$ be defined for each $\mathbf{b} \in B$ as $h_{\mathbf{b}}(\mathbf{a}) = f(\mathbf{a}, \mathbf{b})$.

For any set $S \subset A$ define $S^{\perp} \equiv \{\mathbf{b} \in B : f(\mathbf{s}, \mathbf{b}) = -1 \ \forall \mathbf{s} \in S\}$.

(a) Prove $\mathbf{t} \in S^{\perp} \Rightarrow \alpha\mathbf{t} \in S^{\perp}$ for all $\alpha \geq 0$.

(b) Prove $\mathbf{t} \in S^{\perp}, \mathbf{t}' \in S^{\perp} \Rightarrow (\mathbf{t} + \mathbf{t}') \in S^{\perp}$.

(c) Write the analogous definition of $T^{\perp}$ for $T \subset B$.

(d) Prove that for all $S \subset A$, $S \subseteq (S^{\perp})^{\perp} =$ and that $S \subsetneq (S^{\perp})^{\perp}$ can occur.

(e) Let $n = 2$. Prove that if $S \subset A$ is closed, convex, and a cone (these terms are defined in the Glossary) , then $S = (S^{\perp})^{\perp}$. What does this imply about sets $T \subset B$?

(f) Verify that each property of A – closed, convex, a cone – is necessary for the conclusion of part 23e to be correct.

D Problems

24. Show that for the Riemann integral of continuous function $f(x, y)$,

$$\int_a^b dx \int_c^d f(x, y)\, dy = \int_c^d dy \int_a^b f(x, y)\, dx$$

is a duality. Explain why the interchange in the indefinite integral

$$\int_a^{\infty} dx \int_c^{\infty} f(x, y)\, dy$$

is not necessarily valid.

25. Read the definition of a second order partial derivative in Problem 21. Find a function $f(x, y)$ such that

$$\frac{\partial f}{\partial x \partial y} \neq \frac{\partial f}{\partial y \partial x}.$$

I Problems

26. In Euclidean plane geometry, for any pair of distinct points, there is a unique line through them. Also, for any pair of nonparallel lines, there is a unique point where they intersect. Find the underlying duality between points and lines, and define it precisely.

27. A regular 3-dimensional solid must have these properties: its sides are regular polygons all congruent to each other; for some integer k, exactly k sides meet at any point where more than 2 sides meet; it is convex.

    - (Platonic solids) Use the fact that the sum of the angles where $k \geq 3$ sides meet must be strictly between 180 and 360 degrees to prove that at most 5 such solids (up to geometric similarity) exist.
    - For a regular 3-dimensional solid, let F be the set of sides (also called faces); let V be the set of points where more than two sides meet (also called vertices); and let E be the set of edges, that is, line segments of positive length that are the intersection of two sides. Define a function $f : F \times V \mapsto E \cup \{\{\phi\}\}$ that establishes a duality between F and V.
    - Show that the regular tetrahedron is self-dual.
    - Find the other two pairs of dual Platonic solids.
    - Slice off the corners of a cube to create triangular sides that meet at vertices. The original sides will be reduced to smaller squares rotated 45 degrees. Find the dual solid.

28. The *Domesday Book* (see http://www.nationalarchives.gov.uk/domesday/) is the oldest surviving public record in the Western world. It is a detailed survey of land usage and ownership, animals, and other property and resources in England as of 1086, ordered by King William the Conqueror. It is an invaluable historical resource regarding the economic activity and feudal social structure of its time. Pretend that you are an historian of England. What, for you, would be a dual of the *Domesday Book*?

29. Watch a performance by a group of dancers. This could be a ballet, a Bollywood or Hollywood musical, a country dance, etc. There are two ways to view the choreography, ways that are duals of each other. What are they?

30. Define the duality in a musical score. Analyze the two rounds "Row, row row your boat" and "Hey, ho, nobody home" from both primal and dual viewpoints. Which round is musically superior, and why?

Chapter 2

Introduction to Linear Programming

2.1 What is Linear Programming?

Preview: This section defines the linear programming (*LP*) problem and its two forms, the primal and the dual. It proves the first important theorem of LP theory, called *weak duality*, and several of its consequences.

Linear programming (LP) is all about creating and solving models of the form:

Definition 2.1 (The Linear Programming Problem (LP))

$$\min \mathbf{c} \cdot \mathbf{x} \quad \textit{subject to} \quad A\mathbf{x} \geq \mathbf{b}; \mathbf{x} \geq \mathbf{0}. \tag{2.1}$$

The n-vector $\mathbf{x}$ is the set of variables; the $m \times n$ matrix A, the n-vector $\mathbf{c}$ and the m-vector $\mathbf{b}$ are the data. Linear programming has the form of a general optimization problem that has a linear objective function and linear constraint functions. Geometrically, each constraint $\mathbf{x}_i \geq 0$ and each constraint $A_{j,:}\mathbf{x} \geq \mathbf{b}_j$ defines a halfspace. The intersection of these halfspaces forms the *feasible region*. The problem (2.1) is to find a point in the feasible region which is farthest in the direction $-c$. Figure 2.1 depicts an example in two dimensions.

LP models always come in pairs. For each pair, one is called the primal; the other is called the dual. You will soon see that the dual emerges naturally from the primal via the Lagrangian (see Chapter 1). The relations between the two are symmetric, so it doesn't matter which one you call the primal, as long as you are consistent. Usually the first one you write down, you call the primal; the other you call the dual. The dual to the primal (2.1) is

$$\max\ \mathbf{b} \cdot \boldsymbol{\pi} \quad \textit{subject to} \quad \boldsymbol{\pi}^T A \leq c^T; \boldsymbol{\pi} \geq \mathbf{0}; \tag{2.2}$$

or equivalently

$$\max\ \mathbf{b} \cdot \boldsymbol{\pi} \quad \textit{subject to} \quad A^T \boldsymbol{\pi} \leq \mathbf{c}; \boldsymbol{\pi} \geq \mathbf{0}. \tag{2.3}$$

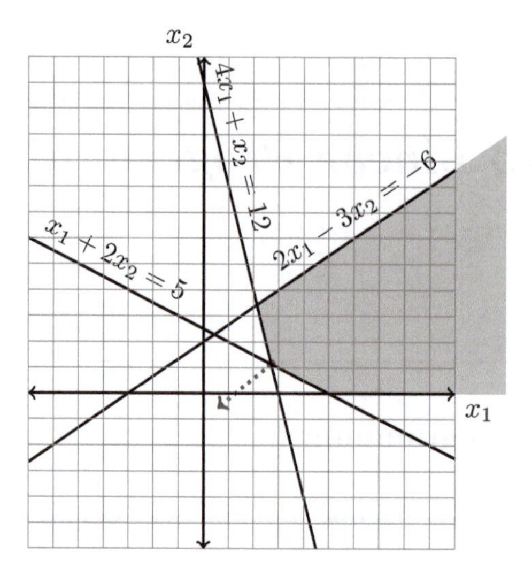

FIGURE 2.1: The feasible region of the LP (2.4-2.8) is shaded. The optimum solution is marked by the heavy dot at $(\frac{19}{7}, \frac{8}{7})$. It is the point farthest in the direction $-\mathbf{c} = (-50, -40)$, indicated by the dotted ray.

Example 2.1

$$\text{minimize } 50x_1 + 40x_2 \qquad \text{subject to} \tag{2.4}$$
$$x_1 + 2x_2 \geq 5 \tag{2.5}$$
$$2x_1 - 3x_2 \geq -6 \tag{2.6}$$
$$4x_1 + x_2 \geq 12 \tag{2.7}$$
$$x_1, x_2 \geq 0 \tag{2.8}$$

is the same as (2.1) with

$$\mathbf{x} = \begin{pmatrix} x_1 \\ x_2 \end{pmatrix}; \qquad \mathbf{c} = \begin{pmatrix} 50 \\ 40 \end{pmatrix};$$
$$A = \begin{bmatrix} 1 & 2 \\ 2 & -3 \\ 4 & 1 \end{bmatrix}; \qquad \mathbf{b} = \begin{pmatrix} 5 \\ -6 \\ 12 \end{pmatrix}.$$

For the matrix A and the vectors $\mathbf{b}, \mathbf{c}$ above, the dual (2.3) is

$$\text{maximize } 5\pi_1 - 6\pi_2 + 12\pi_3 \qquad \text{subject to} \tag{2.9}$$
$$\pi_1 + 2\pi_2 + 4\pi_3 \leq 50 \tag{2.10}$$
$$2\pi_1 - 3\pi_2 + \pi_3 \leq 40 \tag{2.11}$$
$$\pi_1, \pi_2, \pi_3 \geq 0 \tag{2.12}$$

Geometrically, the primal and dual live in different dimensional spaces. Do not expect to see a relationship between the feasible regions of the primal and dual. There are geometric interpretations of the dual, but they will have to wait until later.

There are some beautiful relationships between primal and dual, the first of which is called *weak duality*. In words, the objective value of any feasible solution to the (minimizing) primal is greater than equal to the objective value of any feasible solution to the (maximizing) dual.

Theorem 2.1.1 Weak Duality: *If* $\tilde{\mathbf{x}}$ *satisfies the constraints of* (2.1) *and* $\tilde{\boldsymbol{\pi}}$ *satisfies the constraints of* (2.2) *then*

$$\mathbf{c} \cdot \tilde{\mathbf{x}} \geq \mathbf{b} \cdot \tilde{\boldsymbol{\pi}}.$$

Proof:

$$\mathbf{c} \cdot \tilde{\mathbf{x}} \geq \mathbf{c} \cdot \tilde{\mathbf{x}} - \overbrace{(A\tilde{\mathbf{x}} - \mathbf{b})}^{\geq 0} \cdot \overbrace{\tilde{\boldsymbol{\pi}}}^{\geq 0} = \mathbf{b} \cdot \tilde{\boldsymbol{\pi}} + \tilde{\mathbf{x}}^T \mathbf{c} - (A\tilde{\mathbf{x}})^T \tilde{\boldsymbol{\pi}} = \mathbf{b} \cdot \tilde{\boldsymbol{\pi}} + \overbrace{\tilde{\mathbf{x}}^T}^{\geq 0} \overbrace{(\mathbf{c} - A^T \tilde{\boldsymbol{\pi}})}^{\geq 0} \geq \mathbf{b} \cdot \tilde{\boldsymbol{\pi}}.$$

■

There are several things to notice about this proof. First, it hinges on seeing the scalar value $\tilde{\boldsymbol{\pi}}^T A \tilde{\mathbf{x}}$ in two ways, as $(\tilde{\boldsymbol{\pi}}^T A)\tilde{\mathbf{x}}$ or as $\tilde{\boldsymbol{\pi}}^T (A\tilde{\mathbf{x}})$. This seemingly obvious statement is the essence of duality.

The second thing to notice deserves being stated as a corollary.

Corollary 2.1.2 Complementary Slackness: *If* $\tilde{\mathbf{x}}$ *and* $\tilde{\boldsymbol{\pi}}$ *satisfy the conditions of Theorem 2.1.1, then the following two statements are equivalent:*

1. $\mathbf{c} \cdot \tilde{\mathbf{x}} = \mathbf{b} \cdot \tilde{\boldsymbol{\pi}}$;
2. Complementary slackness:

$$\begin{gathered} \textit{for all } i : \tilde{\mathbf{x}}_i(\mathbf{c}_i - A_{i,:}^T \tilde{\boldsymbol{\pi}}) = 0, \\ \textit{and} \\ \textit{for all } j, (\mathbf{b}_j - A_{j,:}\tilde{\mathbf{x}})\tilde{\boldsymbol{\pi}}_j = 0, \end{gathered} \tag{2.13}$$

and either of these statements implies that $\tilde{\mathbf{x}}$ and $\tilde{\boldsymbol{\pi}}$ are optimal.

The following definitions explain where the term "complementary slackness" comes from.

Definition 2.2 (Slack, Complementary Slackness) *A constraint* $g(\mathbf{x}) \geq 0$ *is slack at* $\tilde{\mathbf{x}}$ *if* $g(\tilde{\mathbf{x}}) > 0$. *Hence* $A_{j,:}\mathbf{x} \geq \mathbf{b}_j$ *is slack at* $\tilde{\mathbf{x}}$ *if* $A_{j,:}\tilde{\mathbf{x}} > \mathbf{b}_j$, *and the constraint* $\boldsymbol{\pi}_j \geq 0$ *is slack at* $\tilde{\boldsymbol{\pi}}$ *if* $\tilde{\boldsymbol{\pi}}_j > 0$. *A pair of primal and dual solutions* $\tilde{\mathbf{x}}, \tilde{\boldsymbol{\pi}}$ *to the pair of LPs* (2.1),(2.2) *satisfies complementary slackness if*

$$\begin{gathered} \forall j \, \{A_{j,:}\tilde{\mathbf{x}} = \mathbf{b}_j \text{ or } \tilde{\boldsymbol{\pi}}_j = 0\} \\ \text{and } \forall i \left\{\tilde{\mathbf{x}}_i = 0 \text{ or } (A^T)_{i,:}\tilde{\boldsymbol{\pi}} = \mathbf{c}_i\right\}. \end{gathered}$$

For each j, the two constraints $A_{j,:}\mathbf{x} \geq \mathbf{b}_j$ and $\boldsymbol{\pi}_j \geq 0$ are called a complementary pair, as are the two constraints $\mathbf{x}_i \geq 0$ and $A_{i,:}^T\boldsymbol{\pi} \leq \mathbf{c}_i$ for each i. In words, complementary slackness means that at most one of each pair is slack.

Example 2.2 Let the primal and dual be as in Example 2.1. The primal solution $\tilde{\mathbf{x}} = (5,0)$ is feasible. In the dual, the solution $\tilde{\boldsymbol{\pi}} = (20,0,0)$ is feasible. Let's see if complementary slackness holds for $\tilde{\mathbf{x}}$ and $\tilde{\boldsymbol{\pi}}$. The first part of the complementary slackness conditions has three requirements:

1. $\tilde{\mathbf{x}}_1 + 2\tilde{\mathbf{x}}_2 = 5$ or $\tilde{\boldsymbol{\pi}}_1 = 0$.
2. $2\tilde{\mathbf{x}}_1 - 3\tilde{\mathbf{x}}_2 = -6$ or $\tilde{\boldsymbol{\pi}}_2 = 0$.
3. $4\tilde{\mathbf{x}}_1 + \tilde{\mathbf{x}}_2 = 8$ or $\tilde{\boldsymbol{\pi}}_3 = 0$.

All three requirements are satisfied because $5+0=5$, $\tilde{\boldsymbol{\pi}}_2 = 0$, and $\tilde{\boldsymbol{\pi}}_3 = 0$, respectively. The second part of the complementary slackness conditions entails two requirements:

1. $\tilde{\boldsymbol{\pi}}_1 + 2\tilde{\boldsymbol{\pi}}_2 + 4\tilde{\boldsymbol{\pi}}_3 = 50$ or $\tilde{\mathbf{x}}_1 = 0$
2. $2\boldsymbol{\pi}_1 - 3\boldsymbol{\pi}_2 + \boldsymbol{\pi}_3 = 40$ or $\tilde{\mathbf{x}}_2 = 0$.

The second requirement is satisfied because $2(20)+0+0 = 40$. It is also satisfied because $\tilde{\mathbf{x}}_2 = 0$. But the first is not because $\tilde{\mathbf{x}}_1 = 5 \neq 0$ and $20+0+0 \neq 50$. Therefore, these two solutions are feasible but not complementary slack. If instead we chose $\tilde{\boldsymbol{\pi}} = (50,0,0)$ complementary slackness would be satisfied because $50+0+0 = 50$ and $\tilde{\mathbf{x}}_2 = 0$. *Question: Why is that pair of values not optimal?*[1]
Question: Verify for yourself that the optimal primal solution $\mathbf{x}^* = (\frac{19}{7}, \frac{8}{7})$ *and the optimal dual solution* $\boldsymbol{\pi}^* = (\frac{110}{7}, 0, \frac{60}{7})$ *are feasible and satisfy complementary slackness.*[2]
By Corollary 2.1.2 their objective values should be equal, and indeed (after factoring out the common term $\frac{1}{7}$), $(50)(19) + (40)(8) = 1270 = (110)(5) + (0)(-6) + (60)(12)$.

The *strong* duality theorem, which we will prove later, says that if $\tilde{\mathbf{x}}$ and $\tilde{\boldsymbol{\pi}}$ are optimal in the primal and dual respectively, then their objective function values are the same.

Theorem 2.1.3 Strong Duality: *If the primal* (2.1) *and dual* (2.2) *both have feasible solutions, then they have optimal solutions* $\tilde{\mathbf{x}}$ *and* $\tilde{\boldsymbol{\pi}}$ *with equal objective function values* $\mathbf{c} \cdot \tilde{\mathbf{x}} = \mathbf{b} \cdot \tilde{\boldsymbol{\pi}}$.

Combining Theorem 2.1.3 with Corollary 2.1.2 tells us that *complementary slackness is equivalent to optimality* for pairs $\tilde{\mathbf{x}}$ and $\tilde{\boldsymbol{\pi}}$ of feasible solutions to the primal and dual, respectively.

[1] Because $(50,0,0)$ is not feasible in the dual.

[2] The first and third primal constraints (2.5),(2.7) are not slack at $\mathbf{x}^*$, which permits $\boldsymbol{\pi}_1^*$ and $\boldsymbol{\pi}_3^*$ to be nonzero. Both dual constraints (2.10),(2.11) are not slack, which permits both $\mathbf{x}_1^*$ and $\mathbf{x}_2^*$ to be nonzero.

Corollary 2.1.4 Strong Duality and Complementary Slackness:
If $\tilde{\mathbf{x}}$ *satisfies the constraints of* (2.1) *and* $\tilde{\boldsymbol{\pi}}$ *satisfies the constraints of* (2.2) *then the following three statements are equivalent:*

1. $\tilde{\mathbf{x}}$ *and* $\tilde{\boldsymbol{\pi}}$ *are optimal in the primal and dual, respectively.*
2. $\mathbf{c} \cdot \tilde{\mathbf{x}} = \mathbf{b} \cdot \tilde{\boldsymbol{\pi}}$.
3. *The pair* $\tilde{\mathbf{x}}, \tilde{\boldsymbol{\pi}}$ *satisfies complementary slackness.*

Third, the proof of Theorem 2.1.1 should remind you of the Lagrangian and of Lagrange multipliers. As discussed in Chapter 1, to solve problems of form $\min f(\mathbf{x}) : g_j(\mathbf{x}) = 0 : i = 1 \ldots m$ Lagrange invented the function $L(\mathbf{x}, \boldsymbol{\pi}) \equiv f(\mathbf{x}) - \sum_{j=1}^{m} \pi_j g_j(\mathbf{x})$. Setting $\nabla L = \mathbf{0}$ and solving for $\mathbf{x}$ and $\boldsymbol{\pi}$ solves the problem. Our problem fits that form except that the $g_j(\mathbf{x})$ are constrained to be ≥ 0 instead of $= 0$. We proved weak duality by showing that the Lagrangian function has a value wedged in between the primal and dual values. The extra variables $\boldsymbol{\pi}$ in the Lagrangian turn out to be the variables $\boldsymbol{\pi}$ in the dual LP. That is why I use the same symbol for them. The correspondence is deep. It will turn out that, just as π_j equals the cost of enforcing constraint $h_j(\mathbf{x}) = 0$ (see Theorem 1.3.4), i.e., the impact on the optimum value of f if $h_j(\mathbf{x})$ is changed from 0, the LP dual variables predict the impact on the optimum value of the objective function if the right-hand-side (RHS) values $\mathbf{b}$ are changed.

There is a geometric meaning to the Lagrangian, as discussed in Chapter 1. Briefly, solving the problem can be thought of as a game between two players. The first player controls the primal $\mathbf{x}$ variables and wants to minimize the Lagrangian. The second player controls the dual $\boldsymbol{\pi}$ variables and wants to maximize the Lagrangian. The solution to the game is found at a *saddlepoint* of the Lagrangian, a point at which any change in the $\mathbf{x}$ variables makes the function larger, and any change in the $\boldsymbol{\pi}$ variables makes the function smaller. The players are stuck at the saddlepoint because neither player can improve her/his position.

Fourth, the weak duality theorem suggests that solving the primal is intimately connected with solving the dual, and vice versa. I shouldn't have to say "and vice versa" because you should know that primal-dual relationships are symmetric.

Fifth, it tells us that if the primal has unbounded optimum, then the dual is infeasible; if the dual has unbounded optimum, then the primal is infeasible. The converse is false. It is possible for both primal and dual to be infeasible. The possibilities are shown in Figure 2.2.

	D feasible	D infeasible
P feasible	equal value optima	P unbounded optimum
P infeasible	D unbounded optimum	possible

FIGURE 2.2: If an LP is infeasible, its dual is infeasible or has unbounded optimum. If both primal and dual are feasible, their optima have the same value.

We can arrive at the dual from the primal by trying systematically to find bounds on the objective value of the primal [236]. The only information we have consists of the primal constraints

$$\begin{aligned} A\mathbf{x} &\geq \mathbf{b}; \\ \mathbf{x} &\geq \mathbf{0}. \end{aligned}$$

Let's take a nonnegative linear combination of the $A\mathbf{x} \geq \mathbf{b}$ constraints,

$$\boldsymbol{\pi}^T A\mathbf{x} \geq \boldsymbol{\pi}^T\mathbf{b},$$

where $\boldsymbol{\pi} \geq \mathbf{0}$. The resulting inequality is satisfied by all feasible solutions. *Question: Why must the linear combination be nonnegative?*[3] If by good fortune $\boldsymbol{\pi}^T A \leq \mathbf{c}^T$ then since $\mathbf{x} \geq \mathbf{0}$ we would know that

$$\boldsymbol{\pi}^T\mathbf{b} \leq \boldsymbol{\pi}^T A\mathbf{x} \leq \mathbf{c}^T\mathbf{x}.$$

Hence $\boldsymbol{\pi}^T\mathbf{b}$ would be a lower bound on the objective value of the primal. The best such bound we could find is the maximum of $\boldsymbol{\pi}^T\mathbf{b}$ subject to the constraints

$$\begin{aligned} \boldsymbol{\pi}^T A &\leq \mathbf{c}^T, \\ \boldsymbol{\pi} &\geq \mathbf{0}, \end{aligned}$$

which is precisely the solution to the dual. (This is really a proof of weak duality in disguise). The remarkable thing is that the best bound we can find in such a way is in fact a tight bound. (This is a statement of the strong duality theorem.)

► *You might wonder if complementary slackness should be defined to require exactly one of each complementary pair to be zero. The immediate answer is "no." There are examples of optimal primal and dual solutions* $\mathbf{x}^*$ *and* $\boldsymbol{\pi}^*$ *containing a complementary pair such that* $\mathbf{x}_i^* = \hat{\boldsymbol{\pi}}_i^* = 0$. *It is easy to construct such examples by setting much of the data to zero (see Problem 19.) Changing the definition would make the statement of Corollary 2.1.2 false.*

On the other hand, we can make a new definition:

Definition 2.3 (Strict Complementary Slackness) *The pair of solutions* $\mathbf{x}$ *and* $\boldsymbol{\pi}$ *of the primal* (2.1) *and dual* (2.3) *respectively satisfy strict complementary slackness iff for all* i *exactly one of* $\mathbf{x}_i, \mathbf{c}_i - A_{:,i} \cdot \boldsymbol{\pi}$ *is nonzero, and for all* j *exactly one of* $A_{j,:}\mathbf{x} - \mathbf{b}_j, \boldsymbol{\pi}_j$ *is nonzero.*

Later we will prove the following stronger version of Corollary 2.1.2:

Theorem 2.1.5 *If primal* (2.1) *and dual* (2.3) *are both feasible, there exists a pair* $\mathbf{x}^*, \boldsymbol{\pi}^*$ *of optimal solutions to the primal and dual respectively that satisfies strict complementary slackness.*

◄

2.1.1 Symmetry and Alternative Forms

Symmetry: The dual of the dual is the primal. If we cast the dual (2.2) in the form of the primal, we get

$$\min(-\mathbf{b}) \cdot \mathbf{x} \text{ subject to } (-A^T)\mathbf{x} \geq (-\mathbf{c}); \quad \mathbf{x} \geq \mathbf{0}.$$

Substituting $(-\mathbf{b})$ for $\mathbf{c}$, $(-A^T)$ for A, and $(-\mathbf{c})$ for $\mathbf{b}$ in (2.2), we find that the dual of the dual is

$$\max(-\mathbf{c}) \cdot \boldsymbol{\pi} \;\text{ subject to } \boldsymbol{\pi}^T(-A^T) \leq (-\mathbf{b})^T; \quad \boldsymbol{\pi} \geq \mathbf{0}; \tag{2.14}$$

which is immediately equivalent to $\min \mathbf{c} \cdot \boldsymbol{\pi}$ subject to $A\boldsymbol{\pi} \geq \mathbf{b}$, $\boldsymbol{\pi} \geq \mathbf{0}$, which is precisely (2.1). So if you apply to a dual the rules that turn a primal into a dual, you get the primal.

[3] Negative multiples reverse the direction of the inequality.

Alternative forms: We could just as well have defined the canonical LP to be any of the following:

minimize $\mathbf{c} \cdot x$ subject to $A\mathbf{x} = \mathbf{b}; x \geq 0$;

minimize $\mathbf{c} \cdot x$ subject to $A\mathbf{x} \geq \mathbf{b}$;

minimize $\mathbf{c} \cdot x$ subject to $A\mathbf{x} \leq \mathbf{b}$,

or other similar forms using maximization. We could even have defined it as find $\mathbf{x}$ such that $A\mathbf{x} \leq \mathbf{b}$. However, LP is never politically correct. It must have some inequality. "Minimize $\mathbf{c} \cdot x$ subject to $A\mathbf{x} = \mathbf{b}$" is not general enough to define LP.

Why are these forms equivalent in expressive power to (2.1)? There are several tricks for transforming from one form to another. The tricks come in primal-dual pairs. The main trick is called "slack variables". How do we convert constraints of form $A\mathbf{x} \leq \mathbf{b}, x \geq 0$ to the form $A\mathbf{x} = \mathbf{b}, x \geq 0$? Let $A \in \Re^{m \times n}$. Introduce a vector $\hat{\mathbf{x}}$ of m additional (so-called *slack*) variables to get the equivalent system

$$[A \quad I] \begin{pmatrix} \mathbf{x} \\ \hat{\mathbf{x}} \end{pmatrix} = b; \begin{pmatrix} \mathbf{x} \\ \hat{\mathbf{x}} \end{pmatrix} \geq \mathbf{0}.$$

Clearly $\exists\mathbf{x}$ that satisfies the first set of constraints iff $\exists(\mathbf{x}, \hat{\mathbf{x}})$ that satisfies the second set of constraints. Nonnegativity in the slack variables takes the place of feasibility in the constraints $A\mathbf{x} \leq \mathbf{b}$. The dual trick converts a system of form $D\boldsymbol{\pi} \geq \mathbf{d}, \boldsymbol{\pi} \geq 0$ to the form $D\boldsymbol{\pi} \geq \mathbf{d}$. Simply write

$$\begin{bmatrix} D \\ I \end{bmatrix} \boldsymbol{\pi} \geq \begin{pmatrix} \mathbf{d} \\ \mathbf{0} \end{pmatrix}.$$

The dual trick is so trivial one can hardly call it a trick. The primal trick, though simple, isn't as obvious. This is one of the reasons to work with duality. Often the something is easier to see in the dual than in the primal. The other main primal trick is to convert the form $A\mathbf{x} = \mathbf{b}, \mathbf{x} \geq \mathbf{0}$ to the form $A\mathbf{x} \leq \mathbf{b}, \mathbf{x} \geq \mathbf{0}$. This one is easy to see in the primal. Write

$$\begin{bmatrix} A \\ -A \end{bmatrix} \mathbf{x} \leq \begin{pmatrix} \mathbf{b} \\ -\mathbf{b} \end{pmatrix}; \quad \mathbf{x} \geq \mathbf{0}.$$

Before reading further try to find the dual of this trick. The dual trick converts $D\boldsymbol{\pi} \geq \mathbf{b}$ to the form $D\boldsymbol{\pi} \geq \mathbf{b}, \boldsymbol{\pi} \geq 0$. The variables in the first form can be positive or negative. In LP parlance they are *unrestricted.* The term "unrestricted" refers to the absence of a restriction on the sign of the variable. The transformation replaces $\boldsymbol{\pi}$ by the difference of two nonnegative vectors, $\boldsymbol{\pi}^+$ and $\boldsymbol{\pi}^-$, to make the system

$$[D \quad -D] \begin{pmatrix} \boldsymbol{\pi}^+ \\ \boldsymbol{\pi}^- \end{pmatrix} \geq \mathbf{b}, \begin{pmatrix} \boldsymbol{\pi}^+ \\ \boldsymbol{\pi}^- \end{pmatrix} \geq \mathbf{0}.$$

This time the primal version of the trick was more obvious.

Example 2.3 Find the dual of

$$\min \mathbf{c} \cdot \mathbf{x} \quad \text{subject to} \quad A\mathbf{x} \geq \mathbf{b}. \tag{2.15}$$

Solution: Convert to

$$\min\ (\mathbf{c}^T \quad -\mathbf{c}^T) \begin{pmatrix} \mathbf{x}^+ \\ \mathbf{x}^- \end{pmatrix} \quad \text{subject to} \quad [A \quad -A] \begin{pmatrix} \mathbf{x}^+ \\ \mathbf{x}^- \end{pmatrix} \geq \mathbf{b}; \begin{pmatrix} \mathbf{x}^+ \\ \mathbf{x}^- \end{pmatrix} \geq \mathbf{0}.$$

This is in the form (2.1). The dual is therefore

$$\max \mathbf{b} \cdot \boldsymbol{\pi} \quad \text{subject to} \quad \begin{bmatrix} A^T \\ -A^T \end{bmatrix} \boldsymbol{\pi} \leq \begin{pmatrix} \mathbf{c} \\ -\mathbf{c} \end{pmatrix}; \boldsymbol{\pi} \geq \mathbf{0}$$

which simplifies to the equivalent

$$\max \mathbf{b} \cdot \boldsymbol{\pi} \quad \text{subject to} \quad A^T \boldsymbol{\pi} = \mathbf{c}; \ \boldsymbol{\pi} \geq \mathbf{0}. \tag{2.16}$$

Corollary 2.1.6 *The weak duality Theorem 2.1.1 applies to the primal-dual pair* (2.15),(2.16).

Proof: The transformation above preserves feasibility and objective function values.
Corollary 2.1.6 suggests a generalization.

Theorem 2.1.7 *Let (P) denote a minimization primal LP and let (D) denote its maximization dual. Then the objective function value of any feasible solution to (P) is greater than or equal to the objective function value of any feasible solution to (D).*

Proof: We only use transformations and simplifications that maintain equivalence of feasibility and objective function values. ■

Slack variables convert general linear inequalities to simple nonnegativity constraints. They are so useful they deserve consistent notation.

Definition 2.4 (Slack Variables) *The vector of slack variables for primal constraints* $A\mathbf{x} \geq \mathbf{b}$ (2.1)*, and the vector of slack variables for dual constraints* $A^T\boldsymbol{\pi} \leq \mathbf{c}$ (2.3) *are, respectively,*

$$\hat{\mathbf{x}} \equiv A\mathbf{x} - \mathbf{b}, \ \textit{making} \ \hat{\mathbf{x}} \geq \mathbf{0} \ \textit{equivalent to} \ A\mathbf{x} \geq \mathbf{b}. \tag{2.17}$$

$$\hat{\boldsymbol{\pi}} \equiv \mathbf{c} - A^T\boldsymbol{\pi}, \ \textit{making} \ \hat{\boldsymbol{\pi}} \geq \mathbf{0} \ \textit{equivalent to} \ A^T\boldsymbol{\pi} \leq \mathbf{c}. \tag{2.18}$$

In these terms, complementary slackness (2.13) *is*

$$\mathbf{x}_i \hat{\boldsymbol{\pi}}_i = 0 \quad \textit{for all } i$$

and

$$\hat{\mathbf{x}}_j \boldsymbol{\pi}_j = 0 \quad \textit{for all } j.$$

The pair of variables $\mathbf{x}_i$ and $\hat{\boldsymbol{\pi}}_i$ is complementary because (at least) one of the pair must equal zero. Symmetrically, each pair of variables $\boldsymbol{\pi}_j$ and $\hat{\mathbf{x}}_j$ is complementary. In words, the ith primal variable is complementary with the ith dual slack variable, and symmetrically the jth dual variable is complementary with the jth primal slack variable.

After you have converted a few primals to their duals, you will see the pattern of relationships between variables in the one and constraints in the other. Equality constraints correspond to unrestricted variables. Natural inequality constraints, i.e., $\leq$ for maximization and $\geq$ for minimization, correspond to nonnegative variables. Unnatural inequality constraints correspond to nonpositive variables. The relationships are summarized in Table 2.1.

TABLE 2.1: Converting Between Primal and Dual

min	max
$A_{j,:}x \geq b_j$	$\pi_j \geq 0$
$A_{j,:}x = b_j$	π_j unrestricted
$A_{j,:}x \leq b_j$	$\pi_j \leq 0$
$x_i \geq 0$	$\pi^T A_{:,i} \leq c_i$
x_i unrestricted	$\pi^T A_{:,i} = c_i$
$x_i \leq 0$	$\pi^T A_{:,i} \geq c_i$
max	**min**
$A_{j,:}x \leq b_j$	$\pi_j \geq 0$
$A_{j,:}x = b_j$	π_j unrestricted
$A_{j,:}x \geq b_j$	$\pi_j \leq 0$
$x_i \geq 0$	$\pi^T A_{:,i} \geq c_i$
x_i unrestricted	$\pi^T A_{:,i} = c_i$
$x_i \leq 0$	$\pi^T A_{:,i} \leq c_i$

Summary: The linear programming (*LP*) problem is to minimize or maximize a linear function of variables subject to linear inequality constraints on the variables. LPs come in pairs, the primal and the dual. Any feasible solution in one gives a bound on the value of the other. A pair of feasible primal and dual solutions satisfies the *complementary slackness* conditions (2.13) if and only if both solutions are optimal. Don't be dismayed by the variety of forms of LPs when minimizing or maximizing, or when writing constraints in the form $A\mathbf{x} = \mathbf{b}, x \geq 0$ or $A\mathbf{x} \geq \mathbf{b}$ or $A\mathbf{x} \leq \mathbf{b}, x \geq 0$, etc. The relationships between primal and dual are valid for all these forms.

2.2 Visualizing LP

Preview: This section shows how to visualize primal LPs with two variables, and how to interpret their duals in the primal geometrical space.

LPs are best visualized when they are of form (2.1) or its dual (2.3), and have two or perhaps three variables. This is because even simple LPs in the other forms take place in too many dimensions to be drawn or visualized easily. We will now focus on the two variable case because it can be drawn easily. LPs with only one variable are not complicated enough to give insight.

Example 2.4 Visualize the following problem in two dimensions:

$$\min 40x_1 - 10x_2 \text{ subject to} \tag{2.19}$$
$$x_1 + 5x_2 \geq 5 \tag{2.20}$$
$$2x_1 + x_2 \geq 6 \tag{2.21}$$
$$x_1 - x_2 \geq -2 \tag{2.22}$$
$$x_1, x_2 \geq 0 \tag{2.23}$$

To draw this problem, first draw the lines corresponding to the equations $A\mathbf{x} = \mathbf{b}$ and $x_i = 0 : i = 1, 2$ as in Figure 2.3 (Left). Then shade the region corresponding to $A\mathbf{x} \geq \mathbf{b}$ and $\mathbf{x} \geq 0$ as in Figure 2.3 (Right). Recall from Definition 1.2 this is the *feasible region*. The shape of the feasible region will always be a polygon if it is bounded. If it is unbounded, as it in this case, it will look like an incomplete drawing of a polygon, missing one or two sides. The places in the polygon where two lines meet are the *vertices* of the polygon.

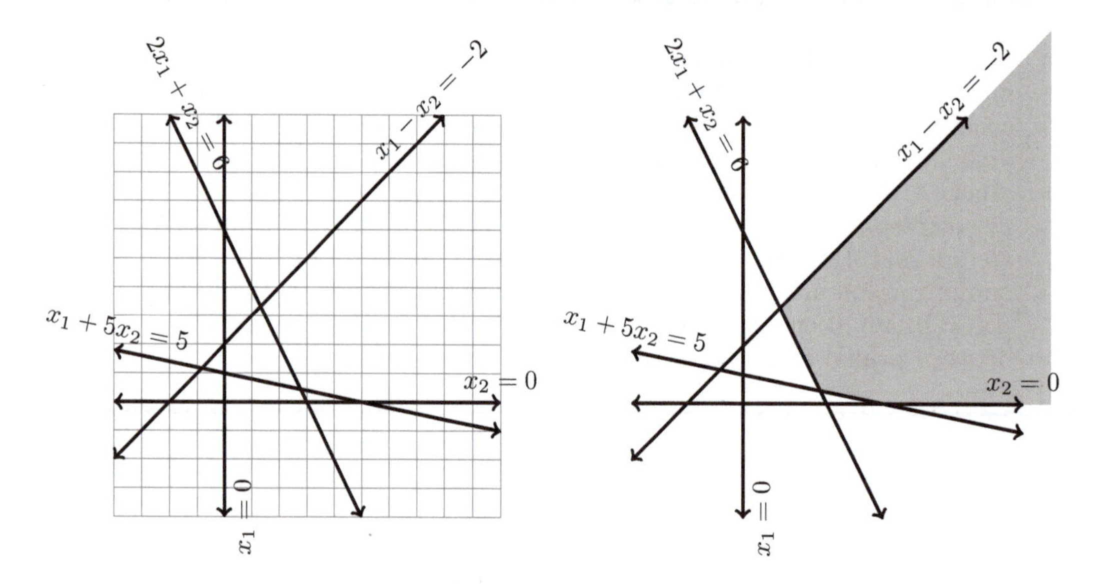

FIGURE 2.3: Left: the lines defined by $A\mathbf{x} = \mathbf{b}, x = 0$ from (2.20-2.23). In higher dimensions, each line would be a hyperplane. Right: segments of the lines form the boundary of the shaded feasible region.

Continuing Example 2.4, $\mathbf{c} \cdot \mathbf{x} = (40, -10) \cdot (x_1, x_2) = 40x_1 - 10x_2$ is the *objective function* or simply the *objective*. To solve the LP graphically, we call upon a geometric meaning of the dot product. In two dimensions, all points on a line orthogonal (perpendicular) to a vector have the same dot product with that vector. Any points further in the direction of the vector have a larger dot product. In Figure 2.4 I've drawn several dashed lines orthogonal to the vector $(40, -10)$. It is visually evident that the point marked x^* is the optimum solution to the LP, because all other points in the feasible region have a larger (worse) objective function value. If your geometric intuition suggests that the optimal solution to an LP will be at a corner point of the feasible region, good for you! Your intuition will be rendered precise in Chapter 4.

Reading Tip: *In general, on your first reading, it is OK to understand either the geometric or the algebraic reasoning in the text. On your second reading, you should aim to understand both. Specifically, for the example above, use the geometry of the dot product to draw the feasible region if you are not certain you have mastery of the concept. Example: Write constraint 2.21 as* $(2,1)\cdot\mathbf{x}\geq 6$. *The boundary of the half plane of points that satisfy the constraint must be a line orthogonal (perpendicular) to* $(2,1)$. *The line must be distance* $\frac{6}{\sqrt{2^2+1^2}}$ *from* $\mathbf{0}$ *in the direction* $(2,1)$, *i.e., distance* $\frac{6}{\sqrt{5}}$ *from the origin in the direction* $(2,1)$. *Moving further in the direction* $(2,1)$ *keeps you feasible with respect to constraint 2.21 because the constraint requires the LHS (left-hand-side) to be* ≥ 6.

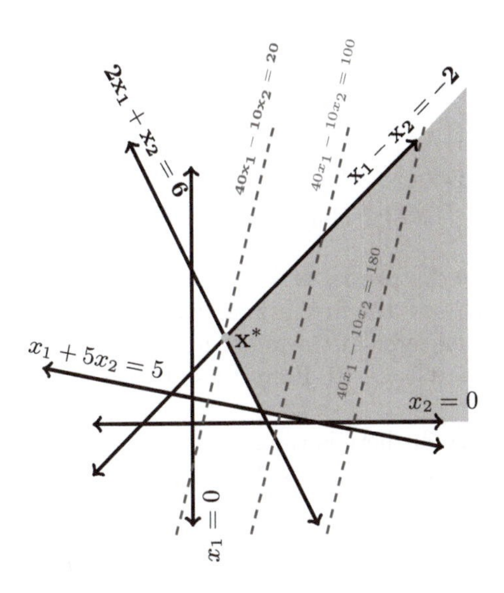

FIGURE 2.4: The objective value $40x_1 - 10x_2$ is constant along each dashed line. Each dashed line is an *isoquant*, meaning "same value." The optimum solution is at $x^* = (1\frac{1}{3}, 3\frac{1}{3})$.

What exactly is $\mathbf{x}^*$? We can estimate it graphically, but we must do some calculations to get its precise value. From Figure 2.4 the point $\mathbf{x}^*$ is at the intersection of the lines $x_1 - x_2 = -2$ and $2x_1 + x_2 = 6$. We say that the constraints $x_1 - x_2 \geq -2$ and $2x_1 + x_2 \geq 6$ are *binding at* $\mathbf{x}^*$. Solving the two equations in two variables, we get $x_1 = \frac{4}{3}; x_2 = \frac{10}{3}$. Thus $\mathbf{x}^* = (\frac{4}{3}, \frac{10}{3})$. (*Question: what would the binding constraints be if the objective were to minimize* $30x_1 + 29x_2$?[4]

The idea of a binding constraint is so important that it needs a formal definition.

Definition 2.5 (Binding, Tight) *A constraint is* binding *or* tight at $\mathbf{y}$ *if it holds with equality at* $\mathbf{y}$. *By definition, an equality constraint is not binding at* $\mathbf{y}$ *iff it is violated at* $\mathbf{y}$.

In terms of this definition, we calculated the optimal solution by visually identifying the constraints that are binding there, and then solving the two linear equations.

Returning now to Example 2.4, what is going on in the dual? By Definition (2.1),(2.3)

[4] $x_1 + 5x_2 \geq 5$ and $2x_1 + x_2 \geq 6$.

the dual is

$$\max 5\pi_1 + 6\pi_2 - 2\pi_3 \text{ subject to} \tag{2.24}$$
$$\pi_1 + 2\pi_2 + \pi_3 \leq 40 \tag{2.25}$$
$$5\pi_1 + \pi_2 - \pi_3 \leq -10 \tag{2.26}$$
$$\pi \geq 0 : i = 1, 2, 3 \tag{2.27}$$

Let's call the optimal dual solution $\boldsymbol{\pi}^*$. We know quite a bit about $\boldsymbol{\pi}^*$. It must be feasible in the dual, and it should be complementary slack with $\mathbf{x}^*$. Plugging in $A, \mathbf{b}, \mathbf{c}$ and $\mathbf{x}^*$ into the complementary slackness equations (2.13) yields

$$\begin{aligned}
x_1^*(c_1 - A_{1,:}^T \boldsymbol{\pi}^*) = {} & \tfrac{4}{3}(40 - \pi_1^* - 2\pi_2^* - \pi_3^*) = {} & 0 \Rightarrow \pi_1^* + 2\pi_2^* + \pi_3^* = 40 \\
x_2^*(c_2 - A_{2,:}^T \boldsymbol{\pi}^*) = {} & \tfrac{10}{3}(-10 - 5\pi_1^* - \pi_2^* + \pi_3^*) = {} & 0 \Rightarrow 5\pi_1^* + \pi_2^* - \pi_3^* = -10 \\
(b_1 - A_{1,:}x^*)\pi_1^* = {} & (5 - (1)(\tfrac{4}{3}) - (5)(\tfrac{10}{3}))\pi_1^* = {} & 0 \Rightarrow (5 - 18)\pi_1^* = 0 \Rightarrow \pi_1^* = 0 \\
(b_2 - A_{2,:}x^*)\pi_2^* = {} & (6 - (2)(\tfrac{4}{3}) - (1)(\tfrac{10}{3}))\pi_2^* = {} & 0 \Rightarrow 0\pi_2^* = 0 \Rightarrow 0 = 0 \\
(b_3 - A_{3,:}x^*)\pi_3^* = {} & (-2 - (1)(\tfrac{4}{3}) + (1)(\tfrac{10}{3}))\pi_3^* = {} & 0 \Rightarrow 0\pi_3^* = 0 \Rightarrow 0 = 0.
\end{aligned}$$

In words, we would say that the two primal variables are nonzero, which forces the two dual constraints to be binding at $\boldsymbol{\pi}^*$, and the first primal constraint is not binding, which forces the first dual variable π_1^* to equal zero. At this point it is easy to calculate $\boldsymbol{\pi}^*$: $\pi_1^* = 0$, from which $2\pi_2^* + \pi_3^* = 40$ and $\pi_2^* - \pi_3^* = -10$. From these two equations in two unknowns, it follows that $\boldsymbol{\pi}^* = (0, 10, 20)$.

The key idea I want to bring out here is the geometric meaning of the dual solution. In matrix form, the two equations we've just solved are

$$\pi_2 \begin{pmatrix} 2 \\ 1 \end{pmatrix} + \pi_3 \begin{pmatrix} 1 \\ -1 \end{pmatrix} = \begin{pmatrix} 40 \\ -10 \end{pmatrix}.$$

In words, the vector $\mathbf{c} = (40, -10)$ is a linear combination of the vectors $A_{2,:} = (2, 1)$ and $A_{3,:} = (1, -1)$. See Figure 2.5. The weights of the linear combination are $\pi_2^* = 10$ and $\pi_3^* = 20$. It is no coincidence that $\boldsymbol{\pi}^* \geq 0$, i.e., that the dual is feasible. The dual should be feasible at optimality.

Definition 2.6 (Cone, generated cone, offset cone) *A set $S \subset \Re^n$ is a* cone *iff $\alpha s \in S$ for all $s \in S$ and $\alpha \geq 0$. The* cone generated *by a set of vectors is the set of nonnegative linear combinations of those vectors. If S is a cone and $\mathbf{w} \in \Re^n$, S* offset *by $\mathbf{w}$ is the set $\{s + \mathbf{w} : s \in S\}$, which is S translated to move $\mathbf{0}$ to $\mathbf{w}$.*

Example 2.5 The nonnegative orthant in $\Re^n$ is the cone generated by the unit vectors $e_1, \ldots, e_n$. The shaded region in Figure 2.5 is the cone generated by the vectors orthogonal to the binding constraints at $\mathbf{x}^*$, offset by $\mathbf{x}^*$.

Geometrically, nonnegativity of the dual variables means that the primal objective is in the cone generated by the vectors that define the binding constraints of the primal.

At other points, the objective may still be expressible as a linear combination of the defining vectors of binding constraints, but the weights might not be nonnegative. Geometrically, a negative weight means that the objective is not in the cone generated by the vectors.

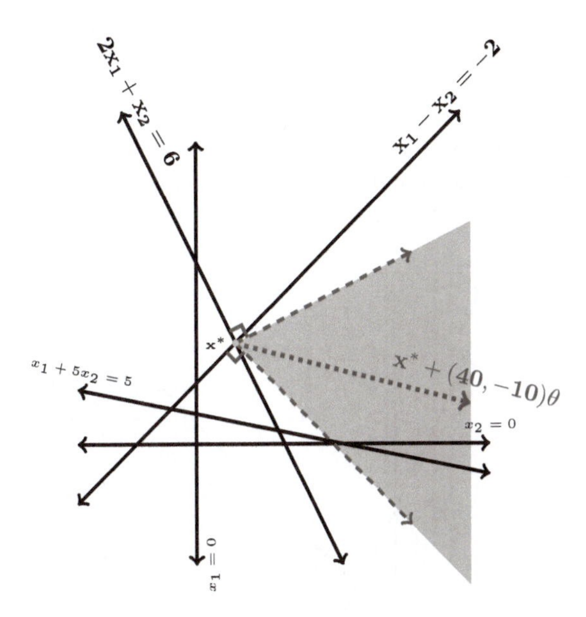

FIGURE 2.5: The vector $\mathbf{c} = (40, -10)$ is a nonnegative linear combination of the orthogonal vectors of the binding constraints at $\mathbf{x}^*$. The shaded region is the cone generated by the orthogonal vectors of the binding constraints, offset by $\mathbf{x}^*$.
The dotted ray within the shaded region, $\mathbf{x}^* + \theta\mathbf{c} : \theta \geq 0$, moves in the direction $\mathbf{c}$ from $\mathbf{x}^*$.

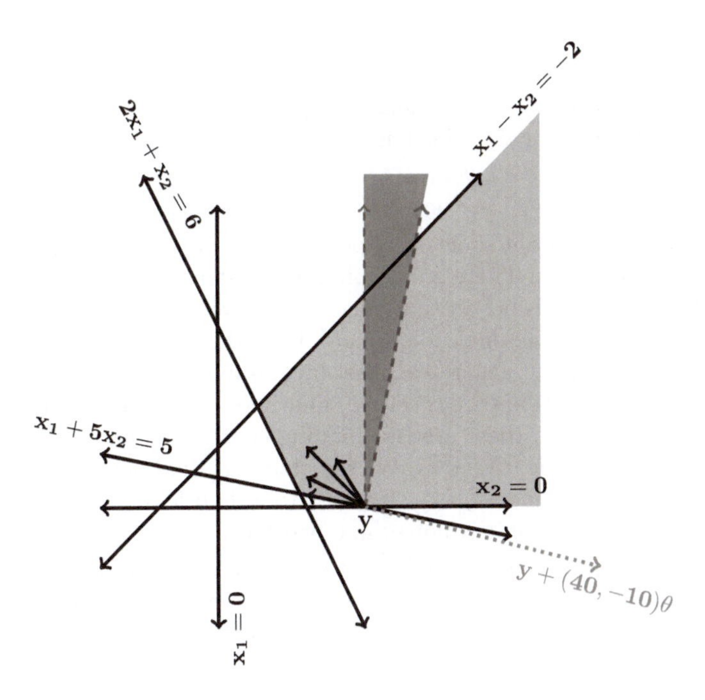

FIGURE 2.6: Because the dotted ray $\mathbf{y} + \theta\mathbf{c} = (5 + 40\theta, -10\theta) : \theta \geq 0$ is not in the shaded cone at $\mathbf{y} = (5, 0)$, there are directions (indicated by solid rays) to move that retain feasibility and improve the objective.

Algebraically a negative weight means infeasibility in the dual (violation of $\pi \geq 0$ constraints), which in turn corresponds to suboptimality in the primal. This suboptimality has a geometric meaning. Consider the point $\mathbf{y} = (5, 0)$ in Figure 2.6. The cone generated by the vectors of the binding constraints is shaded. The vector $\mathbf{c}$ is not in the cone, and there are directions to move from $\mathbf{y}$, indicated by the arrows, that maintain feasibility and improve the objective. On the other hand, in Figure 2.5 at the point $\mathbf{x}^*$ the vector $\mathbf{c}$ is in the cone, and there is no feasible improving direction.

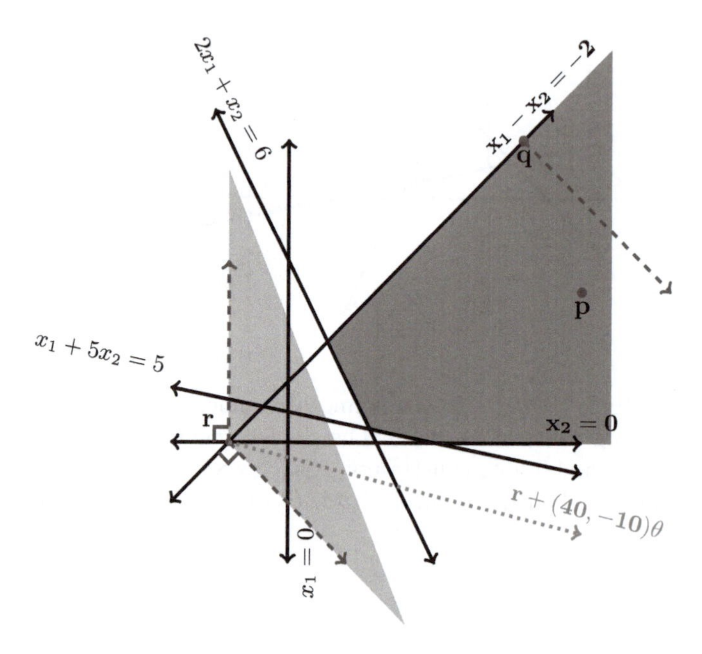

FIGURE 2.7: Only the zero objective makes **p** optimal. Only objectives equivalent to minimizing $x_1 - x_2$ make **q** optimal. Point **r** is primal infeasible but corresponds to a dual feasible point.

Let's check for dual feasibility at some other points. Referring to Figure 2.7, point p is in the interior of the feasible region. There are no binding constraints; the cone they generate is just p itself, and $\mathbf{c} = (30, 40)$ is not in that cone. Therefore, p is not optimal for the LP. *Question: For what objective function is p optimal?*[5] Considering point q now, there is one binding constraint, and the cone it generates, offset by q, is indicated by the arrow at q. The only objectives for which q is optimal are (positive) multiples of that constraint's orthogonal vector. Considering point r now, there are two binding constraints and $\mathbf{c}$ is in the shaded cone generated by their orthogonal vectors. We say that r is dual feasible, although more precisely, r corresponds via complementary slackness to a dual feasible solution. Of course the point r is not optimal, because it is not primal feasible.

Now let's look at a maximization example so you can see what happens when the directions are reversed. The duality may be easier to see in this example because the objective vector points away from the feasible region. The other difference is that we use the *negative* gradients of the nonnegativity constraints.

[5]Minimize $\mathbf{0} \cdot \mathbf{x}$.

Example 2.6 Consider the maximization problem

$$\max 100x_1 + 50x_2 \text{ subject to} \tag{2.28}$$
$$4x_1 + 3x_2 \leq 10 \tag{2.29}$$
$$x_1 + 2x_2 \leq 8 \tag{2.30}$$
$$x_1 \quad\quad \leq 2 \tag{2.31}$$
$$x_1, x_2 \geq 0 \tag{2.32}$$

The feasible region is the shaded trapezoid in Figure 2.8. On each dashed line, all points have the same objective value, since the dashed lines are orthogonal to the objective function vector $(100, 50)$. We call the dashed lines *isoquants* (from *iso*, meaning equal, and *quant* meaning quantity). The isoquants make it clear that the optimum solution is at the point marked $\mathbf{x}^*$.
To find the exact value of $\mathbf{x}^*$, identify the two binding constraints at $\mathbf{x}^*$ as $x_1 \leq 2$ and $4x_1 + 3x_2 \leq 10$. Solving the system $x_1 = 2; 4x_1 + 3x_2 = 10$ yields $\mathbf{x}^* = (2, \frac{2}{3})$.

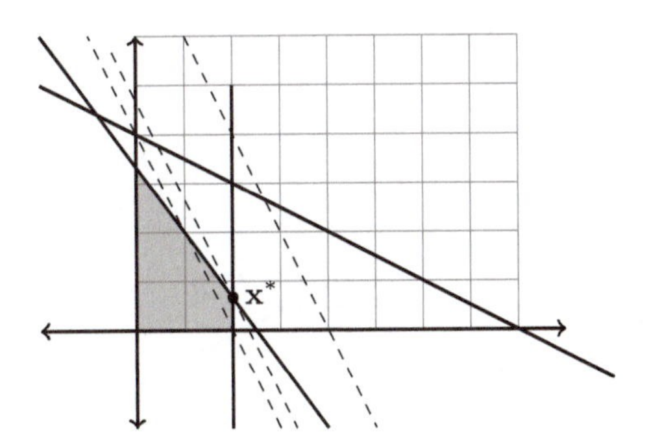

FIGURE 2.8: The dashed isoquant lines of the objective function $100x_1 + 50x_2$ show that $\mathbf{x}^*$ is optimal.

The dual problem to Example 2.6 is

$$\begin{aligned} \min 10\pi_1 + 8\pi_2 + 2\pi_3 &\text{ subject to} \\ 4\pi_1 + \pi_2 + \pi_3 &\geq 100 \\ 3\pi_1 + 2\pi_2 &\geq 50 \\ \pi_1, \pi_2, \pi_3 &\geq 0 \end{aligned}$$

Adding slack variables to the dual model, we get the dual problem

$$\min 10\pi_1 + 8\pi_2 + 2\pi_3 \text{ subject to} \tag{2.33}$$
$$4\pi_1 + \pi_2 + \pi_3 - \pi_4 = 100 \tag{2.34}$$
$$3\pi_1 + 2\pi_2 - \pi_5 = 50 \tag{2.35}$$
$$\pi_1, \pi_2, \pi_3, \pi_4, \pi_5 \geq 0 \tag{2.36}$$

Now let's rewrite the primal in the form $\max c \cdot x$ subject to $A\mathbf{x} \leq \mathbf{b}$.

$$\begin{aligned} \max 100x_1 + 50x_2 & \text{ subject to} \\ 4x_1 + 3x_2 &\leq 10 \\ x_1 + 2x_2 &\leq 8 \\ x_1 &\leq 2 \\ -x_1 &\leq 0 \\ -x_2 &\leq 0 \end{aligned}$$

When the primal is written in this form, the dual feasibility constraints (2.34)-(2.36) require that the primal objective $(100, 50)$ be expressed as a nonnegative linear combination of the coefficients in the rows of the primal constraints. This is the same as for minimization primals, except that the rows for nonnegativity constraints are negative. *Question: what are the coefficients in the 5th row of the primal constraints?*[6] Referring to Figure 2.9, the

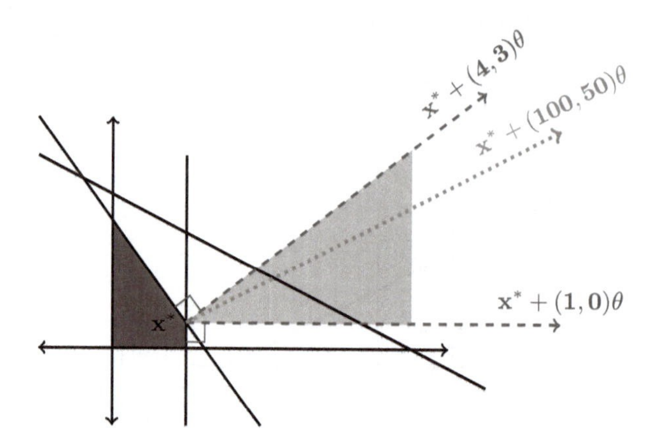

FIGURE 2.9: The dual is feasible at the optimal primal solution $\mathbf{x}^*$ because the dotted ray $\mathbf{x}^* + \theta\mathbf{c} : \theta \geq 0$ is within the shaded offset cone bordered by the dashed rays.

objective $(100, 50)$ is in the cone generated by $(4, 3)$ and $(1, 0)$, offset by the optimal solution $\mathbf{x}^* = (2, \frac{2}{3})$. Hence the corresponding dual solution is feasible, as it should be.

On the other hand, the point $(2, 0)$ is not primal optimal and therefore should not correspond to a dual feasible solution. Referring to Figure 2.10, the objective $(100, 50)$ is outside the cone generated by $(1, 0)$ and $(0, -1)$. *Question: What cone does not contain* $(100, 50)$ *at the point* $0, \frac{10}{3}$?[7]

The heart of the duality here is the reason why dual infeasibility means primal suboptimality. Referring to Figure 2.11, movement in any direction indicated by the arrows will increase (improve) the objective function, because these are the directions which have positive dot product with $(100, 50)$. At the point $(2, 0)$ the binding constraints do not "block off" all of these directions. In particular, the heavy-shaded arrows do not interfere with the constraints that bind at $(2, 0)$. Therefore, $(2, 0)$ is not optimal. There are nearby points that are better, i.e., feasible and have larger objective function value.

In two dimensions, the set $\{\mathbf{x} : \mathbf{c} \cdot \mathbf{x} = c_0\}$ is a line, as are the sets $\{\mathbf{x} : A_{j,:}\mathbf{x} = \mathbf{b}_j\}$ for each j. In n dimensions, this set is a hyperplane. Now let's jump to three dimensions and see how the visualizations from two dimensions extend. Each equation $A_{j,:}\mathbf{x} = \mathbf{b}_j$ now defines

[6] $0, -1$.

[7] The cone generated by $(-1, 0)$ and $(1, 2)$

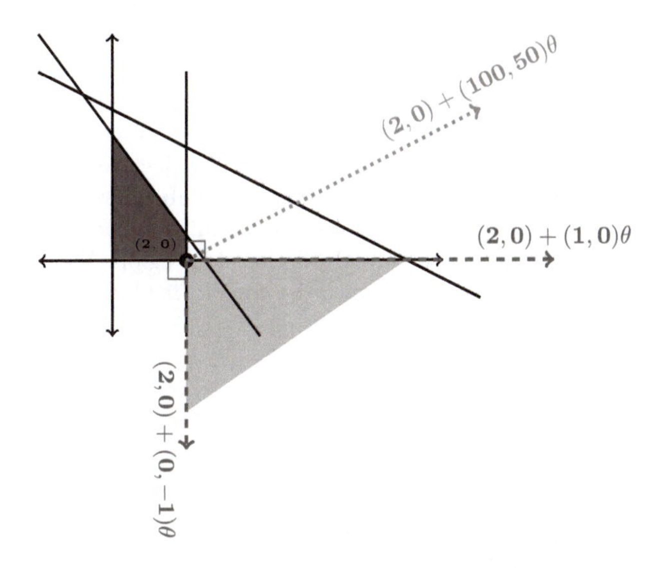

FIGURE 2.10: The offset cone formed by the dashed orthogonal vectors to the binding constraints at $(2,0)$ does not contain the (offset) dotted objective vector.

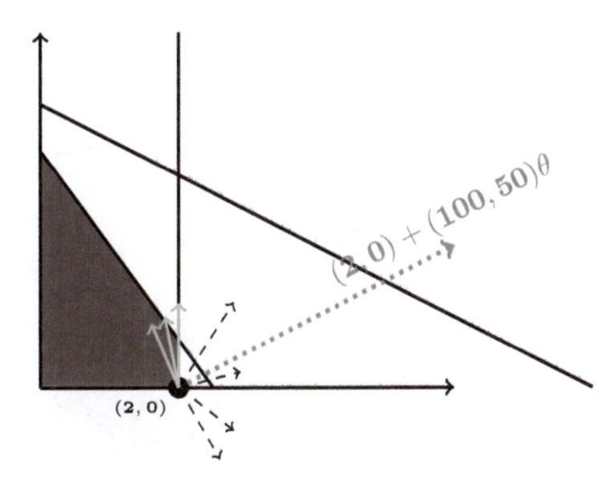

FIGURE 2.11: Movement along any arrow pointing from $(2,0)$ increases the objective function. Move along a solid arrow to maintain primal feasibility, too.

a plane, rather than a line. In the same way, a set of points all with the same objective function value is now a plane, rather than a line. One side of a plane now defines a half-space, rather than a half-plane. The feasible region is now a three-dimensional polyhedron, rather than a polygon. (Polyhedra will be formally defined in Chapter 4).

The plane is a set of points orthogonal to the objective function vector. A vertex of a three-dimensional polyhedron is at the intersection of three planes, rather than at the intersection of two lines. As before, it is visually evident that if there is an optimum solution there is an optimum solution at a vertex. At any vertex, the binding constraints define a cone, this time a three-dimensional cone, which is the set of objective functions for which that vertex is optimal. If you are visualizing a corner of a cube, I recommend that you visualize a maximization problem. For a minimization problem you must be careful, because the cone formed by the constraints coincides with the feasible region. Were you to stretch the cube at that corner, making it a sharper point, the cone formed would get bigger while the feasible region would get smaller.

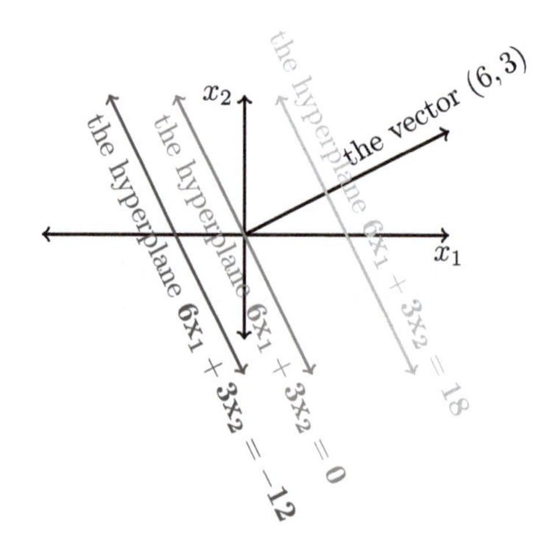

FIGURE 2.12: A hyperplane in 2D is a line orthogonal to a vector.

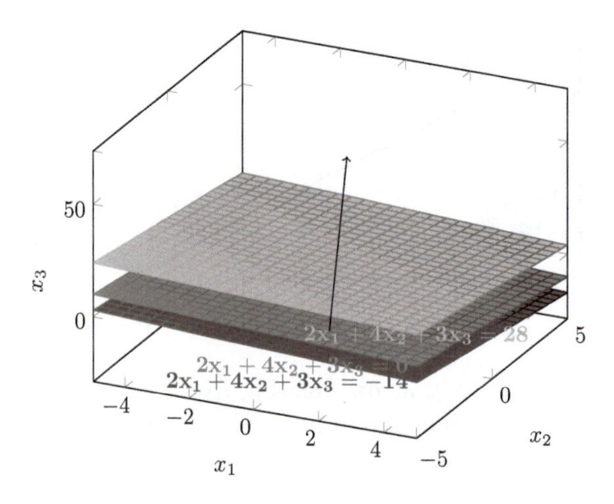

FIGURE 2.13: A hyperplane in 3D is a 2D plane orthogonal to a vector.

Three warnings are as follows:

- In two dimensions, it is hard to appreciate the difference between a vector and a hyperplane, since a hyperplane in 2D is a line. See Figures 2.12 and 2.13. It is better to try to think in 3D when learning LP. I often use the room I am in as an example of a polyhedron.
- LP in two dimensions is easier than it is in higher dimensions. Do not be too quick to generalize your 2D intuition to higher dimensions. See Exercise 32.
- Here is another example of how LP is simpler in 2D than in 3D. In two dimensions, there must be a redundant constraint if more than two constraints are binding at the same point. But in $n \geq 3$ dimensions, any number of constraints can be binding at the same point without any being redundant. See Figures 2.14 and 2.15. When we study degeneracy in later chapters, you will be misled if you visualize in only two dimensions.

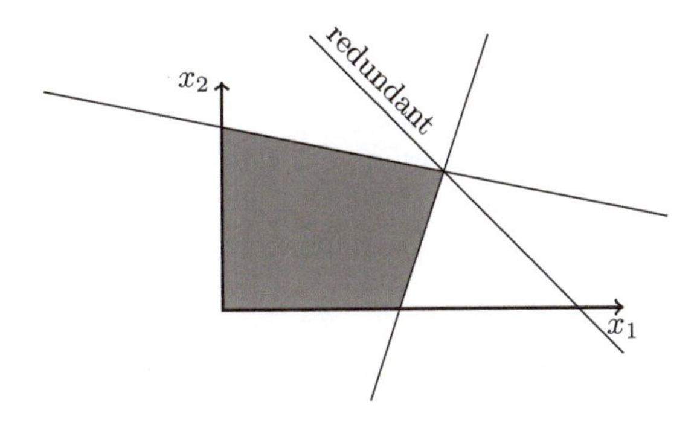

FIGURE 2.14: In two dimensions, if three LP inequality constraints are binding at the same point, one of the three must be redundant and may be removed from the LP without changing the feasible region.

Summary: A two-variable primal's feasible region is a (possibly unbounded) polygon. The sides of the polygon are points at which a constraint is binding and the other constraints are satisfied. The optimal solution is typically at an extreme point of the feasible region, a feasible point at which two constraints are binding. Let the LP primal maximize $\mathbf{c} \cdot x$. The dual is feasible at $\mathbf{x}$ if the $\mathbf{c}$ is in the cone formed by nonnegative linear combinations of the vectors orthogonal to the binding constraints at $\mathbf{x}$. Dual feasibility at $\mathbf{x}$ is therefore equivalent to the objective vector $\mathbf{c}$ pointing "away" from the feasible region at $\mathbf{x}$. Do some of your LP visualization in three dimensions because too many things are true in two dimensions but false in general. In three dimensions, the feasible region is a polyhedron such as a pyramid or cube, which doesn't let you get confused between constraints and objectives. The objective vector $\mathbf{c}$ is still a vector. A constraint defines a plane; to satisfy the constraint, $\mathbf{x}$ has to be in the correct

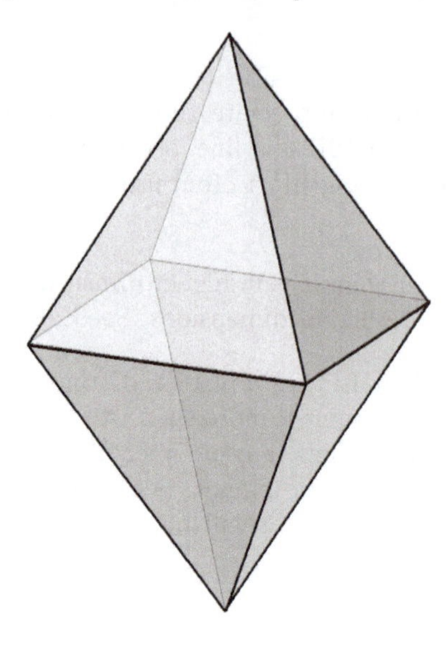

FIGURE 2.15: Four constraints are binding at each vertex of an octahedron, but none of the four is redundant.

half-space defined by the plane. Extreme points have three or more binding constraints.

2.3 Presolving LP

Preview: This section will give you practice with elementary manipulations of LPs, and introduce you to important preliminary steps taken by LP software. Although the rules for each step are simple, taken together they can do wonders.

When it comes to solving LPs, the form (2.1) is not the best. Software will automatically convert your LP from that form to the form $\min \mathbf{c} \cdot \mathbf{x}$ subject to $A\mathbf{x} = \mathbf{b}; \mathbf{l} \leq \mathbf{x} \leq \mathbf{u}$. Here $\mathbf{l}$ and $\mathbf{u}$ are vectors giving lower and upper bounds respectively on the elements of $\mathbf{x}$. Terms in $\mathbf{l}$ are permitted to equal $-\infty$ and terms in $\mathbf{u}$ are permitted to equal ∞, corresponding to the constraint's nonexistence.

What is the dual of this form? Write the primal as

$$\min \mathbf{c}^T\mathbf{x} \quad \text{subject to} \tag{2.37}$$

$$\begin{bmatrix} A \\ -A \\ I \\ -I \end{bmatrix} \mathbf{x} \quad \geq \quad \begin{bmatrix} \mathbf{b} \\ -\mathbf{b} \\ \mathbf{l} \\ -\mathbf{u} \end{bmatrix} \tag{2.38}$$

Using (2.16) the dual is

$$\begin{array}{rcl} \max \mathbf{b}\cdot\boldsymbol{\pi}^+ - \mathbf{b}\cdot\boldsymbol{\pi}^- + \mathbf{l}\cdot\mathbf{z}^+ - \mathbf{u}\cdot\mathbf{z}^- & \text{subject to} & \\ A^T\boldsymbol{\pi}^+ - A^T\boldsymbol{\pi}^- + \mathbf{z}^+ - \mathbf{z}^- & = & \mathbf{c} \\ \boldsymbol{\pi}^+, \boldsymbol{\pi}^-, \mathbf{z}^+, \mathbf{z}^- & \geq & \mathbf{0} \end{array}$$

Replacing $\boldsymbol{\pi}^+ - \boldsymbol{\pi}^-$ by $\boldsymbol{\pi}$ this simplifies to

$$\max \mathbf{b}\cdot\boldsymbol{\pi} + \mathbf{l}\cdot\mathbf{z}^+ - \mathbf{u}\cdot\mathbf{z}^- \quad \text{subject to} \tag{2.39}$$

$$A^T\boldsymbol{\pi} + \mathbf{z}^+ - \mathbf{z}^- = \mathbf{c} \tag{2.40}$$

$$\mathbf{z}^+, \mathbf{z}^- \geq \mathbf{0} \tag{2.41}$$

We can see weak duality in action here when we consider the case $\mathbf{u}_i < \mathbf{l}_i$ for some i. Of course, this renders the primal infeasible. In the dual, if there is a feasible solution, increase the values of $\mathbf{z}_i^+$ and $\mathbf{z}_i^-$ by equal amounts θ. The solution remains feasible and as $\theta \to \infty$ the objective value increases without bound. By the same reasoning, if $\mathbf{u}_i \geq \mathbf{l}_i$ then any feasible solution to the dual will be improved (or retain its objective value in case $\mathbf{u}_i = \mathbf{l}_i$) by reducing $\mathbf{z}_i^+$ and $\mathbf{z}_i^-$ by equal amounts θ until one (or both) reaches zero. We can therefore assume that in a primal-dual optimal pair of solutions, $\mathbf{z}_i^+\mathbf{z}_i^- = 0\ \forall i$.

The complementary slackness conditions between the primal (2.37) and the dual (2.39) are

$$(\mathbf{x}_i - \mathbf{l}_i)\mathbf{z}_i^+ = 0; \tag{2.42}$$

$$(\mathbf{u}_i - \mathbf{x}_i)\mathbf{z}_i^- = 0; \tag{2.43}$$

The first complicated thing most LP solvers do is a so-called presolve step. Presolve simplifies a problem by repeatedly applying a set of simple rules that tighten bounds and eliminate unnecessary constraints and variables. This section will introduce you to several of these rules, some primal, and some dual, so that you will have a basic understanding of the presolve step. The primary early published reference is by Brearley, Mitra, and Williams [32]. Further progress is reported in [7, 132].

zero row If the coefficients in row j are all zero, $A_{ji} = 0\ \forall i$, then the LP is infeasible unless $\mathbf{b}_j = 0$. If $\mathbf{b}_j = 0$ the constraint is redundant and may be removed. *Question: How much CPU time will this step require?*[8]

zero column The dual of a zero row is a zero column. All coefficients in column i are zero, $A_{ji}\ \forall j$. This case is more complicated than the zero row case because of the lower and upper bounds $\mathbf{l}_i \leq \mathbf{x}_i \leq \mathbf{u}_i$. If $\mathbf{c}_i = 0$ the column may be removed. If $\mathbf{c}_i < 0$ and

[8] Since the computer stores only the nonzero coefficients in A, it takes time proportional to the number of rows.

$\mathbf{u}_i = \infty$ the LP has unbounded optimum; if $\mathbf{c}_i < 0$ and $\mathbf{u}_i$ is finite, then x_i may be fixed to the value $\mathbf{u}_i$. If $c_i > 0$ and $\mathbf{l}_i = -\infty$ the LP has unbounded optimum; if $c_i > 0$ and $\mathbf{l}_i$ is finite, then x_i may be fixed to the value $\mathbf{l}_i$. *Question: What is going on in the dual when* $\mathbf{l}_i = -\infty$ *and* $\mathbf{u}_i = \infty$?[9]

fixed variable If $\mathbf{l}_i = \mathbf{u}_i$, variable x_i can be substituted out of the LP.

singleton row If row j has exactly one nonzero coefficient, say $A_{ji} \neq 0$, this forces $x_i = b_j/A_{ji}$. If $\mathbf{l}_i \leq b_j/A_{ji} \leq \mathbf{u}_i$ variable x_i can be substituted out of the LP; if not, the LP is infeasible.

duplicate row If row j is a multiple of row $k \neq j$, it is redundant and may be removed.

free singleton column The dual of a singleton row is a singleton column, $A_{ji} \neq 0$; $A_{ki} = 0\ \forall k \neq j$. Like zero columns, singleton columns are made complicated by the presence of upper and lower bounds. A particularly useful and simple case is when $\mathbf{l}_i = -\infty$ and $\mathbf{u}_i = \infty$. We say that the column is *free*. Then the variable x_i can be replaced by

$$\frac{b_j - \sum_{t \neq i} A_{jt} x_t}{A_{ji}}$$

in the objective function and both row j and column i may be removed. This replacement is highly desirable because it removes both a row and a column without adding any nonzero terms to the A matrix. Moreover, it often adds terms to the objective function which, as we will see later, can be helpful in directing the simplex method.

A presolve routine will make extra efforts to identify free singleton columns because of their desirability. One technique for doing so is called *implied bounds.* For example, suppose x_1 is a column singleton with $\mathbf{l}_1 = 0, \mathbf{u}_1 = \infty$ and constraint $10x_1 - 12x_2 + 5x_3 = 18$. Suppose further $\mathbf{l}_2 = -1$ and $\mathbf{u}_3 = 1$. Then $x_1 = \frac{1}{10}(18 + 12x_2 - 5x_3 \geq \frac{1}{10}(18 + 12(-1) - 5(1) = 0.1$ and the bound $x_1 \geq \mathbf{l}_1$ is redundant. It is *implied* by the other constraints. Therefore, x_1 can be treated as a free column singleton.

In general, we have a column singleton variable x_i with $\mathbf{l}_i > -\infty$ and $\mathbf{u}_i = \infty$, and the constraint $A_{j,:}x = b_j$ where $A_{ji} > 0$. Then the following is an implied lower bound on x_i.

$$x_i \geq \frac{1}{A_{ji}} \left(b_j + \sum_{k:A_{jk}<0} |A_{jk}| \mathbf{l}_k - \sum_{k:A_{jk}>0} A_{jk} \mathbf{u}_k \right)$$

If the implied lower bound is greater than or equal to $\mathbf{l}_j$, the $x_i \geq \mathbf{l}_i$ bound is implied and the variable x_i treated as a free column singleton. Similar implied bounds can be calculated for $A_{ji} < 0$ and by considering the other bounds on the other variables.

bounds For rows j involving variables with non-infinite lower or upper bounds, compute a lower bound L_j on the left-hand side by plugging in the lower bound for each variable with a positive coefficient and the upper bound for each variable with a negative coefficient. If $L_j > b_j$, the LP is infeasible. If $L_j = b_j$ every variable with positive (negative) coefficient may be fixed at its lower (upper) bound. The constraint itself becomes redundant and may be removed. Similarly compute an upper bound U_j on the left-hand side. If $U_j < b_j$ the LP is infeasible. If $U_j = b_j$ every variable with positive (negative) coefficient may be fixed at its upper (lower) bound, and the constraint may be removed. *Question: How much CPU time will this step require?*[10]

[9]The dual has constraint $0\pi_1 + 0\pi_2 + \ldots 0\pi_m = c_i$. So the dual is infeasible unless $c_i = 0$.

[10]Time proportional to the number of nonzero coefficients in the matrix A.

A dual reduction Suppose we could deduce that a dual variable $\mathbf{z}_t^+$ was nonzero. Then by complementary slackness (2.42) we could fix the corresponding primal variable $x_t = \mathbf{l}_t$. Similarly if we knew that $\mathbf{z}_t^- > 0$ we could fix $x_t = \mathbf{u}_t$. How could this come about?

A column singleton in column i of A is a row singleton in row i of A^T. In Equation (2.40) such a row of A^T yields the following constraint:

$$A_{ik}^T \boldsymbol{\pi}_k + \mathbf{z}_i^+ - \mathbf{z}_i^- = c_i.$$

If $\mathbf{l}_i = -\infty$, then by Equation (2.42) $\mathbf{z}_i^+ = 0$. Since $\mathbf{z}_i^- \geq 0$ this implies $A_{ki}\boldsymbol{\pi}_k \geq c_i$. Depending on the sign of A_{ki} this yields an upper or lower bound on $\boldsymbol{\pi}_k$. Similarly, if $\mathbf{u}_i = \infty$, then $\mathbf{z}_i^- = 0$ and $A_{ki}\boldsymbol{\pi}_k \leq c_i$.

Supposing now that by examining all column singletons of A we have found upper and/or lower bounds on several of the $\boldsymbol{\pi}_k$. We perform the same lower and upper bounds computation on A^T that we performed in the previous reduction. That is, we compute a lower bound L_t and upper bound U_t on $A_{t,:}^T \boldsymbol{\pi}$. If $U_t < c_t$, then to keep row t of (2.40) true $\mathbf{z}_t^+$ must be strictly positive. Similarly if $L_t > c_t$, then it must be that $\mathbf{z}_t^- > 0$.

Although each of these rules is simple, taken together they can do wonders [26]. presolve, on a small set of test cases. Part of Karmarkar's [172] reported success [273] with linear programs in the 1980s was due to his use of presolve rather than his interior point algorithm [3].

Summary: Presolving involves a collection of simple reduction rules that in concert often greatly reduce the time and memory requirements of LPs. The dual reductions are good examples of how to think about the dual when looking at the primal. When you solve small LP instances in later chapters, however, do not presolve them or you will be apt to reduce the instances to such triviality that you won't learn anything.

2.4 Notes and References

The linear programming model was invented by George Dantzig in the 1940s. What we now take for granted as natural was novel at the time. Linear production functions and input-output models had been proposed and employed by Leontief [209, 210] to understand equilibrium production levels and prices. Perhaps one can see this leading naturally to systems of linear inequalities. But it took Dantzig about a year to decide that an objective function was needed [68, 74]. Dantzig later discovered [64] that Leonid Kantorovich had previously invented a linear transshipment model and described multiple important applications of it in 1939 [169]. Kantorovich's model was not apparently as general as LP, although it can be proved to be so. Unfortunately, Kantorovich's work was not pursued by his fellow Russian researchers at the time. In the early 1940s, Frank Hitchcock, as well, invented transportation linear programming models [149], and T.C. Koopmans cast several economics problems as LPs [193].

Thus, multiple researchers should be credited with the invention of LP modeling. In hindsight, two more things had to be invented for LP to have real impact: a computationally effective algorithm, and the computer. The simplex method (see Chapter 5) provided the former. LP actually helped develop the latter. Much of the funding for SEAC, BINAC, and the first UNIVAC and IBM computers was instigated by Dantzig and his colleagues at the U.S. Air Force [69].

For a great scholarly study of the early history of linear and combinatorial optimization, see [265].

Cottle [59] gives an interesting and varied list of propositions that are true in all dimensions less than or equal to 4, but are false in all dimensions greater than or equal to 5.

Presolve is even more important for Integer Programming (IP) than for LP. An early example of non-automated IP presolving on a fairly large instance is reported by Gene Woolsey [302]. Done by hand by Charles Krabeck, it reduced an IP with 630 constraints, 25 variables, and approximately 12,000 nonzeros to an IP with two binary variables and one constraint. Automating presolving is not trivial, particularly because it must run quickly enough to be used recursively. A first presolve pass may tighten bounds on some variables. Those new bounds typically imply more bounds that are discovered by the second pass. After multiple passes, the impact of the first pass often has cascaded far beyond what was initially apparent. For IP instances that take more than 10 cpu seconds to solve with presolve, Achterberg and Wunderling [2] find that more than 10 times as much time is required on average without presolve. For an exposition of the present state-of-the-art in IP reductions, see Achterberg *et al.* [1].

2.5 Problems

E Exercises

1. Explain why "$g(x) \geq 0$ is binding at x^*" and "$g(x) \geq 0$ is not slack at x^*" are not logically equivalent statements.

2. Consider the problem of minimizing $2x_1 - 5x_2$ subject to the constraints $3x_1 + x_2 \geq 6$, $-x_1 + x_2 \geq 2$, $9x_1 - 8x_2 \geq 1$, and x_1, x_2 nonnegative. Let $\mathbf{x}$ denote the vector (x_1, x_2). Write matrix A and vectors b and $\mathbf{c}$ to express this problem in the form (2.1).

3. Repeat Exercise 2 for the problem:

$$\begin{aligned} \text{Minimize } x_1 - 6x_2 + 4x_3 & \qquad \text{subject to} \\ x_i \geq 0 : i = 1, 2, 3 & \\ 2x_1 + x_3 & \;\geq\; 10 \\ x_1 \geq x_2 & \\ 3x_1 - x_2 + 2x_3 & \;\geq\; 4 \\ -5x_1 - x_2 + x_3 \geq 0 & \end{aligned}$$

4. For the problem in Exercise 2, write the dual both with and without matrix notation.

5. For the problem in Exercise 3, write the dual both with and without matrix notation.

6. (a) Verify that the weak duality theorem holds for the LP minimize x_1 subject to $x_1 \geq 5$, x_1 nonnegative.
   (b) Prove that the strong duality theorem holds for the LP minimize x_1 subject to $x_1 \geq 5$, x_1 nonnegative.
   (c) Verify that the weak duality theorem holds for the LP minimize $13x_1$ subject to $77x_1 \geq 5$, x_1 nonnegative.
   (d) Prove that the strong duality theorem holds for the LP minimize $13x_1$ subject to $77x_1 \geq 5$, x_1 nonnegative.

7. Find the dual of the form $\min c \cdot x; A\mathbf{x} = \mathbf{b}; x \geq 0$.

8. Convert the following into the form (2.1):

$$\begin{aligned} \text{Minimize } x_1 + 2x_2 & \qquad \text{subject to} \\ 3x_1 + 4x_2 &\geq 5 \\ 6x_1 + 7x_2 &\geq 8 \\ x_1 \text{ unrestricted}, x_2 \geq 0 & \end{aligned}$$

9. Convert the following into the form (2.1):

$$\begin{aligned} \text{Minimize } x_1 - x_2 + x_3 & \qquad \text{subject to} \\ x_1 + 2x_2 - 3x_3 &\leq 10 \\ x_1 \qquad - 2x_3 &= 4 \\ x_i \geq 0 : i = 1, 2, 3 & \end{aligned}$$

10. Convert the following into the form (2.3):

$$\begin{aligned} \text{Maximize } 15x_1 + 16x_2 + 17x_3 & \qquad \text{subject to} \\ x_2, x_3 \text{ unrestricted} & \\ x_1 \geq 0 & \\ 18x_1 - 19x_2 + 20x_3 &\leq 21 \\ 22x_1 + 23x_2 - 24x_3 &\leq 25 \end{aligned}$$

11. Convert the following into the form (2.3):

$$\begin{aligned} \text{Minimize } 4x_1 - x_2 - 2x_3 - 3x_4 & \qquad \text{subject to} \\ -x_1 + x_2 + x_3 + 5x_4 &\leq 12 \\ x_1 - 2x_2 - 3x_3 - 4x_4 &\geq 1 \\ x_i \geq 0 : 1 \leq i \leq 4 & \end{aligned}$$

12. Consider the primal LP

$$\begin{aligned} \text{Minimize } x_1 + x_2 & \qquad \text{subject to} \\ 3x_1 + 4x_2 &\geq 6 \\ x_1, x_2 \geq 0 & \end{aligned}$$

(a) Verify that $\mathbf{x} = (2, 0)$ is feasible.

(b) Write the dual.

(c) Verify that 0.3 is feasible in the dual.

(d) Verify that weak duality holds for the pair $(2, 0)$ (primal) and 0.3 (dual).

(e) Why doesn't the pair $(1, 1)$ (primal), 2 (dual) contradict the weak duality theorem?

(f) Which constraints are binding in the primal at $(2, 0)$? At $(0.1.5)$? At $(6, -3)$?

(g) Which constraints are binding in the dual at 0? At 0.3? At $\frac{1}{3}$? At $\frac{1}{2}$?

(h) Show that complementary slackness does not hold at the pair $(2, 0)$; $\frac{1}{2}$, and does not hold at the pair $(0, 0)$; $\frac{1}{3}$.

(i) Show that complementary slackness does hold at the following pairs:

   i. $(0, 0); 0$

   ii. $(0.1.5); \frac{1}{2}$

   iii. $(2, 0); \frac{1}{3}$

   iv. Draw the primal and geometrically confirm your answers to 12a,12f.

   v. Draw the dual and geometrically confirm your answers to 12c,12g.

13. For the constraints in example (2.19) find the optimal solution if the objective is to minimize $5x_1 - 2x_2$.

14. For the example depicted in Figure 2.8 what is the optimum solution for the objective $\max x_1 + x_2$?

15. For the problem depicted in Figure 2.9, show that at the optimal solution, the values of the dual variables π_1 and π_3 are $\frac{50}{3}, \frac{100}{3}$.

16. Consider the constraints

$$x_1 + x_2 \leq 12 \tag{2.44}$$

$$x_1 + 2x_2 \leq 18 \tag{2.45}$$

$$x_2 \leq 8 \tag{2.46}$$

$$x_1 \geq 0 \tag{2.47}$$

$$x_2 \geq 0 \tag{2.48}$$

(a) Graph the feasible region.

(b) For each of the following pairs of constraints, compute the point at which they are binding. Verify your answers by looking at your graph.

   i. (2.47,2.48)

   ii. (2.44,2.45)

   iii. (2.44,2.46)

   iv. (2.45,2.46)

   v. (2.44,2.47)

   vi. (2.46,2.47)

(c) In your graph, draw the isoquant line $x_1 = 12$ to show that $\max x_1$ is optimized at $(12, 0)$.

(d) Draw the isoquant line $x_1 - x_2 = 12$ to show that $\max x_1 - x_2$ is optimized at $(12, 0)$.

(e) Draw the isoquant line $2x_1+3x_2 = 24$ to show that $\max 2x_1+3x_2$ is not optimized at $(12, 0)$.

(f) Give an objective for which both $(0, 0)$ and $(12, 0)$ are optimal. Repeat for $(0, 0), (0, 8)$. Repeat for $(6, 6), (2, 8)$.

(g) Draw the cones at $(0, 0)$, $(12, 0)$, $(6, 6)$, $(2, 8)$, and $(0, 8)$ that show the set of objectives for which the point is optimal.

(h) For implied bounds on singleton columns, what lower bound can you calculate in the case $A_{ji} < 0$?

(i) For implied bounds on singleton columns, what upper bounds can you calculate for $A_{ji} > 0$?

M problems

17. Find the dual of the following problem, which has no typographical errors.

$$\begin{array}{rcl} \max 3x_1 + 4x_2 & & \text{subject to} \\ 5x_1 - 6x_2 & \leq & 7 \\ -8x_1 + 9x_2 & \geq & 10 \\ 11x_1 + 12x_2 & \leq & 13 \end{array}$$

18. Arrive at the primal (2.1) from the dual (2.2) by trying systematically to find bounds on the objective value of the dual.

19. Solve the LP $\max x_1+x_2$ subject to $2x_1+3x_2 \leq 0$; $x_1, x_2 \geq 0$ by inspection. Let π_1 be the variable in the dual LP. Find an optimal solution π_1^* to the dual that satisfies strict complementary slackness with the optimal primal solution. Find an optimal solution $\pi_1^{'}$ that does not satisfy strict complementary slackness. Why doesn't $\pi_1^{'}$ contradict Theorem 2.1.5?

20. For all minimization LPs in the form (2.1) with one variable and one constraint (not counting nonnegativity), prove explicitly the strong duality theorem.

21. Sketch the feasible region of the LP $\max x_1-9x_2$ subject to $2x_1+x_2 \leq 10; -x_1+2x_2 \leq 4; x_1 \geq 0; x_2 \geq 0$. Determine the exact location of each extreme point of the feasible region.

22. Solve the LP of Problem 21 graphically.

23. For the problem $\max x_1 + x_2$ subject to $x_1 + x_2 \leq 20; x_1 \geq 0; 0 \leq x_2 \leq 15$:

    (a) Identify all primal optimal solutions.

    (b) Write the dual and identify all its optimal solutions.

    (c) Verify that Theorem 2.1.3 holds.

    (d) Verify that Corollary 2.1.4 holds, in the sense that every optimal primal solution has complementary slackness with every optimal dual solution.

24. Does the complementary slackness corollary to the theorem of weak duality apply to other forms? For the system $\min \mathbf{c} \cdot \mathbf{x}; A\mathbf{x} = \mathbf{b}; \mathbf{x} \geq \mathbf{0}$, state and prove a weak duality theorem, including the complementary slackness corollary.

25. Use Corollary 2.1.4 to prove the analogous statement for the the primal $\min \mathbf{c}\cdot\mathbf{x}$; $A\mathbf{x} = \mathbf{b}; \mathbf{x} \geq vzero$ and its dual.

26. Solve the LP to minimize $6x_1 + 6x_2 + 4x_3 + 5x_4$ subject to $3x_1 + x_2 - x_3 + x_4 \geq 20$; $0.5x_1 + x_2 + x_3 - x_4 \geq 10$; $x_i \geq 0$ graphically. Do not presolve the LP.

27. Consider the LPs $\min \mathbf{c} \cdot \mathbf{x}$ subject to $A\mathbf{x} \leq \mathbf{b}; \mathbf{x} \geq \mathbf{0}$ and $\max \mathbf{c} \cdot \mathbf{x}$ subject to $A\mathbf{x} \leq \mathbf{b}; \mathbf{x} \geq \mathbf{0}$. Suppose that the region $\mathbf{x} : A\mathbf{x} \leq \mathbf{b}; \mathbf{x} \geq \mathbf{0}$ is unbounded. Can the following occur? Justify your answers: (a) Both problems have an optimal solution; (b) both problems have an unbounded optimum.

28. Presolve the following LP:

$$\begin{array}{rll} \min 2x_1 + 3x_2 + 10x_3 + x_4 & & \text{subject to} \\ 4x_1 + 6x_2 + 3x_3 - x_4 & \geq & 11 \\ 5x_1 - x_2 - 4x_3 & \geq & 0 \\ x_1 + x_2 + 5x_3 & \geq & 7 \\ 1 \leq x_i \leq 3 \;\; \forall i & & \end{array}$$

29. Construct an example of an LP for which both primal and dual are infeasible.

30. Derive a simple algorithm to solve the problem $\min \mathbf{c} \cdot \mathbf{x}$ subject to $A\mathbf{x} = \mathbf{b}$.

31. Find a point in Figure 2.7 at which two constraints are binding that is infeasible in both the primal and the dual.

D Problems

32. In 2D you can characterize the optimal solution as being formed by the least angle between the objective and the side. Find a counterexample to this idea in 3D.

I Problems

33. A straightforward implementation of the duplicate row rule takes time quadratic in the number of rows, which is too expensive in practice. Figure out how to speed this up to be close to linear in the number of nonzero coefficients, on average, under plausible assumptions. Hint: Reduce the problem to identifying identical rows.

34. (a) Define a random polyhedron P as $\{\mathbf{x}|A\mathbf{x} \leq \mathbf{b}\}$ where the A_{ij} and b_j values are i.i.d. Gaussian (normal) distributed with mean and variance 1. Prove that the probability is zero that P has degeneracy.
    (b) Find weaker conditions under which the probability of degeneracy is zero.
    (c) Define a random polyhedron P as the convex hull of the columns of A, where A is a matrix whose values are i.i.d. Gaussian (normal) distributed with mean and variance 1. Prove that the probability is strictly positive that P has degeneracy.
    (d) **(*)** Prove or disprove: if n points are generated randomly i.i.d. from a uniform distribution on the unit cube in $\Re^3$, then the probability that the convex hull of the points has a degeneracy goes to 1 as $n \to \infty$.

Chapter 3

Formulating and Solving Linear Programs

Preview: This chapter teaches LP formulation, how to build an LP model that accurately describes the critical elements of a problem. The main steps in model building are: understand and visualize the problem; define index sets and data; define the decision variables; write the objective function as a linear combination of the decision variables; and write the constraints as linear inequalities in the decision variables. This chapter also explains how to interpret the results from LP solver software, and points out features of problems that resist LP formulation.

Remember word problems? For example, A and B together can dig a ditch in 20 days. A, working alone, can dig a ditch in 30 fewer days than can B working alone. How long does it take for B to dig a ditch? Linear programming formulation is like word problems. It is a skill different from algebraic manipulation, and it comes naturally to some people. But many others, who are usually good at math, are stumped by formulation problems. If you feel stumped, do not let yourself be fearful or paralyzed. Remember, *formulation is a different skill.* You *can* learn it, and I will help you learn it.

Questions to ask when formulating are as follows:

- What choices do I have? What can I control? This tells you what your decision variables are.
- What is my goal? This tells you your objective function.
- What limits my choices? What is impossible? This tells you your constraints.

Guidelines for formulation are:

- Define index sets first. Use different index variables for different index sets.
- Define the data in full detail, including units.
- Define decision variables in full detail, including units. Choose notation that looks different from the notation for the data. Sometimes I use upper case for data and lower case for variables.
- Separate the data from the model. Some LP modeling languages will force you to make this separation. It may be annoying on small cases, but it is good practice.
- Don't forget nonnegativity constraints and other bounds on variables.

- Write constraints for clarity. Variables may appear on both sides of an equality or inequality; they may appear twice in the same constraint. Permit inequalities in either direction, equalities, and summation and subset notation. It is more important to make the model easy to understand than to put it in some standard form or another.
- Check the units in your constraints from the data and the variables to be sure they match.
- Distinguish auxiliary variables from decision variables. Auxiliary variables are extra variables used in the model for ease or technical reasons, whose values are implied by the values of the decision variables. Use defining (or "linking") constraints to force auxiliary variables to have the values they are supposed to have, and distinguish those constraints from the other constraints.
- Don't use absolute value signs or the terms "min max" or "max min"
- Don't simplify the model using reasoning that relies on specific data values. Leave that to the LP presolver. This rule is important to follow when doing the problems, because otherwise you will often miss the point. This rule is also important in practice because your model is apt to be run many times with data you can't anticipate or oversee.

For real life computation you should follow all of these rules. Possible exception: your software might not permit you to follow the rule about writing constraints for clarity.

3.1 Three Classic Primal/Dual Formulation Pairs

Preview: Here are three classic LPs, the diet problem, the manufacturer's problem, and the 2-person zero-sum game. For each one, it is straightforward to formulate the primal, and the dual can be derived in a natural way by telling a story.

Diet problem and vitamin salesman. Read the label on a bag or can of pet food. It lists minimum amounts of nutrients such as protein, fat, etc. These nutrient requirements do not change, but the pet food itself is made up of different mixtures of ingredients (such as bone meal, beef, rice flour, etc.) at different times of the year. This is because ingredient prices vary seasonally, and the manufacturer uses linear programming to choose the optimal mix of ingredients that minimizes the total cost, while adhering to the nutritional guidelines.

In general, the diet problem involves food choices $i \in I$, unit food prices $c_i : i \in I$, nutrients $j \in J$, and minimum nutrient amounts $b_j : j \in J$. The other data are the values A_{ji} = the number of units of nutrient j contained in one unit of food i. The decision variables are x_i = the number of units of food i to purchase. The constraints, besides $x_i \geq 0$, are $A\mathbf{x} \geq \mathbf{b}$, which say that the food purchased contains at least the minimum nutrient amounts. The objective is to minimize $\sum_{i \in I} c_i x_i$. This is the primal (2.1).

The dual problem arises from a vitamin pill salesman, who sees you about to enter a grocery store, and wants to lure you into his store instead. Why buy carrots, peas, orange juice, and cereal when you could buy his pills instead? The salesman's problem is to choose prices $\pi_j : j \in J$ for the cost of pills containing one unit of nutrient j. The objective is to maximize the net sales, $\sum_{j \in J} \pi_j b_j = \pi \cdot \mathbf{b}$. The constraints, besides $\pi \geq 0$ (*Question: Why?*[1]), are that for each food you could buy, the cost of the pills that contain the same nutrients as that food should not cost more than the food itself. For example, suppose the first food choice $i = 1$ corresponds to carrots in units of one carrot. Then by definition of the data, one carrot supplies A_{j1} units of nutrient j, for each $j \in J$. The combination of pills that have the same nutritional content as a carrot can be thought of as a "synthetic carrot". The term is due to David Luenberger, and we will encounter it later when we derive the simplex method. A synthetic carrot costs $\sum_{j \in J} A_{j1} \pi_j = A_{:,1} \cdot \pi = (A^T)_{1,:} \pi$. The vitamin salesman's constraint is that a real carrot cannot be cheaper than a synthetic carrot, for otherwise you would buy the real carrot. Mathematically the constraint is $A^T_{1,:} \pi \leq c_1$. The set of constraints for all $i \in I$ is $A^T \pi \leq c$. This is precisely the set of dual constraints in (2.3).

If you complain to the salesman that the pills don't taste as good as the foods, he will respond that you have omitted a tastiness constraint from your LP model, and offer to sell you tastiness pills to satisfy the new constraint. This illustrates how constraints in the primal correspond to variables in the dual.

Manufacturer and resource purchaser. A manufacturer owns b_j units of resource $j \in J$. Resources can be thought of as raw materials such as steel, crude oil, etc. or anything else such as floor space, lathe machine time, etc. that contributes towards the manufacture of goods. The manufacturer can choose to use production activities indexed by $i \in I$. Each unit of production activity i requires d_{ij} units of resource j, and produces one unit of product i, which sells for c_i dollars. The decision variables are y_i = the number of units of product i to make. The manufacturer's LP to maximize income is

$$\max \sum_i c_i y_i \quad \text{subject to} \quad \sum_i d_{ij} y_i \leq b_j$$

and $y_i \geq 0$. To convert this to matrix notation, let D be the matrix such that $D_{ij} = d_{ij}$. The problem is to maximize $\mathbf{c} \cdot \mathbf{y}$ subject to $D\mathbf{y} \leq \mathbf{b}, \mathbf{y} \geq 0$. This has the form of the *dual* (2.3) but we are calling it the primal because we wrote it down first. *Question: What in this model corresponds to what is in* (2.3)?[2]

For the dual problem, an agent wants to purchase the manufacturer's resources, for her own mysterious ends. She plans to offer π_j dollars per unit resource $j \in J$. Besides $\pi \geq 0$, her constraints are that, for each production activity available to the manufacturer, she must offer at least as much money for the resources needed by that activity, as the manufacturer would earn by running that activity. That is, for all i,

$$\sum_j a_{ij} \pi_j \geq c_i.$$

In matrix notation this is more compactly written as $D^T \pi \geq \mathbf{c}$. Her objective is to purchase the resources at minimum total cost, i.e., to minimize $\sum_j b_j \pi_j = \mathbf{b} \cdot \pi$. *Question: Why must* $\pi \geq 0$?[3]

Rounding. You may object to the primal manufacturer LP because usually one cannot

[1] A negative price means the salesman pays you.

[2] D corresponds to A^T, $\mathbf{y}$ corresponds to π, $\mathbf{b}$ corresponds to $\mathbf{c}$, and $\mathbf{c}$ corresponds to $\mathbf{b}$.

[3] The manufacturer would keep a resource rather than pay for it to be removed.

sell a fraction of a unit of a product. Therefore, the y_i variables ought to be required to be integers. Similarly, you might object to a primal diet LP because one cannot use a fraction of an egg without paying for a whole egg. In general, if the values of the nonzero variables are large, then rounding fractional values of an LP solution will give an acceptable integer-valued solution. However, if the values are small, e.g. in the single digits, then rounding is apt not to be good enough. *Question: For the primal diet problem, would you round up or down? For the primal manufacturer problem, would you round up or down?*[4]

Two person zero sum games. Two players, I and II, play a game whose data is given in a matrix $A \in \Re^{m \times n}$. Player I has m choices, indexed $i = 1 \ldots m$, corresponding to the rows of A; player II has n choices, indexed $j = 1 \ldots n$, corresponding to the columns of A. If I chooses i and II chooses j, then II must pay A_{ij} dollars to I. If $A_{ij} < 0$ this is interpreted as I paying II. For some matrices A there exists a saddlepoint, a pair of choices α and β that, for each player, is superior to his or her other choices, given what the other player is choosing. For example, the matrix in Figure 3.1 has a saddlepoint in row $\alpha = 2$ and column $\beta = 3$. When such a saddlepoint exists we call $A_{\alpha\beta}$ the *value* of the game. It is what I should pay II for it to be fair to play the game. Morgenstern and von Neumann showed that when no saddlepoint exists, the game still has a value, in a sense of expected value, if the players make choices according to probability distributions [234]. A *mixed strategy* is a probability distribution on a player's set of choices. Morgenstern and von Neumann proved, moreover, that player I has a mixed strategy that guarantees that the expected value of his payoff is at least the value of the game; player II has a mixed strategy that guarantees that the expected amount of her payment to I is at most the value of the game.

1	16	7
29	15	12
3	6	9

FIGURE 3.1: When I chooses row 2, I is sure to receive at least 12. When II chooses column 3, II is sure to pay at most 12.

To find player I's optimal mixed strategy, set up a linear program that finds a mixed strategy with the best guaranteed expected payoff.

$$\max z \quad \text{subject to} \quad \forall j \quad z \leq \sum_{i=1}^{m} x_i A_{ij}; \ \sum_{i=1}^{m} x_i = 1; x \geq 0.$$

The first constraint ensures that z is not more than the expected payoff against each of II's choices. The second and third constraints ensure that the vector $\mathbf{x}$ represents a mixed strategy.

The dual is (try to write this down without looking)

$$\min w \quad \text{subject to} \quad \forall i \quad w \geq \sum_{j=1}^{n} \pi_i A_{ij}; \ \sum_{j=1}^{n} \pi_i = 1; \boldsymbol{\pi} \geq 0. \tag{3.1}$$

In the popular children's game "Roshambo," also called "shoushiling," "rock-paper-scissors," "Burung-batu-air," "Jan-ken," "Baau-jin-dup," etc., each player has three choices, denoted here as rock, paper, and scissors. The two players make their choices simultaneously. Rock wins against scissors, scissors wins against paper, and paper wins against rock. If both

[4]up; down.

players make the same choice, the outcome is a tie. Let the loser pay 5 to the winner. The payoff matrix A is

	Rock	Paper	Scissors
Rock	0	−5	5
Paper	5	0	−5
Scissors	−5	5	0

Consider the (non-mixed) strategy where player I always chooses rock. Player II counters by playing paper. Player I then expects to receive $A_{:,2} \cdot (1,0,0) = -5$. Consider the mixed strategy where I chooses rock with probability 0.6 and scissors with probability 0.4. Player II counters by playing rock. Player I expects to receive $A_{:,1} \cdot (0.6, 0, 0.4) = -2$ on average. Consider the mixed strategy where player I chooses rock, paper, and scissors each with probability $\frac{1}{3}$. Whenever player II chooses rock, player I expects to receive $A_{:,1} \cdot (\frac{1}{3}, \frac{1}{3}, \frac{1}{3}) = 0$ on average. By similar computations, player I expects to receive 0 on average whenever player II chooses scissors or paper. Therefore, no matter what strategy II follows, player I assures himself to do no worse than receiving 0 on average.

Because of the symmetry of the game, a similar computation shows that if player II selects each choice with probability $\frac{1}{3}$, she assures herself to do no worse than paying 0 on average. The value of the game is therefore 0.

The preceding example has too much symmetry to expose the concepts fully, just as equilateral triangles are not representative of triangles in general. Consider now a game with the following payoff matrix. Player I has three choices, a,b,c; player II has two choices, α, β.

	α	β
a	3	5
b	6	2
c	4	1

Reading Tip: *On your second reading, define the variables and formulate each player's LP without peeking at the LPs below.*

Player I's LP has variables $x_i =$ the probability for $i = 1, 2, 3$ of choosing a, b, c, respectively.

$$\begin{array}{llll} \max\ z & & \text{subject to} & \\ & z & \leq 3x_1 + 6x_2 + 4x_3 & \text{versus } \alpha \\ & z & \leq 5x_1 + 2x_2 + 1x_3 & \text{versus } \beta \\ & & x_1 + x_2 + x_3 = 1; x_i \geq 0 & i = 1,2,3 \end{array}$$

The optimal solution is $x_1 = \frac{2}{3}; x_2 = \frac{1}{3}; x_3 = 0$, and $z = 4$.

Player II's LP has variables π_1, π_2 for the probability of choosing α, β respectively. She minimizes the amount she will pay Player I on average.

$$\begin{array}{llll} \min\ w & & \text{subject to} & \\ & w & \geq 3\pi_1 + 5\pi_2 & \text{versus a} \\ & w & \geq 6\pi_1 + 2\pi_2 & \text{versus b} \\ & w & \geq 4\pi_1 + 1\pi_2 & \text{versus c} \\ & & \pi_1 + \pi_2 = 1; \pi_j \geq 0 & j = 1,2 \end{array}$$

The optimal solution is $\pi_1 = \pi_2 = \frac{1}{2}$ and $w = 4$. When players follow their optimal mixed strategies, Player I never chooses c ($x_3 = 0$). Look at the third constraint of Player II's LP.

At $\pi_1 = \pi_2 = \frac{1}{2}$, she would on average pay Player I $4\pi_1 + 2\pi_2 = \frac{4}{2} + \frac{2}{2} = 3$ whenever Player I chose c. Thus if $x_3 > 0$ Player I's expected payoff would fall below his optimal value 4. This illustrates complementary slackness. Player I will never choose a pure strategy for which Player II's LP constraint is not tight.

Summary: We've seen three classic LP formulations for which both the primal and the dual could be written from a verbal description of the problem. More complicated problems rarely offer a natural story that corresponds to the dual.

3.2 Visualizing Problems, Defining Variables, and Formulating Constraints

---

Preview: This section will help you formulate LPs correctly and easily. The main techniques are to visualize the problem, unambiguously identify the decision variables, and convert constraints from words into math.

Most of my students hate formulation. Many stare, baffled, at a formulation problem, but after five minutes their papers and minds are still blank. The only thing they learn is to be traumatized. This bafflement has become even more prevalent among my U.S. students since standardized test preparation crowded word problems out of many primary and secondary school math curricula.

If formulation does not come easily to you, this section will help you. But do not expect a single "aha" moment to suffice. You are acquiring a new skill, like swimming or playing soccer.

Visualize

- *Make the problem your problem.* In any optimization problem, someone has to make a decision. Be that person. For the problem at the beginning of Chapter 1, stand in a field, give yourself a 100-foot roll of chicken-wire fencing, and try to fence off the largest possible amount of rectangular space for your garden. *Question: Read Problem 3.24. Who are you? Read Problem 3.16. Who are you?*[5]
- *Hand-draw a figure that depicts the problem.* This will engage your visual reasoning. Draw on paper because the tactile sensation will activate even more of your brain. For Problem 3.16 I draw two close parallel lines for the road, several stick figures along the road, and a bus stop near but not at the middle of the road. Then I draw an arrow

[5]The pilot deciding where to land; the mayor deciding where to place the bus stop.

from each stick figure to the bus stop and label each arrow with a walking speed and bus arrival time. What do you draw for Problem 3.24?

Example 3.1 Visually depict the following formulation problem:
Next week an oil company will convert certain reserves of five kinds of crude oil into heating oil, jet fuel, and automobile gasoline. Conversion into these products will be performed at its refinery, which can process up to a certain amount of crude per day. Each product is sold by the gallon. For each product and each type of crude oil, there is a specific fraction of a barrel of that type of crude needed per gallon of the product produced. All product must be sold on the day it is produced. Given projected prices for each product on each day of next week, what should the company's production plan be to maximize its revenue?
Figure 3.2 shows my first drawing. The circles are the reserves of each kind of crude oil; the rectangle is the company refinery. Arrows from the circles to the rectangle represent using each kind of crude at the refinery. Arrows leaving the rectangle represent the heating oil, jet fuel, and gasoline produced by the refinery. Why is this figure too simple? If you do not see why, reread the problem description.

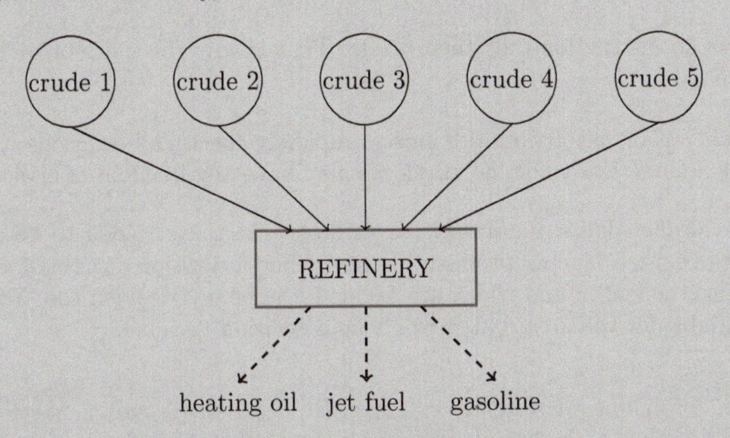

FIGURE 3.2: In this visual depiction of the oil refinery problem, each circle represents a different kind of crude oil to be sent to the refinery, which is represented by a rectangle.

If I am in charge of the company's production, I must decide what the refinery produces on each day of the week. My first figure would only be OK for a one-day problem. But if I replace each arrow in the figure with seven arrows, one for each day, the figure becomes messy and its meaning unclear. Figure 3.3 shows my second drawing. The circles for the crude reserves are the same, but I've drawn a separate rectangle for the refinery on each day for the first three days. The output is depicted separately for each day. I don't draw all seven days because three days are enough to see the pattern, and more days would clutter the figure too much. I will keep this image in my mind as I work through the formulation.

Precisely define a complete set of decision variables

Having made the problem your problem, you are the decision-maker. However, you cannot do anything directly. You can only send instructions to your employees by text message. Define a set of variables such that you can instruct your employees by sending them the values of those variables.

- *No ambiguity* Your variable definitions may not be ambiguous. Employees must know

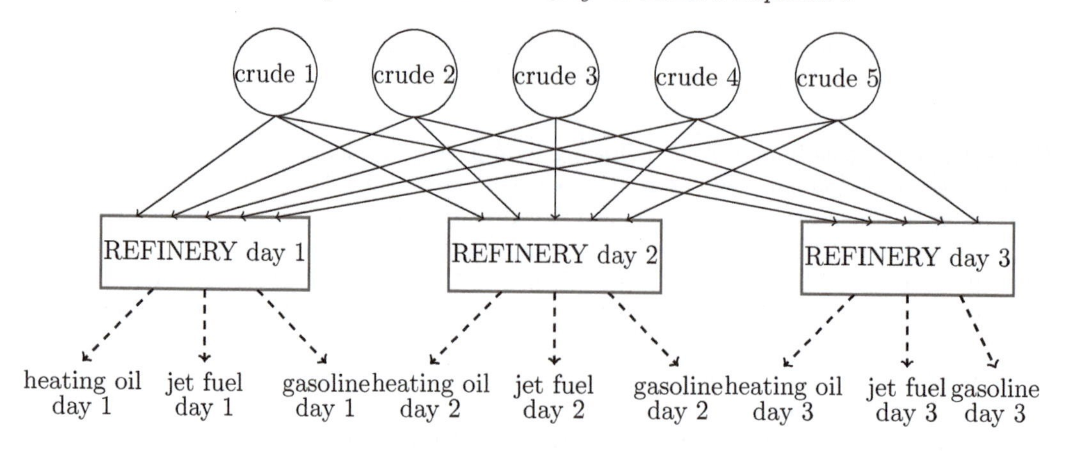

FIGURE 3.3: In this corrected visual depiction of the oil refinery problem, each day of refinery activity is represented by its own rectangle.

exactly what an instruction tells them to do. That is why the units of each variable must be defined.

- *Completeness* Your set of variables must completely specify all decisions. Once each variable has a numerical value, no employee may have any freedom of choice.

As a general rule, do not define more decision variables than are needed to be complete. For example, in the garden fencing problem, do not define a decision variable for the area of the garden. Once the length and width are decided, the area is decided too. Your model may include a variable for the area, but it won't be a decision variable.

Example 3.2 Continuing the refinery example, let's consider some incorrect choices of decision variables. For each choice, try to see what is wrong without reading ahead.

1. x_1 = heating oil; x_2 = jet fuel, x_3 = gasoline.
2. y_j = number of barrels of type j of crude oil, $j = 1, \ldots, 5$
3. x_{jt} = number of barrels of type j crude oil used by the refinery on day t, $j = 1, \ldots, 5; t = 1, \ldots, 7$.
4. x_{jt} = number of barrels of type j crude oil used by the refinery on day t, $j = 1, \ldots, 5; t = 1, \ldots, 7$ and y_{jit} = number of barrels of crude oil type j, $j = 1, \ldots 5$, used to make product i, $i = 1, 2, 3$, on day t, $t = 1, \ldots, 7$.
5. y_{jit} = number of barrels of crude oil type j, $j = 1, \ldots 5$, used to make product i, $i = 1, 2, 3$, on day t, $t = 1, \ldots, 7$.

What's wrong:

1. The x_i don't represent decisions. "Heating oil" is a kind of product, not a decision. No units are defined. If you told your employees "five heating oil, twenty jet fuel, six gasoline" they would not know what to do.

2. Incomplete. The y_j could be taken to mean the total number of barrels of type j crude to use during the week. That isn't specific enough to tell your employees what to do. There could be infinitely many different ways to use those amounts of crude.

3. Incomplete. The x_{jt} are an improvement over the y_j, but they aren't sufficient to tell your employees at the refinery what to do with the crude oils they get each day. They would not know how much of each product to make.

4. The x_{jt} are redundant decision variables. Once the y_{jit} values are decided, the production plan is completely determined. In particular, once you've decided how much of crude type j is used for each product on day t, you've implicitly decided how much crude type j in total is used on day t. In algebra, x_{jt} must equal $\sum_{i=1}^{3} y_{jit}$.

5. The y_{jit} have redundancy. Each product is made of the crude oils in fixed proportion. For example, suppose the formula for 1 barrel of input for jet fuel ($i = 2$) calls for 0.1 barrels type 1, 0.5 barrels type 3, and 0.2 barrels each of types 4 and 5. If $y_{1,2,7} = 50$, then $y_{3,2,7} = 250$ and $y_{4,2,7} = y_{5,2,7} = 100$. Once $y_{1,2,t}$ is decided, $y_{j,2,t}$ are decided for $j = 2, \ldots, 5$.

I would define decision variables like this.

Indices.

$t = 1, \ldots, 7$ index the days of next week.
$j = 1, \ldots, 5$ index the kinds of crude oil.
$i = 1, 2, 3$ index the products, heating oil ($i = 1$), jet fuel, and auto gasoline ($i = 3$).

Units.

All crude oils $j = 1, \ldots, 5$ are measured in barrels.
All products $i = 1, 2, 3$ are measured in gallons.

Decision Variables

y_{it} = # of units (gallons) of product i to make on day t.
Question: why didn't I define y_{it} as the # of barrels of crude oil used to make product i made on day t?[6]
Next I would define my data.
B_j = total units (barrels) of crude type j in reserve.
C_{it} = selling price in dollars on day t of the a unit (gallon) of product i.
U_t = upper bound on # of units (barrels) of crude processed by the refinery on day t. The problem description did not clarify whether the capacity was the same on each day. To play it safe, use a separate term for each t.
A_{ij} = amount of a unit (barrel) of type j crude used when making one unit (gallon) of product i.

Writing the objective function is often a good test of your index, decision variable and data definitions. The total value of refinery production is the sum over each day t and each product i of the sale value $C_{it}y_{it}$, that is,

$$\sum_{t=1}^{7} \sum_{i=1}^{3} C_{it} y_{it}.$$

Convert verbal statements into mathematical constraints

Most constraints sound like common sense when stated in words. The amount of physical inventory cannot be negative; don't exceed a processor's capacity; don't begin step B until step A has been completed. Your job is to translate these statements into math.

- Write the constraint as an inequality or equality with words and index sets. At all times t, inventory at time t is $\geq$ zero; for all k, total work assigned to processor $k \leq$ capacity of processor k; start time of step B $\geq$ finish time of step A.
- Plug in your precisely defined variables and data to convert the constraint into math. Be sure all measurement units (e.g., time, mass, quantity) match, as in a physics equation.

Example 3.3 Continuing the refinery example, one obvious constraint is that you cannot use more of each kind of crude oil than you have in reserve. Write, for all j:

Amount of each kind crude oil used	is at most	Amount each kind crude oil in reserve
# barrels crude type j used	$\leq$	# barrels crude type j in reserve
Sum over days t of # barrels type j used day t	$\leq$	B_j barrels
$\sum_{t=1}^{7}\sum_{i=1}^{3} A_{ij}y_{it}$	$\leq$	B_j

If the left-hand side of the last line is not clear to you, write:

$$\begin{aligned}
&\#\text{ barrels type } j \text{ used on day } t\\
&= \text{sum over } i \ \{\#\text{ barrels type } j \text{ used on day } t \text{ for product } i\}\\
&= \sum_{i=1}^{3} \#\text{ barrels type } j \text{ crude used per unit product } i\\
&\qquad\qquad \times \quad \#\text{ units product } i \text{ made on day } t\\
&= \sum_{i=1}^{3} A_{ij}y_{it}
\end{aligned}$$

Similarly, to convert the common-sense statement that you cannot exceed the refinery capacity on any day, write

On each day don't use more crude oil	than the	refinery can handle that day
amount crude oil used day t	$\leq$	refinery capacity day t $\forall t$
$\sum_{j=1}^{5}$ # barrels crude type j used day t	$\leq$	U_t $\forall t$

$$\sum_{j=1}^{5}\sum_{i=1}^{3} A_{ij}y_{it} \leq U_t \text{ for all } t$$

[6] I don't know how many barrels go into making one gallon. Moreover, the formula might change or be different at some other refinery.

Don't forget nonnegativity constraints. You can't produce a negative amount of any product on any day: $y_{it} \geq 0$ for all i and t.
We don't want to have to change the model if merely the number of days, crude oil types, or product types change. The more adaptable LP model is then:

Indices

$t = 1 \ldots T$ days;
$j = 1 \ldots J$ types of crude oil;
$i = 1 \ldots I$ types of products.

Decision variables

y_{it} = # of units of product i made at the refinery on day t.

Data

B_j = total units of crude type j in reserve.
C_{it} = unit selling price in \$ on day t of product i.
U_t = Maximum # of units of crude the refinery can process on day t.
A_{ij} = # of units of type j crude used when producing one unit product i.

Objective

Maximize $\sum_{t=1}^{T} \sum_{i=1}^{I} C_{it} y_{it}$.

Constraints

$$\sum_{t=1}^{T} \sum_{i=1}^{I} A_{ij} y_{it} \leq B_j \text{ for all } j$$
$$\sum_{i=1}^{I} \sum_{j=1}^{J} A_{ij} y_{it} \leq U_t \text{ for all } t$$
$$y_{it} \geq 0 \text{ for all } i \text{ and } t.$$

Practitioner Tips:
(1) Try to use different letters for different index sets.
(2) Be sure you define all unit quantities and use them consistently, especially since barrels are not gallons.
(3) When a constraint is actually a set of constraints, specify the index sets over which the constraint applies. Often the constraint must be slightly different at the first or last value of an index to capture initialization or termination conditions.
(4) Theoretically, decision variables y_{it} completely determine how much of each crude type should be used on each day towards making each product. But they do not specify *how much. We have implicitly assumed the refinery workers will adhere to the A_{ij} data. It may be wise to post-process the LP output to generate specific production instructions for the refinery, and to assure that the correct amounts of each crude type are sent to the refinery each day. We could instead embed the detail into the LP model. Define additional variables x_{jit}, w_{jt} as the # of barrels of crude type j on day t used to make product i, and used by the refinery, respectively. These are not decision variables; they are* auxiliary *variables (see Section 3.3). To ensure their values are correct, we must add the constraints $x_{jit} = A_{ij} y_{it}$ and $w_{jt} = \sum_{i=1}^{I} x_{ijt}$.*

Summary: The key to modeling is to visualize yourself as the decision maker. Express the choices you have with decision variables, and your goal with your objective function. Describe your requirements and limitations in words; translate them into mathematical constraints on your decision variables.

3.3 LP Modeling Methods

A method is a trick you use more than once. – Richard E. Stone.

Preview: This section will teach you several methods that extend the reach of linear programming formulation.

The first method is for min-max problems. To introduce it, suppose you are given this set of numbers: $10, 16, 85, -33, 91, 70, 21$. Write a linear program to pick the largest number from the set. One answer is: minimize x subject to $x \geq 10, x \geq 16, x \geq 85, \ldots, x \geq 21$. This forces the optimal solution to equal the maximum of these numbers. This method works even if, instead of numbers on the right-hand-side of the constraints, there were linear functions of variables. The value of x will be the value of the largest right-hand side values, and the other variables will take whatever values are needed to make that largest value as small as possible.

By the way, the dual answer is: maximize $10\pi_1 + 16\pi_2 + \cdots + 21\pi_7$ subject to $\pi_j \geq 0\ \forall j; \sum_{j=1}^{7} \pi_j = 1$. In words, choose weights π_j so that the weighted sum of the numbers is as large as possible. The optimal solution places a weight of 1 on the largest number and weights of zero on the other numbers.

You may doubt the validity of this method. It works because linear programming algorithms find optimal solutions. Suppose you had to minimize the maximum of $3+x$ and $5-x$. You model the problem as minimize z subject to $z \geq 3 + x$ and $z \geq 5 - x$. Not only will z equal the maximum of $3 + x$ and $5 - x$ at the optimal solution, but also the algorithm will choose the optimal value $x = 1$ that makes this maximum as small as possible. Convince yourself that you can minimize the maximum of $1 + 2x$ and $5 - 3x$. Now replace the single variable x by a vector of variables, the scalars 2 and 3 by vectors of constants, throw in extra constraints $A\mathbf{x} \leq \mathbf{b}$, and you will see the power of the method.

Example 3.4 The following problem is not an LP because its objective function is not linear.

$$\begin{aligned} \min_{x_1,x_2} \quad \max \quad & \{\{4x_1 + 7x_2\}, \{10x_1 - 3x_2\}, \{-2x_1 + 9x_2\}\} \\ \text{subject to} \quad & \\ & x_1 - x_2 \leq 3.5 \\ & x_1 - x_2 \geq -2.5 \\ & 0 \leq x_i \leq 100 \quad i = 1, 2 \end{aligned}$$

Yet, we can model it with the following LP.

$$
\begin{array}{rcl}
 & & \min z \text{ subject to} \\
z & \geq & 4x_1 + 7x_2 \\
z & \geq & 10x_1 - 3x_2 \\
z & \geq & -2x_1 + 9x_2 \\
x_1 & - & x_2 \leq 3.5 \\
x_1 & - & x_2 \geq -2.5 \\
0 \leq & x_i & \leq 100 \quad i = 1, 2
\end{array}
$$

A related method is for dealing with absolute values. If you want to minimize the absolute value of y_1 but you don't know whether y_1 is going to be positive or negative, you can replace $y_1 = y_1^+ - y_1^-$ where the new variables are nonnegative, and use $y_1^+ + y_1^-$ for $|y_1|$.

Example 3.5 Let A be an m by n matrix and b an m-vector. Let $\mathbf{y} = y_1, \ldots, y_n$ be a vector of variables. The problem

$$\min |y_1| \text{ subject to } A\mathbf{y} \leq \mathbf{b}$$

is not a linear program. Yet it can be modeled as the LP

$$\min y_1^+ + y_1^- \text{ subject to } y_1^+ \geq 0, y_1^- \geq 0, A \begin{pmatrix} (y_1^+ - y_1^-) \\ y_2 \\ y_3 \\ \cdots \\ y_n \end{pmatrix} \leq \mathbf{b}.$$

However, the method fails on the problem, maximize $|y|$ subject to $y = 1$. *Question: What is the value of the LP you would get if you tried to use this method to maximize* $|y_1|$*?*[7]

Reading Tip: *You just read that the method can fail. Did you read passively? Write the analogous LP model for maximizing* $|y|$ *subject to* $y = 1$*. Find precisely why its optimum is unbounded and the method fails. Understand why maximization in this problem is so different from minimization.*

Another method to minimize the absolute value of y is to use both y and $-y$.

Example 3.6 To get x equal to the maximum of $|y|$ and $|w|$, minimize x subject to constraints

$$x \geq y, x \geq -y, x \geq w, x \geq -w.$$

Question: How would you get x *equal to the maximum of* $3|y|$ *and* $4|w|$*?*[8]

---

[7] Unbounded optimum.

[8] Minimize x subject to constraints $x \geq 3y, x \geq -3y, x \geq 4w, x \geq -4w$.

Example 3.7 Given a set of lines in the plane, find the smallest circle that touches all of them. More generally, given a set of hyperplanes in $\Re^m$, find a smallest closed ball that intersects all of them.
Solution: let $1 \le i \le m$ index the dimensions, and let $j \in J$ index the set of hyperplanes. The decision variables are $\mathbf{x} = x_1, \ldots, x_m$, representing the coordinates of the center of the ball, and r, representing the radius of the ball. How about the data? They can be expressed as $Ay = \mathbf{b}$ where A is a $|J|$ by m matrix and $\mathbf{b}$ is an m-vector. Moreover, without loss of generality we can assume $\mathbf{b} \ge 0$ and $||A_{j,:}|| = 1$. The distance between the ball center and the jth hyperplane is $|A_{j,:}x - b_j|$. Combining the two methods for min-max problems and absolute values, the model therefore is

$$\begin{aligned} &\min \quad r \quad \text{subject to} \\ r &\ge A_{j,:}\mathbf{x} - b_j \quad \text{for all } j \\ r &\ge b_j - A_{j,:}\mathbf{x} \quad \text{for all } j \end{aligned}$$

Auxiliary Variables
The most important method for extending the range of LP models is the use of auxiliary variables. These are variables whose values are determined by the decision variables. In that sense the auxiliary variables are not necessary. But they can make it much easier to write a formulation and to do sensitivity analysis. You have to remember to write constraints that force the auxiliary variables to have the values they should have, given the decision variables.

For example, in a production planning problem, our decision variables might be how much to produce and how much to sell in each time period. The inventory at the end of each time period would be good auxiliary variables. To be specific, let $i \in I$ index products; let $t \in T = \{1, 2, \ldots |T|\}$ index time periods. For the data, let c_{it} be the cost of producing one unit of product i in period t; let C_t be the total number of units of product that can be produced in period t; let d_{it} be the demand for product i in period t. Let b be the maximum number of units of product that may be stored from one period to the next. For decision variables, let x_{it} equal the number of units of product i produced in period t. For auxiliary variables, let y_{it} equal the number of units of product i stored in inventory from period t to period $t + 1$. The LP model is

$$\begin{aligned} &\min \sum_{t\in T, i\in I} c_{it}x_{it} \quad \text{subject to} \\ \sum_{i\in I} y_{it} &\le b \;\; \forall t \in T \\ y_{i1} &= x_{i1} - d_{i1} \;\; \forall i \in I \\ y_{it} &= y_{i\ t-1} + x_{it} - d_{it} \;\; \forall i \in I\ \forall t \ge 2 \\ x_{it}, y_{it} &\ge 0 \;\; \forall i \in I, \forall t \in T \end{aligned}$$

The y_{it} are auxiliary in the sense that once the x_{it} values are chosen, the y_{it} values are fixed. The auxiliary variables make it easier to write the model correctly and to handle some real-world complications. In the retail industry, there is a well-known phenomenon called "shrink." Shrink is an unplanned decrease in inventory levels over time, typically due to employee theft, product aging or damage, and mislabeling or physical misplacement of product. Suppose that the shrink rate of product i is approximately such that a fraction a_i of inventory stored from one year to the next is lost. Account for this shrink with the auxiliary variables in the model. Change the definition of y_{it} to the number of units of product i *successfully* stored from period t to period $t+1$. Change the defining constraints to

$$\begin{aligned} y_{i1} &= (1-a_i)(x_{i1}-d_{i1}) && \forall i \in I \\ y_{it} &= (1-a_i)(y_{i\ t-1}+x_{it}-d_{it}) && \forall i \in I\ \forall t \geq 2 \end{aligned}$$

Example 3.8 [Maximizing Expected Profit over a Set of Scenarios]
The Problem: Your bakery makes different types of perishable pastries, indexed by $i = 1, \ldots, n$. You must decide how much of each type to make tonight for sale tomorrow. Any pastries not sold by tomorrow night are discarded. Let x_i be the number of pastries of type i to produce tonight. Your production capability is described by the constraints $A\mathbf{x} \leq \mathbf{b}; \mathbf{x} \geq \mathbf{0}$. The per/unit cost of making pastry type i is c_i, and its unit selling price is s_i.
As described so far, your problem calls for a simple production model with objective to maximize $\sum_{i=1}^{n}(s_i - c_i)x_i$ subject to $A\mathbf{x} \leq \mathbf{b}$, $\mathbf{x} \geq \mathbf{0}$. The complication is that you are quite uncertain as to tomorrow's demand. Based on sales records and weather forecasts, you envision $m = 3$ scenarios, indexed by $j = 1, 2, 3$, depending on tomorrow being hot, cold, or rainy, with probabilities $p_j : j = 1, 2, 3$, respectively. For weather scenario j, you estimate the demand for pastry type i to be D_{ij}. Your problem is to choose $\mathbf{x}$ to maximize your expected profit. *Question: Think of a simple example with* $n = 1, m = 2$ *to show it is incorrect to merely add the constraints* $x_i \leq \sum_{j=1}^{m} p_j D_{ij}$.[9]

The Solution: The decision variables are the same x_i defined above. The constraints $A\mathbf{x} \leq \mathbf{b}$ and $\mathbf{x} \geq \mathbf{0}$ remain. Define auxiliary variables to capture the expected profit. Let $y_{ij} \geq 0$ be the number of pastry type i sold in scenario j. Constrain $y_{ij} \leq x_i$ and $y_{ij} \leq D_{ij}$ for all i and j. The model consists of the indices i, j, data $A, \mathbf{b}, \mathbf{p}, D, \mathbf{c}, \mathbf{s}$, and variables x_i, y_{ij} as defined above, together with the following objective and constraints.

$$\max \sum_{i=1}^{n}\sum_{j=1}^{m} p_j s_i y_{ij} - \sum_{i=1}^{n} c_i x_i \text{ subject to}$$

$$A\mathbf{x} \leq \mathbf{b}$$

$$y_{ij} \leq x_i\ \forall i, j \quad \text{Don't sell more than what you produced}$$

$$y_{ij} \leq D_{ij}\ \forall i, j \quad \text{Don't sell more than the demand in scenario } j$$

$$y_{ij} \geq 0, x_i \geq 0\ \forall i, j \quad \text{Nonnegativity of production and sales}$$

Penalties are another reason to use auxiliary variables. Sometimes we have a set of constraints that we would like, but do not expect, to be fully satisfied. Our objective, or part of our objective, is to minimize the amount by which these constraints are violated, in some sense. Such problems are typical of *multiobjective optimization, goal programming* and *classification* for data mining in statistics. The general idea is to define auxiliary variables

[9] $p_1 = p_2 = \frac{1}{2}$; $D_{1,1} = 0, D_{1,2} = 100$. When $x_1 = 50$ the expected number sold is 25. But if $s_1 > 2c_1$, $x_1 = 100$ is optimal.

that measure constraint violations, and to incorporate minimization of those variables into the objective.

Classification Model for Data Mining

We have data about "good" and "bad" outcomes and we wish to be able to predict a future outcome. For example, we have records of sales prospects whom our sales force pursued, containing data regarding each prospect *at the time the decision was made to pursue the prospect* and telling whether or not the end result was a sale. For another example, we have records of loan recipients, each containing the credit scores, age, income, marital status, years of employment, number of late payments in the past 12 months, etc. *at the time the loan was made* and telling whether or not the recipient defaulted on the loan. We want a prediction method that advises us, in the former case, whether or not to pursue a prospect, and in the latter case, whether or not to extend a loan. This is called a classification problem because we wish to classify prospects into the categories "good" and "bad."

I will build the model in stages. Let $g^i : i = 1, 2, \ldots, m_g$ be the data vector that describes the ith good prospect or loan recipient, and let $b^j : j = 1, 2, \ldots, m_b$ be the data vector that describes the jth bad prospect or loan recipient. Geometrically, the g^i and b^j are good and bad points scattered in $\Re^n$, where n is the number of items of data that describe a prospect or loan recipient. Ideally, we would like to find a hyperplane $\mathbf{x} : \mathbf{v} \cdot \mathbf{x} = \mathbf{v}_0$ that separates all of the good points from all of the bad points. Therefore, only the relative locations of the points matter. Shift the locations of all points to be in the strictly positive orthant by adding a suitably large constant vector to each g^i and each b^j. (You will soon see the reason for this shift.) Let v_{max} equal the largest value of any component of all of the data vectors g^i and b^j. The linear programming model for this does not even require an objective function. There are n variables, $\mathbf{v}_1, \ldots, \mathbf{v}_n$. The constraints are:

$$g^i \cdot \mathbf{v} \geq v_{max}/2; \qquad \forall i = 1, \ldots, m_g; \tag{3.2}$$
$$b^j \cdot \mathbf{v} \leq v_{max}/2; \qquad \forall j = 1, \ldots, m_b; \tag{3.3}$$
$$\mathbf{v}_1, \ldots, \mathbf{v}_n \qquad \text{unrestricted.} \tag{3.4}$$

Question: Why did I fix the right-hand side to $v_{max}/2$ rather than making it a variable $\mathbf{v}_0$*?*[10] The variables $\mathbf{v}_1, \ldots, \mathbf{v}_n$ are unrestricted to permit the hyperplane to be oriented in any possible way. *Question: Before reading further, why is this model apt to fail even though the* $\mathbf{v}$ *variables are unrestricted?*

The LP model (3.2),(3.3),(3.4) may seem good. But as illustrated in Figure 3.4 it could be infeasible. For real world data, it is extraordinarily rare to be able to separate good from bad points perfectly with a hyperplane or with any other simple rule.

To make the model more practical, let's create a "forgiveness zone" on each side of the hyperplane. We will permit points to be on the "wrong" side of the hyperplane by an amount up to α with the following constraints:

$$g^i \cdot \mathbf{v} \geq 1 - \alpha \qquad \forall i = 1, \ldots, G; \tag{3.5}$$
$$b^j \cdot \mathbf{v} \leq 1 + \alpha \qquad \forall j = 1, \ldots, B. \tag{3.6}$$
$$\mathbf{v}_1, \ldots, \mathbf{v}_n \qquad \text{unrestricted.} \tag{3.7}$$

For large enough α feasible values for $\mathbf{v}$ must exist. Moreover, we have a natural objective function:

$$\text{Minimize } \alpha \tag{3.8}$$

[10] To avoid the problem of scaling $\mathbf{v}$. The solution $\mathbf{v} = 0, \mathbf{v}_0 = 0$ would be a feasible but useless solution.

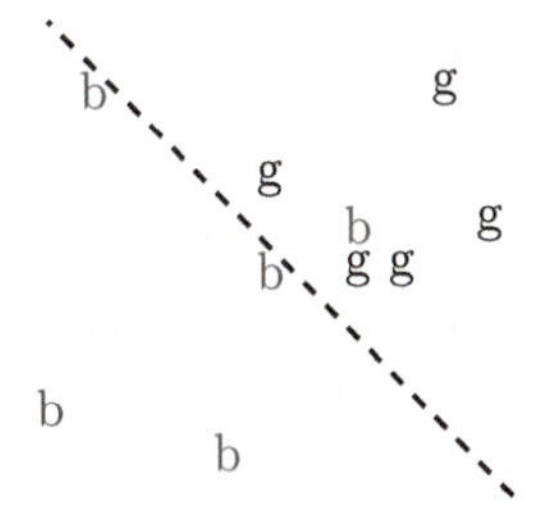

FIGURE 3.4: In this $n = 2$-dimensional example of classification, a hyperplane is a line. However, no line perfectly separates the good points, indicated by the letter g, from the bad points, indicated by the letter b.

What have we done? Instead of trying for a perfect classification, we've sought a classification that has minimum error with respect to our data. In the LP model (3.8)–(3.7), the measure of error is the maximum amount by which a point is on the wrong side of the hyperplane $\mathbf{v} \cdot \mathbf{x} = 1$. *Question: Can you see a weakness in this model.?*[11]

Figure 3.5 illustrates why minimizing the largest error might not be a good choice for the objective. A single data point can dominate the solution and drastically affect the location

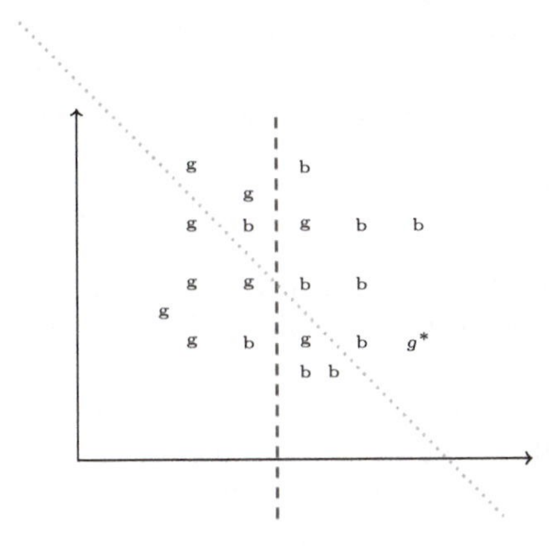

FIGURE 3.5: Without the data point g^* the dashed line is an optimum solution. With g^* the dotted line is optimal. Hence g^* exerts enormous influence on the solution.

and orientation of the hyperplane. Since each prediction is either correct or incorrect, we should not greatly distinguish between "incorrect" and "very incorrect". Fortunately, linear programming offers another natural choice of how to measure error. Define the "error" of a good data point g^i to be

- 0 if $g^i \cdot \mathbf{v} \geq 1$;
- $1 - g^i \cdot \mathbf{v}$ otherwise.

Similarly, define the "error" of a bad data point b^j to be

- 0 if $b^j \cdot \mathbf{v} \leq 1$;
- $b^j \cdot \mathbf{v} - 1$ otherwise.

[11] Read on.

It adds no computational burden, to minimize a weighted sum of the errors. For example, you can represent multiple identical data points by a single weighted point. Let r_i be the weight on the error of g^i and let p_j be the weight on the error of b^j. (I'm using r for the word "regret" and p for the word "penalty.")

The linear program has n decision variables $\mathbf{v}_1, \ldots, \mathbf{v}_n$ and $G+B$ auxiliary variables α_i and β_j, representing the errors of the good and bad points.

$$\begin{array}{lrll} \text{Minimize } \sum_{i=1}^{G} r_i\alpha_i + \sum_{j=1}^{B} p_j\beta_j & & \text{subject to} & \\ & g^i \cdot \mathbf{v} \geq & 1-\alpha_i & \forall i = 1, \ldots, G \\ & b^j \cdot \mathbf{v} \leq & 1+\beta_j & \forall j = 1, \ldots, B. \\ \alpha_i \geq 0 \ \ \forall i; \beta_j \geq 0 \ \ \forall j & & & \end{array}$$

In this model, the error variables are nonnegative. We want to penalize for how much a point is on the wrong side of the hyperplane, but we don't care by how much a point is on the correct side of the hyperplane.

Why might some errors have larger weights than others? In the case of classifying sales prospects as good or bad, a nonzero α_i corresponds to a false negative, that is, a good prospect who is wrongly classified as bad. You would regret missing the opportunity to do business with a good prospect. A nonzero β_j corresponds to a false positive, a bad prospect who is classified as good. If the profit from a sale is much larger than the cost of pursuing a prospect, as is often the case with "big ticket" items such as houses or cars, a false negative is much worse than a false positive. The value of a missed sale is larger than the penalty for pursuing a bad prospect. The r_i would be larger than the p_j in this case.

Banks that loan money are in the opposite situation. A false negative is a customer who would have repaid a loan but was turned down – a missed opportunity to make a profitable loan, but not an important mistake in a world full of loan applicants. However, a false positive is a customer who defaults on a loan. That is an expensive mistake (at least in principle if the bank is not permitted to sell 100% of a loan). In this case, the p_j would be larger than the r_i. As a rule of thumb, put weights on penalty or error terms in your objective function, even if initially the weights are all equal to 1. Weights will not slow the computation, and they add generality to your model.

In the data mining example, there is another more subtle reason to make weights unequal. The data may be a biased sample. A sales department can readily track the outcomes of prospects who were pursued, but not the outcomes of prospects who were not. A bank can easily retain data about all customers who apply for loans. It can also easily determine whether or not a customer who received a loan defaulted. But how can a bank determine whether a customer who was turned down for a loan would or would not have defaulted? At best, the bank can only get partial data based on customers who received loans from other banks, and possibly some sparse data based on experimental loans that it made. In both cases, data corresponding to points that are apt to be classified as "bad" will be underrepresented. False negatives may be indistinguishable from true negatives, and both may be sparsely represented in the data set. To compensate, we may increase the weights on these data points.

Going beyond the scope of LP: You may have thought of choosing a hyperplane that minimizes the number of false positives plus the number of false negatives. Better yet, put a weight on each data point, and minimize the sum of the weights of all data points on the wrong side of the hyperplane. This is a good thought, but it is beyond what linear programming can handle. The objective function requires us to have a yes/no answer to the question of whether a point is on the correct side of the hyperplane. We would need an integer programming model for that, as explained in Chapter 15.

Another idea that may have occurred to you is to use more than one hyperplane to separate the good points from the bad points. We could define a matrix A of variables, and aim to satisfy $Ag^j \leq \mathbf{1}$ and $Ab^i \nleq \mathbf{1}$, or satisfy $Ab^j \leq \mathbf{1}$ and $Ag^j \nleq \mathbf{1}$. *Question: Why are these not the same goal?*[12]

This is a good idea that can permit a more accurate classification than a single hyperplane could provide. However, it requires the model to determine if $A_{1,:}b^j < 1$ or $A_{2,:}b^j < 1$ or ... or $A_{k,:}b^j < 1$. These "or" relationships are beyond the scope of linear programming. Chapter 15 shows you how to model these with integer programming. If you try to wriggle out of this difficulty by constraining $Ag^i \leq \mathbf{1}$ for all i and maximizing the sum of errors of the bad points b^j, you will be trying to maximize the maximum of a set of linear functions. The maximum of a set of linear functions is a convex function. Maximizing a convex function is much more difficult than minimizing a convex function – see Theorem 1.3.1 – and is out of the scope of LP.

Linearization is another method to create LP models of problems that don't initially appear to be LPs. Usually we linearize by substituting new variables in place of products of variables. For example, combine the diet and manufacturing scenarios of Section 3.1 into a blending and manufacturing problem as follows. Let $i \in I$ index ingredients that will be blended together to create a variety of products indexed by $j \in J$. For each product j, there are upper and/or lower bounds U_{ij}, L_{ij} on the fraction of ingredient i that may go into product j. The manufacturer must decide how much of each product j to manufacture, and what mix of ingredients to use for each product. The obvious decision variables would be x_j equal to the number of units, say kilograms, of product j to make, and f_{ij} equal to the number of kilograms of ingredient i to put into each kilogram of product j. Some of the constraints, such as $L_{ij} \leq f_{ij} \leq U_{ij}$ and $\sum_{i \in I} f_{ij} = 1$, are easy to write. Some parts of the objective function, such as $\sum_{j \in J} C_j x_j$ where C_j is the selling price of a kilogram of product j, are easy to write. But the number of units of ingredient i used, which may be constrained by an upper bound and may entail a cost term in the objective function, is not easy to write. In terms of x_j and f_{ij} the number of kilograms of ingredient i used equals $\sum_{j \in J} f_{ij} x_j$. This is nonlinear because it contains products rather than linear combinations of variables. It would seem that this problem cannot be modeled as an LP.

We resolve the nonlinearity by defining new variables $y_{ij} \equiv f_{ij} x_j$, equal to the total number of units of ingredient i used in the creation of product j. Since $x_j = \sum_{i \in I} y_{ij}$, the entire LP model can be written in terms of the y_{ij} variables. In practice, it is more convenient to retain the x_j variables as auxiliary variables. Then the constraint $f_{ij} \leq U_{ij}$ simply becomes $y_{ij} \leq U_{ij} x_j$. The f_{ij} variables cannot be retained, because they are nonlinear functions of the y_{ij} variables, but they are not needed.

For another example of linearization, consider the fractional optimization problem

$$\min \quad \frac{\mathbf{c} \cdot \mathbf{x}}{\mathbf{d} \cdot \mathbf{x}} \quad \text{such that:} \tag{3.9}$$

$$A\mathbf{x} \leq \mathbf{b} \tag{3.10}$$

where $\mathbf{d}, A$ and $\mathbf{b}$ are such that $A\mathbf{x} \leq \mathbf{b} \Rightarrow \mathbf{d} \cdot \mathbf{x} > 0$. (*Question: How could you verify this*

[12]The first encloses the good points in a polyhedron that excludes all bad points; the second encloses the bad points in a polyhedron that excludes all good points. If the good points formed a small cluster surrounded on all sides by bad points, the first but not the second would be feasible.

condition?[13]) Let $r = \frac{1}{\mathbf{d}\cdot\mathbf{x}}$. In terms of r and $\mathbf{x}$ the problem is

$$\begin{aligned} \min \quad & r\mathbf{c}\cdot\mathbf{x} \quad \text{such that:} && (3.11)\\ & r\mathbf{d}\cdot\mathbf{x} = 1 && (3.12)\\ & A\mathbf{x} \le \mathbf{b} && (3.13)\end{aligned}$$

Model (3.11),(3.12),(3.13) is not an LP since it contains products of variables. Create new variables $\mathbf{y} = r\mathbf{x}$. The model now becomes

$$\begin{aligned} \min \quad & \mathbf{c}\cdot\mathbf{y} \quad \text{such that:} && (3.14)\\ & \mathbf{d}\cdot\mathbf{y} = 1 && (3.15)\\ & A\mathbf{y} \le r\mathbf{b}; && (3.16)\end{aligned}$$

which is an LP in variables $\mathbf{y}, r$. I would include the constraint $r \ge \epsilon$ for a small value $\epsilon > 0$ because I don't like to rely on assumptions about my data. Problem 26 asks you to generalize the linearization (3.14), (3.15),(3.16) for cases where (3.9) is the ratio of affine (linear plus a constant) functions.

Unwritten Unwritable Constraints

I call the last method the "unwritten, unwritable constraint." This is a very general method which includes a few of the other methods as special cases. The idea is to count on the objective function to enforce a constraint that is nonlinear. Because the constraint is nonlinear, it can not be and is not written. But it will be enforced at the optimum solution. The substitution of $y = y^+ - y^-, y^{\ge}0, y^- \ge 0$ and use of $y^+ + y^-$ for $|y|$ is a special case of this method. There are infinitely many ways for $y^+ - y^-$ to equal y. Only one of these ways sets $y^+ + y^- = |y|$. That is the one where $y^+ = 0$ or $y^- = 0$. The constraint $y^+ = 0$ or $y^- = 0$ can not be written in an LP because it has an "or" in it. (Alternatively, the constraint $y^+y^- = 0$ can't be written because it is nonlinear.) However, if the coefficients for y^+ and y^- in the minimizing objective function are both strictly positive, the unwritten constraint is automatically satisfied at the optimum. Any solution that violates that constraint could be improved by simultaneously reducing the values of y^+ and y^-.

Another use of the unwritten, unwritable constraint occurs when there are overtime labor costs or other increasing tiers of resource costs. Suppose y_i denotes the amount of hours of labor used in period i, and that up to some amount u_i it costs c_i per hour, but beyond that it costs c_i' per hour, where $c_i' > c_i$, because of overtime labor rates. Or suppose in the diet problem y_i denotes the number of ounces of orange juice in the diet, at cost c_i per ounce. However, your usual supplier only has u_i ounces, and any orange juice in excess of u_i ounces costs $c_i' > c_i$ per ounce. How do we model the problem as an LP? Let y_i denote the amount (of labor hours or orange juice ounces) purchased at price c_i, and let a new variable y_i' denote the amount purchased at price c_i'. Where y_i had appeared in the model, now use $y_i + y_i'$. In the objective function, where c_iy_i had appeared (as a cost to be minimized), now use $c_iy_i + c_i'y_i'$. Include the upper bound $y_i \le u_i$. The unwritten, unwritable constraint is that $y_i' = 0$ if $y_i < u_i$, or equivalently $y_i'(u_i - y_i) = 0$. At the optimum, the unwritten constraint is satisfied because an hour of labor is an hour of labor, or a orange juice is orange juice, and the LP solver will choose the cheaper source as long as it is available.

If we change the story about the orange juice to a bulk discount, so that every ounce of orange juice beyond u_i costs less, we would wish for the same unwritten constraint to be satisfied, but it won't be. The LP solver will use the discounted orange juice and never purchase the more expensive orange juice. An integer program would be needed to model bulk discounts.

[13] Minimize $\mathbf{d}\cdot\mathbf{x}$ subject to $A\mathbf{x} \le \mathbf{b}$ and get a positive objective value.

The same unwritten, unwritable constraint reasoning applies to decreasing tiers of income or other benefits. Suppose you can sell up to u_6 kilograms of chocolate-covered toffee for c_6 \$/kg in your primary market. In a secondary market, you can sell up to u_6' additional kilo for c_6' \$/kg, where $c_6' < c_6$. Let x_6 and x_6' denote the sales in the primary and secondary markets, respectively. The sum $x_6 + x_6'$ denotes the total amount sold, and the objective function, if maximizing, contains the terms $c_6x_6 + c_6'x_6'$. The unwritten constraint, that $x_6' = 0$ if $x_6 < u_6$, will be satisfied at optimality.

Practitioner's warning. If you use an unwritable constraint that depends on properties of the data, such as the $c' > c$ orange juice example, you must state that required property in your data definition documentation. Remember that your model will be used by people whose data you cannot predict, oversee, or control. You can safeguard your LP by including the required data properties in the model. In the juice example, create two auxiliary variables, d_1 and d_2, and the three constraints $d_1 = c$, $d_2 = c'$, and $d_2 - d_1 \geq \epsilon$ where ϵ equals 0 or some suitably small value such as 0.001.

Summary: The main methods for enlarging the scope of LPs are auxiliary variables, minimizing the maximum (or maximizing the minimum) of two or more linear functions, linearization, and unwritten, unwritable constraints that rely on optimization for enforcement. Be sure that you document any data requirements that unwritten constraints rely on. The next section provides more illustrations of these methods.

3.4 More Examples of LP Formulation

Preview: This section contains seven more examples of LP formulation. Each example has several variations or complications to show what can and what can't be handled by LP.

Example: shipping cargo in compartments with balance constraints

A transport vehicle has compartments indexed by $j \in J$. It carries different kinds of cargo types indexed by $i \in I$. Each cargo type has a given volume V_i cubic feet per ton, profit per ton P_i, and total availability of A_i tons. Each compartment has a weight capacity of W_j tons, and a space capacity of S_j cubic feet. To avoid unbalancing the vehicle, the fractions of cargo weight to weight capacity may not vary by more than a factor of $\frac{6}{5}$ from compartment to compartment. For example, if compartment 1 has $W_1 = 100$ and contains 50 tons, then no other compartment may hold more than 60% of its weight capacity, or less than $41\frac{2}{3}$% of its weight capacity. What should the vehicle carry to maximize profit?

The decision variables are x_{ij}= number of tons of cargo type i to carry in compartment

j. The objective is to maximize $\sum_{i\in I, j\in J} P_i x_{ij}$. The simple constraints are

$$\begin{aligned} 0 &\leq x_{ij} \ \forall i \in I, j \in J; \\ \sum_{j\in J} x_{ij} &\leq A_i \ \forall i \in I; \\ \sum_{i\in I} x_{ij} &\leq W_j \ \forall j \in J; \\ \sum_{i\in I} V_i x_{ij} &\leq S_j \ \forall j \in J. \end{aligned}$$

The balancing constraints are not so obviously linear. To make them easier to manage, introduce auxiliary variables representing the total weight of cargo in each compartment, defined by constraints

$$y_j = \sum_{i\in I} x_{ij}.$$

(If we wish we could now rewrite the total weight constraints as $y_j \leq W_j \ \forall j \in J$.) For every ordered pair of compartments $j1 \neq j2$,

$$y_{j1}/W_{j1} \leq 6y_{j2}/5W_{j2}.$$

Question: Why must this constraint be written for each ordered pair, not just for each unordered pair?[14] *Question: There are $(|J|)(|J|-1)$ balancing constraints. If the balancing requirement required exactly the same proportion of weight capacity in each compartment, how many constraints would be needed?*[15]

Complications that can be handled with LP are:

Minimum shipping quantities: At least L_i tons of cargo type i must be carried. Solution: $\sum_{j\in J} x_{ij} \geq L_i \ \forall i$.

Decreasing profit tiers: The profit per ton for cargo type i is not constant. Instead, the profit is P_{i1} per ton for the first A_{i1} tons, P_{i2} per ton for the next A_{i2} tons, and P_{i3} per ton for the next A_{i3} tons. Here the total availability $A_i = \sum_{t=1}^{3} A_{it}$ and $P_{i1} > P_{i2} > P_{i3}$ (see Figure 3.6).

Solution: Replace the index set I by a larger index set I' which contains three indices, $i1, i2, i3$ for each $i \in I$. For example, if cargo type 1 is wheat, then wheat is subdivided into very profitable wheat, profitable wheat, and less profitable wheat, as though these three were different kinds of cargo. The constraints we can't write down in the LP are that we don't ship any profitable wheat unless we ship the maximum possible very profitable wheat, and we don't ship any less profitable wheat unless we ship the maximum possible profitable wheat. These constraints, though unwritten, will be satisfied at the optimum solution, because the three kinds of wheat are the same in all respects except profitability.

These profit tiers are an example of a piecewise linear concave function, which we hereby define.

[14] For example, if $|J| = 2$ there would be only one constraint and y_{j1} could be zero.
[15] $|J| - 1$ equality constraints.

Definition 3.1 *A piecewise linear concave (respectively convex) function is the minimum (respectively maximum) of a finite set of linear functions. If $g^j : \Re \mapsto \Re$ has the form $g^j(x) = \alpha^j x + \beta^j$ for $j = 1, \ldots, m$, then*

$$f(x) \equiv \min_{1 \le j \le m} g^j(x)$$

is a piecewise linear concave function and

$$h(x) \equiv \max_{1 \le j \le m} g^j(x)$$

is a piecewise linear convex function.

Piecewise linear concave benefits and piecewise linear convex costs can be modeled in LPs using the unwritten, unwritable constraint principle. Some LP software will permit you to declare variables as having piecewise linear values, instead of forcing you to reformulate. You can usually approximate concave benefits and convex costs that are not piecewise linear accurately enough with piecewise linear ones.

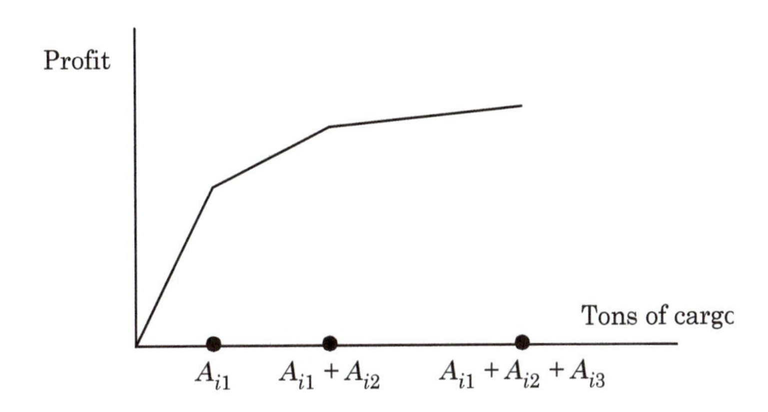

FIGURE 3.6: Handle concave profits in an LP by dividing the variable into several variables.

Complications that cannot be handled with LP:

- Mutually exclusive cargo: compartment j may not contain nonzero amounts of both cargo type $i1$ and $i2$.
- Minimum shipping lot size: if $x_{ij} > 0$, then $x_{ij} \ge L_{ij}$.
- Convex profits: a higher profit is earned per ton for cargo type i over a certain amount. If we try the same trick we used above for piecewise linear concave profits, the constraints we can't write down in the LP are that we don't ship any very profitable wheat unless we ship the maximum possible profitable wheat, and we don't ship any profitable wheat unless we ship the maximum possible less profitable wheat. But these constraints will be violated by the LP solver as it maximizes profit.

- Three-dimensional packing: the shapes of the different cargo types, and the shapes of the compartments, must be taken into account to see if they can fit. Unlike the other three complications, this complication is even beyond the reach of integer programming.

Example: Linear regression to minimize the maximum or sum of absolute deviations. In linear regression problems, we seek a linear predictive formula of form $p = \sum_{i \in I} w_i m_i$ that predicts the value of p given the values of the m_i. The weights w_i are real numbers. The m_i are called "independent variables"; p is called the "dependent variable". For example, we might want to predict the selling price at auction of a certain manufacturer's used cars as a function of the car's original sales price, the age of the vehicle, and the mileage on the vehicle. We choose values for the w_i that optimize the accuracy of the predictions on a given set of data. The data consist of a set of observations indexed by $j \in J$. Observation j consists of values $M_{ij} : i \in I; P_j$, which are the values of the m_i and of p respectively for that observation. In the case of the used car prediction, the data would consist of past auction data giving, for each vehicle sold (each observation), the original sales price, age, and mileage, and the auction sales price.

The most commonly used accuracy criterion is to minimize the sum of squares of the deviations of the predictions from the observed values of the dependent variable, i.e., to minimize $\sum_{j \in J}(P_j - \sum_{i \in I} w_i M_{ij})^2$. That would be the subject of another book about multiple regression. Linear programming can be used to optimize the w_i values with respect to two other accuracy criteria. The first is to minimize the maximum absolute deviation. This is an application of the min-max method described earlier. The model is

$$\begin{array}{rll} \min z & \text{subject to} & \\ z \geq & P_j - \sum_{i \in I} w_i M_{ij} & \forall j \\ z \geq & \sum_{i \in I} w_i M_{ij} - P_j & \forall j. \end{array}$$

The w_i are unrestricted in sign. It is quite possible, for example, for the selling price at auction to tend to be smaller the greater the age of the vehicle.

The second accuracy criterion that LP can handle is to minimize the sum of the absolute deviations. The model is

$$\begin{array}{rll} \min \sum_{j \in J} d_j & \text{subject to} & \\ d_j \geq & P_j - \sum_{i \in I} w_i M_{ij} & \forall j \\ d_j \geq & \sum_{i \in I} w_i M_{ij} - P_j & \forall j. \end{array}$$

The unwritten constraint is that d_j should be no larger than $|P_j - \sum_{i \in I} w_i M_{ij}|$. This can't be explicitly written as an LP constraint, but it is automatically satisfied at the optimum because the objective is to minimize the sum of the d_j terms. One can put arbitrary nonnegative weights on the deviations d_j if desired.

Example: Multicommodity Flow A graph is a set V of points, called vertices, and a set E of ordered pairs of vertices, called edges. For vertices i and j, the edge (i, j) represents a capability of sending commodities from i directly to j. Suppose for each edge (i, j) there is an upper bound U_{ij}, and a unit cost c_{ij}. Also suppose for each of several commodities indexed by $k \in K$ there is a supply amount for each vertex v, denoted $S_{v,k}$. A value $S_{v,k} < 0$ denotes a demand at vertex v for commodity k. The problem is to send flows of each commodity type along the edges of the graph to minimize total cost while meeting all supply (and demand) requirements, without exceeding the upper bound on total flow along each edge. The LP formulation defines variables x_{ijk} = the amount of flow of commodity k along edge i, j. The objective is

$$\min \sum_{k \in K} \sum_{(i,j) \in E} c_{ij} x_{ijk}.$$

The constraints are

$$\begin{aligned} 0 \leq \; & x_{ijk} && \forall (i,j) \in E, \; \forall k \in K; \\ \sum_{k \in K} x_{ijk} \leq \; & U_{ij} && \forall (i,j) \in E, \\ \sum_{i:(i,v) \in E} x_{ivk} - \sum_{j:(v,j) \in E} x_{vjk} = \; & d_{vk} && \forall v \in V, \; \forall k \in K. \end{aligned}$$

This is a kind of multicommodity flow problem. If $|K| = 1$ it is a special kind of LP called a network flow problem, to be discussed in Chapter 11. If edges are not permitted to carry more than one commodity, so that for each edge we must decide which commodity it will carry, the problem cannot be modeled as an LP.

Example: Blending Blending is a generalization of the diet problem. Raw materials must be blended together to form a variety of products. Let $i \in I$ index a set of raw materials such as light sweet crude oil, corn, etc. Let $j \in J$ index a set of substances such as water, hexane, lysene, etc. Let A_{ij} denote the number of units of substance j contained in one unit of material i. Let $k \in K$ index a set of products. Let u_{kj} and $\mathbf{l}_{kj}$ respectively denote the maximum (respectively minimum) number of units of substance j permitted to be in a unit of product k. Let w_i and v_k denote the mass of one unit of material i and product k, respectively. The w_i and v_k data must all be in the same units, such as kilograms. In a classical blending problem, the amount of substance j in a blended product equals the sum of the amounts of substance j contained in the raw materials that compose it.

This example is an example of the linearization method. The natural decision variables are f_{ik} = the number of units of raw material i that go into making one unit of product k, and y_k = the number of units of product k to make. However, these decision variables would lead to a nonlinear model which has terms $f_{ik} y_k$.

Instead, define x_{ik} = the number of units of raw material i that will be blended (with other materials) to form product k. These are the new decision variables. Let y_k = the number of units of product k produced, as before. The y_k are now auxiliary variables, rather than decision variables. Their values are implied by the values of the x_{ik} decision variables. By conservation of mass, $\sum i \in I w_i x_{ik} = v_k y_k \; \forall k \in K$. These are the defining constraints for the auxiliary y_k variables.

Let c_i = the cost of one unit of material i. All c_i must be measured in the same units, such as dollars. The classic blending problem formulation is then as follows.

$$\begin{aligned} \min \sum_{i \in I} \sum_{k \in K} c_i x_{ik} \quad & \text{subject to} \\ \sum i \in I w_i x_{ik} = v_k y_k \; & \forall k \in K \\ \mathbf{l}_{kj} y_k \leq \sum_{i \in I} A_{ij} x_{ik} \leq & \; u_{kj} y_k \\ x_{ik} \geq 0 \; & \forall i \in I, k \in K \end{aligned}$$

Complications:

- The blending process need not conserve the amounts of substances in the resources, nor their mass. For example, a fixed percentage of water might evaporate during blending. All that is needed is a linear relationship between inputs and outputs.
- More than one blending process might be available to produce product k. Let $p \in P$ index blending processes. For each $p \in P$ there is one $k \in K$ denoting the the product that results from process p. Each process has its own cost and its own linear relationship between inputs and output, and possibly its own upper bound on production capacity. To handle this complication, define decision variables x_{ip} = the number of units of material i to be blended by process p. This set of decision variables also applies to the more general situation where process p produces more than one product, as long as the ratios of the amounts of products produced by each process are fixed.
- Some characteristics of a blend may not behave linearly. For example, the octane level of gasoline is a nonlinear function of lead content. This complication usually cannot be handled by linear programming.

Example: Production Scheduling with Inventory

For this example, read the model and figure out the problem.

1. Index sets
   - $t \in T$: time periods, such as months.
   - $p \in P$: products.
   - $m \in M$: machines.
2. Data
   - d_{pt}: maximum number of units of product p demanded during period t (due by the end of period t).
   - A_{mp}: fraction of machine m's one-period capacity needed to produce one unit of product p.
   - c_{mpt}: cost to make one unit of product p during period t on machine i, in dollars.
   - s_{pt}: selling price of one unit of product p during period t.
   - h_{pt}: cost of holding one unit of product p from time period t to time period $t+1$.
3. Decision Variables
   - x_{mpt}: number of units of product p made during period t on machine m.
   - y_{pt}: number of units of product p sold during period t.
4. Auxiliary Variables
   - $I_{p,t}$: number of units product p held in inventory from period t to period $t+1$, $p \in P, t \in T$.
   - $I_{p,0}$: number of units product p on hand in inventory before the start of period 1, $p \in P$.
5. Objective Function

$$\max \sum_{p \in P} \sum_{t \in T} \left(s_{pt} y_{pt} - h_{pt} I_{pt} - \sum_{m \in M} c_{mpt} x_{mpt} \right)$$

6. Constraints

   - Nonnegativity: $x_{mpt} \geq 0 \ \ \forall m \in M, p \in P, t \in T$; $y_{pt} \geq 0$ and $I_{pt} \geq 0 \ \ \forall p \in P, t \in T$
   - Define I_{pt}: $I_{p,t} = I_{p,t-1} + \sum_{m \in M} x_{mpt} - y_{pt}$.
   - Demand: $y_{pt} \leq d_{pt}$.
   - Production Capacity: $\sum_{p \in P} A_{mp} x_{mpt} \leq 1 \ \ \forall i \in I, t \in T$.
   - No initial inventory: $I_{p,0} = 0 \ \forall p \in P$.

Question: Why is $I_{pt} \geq 0$ important? What would it mean for I_{pt} to be less than zero?[16]

Modeling Tip: Multi-stage models usually require specific attention to be paid to the initial and final stages to account for startup and termination conditions. When you create a multi-stage model, focus first on the general logic that carries things from generic stage t to generic stage t+1. After your generic logic is established, tweak it as necessary to handle startup and termination.

Complications include:

- Each machine can be utilized to manufacture only one product type in any one time period. This complication calls for an integer programming model.

- There is a setup cost for switching from one product to another during a time period, or between time periods. This complication also calls for an integer programming model.

Example: Work Scheduling Nurses work nine-hour shifts at Holy Cross Hospital. Each shift includes a one-hour break after the first four hours on duty. The staffing requirements at Holy Cross call for a certain minimum number of nurses on duty during each hour of each day during the week. (The number varies by day and hour). Nurse pay varies depending on the time of day and the day of week of the shift. How should the nurses be scheduled to meet staffing requirements at minimum cost?

Let $1 \leq i \leq 168$ index the $7 \times 24 = 168$ hours of the week, starting at a particular time such as $00:00$ Monday. For this model we want the index i to have *wraparound.* This means that the value $i = 169$ means $i = 1$, the value $i = 0$ means $i = 168$, and so on. Most modeling languages permit you to specify wraparound for an index set.

Let x_i = the number of nurses scheduled to begin their shifts at hour i, let c_i denote the cost of such a shift, and let L_i denote the minimum number of nurses on duty during hour i. The model is as follows.

$$\min \sum_{i=1}^{168} c_i x_i \qquad \text{subject to}$$

$$\sum_{\substack{i-8 \leq j \leq i \\ j \neq i-4}} x_j \quad \geq \quad L_i \ \forall i$$

$$x_i \quad \geq \quad 0 \ \ \forall i$$

Variation: Nurses work between one and eight consecutive hours, staying on duty during their entire shift. Let $x_{ik} \geq 0$ equal the number of nurses who start a k hour shift at hour

[16]Carrying negative inventory from period t to period $t+1$ would correspond to selling units of the product before they were manufactured, and receiving an additional h_{pt} dollars per unit as well.

i, $1 \le k \le 8; 1 \le i \le 168$. It is obvious how to modify the objective function. The staffing requirement constraint becomes

$$\sum_{j=i-7}^{i} \sum_{k=i-j+1}^{8} x_{jk} \ge L_i \ \ \forall i.$$

Work Scheduling with Specialized Tasks In this more complex scenario, minimum staffing levels are required for each of several task categories, but not every nurse is qualified to perform each kind of task. Each nurse, if hired, must work at least four and at most eight consecutive hours. Nurses may switch from one work task to another each hour.

Let $1 \le i \le 168$ as before index hours of the week. Let $t \in T$ index the set of task categories. Let $s \in S$ index subsets of task categories that define a nurse's qualifications. Let $k \in K$ index the shift lengths, so $4 \le k \le 8$. The data are now L_{it} = the minimum number of nurses doing task t during hour i; c_{iks} = wages paid to a nurse with qualification set s for a k-hour shift beginning at time i.

The decision variables are

- x_{isk}: number of nurses with qualification set s who begin a k-hour shift at hour i.
- w_{ist}: number of nurses with qualification set s assigned to task category t during hour i.

There are also auxiliary variables y_{is} = number of nurses with qualification set s at work during hour i. These auxiliary variables are defined by the constraints

$$y_{is} = \sum_{k \in K} \sum_{j=i-k+1}^{i} x_{jsk}.$$

The other constraints are

$$\begin{array}{lll} x_{iks} \ge 0; & w_{ist} \ge 0; & y_{is} \ge 0; \\ \sum_{t \in T} w_{ist} & \le & y_{is}; \\ w_{ist} & = 0 & \forall i \ \forall t \notin s; \\ \sum_{s \in S} w_{ist} \ge L_{it}. & & \end{array}$$

The objective is to minimize

$$\sum_{i \in I} \sum_{k \in K} \sum_{s \in S} c_{iks} x_{iks}.$$

Work Scheduling Pitfalls.

Pitfall 1: I've written this model as an LP, but a solution only makes sense if the x_i, x_{ik}, or x_{iks} variables have integer values. The following example shows that the optimal LP solution could have fractional values even in a very simple work scheduling model.

Example: There are three time periods, indexed by $i = 1, 2, 3$. The staffing requirements are one nurse in each period. The length of a shift is two. The hiring costs are $c_i = 20 \ \forall i$. The optimal LP solution is $x_i = \frac{1}{2} \ \forall i$, at cost 30. Rounding up this solution gives $x_i = 1 \ \forall i$, at cost 60. However, the optimal integer solution costs only 40.

In practice, the simple work scheduling LP models do not have a lot of fractional values. Unless the nonzero values of the x_i are small, rounding up is very likely to give a good solution. However, the more variables compared to the total staffing requirements, the more likely it becomes that fractional values will occur.

Pitfall 2: In the third model with x_{iks} variables, it may be unreasonably costly to acquire the c_{iks} data. Create a fairly accurate formula in terms of i, k and s, and use it to generate the data.

Pitfall 3: The size of the third model grows exponentially in $|T|$. *Question: Estimate the number of variables and the number of nonzero coefficients when* $|T| = 10$.[17]

Pitfall 4: The model assumes that any desired number of nurses with any particular qualifications set can be hired to work at any desired time. In practice the pool of available qualified nurses is severely limited. It may be better to model the assignment of individual nurses to shifts. Let P denote the pool of available nurses to hire. Let p index the set P. Let the possible shifts be indexed by $i \in I$. Let the variable x_{pi} represent the fraction of person p who is hired to take shift i. This variable makes sense only if it equals 0 or 1. This model has now moved to the realm of integer programming, a topic for a later chapter.

Non-convexity in Feasibility or Costs A set is *convex* if every weighted average of points in the set is also in the set. As you will see in Chapter 4, the set of feasible solutions to a linear program is always *convex*. As a general rule, non-convexity of the set of feasible solutions makes it difficult to model a problem as an LP. Similarly, costs that are not convex make it difficult to model a problem as an LP. For example, one of the problems in this chapter asks you to construct an LP model that maximizes the number of nonzero components in a vector $\mathbf{x}$ subject to the constraints $A\mathbf{x} = \mathbf{b}; \mathbf{x} \geq 0$. It does not ask you to minimize the number of nonozero components. Why not? Suppose $A\mathbf{x} = \mathbf{b}, \mathbf{x} \geq \mathbf{0}$ and $A\mathbf{y} = \mathbf{b}, \mathbf{y} \geq \mathbf{0}$. Suppose further that $x_i = 0 : 1 \leq i \leq n/2$, $x_i > 0 : n/2 < i \leq n$, $y_i > 0 : 1 \leq i \leq n/2$, and $y_i = 0 : n/2 < i \leq n$. Then $\frac{\mathbf{x}+\mathbf{y}}{2}$ is a strictly worse solution than either $\mathbf{x}$ or $\mathbf{y}$ for the objective of minimizing the number of nonzero components. Hence that objective is non-convex. Suppose that, instead of a non-convex objective, we had a constraint that the vector could not have more than $\frac{n}{2} + 1$ nonzero components. Then $\mathbf{x}$ and $\mathbf{y}$ would be feasible, but their average $\frac{x+y}{2}$ would be infeasible. Therefore, the feasible region would be non-convex. In either case the problem would not be suitable to be modeled as an LP. If you have trouble modeling a problem as an LP, check for non-convexity. That might help you understand your trouble.

Network models with integrality constraints constitute the main exception to the general rule that non-convexity of the feasible region interferes with LP modeling. In many situations, a decision variable must take an integer value, because violation of integrality reflects a physical impossibility. For example, suppose setting variable $x_{ij} = 1$ means that a delivery truck will travel directly from point i to point j, and setting $x_{ij} = 0$ means that it won't. A value $x_{ij} = 0.2$ has no physical meaning and renders the solution infeasible. But if a variable is required to take an integer value, the feasible region will be non-convex (unless at most one integer value is possible in a feasible solution). This non-convexity typically takes the problem out of the realm of LP into the more computationally difficult realm of IP (integer programming). However, there are special LP models, called network models, that can be solved as LPs while assuring integer values of variables. Chapter 11 describes the main network models; Chapter 15 introduces integer programming.

[17] 1.9×10^6 variables and 5.5×10^6 nonzeros. This size is feasible on commercial software for LPs but is apt to be difficult to solve as an IP.

Summary: Follow the rules stated in this chapter when formulating LPs. Define index sets, data, and variables, separate the data from the model, and don't make data-based simplifications to the model. Linearization, auxiliary variables and unwritten constraints are three major methods to extend the power of LP modeling. When using auxiliary variables, be sure to include the defining constraints in your model. When using unwritten constraints be sure that the objective function enforces the constraints. If some of your variables may take only integer values, rounding an LP solution might be good enough, but otherwise you will in general have to use integer programming, which is introduced in Chapter 15. The only common exception to this general rule occurs when your LP takes the special form of a network LP, as explained in Chapter 11. Non-convex feasibility or costs, non-concave profits, "or" constraints, and yes/no decisions are other common problem characteristics that call for integer programming.

3.5 Using Software to Solve LPs

"You never run a model once — if it is going to be successful" (R.E.S.)

Preview: This section gives some tips about what to do when solving LPs with software.

Get to know your software.

When you are learning to use your solver software, it helps to run a small example for which you know the optimal primal and dual solutions. Here is an example I have used. It has the virtue that all of the variables have different values, except the ones that are zero, which is unavoidable. That way it is easy to be sure you are interpreting the output from the solver correctly.

Maximize $100x_1 + 200x_2$ subject to constraints $x_1 + 2x_2 \leq 6; 2x_1 + 3x_2 \leq 4, x \geq 0$. The solution has objective value $266.66\ldots$ and $x_2 = 1.33\ldots$ and $x_3 = 3.33\ldots$. x_1 is nonbasic with reduced cost (dual variable value) $33.33\ldots$ and x_4 is nonbasic with reduced cost (dual variable value) $66.66\ldots$.

When you are getting to know your LP solver, you should run an example like the one above, and also an example of an infeasible LP and an example of an LP with unbounded optimum.

An LP solver makes it easier to learn how to model, because it provides rapid, concrete feedback. An expert might look at your model and see that it is incorrect, but find it difficult to explain to you why it is incorrect. If you run the model and get an incorrect answer, you will often be able to see for yourself what is wrong. This is especially true if you get a nonsensical answer such as an unbounded optimum.

Debug your model before solving real cases.

The first thing you should do before trying to solve a real-life case of an LP model is to debug it on small cases. If you have done your modeling properly, you will have separated the data from the model. That should make it fairly easy to create and solve toy size cases. If you can, construct at least one of your cases such that you know what the optimum solution should be.

Diagnosing Infeasibility.

What do you do when the solver tells you that your LP is infeasible? The first thing you need to do is figure out whether there is a bug in your LP model, or whether the LP really ought to be infeasible. If you think that your LP should be feasible, first try finding a solution that ought to be feasible, and plug it into your model. This will help debug your model. If you think that the LP ought to be infeasible, but you don't understand why, several software solvers will on request provide a minimal set of contradictory constraints. Studying that subset of constraints can be a lot easier than studying the whole model. If you suspect a few of the constraints of causing the infeasibility, test your suspicion by relaxing those constraints, i.e., change the right hand sides. If a suspected constraint is an equality constraint, replace it by a pair of inequality constraints with a gap between them. For example, replace the constraint $\mathbf{a} \cdot \mathbf{x} = a_0$ with the pair of constraints $\mathbf{a} \cdot \mathbf{x} \leq 1.25a_0; \mathbf{a} \cdot \mathbf{x} \geq 0.8a_0$. *Question: What if $a_0 < 0$? What if $a_0 = 0$?*[18]

Once you have thoroughly debugged your model, and it still is infeasible, but you want a solution, what do you do? Replace the troublesome constraints with penalty functions. The general form of penalty functions is as follows. Suppose the objective is minimization and the constraints are $A\mathbf{x} \geq \mathbf{b}, \mathbf{x} \geq 0$. Change the constraints to $A\mathbf{x}+\mathbf{y} \geq \mathbf{b}; \mathbf{x} \geq 0; \mathbf{y} \geq 0$. Add the term $\sum_i w_i y_i$ to the objective function. The coefficients w_i are positive penalty weights. Play around with the weights until you have a satisfactory solution. You can also put upper bounds or convex costs on the y_i.

Diagnosing Unboundedness

It is very rare for a correct model to have unbounded optimum, because infinite production levels, revenues, etc. are not possible in real life except in financial markets where arbitrage can occur. The solver will provide a ray of unboundedness. Interpret the ray using your definitions of your variables and you will find the error in your model. The values of the variables along the ray should correspond to a set of decisions impossible to follow in the real situation you have modeled. If they don't you probably have not defined your variables clearly. The most common reason for unboundedness, in my experience, is a failure to link auxiliary variables to decision variables. Another common reason is simply the accidental omission of a constraint.

Size and speed issues when solving large cases.

OK, you've debugged your model and you try it on a large case. What do you do if the software fails to solve it at all or can't solve it in a reasonable amount of time?

The first thing to do is to estimate the size of your problem. Calculate the approximate number of variables, constraints, and nonzero coefficients in the constraint matrix A. Compare your values with the restrictions on your software. Student and trial software licenses often limit the number of constraints and/or variables. If there are no such limits, you must

[18] $1.25a_0 \leq \mathbf{a} \cdot \mathbf{x} \leq 0.8a_0$; separate the positive and negative components of $\mathbf{a} \cdot \mathbf{x}$ and bound their ratios.

still check to be sure that your computer has enough memory (onboard memory, not slower storage such as a hard drive) to hold all the nonzeros, times a cushion factor of between 1.2 and 2. If you don't have enough core memory, you are apt to be clobbered by page fault time as the computer swaps data between the slow storage and the core. Remember that a double float variable in C or JAVA takes 64 bits = 8 bytes of memory with most compilers on 16, 32, and 64 bit machines. If you are short of memory by only a factor of two, it may be easiest to buy more memory or switch to a different machine that has more. Because swapping data in and out of core memory takes orders of magnitude more time than it takes to retrieve and store data in core, doubling your memory can increase speed by a factor of 100 or more. But if the additional memory does not eliminate swapping, the speed increase might be negligible. If you are having trouble estimating how much memory you need, run some smaller cases to determine the breakpoint at which page faults occur.

The number of nonzero coefficients is important because your software expects your A matrix data to be *sparse*, that is, to consist mainly of zeros. Matrices containing a high proportion of nonzeros are called *dense*. Software doesn't waste space storing each zero explicitly. Instead, it stores the nonzero values together with their locations in A. *Question: What takes more space to store in memory: a vector of length* 100 *with* 90 *nonzero entries, or a vector of length* 10000 *with* 30 *nonzero entries?*[19] The great majority of real-world A matrices contain less than 1% nonzeroes. Larger matrices with, say, 10,000 rows and 100,000 columns typically contain significantly less than 1% nonzeros. Indeed, sparseness in vectors is often assessed by the absolute number rather than the percentage of nonzeroes. A column of an A matrix with more than 30 nonzeroes is usually considered dense. In general, dense columns are worse than dense rows. This is particularly true for interior methods (see Chapter 18) because a few dense columns can make the matrix AA^T dense throughout.

The best LP solvers offer dual simplex (see Chapter 6), primal simplex (see Chapter 5), and an interior method (see Chapter 18) such as log-barrier as algorithm choices. Whichever one you've been using, try the other two. In general, barrier algorithms lack the numerical stability of simplex algorithms but are faster on average on large instances. Unlike simplex algorithms, interior algorithms are not slowed by degeneracy. Also, log-barrier algorithms parallelize readily, but simplex algorithms do not. This can make the former more effective on multi-core CPUs. Some LP models have a lot of primal degeneracy but little dual degeneracy, thus favoring dual simplex.

Sometimes (but not often), primal simplex is significantly faster than dual simplex. Models with many more columns than rows, or with much dual degeneracy, can make dual simplex run slowly compared with primal simplex. If both primal and dual degeneracy slow simplex algorithms, either use barrier or try perturbing the problem (both options are available in commercial software). Perturbing the objective function vector in the primal will usually greatly reduce the amount of degeneracy in the dual.

If your software has trouble solving your model in a reasonable amount of time, you should also check to see if your model has any dense rows or columns. You should also check the number of nonzeros before and after presolve. Presolve usually reduces the number of nonzeros, but sometimes constraint aggregation increases the number instead. Dense columns are also well known to slow down barrier algorithms. Commercial software may handle dense columns separately, if there are not too many of them. Other density issues are not as easy to anticipate for barrier algorithms, because they work with the matrix product AA^T. Suppose your software tells you that Cholesky factorization has a huge number of nonzeros. To understand why AA^T is much more dense than A, create a much smaller instance of your model and look at the patterns of nonzeros in A, A^T, and AA^T.

Good software will offer you many other parameter settings to try. The default parameter

[19]The former needs three times the space.

values are chosen based on their performance on a large testbed. Your model might be atypical. Many of the alternative parameter settings for simplex algorithms make a tradeoff between the number of iterations and the time required per iteration. Others trade off higher precision or numerical stability for faster runtime.

Numerical issues when solving large cases.
If the solver is giving strange answers such as variable values exceeding their upper bounds, or, more generally, constraints being violated by a large amount, you may have a numerical instability problem. Numerical instability is often caused by what is called an "ill-conditioned" A matrix, which can occur if A contains submatrices that have small but nonzero determinants. Other causes of instability are unnecessarily large coefficient values to represent penalties, and rounding that changes a zero determinant into a small but nonzero determinant. For example, the matrix

$$\begin{bmatrix} \frac{1}{3} & 1 \\ \frac{2}{3} & 2 \end{bmatrix}$$

is singular, which is not a bad thing. But the rounded values give the matrix

$$\begin{bmatrix} 0.333 & 1 \\ 0.667 & 2 \end{bmatrix}$$

which is numerically nasty. You'd be better off multiplying the data by 3 to get the integer matrix

$$\begin{bmatrix} 1 & 3 \\ 2 & 6 \end{bmatrix}$$

or, believe it or not, keeping the full precision of the data with the matrix

$$\begin{bmatrix} 0.333333333333333 & 1.0 \\ 0.666666666666667 & 2.0 \end{bmatrix}$$

If you can't fix the problem by changing your model, check your software manual for adjustable tolerances or other optional parameter settings to help deal with ill-conditioning.

A badly scaled A matrix contains numbers of very different magnitudes. This can arise when different variables or different constraints are defined in different units. For example, if some energy requirements were stated in kilowatt-hours and others were stated in gigawatt-hours, the numerical values in one constraint could be 10^6 larger than in another one. That disparity could be eliminated simply by multiplying the first constraint by 10^{-6}.

You might think that poor scaling is a frequent cause of numerical instability. And indeed, a badly scaled matrix is apt to be ill-conditioned. However, all good LP software rescales the data, both rows and columns, even before it runs the presolve [278]. That rescaling usually repairs poor scaling, so that scaling doesn't cause instability all that frequently after all. Nonetheless, it is better for you to avoid scaling your model badly than to hope your software will compensate.

Don't celebrate once you've debugged your model and found a numerically sensible optimal solution. You still have a lot more to do.

Check the optimal solution against reality for feasibility.
After you think you have debugged your model and have an optimum solution, ask your users to examine that solution. Often the LP solution will be infeasible with respect to a real-world constraint that the users never thought to tell you about because it was obvious

to them. "Of course the solution can't be like *that*," they will say. Don't be peeved, but don't let yourself feel at fault, either. Thank them politely for their domain expertise and move ahead.

Check the optimal solution against reality for optimality.
Users often don't understand that a computer has no common sense. It only does what you tell it to do. An optimization algorithm will sacrifice anything that isn't included in the objective function in order to get a tiny improvement in the objective function value (as long as feasibility is maintained). The user frequently has secondary considerations, not included in the objective function, and would be delighted to sacrifice a small fraction of the primary objective in order to get a significant gain with respect to a secondary one. For example, cost could be the primary objective of a daily diet problem, but the user might prefer to spend 0.2% more to get a solution that is not drastically different from the previous day's solution. A user whose primary objective is to balance workload might love to trade a little bit of imbalance for a big reduction in average job completion time.

Using a modeling language
If you are going to solve real problems, I highly recommend that you use a modeling language such as AMPL, LINGO, or GAMS, or a programming language for which your software has a well-supported interface, such as Python with Gurobi. You will appreciate the power of a modeling language when you have to alter your data, or, even more so, when you have to alter your model. You won't experience this in textbook assignments because textbook problems don't keep changing on you. A modeling language lets you use algebraic expressions such as $\sum_{1\leq i<j-2\leq n} x_{ij}$ and other summations over index sets with logical relationships. You've already learned the most important things to know when modeling so as to be able to use a modeling language, which are to define your index sets, define your variables, define your data, and separate the data from the model. Some modeling languages use the term "parameter" to mean "data". Some modeling languages require index sets to be a range of integers, while others permit any finite ordered set of distinct symbols such as January, February, March, ..., December. If you have to learn a modeling language but it seems too difficult, try writing your own LP matrix generator. You will probably find that the latter task is more time-consuming.

Other tips

- There may be two ways to generate the variables in your model, sparse as-needed, or all-possible. The latter can use up too much memory. If you run out of memory during presolve, switch to the sparse way. If you don't have trouble, don't worry. Presolve will get rid of the unneeded variables.

- Multi-core CPUs have become standard. This favors barrier algorithms over simplex algorithms. Since barrier algorithms usually end with a simplex algorithm-like crashing procedure (see Chapter 4), there may be little benefit to having more than four cores.

- CPU time is the time the computer spends on actual computation, as opposed to run time, which insufficient memory can cause to be large as discussed previously. CPU time is not often an issue with LPs. When it is, keep in mind that as a general rule larger LPs take longer to solve. Larger means more rows, nonzero coefficients, and columns, in that order of importance. This rule does *not* hold for IPs, as will be discussed in Chapter 15.

- If your LP model has a huge number of variables you might be able to employ column generation, which only keeps a small fraction of the columns explicitly in the model. The rest of the columns are only brought in as needed. Consult Section 7.5 to learn more. The dual situation is a huge number of constraints, and the dual technique is called constraint generation.

- Dual degeneracy means multiple primal optima. If this is causing computational problems, try a barrier algorithm. Also, dual degeneracy may be an opportunity to discriminate among different optimal solutions. Consider adding a small tie-breaking term to the objective function. The LP may solve faster and give a more desirable answer.
- Often getting the data set up, clean, and correct is a much bigger project than the operations research problem. On projects whose budgets I've seen, for which the data were not readily available, the data part has accounted for 4 to 20 times the operations research part.
- Beware people who expect the LP model to do everything, i.e., to solve the problem in all detail. They will be disappointed when it doesn't and may reject it.
- Learn to work with people who are afraid the LP will do everything. These are people who know that the model has not captured all of the complications of the real problem. They will want to reject the model unless they realize that there is still room for them to modify the solution or handle exceptional cases separately. If you can help with the 90% of the cases that are typical, you have a success. Make sure your users know they can still take care of the other 10% separately.

Summary: Be prepared to learn a modeling language if you are going to solve real problems with LP. The language will make it easier for you to handle the many changes to the model and data you will be obliged to make. The tips in this section will help should you encounter difficulty with time or memory constraints. However, it is more likely that you will encounter difficulty with acquiring the necessary data.

3.6 Shadow Prices

Preview: This section explains how the dual variables provide useful information about the cost of enforcing constraints at the optimal solution.

When an LP is solved, both the optimal primal and dual values are found. The meaning of the primal variables had better be clear. If it isn't, you didn't define your variables properly in your formulation. The dual variables have meaning, too. Recall that there is a dual

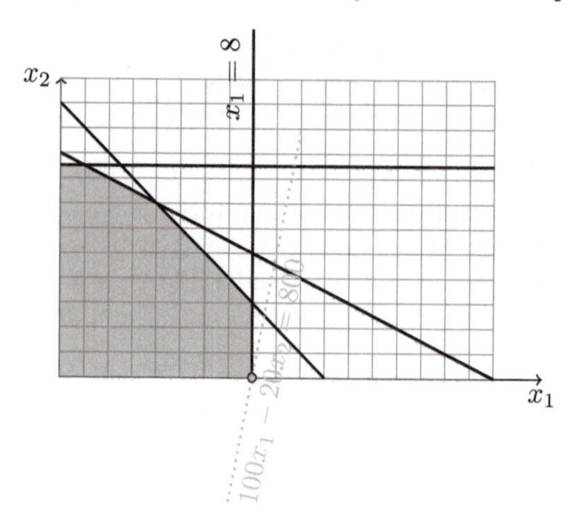

FIGURE 3.7: The feasible region defined by constraints (3.17) through (3.22) is shaded. The dotted isoquant line $100x_1 - 20x_2 = 800$ intersects the feasible region at the optimum solution $(8, 0)$.

variable for each constraint in an LP. The value of the dual variable is the cost of enforcing its corresponding constraint (see Definition 1.8), that is, the effect on the objective function of changing the right-hand side of the constraint. When the constraint is an upper bound on the availability of a resource, the dual variable equals the value to you of an additional unit of the resource. That is why it is called the *shadow price* of the resource. It is what you would be willing to pay for an additional unit. If the market price is lower than the shadow price, it would be advantageous to purchase more of that resource. If the market price is higher, it would be advantageous (in the LP model) to sell a unit of the resource. In general, the shadow price of a constraint is the rate at which the objective function gets better as the constraint is relaxed.

For a geometric view of shadow prices, consider the following set of constraints on the pair of variables x_1, x_2, depicted in Figure 3.7.

$$x_1 \leq 8 \tag{3.17}$$
$$x_2 \leq 8.5 \tag{3.18}$$
$$x_1 + x_2 \leq 11 \tag{3.19}$$
$$x_1 + 2x_2 \leq 18 \tag{3.20}$$
$$x_1 \geq 0 \tag{3.21}$$
$$x_2 \geq 0 \tag{3.22}$$

- For the objective maximize $100x_1 - 20x_2$, the optimum solution is $(8, 0)$ with value 800. The shadow price of constraint (3.17) is the change in the objective if the constraint is relaxed by one unit, from $x_1 \leq 8$ to $x_1 \leq 9$, as shown in Figure 3.8. The optimal solution would shift to $(9, 0)$, with value 900. The shadow price is $900 - 800 = 100$. *Question: If the objective were a minimization, what impact would raising an upper bound have?*[20]

[20] It would decrease the objective or leave it unchanged, because expanding the feasible region can not worsen the optimal objective value.

The shadow price 100 accurately predicts the change in optimal objective value up to the relaxation $x_1 \leq 11$. For upper bounds on x_1 greater than 11, the shadow price is inaccurate, because the constraint (3.19) becomes binding. This is an example of the rule of optimism, Theorem 7.1.1, which we will prove in a later chapter.

The shadow price also predicts the change in objective value caused by tightening constraint (3.17). For example, if the upper bound on x_1 were reduced from 8 to 5, the optimal objective value (at $(5, 0)$) would equal $800 - 3(100) = 500$.

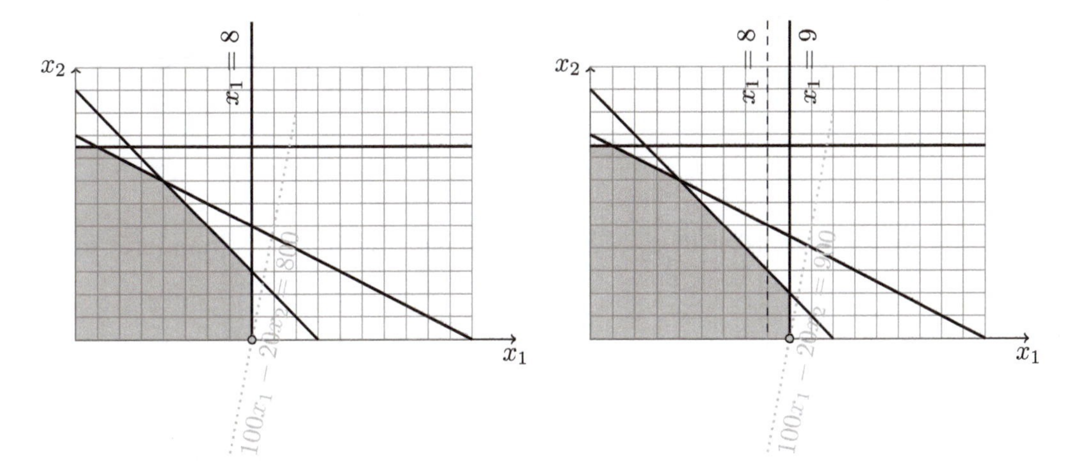

FIGURE 3.8: When maximizing $100x_1 - 20x_2$, the optimal solution shifts from $(8, 0)$ to $(9, 0)$ if the upper bound on x_1 is increased by 1. The shadow price of the upper bound constraint is $900 - 800 = 100$.

- For the objective maximize $100x_1 - 20x_2$, what is the shadow price of (3.22), the nonnegativity constraint on x_2? Usually there is no physically feasible meaning of a nonnegative variable being less than zero. The interpretation of this shadow price is the amount by which the objective value worsens if the constraint is tightened by one unit. As depicted in Figure 3.9, if $x_2 \geq 1$ were required, the objective would decrease by 20 because the optimal solution would shift to $(8, 1)$. Hence the shadow price is 20. *Question: Why might a company want to provide a service or product that is not profitable compared to other ones?*[21]

- The preceding examples are very simple because the constraint that changes is orthogonal to the other binding constraints. In general, the shadow price depends on how the changing constraint interacts with the other binding constraints to shift the optimal solution. For the objective maximize $200x_1 + 160x_2$, the point $(8, 3)$ is optimal. Were the constraint (3.17) relaxed to $x_1 \leq 9$, as depicted in Figure 3.10, the optimal solution would shift to $(9, 2)$. The resulting increase of the objective function by 40 equals the shadow price of constraint (3.17). The shadow price of constraint (3.19) is 160. The optimal solution would shift from $(8, 3)$ to $(8, 4)$ were the right-hand-side increased from 11 to 12, as depicted in Figure 3.11. *Question: What are the shadow prices of constraints* (3.18) *and* (3.22)*?*[22]

- For the objective maximize $222x_1 + 400x_2$, the optimal solution is $(4, 7)$. If constraint

[21] To preserve a market presence or corporate image, for example.

[22] Zero, since they are not binding at $(8, 3)$. The same is true for (3.20) and (3.21).

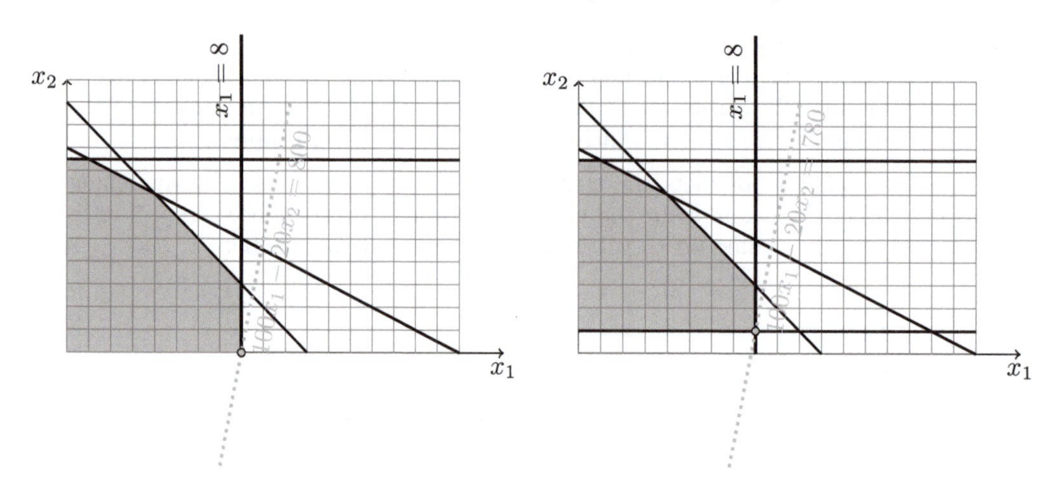

FIGURE 3.9: When maximizing $100x_1 - 20x_2$, the optimal solution shifts from $(8, 0)$ to $(8, 1)$ if the lower bound on x_2 is increased by 1.

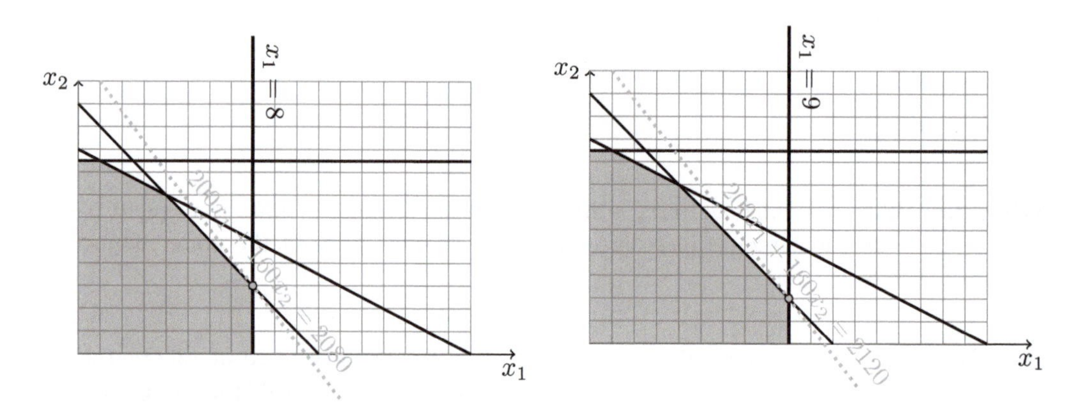

FIGURE 3.10: When maximizing $200x_1 + 160x_2$, the optimal solution shifts from $(8, 3)$ to $(9, 2)$ if the upper bound on x_1 is increased by 1. If the objective function were $2x_1 + 1.6x_2$ the shadow price for the x_1 upper bound would be $\frac{2120-2080}{100} = 0.4$.

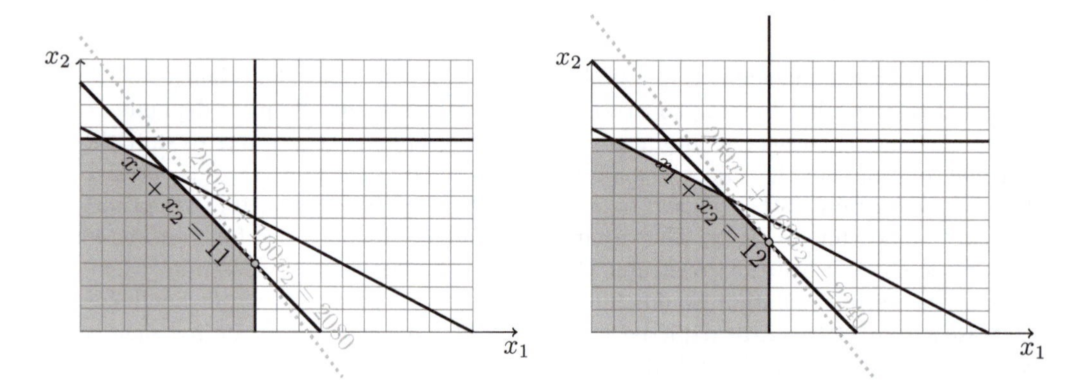

FIGURE 3.11: When maximizing $200x_1 + 160x_2$, the optimal solution shifts from $(8, 3)$ to $(8, 4)$ if constraint (3.19) is relaxed by 1. The shadow price for that constraint is $2240 - 2080 = 160$.

(3.19) were relaxed to $x_1 + x_2 \leq 12$, the optimal solution would shift to $(6, 6)$, as shown in Figure 3.12. The objective would increase from 3688 to 3732, indicating a shadow price of 44.

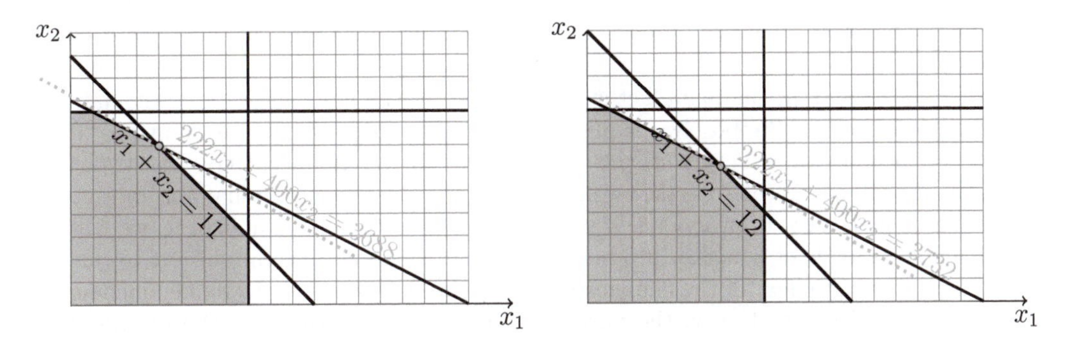

FIGURE 3.12: When maximizing $222x_1 + 400x_2$, the optimal solution shifts from $(4, 7)$ to $(6, 6)$ if constraint $x_1 + x_2 \leq 11$ is relaxed to $x_1 + x_2 \leq 12$. Hence the shadow price is $(6 - 4)(222) + (6 - 7)(400) = 44$.

- Shadow prices are very rarely accurate for large changes in a constraint's right-hand-side. Often they are inaccurate even for modest changes. For the objective maximize $60x_1 + 200x_2$, the optimal solution is $(1, 8.5)$ with value 1760. If constraint (3.18) were relaxed to $x_2 \leq 9$, the optimal solution would shift to $(0, 9)$, consistent with the shadow price of $80 = \frac{1800-1760}{9-8.5}$, as shown in Figure 3.13. But if the constraint were relaxed any further, to $x_2 \leq 9 + \epsilon$ for any $\epsilon > 0$, the optimal solution would remain at $(0, 9)$. You should see visually in Figure 3.13 that the shadow price becomes inaccurate because another constraint that was not previously binding comes into play as the optimal solution shifts. *Question: which constraint?*[23] Intuitively, this is the reason why, when a shadow price fails to be accurate, it is overly optimistic. It does not take into account constraints that are not binding at the optimal solution, but become binding as the solution shifts.

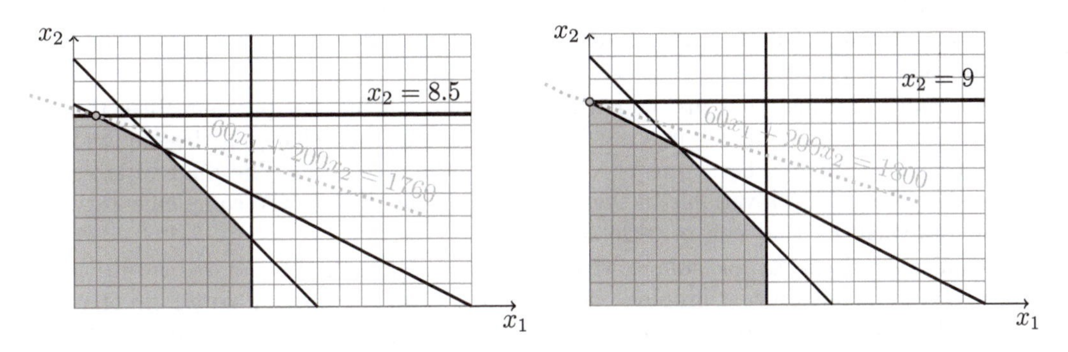

FIGURE 3.13: When maximizing $60x_1 + 200x_2$, the optimal solution shifts from $(1, 8.5)$ to $(0, 9)$ if the upper bound $x_2 \leq 8.5$ is relaxed by 0.5. The shadow price $(1800-1760)/0.5 = 80$ does not provide an accurate prediction if the upper bound is relaxed to $x_2 \leq 9.5$, because the optimal solution does not shift further if the bound is relaxed by more than 0.5.

[23]The constraint $x_1 \geq 0$.

Now that you've seen their geometric meaning, let's see what shadow prices mean in the first two classic problems of Section 3.1. In the manufacturing problem, the shadow price of a resource upper bound constraint is the rate of increase in the objective if the upper bound were increased. Therefore, it is the value to the manufacturer of an additional unit of that resource. This is precisely what the purchaser in the dual story must offer the manufacturer for that resource.

The other constraints in the manufacturing problem are the nonnegativity constraints on the production levels x_i. Suppose we changed the right-hand side of the constraint $x_i \geq 0$ to 1. We'd get the constraint $x_i \geq 1$. If the dual variable for the nonnegativity constraint equals 50, then it will cost us 50 to make a unit of product i. Let's be clear what this means. At optimality we make zero of product i. If we force ourselves to make 1 unit of it, we will change the amounts made of the other products in order to maximize income subject to this new constraint. The net effect of the changes in production levels will be to reduce income by 50. From the dual point of view, the purchaser must offer at least c_i for the resources needed to make a unit of product i. A shadow price of 50 in the primal means that the purchaser offers $c_i + 50$ for that resource bundle.

In the diet problem, each nutrient constraint has a shadow price, which is what the vitamin salesman will charge for a pill containing one unit of that nutrient. This is the cost of enforcing that nutrient constraint. It predicts the additional cost that would be incurred if the nutritional requirement were increased by one unit, or the reduction in cost that would occur if the requirement were decreased by one unit. *Question: Reason by analogy with the manufacturing problem to see the meaning of the shadow price of a nonnegativity constraint in the diet problem.*[24]

Shadow prices provide a charmingly intuitive explanation of complementary slackness at optimal solutions. In the manufacturing problem, what is the shadow price of a resource constraint that is not binding in the optimal solution? From an economics point of view, the shadow price of the jth resource must be zero. You wouldn't pay anything to get another unit of resource j. You are not using all that you have. More generally, if a constraint $A_{j,:}x \leq b_j$ is not binding at the optimal solution, the corresponding shadow price must be zero because an increase in b_j will have zero impact on the optimal objective value. Referring to Definition 2.2, complementary slackness requires that at optimality either the slack variable that equals $b_j - A_{j,:}x$ or the dual variable π_j be zero. If the constraint is not binding, the slack variable is strictly positive, whence $\pi_j = 0$. Since π_j is the shadow price, complementary slackness has the natural meaning that the shadow price of a non-binding constraint must be zero.

Complementary slackness has just as natural a meaning in the diet problem. If in the optimal solution the constraint for nutrient j is not binding, then changing the required amount of that nutrient will have zero effect on the optimal solution. The dual variable π_j is zero when the jth slack variable in the primal is nonzero. This is the complementary slackness requirement for that pair of variables.

Complementary slackness for nonnegative decision variables in LPs makes intuitive sense, too. In the manufacturing problem, suppose $x_j > 0$ in the optimal solution. That means we make a nonzero amount of product j. It would cost us zero to be required to make a nonzero amount of product j, since we are already doing so at optimality. That is the complementary slackness condition between x_j and the jth dual slack variable. Similarly, in the diet problem, it costs us nothing to be required to eat some nonzero amount of food i if we are already doing so at optimality. That is the complementary slackness condition for the decision variable x_i and its corresponding dual variable.

---

[24] At optimality we eat zero of food j. The shadow price predicts how much more our meal would cost if we required ourselves to eat at least one unit of food j.

The mathematical underpinnings of shadow prices will be given in Chapter 7. In practice, look for large dual values. Most real constraints are soft. If the shadow price is high enough, often the constraint itself can be nudged.

Example of the use of shadow prices. An airline has a large set of legs. Each leg is a direct flight from one city to another. Each leg has its own capacity, which is the number of seats on the plane making that flight. Potential passengers want to fly routes, which are collections of legs. Different fare classes for the same route are modeled as different routes that happen to use the same legs. First class seats on a leg are modeled as a different leg that has its own capacity and happens to have the same schedule as another leg. The airline has a forecast demand, giving the number of passengers who want to fly each route. Each route has a given price charged to the passenger. A simple yield management model is to maximize the total prices collected subject to not exceeding demand, and not exceeding leg capacity. The model, more precisely is as follows. Let $i \in I$ index the legs. Let $j \in J$ index the routes, and let U_j denote the forecast demand for route j (in passengers). Let $A_{ij} = 1$ if route j includes leg i, and 0 otherwise. Let L_i denote the capacity (in passengers) of leg i. Let P_j denote the price charged for route j. Let $x_j =$ the number of passengers who are sold tickets for route j. Then the model is

$$\begin{aligned} \max P \cdot x &= \sum_{j \in J} P_j x_j \text{ subject to} \\ \mathbf{A}\mathbf{x} &\leq L; \\ 0 &\leq x \leq U. \end{aligned}$$

The forecast demands are changing several times a day. Decisions as to whether to sell tickets to customers must be made minute-by-minute. How do we use the basic LP model to make decisions? Keep in mind that the idea of this simplified version of yield management is that the prices for routes are given, but the number of seats available for each route can change at the discretion of the airline. If it does not seem worth it to sell a customer a ticket, because the airline expects other more valuable customers to buy the same legs, then the airline says, sorry, the seat is not available. This decision of seat availability has to be made in real time. We can't run a whole LP every time a customer asks if a seat is available. So what do we do?

Reading Tip: *Think about this before reading ahead. Force yourself to make a guess, even if you expect it to be wrong. What are you afraid of? Guessing will engage your brain and give you a bigger stake in what you read next.*

We use the dual variables, the shadow prices. Run the basic model. It tells us, for each leg, how valuable one more or one less seat would be on that leg. When a potential customer asks if the route using legs 15, 244, and 187 is available for purchase, we add up the shadow prices of legs 15, 244, and 187, and compare the sum with the price of the ticket for that route. If the ticket price is higher, we sell the seat. This simple computation can be run in real time. The basic LP model might only have to be run a few times a day. This example is simpler than actual yield management for airlines, hotels, etc. but is a good illustration of the meaning of dual variables.

In a later chapter, we will see that predictions based on dual variables are not always accurate. Memorize the following two guidelines:

(1) Predictions based on dual variables, when not accurate, are always optimistic.
(2) The larger the change in the right-hand-side value, the less likely is the dual variable's prediction to be accurate.

Summary: At the optimum, primal variable values tell you what to do to attain the best possible objective value for the given data. Dual variables tell you what impact changes in the right-hand-side data would have on the objective value. This dual information can be very important because in real problems the amount of a resource available or a bound on the level of an activity can often be changed, at some cost. The dual variables tell you how much it would be worth paying for such a change.

3.7 Notes and References

See Section 2.4 for origins of the linear programming model. In addition to those who developed the LP model, Abraham Charnes and his colleagues, notably William Cooper, contributed a great deal to expanding the range of LP applications. These include the first oil blending application [42], chance-constrained programming [40], and many others in both management and production [41, 43, 44].

3.8 Problems

E Exercises

1. Find government-recommended minimum daily requirements for calories, protein, vitamin C, and vitamin D. Find price and nutritional information for bread, orange juice, rice, peanuts, and milk. Formulate an LP to minimize cost and meet the requirements. Be sure to specify the units of each nutrient and each food.

2. Verify that the LP (3.1) is the dual to the 2-person 0-sum primal.

3. In the game Even-Odd, players I and II each choose either the number 1 or the number 2. They reveal their choices simultaneously. If the sum of the two numbers is even, player II gives player I one dollar. If the sum of the two numbers is odd, player I gives player II one dollar. Write the LP for player I and the LP for player II. Guess optimal solutions and verify by the weak duality theorem that your guesses are optimal.

4. Write an LP to find the minimum of the numbers $5, 18, 3, 10, 22$. Write its dual. Interpret the dual so that it is clear why it too finds the minimum.

5. In the preceding exercise, why is a constraint $\sum_{i=1}^{5} \pi_i \leq 1$ incorrect? If this constraint is used instead of $\sum_{i=1}^{5} \pi_i = 1$, what happens to the value of the linear program?

6. Find two or three ways to get x equal to $|y| + |w|$ in a linear program.

7. Show why $\max u^+ + u^-$ subject to $-6 \leq u^+ - u^- \leq 5; u^+ \geq 0; u^- \geq 0$ does not solve the problem of maximizing $|u|$ subject to $-6 \leq u \leq 5$.

8. Model the following problem as an LP. You can buy ordinary crude oil for \$100/barrel, low sulfur crude oil for \$130/barrel, and plastic-derived oil for \$110/barrel. The maximum amounts you can buy are $5000, 2000, and 500$, respectively. You can produce and sell a regular blend, a high-octane blend, and a green blend, at prices $c_1, c_2, and c_3$ dollars per barrel, respectively. The regular blend must be between 80% and 90% ordinary crude. The high-octane blend must be less than 60% ordinary and at least 30% low-sulfur. The green blend must be at least 10% plastic-derived and at most 40% ordinary. How much of each blend should you produce, and what should be the composition of each blend?

9. For the matrix in Figure 3.1 show how to find the saddlepoint by repeatedly removing rows and columns that represent strategies sure to be inferior to other strategies.

10. Prove that a matrix can not have more than one saddlepoint, i.e., an element that is minimal in its row and maximal in its column.

11. For the objective maximize $222x_1 + 400x_2$ subject to constraints (3.17) ... (3.22), find the shadow prices of constraints (3.20) and (3.21).

12. Interpret complementary slackness in terms of shadow prices in the two-person-zero-sum game LP. In particular, if player I's constraint about playing against II's strategy j is not binding, what does that tell you about how often II chooses j when she plays? How about player II's constraint against I's strategy i?

13. Suppose the dual variable π_j corresponding to constraint $A_{j,:}x \geq b_j$ is strictly positive at the optimal solution. Assuming that π_j is the cost of enforcing that constraint, show that complementary slackness holds for the pair of variables π_j and its associated primal variable.

M Problems

14. Repeat problem 3 for the game that is identical except that player II gives player I two dollars if the sum is even.

15. A vehicle distributor's market is split into 2 regions. It costs 50 per car and 70 per truck to deliver from her distribution center to region 1, and 30 per car and 45 per truck to deliver to region 2. Sales limits on trucks are 5 in region 1 and 8 in region 2 in the first week. In the second week the sales limits are 10 and 11, respectively. For cars, the sales limits are 8 in each region in week 1 and 12 in each region in week 2. Up to 9 vehicles can be stored at the distribution center from one week to the next. It costs 20 per car and 30 per truck to order and bring vehicles from the manufacturer to the distribution center at the start of week 1. For the start of week 2, the costs are 100 for cars and 80 for trucks. The distributor is paid 120 per car and 125 per truck delivered to region 1. She is paid 95 per car and 115 per truck for delivery to region 2. These amounts are the same each week. (i) Formulate the problem of determining the ordering, storage, and distribution policy that maximizes the distributor's net income. Remember to separate the model from the data.

    (ii) Formulate the more general problem for a larger number of regions, vehicle types, and time periods. Should you assume that payments for deliveries are the same in each time period?

16. The citizens of Thurber all live on a straight road 3 miles long. This strange town has but one alarm clock, which is loud enough to awaken all of the citizens each morning. When the alarm goes off, each citizen gets dressed, eats breakfast, and walks to the town bus stop. Each citizen takes a fixed amount of time to dress and eat, though this amount of time differs from person to person. Likewise, each citizen walks at a fixed rate that differs from person to person, and each citizen has to catch a bus that leaves at a certain time, that time differing from person to person. Where should the bus stop be located to allow the citizens to sleep as late as possible? Formulate this problem as a linear program. For simplicity, treat the speed of sound as infinite.

17. Do Problem 8 with all of the following complications: Let i index the types of oil used as raw material and let j index the types of blends to be produced and sold. It costs d_j to produce a barrel of blend j (in addition to the cost of the raw materials). The green blend may not have a higher percentage of regular than of plastic-derived, nor a higher percentage of low-sulfur than of plastic-derived. You can buy up to f_i additional barrels of oil type i for 20% additional cost per barrel.

18. Given a finite set of open half-spaces in $\Re^n$, formulate an LP to find a point in $\Re^n$ that is contained in all the half-spaces, and such that the point closest to it that is not in all the half-spaces is as far away as possible. Use the Euclidean distance.

19. The Euclidean distance from a point $\mathbf{x} \in \Re^n$ to a set $S \subset \Re^n$ is

$$\inf_{\mathbf{s} \in S} ||\mathbf{x} - \mathbf{s}||.$$

If S is closed the "inf" can be replaced by "min" and the distance is zero iff $\mathbf{x} \in S$. Given a finite set of hyperplanes in $\Re^n$, formulate an LP to find a point in $\Re^n$ such that the sum of its Euclidean distances to the hyperplanes is as small as possible. Complications: How would you change your model if a nonnegative weighted sum of distances were to be minimized? What is the solution to the problem of maximizing the sum of distances? Explain why it is unlikely that you could write an LP to maximize the sum of distances to $\mathbf{x}$ subject to upper bounds $|\mathbf{x}_i| \leq K \; \forall i$.

20. ([223]) Given a finite set of hyperplanes in $\Re^n$, formulate an LP to find the smallest Euclidean ball that intersects all the hyperplanes. (A Euclidean ball is a set of form $\{x : ||x - y|| \leq \epsilon\}$ for some $\epsilon \geq 0$ and some $\mathbf{y} \in \Re^n$. $||w||$ denotes the Euclidean norm of w, $\sqrt{\sum_{i=1}^{n} w_i^2}$.)

21. Quimby's Real Sausages (QRS), a meat-packing company, has purchased 8,000 cows for the coming year of production. Each cow yields 9,000 sausages. Sausages will sell at 2 per dollar in each month except July, when barbecue contests allow a higher price of \$0.75 each. The marketing department at QRS estimates maximum sales of five million sausages in January and ten million each month thereafter. Sausages may only be sold fresh, that is, in the month they are produced. QRS incurs a cost of \$140.00 each month for the pasturage of each cow on hand not converted to sausage that month.

    Formulate the problem of determining QRS's optimal sausage production strategy as a linear program. Remember not to use reasoning based on the specific data given to simplify your model.

    Complications

    Marketing reports that a 25% higher price can be charged for sausages made from cows that have been grass fed, as consumers request more "natural" food. A cow counts as being "grass fed" if it has been in pasture for at least one month.

Sausages may be frozen in the month they are produced and stored at cost 1/10 dollar/month until sold. The refrigerated storage capacity is three million sausages.

Cows that are in pasturage at the end of the month beget additional cows at a rate of 1/12 cow per cow.

22. Model the following as an LP. Remember to separate the data from the model, and not to use reasoning based on the specific data given to simplify your model. You have production capacity to manufacture 500, 600, 200, 700, and 1000 silver-plated left-handed tweezers in 2020, 2021, 2022, 2023, and 2024, respectively. The production costs (all in present value) are 2, 4, 6, 5, and 4 dollars, in each year, respectively. By paying overtime, you can produce up to 200 additional tweezers per year (above the capacity given), at a cost of 1.5 times the usual cost that year. Tweezers sell for 5, 7, 8, 8, and 5 dollars, in each year, respectively, and maximum sales amounts are 500, 800, 400, 500, and 300. Tweezers may be stored from one year to the next, but storage costs \$.50 per tweezer per year. Also, goblins remove the silver from 25% of whatever is stored at year's end.

    In the year 2025, the silver from stored tweezers may be melted down and sold for 2 dollars each. Find a production, sales, and inventory plan to maximize your net profits.

23. Let $\mathbf{w} \in \Re^n$ be an extreme point of polyhedron $P = \{\mathbf{x} : A\mathbf{x} \leq \mathbf{b}\}$. Define the distance between two points $\mathbf{x}, \mathbf{y} \in \Re^n$ by the L_1 norm to be $\sum_{i=1}^{n} |x_i - y_i|$. Let $\mathbf{q} \in \Re^n$. Formulate a linear program that finds a vector $\mathbf{s} \in \Re^n$ that has minimum distance from $\mathbf{q}$ subject to $\mathbf{w}$ being an optimal solution to $\max \mathbf{s}\cdot\mathbf{x}$ subject to $\mathbf{x} \in P$.

24. $|J|$ robots, indexed by $j \in J$, are all to be dropped by spaceship at a point in a flat terrain on Mars. Points in the terrain are mapped as x, y where $0, 0$ is the center of the terrain, $x > 0$ means E, $y < 0$ means S, and so on. After landing, robot $j : j \in J$ must perform a task at coordinates x_j, y_j, which takes t_j minutes to complete. Each robot can travel in any one of the four directions, N, S, E, W, at any time. Robot j travels at r_j meters per minute. After a robot completes its task, it returns to the drop point. When all robots have returned, the spaceship lifts off.

    (a) Formulate a linear program to determine the drop point that permits the spaceship to lift off as soon as possible.

    (b) Formulate an LP that determines the drop point that minimizes the total fuel expended by the robots. Assume that a robot only uses fuel when traveling, and that robot j expends f_j grams of fuel to travel 1 meter.

25. Raw materials are indexed by $i \in I$, each measured in kilograms, costing r_i per kilogram. Products are indexed by $j \in J$, also each measured in kilograms, and selling for s_j dollars per kilogram. Processes are indexed by $k \in K$.

    Each process k takes as input a variable mix of raw materials with total weight measured in kilograms, and ouputs a known fixed mix of products with total weight equal to p_k times the weight of the input materials, where $0 < p_k < 1$ is known. For example, for every kilogram of input, process 9 might output $.5p_9$ kilograms of product $j = 19$, $.4p_9$ kilograms of product $j = 2$, and $.1p_9$ kilograms of product $j = 5$. The constant p_9 might be 0.8838. The constraints on the variable mix are upper and lower bounds on the fraction of the input that is material i, and upper and lower bounds on the fraction of the input that is material i or i', for each i and each pair i, i'. For example, process 6 might require between 10% and 15% input to be material $i = 3$, between 5% and 20% to be material $i = 7$, and that materials 3 and 7 together be between 16% and 35% of the input. Process k costs c_k dollars per kilogram of input. There are lower and upper bounds on the sales of each product. Excess product j may be disposed of at cost $d_j > 0$ per kilogram. There are upper bounds on the amount of each raw material available.

(a) Formulate the production planning problem to maximize profit as an LP. Begin by defining any necessary notation (notation needed to describe the problem data) not given in the problem statement.

(b) Show how to modify your LP if the output of a process may be a mix of raw materials and products, and the processes are run in sequence $k = 1, 2, \ldots, |K|$ so that some or all of the output of process k may be used as input for process $k' > k$.

(c) Could you modify your LP to model the problem as in part 25b, but if the processes may be run in any sequence? That is, each process may be run once, but the sequence is not fixed. Explain.

(d) Could you modify your LP to model the problem as in part 25b, but if the processes can be run up to $|K|$ times, each time in the sequence $1, 2, \ldots, |K|$? That is, each process may be run many times, but the sequence is fixed. Explain.

26. Generalize the fractional linear objective function in (3.9) to the fractional affine objective function

$$\min \frac{\mathbf{c} \cdot \mathbf{x} + c_0}{\mathbf{d} \cdot \mathbf{x} + d_0}.$$

Linearize to get an equivalent LP that generalizes (3.14), (3.15),(3.16).

27. Construct an example of an LP that contains a constraint whose shadow price remains strictly positive no matter how much the constraint is relaxed. Interpret the meaning of this phenomenon in the dual.

28. Suppose at an optimal solution to an LP with constraints $A\mathbf{x} = \mathbf{b}, x \geq 0$, the optimum basic feasible solution is nondegenerate. Argue that if one of the b_j values is changed by small enough amount $\epsilon > 0$, then the corresponding dual variable (shadow price) correctly predicts the effect on the objective function value. Illustrate for the case:

$$\begin{aligned} \max 100x_1 + 100x_2 & \quad \text{subject to} \\ x_1 + x_2 \leq & \quad \beta \\ 2x_1 + x_2 \leq & \quad 12 \\ x_1 + 2x_2 \leq & \quad 12 \\ x_i \geq 0 & \end{aligned}$$

by comparing $\beta = 9$ with $\beta = 8$. At the value $\beta = 8$ how do the different choices of optimal basis give accurate or inaccurate predictions? (Remember β could increase or decrease slightly.)

29. Let $P = \{\mathbf{x} : A\mathbf{x} = \mathbf{0}; \mathbf{x} \geq \mathbf{0}\} \subset \Re^n$, a polyhedron in $\Re^n$. Formulate an LP to find a point $\mathbf{x} \in P$ such that the number of strictly positive components of $\mathbf{x}$ is as large as possible.

30. *Convex penalties for constraint violation.* Your LP to minimize $\mathbf{c} \cdot x$ subject to $A\mathbf{x} \leq \mathbf{b}; Fx \leq g; x \geq 0$ is infeasible. After discussion with domain experts, you determine that only the $A\mathbf{x} \leq \mathbf{b}$ constraints are soft, and that the polyhedron $Fx \leq g; x \geq 0$ is not empty. Also, the penalty for exceeding b_i should be α_i per unit up to $b_i + \beta_i$, and ρ_i per unit exceeding $b_i + \beta_i$. Formulate an LP that minimizes a weighted sum of $\mathbf{c} \cdot x$ and the sum of penalties for violating $A\mathbf{x} \leq \mathbf{b}$. Also, explain why $\rho_i \geq \beta_i$ is required.

31. The standard hypercube in $\Re^n$ with center w and edge length L is the polytope $\{x \in \Re^n : w_i - L/2 \leq x_i \leq w_i + L/2 \ \forall i\}$. Given $m \times n$ matrix A and m-vector b, formulate an LP to find a standard hypercube of minimum possible edge length that contains the polyhedron $\{y \geq 0 : Ay \leq b\}$.

D problems

32. ([277]) You must choose to enter door 1 or door 2. One or two leopards are behind door 1. One or two bears are behind door 2. There are three animals total and the probability there are more leopards is 1/2.

    Leptons (respectively baryons) emitted by leopards (respectively bears) pass through door 1 (respectively door 2) at rate 1 per minute for each animal, as independent Poisson (memoryless) processes. You have a 75% chance to live if you choose a door behind which there is only one animal. Otherwise you will certainly die.

    You have \$500 to pay laboratory technicians to monitor the doors with subatomic particle detectors for both leptons and baryons. They agree to follow a plan you will provide. The plan specifies, given the particles that have been detected so far, whether to stop or continue monitoring. The cost of the plan would ordinarily be \$100 per particle detected, but because you are likely to die you agree to pay in advance. In exchange, they agree that you pay for the expected number of particles detected. However, you know that regardless of your plan they will not detect more than 8 particles. Formulate an LP that maximizes your chance of survival. *Hint: your optimal plan might be randomized.*

33. Formulate the following as an linear program. State clearly any simplifying assumptions you make to keep your model linear. Electricity must be produced and delivered to customers to meet demand at minimum cost, over a 24 hour period. A set of generators is available. Each generator is characterized by a convex piecewise linear cost function that maps power output (in MW) to hourly cost. The functions are different for each generator. Each hour the power output of each generator can be set to a new level; however, the new level can't differ from the previous level by more than some specified amount (in MW). For each generator-customer pair there is a specified transmission efficiency, which is strictly positive and strictly less than 1. If the efficiency is p, the customer receives p units of electric energy for every 1 unit of energy sent from the generator. Each customer has a known demand pattern specified in the following way: for every $i, j : 1 \leq i < j \leq 24$ there is a certain number of MW-hours of energy that must be received during the time interval from the beginning of hour i to the beginning of hour j, independent of the demand for any other pair i, j. For example, a customer might need 0.4 MW-hour between 7am and 8am, 0.5 MW-hour between 8am and 9am, and an additional 0.8 MW-hour between 7am and 9am. That customer would be satisfied to receive 0.6 MW-hour between 7am and 8am and 1.1 MW-hour between 8am and 9am. The customer would be equally satisfied to receive 1.2 MW-hour between 7am and 8am, and 0.5 MW-hour between 8am and 9am.

Follow the usual rules for formulating LPs. Define index sets, data, and variables first. Then state the model (objective function and constraints).

34. Model the following problem as an LP. Treat amounts of traffic as continuous variables rather than as discrete variables. Treat time as occurring in discrete steps. A network contains three kinds of nodes: sources, bottlenecks, and destinations. Traffic in the network originates at $|S|$ source points indexed by $s \in S$ and travels to destination points $d \in D$. In particular, for each time step $t = 1 \ldots |T|$, V_{sdt} vehicles with destination $d \in D$ originate at s. For each source-destination pair s, d, traffic will always follow a certain prescribed route from s to d. The route is described as a set of times after a vehicle departs from s to traverse a set of bottleneck points indexed by $b \in B$ on the way to d. For example, a route might traverse b_1 three time units after departure from s, traverse b_6 five time units after departure from s, and traverse b_5 nine time units after departure from s. A route never traverses more than one bottleneck point during a single time step.

    The capacity of bottleneck b in time step t is C_{bt}. If more than that number of vehicles attempts to traverse the bottleneck during that time step, traffic flow becomes congested. To avoid congestion, vehicles at source points may be held back from departure into the network until times later than their origination times. The time step at which a vehicle is allowed to depart into the network is called the *release time.*

    The problem is to decide when to release vehicles so as to avoid congestion but permit all vehicles to arrive at their destinations. The objective is to minimize the sum (taken over all vehicles) of the release times. (This problem has been solved as an LP since the 1960s.)

    Hints:

    - *If you add a fixed amount to the objective of a linear program, you don't change the optimal solution. There are two fixed amounts in the problem. The first fixed amount is the total time spent driving rather than waiting to be released (the sum over all vehicles). The other fixed amount is the sum of* tV_{sdt}*, the time between time step zero and the vehicle origination time, i.e., when it is ready to leave (though it might be forced to wait).*
    - *The hardest part of this formulation problem is to define the data so that the constraints are linear.*

35. For two $m \times n$ matrices D and F, define $D \leq F \Leftrightarrow D_{i,j} \leq F_{i,j}\ \forall 1 \leq i \leq m; \forall 1 \leq j \leq n$. Let $D, F, \mathbf{c}, \mathbf{b}$ be given data. Model the following problem as a linear program without using exponentially many constraints:

$$\begin{aligned} \max \mathbf{c}^T\mathbf{x} &\text{ subject to} \\ E\mathbf{x} \leq \mathbf{b} &\text{ for all } D \leq E \leq F \\ \mathbf{x} \geq &\mathbf{0} \end{aligned}$$

    Hint: First model the special case $m = 1$.

36. Let $P = \{\mathbf{x} : A\mathbf{x} = \mathbf{b}; x \geq \mathbf{0}\} \subset \Re^n$, a polyhedron in $\Re^n$. Formulate an LP to find a point $\mathbf{x} \in P$ such that the number of strictly positive components of $\mathbf{x}$ is as large as possible. (Hint: Do Problem 29 first.)

I Problems

37. A simplex is a polytope in $\Re^n$ with $n+1$ extreme points and nonzero n-dimensional volume (or equivalently with nonempty interior). A triangulation of the n-cube $[0,1]^n$ is a collection of simplices whose union is the n-cube and whose pairwise intersections are all faces of those simplices. Formulate and solve a linear program that proves that all triangulations of the 4-cube have cardinality at least 16 [257].

38. Suppose you need to solve $\min \mathbf{c} \cdot \mathbf{x}$ subject to $A\mathbf{x} \geq \mathbf{b}; \mathbf{x} \geq 0$ but you have software that only solves feasibility problems. That is, your software takes as input a description $A\mathbf{x} \geq \mathbf{b}; \mathbf{x} \geq 0$ of a polyhedron and outputs a point in that polyhedron, if one exists; otherwise, it outputs a message that the polyhedron is empty.

    (a) How could you get a close approximation to the optimal solution (if one exists) by solving a sequence of feasibility problems?

    (b) How could you find an optimal solution (if one exists) by solving one feasibility problem? (Hint: Use duality.)

39. Assume that an $n \times n$ matrix A is stored such that A_{ij} can be retrieved in a single step. Devise an algorithm that either finds a saddlepoint of A or determines that none exists, in

    (a) time $O(n^{1.99})$ [212];

    (b) time $O(n \log n)$[24].

    For the definition of $O()$ see Definition 9.2.

40. How would you compute the dot product $\mathbf{c} \cdot \mathbf{d}$ of two vectors $\mathbf{c}, \mathbf{d}$ that are stored as sparse vectors? If you doubled the length of $\mathbf{c}$ and $\mathbf{d}$ but kept the number of nonzeros the same, how much longer would it take to compute $\mathbf{c} \cdot \mathbf{d}$? If you doubled the number of nonzeroes in both $\mathbf{c}$ and $\mathbf{d}$, but kept their length the same, how much longer would it take to compute $\mathbf{c} \cdot \mathbf{d}$?

Chapter 4

Polyhedra

Preview: Polyhedra are the feasible regions of LPs. This chapter derives properties of polyhedra, which underly algorithms for linear optimization.

A polyhedron is a generalization of a polygon. Polyhedra live in $\Re^n$, not just $\Re^2$; polyhedra need not be bounded. The cube in $\Re^3$ is a bounded polyhedron, and the nonnegative orthant in $\Re^3$ is an unbounded polyhedron.

Definition 4.1 (Polyhedron) *A* **polyhedron** *is a set of form* $\{\mathbf{x} : A\mathbf{x} \leq \mathbf{b}\}$*, where* $\mathbf{x} \in \Re^n$*, A is an m by n matrix, and* $\mathbf{b} \in \Re^m$*. Geometrically, each linear inequality* $A_{j,:}\mathbf{x} \leq b_j$ *defines a halfspace, and the polyhedron is the intersection of these m halfspaces.*

Definition 4.2 (Bounded set, Polytope) *A set $S \subset \Re^n$ is* **bounded** *iff there exists* $\alpha \in \Re$ *such that* $||s|| \leq \alpha$ *for all* $s \in S$*. A bounded polyhedron is called a* **polytope**.

Polytopes in two and three dimensions are depicted in Figures 4.1 and 4.2, respectively.

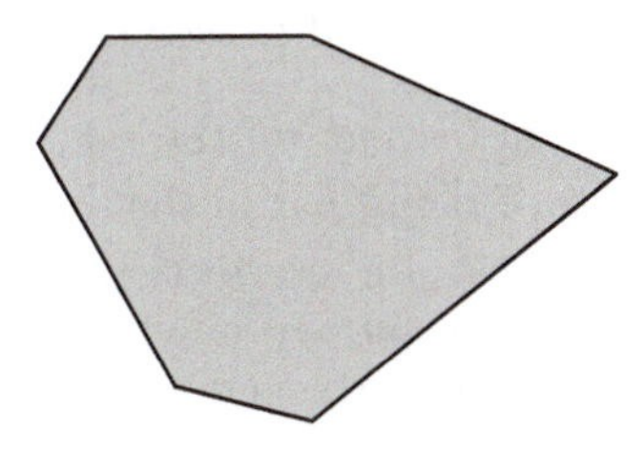

FIGURE 4.1: In two dimensions, a polytope is a convex polygon.

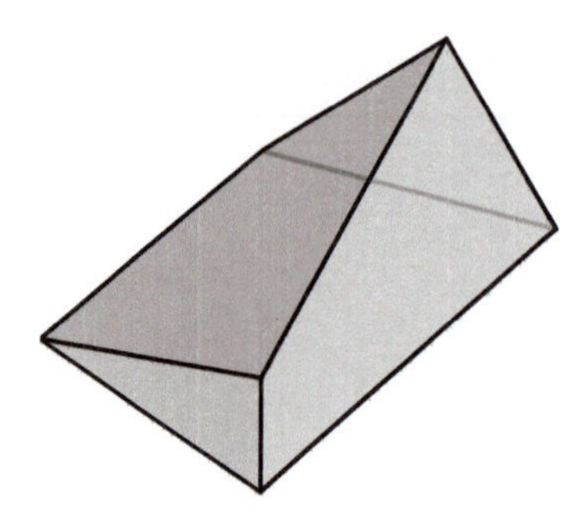

FIGURE 4.2: In three dimensions, a polytope is convex, has flat sides, and is bounded.

Definition 4.3 (Convex Combination, Affine Combination) *For any two points* $\mathbf{x}, \mathbf{y} \in \Re^n$, *and any* $0 \leq \lambda \leq 1$, *the point* $\lambda\mathbf{x} + (1-\lambda)\mathbf{y}$ *is a* **convex combination** *of* $\mathbf{x}$ *and* $\mathbf{y}$. *For any* $\lambda \in \Re$, *the point* $\lambda\mathbf{x} + (1-\lambda)\mathbf{y}$ *is an* **affine combination** *of* $\mathbf{x}$ *and* $\mathbf{y}$. *More generally, for any set of points* $\mathbf{x}^1, \mathbf{x}^2, \ldots, \mathbf{x}^N$ *in* $\Re^n$, *a point* $\mathbf{y}$ *is a* **convex** *(respectively* **affine***)* **combination** *of these points iff there exist* $\lambda^1 \ldots \lambda^N$ *such that* $\lambda \geq 0$ *(respectively nothing) and* $\sum_{j=1}^N \lambda^j = 1$ *such that* $\mathbf{y} = \sum_{j=1}^N \lambda^j \mathbf{x}^j$.

Definition 4.4 (Convex Hull, Affine Hull) *The convex (respectively affine) hull of points* $\mathbf{x}^1, \mathbf{x}^2, \ldots, \mathbf{x}^N$ *in* $\Re^n$ *is the set of all their convex (respectively affine) combinations.*

Geometrically, the line segment between $\mathbf{x}$ and $\mathbf{y}$ is their convex hull, and the line that they define (if they are distinct) is their affine hull. Figure 4.3 illustrates convex and affine combinations of points.

Reading Tip: *Be sure you grasp the equivalence between the geometry and the algebra. As you move to the right from* y *on the line through* x *and* y *in Figure 4.3, is* $\lambda < 0$ *or is* $1-\lambda < 0$*?*

Definition 4.5 (Convex Set) *A set* $S \subseteq \Re^n$ *is* **convex** *iff for all* $\mathbf{s} \in S, \mathbf{t} \in S$, *all convex combinations of* $\mathbf{s}$ *and* $\mathbf{t}$ *are in* S.

Figure 4.4 depicts a convex set and a non-convex set.

Proposition 4.0.1 *A polyhedron is closed and convex.*

Proof: For any vector $\boldsymbol{\pi} \in \Re^n$ and scalar $\boldsymbol{\pi}_0$ the set $\{x : \boldsymbol{\pi} \cdot x < \boldsymbol{\pi}_0\}$ is open. (Set $\epsilon = \frac{|\boldsymbol{\pi}_0 - \boldsymbol{\pi} \cdot x|}{2||\boldsymbol{\pi}||}$; then the ball around $\mathbf{x}$ of radius ϵ is in the set.) By definition its complement is closed. A polyhedron is therefore the finite intersection of closed sets, which is closed.

Now let $P = \{\mathbf{x} : A\mathbf{x} \leq \mathbf{b}\}$ be an arbitrary polyhedron and let $\mathbf{u} \in P, \mathbf{v} \in P$, and $0 \leq \lambda \leq 1$. Then $A\mathbf{u} \leq \mathbf{b}; A\mathbf{v} \leq \mathbf{b}; \lambda A\mathbf{u} \leq \lambda\mathbf{b}$, and (since $\lambda \leq 1$) $(1-\lambda)A\mathbf{v} \leq (1-\lambda)\mathbf{b}$. Hence

$$A(\lambda\mathbf{u} + (1-\lambda)\mathbf{v}) = \lambda A\mathbf{u} + (1-\lambda)A\mathbf{v} \leq \lambda\mathbf{b} + (1-\lambda)\mathbf{b} = \mathbf{b} \tag{4.1}$$

and P is convex. ■

That the feasible region of an LP is always convex helps explain why the method in

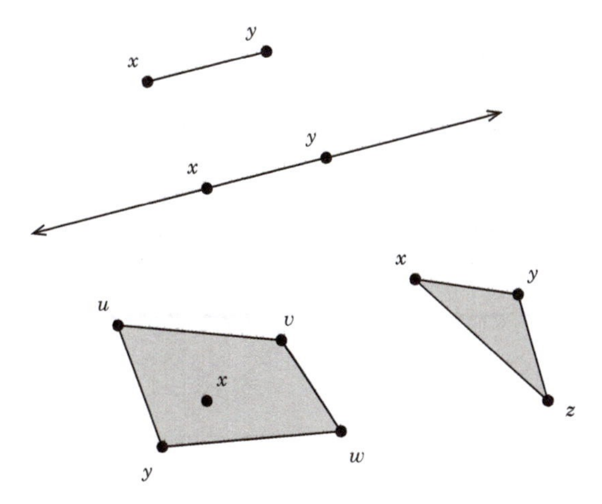

FIGURE 4.3: The convex combinations of x and y form a line segment; their affine combinations form a line. The sets of convex combinations of three and five points are depicted.

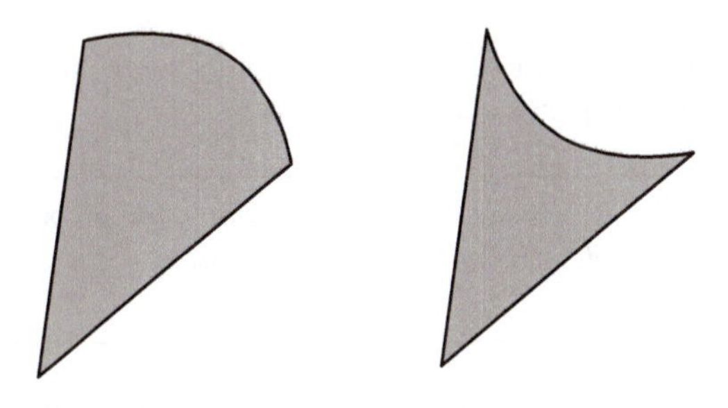

FIGURE 4.4: The set on the left is convex; the set on the right is not convex.

Section 3.3 of replacing $|y|$ with $y^+ + y^-$ can fail. We want only one of y^+, y^- to be nonzero. Geometrically, we want the feasible region to comprise only the boundary of the positive quadrant, the positive parts of the horizontal and vertical axes. But that region is not convex, so we cannot force it to be the feasible region. To be convex, the feasible region must include the entire positive quadrant.

Later in this chapter we will prove that for any finite set of points in $\Re^n$, the set of all of their convex combinations is a polytope, and conversely for any polytope there exists a finite set of points whose convex combinations form that polytope. This equivalence is an aspect of duality. For now, observe how each definition is useful for different things. If I give you a point $\mathbf{s}$ and ask you, "Is $\mathbf{s}$ in this polytope?", you would like to use the first definition. Simply compute $A\mathbf{s}$ and see if it is $\leq \mathbf{b}$. But if I ask you to give me merely one point in the polytope, you have to solve a linear program. For the dual description of the polytope, the opposite is true. If I ask you whether $\mathbf{s}$ is in the polytope, you have to solve a linear program to know the answer. But if I ask you to give me a point in the polytope, you simply take one of the points from the dual description. If I ask you to give me many points in the polytope, you pick many sets of values $\pi_1 \ldots \pi_M$, $\pi_i \geq 0; \sum_i^M \pi_i = 1$ and compute the convex combinations of the points given in the dual description.

Summary: Polyhedra are the feasible regions of LPs. Memorize the definitions of convex combinations and convexity of a set. All polyhedra are closed and convex. Polytopes are bounded polyhedra.

4.1 Separation

Preview: This section proves the separation theorem, an important duality property of polyhedra which is geometrically intuitive.

We proved in the previous section that polyhedra are closed and convex subsets of $\Re^n$. Such subsets enjoy a duality property called *separation*.

Theorem 4.1.1 *Let $S \subset \Re^n$ be closed and convex. Then for all $\mathbf{x} \in \Re^n$ either $\mathbf{x} \in S$ or there exist $\boldsymbol{\pi} \in \Re^n$, $\pi_0 \in \Re$ (depending on $\mathbf{x}$) such that $\boldsymbol{\pi} \cdot \mathbf{x} < \pi_0$ and for all $\mathbf{s} \in S$, $\boldsymbol{\pi} \cdot \mathbf{s} > \pi_0$.*

The geometric meaning of $\mathbf{x} \in S$ is clear, but what is the meaning of the theorem when $\mathbf{x} \notin S$? Geometrically the vector $\boldsymbol{\pi}$ is orthogonal to the hyperplane $\{\mathbf{y} : \boldsymbol{\pi} \cdot \mathbf{y} = \pi_0\}$. The point $\mathbf{x}$ lies in the open halfspace on one side of this hyperplane, where all points have dot product with $\boldsymbol{\pi}$ less than π_0. The set S lies on the other side of this hyperplane, so all points in S have dot product with $\boldsymbol{\pi}$ more than π_0. The hyperplane separates $\mathbf{x}$ from S. If we wanted to minimize $\boldsymbol{\pi} \cdot \mathbf{y}$, there would exist an $\epsilon > 0$ such that the point $\mathbf{x}$ would have an objective value better by at least ϵ than any point in S. In some texts, this kind of separation is called *strict separation.*

Proof: First we prove the theorem for the case of S bounded. Suppose $\mathbf{x} \notin S$. Define the function $f(\mathbf{s}) \equiv ||\mathbf{s} - \mathbf{x}||$, the Euclidean distance from $\mathbf{s}$ to $\mathbf{x}$. Since $f(\mathbf{s})$ is continuous, and S is compact, $f()$ attains its minimum on S.

Aside: We claim that the minimum is unique. The proof is by contradiction. If $||\mathbf{s}-\mathbf{x}|| = ||\mathbf{t} - \mathbf{x}||$ and $\mathbf{s} \neq \mathbf{t}$, the three points $\mathbf{s}, \mathbf{t}, \mathbf{x}$ form an isosceles triangle.

Reading Tip: *This illustrates a proof technique. S lives in high dimensional space $\Re^n$, but the three points determine a triangle, for which we can use elementary geometry.*

Then the midpoint of $\mathbf{s}$ and $\mathbf{t}$ is strictly closer to $\mathbf{x}$ than are $\mathbf{s}$ and $\mathbf{t}$, and is in S by convexity. Let $\mathbf{s}^*$ denote the (unique) point at which $f()$ attains its minimum on S.

We now want to construct a hyperplane separating S from $\mathbf{x}$. Geometrically, it is plausible to bisect the line segment connecting $\mathbf{s}^*$ and $\mathbf{x}$ with a hyperplane (see Figure 4.5). That hyperplane is defined by $\{\mathbf{y} : (\mathbf{s}^* - \mathbf{x}) \cdot \mathbf{y} = (\mathbf{s}^* - \mathbf{x}) \cdot (\mathbf{s}^*/2 + \mathbf{x}/2)\}$. (Why? The

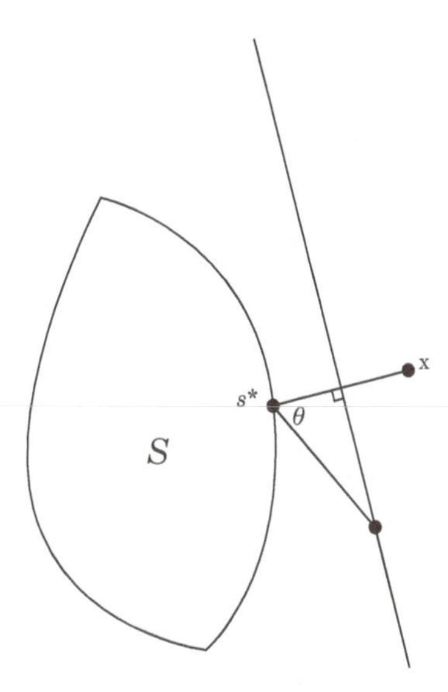

FIGURE 4.5: Separate S from $\mathbf{x}$ with the hyperplane bisecting the line segment between s^* and $\mathbf{x}$.

hyperplane must be defined by an equation of form $\boldsymbol{\pi} \cdot \mathbf{y} = \boldsymbol{\pi}_0$. The vector $\boldsymbol{\pi}$ is $\mathbf{s}^* - \mathbf{x}$. At value $\boldsymbol{\pi}_0 = (\mathbf{s}^* - \mathbf{x}) \cdot \mathbf{x}$ the hyperplane would pass through $\mathbf{x}$; at value $\boldsymbol{\pi}_0 = (\mathbf{s}^* - \mathbf{x}) \cdot \mathbf{s}^*$ the hyperplane would pass through $\mathbf{s}^*$. The given value of $\boldsymbol{\pi}_0$ is halfway between.) Everything on one side of the hyperplane has dot product with $\boldsymbol{\pi}$ greater than $\boldsymbol{\pi}_0$; everything on the other side (including $\mathbf{x}$) has smaller dot product.

We must show that $\boldsymbol{\pi} \cdot \mathbf{s} > \boldsymbol{\pi}_0 \ \forall \mathbf{s} \in S$. Suppose to the contrary $\mathbf{t} \in S$ but $\boldsymbol{\pi} \cdot \mathbf{t} \leq \boldsymbol{\pi}_0$. Our geometric intuition says that (see Figure 4.5) by convexity, there would be points very close to $\mathbf{s}^*$ that are closer to $\mathbf{x}$ than $\mathbf{s}^*$, contradicting the minimality of $\mathbf{s}^*$. To be precise, consider the three points $\mathbf{x}, \mathbf{s}^*, \mathbf{t}$. As in Figure 4.5, let θ be the angle these points define at $\mathbf{s}^*$. Then $(\cos\theta)||\mathbf{s}^* - \mathbf{x}||||\mathbf{s}^* - \mathbf{t}|| = (\mathbf{s}^* - x) \cdot (\mathbf{s}^* - \mathbf{t}) = \boldsymbol{\pi} \cdot (\mathbf{s}^* - \mathbf{t}) \geq \boldsymbol{\pi} \cdot \mathbf{s}^* - \boldsymbol{\pi}_0 > 0$. Hence $\theta < \pi/2$. Since the angle θ at s^* is acute, there exist points on the line segment between $\mathbf{t}$ and $\mathbf{s}^*$ that are closer to $\mathbf{x}$ than is $\mathbf{s}^*$. This contradicts the minimality of $\mathbf{s}^*$. Hence the theorem is proved for bounded S.

If S is not bounded, let $\mathbf{s}$ be any point in S. Define $\overline{B(\mathbf{x}, \alpha)} \equiv \{\mathbf{y} : ||\mathbf{x} - \mathbf{y}|| \leq \alpha\}$, the closed ball around $\mathbf{x}$ of radius α. Let $\hat{S} = S \bigcap \overline{B(\mathbf{x}, ||\mathbf{x} - \mathbf{s}||)}$. Now $\hat{S}$ is compact and there is a closest point $\mathbf{s}^*$ to $\mathbf{x}$ in $\hat{S}$. Any point in $S - \hat{S}$ must be farther from $\mathbf{x}$ than $\mathbf{s}^*$. Hence $\mathbf{s}^*$ is the (unique) point at which $f()$ attains its minimum on S and the rest of the proof follows without change. ■

Reading Tip: *This proof illustrates how to let your geometric intuition guide your proofs. Spatial reasoning often suggests to you why something is true. Translate your geometric idea into algebra to get a rigorous proof.*

Try to determine whether all of the hypotheses of the theorem are necessary. Find an example of a closed non-convex set S and a point $\mathbf{y} \notin S$ that cannot be separated from S. Find an example of a convex set S that is not closed and a point $\mathbf{y} \notin S$ that cannot be strictly separated from S.

Question: What would it mean for the separation property to characterize convex closed sets? Do you think that it does?[1]

Summary: If a point is not in a polyhedron or some other closed convex set S, there exists a hyperplane strictly between the point and S. This geometrically plausible fact is our first major duality theorem. Its proof appears to be elementary, but actually relies on a deep theorem from real analysis.

4.2 Fourier-Motzkin Elimination

Preview: This section presents Fourier-Motzkin elimination, a way to solve systems of linear inequalities by reducing the number of variables by one at a time. Though not ordinarily useful in practice, it provides an important property of polyhedra: projections and other linear transformations of polyhedra are polyhedra.

Fourier-Motzkin elimination is a way to solve systems of linear inequalities by systematically removing one variable at a time. It is not computationally efficient but it is theoretically powerful. There is an extension to integer programming which, too, is not efficient but is theoretically useful. Let's start with an example.

$$x_1 + 2x_2 - 4x_3 \leq 4 \tag{4.2}$$
$$-5x_1 - x_2 + 2x_3 \leq 6 \tag{4.3}$$
$$2x_1 - x_2 + 10x_3 \geq 4 \tag{4.4}$$
$$x_1 \leq 12 \tag{4.5}$$
$$x_2 \geq -5 \tag{4.6}$$
$$0 \leq x_3 \leq 20 \tag{4.7}$$

We want to eliminate x_3 from this system, reducing it to a set of inequalities on x_1 and x_2. Group the constraints into three sets, those that do not involve x_3, those that are upper bounds on x_3, and those that are lower bounds on x_3. The first set stays the same. Create all combinations of a bound from the second set with a bound from the third set.

In the example, inequalities (4.5) and (4.6) compose the first set. The second set comprises (4.3) and (4.7):

---

[1] It would mean that the converse of Theorem 4.1.1 was also true, i.e., that every set that possesses the separation property must be closed and convex.

$$\begin{aligned} x_3 &\leq 3 + 2.5x_1 + 0.5x_2 \\ x_3 &\leq 20 \end{aligned}$$

The third set comprises (4.2), (4.4), and (4.7):

$$\begin{aligned} 0.25x_1 + 0.5x_2 - 1 &\leq x_3 \\ -0.2x_1 + 0.1x_2 + 0.4 &\leq x_3 \\ 0 &\leq x_3 \end{aligned}$$

We have two upper bounds and three lower bounds on x_3. That will give us six inequalities, as follows:

$$\begin{aligned} 0.25x_1 + 0.5x_2 - 1 &\leq 3 + 2.5x_1 + 0.5x_2 \\ 0.25x_1 + 0.5x_2 - 1 &\leq 20 \\ -0.2x_1 + 0.1x_2 + 0.4 &\leq 3 + 2.5x_1 + 0.5x_2 \\ -0.2x_1 + 0.1x_2 + 0.4 &\leq 20 \\ 0 &\leq 3 + 2.5x_1 + 0.5x_2 \\ 0 &\leq 20 \end{aligned}$$

The last of these involves only constants. It *must* be checked for validity. If it is not valid, the original system of inequalities has no solution. If it is valid, it can be discarded.

After collecting terms, the new system is:

$$\begin{aligned} -4 &\leq 2.25x_1 \\ 0.25x_1 + 0.5x_2 - 1 &\leq 20 \\ -2.6 &\leq +2.7x_1 + 0.4x_2 \\ -0.2x_1 + 0.1x_2 &\leq 19.6 \\ -3 &\leq +2.5x_1 + 0.5x_2 \\ x_1 &\leq 12 \\ x_2 &\geq -5 \end{aligned}$$

► *Suppose the variable x_3 were constrained to be integer. Then every upper bound on x_3 could be reduced to its floor, and every lower bound could be increased to its ceiling. This gives an inefficient but theoretically powerful algorithm for integer programming.* ◄

In general, to eliminate the variable x_i from a system of linear inequalities of form $A\mathbf{x} \leq \mathbf{b}$, follow these steps:

1. Scale all constraints so that x_i has coefficient $1, -1$ or 0 in each inequality.

2. Each constraint in which x_i has coefficient 1 (respectively -1) is an upper (respectively lower) bound on x_i. For each pair of upper and lower bounds, create a new constraint not involving x_i that sets the lower bound less than or equal to the upper bound.

3. Discard the constraints in which x_i has nonzero coefficient. Convert the remaining constraints into $A\mathbf{x} \leq \mathbf{b}$ form.

If for some $b_j < 0$ a contradiction of form $0 \leq \mathbf{b}_j$ appears, the system has no solution and additional variables need not be eliminated from the system.

Reading Tip: *I hope that you have wondered if there is a dual to FM elimination. What would one step of such a procedure do? See the exercises.*

Now we make an observation about FM elimination that will be very useful.

Remark: *One step of FM-elimination consists of multiplying the system $A\mathbf{x} \leq \mathbf{b}$ by a nonnegative matrix M.*

Proof: First scale the rows by a positive diagonal matrix M_1 so that the nth column consists of 1's, -1's, and 0's. Second, permute the rows by permutation matrix M_2 so that the first k_1 rows have 1 in the nth column, the next k_2 have -1, and the last k_3 have 0 in the nth column. Third, multiply by a $k_1k_2 + k_3$ by n matrix M_3 defined as: for each $1 \leq i \leq k_1$ and each $k_1 + 1 \leq j \leq k_1 + k_2$, M_3 has a row consisting of $e_i + e_j$; for each $k_1 + k_2 + 1 \leq i \leq k_1 + k_2 + k_3$, M_3 has a row consisting of e_i. The order of the rows is immaterial since we re-permute them each step. *(Remember, when you multiply a row vector times the matrix A, the row vector tells you which multiples of each row of M to add together. The matrix M_3 is a collection of formulas for making linear combinations out of A. If the row is $e_i + e_j$ the formula is to add one of row i to one of row j. This follows directly from the definition of matrix multiplication but people sometimes forget it.)* Let $M = M_3M_2M_1$. One step of FM-elimination consists of replacing $A\mathbf{x} \leq \mathbf{b}$ by $MA\mathbf{x} \leq Mb$. The nth column of MA is all zeros.

The theorem about FM elimination is that it works. To be precise:

Theorem 4.2.1 *A single step of FM elimination on the system $Ay \leq \mathbf{b}$ multiplies the system by a nonnegative matrix M such that the nth column of MA is all zeros, and $x_1 \dots x_{n-1}, 0$ is a solution to $MAy \leq Mb$ iff $\exists x_n$ such that $x_1 \dots x_{n-1}, x_n$ is a solution to $Ay \leq \mathbf{b}$.*

Proof: On the one hand, if $A\mathbf{x} \leq \mathbf{b}$, then $MA\mathbf{x} \leq Mb$ since M is nonnegative. Now $MA\mathbf{x} = MA(x_1 \dots x_{n-1}, 0)$ because the nth column of MA is all zeros. Therefore, $x_1 \dots x_{n-1}, 0$ is a solution to $MA\mathbf{x} \leq Mb$. On the other hand, let $x_1 \dots x_{n-1}, 0$ solve $MAy \leq Mb$. Let x_n^+ equal the minimum of all upper bounds formed by the first k_1 rows of $M_2M_1A\mathbf{x} \leq \mathbf{b}$. In other words, let x_n^+ equal

$$\min_{1 \leq j \leq k_1} b_j - \sum_{i=1}^{n-1} M_2M_1A_{ji}x_i.$$

Letting $\mathbf{x}^+ = x_1 \dots x_{n-1}, x_n^+$, by definition of x_n^+ we have $M_2M_1A_{j,:}x^+ \leq b_j$ for $1 \leq j \leq k_1$. Similarly letting

$$x_n^- = \max_{k_1+1 \leq j \leq k_1+k_2} b_j - \sum_{i=1}^{n-1} M_2M_1A_{ji}x_i,$$

we have $M_2M_1A_{j,:}x^- \leq b_j$ for $k_1 + 1 \leq j \leq k_1 + k_2$. Since FM elimination compares every pair of upper bound and lower bound, it must be that $x_n^+ \geq x_n^-$. Therefore, for any value of x_n in the range $x_n^- \leq x_n \leq x_n^+$, the inequalities $A_{j,:}(x_1 \dots x_{n-1}, x_n) \leq b_j$ hold for $1 \leq j \leq k_1 + k_2$. Finally, since x_n does not appear in the last k_3 inequalities, and $MA(x_1 \dots x_{n-1}, 0) \leq Mb$, the last k_3 inequalities are satisfied by $\mathbf{x}$. Therefore, $A\mathbf{x} \leq \mathbf{b}$ as desired. ■

Example 4.1 Let's construct the matrices M_1, M_2, M_3 such that $M_3M_2M_1 = M$ in the statement of Theorem 4.2.1 when eliminating x_3 from the following system:

$$
\begin{aligned}
11x_1 - 12x_2 + 13x_3 &\leq 200 && (4.8)\\
15x_1 + 16x_2 - 17x_3 &\leq 300 && (4.9)\\
-x_1 &\leq 0 && (4.10)\\
-x_2 &\leq 0 && (4.11)\\
-x_3 &\leq 0 && (4.12)\\
-21x_1 - 22x_2 + 23x_3 &\leq 400 && (4.13)\\
27x_1 + 28x_2 - 29x_3 &\leq 500 && (4.14)
\end{aligned}
$$

This system has the form $A\mathbf{x} \leq \mathbf{b}$ where $b = (200, 300, 0, 0, 0, 400, 500)$ and

$$
A = \begin{bmatrix} 11 & -12 & 13 \\ 15 & 16 & -17 \\ -1 & 0 & 0 \\ 0 & -1 & 0 \\ 0 & 0 & -1 \\ -21 & -22 & 23 \\ 27 & 28 & -29 \end{bmatrix}.
$$

The matrix M_1 scales the rows so that all entries in the last column of A are $0, -1$, or 1. Therefore,

$$
M_1 = \begin{bmatrix} \frac{1}{13} & 0 & 0 & 0 & 0 & 0 & 0 \\ 0 & \frac{1}{17} & 0 & 0 & 0 & 0 & 0 \\ 0 & 0 & 1 & 0 & 0 & 0 & 0 \\ 0 & 0 & 0 & 1 & 0 & 0 & 0 \\ 0 & 0 & 0 & 0 & 1 & 0 & 0 \\ 0 & 0 & 0 & 0 & 0 & \frac{1}{23} & 0 \\ 0 & 0 & 0 & 0 & 0 & 0 & \frac{1}{29} \end{bmatrix}.
$$

The matrix M_2 rearranges the rows so that the coefficients in the last column appear in the order 1's, -1's, and 0's. Row 1 remains row 1; row 2 moves to row 3; rows 3 and 4 move to rows 6 and 7; row 5 remains row 5; row 6 moves to row 2; row 7 moves to row 4. Remember that the kth row of M_2 tells what linear combination of the rows of M_1A to put into row k of the product M_2M_1A. Therefore,

$$
M_2 = \begin{bmatrix} 1 & 0 & 0 & 0 & 0 & 0 & 0 \\ 0 & 0 & 0 & 0 & 0 & 1 & 0 \\ 0 & 1 & 0 & 0 & 0 & 0 & 0 \\ 0 & 0 & 0 & 0 & 0 & 0 & 1 \\ 0 & 0 & 0 & 0 & 1 & 0 & 0 \\ 0 & 0 & 1 & 0 & 0 & 0 & 0 \\ 0 & 0 & 0 & 1 & 0 & 0 & 0 \end{bmatrix}.
$$

The M_2 matrix is always a *permutation matrix*; it has exactly one 1 in each row and in each column, and the rest of the entries equal 0. The matrix M_3 creates a row for each pair of rows of M_2M_1A with one "1" and one "-1" in the last column. That row is simply the sum of the rows the compose the pair. Since M_2 has conveniently placed all the "1"s first, followed by all the "−1"s, M_3 has the following form:

$$M_3 = \begin{bmatrix} 1 & 0 & 1 & 0 & 0 & 0 & 0 \\ 1 & 0 & 0 & 1 & 0 & 0 & 0 \\ 1 & 0 & 0 & 0 & 1 & 0 & 0 \\ 0 & 1 & 1 & 0 & 0 & 0 & 0 \\ 0 & 1 & 0 & 1 & 0 & 0 & 0 \\ 0 & 1 & 0 & 0 & 1 & 0 & 0 \\ 0 & 0 & 0 & 0 & 0 & 1 & 0 \\ 0 & 0 & 0 & 0 & 0 & 0 & 1 \end{bmatrix}.$$

Definition 4.6 (Projection of a polyhedron) *Let P be a polyhedron in $\Re^n$. Let I denote a nonempty strict subset of the indices $1 \ldots n$ and let $\bar{I}$ denote its complement. The projection Q of P onto I is the set*

$$\left\{ x_I \in \Re^{|I|} | \exists x_{\bar{I}} \in \Re^{\bar{I}} \text{ such that } x_{I \cup \bar{I}} \in P \right\}. \tag{4.15}$$

Figure 4.6 shows an example of projections of a polygon. *Question: The cube $0 \le x_i \le 1; i = 1, 2, 3$ has a projection in the shape of a square onto the indices $1, 2$. If the cube were rotated randomly, what shape would the projection almost surely have?*[2]

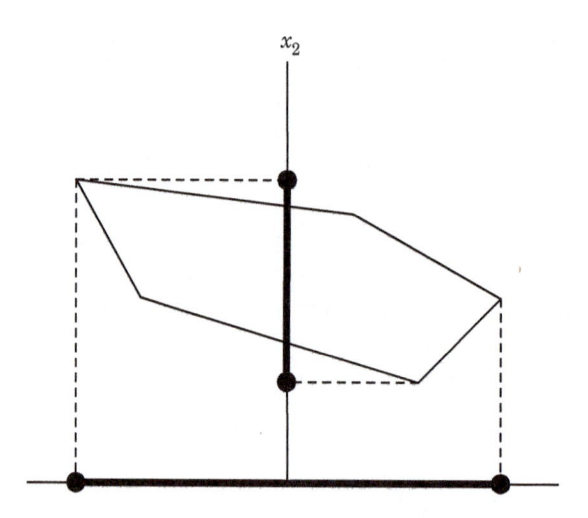

FIGURE 4.6: Two projections of a polygon, onto index set $\{1\}$ and index set $\{2\}$.

Theorem 4.2.2 *Let $P \subset \Re^n$ be a polyhedron. Let Q be the projection of P onto index set $\phi \ne I \subsetneq \{1 \ldots n\}$. Then Q is a polyhedron.*

[2]A hexagon.

Proof: Perform Fourier-Motzkin elimination on the variables $x_i, i \in \overline{I}$. By Theorem 4.2.1 the result is the projection Q. By the definition of Fourier-Motzkin elimination, the result is defined by a finite set of linear inequalities, and hence by definition is a polyhedron. *Question: Think of a counterexample to the converse of Theorem 4.2.2.*[3]

Corollary 4.2.3 *Let P be a polyhedron. Let Q be a linear transformation of P. Then Q is a polyhedron.*

Proof: By assumption $Q = \{Tx : x \in P\}$ for some matrix T. Let P be defined by inequalities $A\mathbf{x} \leq \mathbf{b}$. Then $R = \{(\mathbf{x}, y) : A\mathbf{x} \leq \mathbf{b}, y = Tx\}$ is a polyhedron, Q is a projection of R, and the result follows from Theorem 4.2.2.

Summary: Fourier-Motzkin elimination is a way to recursively solve systems of linear inequalities by reducing the dimension. Hence it is an inductive mechanism to analyze polyhedra. We used it to prove that linear transformations of polyhedra are polyhedra, and in the next section we will use it to prove the very important theorem of the alternative for polyhedra. The FM algorithm works by pairing every upper bound constraint on a variable with every lower bound constraint on that variable. The number of constraints can increase quadratically in a single step, and thus exponentially overall. That is why the algorithm is not computationally useful.

4.3 Theorems of the Alternative

Preview: Theorems of the alternative state that, of two systems of constraints, exactly one system has a feasible solution. These are among the most useful of the important duality theorems. We start with a theorem of the alternative for a system of linear equalities, which is related to Lagrange's method, and then move to theorems for polyhedra.

Much of linear programming can be thought of as extending the theory of linear equalities to linear inequalities. Theorems of the alternative compose one of the prettiest and most powerful of these extensions. Let's begin with a theorem about linear equalities. Remember the Lagrangean function, which we used to prove weak duality? Why does it work to solve optimization problems subject to equality constraints?

[3]The projection of a circle or annulus is a line segment, which is a polyhedron.

Theorem 4.3.1 *If $A \in \Re^{m\times n}$ and $\mathbf{b} \in \Re^m$, then exactly one of the following two systems has a feasible solution: $A\mathbf{x} = \mathbf{b}$ (system I); $\boldsymbol{\pi}^T A = \mathbf{0}, \boldsymbol{\pi}^T\mathbf{b} \neq 0$ (system II).*

Proof: If both systems had feasible solutions, there would be a contradiction: $\mathbf{0} = \mathbf{0x} = \boldsymbol{\pi}^T A\mathbf{x} = \boldsymbol{\pi}^T\mathbf{b} \neq 0$. Suppose that system I has no feasible solution. Then $\mathbf{b}$ is not in the column space of A. Let J be a subset of the column indices such that $A_{:,J}$ is a basis for the column space of A. Note $|J| < m$. Let $\boldsymbol{\pi}^1, \dots \boldsymbol{\pi}^{m-|J|}$ be additional vectors, orthogonal to $A_{:,j}$ for all $j \in J$ and to each other, that extend the columns to a basis for $\Re^m$. If we express $\mathbf{b}$ as a linear combination of these basis vectors, at least one $\boldsymbol{\pi}^i$ has a nonzero coefficient, for otherwise $\mathbf{b}$ would be in the column space of A. Then $\boldsymbol{\pi}^i$ satisfies system II. ∎

Let $f()$ be a differentiable real-valued function on $\Re^m$. We wish to minimize $f(\mathbf{z})$ subject to the equality constraints $g^j(\mathbf{z}) = 0$, $j = 1 \dots m$. Let $\mathbf{y}$ satisfy $g^j(\mathbf{y}) = 0 : j = 1 \dots m$. If there exists a direction $\mathbf{d} \in \Re^m$ such that $\nabla g^j(\mathbf{y}) \cdot \mathbf{d} = 0\ \forall j$, then moving infinitesimally in that direction from $\mathbf{y}$ keeps the constraints satisfied. If in addition $\mathbf{d} \cdot \nabla f(\mathbf{y}) \neq 0$, moving in either direction $\mathbf{d}$ or $-\mathbf{d}$ will decrease $f()$. Letting $\boldsymbol{\pi}$ denote $\mathbf{d}$, $\mathbf{b}$ denote $\nabla f(\mathbf{y})$, and letting A denote the matrix whose jth column is $\nabla g^j(\mathbf{y})$, system II says that there exists a direction to travel from $\mathbf{y}$ that retains feasibility and improves the objective function. If no such direction exists, system I has a feasible solution. But what is system I? It is the Lagrangian condition $\nabla f(\mathbf{y}) = \sum x_j \nabla g^j(\mathbf{y})$.

A theorem of the alternative lets you switch between a "there doesn't exist" statement and a "there exists" statement. The Lagrangian condition is simply the statement, "$\mathbf{y}$ can only be a solution if $\mathbf{y}$ is feasible and there does not exist an improving direction $\mathbf{d}$ that keeps the constraints satisfied", converted into a "there exists" form. From now on there should not be anything mysterious about the Lagrangian function.

Looking ahead for a moment, the first order optimality conditions for inequality-constrained optimization are based on the same idea. There are two main differences. The first is that if an inequality constraint holds strictly, it is not relevant because no infinitesimal movement will violate that constraint. The second is that we don't need $\mathbf{d} \cdot \nabla g^j(\mathbf{y}) = 0$, we only need $\mathbf{d} \cdot \nabla g^j(\mathbf{y}) \geq 0$ (for a constraint of form $g^j(\mathbf{y}) \geq 0$).

The analogue to Theorem 4.3.1 for inequalities ought to be similar. Either there is a feasible solution to $A\mathbf{x} \leq \mathbf{b}, \mathbf{x} \geq \mathbf{0}$ or there is a reason why not, a contradiction. To get a contradiction with inequalities, we can only take nonnegative linear combinations of the inequalities. A contradiction would be the existence of $\boldsymbol{\pi} \geq \mathbf{0}$ such that $\boldsymbol{\pi}^T A \geq \mathbf{0}$ and $\boldsymbol{\pi}^T\mathbf{b} < 0$. For the system $A\mathbf{x} \leq \mathbf{b}$, a contradiction would be the existence of $\boldsymbol{\pi} \geq \mathbf{0}$ such that $\boldsymbol{\pi}^T A = \mathbf{0}$ and $\boldsymbol{\pi}^T\mathbf{b} < 0$.

Theorem 4.3.2 *For any $A \in \Re^{m\times n}$ and $\mathbf{b} \in \Re^m$ exactly one of the following two systems has a feasible solution: $A\mathbf{x} \leq \mathbf{b}$ (system I); $\boldsymbol{\pi} \geq \mathbf{0}$, $\boldsymbol{\pi}^T A = \mathbf{0}, \boldsymbol{\pi}^T\mathbf{b} < 0$ (system II).*

Proof: If $\mathbf{x}$ and $\boldsymbol{\pi}$ solved systems I and II, respectively, then $0 = 0x = \boldsymbol{\pi}^T A\mathbf{x} \leq \boldsymbol{\pi}^T\mathbf{b} < 0$ would be a contradiction. Therefore, at most one of the two systems has a feasible solution. Perform FM-elimination on system I n times. By Theorem 4.2.1 this is equivalent to multiplying by the product of n nonnegative matrices $M^n M^{n-1} \dots M^1 = M^*$ such that $M^* A = 0$. If $M^*\mathbf{b} \geq \mathbf{0}$, then by Theorem 4.2.1 system I has a feasible solution. If $M^*\mathbf{b} \ngeq \mathbf{0}$, then there exists i such that $M^*_{i,:}\mathbf{b} < 0$. Set $\boldsymbol{\pi}^T = M^*_{i,:}$. Then $\boldsymbol{\pi}$ is a feasible solution to system II. ∎

Once we have one theorem of the alternative, it is fairly easy to derive other such theorems. The methods are mainly the same as the ones we use to transform from one LP form to another. For example, the system $A\mathbf{x} \leq \mathbf{b}; \mathbf{x} \geq 0$ has a feasible solution iff the system $[A^T \quad -I]^T\mathbf{x} \leq (\mathbf{b}^T, 0)^T$ has a feasible solution. By Theorem 4.3.2 that system has a feasible solution iff the system $\boldsymbol{\pi} \geq \mathbf{0}, \boldsymbol{\pi}^T[A^T \quad -I]^T = 0$, $\boldsymbol{\pi} \cdot (\mathbf{b}, 0) < 0$ has no feasible solution. Separating $\boldsymbol{\pi}$ into $\boldsymbol{\pi}^O, \boldsymbol{\pi}^S$, where $\boldsymbol{\pi}^S$ play the role of slack variables, this system is equivalent to $\boldsymbol{\pi}^O \geq \mathbf{0}$; $A^T\boldsymbol{\pi}^O \geq \mathbf{0}$; $\boldsymbol{\pi}^O \cdot \mathbf{b} < 0$. We've derived another theorem of the alternative.

Corollary 4.3.3 *For any $A \in \Re^{m\times n}$ and $\mathbf{b} \in \Re^m$ exactly one of the following two systems has a feasible solution: $A\mathbf{x} \leq \mathbf{b}; \mathbf{x} \geq 0$ (system I); $\boldsymbol{\pi} \geq \mathbf{0}$, $\boldsymbol{\pi}^T A \geq \mathbf{0}, \boldsymbol{\pi}^T\mathbf{b} < 0$ (system II).*

Reading Tip: *On a first reading, identify the LP transformation methods embedded in the proof of Corollary 4.3.3. Hint: There are two, and each is the other's dual method. Also, find a very simple proof that at most one of the two systems in the corollary can be feasible. On a second reading, try to derive your own theorem, for example, with $A\mathbf{x} \geq \mathbf{b}$ constraints.*

Section 8.1 of Chapter 8 provides more of these theorems.

Summary: Theorems of the alternative tell us that exactly one of two systems of linear inequalities is feasible. The two systems are related to each other very much as primal and dual LPs are related. To prove a theorem of the alternative, it is easy to show that if one system is feasible, then the other is infeasible. Indeed, a feasible solution to one system is a recipe for deriving a contradiction in the other system. The hard part is to prove that at least one system is feasible. This echoes the easy proof of the weak duality theorem 2.1.1 and the harder proof, coming in Chapter 5, of the strong duality theorem 2.1.3.

4.4 Extreme Points and Optimal Solutions

Preview: This section defines extreme points, a generalization of vertices of polygons. Extreme points can be defined both geometrically and algebraically. Their most important property is this: every LP (except for some unimportant exceptions) that has an optimal solution has an optimal solution which is an extreme point. This property reduces the search for an optimal solution from an infinite set of points (the feasible region) to a finite set of points.

The idea of a vertex of a polygon generalizes to the idea of an "extreme point" of a polyhedron.

Definition 4.7 (Extreme Point) *An extreme point of a set $S \subset \Re^n$ is a point in S that is not the convex combination of other points in S.*

For example, the cube in $\Re^3$ has 8 extreme points, and the nonnegative orthant has only one extreme point, the origin. Geometrically, $\mathbf{x}$ is an extreme point of polyhedron P if $\mathbf{x}$ is the farthest point in P in some direction. Unbounded polyhedra also have "extreme directions," rays along which there is no farthest point and which are not combinations of other such rays. For example, the n rays emanating from the origin along the n positive axes are the extreme directions of the nonnegative orthant.

Polyhedra usually contain an uncountably infinite number of points but can be described finitely in two fundamentally different ways, by a finite set of constraints, or by a finite set of extreme points and a finite set of extreme directions. The latter description is constructive: every point in the polyhedron is a weighted average of the vertices, plus nonnegative multiples of the extreme directions. Therefore, it is sufficient to consider only vertices and extreme directions when solving an LP. Restricting our attention from an infinite set to a finite set is an example of the general approach to optimization described in Section 1.3.

If S is convex, a slightly more restrictive definition of extreme point suffices.

Proposition 4.4.1 *If S is convex, $\mathbf{x}$ is an extreme point of S iff $\mathbf{x} \in S$ and there do not exist $\mathbf{u} \in S, \mathbf{v} \in S, \mathbf{u} \neq \mathbf{x}, \mathbf{v} \neq \mathbf{x}$ and $0 \leq \lambda \leq 1$ such that $\mathbf{x} = \lambda\mathbf{u} + (1-\lambda)\mathbf{v}$.*

Proof: See Figure 4.7. If $\mathbf{x}$ is an extreme point of S, then by definition there do not exist $\mathbf{u}, \mathbf{v}, \lambda$ as stated. Suppose on the other hand $\mathbf{x} \in S$ is not an extreme point of S. By definition there exist $\mathbf{y}^1, \dots, \mathbf{y}^m$ all in S and different from $\mathbf{x}$, and strictly positive weights $\lambda_1, \dots, \lambda_m$ such that $\mathbf{x} = \sum_{k=1}^m \lambda_k \mathbf{y}^k$ and $\sum_{k=1}^m \lambda_k = 1$. Set $\lambda = \lambda_1 > 0$, $\mathbf{u} = \mathbf{y}^1$ and $\mathbf{v} = \frac{1}{1-\lambda_1}\sum_{k=2}^m \lambda_k \mathbf{y}^k$. By convexity of S, $\mathbf{v} \in S$. Because $\mathbf{u} = \mathbf{y}^1 \neq \mathbf{x}$, also $\mathbf{v} \neq \mathbf{x}$. Hence there do exist $\mathbf{u}, \mathbf{v}, \lambda$ as stated. ■

Reading Tip: *In what way is this definition more restrictive? Why doesn't the more restrictive definition apply to non-convex sets? You should find a counterexample.*

There are other equivalent definitions (e.g. require $u \neq v$ instead of $u \neq x, v \neq x$, etc.) but we won't need them. The Krein-Milman theorem, which we will prove later, says that a polytope (actually any compact convex set) is the set of convex combinations of its extreme points. This isn't true of polyhedra in general – consider the nonnegative orthant, for example, which has only one extreme point, or a line, which has none.

For now, we would like to have a theorem which says that if a linear program has an optimal solution, it has an optimal solution that is an extreme point. Unfortunately, we can't have such a theorem. It would be false. Consider the LP: minimize $0x_1 + x_2$ subject to $5 \leq x_2 \leq 6$. Any point on the line $x_2 = 5$ is an optimal solution, but the feasible region has no extreme points. If we want a correct theorem, at the very least we must rule out feasible regions that have no extreme points. By good fortune, those are the only cases we must exclude. We will restrict ourselves to feasible regions that are *pointed polyhedra*, polyhedra that have at least one extreme point.

The definition of extreme point can be awkward to use because it tells what can not happen instead of what does happen. We now seek a more positive characterization of extreme points. Let's look at the geometry to get a clue as to what is true. In Figure 4.8 points a and b are extreme points of the polyhedron, but points c, d, and e are not. What happens at these extreme points that doesn't happen elsewhere? The extreme points are at the intersection of two lines that are part of the boundary of the polyhedron. Look at a corner of the room you are in. It is at the intersection of three planes that are part of the boundary of the room, and only corners have this property. In general, if we are in

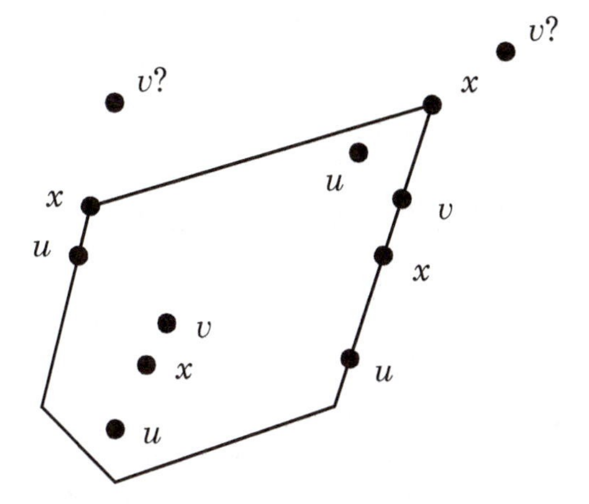

FIGURE 4.7: Extreme points of a convex set cannot be in the line segment defined by other points in the set.

m dimensions, our extreme points will be at the intersection of m hyperplanes that define boundaries of the polyhedron. Now algebraically, what does it mean to be on a hyperplane that defines the boundary of a polyhedron? The answer is given in Definition 2.5, which I repeat here.

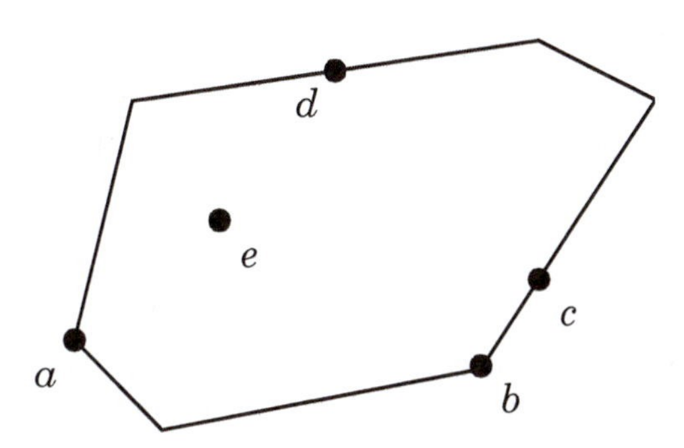

FIGURE 4.8: Points a and b are at the intersections of two defining boundary hyperplanes of the polyhedron.

Definition 2.5 *A constraint is* binding *or* tight *at* $\mathbf{y}$ *if it holds with equality at* $\mathbf{y}$.

Remember, you may not say that a constraint is binding. You must say that it is binding at a particular point. Let A be an $m \times n$ matrix, and b be an m-vector. Of the m inequalities defined by $A\mathbf{x} \leq \mathbf{b}$, zero to m of them might be binding at a point $\mathbf{y}$. If the rows of A that form binding constraints at $\mathbf{y}$ have rank n, we say that $\mathbf{y}$ is a *basic solution* to $A\mathbf{x} \leq \mathbf{b}$. If in addition $A\mathbf{y} \leq \mathbf{b}$, we say that $\mathbf{y}$ is a *basic feasible solution*, abbreviated BFS. For clarity in the formal definition, we treat equality constraints explicitly.

Definition 4.8 (Basic Solution, Basic Feasible Solution (BFS)) *The point $\mathbf{y} \in \Re^n$ is a basic solution of the system $\{\bar{A}\mathbf{x} = \bar{\mathbf{b}}; A\mathbf{x} \leq \mathbf{b}\}$ if the rows of $\bar{A}$ and A that are binding at $\mathbf{y}$ have rank n. If in addition $\bar{A}\mathbf{y} = \bar{\mathbf{b}}$ and $A\mathbf{y} \leq \mathbf{b}$, then $\mathbf{y}$ is a basic feasible solution, abbreviated BFS.*

Proposition 4.4.2 *$\mathbf{y}$ is a BFS of $P = \{\mathbf{x} : A\mathbf{x} \leq \mathbf{b}\}$ iff $\mathbf{y}$ is an extreme point of P.*

Proof: Suppose $\mathbf{y}$ is a BFS. Let n linearly independent constraints that bind at $\mathbf{y}$ be indexed by I, so $A_{i,:}y = b_i \; \forall i \in I$, and $\mathbf{y}$ is the unique solution to $A_{I,:}x = b$. Let $\mathbf{c}^T = \sum_{i \in I} A_{i,:}$. Then $\mathbf{y}$ is the unique maximizer of $\mathbf{c}{\cdot}x : x \in P$. Therefore, $\mathbf{y}$ can't be the convex combination of other points in P.

Conversely, suppose $\mathbf{y}$ is not a BFS. Let the binding constraints at $\mathbf{y}$ be indexed by I. Since they are not full rank, there exists $\boldsymbol{\pi} \neq 0$ such that $A_{I,:}\boldsymbol{\pi} = 0$. For any ϵ, the point $\mathbf{y} + \epsilon\boldsymbol{\pi}$ satisfies the binding constraints at equality, i.e., $A_{I,:}(\mathbf{y} + \epsilon\boldsymbol{\pi}) = b_I$. For all nonbinding constraints, for sufficiently small $\epsilon > 0$ the points $\mathbf{y} \pm \epsilon\boldsymbol{\pi}$ are feasible, *i.e.*, $A_{\bar{I},:}(\mathbf{y} \pm \epsilon\boldsymbol{\pi}) \leq \mathbf{b}_{\bar{I}}$. Therefore, $\mathbf{y}$ is not an extreme point of P. ∎

Definition 4.9 (Direction of a polyhedron) *A vector $\mathbf{d}$ is a direction of polyhedron P iff $\mathbf{x} + \theta\mathbf{d} \in P$ for all $\mathbf{x} \in P, \theta \geq 0$.*

Proposition 4.4.3 *If a polyhedron P contains an infinite ray, then that ray defines a direction of P.*

Corollary 4.4.4 *If a polyhedron P contains a line (infinite in both directions), then it does not contain an extreme point.*

Proof 1 (algebraic): Let $P = \{\mathbf{x} | A\mathbf{x} \leq \mathbf{b}\}$ and let $\{\mathbf{w} + \theta\mathbf{d} : \theta \geq 0\} \subset P$ be the ray contained in P. Therefore, $A\mathbf{d} \leq 0$. For any $\mathbf{v} \in P$, $A\mathbf{v} \leq \mathbf{b}$. Therefore, $A(\mathbf{v} + \theta\mathbf{d}) = Av + \theta A\mathbf{d} \leq A\mathbf{v} \leq \mathbf{b}$ for all $\theta \geq 0$. Since $\mathbf{v}$ is an arbitrary element of P, $\mathbf{d}$ is a direction of P.

Proof 2 (geometric): Suppose $\mathbf{w} \in P$. Let the ray be defined by $\mathbf{v} + \theta\mathbf{d} : 0 \leq \theta < \infty$. Consider the sequence of points $\mathbf{v}, \frac{\mathbf{v}+\mathbf{w}+\mathbf{d}}{2}, \frac{\mathbf{v}+3(\mathbf{w}+\mathbf{d})}{4}, \frac{\mathbf{v}+7(\mathbf{w}+\mathbf{d})}{8}, \ldots$ (see Figure 4.9). Each of these points is in P because (P is convex and) each is the convex combination of a $\mathbf{w}$ and a point on the ray. The sequence converges to $\mathbf{w} + \mathbf{d}$. Since P is closed, $\mathbf{w} + \mathbf{d} \in P$. ∎

The difference between a polytope and an unbounded polyhedron is that the latter has at least one direction. We've already seen (though not yet proved) that a polytope can be characterized as the set of convex combinations of its (finitely many) extreme points. It will turn out that a polyhedron can be characterized by the (finite) set of its extreme points together with a finite set of directions called the extreme directions.

Reading Tip: *On a second reading, ask yourself questions about details that at first glance seem extraneous. Why does the preceding paragraph contain the parenthetical words "finitely many" and "finite"?*

Now we give the converse to Corollary 4.4.4.

Theorem 4.4.5 *If $P \subset \Re^n$ is a nonempty polyhedron containing no line, then P contains at least one extreme point.*

Proof: Let P be defined by $\{\mathbf{x} : A\mathbf{x} \leq \mathbf{b}\}$. Since P is nonempty, it contains a point $\mathbf{w}$.

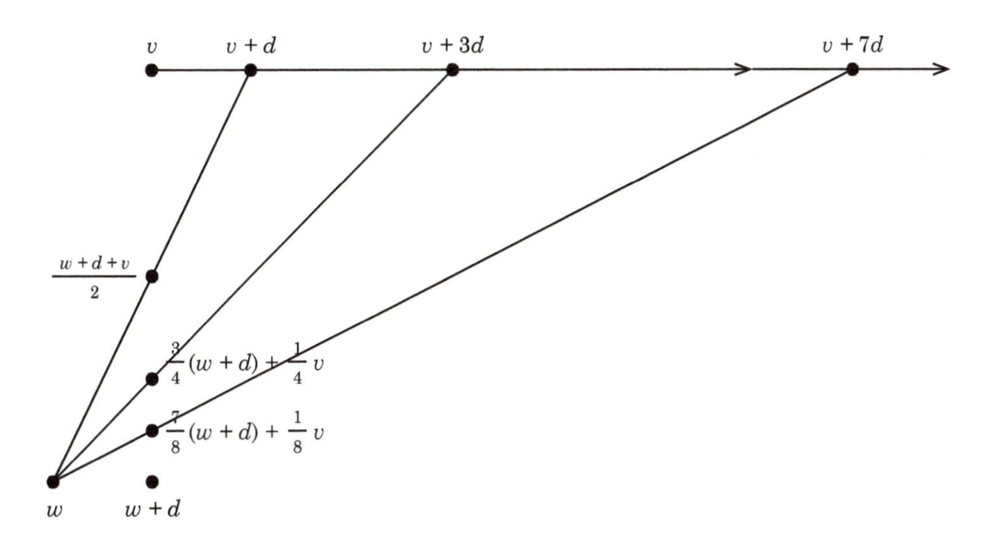

FIGURE 4.9: $\mathbf{w}+\mathbf{d}$ is the limit of the sequence, and is in P since P is closed.

Let I index the rows of A that define constraints binding at $\mathbf{w}$, so $A_{i,:}\mathbf{w} = \mathbf{b}_i \; \forall i \in I$. If the submatrix $A_{I,:}$ has rank n, then $\mathbf{w}$ is a basic feasible solution and hence is extreme by Proposition 4.4.2.

Otherwise, we repeatedly increase its rank as follows. There exists a nonzero vector $\boldsymbol{\pi}$ such that $A_{I,:}\boldsymbol{\pi} = 0$. Since P has no line, either $\boldsymbol{\pi}$ or $-\boldsymbol{\pi}$ (or both) is not a direction of P. Without loss of generality let $\boldsymbol{\pi}$ not be a direction of P. The ray $\mathbf{w} + \theta\boldsymbol{\pi} : \theta \geq 0$ is inside P at $\theta = 0$, exits P for sufficiently large θ, and by convexity of P cannot reenter once it has exited P. Let θ^* be the largest value of θ at which the ray is within P. There must be at least one constraint $A_{j,:}$ that is satisfied at $\mathbf{w} + \theta^*\boldsymbol{\pi}$ but is violated for $\theta > \theta^*$. This constraint must therefore be binding at $\mathbf{w} + \theta^*\boldsymbol{\pi}$. Since $A_{I,:}\boldsymbol{\pi} = 0$, $j \notin I$. (This is because $\boldsymbol{\pi}$ moves us in a direction orthogonal to the gradients of the constraints in I, i.e., a direction that is neutral with respect to those constraints). Add j to the set I and repeat from $\mathbf{w} + \theta^*\boldsymbol{\pi}$. ∎

Definition 4.10 (Pointed Polyhedron) *A polyhedron that contains at least one extreme point is called a pointed polyhedron. By Corollary 4.4.4 and Theorem 4.4.5, a polyhedron is pointed iff it is nonempty and contains no line.*

The proof to Theorem 4.4.5 is closely related to an algorithmic procedure called *crashing*. The procedure repeatedly crashes into constraints to increase the rank of the set of binding constraints.

Algorithm 4.4.6 (Crashing) *Input: A polyhedron* $P = \{\mathbf{x} | A\mathbf{x} \leq \mathbf{b}\} \subset \Re^n$, *a point* $\mathbf{x} \in P$, *an objective* $\min \mathbf{c} \cdot \mathbf{x}$.
Output: A basic feasible solution that is as good or better than $\mathbf{x}$, *if one exists. If no such solution exists, the output is a ray of unboundedness, or a line in* P.

1. *If* $\mathbf{x}$ *itself is not basic, use orthogonalization to construct a nonzero vector* $\boldsymbol{\pi}$ *such that* $A_{i,:}\boldsymbol{\pi} = 0$ *for all constraints* i *binding at* $\mathbf{x}$. *If* $\mathbf{c} \cdot \boldsymbol{\pi} > 0$ *replace* $\boldsymbol{\pi}$ *by* $-\boldsymbol{\pi}$, *so that now* $\mathbf{c} \cdot \boldsymbol{\pi} \leq 0$.

2. *Determine* k, *the index of the first constraint of* P, $A_{k,:}\mathbf{x} \leq \mathbf{b}_k$, *that would be violated if one moved along the ray* $\mathbf{x} + \theta\boldsymbol{\pi} : \theta \geq 0$.
Case 1: The index k *exists. Set* $\mathbf{y} = \{\mathbf{x} + \theta\boldsymbol{\pi} : \theta \geq 0\} \bigcap \{\mathbf{x} : A_{k,:}\mathbf{x} = \mathbf{b}_k\}$. *The point* $\mathbf{y}$ *is where the ray exits* P. *Replace* $\mathbf{x}$ *by* $\mathbf{y}$ *and return to Step 1, having increased the rank of the binding constraints.*
Case 2: No such k *exists, i.e., the ray never leaves* P. *If* $\mathbf{c} \cdot \boldsymbol{\pi} < 0$ *we have found a ray of unboundedness. The optimal value of the linear program is unbounded. If instead* $\mathbf{c} \cdot \boldsymbol{\pi} = 0$, *travel on the ray* $\mathbf{x} - \theta\boldsymbol{\pi} : \theta \geq 0$. *Either the ray exits* P *at the point* $\mathbf{y}$, *which replaces* $\mathbf{x}$ *as in Case 1, or the ray does not exit* P. *In the latter case, the two rays* $\mathbf{x} \pm \theta\boldsymbol{\pi} : \theta \geq 0$ *form a line of* P, *which implies that no basic feasible solution exists.*

Figure 4.10 illustrates the crashing algorithm. In three dimensions, the algorithm moves from an initial point until it crashes into a constraint. It keeps that constraint binding and moves in a different direction until it crashes into another constraint, which it also maintains as binding. Then it moves again until it crashes into a third constraint at an extreme point.

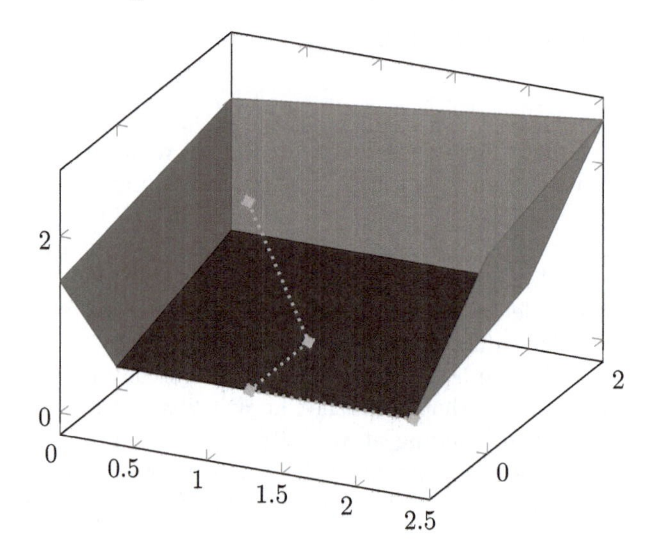

FIGURE 4.10: Each time you crash into a constraint, maintain it as a binding constraint thereafter.

The crashing algorithm is valid because $\boldsymbol{\pi}^T A_{i,:} = 0$ for all indices i of constraints binding at $\mathbf{x}$, but $\boldsymbol{\pi}^T A_{k,:} > 0$ since movement in the direction $\boldsymbol{\pi}$ from $\mathbf{x}$ eventually violates constraint k. This implies that $A_{k,:}$ is not in the space spanned by the rows $A_{i,:}$, so that Case 1 strictly increases the rank of the binding constraints. Crashing cannot take more than n iterations, and each iteration requires only a limited number of straightforward linear algebra computations.

Theorem 4.4.7 *Algorithm 4.4.6 terminates correctly in at most* n *iterations.*

Crashing proves that for any pointed polyhedron P, and any vector $\mathbf{c}$, either the linear program $\min \mathbf{c}\cdot\mathbf{y}$ subject to $\mathbf{y} \in P$ has unbounded optimum, or for all $\mathbf{x} \in P$ there exists a basic feasible $\mathbf{w} \in P$ such that $\mathbf{c}\cdot\mathbf{w} \leq \mathbf{c}\cdot\mathbf{x}$. Now we have the fundamental theorem of linear programming.

Theorem 4.4.8 *Let $P \subset \Re^n$ be a pointed nonempty polyhedron, and let $\mathbf{c} \in \Re^n$. Then the linear program* $\min \mathbf{c}\cdot\mathbf{x}$ *subject to* $\mathbf{x} \in P$ *either:*

- *Has unbounded optimum.*
- *Has an optimal solution that is basic, and hence is an extreme point of P.*

Proof: Let $\mathbf{x}^*$ be a basic feasible solution that has minimum objective value over all basic feasible solutions. (Note that there are finitely many, in fact at most $\binom{m}{n}$ basic solutions, where m is the number of rows in a matrix defining P, so $\mathbf{x}^*$ exists). We claim that either the LP has unbounded optimum or $\mathbf{x}^*$ is optimal. Suppose to the contrary the LP has no ray of unboundedness but for some $\mathbf{x} \in P$ we have $\mathbf{c}\cdot\mathbf{x} < \mathbf{c}\cdot\mathbf{x}^*$. Crash from $\mathbf{x}$; Algorithm 4.4.6 terminates correctly by Theorem 4.4.7. Since there is no ray of unboundedness, crashing finds a basic feasible $\mathbf{w}$ such that $\mathbf{c}\cdot\mathbf{x} \geq \mathbf{c}\cdot\mathbf{w}$. But $\mathbf{c}\cdot\mathbf{w} \geq \mathbf{c}\cdot\mathbf{x}^*$, whence $\mathbf{c}\cdot\mathbf{x} \geq \mathbf{c}\cdot\mathbf{x}^*$, a contradiction. *Question: Why can't the proof assume to the contrary that there is no unbounded optimum and no basic optimal solution, let $\mathbf{x}$ be an optimal nonbasic solution, and crash from $\mathbf{x}$?*[4]

Reading Tip: *Where does the proof of Theorem 4.4.8 rely on P being nonempty? Pointed? Construct an example to show the theorem would be false were P not required to be pointed.*

Theorem 4.4.8 gives a finite procedure to solve LPs on pointed polyhedra. Enumerate all the basic solutions, and choose, among those that are feasible, one with best objective function value. If no basic solutions are feasible, stop, the LP is infeasible. Finally, check to see if there are any directions that make the LP unbounded. For an LP in form (2.1), solve the associated LP

$$\begin{aligned} \min \quad & \mathbf{c}\cdot\mathbf{y} \quad \text{subject to} \\ & A\mathbf{y} \geq \mathbf{0}; \\ & \mathbf{y} \geq \mathbf{0}; \\ & -1 \leq \mathbf{y}_i \leq 1 \ \forall i \end{aligned}$$

The associated LP is over a nonempty polytope, so it has no directions and can be solved by examining all basic solutions. The optimal objective value of the associated LP is < 0 iff the original LP has unbounded optimum. What if the polyhedron is not pointed? Re-express the LP in a form with nonnegativity constraints. The resulting equivalent LP will be over a polyhedron with no lines.

The procedure sketched above terminates in finitely many steps and is guaranteed to solve the LP correctly. In the next chapter we will derive a much faster procedure, the simplex method.

[4]It may not assume the existence of an optimal solution. There are optimization problems such as $\min e^{-x} : x \geq 0$ that have no unbounded optimum and no optimum solution.

4.4.1 Degeneracy

A basic solution $\mathbf{x} \in \Re^n$ is *degenerate* if more than n constraints are binding at $\mathbf{x}$. You can make any basic solution degenerate by adding a redundant constraint to the polyhedron's description. However, don't be misled by what happens in two dimensions, where degeneracy always implies redundancy. See Figure 4.11. The Egyptian pyramids (square base, triangular sides) are degenerate at their top, but none of the defining constraints is redundant. Two of the Platonic solids are degenerate at their vertices. *Question: Which two?*[5] Degeneracy can be a pain in the neck, both theoretically and computationally. It makes algorithm correctness harder to prove, it can slow down the simplex method considerably, and it can inhibit the usefulness of predictions that are based on dual variables.

FIGURE 4.11: In three dimensions, a vertex is degenerate if it is incident on more than three facets.

Degeneracy in the dual corresponds to the existence of multiple optimal solutions in the primal. You can see this intuitively by looking at the complementary slackness conditions (2.13). Extra binding constraints in the dual mean more zero terms contributed by the dual to the complementary slackness conditions. These extra zero terms give extra degrees of freedom in the primal variables, so there are multiple primal solutions that are complementary slack with a degenerate dual solution. This intuition will be made rigorous by the simplex method in Chapter 5.

Theorem 4.4.8 states that when there is an optimal solution there is an optimal basic feasible solution. The simplex method, which is derived in the next chapter, uses that fact by limiting its search to basic feasible solutions. Central path following methods such as the barrier algorithm derived in Chapter 18, on the other hand, never converge to a basic feasible solution unless it is the unique optimal solution. When dual degeneracy creates multiple optimal solutions, central path methods converge to the center of the region of optimal

[5]The octahedron and icosahedron.

solutions. To get from the center to a basic feasible solution, one must then perform a procedure like crashing.

In practice most large LPs have degeneracy in both primal and dual. As best I know, no fully satisfactory explanation of this phenomenon has been discovered.

Summary: Memorize the definitions of extreme points, binding constraints, basic feasible solutions, and degeneracy. Remember the equivalence of extreme points and basic feasible solutions, and that the former is a statement about nonexistence but the latter is a statement about existence. Crashing is an efficient algorithm to move from any LP feasible solution to an equally good or better extreme point, or find a ray of unboundedness. The consequent fundamental theorem of LP tells us that to search for an optimal solution, it suffices to search among the extreme points.

4.5 Two Ways to Represent Polyhedra

Preview: At the beginning of this chapter, we described polyhedra as the feasible regions of LPs. In the middle of this chapter, we found that polyhedra have special points called extreme points, and special vectors called directions. Here, at the end of the chapter, we will describe polyhedra as constructed from extreme points and directions.

A polyhedron can be defined as the set of solutions to finitely many linear inequalities, or, in a dual way, as the set of convex combinations of finitely many points plus nonnegative combinations of finitely many directions. These two kinds of definitions are sometimes called "outer" and "inner" characterizations, respectively.

Theorem 4.5.1 (Representation of Polyhedra) *Let $N \geq 1$ be an integer and let $M \geq 0$ be an integer. Let $\mathbf{y}^1 \ldots \mathbf{y}^N$ and $\mathbf{d}^1 \ldots \mathbf{d}^M$ be elements of $\Re^n$. Let P be the set of all of sums of a convex combination of the $\mathbf{y}^i$ and a nonnegative linear combination of the $\mathbf{d}^j$. That is, let*

$$P = \left\{ \sum_{i=1}^{N} \lambda_i \mathbf{y}^i + \sum_{j=1}^{M} \mu_j \mathbf{d}^j | \lambda_i \geq 0 \ \forall i; \sum_{i=1}^{N} \lambda_i = 1; \mu_j \geq 0 \ \forall j \right\}.$$

Then P is a polyhedron. That is, there exists a matrix A and vector $\mathbf{b}$ such that $P = \{\mathbf{x} | A\mathbf{x} \leq \mathbf{b}\}$.

Proof: Let Q be the polyhedron

$$Q = \left\{ (\mathbf{x}, \lambda, \mu) | \mathbf{x} - \sum_{i=1}^{N} \lambda_i \mathbf{y}^i + \sum_{j=1}^{M} \mu_i \mathbf{d}^i = 0; \lambda \geq 0; \sum_{i=1}^{N} \lambda_i = 1; \mu \geq 0 \right\}.$$

Then P is the projection of Q onto the $\mathbf{x}$ variables and is a polyhedron by Theorem 4.2.2. ■

The next theorems give converses to Theorem 4.5.1. They state that polyhedra can be described as sets of all convex combinations of finitely many points added to nonnegative multiples of vectors. These theorems are not exact converses because they apply only to pointed polyhedra, whereas the polyhedron P in Theorem 4.5.1 could contain a line. First we restrict our attention to polytopes.

Theorem 4.5.2 (Converse Representation of Polytopes) *Let P be a nonempty polytope in $\Re^n$. Then there exist integer $N \geq 1$ and vectors $\mathbf{y}^1 \ldots \mathbf{y}^N$ such that*

$$P = \left\{ \sum_{i=1}^{N} \lambda_i \mathbf{y}^i | \lambda_i \geq 0 \ \forall i; \sum_{i=1}^{N} \lambda_i = 1 \right\}.$$

Moreover, $\mathbf{y}^1 \ldots \mathbf{y}^N$ may be taken to be the extreme points of P.

Proof: Let $\mathbf{y}^1 \ldots \mathbf{y}^N$ be the basic feasible solutions of P. *Question: What fact is implicitly used in the previous sentence?*[6] Let Y denote the matrix whose columns are $\mathbf{y}^1, \mathbf{y}^2, \ldots \mathbf{y}^N$. Since P has no line it has at least one extreme point, hence $N \geq 1$. We claim that this choice of $\mathbf{y}^i$ fulfils the theorem. If not, there would exist a point $\mathbf{p} \in P$ such that there did not exist a solution to

$$Y\lambda = \mathbf{p}; \lambda \geq \mathbf{0}; \mathbf{1} \cdot \lambda = 1.$$

(Here $\mathbf{1}$ denotes an N-vector of 1s.) By Theorem 4.3.2 there exists $\boldsymbol{\pi}, \pi_0$ such that $\boldsymbol{\pi}^T Y + \pi_0 \mathbf{1} \geq \mathbf{0}; \boldsymbol{\pi}^T p + \boldsymbol{\pi}_0 < 0$. Consider the problem of minimizing $\boldsymbol{\pi} \cdot \mathbf{z}$ subject to $\mathbf{z} \in P$. Crash from $\mathbf{p}$ (Algorithm 4.4.6). Since P is a polytope, we will get an extreme point with value $\leq \mathbf{p} \cdot \boldsymbol{\pi}$. Now, the point $\mathbf{p}$ has strictly better objective function value than $-\boldsymbol{\pi}_0$, but every extreme point $\mathbf{y}^i$ has objective value $\geq -\pi_0$. This is a contradiction. ■

Another way to prove Theorem 4.5.2 is to define Q to be the set of convex combinations of the basic feasible solutions of P. By Theorem 4.5.1 Q is a polyhedron, hence by Proposition 4.0.1 Q is closed. Also, the set Q is bounded because there are finitely many basic feasible solutions of P. If there exists a point $\mathbf{p} \in P \setminus Q$, then by Theorem 4.1.1 there exists $\boldsymbol{\pi}, \pi_0$ such that $\boldsymbol{\pi} \cdot \mathbf{p} < \boldsymbol{\pi}_0$ and $\boldsymbol{\pi} \cdot \mathbf{x} > \boldsymbol{\pi}_0 \ \forall \mathbf{x} \in Q$. Crash (Algorithm 4.4.6) from $\mathbf{p}$ in polyhedron P with respect to the objective function $\boldsymbol{\pi} \cdot \mathbf{x}$. Crashing will arrive at a basic feasible solution $\mathbf{y} \in P$ such that $\boldsymbol{\pi} \cdot \mathbf{y} \leq \boldsymbol{\pi} \cdot \mathbf{p} < \boldsymbol{\pi} \cdot \mathbf{x} \ \forall \mathbf{x} \in Q$, a contradiction since $\mathbf{y} \in Q$.

[6] The number of BFSs is finite.

Theorem 4.5.3 (Converse Representation of Pointed Polyhedra) *Let P be a polyhedron in $\Re^n$ containing at least one extreme point. Then there exist integers $N \geq 1$ and $M \geq 0$ and vectors $\mathbf{y}^1 \ldots \mathbf{y}^N$ and $\mathbf{d}^1 \ldots \mathbf{d}^M$ in $\Re^n$ such that*

$$P = \left\{ \sum_{i=1}^{N} \lambda_i \mathbf{y}^i + \sum_{j=1}^{M} \mu_i \mathbf{d}^i | \lambda_i \geq 0 \ \forall i; \sum_{i=1}^{N} \lambda_i = 1; \mu_j \geq 0 \ \forall j \right\}.$$

Proof: Let $\mathbf{y}^1 \ldots \mathbf{y}^N$ be the basic feasible solutions of P. Since P has at least one extreme point $N \geq 1$. Suppose $P = \{\mathbf{x} | A\mathbf{x} \leq \mathbf{b}\}$. Let $\mathbf{d}^1 \ldots \mathbf{d}^M$ be the basic feasible solutions to the polyhedron $P^R \equiv \{\mathbf{d} : A\mathbf{d} \leq \mathbf{0}; -\mathbf{1} \leq \mathbf{d} \leq \mathbf{1}\}$. By Theorem 4.5.2 every element of P^R is a convex combination of $\mathbf{d}^1 \ldots \mathbf{d}^M$.

We claim that this choice of $\mathbf{y}^i$ and $\mathbf{d}^j$ fulfills the theorem. If not, there would exist a point $\mathbf{p} \in P$ such that there did not exist a solution to

$$Y\lambda + D\mu = \mathbf{p}; (\lambda, \mu) \geq \mathbf{0}; (\mathbf{1}, \mathbf{0}) \cdot (\lambda, \mu) = 1.$$

The columns of the matrix Y are $\mathbf{y}^1 \ldots \mathbf{y}^N$; the columns of the matrix D, if it is nonempty, are $\mathbf{d}^1 \ldots \mathbf{d}^M$. By Theorem 4.3.2 there exists $\boldsymbol{\pi}, \pi_0$ such that $\boldsymbol{\pi}^T Y + \pi_0 \mathbf{1} \geq \mathbf{0}; \boldsymbol{\pi}^T D \geq \mathbf{0}; \boldsymbol{\pi}^T \mathbf{p} + \pi_0 < 0$. Consider the problem of minimizing $\boldsymbol{\pi} \cdot \mathbf{z}$ subject to $\mathbf{z} \in P$. Crash from $\mathbf{p}$ (Algorithm 4.4.6). Since P has an extreme point, we will either get a ray of unboundedness or an extreme point with value $\leq \mathbf{p} \cdot \boldsymbol{\pi}$. Now, the point $\mathbf{p}$ has strictly better objective function value than $-\pi_0$, but every extreme point $\mathbf{y}^i$ has objective value $\geq -\pi_0$. Therefore, when we crash we can't arrive at an extreme point. Instead we must find a ray $\mathbf{y} + \theta\mathbf{r} : \theta \geq 0$ of unboundedness along which $\boldsymbol{\pi} \cdot (\mathbf{y} + \theta\mathbf{r}) \to -\infty$ as $\theta \to \infty$. Therefore, $\boldsymbol{\pi} \cdot \mathbf{r} < 0$. Also, $A\mathbf{r} \leq 0$ because by Proposition 4.4.3 $\mathbf{r}$ is a direction of P. Without loss of generality, scale $\mathbf{r}$ so that $||\mathbf{r}|| = 1$. Hence $\mathbf{r} \in P^R$, and so $\mathbf{r} = D\mu$ for some $\mu \geq 0$. Now $0 > \boldsymbol{\pi}^T \frac{\mathbf{r}}{||\mathbf{r}||} = \boldsymbol{\pi}^T D\mu \geq 0$, a contradiction. The point $\mathbf{p}$ cannot exist and the theorem is proved. ■

As with Theorem 4.5.2, Theorem 4.5.3 could instead be proved with Theorems 4.5.1 and 4.1.1, and Proposition 4.0.1. Define Q to be the set of all sums of a convex combination of the extreme points of P plus a nonnegative combination of the columns of D. Let $(\boldsymbol{\pi}, \pi_0)$ define the hyperplane that separates the polyhedron Q from the point $\mathbf{p} \in P \setminus Q$. When we crash from $\mathbf{p}$ we must find a ray $\mathbf{r}$ of unboundedness such that $\mathbf{r} \cdot \boldsymbol{\pi} < 0$. By Proposition 4.4.3 $\mathbf{r}$ is a direction of P. Hence $\mathbf{r}/||\mathbf{r}|| \in P^R$ and so $\mathbf{r} = D\mu$ for some $\mu \geq \mathbf{0}$. Therefore, $\mathbf{r}$ is a direction of Q. Since Q is nonempty, Q must contain points $\mathbf{x}$ with arbitrarily low values $\boldsymbol{\pi} \cdot \mathbf{x}$, which contradicts the hyperplane separating Q from $\mathbf{p}$.

Definition 4.11 (Extreme Direction) *Let D be the set of directions of a polyhedron. A direction $\mathbf{d} \in D$ is* extreme *if $vd \neq \mathbf{0}$ and $\mathbf{d} = \sum_j \mu_j \mathbf{d}^j; \mu_j > 0 \forall j; \mathbf{d}^j \in D \ \forall j$ implies $\mathbf{d}^j = \alpha_j \mathbf{d} \ \forall j$ where the α_j are scalars. That is, $\mathbf{d}$ cannot be expressed as a positive linear combination of other directions unless those directions are scalar multiples of $\mathbf{d}$ itself.*

Corollary 4.5.4 *In Theorem 4.5.3 the vectors $\mathbf{y}^1 \ldots \mathbf{y}^N$ may be taken to be the extreme points of P, and the vectors $\mathbf{d}^1 \ldots \mathbf{d}^M$ may be taken to be the extreme directions of P.*

Proof: since N is finite, simply remove any points that are convex combinations of other points. Since M is finite, remove any directions that are nonnegative linear combinations of other directions. ■

The sum of two sets $S \subset \Re^n$ and $T \subset \Re^n$ is defined as $S + T \equiv \{s + t | s \in S, t \in T\}$. See Figure 4.12.

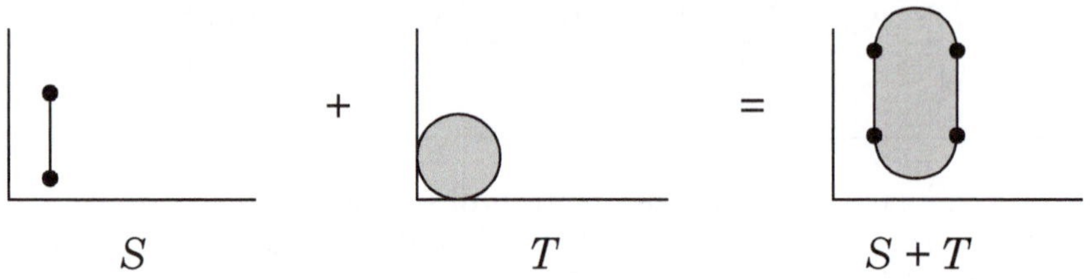

FIGURE 4.12: The sum of two sets S and T is the set of all sums of pairs of points, one each from S and T.

Definition 4.12 (Cone, Recession Cone, Finitely Generated Cone) *A set $C \subset \Re^n$ is a* cone *if it is closed under multiplication by nonnegative scalars. That is, C is a cone if $\mathbf{x} \in C$ and $\mu \geq 0$ imply $\mu\mathbf{x} \in C$. The set of directions of a polyhedron is a cone; it is often called the* recession cone *of the polyhedron.*
A cone C is finitely generated *if there exists a finite set $d^1 \ldots d^K$ of points in $\Re^n$ whose nonnegative linear combinations equal C, that is, $C = \{\sum_{j=1}^{K} \mu_j \mathbf{d}^j : \mu_j \geq 0 \forall j\}$.*

With these definitions, Theorem 4.5.3 can be stated as follows: *Every pointed polyhedron is the sum of a polytope and a finitely generated cone. Question: Why is the recession cone a cone? Is it finitely generated?*[7]

Summary: An extreme point of a polyhedron is a basic feasible solution, and vice versa. The definition of extreme point says what cannot happen; the definition of basic feasible solution says what must happen. Similarly, a theorem of the alternative is an equivalence between a "there does not exist" statement and a "there exists" statement. Also similarly, polyhedra can be described in terms of what constraints cannot be violated, or in terms of a construction based on extreme points and directions. For all polyhedra containing at least one extreme point, every LP that has an optimal solution has an optimal extreme point solution. This property guides us to the simplex algorithm of the next chapter.

---

[7]If $\mathbf{x} + \theta\mathbf{d} \in S \ \forall\theta > 0$, then for any $\alpha \geq 0$, $\mathbf{x} + \theta(\alpha\mathbf{d}) = \mathbf{x} + (\theta\alpha)\mathbf{d} \in S$. Yes.

4.6 Notes and References

Theorem 4.5.2 was originally proved by Steinitz in 1916 for arbitary convex sets in $\Re^n$ [271]. Steinitz was generalizing a theorem of Minkowski's from $\Re^3$ to $\Re^n$. That a closed compact convex set in $\Re^n$ is the convex combinations of its extreme points is nowadays often attributed to Mark Krein and David Milman, who proved a more general theorem about sets of linear functionals adjoint to a Banach space [195].

4.7 Problems

E Exercises

1. Draw the set of convex combinations of the following sets of points:
   (a) $(0, 0)$ and $(1, 1)$
   (b) $(1, 3)$ and $(-2, -1)$
   (c) $(1, 2)$, $(2, 1)$, and $(0, 3)$
   (d) $(1, 2)$, $(2, 1)$, and $(1, 1)$
2. Draw the set of affine combinations of the sets of points in Exercise 1.
3. Which of the following constraints define a set in $\Re^3$ that is convex?
   (a) $x_1 \leq x_2 + x_3$
   (b) $x_1 - x_2 \leq |x_3|$
   (c) $x_1 = |x_2|$
   (d) $x_1 + x_2 \leq |x_3|$
   (e) $x_1 + x_2 \geq |x_3|$
   (f) $|x_1||x_2| \geq x_3$
   (g) $|x_1||x_2| \leq x_3$
   (h) $|x_1| + |x_2| + |x_3| \leq 1$
4. For each of the following sets, determine whether or not it is convex, and find all of its extreme points. (Note: The definition of extreme point applies equally well to arbitrary sets in $\Re^n$ as to polyhedra.)
   (a) $(x_1, x_2) \in \Re^2 : x_1 \geq 0, x_2 \geq 0, x_1 + 2x_2 \geq 6.$
   (b) $(x_1, x_2) \in \Re^2 : x_2 \leq x_1^2.$
   (c) $\mathbf{x} \in \Re^3 : ||x|| \leq 1$ (Euclidean norm).
5. Using the definition of an open set, prove explicitly that the set $\{x \in \Re^3 : x_1 + 8x_2 + 27x_3 < 13\}$ is open.
6. Give an example of an infinite set of linear inequalities whose feasible region is not a polyhedron.

7. Can you find an infinite set of linear inequalities whose feasible region is not convex?

8. Suppose P and Q are polytopes such that every extreme point of P is an extreme point of Q. Prove that $P \subset Q$.

9. Suppose P and Q are polytopes such that every extreme point of P is an extreme point of Q, and vice versa. Prove that $P = Q$.

10. Suppose P and Q are polyhedra such that every extreme point of P is an extreme point of Q. Give an example to demonstrate that P is not necessarily contained in Q.

11. Restate Theorem 4.1.1 in terms of LPs, infeasible solutions, and superoptimal solutions.

12. Find infinitely many hyperplanes that separate $(10, -1)$ from the positive quadrant $\{x \in \Re^2 | x_1 \geq 0; x_2 \geq 0\}$.

13. Find infinitely many hyperplanes that separate $(-1, -1, -1)$ from the positive orthant $\{x \in \Re^3 | x_i \geq 0 : i = 1, 2, 3\}$.

14. Find infinitely many hyperplanes that separate $(-1, -1, -1, -1, -1)$ from the positive orthant $\{x \in \Re^5 | x_i \geq 0 : i = 1, 2, 3, 4, 5\}$.

15. George says that he has 10 distinct linear inequality constraints on $\Re^3$ such that the resulting polyhedron is unbounded. Could he be telling the truth?

16. Sally says she has two constraints on $\Re^3$ such that the resulting polyhedron is not unbounded. Could she be telling the truth?

17. Find the extreme points and extreme directions of $P =$

$$\{(x_1, x_2) | x_1 \geq 1; 5x_1 - x_2 \geq 10; -2x_1 + 3x_2 \leq 8; x_1 \geq 0; x_2 \geq 0; x_1 + x_2 \geq 2\}.$$

18. Use FM elimination to prove that the system of inequalities

$$x_i \geq 0 : i = 1, 2, 3; x_1 + x_2 + x_3 \leq 5; 2x + 1 + 3x_2 - 2x_3 \geq 8; 4x_1 - 3x_2 \geq 1; -x_1 + 2x_2 + 4x_3 \geq 16$$

has or does not have a feasible solution.

19. Use FM elimination to construct a feasible solution to the system

$$\begin{aligned} x_1 &\geq 0 \\ x_2 &\geq 0 \\ x_1 + 2x_2 &\leq 5 \\ x_1 - x_2 &\leq 4 \\ x_1 - x_2 &\geq 2 \end{aligned}$$

20. Give an example of a 2-dimensional polyhedron, strictly contained in $\Re^2$, but containing a line. For some point on the boundary of your polyhedron, specify the binding constraints (at that point).

21. Give an example of a 2-dimensional polyhedron that does not contain a line but whose projection onto 1 dimension does contain a line.

22. Give an example of a 2-dimensional polyhedron in $\Re^2$ containing exactly 2 extreme points. At each extreme point, specify the binding constraints and show that they span $\Re^2$. Modify your example so that there are more than 2 binding constraints at one of the extreme points, such that not every pair of binding constraints is linearly independent.

23. One version of the theorem of the alternative states that exactly one of the systems

$A\mathbf{x} = b$	$A^T\pi \geq 0$
$\mathbf{x} \geq 0$	$b^T\pi < 0$

has a (feasible) solution. A 2nd version states that exactly one of the systems

$A\mathbf{x} \leq \mathbf{b}$	$A^T\pi \geq 0$
$\mathbf{x} \geq 0$	$b^T\pi < 0$
	$\pi \geq 0$

has a (feasible) solution. Assuming the **2nd** version, prove the **1st** version.

24. Let $P = \{x | A\mathbf{x} \leq \mathbf{b}\}$ where $A =$

$$\begin{bmatrix} 1 & 1 \\ -1 & 0 \\ 0 & -1 \end{bmatrix}$$

and $b^T = (4, 0, 0)$.

(a) Draw P.

(b) Show that $\mathbf{x} = (0, \frac{1}{2})$ is not an extreme point of P. Use the definition of extreme point. You may show this graphically (with a picture) or algebraically.

(c) Show that $\mathbf{x} = (0, 0)$ is an extreme point of P by giving a vector $\mathbf{c}$ such that $\mathbf{c} \cdot \mathbf{y} < c \cdot x \; \forall y \in P, y \neq x$. You may draw $\mathbf{c}$ with a picture or write $\mathbf{c}$ numerically. Be sure not to point $\mathbf{c}$ the wrong way.

(d) Show that $\mathbf{x} = (0, 0)$ is a basic solution of P. For this question, don't use a picture. Write the binding constraints algebraically and show that they are linearly independent.

25. Prove: if the projection of hyperplane H separates the projection of point $\mathbf{y}$ from the projection of polyhedron P, then H separates $\mathbf{y}$ from P.

26. Disprove each of the following statements by finding a counterexample:

(a) If P is a polyhedron and $\mathbf{y} \in P$ and $\mathbf{y}$ is the unique minimizer of $\mathbf{c} \cdot x$ subject to $\mathbf{x} \in P$, then the projection of $\mathbf{y}$ is the unique minimizer of $\mathbf{c}' \cdot x'$ subject to $\mathbf{x}' \in P'$ where $\mathbf{c}'$ is the projection of $\mathbf{c}$ and P' is the projection of P.

(b) If P' is the projection of polyhedron P, and $\mathbf{y}' \in P'$ is the unique minimizer of $\mathbf{c}' \cdot x'$ subject to $\mathbf{x}' \in P'$, then must there exist a point $\mathbf{y} \in P$ and vector $\mathbf{c}$ such that $\mathbf{y}'$ is the projection of $\mathbf{y}$, $\mathbf{c}'$ is the projection of $\mathbf{c}$, and $\mathbf{y}$ is the unique minimizer of $\mathbf{c} \cdot x$ subject to $\mathbf{x} \in P$.

27. If $A =$

$$\begin{bmatrix} -1 & -1 \\ -1 & 0 \\ 0 & -1 \end{bmatrix}$$

and $b^T = (-4, 0, 0)$, determine whether or not $\mathbf{x} = (0, 0)$ is a basic solution.

28. Write in detail the alternative proof of Theorem 4.5.3 sketched in the paragraph following the proof.

29. Construct a matrix A and vector b such that $\{x | A\mathbf{x} \geq \mathbf{b}\}$ is empty and there exists $d \neq 0$ such that $Ad \geq 0$. Conclude that a polyhedron could seem to be unbounded but actually be empty.

30. For the polyhedron $P = \{\mathbf{x} : \mathbf{x} \geq 0; -4x_1 + x_2 \leq 10\}$ show that $(1, 1) \in P^R$ as defined in the proof of Theorem 4.5.3, but is not an extreme direction of P.

M Problems

31. Suppose P and Q are polytopes in $\Re^n$ such that for every vector $\mathbf{c} \in \Re^n$,

$$\arg \max_{\mathbf{x} \in P} \mathbf{c} \cdot \mathbf{x} = \arg \max_{\mathbf{x} \in Q} \mathbf{c} \cdot \mathbf{x}.$$

- Prove or disprove: $P = Q$.
- Repeat the question under the weaker assumption that $\forall \mathbf{c} \neq \mathbf{0} \arg \max_{\mathbf{x} \in P} \mathbf{c} \cdot \mathbf{x} = \arg \max_{\mathbf{x} \in Q} \mathbf{c} \cdot \mathbf{x}$.

32. Suppose P and Q are polytopes in $\Re^n$ such that for every vector $\mathbf{c} \in \Re^n$,

$$\max_{x \in P} c \cdot x = \max_{x \in Q} c \cdot x.$$

Prove or disprove: $P = Q$.

33. Suppose P and Q are polyhedra in $\Re^n$ such that for every vector $\mathbf{c} \neq \mathbf{0} \in \Re^n$,

$$\sup_{\mathbf{x} \in P} \mathbf{c} \cdot \mathbf{x} = \sup_{\mathbf{x} \in Q} \mathbf{c} \cdot \mathbf{x},$$

where the supremum equals ∞ if the optimum is unbounded.

Prove or disprove: $P = Q$.

34. Can you find an infinite set of linear inequalities whose feasible region is not closed? What if every inequality is necessary to the definition of the feasible region?

35. Prove that the open orthant $\{x \in \Re^n | x > 0\}$ is convex but not closed. Prove that the open orthant does not possess the separation property of Theorem 4.1.1. Define a weaker kind of separation property that the open orthant does possess.

36. Prove or disprove: If P is a polyhedron and $t \notin P$, then there exist infinitely many distinct vectors $\boldsymbol{\pi}$ with $||\boldsymbol{\pi}|| = 1$ such that for some $\boldsymbol{\pi}_0$ (depending on $\boldsymbol{\pi}$) the hyperplane $\boldsymbol{\pi} \cdot \mathbf{y} = \boldsymbol{\pi}_0$ separates P from t.

37. (a) Prove that, if v is not an extreme point of a polyhedron P, then there does not exist an objective vector $\mathbf{c}$ such that v is the unique optimal solution to the LP $\max c \cdot x : x \in P$.

    (b) From the proof of Proposition 4.4.2 extract a proof that if v is a basic feasible solution of P, then there exists an objective $\mathbf{c}$ such that v is the unique maximizer of $\mathbf{c} \cdot x$ over $\mathbf{x} \in P$.

(c) Conclude that the property of being a unique maximizer is equivalent to being an extreme point.

38. Let P be a polyhedron and let Q be a projection of P. Prove or disprove each of the following statements:

    (a) If P has a nonzero direction, then Q must contain a nonzero direction.
    (b) If Q has a nonzero direction, then P must contain a nonzero direction.
    (c) If P contains a line, then Q must contain a line.
    (d) If Q contains a line, then P must contain a line.

39. Let P be a full-dimensional polyhedron in $\Re^{2n}$ and let P' be its projection onto $\Re^n$. Prove or disprove:

    (a) If x' is an extreme point of P' there exists an extreme point x of P such that x' is the projection of x.
    (b) If P is bounded, then P' does not have more extreme points than P.
    (c) If P can be defined by m inequality constraints, then P' can be defined by 2^m or fewer constraints.

40. Let polyhedron $P = \{x : A\mathbf{x} \leq \mathbf{b}\}$ and let $v \in P$.

    (a) Let $C(v)$ be the set of vectors $\mathbf{c}$ such that v is a (not necessarily unique) maximizer of $\mathbf{c} \cdot x : x \in P$. Prove that $C(v)$ is a convex cone, that is, $C(v)$ is convex and closed under multiplication by nonnegative scalars.
    (b) Let $I \subset \{1, 2, \ldots, m\}$ be the set of indices of binding constraints at v. Prove $C(v)$ consists of the zero vector when $I = \phi$. Prove that $C(v) = \{\sum_{i\in I} \alpha_i A_{i,:} : \alpha_i \geq 0 \ \forall i \in I\}$ when $I \neq \phi$. Interpret this formula in terms of the dual LP.
    (c) Let $C^\circ(v)$ be the set of vectors $\mathbf{c}$ such that v is the unique maximizer of $\mathbf{c} \cdot x : x \in P$. Prove or disprove that $C^\circ(v)$ is the interior of $C(v)$.
    (d) Assume that v is a nondegenerate extreme point of P. Prove that $C^\circ(v) = \{\sum_{i\in I} \alpha_i A_{i,:} : \alpha_i > 0 \ \forall i \in I\}$.
    (e) Assume that v is a degenerate extreme point of P. Prove or disprove: $C^\circ(v) = \{\sum_{i\in I} \alpha_i A_{i,:} : \alpha_i > 0 \ \forall i \in I\}$.
    (f) Assume that v is a degenerate extreme point of $P \subset \Re^n$. Assume further that every subset $J \subset I$ with $|J| = n$ corresponds to a set of linearly independent rows of A. Prove or disprove:

$$C^\circ(v) = \bigcup_{J \subset I; |J|=n} \{\sum_{i\in J} \alpha_i A_{i,:} : \alpha_i > 0 \ \forall i \in J\}.$$

41. Add one word to Sally's statement in Exercise 16 so that it is certain she is lying.

42. Prove: if S is convex, then the set of directions of S is convex. Be sure your proof handles the case S empty.

43. Let $S \subseteq \Re^n$ be a convex set. Suppose there exists $\mathbf{y}^* \in S$ which is not an extreme point of S, and which is an optimal solution to the optimization problem

$$\max_{x\in S} c \cdot x.$$

Prove there exist infinitely many optimal solutions.

44. If S is nonempty and has a nonzero direction, must S be convex? Prove: the set of directions of any set S is convex.

45. Let $S \subseteq \Re^n$ be a convex set. Let $E(S)$ denote the set of extreme points of S. Prove: the set $S \setminus E(S)$ is convex. That is, if you remove the extreme points from S, what remains is still convex.

46. Use FM elimination to find the projection of the polyhedron defined by the following constraints onto x_1, x_2, onto x_1, x_3, and onto x_2, x_3. Draw each projection in two dimensions.

$$\begin{aligned} x_1 &\geq 0 \\ x_2 &\geq 2 \\ x_3 &\geq 4 \\ x_1 + 2x_2 - 3x_3 &\leq 5 \\ -x_1 + 2x_2 + 3x_3 &\leq 18 \\ x_1 - 2x_2 + 3x_3 &\leq 14 \\ x_1 + x_2 + x_3 &\leq 12 \end{aligned}$$

47. Prove Theorem 2.1.1 using the easy part of Theorem 4.3.2. (The easy part states that it is impossible for both systems to have a feasible solution.)

48. Prove the easy part of Theorem 4.3.2 using the weak duality theorem 2.1.1. (The easy part states that it is impossible for both systems to have a feasible solution.)

49. (a) Let $S \subset \Re^n$. Prove that a convex combination of convex combinations of elements of S is a convex combination of elements of S.

    (b) Let S be the set of extreme points of a polytope P. Let $p \in P$. Use part 49a and reasoning similar to that in the proof of Theorem 4.4.5 to prove that p is a convex combination of members of S.

    (c) Part 49b proves Theorem 4.5.2 without relying on a theorem of the alternative, separating hyperplane, or other duality theorem. What doesn't it prove about the set of convex combinations of S?

    (d) Let D be a finite set of directions of polyhedron P such that every direction of P is a convex combination of members of D. State and prove a generalization of part 49b that applies to P.

50. Devise an analogue to FM elimination for linear inequalities on integer-valued variables. Prove that your method works. Hint: Use the ceiling ($\lceil\ \rceil$) and floor ($\lfloor\ \rfloor$) operators. Does your method work for systems involving some integer and some continuous variables?

51. Suppose the polyhedron $P = \{\mathbf{x} | A\mathbf{x} \geq \mathbf{b};\ \mathbf{x} \leq \mathbf{0}\}$ is nonempty. Prove that $\mathbf{d}$ is a direction of P if and only if $\mathbf{d}$ is a solution to

$$A\mathbf{y} \geq \mathbf{0}; \mathbf{y} \leq \mathbf{0}.$$

Why do you need to use the fact that P is nonempty?

52. Working only from the definitions, prove that if the LP feasible region $F = \{\mathbf{x} \in \Re^n | A\mathbf{x} \leq \mathbf{b}, \mathbf{x} \geq \mathbf{0}\}$ is unbounded, then there exists an integer k between 1 and n such that the linear program

$$\max_{x \in F} x_k$$

is unbounded (has unbounded optimum). Note: a region T is unbounded iff for all $M \in \Re$ there exists $\mathbf{z} \in T$ such that $||\mathbf{z}|| \geq M$. Why does this question explicitly include the property of nonnegativity ($\mathbf{x} \geq \mathbf{0}$) in the description of F?

53. Let P be a polyhedron in $\Re^n$. Define $T(P) \subset \Re^n$ as $T(P) = \{\boldsymbol{\pi} : \boldsymbol{\pi}^T \mathbf{x} \leq 1 \; \forall \mathbf{x} \in P\}$. Consider now the polyhedron C^n in $\Re^n$ defined by $-1 \leq x_i \leq 1 : i = 1 \ldots, n$. For $n = 2$, find $T(C^n)$ and $T(T(C^n))$. If P were a regular pentagon centered at the origin, what do you suppose $T(P)$ and $T(T(P))$ would be? Now find $T(C^3)$. This is another example of how thinking in two dimensions can be misleading.

54. A *polyhedral cone* is a set of form $\{\mathbf{x} : A\mathbf{x} \leq 0\}$. (a) Prove that the set of directions of a polyhedron is a polyhedral cone.
(b) Prove that a polyhedral cone is finitely generated. Hint: Use the set P^R as in the proof of theorem 4.5.3.
(c) Conclude that a polyhedron P containing at least one extreme point is the sum of the polytope generated by the extreme points of P and a polyhedral cone generated by the extreme directions of P.

55. Using only Theorem 4.3.2 and elementary mathematical reasoning, derive theorems of the alternative for each the following systems:

    (a) $A\mathbf{x} \geq \mathbf{b}$
    (b) $A\mathbf{x} = \mathbf{b}; x \geq 0$
    (c) $A\mathbf{x} < 0$ (Hint: This system is equivalent to $A\mathbf{x} + 1\epsilon \leq 0; (\mathbf{x}, \epsilon) \cdot (0, 0, \ldots, 0, 1) > 0$.)

56. Prove that FM elimination can require at least exponential time in $m + n$ to solve $A\mathbf{x} \leq \mathbf{b}$ where $A \in Z^{m \times n}$. To be precise, prove that $\exists c_1 > 0, c_2 > 0$ such that $\forall N \exists m, n, A \in Z^{m \times n}, b \in Z^m$ such that $m + n \geq N$ and FM elimination requires at least $c_1 2^{c_2(m+n)}$ steps to solve $A\mathbf{x} \leq \mathbf{b}$.

57. Employ Corollary 4.5.4 and elementary reasoning to give a quick proof that every LP of form $\max c \cdot x$ subject to $A\mathbf{x} = \mathbf{b}; x \geq 0$ that is feasible has either an optimum solution that is an extreme point, or an extreme direction that makes the LP have unbounded optimum.

58. The n-dimensional cube can be defined with the $2n$ constraints $0 \leq x_i \leq 1 : i = 1 \ldots n$ and has 2^n extreme points.

    (a) Construct a polyhedron with exponentially many extreme directions compared with the number of constraints that define it.
    (b) Find a polytope that has exponentially many more constraints in its definition than it has extreme points.

59. How can you solve an LP with FM elimination? Solve only one system of linear inequalities, not a sequence.

60. For the primal system of linear inequalities $x_1 + 2x_2 \leq 35; 4x_1 + 2x_2 \geq 80; x_1 - x_2 \leq 10; x_1 \geq 0, x_2 \geq 0$:

    (a) Write the constraints of the dual if the primal LP objective were to maximize $0x_1 + 0x_2$.

(b) Remove x_1 from the primal constraints by FM elimination.

(c) Write the constraints of the dual of the new primal system you obtained in part 60b, assuming the primal objective coefficient is zero.

(d) Obtain a feasible solution other than all variables equal to zero for the dual system you obtained in part 60c.

(e) Remove x_2 by FM elimination from the primal system you obtained in 60b. Obtain a feasible solution to the original primal.

61. Suppose at least one of the problems, min $c^T x$ subject to $Ax \geq b, x \geq 0$; max $\pi^T b$ subject to $\pi^T A \leq c^T, \pi \geq 0$ has a feasible solution. Prove that the set of feasible solutions to at least one of the two problems is unbounded.

62. Prove that if v is an extreme point of the polyhedron $P + \hat{P}$, then v equals the sum of an extreme point of P and an extreme point of $\hat{P}$.

63. Derive a theorem of the alternative for the system $A\mathbf{x} < b$.

64. Prove that the sum of two polytopes is a polytope.

65. Prove that the sum of two polyhedra is a polyhedron.

66. You have two polyhedra, $P \subset \Re^n$ and $Q \subset \Re^n$, defined by constraints $P\mathbf{x} \leq \mathbf{p}$ and $Q\mathbf{x} \leq \mathbf{q}$ respectively. You want to know whether P contains a point that is not in Q. One approach to your question would begin by finding all the extreme points and extreme directions of P and Q, and work with those. What is likely to be impractical about that approach?

    Use duality to create a linear program that, if solved, answers your question. If you can't find a single LP that answers the question, find a set of LPs instead (but not an exponentially large set!).

67. Let $P^i = \{x | A^i x \leq b^i\} \subset \Re^n$ for $i = 1, 2, 3$. Let $d \in \Re^n$ and $d \in \Re$. Suppose that P^i is nonempty for $i = 1, 2, 3$. Formulate a set of strict and/or nonstrict linear inequalties that has a feasible solution iff:

    (a) $d \cdot \mathbf{y} \geq d^0 \ \forall y \in P^1$.

    (b) $P^1 \neq P^2$.

    (c) $P^1 = P^2$ and P^3 is a polytope.

D Problems

68. State and prove a version of Theorem 4.1.1 for convex sets that are not necessarily closed.

69. Invent a dual version of FM elimination. Hint: Study a numerical example as in Exercise 60.

70. Prove that if the primal LP (2.1) has a feasible solution, then either its feasible region is unbounded, or its dual's feasible region is unbounded, or both.

71. Let $u^0, u^1, \ldots u^N$ be elements of $\Re^n$. Define $C = \{u^0 + \sum_{i=1}^N \lambda_i u^i | \lambda_i \in \Re; \lambda_i \geq 0 \ \forall i = 1 \ldots N\}$. Use the following theorem of the alternative to prove that C is closed. Theorem: Exactly one of these systems has a feasible solution: $A\mathbf{x} = \mathbf{b}; x \geq 0$ or $\boldsymbol{\pi}^T A \geq 0^T; \boldsymbol{\pi}^T b < 0$.

72. Suppose that A is a real m by n matrix with $m > n$. The system

$$A\mathbf{x} \leq \mathbf{b}$$

with $b \in \Re^m$ represents a set of m inequalities in the vector variable $\mathbf{x} \in \Re^n$. Show that if each subset of $n+1$ of these m inequalities has a solution $\mathbf{x} \in \Re^n$ satisfying all $n+1$ inequalities simultaneously, then there exists an $\mathbf{x} \in \Re^n$ satisfying all m inequalities simultaneously.

73. (a) Prove Carathéodory's Theorem: Every point v in polytope $P \subset \Re^n$ equals the convex combination of at most $n+1$ extreme points of P.
    (b) State and prove a generalization of Carathéodory's Theorem for pointed polyhedra.

74. Let A be an $n \times n$ matrix and b be an n-vector. Let $P = \{\mathbf{x} | A\mathbf{x} \leq \mathbf{b}\}$. Prove or disprove: if P is nonempty, then P is unbounded.

75. Prove or disprove: If $P \subset \Re^n : n \geq 2$ is a pointed polyhedron and $t \notin P$, then there exist infinitely many distinct vectors $\boldsymbol{\pi}$ with $||\boldsymbol{\pi}|| = 1$ such that for some $\boldsymbol{\pi}_0$ (depending on $\boldsymbol{\pi}$) the hyperplane $\boldsymbol{\pi} \cdot \mathbf{y} = \boldsymbol{\pi}_0$ separates P from t.

76. State and prove a more general version of Theorem 4.5.3 that does not require P to be pointed.

77. Let $P^i = \{\mathbf{x} | A^i\mathbf{x} \leq \mathbf{b}^i\} \subset \Re^n$ for $i = 1, 2, 3$. Let $\mathbf{d} \in \Re^n$ and $d^0 \in \Re$. Formulate a set of strict and/or nonstrict linear inequalities that has a feasible solution iff:
    (a) $\mathbf{d} \cdot \mathbf{y} \geq d^0 \; \forall \mathbf{y} \in P^1$.
    (b) $P^1 \neq P^2$.
    (c) $P^1 = P^2 \neq \phi$ and P^3 is a polytope.
    (d) $P^1 = P^2$ and P^3 is a nonempty polytope.
    (e) $P^1 = P^2$ and P^3 is a polytope.

78. Let Q^i, $i = 1, 2, 3$ be polyhedra in $\Re^m$ defined by the constraints $A^i\mathbf{x} \leq \mathbf{b}^i$ for $i = 1, 2, 3$ respectively.
    (a) Formulate an LP that if solved determines whether there exists $\mathbf{w} \in \Re^m$ such that $\mathbf{w} \in Q^1, \mathbf{w} \notin Q^2$, and $\mathbf{w} \in Q^3$.
    (b) Formulate an LP that if solved determines whether there exists $\mathbf{w}$ such that $\mathbf{w} \in Q^1 \bigcup Q^2$ and $\mathbf{w} \notin Q^3$.
    (c) Formulate an LP that if solved determines whether there exists $\mathbf{w}$ such that $\mathbf{w}$ is in the convex hull of $Q^1 \bigcup Q^2$ and $\mathbf{w} \notin Q^3$.
    (d) Formulate an LP that if solved finds a vector $\mathbf{c}$ and point $\mathbf{y}$ such that $\mathbf{y}$ is a minimizer of $\mathbf{c} \cdot \mathbf{x}$ over Q^1; $\mathbf{y}$ is a minimizer of $\mathbf{c} \cdot \mathbf{x}$ over Q^2; and $\mathbf{y}$ is not a minimizer of $\mathbf{c} \cdot \mathbf{x}$ over Q^3.

I Problems

79. Given the LP $\max \mathbf{c} \cdot \mathbf{x}$ subject to $A\mathbf{x} \leq \mathbf{b}; \mathbf{x} \geq \mathbf{0}$ and the LP $\min \mathbf{g} \cdot \mathbf{x}$ subject to $E\mathbf{x} = \mathbf{f}, \mathbf{x} \geq \mathbf{0}$, formulate an LP that determines whether or not these two LPs have optimal solutions with equal objective function value. Justify your answer.

80. Given two polyhedra, $P = \{\mathbf{x} : A\mathbf{x} \leq \mathbf{b}; \mathbf{x} \geq \mathbf{0}\}$ and $Q = \{\mathbf{y} : D\mathbf{y} \leq \mathbf{h}; \mathbf{y} \geq \mathbf{0}\}$, where matrices A and D have the same number of columns, formulate an LP that if solved determines whether or not $P \subsetneq Q$. Justify your answer.

81. Let $P = \{\mathbf{x} \in \Re^n : A\mathbf{x} \leq \mathbf{b}; \mathbf{x} \geq \mathbf{0}\}$. Formulate a system of linear inequalities that has a feasible solution iff P is an unbounded polyhedron. Find an alternative system that is feasible iff P is not an unbounded polyhedron.

82. Let $P = \{\mathbf{x} \in \Re^n : A\mathbf{x} \leq \mathbf{b}; \mathbf{x} \geq \mathbf{0}\}$. Formulate a system of linear inequalities that has a feasible solution iff P is a nonempty polytope. Find an alternative system that is feasible iff P is not a nonempty polytope.

83. Let A and $\mathbf{b}$ be given. Formulate a system of linear inequalities that has a feasible solution if and only if $P = \{\mathbf{x} : A\mathbf{x} \leq \mathbf{b}\}$ is a nonempty polytope.

84. Membership, separation, and boundedness in polyhedra. Solve as many of the following problems as you can. Determine whether there are any duality relationships between any of your answers.

    (a) Let A and $\mathbf{b}$ be given. Suppose $P = \{\mathbf{x} : A\mathbf{x} \leq \mathbf{b}\}$ is a polytope. Write an LP that, if solved, provides a point in P iff P is not empty.

    (b) Let A and $\mathbf{b}$ be given. Let $P = \{\mathbf{x} : A\mathbf{x} \leq \mathbf{b}\}$. Write an LP that, if solved, provides a point in P iff P is not empty.

    (c) Let vectors $\mathbf{y}^i \in \Re^n : i = 1, \ldots, N$ be given. Let $\mathbf{p} \in \Re^n$ be given. Write an LP that, if solved, expresses $\mathbf{p}$ as a convex combination of the $\mathbf{y}^i$ iff that is possible, and otherwise shows that it is not possible.

    (d) Let vectors $\mathbf{y}^i \in \Re^n : i = 1, \ldots, N$ and $\mathbf{d}^j : j = 1, \ldots M$ be given. Let $\mathbf{p} \in \Re^n$ be given. Write an LP that, if solved, expresses $\mathbf{p}$ as the sum of a convex combination of the $\mathbf{y}^i$ plus a nonnegative linear combination of the $\mathbf{d}^j$, iff that is possible.

    (e) Let vectors $\mathbf{y}^i \in \Re^n : i = 1, \ldots, N$ be given. Let P denote the set of convex combinations of the $\mathbf{y}^i$. Let $\mathbf{w} \in \Re^n$ be given. Write an LP that, if solved, produces a hyperplane that separates $\mathbf{w}$ from P iff that is possible, and otherwise shows that it is not possible.

    (f) Let vectors $\mathbf{y}^i \in \Re^n : i = 1, \ldots, N$ and $\mathbf{d}^j : d = 1, \ldots, M$ be given. Let P denote the set sums of convex combinations of the $\mathbf{y}^i$ plus nonnegative linear combinations of the $\mathbf{d}^j$. Let $\mathbf{w} \in \Re^n$ be given. Write an LP that, if solved, produces a hyperplane that separates $\mathbf{w}$ from P iff that is possible, and otherwise shows that it is not possible.

    (g) Let $P = \{\mathbf{x} : A\mathbf{x} \leq \mathbf{b}\}$ and let $Q = \{\mathbf{x} : Dx \leq \mathbf{d}\}$ where A and D have the same number of columns. Write an LP that, if solved, determines whether or not $P = Q$ and, if $P \neq Q$, exhibits a point in $(P \bigcup Q) \setminus (P \bigcap Q)$.

    (h) Let A and $\mathbf{b}$ be given. Let $P = \{\mathbf{x} : A\mathbf{x} \leq \mathbf{b}\} \subset \Re^n$. Let $\mathbf{w} \in \Re^n$ be given. Write an LP that, if solved, finds a hyperplane that separates $\mathbf{w}$ from P iff $\mathbf{w} \notin P$, and otherwise determines that $\mathbf{w} \in P$.

85. (*) Find a way to crash that is computationally parallelizable for a 4 to 16 processor CPU.

86. Use Theorem 4.3.2 to prove Theorem 2.1.3. Hint: To avoid an incorrect proof, remember that an LP and its dual can both be infeasible.

87. Prove rigorously that if the entries of $\mathbf{b}$ are independently identically Gaussian (normally) distributed, then regardless of A the polyhedron $\{x|A\mathbf{x} \leq \mathbf{b}\}$ is degenerate with probability 0. Prove rigorously that if the entries of $m \geq 5$ vectors $\mathbf{y}^1, \ldots, \mathbf{y}^n$ in $\Re^3$ are independently identically Gaussian (normally) distributed, then the polytope consisting of the convex combinations of $\mathbf{y}^1, \ldots, \mathbf{y}^m$ is degenerate with probability strictly greater than 0.

Chapter 5

The Simplex Method

Preview: The simplex method, invented by George Dantzig in 1947 [64, 66], is still the most widely used algorithm for solving LPs.

5.1 The Simplex Method

Preview: This section derives the core version of the simplex method, an algorithm that starts at a basic feasible solution and terminates at an optimal basic feasible solution if one exists, and at a ray of unboundedness otherwise. I will describe both the geometry and algebra of the algorithm from both the primal and dual points of view.

Geometric overview: We know from the previous chapter that if the LP has an optimum solution, it has an optimum extreme point solution. Instead of enumerating all the extreme points, and picking the best, start at one extreme point and check the adjacent extreme points, which are the ones you can reach along an edge of the polyhedron. If an adjacent one is better, move to it. If not, then intuitively we should be optimal. Consider Figure 5.1. Suppose the polyhedron has been rotated so that the optimal point is the highest point. At any extreme point but the highest there is a higher adjacent extreme point.

Algebraic overview: Start at a BFS. Check to see if relaxing one of the binding constraints improves the objective value. If there isn't such a constraint, you are optimal. If there is, relax that one constraint as much as possible without violating feasibility. If the constraint can be relaxed indefinitely, there is an unbounded optimum. Otherwise, a new constraint will become binding. That newly binding constraint, together with the already binding constraints, defines a new starting BFS. Repeat until solved.

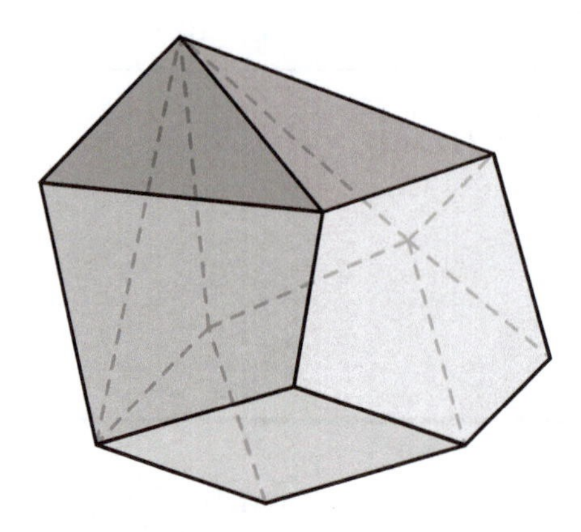

FIGURE 5.1: Every extreme point except the highest has a higher adjacent extreme point.

The simplex method solves the following problem: given a basic feasible solution to an LP, find an optimal basic feasible solution, or a ray of unboundedness. The simplex method operates on an LP in the form min $\mathbf{c} \cdot \mathbf{x}$ subject to $A\mathbf{x} = \mathbf{b}; \mathbf{x} \geq 0$. Here A is an m by n matrix and $\mathbf{b}$ is an m-vector. There is no loss of generality in considering only the minimization case. The dual LP is then max $\mathbf{b} \cdot \boldsymbol{\pi}$ subject to $\boldsymbol{\pi}^T A \leq \mathbf{c}^T$.

We will assume that A has m linearly independent columns. For the LP format at hand, therefore, a basic solution $\mathbf{x}$ must satisfy $A\mathbf{x} = \mathbf{b}$ and $x_j = 0 \;\; \forall j \in \mathbb{N}$ where $|\mathbb{N}| = n - m$. We often think of a basic solution as follows: all but m of the variables $x_1 \ldots x_n$ are fixed at zero. We must satisfy the constraints $A\mathbf{x} = \mathbf{b}$. We have $n - (n - m) = m$ free variables, so we solve m equations in those m unknowns (the free variables) to attain $A\mathbf{x} = \mathbf{b}$. Let $\mathbb{B}$ denote the complement of $\mathbb{N}$. Rewrite $A\mathbf{x} = \mathbf{b}$ as $A_{:,\mathbb{N}}\mathbf{x}_{\mathbb{N}} + A_{:,\mathbb{B}}\mathbf{x}_{\mathbb{B}} = b$. Then we solve the equation $A_{:,\mathbb{B}}\mathbf{x}_{\mathbb{B}} = \mathbf{b}$.

Geometry: We visualize the LP in $n-m$ dimensional space, not in n dimensional space. Imagine that the original form of the problem was to minimize $\mathbf{c}' \cdot \mathbf{x}'$ subject to $\mathbf{x}' \in P$ where $P = \{\mathbf{x}' | A'\mathbf{x}' \geq \mathbf{b}; \mathbf{x}' \geq \mathbf{0}\}$. In $n-m$ dimensional space, a basic solution $\mathbf{x}'$ has $n-m$ linearly independent binding constraints, which are a mix of rows of $A'\mathbf{x}' \geq \mathbf{b}$ and $\mathbf{x}' \geq \mathbf{0}$. A basic feasible solution $\mathbf{x}'$ is an extreme point of the polyhedron P.

Imagine that slack variables $\hat{\mathbf{x}}$ were later introduced to create the reformulation, minimize $\mathbf{c}' \cdot \mathbf{x}' + \mathbf{0} \cdot \hat{\mathbf{x}}$ subject to $A'\mathbf{x}' - I\hat{\mathbf{x}} = \mathbf{b}; \mathbf{x}' \geq \mathbf{0}, \hat{\mathbf{x}} \geq \mathbf{0}$. In this reformulation the constraints of P that are binding at an extreme point of P are a mix of rows of $\hat{\mathbf{x}} \geq \mathbf{0}$ and $\mathbf{x}' \geq \mathbf{0}$. Figure 5.2 labels the binding constraints at an extreme point in terms of $\mathbf{x}'$ and $\hat{\mathbf{x}}$. The constraints $[A - I](\mathbf{x}, \hat{\mathbf{x}}) = \mathbf{b}$ also are binding at any feasible solution in this higher dimensional space, but they are only implicit in the $n - m$ dimensional space we visualize.

It's time to introduce the most useful notational shortcut in LP theory. We let B denote $A_{:,\mathbb{B}}$. The letter B is a mnemonic from the word "basis," but so is the specially typeset symbol $\mathbb{B}$ that denotes the set of indices of a basis. When you read this text, the visual difference between B and $\mathbb{B}$ and the location of $\mathbb{B}$ as a subscript help you distinguish between the two. However, when the text is read aloud the difference between B and $\mathbb{B}$, not to mention $\mathbf{b}$, is not clear. For example, we will often solve the equation $Bx_{\mathbb{B}} = b$. *Question: Say that aloud and remember the three meanings of the "b" sound. Then do the same for*

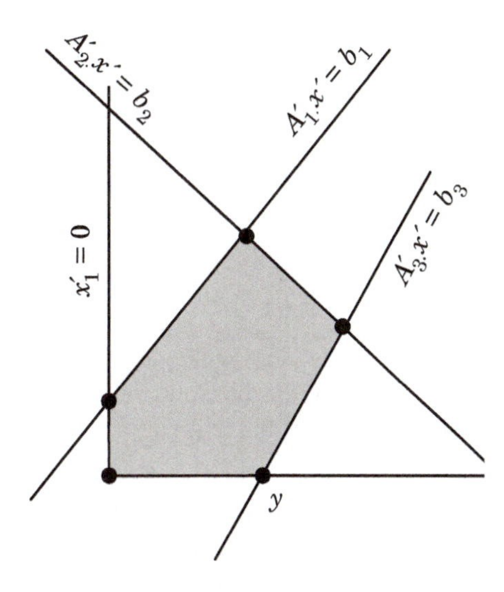

FIGURE 5.2: The binding constraints at y are $x_2 \geq 0$ and $x_5 \geq 0$ where $\mathbf{x}_S = \{3, 4, 5\}$.

the equation's solution, $\mathbf{x}_{\mathbb{B}} = B^{-1}b$.[1] Notice that we've implicitly required that B have full rank. If the basic solution satisfies $\mathbf{x}_{\mathbb{B}} \geq 0$, it is feasible. *Question: If data are random, what is the chance that a basic solution is feasible?*[2]

Example 5.1

$$\begin{aligned} \min \quad & 4x_1 + 5x_2 - 6x_3 && \text{subject to} \\ & x_1 - x_2 + 7x_3 + x_4 && = 10 \\ & 2x_1 + x_2 - 5x_3 + x_5 && = 15 \\ & x_i && \geq 0 \ \forall i \end{aligned}$$

In this example, $n = 5, m = 2$. Hence $|\mathbb{N}| = 5 - 2 = 3$. Let $\mathbb{N} = \{1, 2, 3\}$ and $\mathbb{B} = \{4, 5\}$. Then

$$B = \begin{bmatrix} 1 & 0 \\ 0 & 1 \end{bmatrix}. \tag{5.1}$$

The basic solution is $\mathbf{x}_{\mathbb{B}} = (x_4, x_5) = (10, 15)$, a feasible solution. Now let $\mathbb{N} = \{3, 4, 5\}$, so $\mathbb{B} = \{1, 2\}$. Then

$$B = \begin{bmatrix} 1 & -1 \\ 2 & 1 \end{bmatrix}. \tag{5.2}$$

The basic solution is the solution to

$$\begin{aligned} x_1 - x_2 &= 10 \\ 2x_1 + x_2 &= 15 \end{aligned}$$

[1] I used to employ the same symbol B for both $\mathbb{B}$ and $A_{:,\mathbb{B}}$, but my students condemned my callous abuse of notation.

[2] 2^{-m}.

which is $(x_1, x_2) = (25/3, -5/3)$, an infeasible solution because $x_2 < 0$. Now let $\mathbb{N} = \{1, 2, 4\}$ and $\mathbb{B} = \{5, 3\}$. Then

$$B = \begin{bmatrix} 0 & 7 \\ 1 & -5 \end{bmatrix}. \tag{5.3}$$

The basic solution is the solution to

$$\begin{aligned} 7x_3 &= 10 \\ x_5 \quad -5x_3 &= 15 \end{aligned}$$

which is $\mathbf{x}_{\mathbb{B}} = (x_5, x_3) = (\frac{10}{7}, 22\frac{1}{7})$, a feasible solution. This example shows that actually $\mathbb{B}$ is an ordered set. The ordering may be arbitrary as long as it is consistent.

The simplex method asks, what happens if we take a variable $x_j : j \in \mathbb{N}$, which now equals zero, and increase it by $\theta \geq 0$? We keep all of the other $x_k : k \in \mathbb{N}, k \neq j$ fixed at zero. We must keep the equations $A\mathbf{x} = \mathbf{b}$ true, so as we *increase* the left-hand side (LHS) by $\theta A_{:,j}$ we must alter the values of $\mathbf{x}_{\mathbb{B}}$ to offset the change in the LHS. So the change in $\mathbf{x}_{\mathbb{B}}$ must be to *decrease* it from $B^{-1}b$ by the amount $\theta B^{-1}A_{:,j}$. For the diet problem, David Luenberger gives a beautiful interpretation of this formula [217]. Imagine that x_j denotes the number of carrots in the diet. Right now $j \in \mathbb{N}$ so we eat no carrots. If we add a carrot to our diet, we increase the LHS of our constraints by the nutritional composition of one carrot. To keep $A\mathbf{x} = \mathbf{b}$ true, we have to change the amounts of the other foods in the diet to offset, exactly, the nutritional contribution of a carrot. The formula $B^{-1}A{\cdot}j$ is the mix of foods that compose a synthetic carrot, a mix of foods with exactly the same nutritional contribution as a real carrot.

Geometry: Starting from an extreme point, we move off one binding constraint $x_j \geq 0$, while we keep all other binding constraints binding. Hence we move along an edge of the polyhedron.

What is the effect on the objective function of substituting a real carrot for a synthetic carrot? It costs c_j to buy one carrot. On the other hand, we save $\mathbf{c}_{\mathbb{B}}^T B^{-1} A_{:,j}$ because we remove a synthetic carrot from our shopping list. The net effect on the objective function is to increase by $c_j - c_{\mathbb{B}}^T B^{-1} A_{:,j}$. This value is actually a dual variable value. It is also called the "reduced cost" because the cost c_j of the real carrot is reduced by the cost of a synthetic carrot.

Geometry: The effect on the objective function is the dot product of $\mathbf{c}$ *with the direction of movement along the edge.*

If there is a nonbasic variable $x_j : j \in \mathbb{N}$ for which the reduced cost $c_j - c_{\mathbb{B}}^T B^{-1} A_{:,j} < 0$, then we increase x_j (from its present value of zero) by the largest possible value of θ that keeps us feasible.

Geometry: We've found a direction that is good to travel in, so we move in that direction until we are about to exit the polyhedron.

The ith basic variable decreases by $\theta B_{i,:}^{-1} A_{:,j}$ as x_j is set to θ. If $B_{i,:}^{-1} A_{:,j} < 0$, the ith basic variable is increasing with θ. Its nonnegativity constraint becomes looser. If $B^{-1}_{i,:}A_{:,j} = 0$, that variable doesn't change. Its nonnegativity constraint neither loosens nor tightens. In both these cases, θ can increase indefinitely, as far as that variable is concerned.

However, if $B^{-1}_{i,:}A_{:,j} > 0$, the variable will go negative if θ is too large.

Geometry: $B^{-1}_{i,:}A_{:,j} < 0$ *means we are moving away from the variable's nonnegativity constraint.*
$B^{-1}_{i,:}A_{:,j} = 0$ *means we are moving parallel to the variable's nonnegativity constraint.*
$B^{-1}_{i,:}A_{:,j} > 0$ *means we are moving towards the variable's nonnegativity constraint.*
Example: Draw the feasible region defined by the constraints

$$\begin{aligned} x_1 + 2x_2 &\leq 12 \\ x_1 &\leq 6.5 \\ x_1 + x_2 &\leq 8 \\ x_1, x_2 &\geq 0 \end{aligned}$$

Add slack variables to get the equivalent system

$$\begin{aligned} x_1 + 2x_2 + x_3 &= 12 \\ x_1 \qquad + x_4 &= 6.5 \\ x_1 + x_2 \qquad + x_5 &= 8 \\ x_i \geq 0 : i = 1, \ldots, 5 & \end{aligned}$$

but continue to visualize the feasible region in two dimensions. As depicted in Figure 5.3, each of the five original constraints is now a nonnegativity constraint.

At the extreme point $\mathbf{y}$*, inequalities* $x_1 \geq 0$ *and* $x_3 \geq 0$ *are binding. Hence variables* x_1, x_3 *are nonbasic, equal to* 0*. As we travel in the direction of the arrow we keep* $x_3 = 0$ *(it stays nonbasic), move away from the line* $x_1 = 0$ *(increase* x_1*), and move towards the lines* $x_2 = 0, x_4 = 0$*, and* $x_5 = 0$ *(decrease all three basic variables.) To stay feasible, we must not cross the line* $x_5 = 0$*. Thus* x_5 *is the first of the basic variables to drop to* 0 *as* θ *increases.*

At the extreme point $\mathbf{w}$ *in Figure 5.3, moving in the direction of the arrow brings us away from* $x_2 = 0$*, but closer to* $x_3 = 0$ *and* $x_1 = 0$*. We reach the* $x_3 = 0$ *line first. Therefore, basic variable* x_3 *leaves the basis. Nonbasic variable* x_4 *enters the basis, and nonbasic variable* x_5 *remains at* 0*. Question: What happens to the basic variable* x_1*?*[3]

It's visually obvious in the example of Figure 5.3 which constraint boundary we will reach first as we move. Now let's derive the general algebraic equivalent of determining how far we can move without exiting the feasible region. Let i be any index such that $B^{-1}_{i,:}A_{:,j} > 0$. Right now the ith basic variable equals $B^{-1}_{i,:}b$. To stay feasible, we must have $\theta \leq B^{-1}_{i,:}b / B^{-1}_{i,:}A_{:,j}$. The minimum of these bounds over all such i determines the largest possible value of θ that keeps us feasible. The variable that supplied the minimum bound drops to zero and becomes nonbasic. We then have a new basic feasible solution with better objective value. This would complete an iteration of the algorithm. If no basic variable decreases, there are no such bounds on θ, and so the simplex method has found a ray of unboundedness. *Question: What is the formula for the ray?*[4]

If no reduced costs are good, we have found a choice of basis such that $\mathbf{c}^T_{\mathbb{N}} - c^T_{\mathbb{B}}B^{-1}A_{:,\mathbb{N}} \geq$

[3]It decreases but remains basic because we move towards but do not reach $x_1 = 0$.
[4]$x_j = \theta$; $x_k = 0\ \forall k \neq j, k \in N$; $\mathbf{x}_{\mathbb{B}} = B^{-1}\mathbf{b} - \theta B^{-1}A_{:,j} : 0 \leq \theta < \infty$.

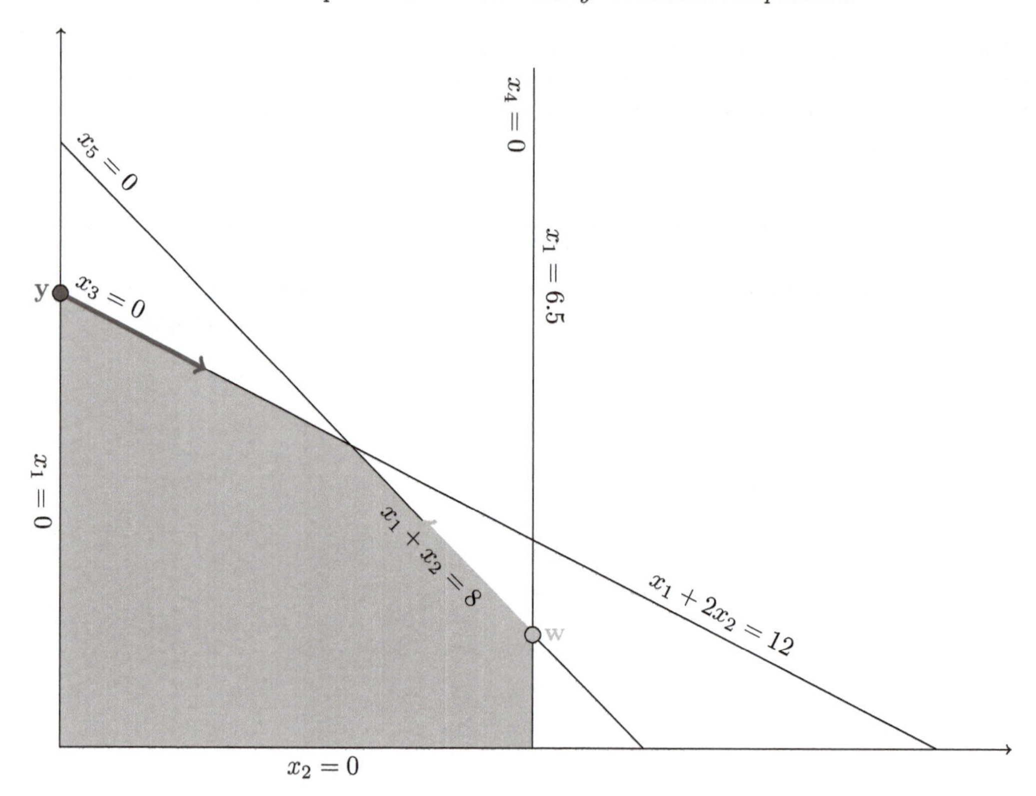

FIGURE 5.3: As we move from y, nonbasic variable x_1 increases, nonbasic variable x_3 remains 0, basic variables x_2, x_4, and x_5 decrease. Of the basic variables, x_5 drops to 0 first. As we move from $\mathbf{w}$ by increasing x_4 from 0 and keeping $x_5 = 0$, x_2 increases while x_3 decreases to 0 before x_1 does.

$\mathbf{0}$. Now, by definition $\mathbf{c}_{\mathbb{B}}^T - c_{\mathbb{B}}^T B^{-1} A_{:,\mathbb{B}} = \mathbf{0}$. Therefore, $\mathbf{c}^T - \mathbf{c}_{\mathbb{B}}^T B^{-1} A \geq \mathbf{0}$. Set $\boldsymbol{\pi}^T = \mathbf{c}_{\mathbb{B}}^T B^{-1}$. Then $\boldsymbol{\pi}^T A \leq c^T$. Saying that there are no good reduced costs is the same as saying that $\boldsymbol{\pi}$ is feasible in the dual! And the objective value of this feasible dual solution equals $\boldsymbol{\pi} \cdot \mathbf{b} = (\mathbf{c}_{\mathbb{B}}^T B^{-1})\mathbf{b} = \mathbf{c}_{\mathbb{B}}^T (B^{-1}\mathbf{b}) = \mathbf{c}_{\mathbb{B}}^T \mathbf{x}_{\mathbb{B}}$ which is the value of our current primal solution. By the weak duality Theorem, they are both optimal. Therefore, if the simplex method gets stuck because it can't find a way to improve the primal solution, it has found a pair of optimal primal and dual solutions. We have proved the main relationship between the termination of the simplex method and duality.

Theorem 5.1.1 *If the simplex algorithm terminates at a basic feasible solution* $\mathbf{x}$ *for which no reduced costs* $\boldsymbol{\pi}$ *are good, i.e., for which* $\boldsymbol{\pi} \geq 0$ *if minimizing, then* $\mathbf{x}$ *is optimal in the primal,* $\boldsymbol{\pi}$ *is optimal in the dual, and the objective values of these solutions are equal.*

Table 5.1 summarizes the simplex algorithm.

We can now see all of the algebra underlying the geometric meaning to optimality that was given in Section 2.2. Let the original form of the LP be to maximize $\mathbf{c}^T\mathbf{x}$ subject to constraints $A\mathbf{x} \leq \mathbf{b}$. The dual is to minimize $\mathbf{b} \cdot \boldsymbol{\pi}$ subject to the constraints $\boldsymbol{\pi}^T A = \mathbf{c}^T; \boldsymbol{\pi} \geq 0$. At each extreme point of the polyhedron of feasible primal solutions, there is a corresponding complementary slack dual solution. What is it? It is $\boldsymbol{\pi}$ such that $\boldsymbol{\pi}^T A = \mathbf{c}^T$,

TABLE 5.1: Simplex Algorithm to minimize $\mathbf{c} \cdot \mathbf{x}$ subject to $A\mathbf{x} = \mathbf{b}; \mathbf{x} \geq \mathbf{0}$

Step 1: (Initialize). We are given $\mathbb{B} \subset \{1, 2, \ldots n\}$ such that $B = A_{:,\mathbb{B}}$ is full rank and $B^{-1}\mathbf{b} \geq \mathbf{0}$.
The current basic feasible solution is $\mathbf{x}_{\mathbb{B}} = B^{-1}\mathbf{b}; x_j = 0\ \forall j \notin \mathbb{B}$.
Step 2: Calculate reduced costs $\boldsymbol{\pi}_j = c_j - \mathbf{c}_{\mathbb{B}}^T B^{-1} A_{:,j}$ for all $j \notin \mathbb{B}$.
Step 3: Select k to minimize $\{\boldsymbol{\pi}_j | j \notin \mathbb{B}\}$. If $\boldsymbol{\pi}_k \geq 0$ go to Step 6.
Step 4: (Determine which basic variables are decreasing).
Compute the values $\mathbf{h}_{\mathbb{B}} = B^{-1} A_{:,k}$. Let $\mathbb{B}^+ = \{i \in \mathbb{B} | h_i > 0\}$.
If $\mathbb{B}^+ = \phi$ go to Step 7.
Step 5: (Minimum Ratio Test: Determine the variable to remove from $\mathbb{B}$).
Select r to minimize $x_r/h_r : r \in \mathbb{B}^+$.
Replace index r in $\mathbb{B}$ with index k. Go to Step 1.
Step 6: The current solution is optimal. Stop.
Step 7: The optimal solution is unbounded along the ray
$x_k = \theta; \mathbf{x}_{\mathbb{B}} = B^{-1}(\mathbf{b} - \theta A_{:,k}); x_j = 0\ \forall j \notin \mathbb{B} \bigcup \{k\}, \theta \geq 0$.

but not necessarily $\boldsymbol{\pi} \geq \mathbf{0}$ (i.e., not necessarily dual feasible). This means that the objective function vector of the primal $\mathbf{c}$ is a linear combination of the constraints (the rows) of the primal, with $\boldsymbol{\pi}$ supplying the multiplier values for the linear combination. See Figure 5.4. At each extreme point, $\mathbf{c}$ is a linear combination of primal rows. Moreover, by complementary slackness, the only rows whose multipliers are nonzero are the constraints that are binding there. Only at the optimal solution are the multipliers all nonnegative, i.e., is the dual feasible. *Question: What does it mean to make this the "original form" of the LP?*[5]

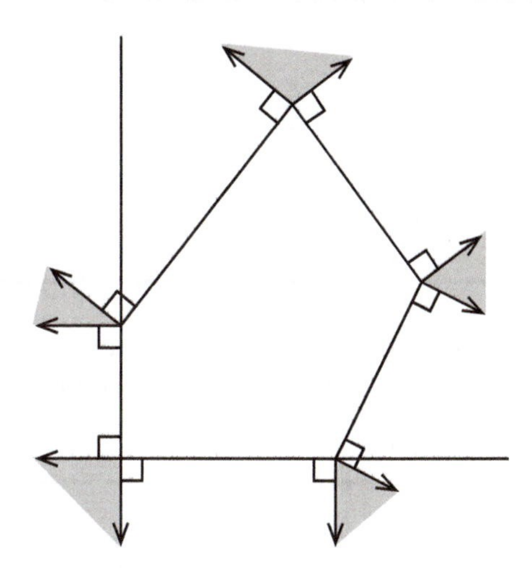

FIGURE 5.4: At each extreme point the shaded area indicates objectives which make that point optimal.

The inverse way of looking at it is to ask, given a particular extreme point $\mathbf{x}$ of a polyhedron, for which objective functions is that extreme point optimal? The answer is, any vector that is a nonnegative linear combination of the binding constraints at $\mathbf{x}$. In Figure 5.4, the sets of vectors for which each extreme point is optimal are indicated by the shaded areas.

[5]The polyhedron is completely described by the $A\mathbf{x} \leq \mathbf{b}$ constraints. If there are nonnegativity constraints they are part of the explicit description of the polyhedron.

Theorem 5.1.2 *If all basic feasible solutions of a pointed polyhedron are not degenerate, then the simplex method will terminate in a finite number of steps at either an optimal basic feasible solution or a ray of unboundedness, regardless of the starting basic feasible solution.*

Proof: Assuming nondegeneracy of all basic feasible solutions, the values $\mathbf{x}_{\mathbb{B}} = B^{-1}\mathbf{b} > \mathbf{0}$. For if $x_i = 0$ for some $i \in \mathbb{B}$ we'd have $n+1$ binding constraints. Therefore, the value of θ at each iteration is > 0, and so the objective function value strictly decreases at each iteration. No choice of basis can be repeated. The number of choices of basis is finite because it is bounded by $\binom{m+n}{m}$. Hence the algorithm must terminate in a finite number of steps. ■

In the absence of degeneracy, Theorem 4.4.8 is an immediate corollary to Theorem 5.1.2. This is an example of an algorithmic proof. We've proved a fundamental fact about LP by inventing an algorithm and analyzing what it does.

Summary: If the feasible region has no degeneracy, the simplex method can be started at any extreme point and will traverse a sequence of adjacent extreme points until it terminates, in a finite number of steps, at a ray of undboundedness or an extreme point optimal solution. The next section deals with the evident weakness of the algorithm as stated. *Question: What is the weakness?*[6]

5.2 Degeneracy

Preview: This section completes the rigorous derivation of the simplex method by showing two different ways that make it work correctly when there is degeneracy.

Recall that a BFS $\mathbf{x} \in \Re^n$ is degenerate if there are more than n constraints binding at $\mathbf{x}$. In the simplex method, because the LP constraints have been cast in the form $A\mathbf{x} = \mathbf{b}; \mathbf{x} \geq \mathbf{0}$, degeneracy manifests as one or more basic variables equal to 0. We say that these basic variables are degenerate. In general, degeneracy arises when there is a tie for the leaving basic variable, during Step 5 in Table 5.1. Two or more basic variables hit 0 simultaneously as the entering nonbasic variable is increased. Geometrically, degeneracy arises when we hit two or more constraints simultaneously as we traverse the edge. Once we are degenerate, it is easy to remain so. All we need is for a degenerate basic variable x_i to have $h_i \leq 0$ in Step 4 of the algorithm. *Question: Why did I write $h_i \leq 0$ rather than $h_i < 0$?*[7] If $h_i = 0$, which is likely if the data are sparse, the basic variable remains in the basis at value zero,

[6] Read the next section.
[7] Read on.

even if the entering variable becomes strictly positive. If $h_i < 0$ the entering variable takes value zero.

Theoretically speaking, we care about degeneracy because we want the simplex method to terminate. The proof of termination depends on the objective function strictly improving at each step, whence no choice of basis can be repeated. There are finitely many choices of basis, hence the algorithm terminates finitely. If there is degeneracy, it is possible for the simplex method to traverse a sequence of basic feasible solutions that loops back on itself. The simplex method gets stuck in an infinitely repeating cycle. This *cycling* phenomenon can occur. Alan Hoffman constructed the first example of cycling [151]; a nice geometric explanation of Hoffman's example is given by Jon Lee [204] and precise sufficient ranges on some parameter values are given by Guerrero-Garcia and Santos-Palomo [139]. Beale later constructed a smaller example based on the dual simplex method [21]. Kotiah and Steinberg report that cycling occurred in practice in instances of dimension roughly 20 by 15 [194]; they further suggest that numerical errors due to the limited precision of computer calculation prevent cycling from occurring more often; their papers met with some criticism (see e.g., [115]). Now, decades later, their research is still somewhat controversial.

5.2.1 The Perturbation/Lexicographic Method

There are two main ways to resolve the problem of potential cycling. The first is the perturbation/lexicographic method, and the second is Bland's rule.

The perturbation method considers a slightly perturbed problem, changing the original right hand side $\mathbf{b}$ so that at the initial basis, $\mathbf{x}_{\mathbb{B}} = B^{-1}\mathbf{b} + (\epsilon, \epsilon^2, \ldots \epsilon^m)$. Imagine $\epsilon > 0$ to be infinitesimally small, like a dx term in calculus. As a first-order approximation, $\mathbf{x}_{\mathbb{B}} = B^{-1}\mathbf{b}$, which might contain values equal to 0. However, each term in $\mathbf{x}_{\mathbb{B}}$ has been increased by a power of $\epsilon > 0$. Each term in $\mathbf{x}_{\mathbb{B}}$ is strictly positive. Hence, the initial basic feasible solution is not degenerate. We will prove that the basic solution is not degenerate in any subsequent iterations, as well.

Figure 5.5 depicts an intuitive explanation of the perturbation method. Before the LP is perturbed, the point $\mathbf{y}$ is degenerate because constraints i, ii, iii are all binding at $\mathbf{y}$. Let x_i, x_{ii}, x_{iii} be the slack variables for these constraints, respectively. Then the three choices of basis, (x_1, x_2, x_i); (x_1, x_2, x_{ii}); (x_1, x_2, x_{iii}) all yield the same point $\mathbf{y}$. Perturbing the right-hand sides of constraints i, ii, iii splits $\mathbf{y}$ into three distinct points, one for each of the three choices of basis, and none of them degenerate. In Figure 5.5 you should see that $\mathbf{y}_i$ is the point for which $\mathbf{x}_i$ is basic, because constraint i is not binding there.

Before we consider subsequent basic feasible solutions, let's define $\hat{\mathbb{B}}$ to be the initial basis indices. Let $\hat{B}$ denote $A_{:,\hat{\mathbb{B}}}$, the initial basis matrix. The perturbed right-hand side values are therefore $\hat{\mathbf{b}} \equiv \mathbf{b} + \hat{B}(\epsilon, \epsilon^2, \ldots \epsilon^m)$. *Question: Why?*[8]

If the current basis during the simplex algorithm is not degenerate, the only way that the next basis can be degenerate is if there is a tie in the minimum ratio test to determine the exiting basic variable. *Question: Why?*[9] Therefore, all we must prove is that there is no tie.

The current basic feasible solution is

$$\mathbf{x}_{\mathbb{B}} = B^{-1}\hat{\mathbf{b}} = B^{-1}\mathbf{b} + \epsilon B^{-1}\hat{B}_{:,1} + \epsilon^2 B^{-1}\hat{B}_{:,2} + \ldots + \epsilon^m B^{-1}\hat{B}_{:,m}. \tag{5.4}$$

If there is no tie in the first-order approximation of the minimum ratio test, the ϵ terms are too small to create a tie. If there is a first-order tie, we look to the terms of order

[8]Because $\hat{B}^{-1}\hat{\mathbf{b}} = \hat{B}^{-1}\mathbf{b} + \hat{B}^{-1}\hat{B}(\epsilon, \epsilon^2, \ldots \epsilon^m) = \hat{B}^{-1}\mathbf{b} + (\epsilon, \epsilon^2, \ldots \epsilon^m) = x_{\mathbb{B}}$.

[9]The entering variable equals $\theta > 0$ because $\mathbf{x}_{\mathbb{B}} > 0$. Since there is no tie for the exiting variable, all other basic variables are not 0 at the value of θ for which the exiting variable reaches 0.

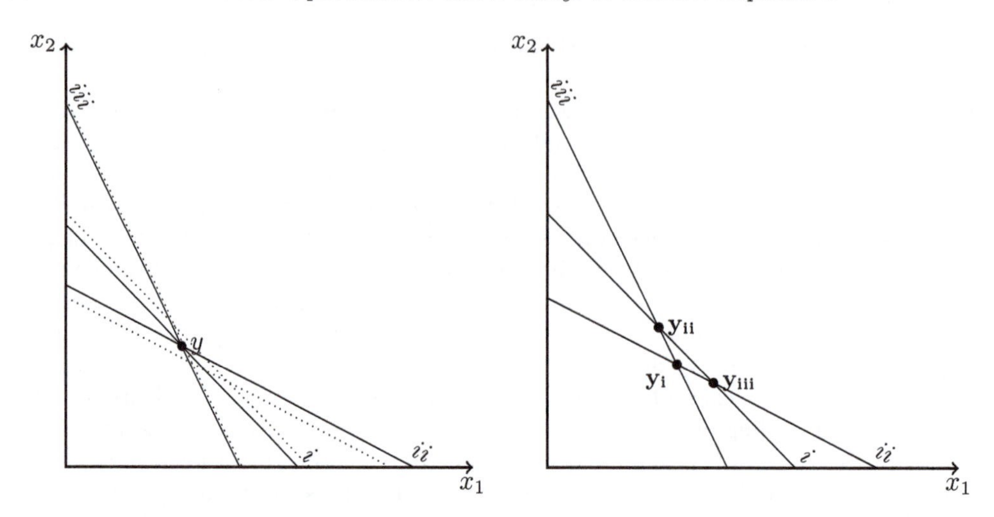

FIGURE 5.5: Perturb the binding constraints at the degenerate point y to split it into distinct nondegenerate points.

ϵ to break the tie. That is, if two or more basic variables have equal first-order ratios $B^{-1}_{i,:}b/B^{-1}_{i,:}A_{:,r}$, then we compare the ratios

$$B^{-1}{}_{i,:}\hat{B}_{:,1}/B^{-1}_{i,:}A_{:,r}. \tag{5.5}$$

This comparison appears to be computationally elegant. We already have computed the denominator terms $B^{-1}_{i,:}A_{:,r}$ to perform the first-order approximation, and the numerator terms are simply dot products of rows of B^{-1}, which presumably we already have computed, times the first column of $\hat{B}$, which we have on hand. Whichever of the tied variables has the least ratio – note that a ratio can be < 0 – is the first to reach zero, hence is the variable to leave the basis. In case there is a tie for the least ratio of the ϵ terms, we consider only those that are tied, and look to the terms of order ϵ^2 to break the tie. We continue in this way until the tie is broken. To compare tied variables at order ϵ^k we use the ratios

$$B^{-1}{}_{i,:}\hat{B}_{:,k}/B^{-1}_{i,:}A_{:,r}. \tag{5.6}$$

The tie must be broken eventually because otherwise for two rows i, j these ratios are all equal:

$$B^{-1}_{i,:}\hat{B}_{:,k}/B^{-1}_{i,:}A_{:,r} = B^{-1}_{j,:}\hat{B}_{:,k}/B^{-1}_{j,:}A_{:,r} \qquad \forall\, 1 \le k \le m. \tag{5.7}$$

Because the columns of $\hat{B}$ form a basis, this would imply that two rows of B^{-1} are multiples of each other, contradicting the nonsingularity of B^{-1}.

We have proved that the perturbed LP never experiences degeneracy. Therefore, Theorem 5.1.2 will apply to the perturbed problem. From the point of view of computation, we do not actually perturb the values of b. We just perform the thought experiment of perturbing the values. Then we make the same choices for the leaving basic variable that the simplex method would have made on the perturbed problem. On the perturbed problem we would have encountered no degeneracy, hence we would have terminated finitely. Since we make the same choices of basis as would the simplex method on the perturbed problem, we too terminate finitely.

Theorem 5.2.1 *If the perturbation method is used to select the exiting basic variable, the simplex algorithm will terminate in a finite number of steps at either an optimal basic feasible solution or a ray of unboundedness.*

Together with Theorem 5.1.1, Theorem 5.2.1 gives a proof of the strong duality theorem, which was stated without proof in Chapter 2. Indeed, we have the stronger result that there are optimal basic feasible solutions. The following version of the strong duality theorem combines Theorems 2.1.3 and 4.4.8.

Theorem 5.2.2 *If both primal and dual LPs* (2.1) *and* (2.2) *are feasible, they both have optimal basic feasible solutions, and those solutions have the same objective function value.*

Proof: The primal feasible region contains no lines because of the constraints $\mathbf{x} \geq 0$. By Theorem 4.4.8 the primal has at least one basic feasible solution. Apply the simplex algorithm with the perturbation method. Since the dual is feasible, Theorem 2.1.1 implies that the primal cannot have a ray of unboundedness. By Theorem 5.2.1 the simplex algorithm terminates at an optimal primal BFS. In the perturbed primal, all basic variables are nonzero. By complementary slackness and the nonsingularity of B, the corresponding dual solution is basic. The result now follows from Theorem 5.1.1. ∎

Lexicography: Since we don't actually perturb the values of $\mathbf{b}$, we can think of the perturbation method as replacing the right-hand-side vector $\mathbf{b}$ by the matrix $[\mathbf{b}|I]$. Each $\mathbf{b}_i$ has been replaced by a vector. We attempt to break ties in the first component of the vectors by considering the second component; we attempt to break remaining ties by considering the third component, and so forth. Think of words as vectors of letters. This tie-breaking rule is the same by which words are ordered alphabetically. That is why Dantzig, Orden, and Wolfe called it *lexicographic* ordering [71]. Breaking ties lexicographically is the same algorithm as the perturbation method. It is a different interpretation of the same algorithm.

Later, when we discuss computational issues, we will see that we might not have the coefficients of B^{-1} at hand explicitly. Moreover, even if we did, we might have to make on the order of m iterations each requiring on the order of m comparisons, where m is the number of rows. The perturbation method then loses some of its attractiveness. We turn now to Bland's rule, which is remarkably simple to implement.

5.2.2 Bland's Rule

Bland invented the following rule to avoid cycling. For the entering variable, pick the least index that has "good" reduced cost. For the leaving variable, if there is a tie, pick the least index.

Theorem 5.2.3 *If Bland's rule is used when the current basic feasible solution is degenerate, the simplex method will terminate in a finite number of steps at either an optimal basic feasible solution or a ray of unboundedness.*

Proof: This is Bland's original proof translated into revised simplex method notation. When the current basic feasible solution is not degenerate, the objective function will strictly improve in the ensuing iteration. Since the simplex method never worsens the objective function, there is no possibility of cycling. Therefore, we only must examine the situation where there is degeneracy and Bland's rule is invoked.

We will obtain a contradiction from the assumption that a cycle occurs. Let m be the number of rows and n the number of variables in the problem $\max \mathbf{c} \cdot \mathbf{x} : A\mathbf{x} = \mathbf{b}, \mathbf{x} \geq \mathbf{0}$. Let $T \subseteq \{1 \ldots n\}$ be the index set of all variables that enter the basis during the cycle. If $j \notin T$, then without loss of generality x_j is basic throughout the cycle, otherwise we could discard the column altogether. Let $q = \max j : j \in T$ be the largest index of any variable that enters during the cycle. Let $\mathbb{B}$ be the choice of basis when x_q enters. Let $Y = y_0, \ldots y_n$ be defined by $y_0 = 1; y_j = \mathbf{c}_{\mathbb{B}}^T B^{-1} A_{:,j} - c_j : j = 1 \ldots n$. (These are the dual variables). Since we are maximizing and x_q enters, then $y_q < 0$ and $y_j \geq 0 \ \forall j < q$ (otherwise y_j would have been chosen to enter). Also, Y is a linear combination of the rows of the matrix

$$A' = \begin{pmatrix} 1 & -\mathbf{c}^T \\ 0 & A \end{pmatrix} \tag{5.8}$$

Let $\bar{\mathbb{B}}$ be the basis indices when x_q leaves the basis (since it enters during the cycle it must also exit). Let $\bar{B}$ be the corresponding submatrix of A. Say x_t enters and say $q = \bar{\mathbb{B}}_r$ (x_q is the rth basic variable). Define $Z = z_0, \ldots, z_n$ with $z_0 = c_{\bar{\mathbb{B}}} \bar{B}^{-1} A_{:,t} - c_t$, $z_{\bar{\mathbb{B}}_i} = \bar{B}^{-1}_{i,:} A_{:,t}$ for the basic variables $\bar{\mathbb{B}}_1 \ldots \bar{\mathbb{B}}_m$, with $z_t = -1$ and $z_j = 0$ otherwise. (This vector is the synthetic carrot minus the carrot (see Section 5.1), with the cost difference between the real and synthetic carrot at the beginning of the vector). This cost difference $z_0 < 0$ since x_t enters the basis.

We claim that $A'Z = 0$. The top row of A' dotted with Z equals $(1)(\mathbf{c}_{\bar{\mathbb{B}}} \bar{B}^{-1} A_{:,t} - c_t) - \mathbf{c}_{\bar{\mathbb{B}}} \bar{B}^{-1} A_{:,t} + (-c_t)(-1) = 0$; the other rows of A' times Z are $\bar{B}\bar{B}^{-1} A_{:,t} + (-1) A_{:,t} = 0$. This proves the claim.

Now, since $Y = \boldsymbol{\pi}^T A'$ for some $\boldsymbol{\pi}$, $Y \cdot Z = 0$. From $y_0 = 1$ and $z_0 < 0$ there must be $y_j z_j > 0$ for some $j, 1 \leq j \leq n$. From $y_j \neq 0$ we have $j \notin \mathbb{B}$; from $z_j \neq 0$ we have $j \in \bar{\mathbb{B}}$ or $j = t$, whence $j \in T$ and $j \leq q$. But j can't equal q because $y_q z_q < 0$. Hence $j < q$, so from the definition of Y, $y_j > 0$ whence $z_j > 0$. Since $z_t = -1$, $j \neq t$. Therefore, $j \in \bar{\mathbb{B}}$. Let p be such that $j = \bar{\mathbb{B}}_p$, that is, j is the pth basic variable when x_q leaves the basis.

We claim that when x_q leaves the basis $\bar{\mathbb{B}}$, variable x_j is eligible to leave the basis, which will contradict the least index choice rule as $j < q$. Proof of claim: every pivot in the cycle must be degenerate, for otherwise the objective value would increase. Therefore, every variable that enters the basis during the cycle does so at value 0, and remains at value 0 throughout the cycle. Therefore, $(\bar{B}^{-1}\mathbf{b})_p = 0$. We already know that $\bar{B}^{-1} A_{:,t_p} = Z_{\bar{\mathbb{B}}_p} = z_j > 0$. This proves the claim and yields our contradiction. ■

As a corollary to Theorem 5.2.1 or Theorem 5.2.3, we have obtained an alternative proof of Theorem 4.4.8.

Summary: Degeneracy only endangers the correctness and finite termination of the simplex method because it makes cycling possible. The perturbation method prevents cycling by following the sequence of choices of basic variables that the simplex method would take on a nearly identical polyhedron that is not degenerate. The lexicographic method is conceptually different but algorithmically identical to the perturbation method. Bland's rule is simpler and requires much less computation per step.

5.3 The Full Tableau Method

Preview: This section illustrates the simplex method with a computationally useless but conceptually useful method.

The full tableau simplex method uses more computer time and memory to make the same choices as the regular simplex method, often with less numerical stability. You might wonder why I include this computationally obsolete method in this book. I do so because it is a great way to understand the simplex algorithm.

In matrix algebraic terms, the method works as follows. Suppose the LP originally has the form $\min \mathbf{c}' \cdot \mathbf{x}'$ subject to $A'\mathbf{x}' = \mathbf{b}', \mathbf{x}' \geq \mathbf{0}$. Add a new unrestricted variable x_0 giving $\mathbf{x} = (x_0, \mathbf{x}')$ as the set of variables. Add a new constraint $x_0 - c' \cdot x' = 0$ to the constraint set, and call the resulting constraints $A\mathbf{x} = \mathbf{b}$. By convention, the new constraint is the first row, $A_{0,:}\mathbf{x} = b_0$. The new objective is to minimize x_0, which we have just forced to equal $\mathbf{c}' \cdot \mathbf{x}'$. At each iteration, the tableau is a table of the coefficients of the system

$$B^{-1}A\mathbf{x} = B^{-1}\mathbf{b}. \tag{5.9}$$

The tableau (5.9) is filled with helpful features. The leftmost column always has a 1 in row 0 and is 0 everywhere else. Sometimes it is omitted, because it is always the same, but I like to keep it as a reminder. Since B is a submatrix of A, the tableau always contains the columns of an identity matrix, in exactly the columns indexed by $\mathbb{B}$. One glance at the tableau tells you the index set $\mathbb{B}$. The rightmost column contains the current basic solution values $\mathbf{x}_{\mathbb{B}} = B^{-1}b$. The values $B^{-1}A_{:,k}$ needed for the minimum ratio test are ready for us in column k.

The prettiest feature of the tableau is where to find the reduced costs that we need to determine the entering variable. The cost vector has the simple form $\mathbf{c} = (1, 0, 0, \ldots 0)$. The first index in $\mathbb{B}$ is 0. Hence $\mathbf{c}_{\mathbb{B}} = (1, 0, \ldots, 0)$. Let $j \notin \mathbb{B}$. Then $j \geq 1$ hence $c_j = 0$. The reduced cost of x_j is therefore $c_j - c_{\mathbb{B}}^T B^{-1}A_{:,j} = 0 - (1, 0, \ldots, 0)^T B^{-1}A_{:,j} = (-B^{-1}A_{:,j})_0$. That is, the reduced cost of x_j is the number in row 0, column j of the current tableau, times -1.

This reduced cost has a natural interpretation in terms of the original LP $\min \mathbf{c}' \cdot \mathbf{x}'$ subject to $A'\mathbf{x}' = \mathbf{b}', \mathbf{x}' \geq 0$. Let $\mathbb{B}'$ denote the index set $\mathbb{B} \setminus \{0\}$, and let B' denote the submatrix $A'_{\mathbb{B}'}$. The reduced cost of x_j in the original LP is $c'_j - \mathbf{c}'^T_{\mathbb{B}'}(B')^{-1}A'_{:,j}$. A simple calculation shows that since

$$[B] = \begin{bmatrix} 1 & -(\mathbf{c}'_{\mathbb{B}'})^T \\ 0 & B' \end{bmatrix}, \tag{5.10}$$

then

$$[B^{-1}] = \begin{bmatrix} 1 & (\mathbf{c}'_{\mathbb{B}'})^T(B')^{-1} \\ 0 & (B')^{-1} \end{bmatrix}. \tag{5.11}$$

Therefore,

$$[B^{-1}]\, A_{:,j} = \begin{bmatrix} 1 & (\mathbf{c}'_{\mathbb{B}'})^T(B')^{-1} \\ 0 & (B')^{-1} \end{bmatrix} \begin{pmatrix} -\mathbf{c}'_j \\ A'_{:,j} \end{pmatrix} = \begin{pmatrix} -\mathbf{c}'_j + \mathbf{c}'^T_{\mathbb{B}'}(B')^{-1}A'_{:,j} \\ (B')^{-1}A'_{:,j} \end{pmatrix}. \tag{5.12}$$

In words, column j of the full tableau consists of what would be the reduced cost in the original LP of x_j times -1 placed above the column vector that could be computed in the original LP for x_j, namely $(B')^{-1}A'_{:,j}$. All of the numbers needed to choose the entering and exiting variable are in the full tableau.

Once the entering and exiting variables x_k and x_r have been selected, the full tableau needs to be updated to keep Equation (5.9) true. This consists simply of the elementary row operations needed to create the columns of an identity matrix in the columns corresponding to the new basis B. All but one of these columns are already in place. The missing column is column k, and the column we don't mind changing is column r. Let t denote the row of the tableau where column r has a 1. For reasons that will become clear in a later chapter, this row is called the "pivot" row. (The value t is the place r holds in the index set $\mathbb{B}$.) Divide row t by the number in row t, column k. By Step 4 in Table 5.1 this number, called the "pivot term," is strictly positive. This creates a 1 in row t of column k. Next, add appropriate multiples of row t to all other rows, including row 0, to make the rest of the terms in column k equal 0. The iteration is now complete.

Example 5.2 The original LP is

$$
\begin{array}{rcl}
\min 25x_1 - 40x_2 & \text{subject to} & \\
3x_1 + x_2 + x_3 & = & 200 \\
-2x_1 + 5x_2 + \quad x_4 & = & 500 \\
-4x_1 + 2x_2 + \qquad x_5 & = & 100
\end{array}
$$

Adding the variable $x_0 = 25x_1 - 40x_2$, the following tableau contains the coefficients of the LP constraints. The objective is to minimize x_0.

x_0	x_1	x_2	x_3	x_4	x_5	*RHS*
1	−25	40	0	0	0	0
0	3	1	1	0	0	200
0	−2	5	0	1	0	500
0	−4	**2**	0	0	1	100

(5.13)

The tableau above represents the basic feasible solution $B = \{0, 3, 4, 5\}$, $\mathbf{x}_{\mathbb{B}} = (x_0, x_3, x_4, x_5) = (0, 200, 500, 100)$. The nonbasic variable x_2 has favorable reduced cost -40; the nonbasic variable x_1 has unfavorable reduced cost 25. Therefore, x_2 is the entering variable. Since 1, 5 and 2 are all positive, the exiting variable is selected by finding the minimum of $\frac{200}{1}$, $\frac{500}{5}$ and $\frac{100}{2}$. Therefore, x_5 exits the basis. The new set of basic variables will be $\mathbf{x}_{\mathbb{B}} = (x_3, x_4, x_2)$. The pivot row is the bottom row, row 3. The pivot term equals 2, the number in the x_2 column of the pivot row.

Update the tableau with elementary row operations so that the columns of the identity matrix are found in the columns of x_3, x_4, x_2. Before performing these operations, let's predict the values of the new basic solution. The nonbasic variables x_1 and x_5 will remain 0. The value of θ is $\frac{100}{2} = 50$. Therefore, the incoming basic variable x_2 will equal $\theta = 50$. The objective function value will decrease from 0 to $0 - 40\theta = -2000$. From the other values in the x_2 column, the basic variable x_3 will decrease from 200 to $200 - (1)\theta = 150$, while the basic variable x_4 will decrease from 500 to $500 - 5\theta = 250$. The first elementary row operation divides the pivot row by the pivot value 2.

x_0	x_1	x_2	x_3	x_4	x_5	*RHS*
1	−25	40	0	0	0	0
0	3	1	1	0	0	200
0	−2	5	0	1	0	500
0	**−2**	**1**	**0**	**0**	**0.5**	**50**

(5.14)

The next elementary row operation subtracts 40 times the pivot row from row 0.

x_0	x_1	x_2	x_3	x_4	x_5	*RHS*
1	55	0	0	0	−20	−2000
0	3	1	1	0	0	200
0	**−2**	**5**	**0**	**1**	**0**	**500**
0	−2	1	0	0	0.5	50

(5.15)

The last two row operations subtract 1 times the pivot row from row 1 and 5 times the pivot row from row 2.

x_0	x_1	x_2	x_3	x_4	x_5	*RHS*
1	**55**	**0**	**0**	**0**	**−20**	**−2000**
0	**5**	**0**	**1**	**0**	**−0.5**	**150**
0	8	0	0	1	−2.5	250
0	−2	1	0	0	0.5	50

(5.16)

The identity matrix is now located in the columns of x_0, x_3, x_4, x_2, in that order. The basic solution is now $\mathbf{x}_{\mathbb{B}} = (x_0, x_3, x_4, x_2) = (-2000, 150, 250, 50)$. The variable x_1, which previously had an unfavorable reduced cost 25, now has a favorable reduced cost -55. The formerly basic variable x_5 has an unfavorable reduced cost 20. *Question: Why is this reduced cost unfavorable?*[10] The value 20 equals $0 - (40)(1)/2$. In general, the reduced cost of the variable that exited the basis equals the negative of the previous reduced cost of the variable that entered the basis, divided by the pivot term. Since the pivot term is always strictly positive, the reduced cost of the exited variable is always unfavorable. Therefore, a variable can not leave the basis in one iteration and reenter the basis in the next iteration. However, it is possible for the variable to reenter the basis later in the future.

In the next iteration of the simplex algorithm, x_1 enters the basis, and x_3 exits the basis at $\theta = 30$, because $\frac{150}{5} < \frac{250}{8}$. *Question: What will be the values of x_4 and x_0?*[11] *What will the value of x_2 be?*[12] After performing elementary row operations, the new tableau is as follows.

x_0	x_1	x_2	x_3	x_4	x_5	*RHS*
1	0	0	−11	0	−14.5	−3650
0	1	0	0.2	0	−0.1	30
0	0	0	−1.6	1	−1.7	10
0	0	1	0.4	0	0.3	110

(5.17)

All reduced costs are nonnegative. *Question: Why are the reduced costs of basic variables of the original LP all zero?*[13] The tableau displays the optimal solution.

Summary: The simplex method for optimizing $\mathbf{c} \cdot x$ subject to $A\mathbf{x} =$

[10] Read on.
[11] $250 - 8\theta = 10$ and $-2000 - 55\theta = -3650$.
[12] $50 + 2\theta = 110$.
[13] $c'^T_{\mathbb{B}'} - c'^T_{\mathbb{B}'}(B')^{-1}B' = 0$.

$\mathbf{b}; x \geq 0$ can be seen as consisting entirely of elementary row operations on the system of equations $A\mathbf{x} = \mathbf{b}; x_0 - c \cdot x = 0$. Doing a few of these operations by hand, or coding them in a programming language, will help you to understand the algorithm well.

5.4 Phase I: Getting an Initial Basic Feasible Solution

Preview: Although we've proved the correctness of the simplex algorithm, and we've proved the strong duality theorem, we are still missing one component of the algorithm itself. How do we start it? How do we determine whether or not the feasible region is empty, and if it is not empty, how do we find an initial extreme point?

Usually no initial basic feasible solution is immediately available. In fact, often an LP model is infeasible and you don't know it. To get around this problem, we use the simplex method to start up the simplex method. We temporarily expand the feasible region so that there is an obvious BFS. Then we use a temporary objective to force us out of the expanded region into the true feasible region. The use of temporary variables and objective is called *Phase I*; optimizing with respect to the true objective after Phase I has found an initial basic feasible solution is called *Phase II* of the simplex method. *Question: Why do we not also worry about whether or not the feasible region is pointed?*[14]

The idea is simple but the notation is a little ugly, so let's do an example first.

Example 5.3 The initial system of constraints is:

$$\begin{aligned} 2x_1 + 3x_2 &\leq 4 \\ 4x_1 - 5x_2 &\leq -6 \\ -7x_1 + 8x_2 &\leq -9 \\ 10x_1 + 11x_2 &= 12 \\ x_1, x_2 &\geq 0 \end{aligned}$$

Add nonnegative slack variables x_3, x_4, x_5 to get the following equivalent system.

$$2x_1 + 3x_2 + x_3 = 4 \tag{5.18}$$

$$4x_1 - 5x_2 + x_4 = -6 \tag{5.19}$$

$$-7x_1 + 8x_2 + x_5 = -9 \tag{5.20}$$

$$10x_1 + 11x_2 = 12 \tag{5.21}$$

$$x_i \geq 0 : i = 1 \ldots 5 \tag{5.22}$$

[14]Because we've converted the LP constraints to the form $A\mathbf{x} = \mathbf{b}; x \geq 0$, which admit no line.

Add temporary (*artificial*) variables y_1, y_2 to constraints (5.19, 5.20) because their slack variables would be negative if we made them basic. Add temporary (artificial) variable y_3 to constraint (5.21), which has no slack variable. The temporary variables expand the feasible region. They allow points to be on the "wrong" side of their constraints.

$$\begin{aligned}
2x_1 + 3x_2 + x_3 &= 4\\
4x_1 - 5x_2 + x_4 - y_1 &= -6\\
-7x_1 + 8x_2 + x_5 - y_2 &= -9\\
10x_1 + 11x_2 + y_3 = 12&\\
x_i \geq 0 : i = 1 \ldots 5; \quad & y_1, y_2, y_3 \geq 0
\end{aligned}$$

The new system possesses the basic feasible solution

$$\begin{aligned}
x_1 = x_2 = x_4 = x_5 &= 0\\
x_3 &= 4\\
y_1 &= 6\\
y_2 &= 9\\
y_3 &= 12
\end{aligned}$$

In this new system, starting with that basic feasible solution, we apply the simplex method to minimize $y_1 + y_2 + y_3$. If a feasible solution to the original system exists, then a feasible solution to the new system with $y_1 = y_2 = y_3 = 0$ exists. Because the y_i are constrained to be nonnegative, that feasible solution must minimize $y_1 + y_2 + y_3$. Conversely, if a feasible solution to the new system with objective value 0 exists, it must have $y_1 = y_2 = y_3 = 0$, and the values of $x_i : i = 1 \ldots 5$ must be feasible in the original system.

In general, set up Phase I as follows:

Simplex Algorithm Phase I Setup

1. Add slack variables to all inequality constraints to convert them to equality constraints.
2. Multiply all constraints with negative right-hand-side values by -1. The constraints should now be in form $A\mathbf{x} = \mathbf{b}; \mathbf{x} \geq \mathbf{0}$ where $\mathbf{b} \geq \mathbf{0}$.
3. For each constraint whose slack variable has coefficient -1, and for each constraint which has no slack variable, add a temporary variable $y_j \geq 0$ with coefficient 1 to the left-hand-side. Set a different value of j for each such constraint.
4. In Phase I, ignore the true objective function. Instead, minimize $\sum_j y_j$.
5. The initial basic feasible solution for Phase I comprises all the y_j and the slack variables whose coefficients equal 1.

If you are using the full tableau method, you must first perform elementary row operations to force the basic variables to have reduced costs equal to zero. These operations are particularly simple: just add each row containing a temporary variable to the top row.

Example 5.4 Set up Phase I from the following constraints.

$$\begin{array}{rrcr} 3x_1 & +4x_2 & = & 100 \\ 5x_1 & -6x_2+7x_3 & \geq & 200 \\ 8x_1 & +9x_2-10x_3 & \leq & 300 \\ -11x_1 & +12x_2+13x_3 & \leq & -400 \\ 14x_1 & +15x_2-16x_3 & \geq & -500 \\ -17x_1 & -18x_2+19x_3 & = & -600 \\ & x_i \geq 0: \ i=1,2,3 & & \end{array}$$

Add slack variables to get an equivalent system.

$$\begin{array}{rllllll} 3x_1 & +4x_2 & & & & & = 100 \\ 5x_1 & -6x_2+7x_3-x_4 & & & & & = 200 \\ 8x_1 & +9x_2-10x_3 & +x_5 & & & & = 300 \\ 11x_1 & -12x_2-13x_3 & & +x_6 & & & = -400 \\ 14x_1 & +15x_2-16x_3 & & & -x_7 & & = -500 \\ -17x_1 & -18x_2+19x_3 & & & & & = -600 \\ & x_i \geq 0: \ i=1,\ldots,7 & & & & & \end{array}$$

Multiply the equality constraints that have negative right-hand-side values by -1 to make them nonnegative.

$$3x_1 \quad +4x_2 \quad = 100 \tag{5.23}$$
$$5x_1 \quad -6x_2+7x_3-x_4 \quad = 200 \tag{5.24}$$
$$8x_1 \quad +9x_2-10x_3 \quad +x_5 \quad = 300 \tag{5.25}$$
$$-11x_1 \quad +12x_2+13x_3 \quad -x_6 \quad = 400 \tag{5.26}$$
$$-14x_1 \quad -15x_2+16x_3 \quad +x_7 \quad = 500 \tag{5.27}$$
$$17x_1 \quad +18x_2-19x_3 \quad = 600 \tag{5.28}$$
$$x_i \geq 0: \ i=1,\ldots,7 \tag{5.29}$$

Add temporary variables to equations (5.23, 5.24, 5.26, 5.28) to get the equivalent system

$$3x_1 \quad +4x_2 \quad +y_1 \quad = 100 \tag{5.30}$$
$$5x_1 \quad -6x_2+7x_3-x_4 \quad +y_2 \quad = 200 \tag{5.31}$$
$$8x_1 \quad +9x_2-10x_3 \quad +x_5 \quad = 300 \tag{5.32}$$
$$-11x_1 \quad +12x_2+13x_3 \quad -x_6 \quad +y_3 \quad = 400 \tag{5.33}$$
$$-14x_1 \quad -15x_2+16x_3 \quad +x_7 \quad = 500 \tag{5.34}$$
$$17x_1 \quad +18x_2-19x_3 \quad +y_4 \quad = 600 \tag{5.35}$$
$$x_i \geq 0, y_j \geq 0: \ i=1,\ldots,7; \ j=1,\ldots,4 \tag{5.36}$$

In full tableau format these equations, together with objective function $x_0 = y_1 + y_2 + y_3 + y_4$, appear as follows.

x_0	x_1	x_2	x_3	x_4	x_5	x_6	x_7	y_1	y_2	y_3	y_4	RHS
1	0	0	0	0	0	0	0	−1	−1	−1	−1	0
0	3	4	0	0	0	0	0	1	0	0	0	100
0	5	−6	7	−1	0	0	0	0	1	0	0	200
0	8	9	−10	0	1	0	0	0	0	0	0	300
0	−11	12	13	0	0	−1	0	0	0	1	0	400
0	−14	−15	16	0	0	0	1	0	0	0	0	500
0	17	18	−19	0	0	0	0	0	0	0	1	600

(5.37)

Adding the rows for y_1, y_2, y_3, y_4 to the row for x_0 gives the new top row:

x_0	x_1	x_2	x_3	x_4	x_5	x_6	x_7	y_1	y_2	y_3	y_4	RHS
1	14	28	1	−1	0	−1	0	0	0	0	0	1300

(5.38)

The optimal solution to Phase I has strictly positive value iff the temporary variables cannot all be driven down to zero, meaning that there does not exist a feasible solution to the original system. In this case, your likely next step will be to diagnose the infeasibility (see Section 3.5).

If the optimal value is 0, you have a basic feasible solution that is feasible in the original system. Moreover, all temporary variables equal zero. Throw out the temporary variables and temporary objective, bring in the original objective (which has been irrelevant until now), and apply the simplex method.

5.4.1 Degeneracy at the End of Phase I

There is one complication that seems picayune, but actually occurs frequently in practice. If there is no degeneracy, a Phase I optimal value of zero forces all temporary variables y_i to be nonbasic. Therefore, we have a BFS for the original system. But if there is degeneracy there could be temporary variables that are in the basis but equal to zero. These degenerate temporary variables must be driven out of the basis.

The procedure is as follows.

Algorithm 5.4.1 Procedure to remove degenerate temporary variables from the basis

Step 1 *Remove all nonbasic temporary variables from the system.*

Step 2 *While there are temporary basic variables:*

1. *Choose an arbitrary basic temporary variable y_i and a nonbasic non-temporary variable x_j such that the value in y_i's row of $B^{-1}A_{:,j}$ is nonzero.*
2. *Swap the latter for the former into the basis.*
3. *Remove the newly nonbasic temporary variable from the system.*

The only difficulty that can arise is if all nonbasic variables have a 0 term in the row of $B^{-1}A$ that corresponds to the basic temporary variable. But in that case the constraint represented by that row is vacuous and may be removed.

Example 5.5 Here is the final tableau of Phase I, where the temporary variables are y_1, y_2, y_3. Two of the temporary variables are equal to zero, but basic.

x_0	x_1	x_2	x_3	x_4	x_5	y_1	y_2	y_3	RHS
1	0	−2	−6	3	−1	0	−4	0	0
0	0	−1	0	1/2	4	1	3/2	0	0
0	1	3	0	1/2	2	0	−4	0	11/2
0	0	−2	−1/2	0	**−1**	0	−3	1	0

(5.39)

Remove nonbasic temporary variable y_2 from the system. To drive temporary variable y_3 out of the basis, any nonzero coefficient on a non-temporary variable suffices, regardless of its sign and regardless of the reduced cost of the variable. For the purpose of illustration, I put variable x_5 into the basis. The pivot term −1 is in boldface. The new tableau is shown next.

x_0	x_1	x_2	x_3	x_4	x_5	y_1	y_3	RHS
1	0	0	$-5\frac{1}{2}$	3	0	0	−1	0
0	0	−9	−2	**1/2**	0	1	4	0
0	1	−1	−1	1/2	0	0	2	11/2
0	0	2	1/2	0	1	0	−1	0

(5.40)

Remove nonbasic temporary variable y_3 from the system. To drive temporary variable y_1 out of the basis, we could put any of x_2, x_3, or x_4 into the basis. For illustration, I put x_4 into the basis. The pivot value 1/2 is in boldface. The new tableau is below. It was not necessary to compute the values in the y_1 column since we will remove that column from the system of equations.

x_0	x_1	x_2	x_3	x_4	x_5	y_1	RHS
1	0	54	$6\frac{1}{2}$	0	0	−6	0
0	0	−18	−4	1	0	2	0
0	1	8	1	0	0	−1	11/2
0	0	2	1/2	0	1	0	0

(5.41)

5.4.2 Phase II: Bringing in the Original Objective

Once the basis has no temporary variables, we bring in the original objective function, which we had ignored when we sought a basic feasible solution. From a theoretical point of view, this step is trivial. We have the basis B, so the reduced costs follow the usual formula $\mathbf{c}^T - \mathbf{c}_{\mathbb{B}}^T B^{-1}A$. But from a computational point of view this step is not trivial. When using the usual simplex method, the dual variable vector $\boldsymbol{\pi}^T = \mathbf{c}_{\mathbb{B}}^T B^{-1}$ must be computed. When using the full tableau method, you must perform elementary row operations to force all basic variables to have reduced costs of zero.

Example 5.6 Continuing the previous example, suppose that the true objective function vector is $\mathbf{c} = (4, -2, 1, 2, 0)$. Remove the y_1 column and replace the top row with the equation $x_0 - \mathbf{c} \cdot \mathbf{x} = 0$.

x_0	x_1	x_2	x_3	x_4	x_5	RHS
1	−4	2	−1	−2	0	0
0	0	−18	−4	1	0	0
0	1	8	1	0	0	11/2
0	0	2	1/2	0	1	0

(5.42)

Add 4 times the 3rd row and 2 times the 2nd row to make the tableau represent $B^{-1}A\mathbf{x} = B^{-1}b$.

x_0	x_1	x_2	x_3	x_4	x_5	RHS
1	0	−2	−5	0	0	22
0	0	−18	−4	1	0	0
0	1	8	1	0	0	11/2
0	0	2	**1/2**	0	1	0

(5.43)

Phase II of the simplex method may now proceed. If minimizing, the current solution is optimal. If maximizing, x_3 will enter the basis and x_5 will leave the basis. The new basic feasible solution will be degenerate. Both x_3 and x_4 will equal 0.

Big-M Method: Another technique to start the simplex algorithm, known as the big-M method, is to solve a single LP with a weighted sum of the temporary and the original objectives. The weight on the temporary objective is a large number M. This method is obsolete. It is bad computationally because it mixes numbers of greatly different magnitudes on a finite precision computer, and does not force degenerate temporary variables to leave the basis.

Summary: You should now understand the simplex method, from determining whether the feasible region is empty or not through finding an optimal extreme point solutions or a ray of unboundedness. You should be able to prove its correctness, and use its correctness to prove the strong duality theorem.

5.5 Notes and References

For an account of the development of the simplex method, see [70] or Chapter 2 of [64]. Appropriate credit for the strong duality theorem is obscured in the published literature. Dantzig [65] wrote and circulated the first statement and proof a few months after his October 1947 meeting with von Neumann. Dantzig did not publish it because he considered it to be von Neumann's result. That makes sense to me, because in my experience with really powerful mathematicians, when they say "A is true by applying B to C to get D, from which E gives you A," it may take me weeks to understand and verify, but they saw all of that in real time. Von Neumann had not just formulated the dual and the strong duality theorem; he had cited Farkas's Lemma, and also correctly conjectured LP duality to be mathematically equivalent to his and Morgenstern's minimax theorem for 2-person zero sum games [234], which Dantzig later verified [67]. However, the consequence was that the strong duality theorem was not published prior to 1951. The 1951 paper by Gale, Kuhn, and Tucker [111] is therefore often cited, particularly in the economics literature. On the other hand, for some pure mathematicians, strong duality is a direct consequence of Minkowski's 1896 theorem [229, 230], or Farkas's 1902 Lemma [91], Weyl's convex polyhedral theory [293], or Gordan's theorem [133].

Dantzig first proposed perturbation in a course he taught in 1950-1951 (see references in [64]). Abe Charnes [39] independently invented the same perturbation method. Dantzig later re-interpreted the tie-breaking method in terms of lexicographic ordering. Most of my students find the geometric perturbation description more intuitive.

5.6 Problems

E Exercises

1. Let $A \in R^{5 \times 6}$. Express, as a product of two matrices, the matrix resulting from A by:
   (a) multiplying the 4th row by 9.
   (b) adding the 2nd row to the 4th row.
   (c) subtracting $\frac{1}{2}$ times the 3rd row from the 1st row.
   (d) adding i times the 5th row to row i for all $i = 1, 2, 3, 4$ (resulting in one matrix, not four).
   (e) subtracting j times column 3 from column j for all $j = 1, 2, 3, 4, 5, 6$.

2. For the feasible region defined by the constraints $0 \leq x_1 \leq 3; 0 \leq x_2 \leq 5$:
   (a) Draw the feasible region.
   (b) Find all basic solutions and determine which are feasible.
   (c) Add slack variables and write the constraints in the form $Ax = b; x \geq 0$.
   (d) For each basic solution, find $\mathbb{B}$, B, and B^{-1}. Verify that $\mathbf{x}_{\mathbb{B}} = B^{-1}\mathbf{b}$.

3. For the feasible region in Exercise 2:
   (a) Write the constraints $Ax = b$ in tableau form. Verify that the tableau represents the basic solution $x_1 = x_2 = 0; x_3 = 3; x_4 = 5$ where x_3, x_4 are the slack variables.
   (b) From a drawing of the feasible region, identify which variable would leave the basis if x_2 were increased from zero, and identify what the new basic feasible solution would be.
   (c) Perform elementary row operations to make x_2 enter the basis. Verify that the new tableau represents the new basic feasible solution you identified.
   (d) Repeat 3b and 3c, increasing x_1 starting from the new solution.

(e) From the solution 3, 5, 0, 0 repeat 3b and 3c, increasing x_4.

(f) From the solution 3, 0, 0, 5 repeat 3b and 3c, increasing x_3. Verify that you have returned to $x_1 = x_2 = 0; x_3 = 3; x_4 = 5$.

4. Add the constraint $x_1+x_2 \leq 6$ and redo Exercises 2 and 3. Let x_5 be the slack variable for the added constraint. Start at $x_1 = x_2 = 0; x_3 = 3; x_4 = 5, x_5 = 6$. Bring variables into the basis in order x_2, x_1, x_4, x_5, x_3.

5. For the system $x_1 \geq 0, x_2 \geq 0, x_1 \leq 3, x_2 \leq 5, x_1 + x_2 \leq 6$:

   (a) Add slack variables x_3, x_4, x_5 and write the tableau for the basic solution $(0, 0, 3, 5, 6)$.

   (b) Identify geometrically which variable is negative at the basic solution with x_2, x_3, x_4 basic.

   (c) Do elementary row operations to bring x_2 into the basis and remove x_5 from the basis. Verify which basic variable is negative.

   (d) Do elementary row operations to bring x_1 into the basis and remove x_4. Verify geometrically that you have arrived at a basic feasible solution.

6. For the system $x_1 \geq 0, x_2 \geq 0, x_1 \leq 3, x_2 \leq 5, x_1+x_2 \leq 6$ and the objective $\max x_0 = 100x_1 + 120x_2$:

   (a) Draw the feasible region and compute the objective value of each basic feasible solution.

   (b) Write the initial simplex tableau.

   (c) Calculate the rate at which x_0 increases as you move from $x_1 = 0, x_2 = 0$ to $x_1 = 3, x_2 = 0$. Identify where the rate appears in the tableau.

   (d) Calculate the rate at which x_0 changes as x_2 is increased to move from $(3, 0)$ to $(3, 3)$. Repeat for x_3 increased to move from $(3, 3)$ to $(1, 5)$.

   (e) Perform two iterations of the simplex algorithm, first moving to $x_1 = 3, x_2 = 0$ and next moving to $x_1 = 3, x_2 = 3$. Identify where the rates you calculated appear in the tableaux.

7. If you think that $B^{-1}A_{:,j} \geq 0$ invent a small counterexample to your belief.

8. Memorize the reduced cost formula, $c_j - c_{\mathbb{B}}^T B^{-1} A_{:,j}$.

9. Let $\mathbf{c} = (10, 20, 30)$. Let $b = (190, 150, 120)$. Let

$$A = \begin{bmatrix} 1 & 1 & 1 \\ 1 & 2 & 1 \\ 1 & 1 & 2 \end{bmatrix}.$$

Compute the tableau for $\max vc \cdot \mathbf{x}$ subject to $A\mathbf{x} \leq \mathbf{b}, \mathbf{x} \geq \mathbf{0}$ for use in the simplex method for the basic solution where x_1 and x_2 are basic, and the constraint $A_{1,:}\mathbf{x} \leq \mathbf{b}_1$ is not binding. Do not use iterations of the simplex method to compute the tableau. Instead, find the 3×3 matrices B, B^{-1} and use the $B^{-1}A\mathbf{x} = B^{-1}\mathbf{b}$ definition. To compute the top row of the tableau, use the formula for reduced costs. Verify that your top row is correct by finding the 4×4 matrix $\tilde{B}$ when the equation $x_0 - \mathbf{c} \cdot \mathbf{x} = 0$ is added to the $A\mathbf{x} = \mathbf{b}$ constraints; finding the inverse of $\tilde{B}$ from the tableau you have already computed; verifying that it really is the inverse by multiplication by $\tilde{B}$

(to get the identity matrix); and finally multiplying the top row of the inverse of $\tilde{B}$ (i.e.,$((\tilde{B})^{-1})_{1,:}$) by the enlarged 4×7 matrix $\tilde{A}$.

M Problems

10. Solve the following problem with the simplex method in matrix form. Minimize $100x_1 + 100x_2$ subject to $x_1 - x_2 \geq 10; x_i \geq 0 : i = 1, 2$. Start with the basic feasible solution in which x_1 is basic.

11. Find the reduced costs with respect to the objective min $6000x_1 - 1200x_2$ at the basic feasible solution you found to be optimal in Exercise 10. Interpret them geometrically.

12. Solve Exercise 10 geometrically and match each step with the simplex method solution. Indicate at each step which constraints are binding.

13. Repeat Exercise 10 with the full tableau form.

14. Solve the following problem with the simplex method in matrix form. Maximize $99x_1 + 99x_2$ subject to $x_1 - x_2 \geq 10; x_i \geq 0 : i = 1, 2$. Start with the basic feasible solution in which x_1 is basic.

15. Solve Exercise 14 geometrically and match each step with the simplex method solution. Indicate at each step which constraints are binding.

16. Repeat Exercise 14 with the full tableau form.

17. (a) Find a basic feasible solution for the polyhedron

$$20x_1 + 30x_2 \geq 100; x_i \geq 0 : i = 1, 2.$$

Start with the basic solution in which $x_1 = x_2 = 0$. You may use the algebraic, revised, or a (primal) tableau simplex method.

(b) With a 2-D diagram, show the progress of your algorithm. Indicate clearly which constraints are active at which steps in the algorithm.

(c) Find the reduced costs with respect to the objective

$$\min\ 6000x_1 - 1200x_2$$

at the basic feasible solution you found in part 17a.

18. Explain how to recognize the case of an unbounded optimum in the simplex method. Concoct your own example and show explicitly how to extract a ray along which the objective function value is unbounded.

19. Consider a diet problem with 3 foods – juice, beer, and wine – and 2 nutrients, sugar and alcohol. Each ounce of juice contains 4 grams of sugar but no alcohol; each pint of beer contains 1 ounce of alcohol but no sugar. Each glass of wine contains 0.75 ounces of alcohol and 8 grams of sugar. The minimum nutritional requirements are 40 grams of sugar and 3 ounces alcohol.

(a) Let x_i denote the number of ounces of juice ($i = 1$), pints of beer ($i = 2$), and glasses of wine ($i = 3$) served. Invent your own prices and write the corresponding diet problem linear program.

(b) What is the basic feasible solution that serves juice and beer but no wine, and supplies the minimum nutritional amounts?

(c) Starting from the basic feasible solution you just computed, consider increasing the nonbasic variable x_3 from 0 to 1. How must the basic variables change to keep the nutritional content of the meal the same?

(d) Write the matrix B, and the matrix B^{-1}. Write the vector $B^{-1}A_3$. What does this vector say about a glass of wine? Compute $\mathbf{c}_B B^{-1} A_3$. What does this scalar say about juice, beer, and a glass of wine?

20. Suppose that the original LP is to minimize $\mathbf{d} \cdot \mathbf{y}$ subject to $D\mathbf{y} \geq \mathbf{b}; \mathbf{y} \geq \mathbf{0}$. Let $A = [D|-I]$, let $\mathbf{x} = \mathbf{y}, \mathbf{y}_S$ where $\mathbf{y}_S$ is a vector of slack variables, and let $\mathbf{c}^T = [\mathbf{d}^T|\mathbf{0}^T]$. Show that $\boldsymbol{\pi}^T A \leq \mathbf{c}$ means that we have a feasible solution to the dual of the original LP.

21. Set up the following problem for Phase I of the 2-phase method. Find the initial tableau, the entering variable, and the leaving variable.

$\max 7x_1 + 3x_2 - x_3$ subject to

$$\begin{aligned} 2x_1 + 3x_2 + 5x_3 &\geq 8 \\ 5x_1 - 6x_2 + 7x_3 &= 9 \\ -10x_1 + 11x_2 + 13x_3 &\leq -14 \\ x_1 + x_2 - x_3 &= -5 \\ 2x_1 - x_2 + x_3 &\geq -7 \\ x_1 + x_2 &\leq 20 \end{aligned}$$

22. Consider the LP $\max x_1 - x_2$ subject to $x_1 - 2x_2 <= 10$, $x_1 \geq 0, x_2 \geq 0$, $-3x_1 + 2x_2 <= 12$. Solve this LP using the simplex algorithm, starting at the basic solution $x_1 = x_2 = 0$. Show why the initial primal solution is feasible and the initial dual solution is infeasible, both geometrically and algebraically. Explain how the primal solution changes in the first iteration. State the dual solution that is at hand at the end of the first iteration. For the second iteration, state exactly the ray of unboundedness that you find in the primal, and state the contradiction that you obtain in the dual.

23. The equations below arise from a maximization problem with $\leq$ constraints which was converted to standard form by adding the usual I of slacks. The original RHS vector is $\mathbf{b} = (2, 4, 5)$; the objective function coefficient vector is $\mathbf{c} = (8, 6, 3)$. Find the missing numbers. Do not use elementary row operations to reverse the simplex method and reconstruct the initial tableau. Instead, use properties of the tableau (e.g., that it represents $B^{-1}A\mathbf{x} = B^{-1}\mathbf{b}$).

$$\begin{aligned} x_0 - \bigcirc x_1 &+ \bigcirc x_2 + \bigcirc x_3 + \bigcirc x_4 + \bigcirc x_5 + \bigcirc x_6 &= \bigcirc \\ 2x_1 &+ \bigcirc x_2 + x_3 + 3x_4 + \bigcirc x_5 - x_6 &= \bigcirc \\ 3x_1 &+ \bigcirc x_2 + 0x_3 - 7x_4 + x_5 + \bigcirc x_6 &= 10 \\ 3x_1 &+ x_2 + 0x_3 + x_4 + \bigcirc x_5 + 2x_6 &= \bigcirc \end{aligned}$$

24. Below is a set of equations occurring at the end of Phase I for a problem with objective Max $z = 9x_1 + 8x_2 - 7x_3 + 6x_4$. Temporary variables are indicated by a horizontal

bar, e.g., $\bar{x}_i$. Set up the initial tableau for Phase II. Find the entering and leaving variables.

$$\begin{aligned} z = \bar{x}_8 + \bar{x}_9 \quad & +0 \\ x_1 - x_3 + x_4 + 2x_6 + 3x_7 - \bar{x}_9 \quad & = 20 \\ 2x_1 - 4x_3 + x_5 + 6x_6 - \bar{x}_8 \quad & = 30 \\ -3x_1 + x_2 + 9x_6 - 2x_7 + 2\bar{x}_9 \quad & = 40 \end{aligned}$$

25. Let $\mathbf{x}, \mathbf{y}$ be any two extreme points of polyhedron P. Prove that you can go from $\mathbf{x}$ to $\mathbf{y}$ by repeatedly traveling from extreme point to adjacent extreme point. *Note: this seemingly obvious statement is quite difficult to prove from scratch.*

26. Below is an optimum tableau.

$$\begin{aligned} z + 3x_3 + 2x_4 \quad & = 13 \\ x_2 + x_3 \quad & = 3 \\ x_1 + x_3 + x_4 \quad & = 5 \\ -2x_3 - x_4 + x_5 \quad & = 2 \end{aligned}$$

The original problem was

$$\begin{aligned} \max z = 2x_1 + x_2 & \\ x_2 \quad & \leq 3 \\ x_1 - x_2 \quad & \leq 2 \\ x_1 + x_2 \quad & \leq 10 \\ x_i \quad & \geq 0 \end{aligned}$$

Write the dual problem, and its optimal solution including values of dual slack variables and the objective function.

27. Consider the Phase I problem

$$\begin{aligned} \min x_5 + x_6 \quad & = \quad z_0 \\ x_1 - x_2 - x_3 + x_5 \quad & = \quad 100 \\ x_1 + x_2 - x_4 + x_6 \quad & = \quad 60 \\ x_i \geq 0; i = 1 \ldots 6 & \end{aligned}$$

(a) What system of linear inequalities on the original variables x_1, x_2 describes the feasible region? (not a trick question).

(b) Starting from the basic solution in which x_5 and x_6 are basic, use the simplex method to find a basic feasible solution (i.e., do phase I).

(c) Identify the (2 by 2) matrix B, the vector $\mathbf{c}_B$, the matrix B^{-1}, the vector $\mathbf{c}_B B^{-1}$, the vector $B^{-1}b$, the scalar $\mathbf{c}_B B^{-1}b$, and the matrix $B^{-1}A$ at each iteration of your phase I computation (including the initial and final tableaux).

(d) Starting from the basic solution in which x_1 and x_3 are basic, solve the problem $\max 5000x_1 - 1000x_2$. Obtain the ray of unboundedness explicitly.

(e) Starting from the basic solution in which x_1 and x_3 are basic, solve the problem $\min 1000x_1 + 5000x_2$.

28. Prove or disprove each of the following:

(a) The dual of a phase I problem can never be unbounded.

(b) The dual of a phase I problem can never be infeasible.

(c) If the current tableau is degenerate, the objective function value will not improve in the next iteration of the simplex method.

(d) If there is a tie for the leaving variable in an iteration of the simplex method, then the next tableau must be degenerate.

29. Let $\mathbf{c} = (300, 400)$. Let $b = (4, 12, 16)$. Let

$$A = \begin{bmatrix} 1 & -1 \\ 1 & 1 \\ 1 & 2 \end{bmatrix}.$$

(a) Solve $\max c \cdot \mathbf{x}$ subject to $A\mathbf{x} \leq \mathbf{b}, \mathbf{x} \geq \mathbf{0}$ with the simplex method (either the matrix form or the tableau form) starting from $x_1 = x_2 = 0$.

(b) Identify the three choices of basis that correspond to an optimal solution. Compute the reduced costs for each choice of basis. Which choice of basis does not correspond to a dual basic feasible solution via complementary slackness?

(c) If the vector b is changed to $(5, 12, 16)$, which choices of basis, if any, have dual values that correctly predict the change in the optimal objective value?

(d) Same as the previous question for changing b to $(3, 12, 16)$.

(e) Same as the previous question for changing b to $(4, 11, 16)$.

(f) Same as the previous question for changing b to $(4, 12, 15)$.

30. Suppose you have a full tableau at an optimal solution to an LP. Suppose the solution is not degenerate (in the primal). Explain how to detect whether there are other optimal solutions. Prove that your detection rule is both necessary and sufficient. Explain why your rule is not necessary and sufficient if the solution is degenerate.

31. You start the simplex method at a BFS. After 20 iterations it detects that an LP has an unbounded optimum. Why it is possible that unboundedness could have been detected readily with the full tableau method after 6 iterations?

32. You start the simplex method at a BFS. After 20 iterations it detects that an LP has an unbounded optimum. From Proposition 4.4.3 the ray of unboundedness found by the simplex method defines a direction of the feasible region. Therefore, there was a ray emanating from the initial BFS along which the objective value is unbounded. Why is it possible, indeed likely, that this ray could not be detected by the simplex method at initial BFS?

D problems

33. Let A be a matrix with m columns and n rows. Consider the problem

$$\begin{array}{ccc} \min & \sum_{i=1}^{n} f_i(x_i) & \text{subject to} \\ A\mathbf{x} & = & \mathbf{b} \\ \mathbf{x} & \geq & \mathbf{0} \end{array}$$

where for each $i = 1 \ldots n$ the function $f_i(x_i)$ is piecewise linear, defined in the range $[0, \infty)$.

(a) Derive a variation of the simplex method similar to the usual primal method which operates directly on this problem, not on a reformulation. Begin by defining basic solution and basic feasible solution. Assume that you will be given an initial basic feasible solution according to your definition of BFS.

(b) Give a simple example for which your algorithm fails to find an optimum solution, even though one exists. Your example must be feasible. You may choose the starting BFS.

(c) Under what natural condition is your algorithm guaranteed to find an optimal solution? Prove that you are correct. You may assume that cycling does not occur.

(d) Does complementary slackness have a meaning in your algorithm?

34. Instead of creating a separate temporary variable for each $\leq$ constraint with negative RHS, you could create a single temporary variable that appeared with a coefficient of -1 in each such constraint. Its initial value would be the maximum absolute value of the negative RHS terms. Analyze this method, and explain why it is not commonly used, despite its reduction in the number of variables.

35. (a) For the LP $\max 333x_1$ subject to $0 \leq x_1 \leq 10; x_1 \leq 12$, guess the values of the dual variables at the primal's optimum. Do not discard the constraint $x_1 \leq 12$.

(b) Solve the LP with the simplex method, starting at the BFS $x_1 = 0$. Confirm that your guess is correct.

(c) What happens to the optimal objective value if the lower bound on x_1 is increased from 0 to 0.5? Compare the shadow price predictions with the actual values.

(d) What happens to the optimal objective value if the upper bound is increased from 10 to 10.5? Compare the shadow price predictions with the actual values.

(e) What happens to the optimal objective value if the upper bound is increased from 10 to 20? Compare the shadow price predictions with the actual values.

I Problems

36. (*)Find a geometric interpretation of Bland's rule.

Chapter 6

Variants of the Simplex Method

Preview: There are several variants of the simplex method. All of them move from basic solution to adjacent basic solution until arriving at an optimum. Here "adjacent" is a combinatorial notion, not a geometric one. It means that the choice of basic variables differs by one swap. The basic solution might not be feasible, so there may not be a geometric meaning in terms of movement along edges of a polyhedron.
Simplex method variants enforce complementary slackness. At each step they maintain a pair of corresponding primal and dual solutions.

6.1 Lower and Upper Bounds

Preview: The simplex method in Chapter 5 solves LP models in which all variables have no upper bounds and have lower bound zero. Every LP can be converted into an equivalent one with such variables, but it is computationally more efficient to generalize the simplex method to solve models in which variables may have lower and upper bounds, than to express upper bounds as additional constraints.

I have presented the simplex method to solve LPs of form min $\mathbf{c} \cdot \mathbf{x}$ subject to $A\mathbf{x} = \mathbf{b}; \mathbf{x} \geq \mathbf{0}$. In software implementations, the simplex method operates on an LP of the following form:

Linear Program with Lower and Upper Bounds

$$\begin{aligned} \min \quad & \mathbf{c} \cdot \mathbf{x} \quad \text{subject to} \\ & A\mathbf{x} = \mathbf{b} \\ & \mathbf{l} \leq \mathbf{x} \leq \mathbf{u} \end{aligned}$$

where $\mathbf{l}$ and $\mathbf{u}$ are vectors in $\{\Re \bigcup -\infty\}^n$ and $\{\Re \bigcup \infty\}^n$, respectively.

For example, in this form a nonnegative variable $\mathbf{x}_j$ has lower bound $\mathbf{l}_j = 0$ and upper bound $\mathbf{u}_j = \infty$.

Before getting into the details of the algorithm, let's see the geometry of this more general concept of basic solutions. Consider the problem

$$\begin{array}{lcr} \min -x_1 - 10x_2 & \text{subject to} & \\ x_1 + x_2 + x_3 & = & 12 \\ x_1 + x_2 \quad - x_4 & = & -2 \\ -x_1 + 3x_2 \quad + x_5 & = & 15 \\ -6 \le x_1 \le 8 & & \\ 2 \le x_2 \le 6 & & \\ 0 \le x_i : i = 3,4,5 & & \end{array}$$

Figure 6.1 depicts the feasible region and its six extreme points. Geometrically, the upper and lower bounds $-6 \le x_1 \le 8; 2 \le x_2 \le 6$ behave just like any other constraints. An extreme point is a feasible point where $n = 2$ constraints are binding. The arrows mark a path the simplex algorithm could take from an initial BFS at $(-4, 2)$ to $(-5.25, 3.25)$ to $(3, 6)$ to the optimum, $(6, 6)$. Thus, from a geometric point of view, the simplex algorithm needs no alteration to handle lower and upper bounds on variables.

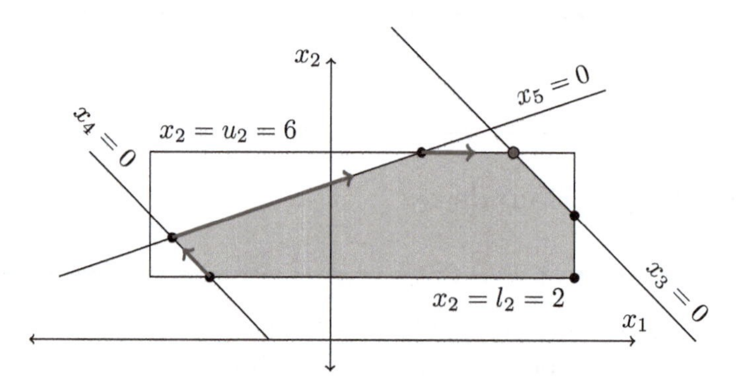

FIGURE 6.1: Lower bounds may be nonzero and upper bounds may be finite without affecting the definition of a BFS or the logic of the simplex algorithm.

However, the algebra of the algorithm requires alteration. In this more general form, each nonbasic variable $\mathbf{x}_j : j \in \mathbb{N}$ must be set to a finite value that is either its lower bound $\mathbf{l}_j$ or upper bound $\mathbf{u}_j$. As usual, a basic solution $\mathbf{x}$ must also satisfy $A\mathbf{x} = \mathbf{b}$. Think of a basic solution as follows: all but m of the variables $x_1 \ldots x_n$ are fixed at their upper or lower bounds. Plug in those values into the equation $A\mathbf{x} = \mathbf{b}$ and solve for the remaining m basic variables. If the basic variables all take values within their lower and upper bounds, the basic solution is feasible. Therefore, the last part of Step 1 in Table 5.1 must be altered to the following:

> *Computing the Basic Solution with Lower and Upper Bounds:*
> Let $L \subseteq \mathbb{N}$ and $U \subseteq \mathbb{N}$ denote, respectively, the nonbasic variables that are at their lower and upper bounds. The current basic solution is
> $\mathbf{x}_{\mathbb{L}} = \mathbf{l}_{\mathbb{L}}; \mathbf{x}_{\mathbb{U}} = \mathbf{u}_{\mathbb{U}}; \mathbf{x}_{\mathbb{B}} = B^{-1}(\mathbf{b} - A_{:,\mathbb{L}}\mathbf{l}_{Ll} - A_{:,\mathbb{U}}\mathbf{u}_{\mathbb{U}})$.
> The basic solution is feasible if $\mathbf{l}_{\mathbb{B}} \le \mathbf{x}_{\mathbb{B}} \le \mathbf{u}_{\mathbb{B}}$.

Any variable $\mathbf{x}_j$ for which $\mathbf{l}_j = \mathbf{u}_j$ can be substituted out of the LP. Hence we can assume $\mathbf{l} < \mathbf{u}$, so that there is no ambiguity about the status of a nonbasic variable.

The reasoning behind the simplex algorithm remains the same: Change the value of a nonbasic variable by θ if the net effect is good for the objective function. Make θ as large as possible without violating feasibility, at which point some basic variable reaches its (upper or lower) bound, and becomes nonbasic. Iterate from the new choice of basis. Stop if no nonbasic variables can improve the objective function, or if a ray of unboundedness is found.

The specifics of an algorithm iteration are more complicated. If a nonbasic variable $\mathbf{x}_j$ is at its upper bound, i.e., $j \in \mathbb{U}$, changing $\mathbf{x}_j$ by θ means *subtracting* θ from $\mathbf{x}_j$. Therefore, for nonbasic variables at their upper bounds $\mathbf{x}_j = \mathbf{u}_j$, a *positive* dual variable value $\boldsymbol{\pi}_j > 0$ signifies improvement in the objective function. Step 2 remains the same, but Step 3 in Table 5.1 must be altered to the following:

Step 3 with Lower and Upper Bounds: Select k to minimize $\{\boldsymbol{\pi}_j \vert j \notin \mathbb{B}, \mathbf{x}_j = \mathbf{l}_j\} \bigcup \{-\boldsymbol{\pi}_j \vert j \notin \mathbb{B}, \mathbf{x}_j = \mathbf{u}_j\}$.

Finding which variable leaves the basis, too, is more complicated with lower and upper bounds. The first complication is this: In the usual simplex algorithm, we ignore basic variables that increase as the nonbasic variable enters, because they will remain basic. They have upper bounds of infinity. With lower and upper bounds, a basic variable that is increasing might hit its *upper* bound and leave the basis. The only basic variables we ignore are those with $B^{-1}_{i,:} A_j = 0$, the ones that do not change as θ increases.

The second complication is this: In the usual simplex algorithm, an entering nonbasic variable increases by θ from its *lower* bound. If $B^{-1}_{i,:} A_j > 0$, the ith basic variable *decreases* as θ increases, so as to keep the equations $A\mathbf{x} = \mathbf{b}$ true. With lower and upper bounds, an entering nonbasic variable might instead *decrease* by θ from its *upper* bound. In that case, $B^{-1}_{i,:} A_j > 0$ implies that the ith basic variable *increases* as θ increases, to keep $A\mathbf{x} = \mathbf{b}$. Alter Steps 4 and 5 in Table 5.1 as follows:

Step 4 with Lower and Upper Bounds. Set $h_{\mathbb{B}} = B^{-1} A_{:,k}$. If $\mathbf{x}_k = \mathbf{u}_k$ set $h_{\mathbb{B}} = -h_{\mathbb{B}}$. Let $\overset{+}{\mathbb{B}} = \{i \in \mathbb{B} \vert h_i < 0\}$; let $\overset{-}{\mathbb{B}} = \{i \in \mathbb{B} \vert h_i > 0\}$. Now $\overset{+}{\mathbb{B}}$ is the set of indices of basic variables that increase as $\mathbf{x}_k$ enters the basis; $\overset{-}{\mathbb{B}}$ is the set of indices of those that decrease. If $\overset{+}{\mathbb{B}} \bigcup \overset{-}{\mathbb{B}} = \phi$ go to Step 7.

Step 5 with Lower and Upper Bounds:
Determine the variable to remove from $\mathbb{B}$.
Select r to attain the smallest value in the set $\{(\mathbf{u}_r - \mathbf{x}_r)/|h_r| : r \in \overset{+}{\mathbb{B}}\} \bigcup \{(\mathbf{x}_r - \mathbf{l}_r)/|h_r| : r \in \overset{-}{\mathbb{B}}\} \bigcup \{\mathbf{u}_r - \mathbf{l}_r : r = k\}$. If the smallest value attained is ∞ go to Step 7. Otherwise,

- If $r \in \overset{+}{\mathbb{B}}$, replace index r in $\mathbb{B}$ with index k. Add r to the set $\mathbb{U}$ (and the set $\mathbb{N}$).
- If $r \in \overset{-}{\mathbb{B}}$, replace index r in $\mathbb{B}$ with index k. Add r to the set $\mathbb{L}$ (and the set $\mathbb{N}$).
- If $r = k$, then if $k \in \mathbb{L}$ move k from $\mathbb{L}$ to $\mathbb{U}$; if $k \in \mathbb{U}$ move k from $\mathbb{U}$ to $\mathbb{L}$.

Calculate the new basic feasible solution by the formula in Step 1. Go to Step 2.

The calculation of the ray of unboundedness in Step 7 changes slightly, as follows:

Step 7 with Lower and Upper Bounds:
Let $\hat{\mathbf{x}}_{\mathbb{B}}$ denote the current values of the basic variables as were calculated in Step 1. If $k \in \mathbb{L}$ the optimal solution is unbounded along the ray $\mathbf{x}_k = \mathbf{l}_k + \theta$; $\mathbf{x}_{\mathbb{B}} = \hat{\mathbf{x}}_{\mathbb{B}} - \theta B^{-1}A_{:,k}$; $\mathbf{x}_j = \mathbf{l}_j \ \forall j \in L \setminus \{k\}$,$\mathbf{x}_j = \mathbf{u}_j \forall j \in \mathbb{U}$, $\theta \geq 0$. If $k \in \mathbb{U}$ the optimal solution is unbounded along the ray $\mathbf{x}_k = \mathbf{u}_k - \theta$; $\mathbf{x}_{\mathbb{B}} = \hat{\mathbf{x}}_{\mathbb{B}} + \theta B^{-1}A_{:,k}$; $\mathbf{x}_j = \mathbf{l}_j \ \forall j \in \mathbb{L}$, $\mathbf{x}_j = \mathbf{u}_j \forall j \in \mathbb{U} \setminus \{k\}$, $\theta \geq 0$.

There is one special, new kind of iteration. It is weird, but it is your friend. It is the computationally easiest kind of iteration. It can happen that the variable $\mathbf{x}_k$ that enters the basis is also the variable that leaves the basis! To keep $\mathbf{x}_k$ within its bounds, it must be that $\theta \leq \mathbf{u}_k - \mathbf{l}_k$. Should that bound be the tightest bound on θ, we get the particularly simple iteration in which $\mathbb{B}$ does not change. This is the case $r = k$ in Step 5. The only change is that index k shifts its membership from $\mathbb{L}$ to $\mathbb{U}$ or vice versa. Of course the values of the basic variables must change to compensate for the change in $\mathbf{x}_k$. If you ever have the choice to perform this kind of iteration, take it because it is computationally the least expensive.

The simplex method variant for lower and upper bounds was first developed when computer memory was severely limited. Treating upper bounds implicitly instead of as part of the system $A\mathbf{x} \leq \mathbf{b}$ often made the difference between being able to solve a model or not. Nowadays memory limitations seldom affect solvability of LPs by the simplex method. Nonetheless, the variant presented here is universally used in simplex method implementations. There is no reason not to use it, and it reduces computation time. It does not matter that the specifics of the algorithm are a bit more complicated, since you only write the extra few lines of code once.

Example 6.1 *Simplex Algorithm with Lower and Upper Bounds*

The problem is

$$\begin{aligned}
\min x_0 = 50x_1 + 100x_2 + 150x_3 - 50x_4 \quad & \text{subject to} \\
x_1 - x_2 + x_3 - 2x_4 \quad & \leq \quad 5 \\
-x_1 + x_2 - x_3 + x_4 \quad & \leq \quad 9 \\
-2x_1 + x_2 + x_3 \quad & \leq 0 \\
-1 \leq x_1 \leq 2 & \\
1 \leq x_2 \leq 4 & \\
0 \leq x_3 \leq 10 & \\
2 \leq x_4 \leq 5 &
\end{aligned}$$

Adding slack variables to the problem gives

$$\begin{aligned}
\min x_0 = 50x_1 + 100x_2 + 150x_3 - 50x_4 \quad & \text{subject to} \\
x_1 - x_2 + x_3 - 2x_4 + x_5 \quad & = \quad 5 \\
-x_1 + x_2 - x_3 + x_4 + \quad x_6 \quad & = \quad 9 \\
-2x_1 + x_2 + x_3 \quad + x_7 \quad & = \quad 0 \\
-1 \leq x_1 \leq 2 & \\
1 \leq x_2 \leq 4 & \\
0 \leq x_3 \leq 10 & \\
2 \leq x_4 \leq 5 & \\
0 \leq x_i \leq \infty \quad & i = 5, 6, 7
\end{aligned}$$

We will start with the basic feasible solution defined by $x_1 = u_1 = 2, x_2 = l_2 = 1$, $x_3 = l_3 = 0$, and $x_4 = l_4 = 2$. The slack variables x_5, x_6, x_7 along with the objective function value variable x_0 will be basic. To find the values of the basic variables we must plug in the values of the nonbasic variables to the equations. Variable x_5 has value determined by the equation

$$x_1 - x_2 + x_3 - 2x_4 + x_5 = 5 \Rightarrow 2 - 1 + 0 - 2(2) + x_5 = 5 \Rightarrow x_5 = 8.$$

Similarly, $x_6 = 8, x_7 = 5$, and $x_0 = 100$. We alter the content of the full tableau in two ways. First, for each nonbasic variable, we record whether it is at its lower or upper bound, denoted L or U respectively. Second, in the rightmost column we keep the values of the basic variables rather than the values $B^{-1}b$. The initial tableau is shown below.

	U	L	L	L				
x_0	x_1	x_2	x_3	x_4	x_5	x_6	x_7	RHS
1	-50	-100	-150	50	0	0	0	100
0	1	-1	1	-2	1	0	0	8
0	-1	1	-1	1	0	1	0	8
0	-2	-1	1	0	0	0	1	5

In the simplex algorithm as presented in Section 5.1, bringing a nonbasic variable into the basis improves a minimization objective if its reduced cost is negative. In this example, increasing the nonbasic variable x_4 from its lower bound $l_4 = 2$ would decrease the objective value at a rate of 50. Increasing the nonbasic variable x_2 or x_3 would not be beneficial as it would increase the objective value. However, bringing x_1 into the basis would decrease the objective value at a rate of 50 because x_1 is now at its upper bound. The reduced costs of x_1 and x_2 have the same sign, but only x_1 would improve the objective by entering the basis.

In this example, both x_1 and x_4 would improve the objective by entering the basis. We will bring x_4 into the basis because that choice illustrates the special type of iteration that is a feature of this variant. Let x_4 increase by θ. Basic variable x_5 decreases by -2θ. Since x_5 has upper bound ∞, it does not limit the increase in θ. Basic variable x_6 decreases by θ from its current value of 8 towards its lower bound 0. Therefore, $\theta \leq 8$. Basic variable x_7 does not change from its current value of 5, so it does not limit θ. So far, the most stringent bound on θ of 8 is due to x_6 approaching its lower bound. We must also check the entering variable x_4 which increases to $l_4 + \theta = 2 + \theta$. At $\theta = 3$, the variable x_4 reaches its upper bound $u_4 = 5$. Thus the most stringent bound $\theta \leq 3$ is due to x_4. The variable x_4 both enters and leaves the basis. *Question: What are the only changes that will occur in the tableau?*[1]

The new tableau is the same as the old except that x_4 is now recorded to be at its upper bound, and the values of the basic variables change by multiples of $\theta = 3$. The objective value decreases from 100 to $100 - 3(50) = -50$. Basic variable x_5 increases by $2\theta = 6$ from 8 to 14; x_6 decreases by $\theta = 3$ from 8 to 5; x_7 remains at 5. The new tableau is shown below.

	U	L	L	U				
x_0	x_1	x_2	x_3	x_4	x_5	x_6	x_7	RHS
1	-50	-100	-150	50	0	0	0	-50
0	1	-1	1	-2	1	0	0	14
0	-1	1	-1	1	0	1	0	5
0	**−2**	-1	1	0	0	0	1	5

Only x_1 has a desirable reduced cost. We bring x_1 into the basis, decreasing it by θ. The variable x_5 increases by θ and has an upper bound of ∞. That does not restrict θ. The basic variable x_6 decreases by θ from its value 5 towards its lower bound of 0. That restricts θ to be ≤ 5. The basic variable x_7 decreases by 2θ from its value 5 towards its lower bound of 0. That restricts θ to be $\leq 5/2$. Finally, x_1 decreases from 2 by θ towards its lower bound -1. That restricts θ to be $\leq 2 - (-1) = 3$. The bound 5/2 is the most stringent. Therefore, x_7 leaves the basis at its lower bound of 0. The new value of x_1 is $u_1 - \theta = 2 - 5/2 = -1/2$. The pivot term -2 is in boldface in the tableau above. The new tableau is given below.

		L	L	U			L	
x_0	x_1	x_2	x_3	x_4	x_5	x_6	x_7	RHS
0	0	-75	-175	50	0	0	-25	-175
0	0	−1.5	1.5	-2	1	0	0.5	16.5
0	0	1.5	−1.5	1	0	1	−0.5	2.5
0	1	0.5	-0.5	0	0	0	−0.5	-0.5

All reduced costs are now unfavorable. The new solution is optimal.

[1] Read on.

Summary: Generalize the idea that each nonbasic variable equals zero to the idea that each nonbasic variable equals its lower or upper bound. The resulting generalization of the simplex method handles bounds on individual variables with less memory and less cpu time than would the "plain vanilla" version. It requires nonbasic variables to be labelled as being at a lower or upper bound. It also permits the entering variable to be the leaving variable, a peculiar but very inexpensive type of iteration.

6.2 Dual Simplex Method

Preview: The dual simplex method starts from a dual feasible but primal infeasible basic solution. It works its way to primal feasibility while maintaining dual feasibility.

The usual simplex method starts with a basic feasible solution in the primal with a complementary basis for an infeasible solution in the dual, and works its way towards dual feasibility. Because it moves within the feasible region of the primal, it is often called a *primal simplex algorithm.* The dual simplex method, introduced by Carlton Lemke [205] does the dual of this procedure. It starts with a basic feasible solution in the dual, whose complementary basis in the primal is infeasible, and works its way towards primal feasibility. Like the primal simplex algorithm, it terminates when it attains both primal and dual feasibility. You may be wondering why the dual simplex method is an algorithm in its own right. Conceptually, it is the same as running the primal simplex method on the dual of the LP. Computationally speaking, however, it is not precisely the same.

In practice, the dual simplex method is used far more often than the primal simplex method. There are two reasons. First, the dual simplex is usually faster than the primal simplex at solving an LP from scratch. Researchers have not yet found a satisfactory explanation of this phenomenon. Second, algorithms that solve integer programs repeatedly add a constraint to a solved LP, then re-solve. The constraint that is added always renders the optimum solution infeasible. This is a perfect scenario for the dual simplex method. The basic solution on hand from the solved LP is dual feasible (since in the dual a variable has been added) but primal infeasible. *Question: What isn't quite right in theory about saying that the solved LP's basis is dual feasible?*[2] *Question: Why is the issue raised in the preceding question not a difficulty in reality?*[3]

Let's look at an example in full tableau format. Consider the following problem:

[2]Since a new constraint has been added, the number of variables in that dual basis is one less than it should be. The simple resolution to this difficulty is explained in Chapter 7.

[3]In practice the method with upper and lower bounds is used. The new constraint is a change in a variable's upper or lower bound. Hence it is a change in constraints rather than a new constraint, and does not add a new variable to the dual.

$$\begin{array}{rcl} \min 40x_1 + 50x_2 & \text{subject to} & \\ x_1 + 2x_2 & \geq & 60 \\ -3x_1 + 5x_2 & \geq & 15 \\ 4x_1 + 3x_2 & \geq & 90 \\ x_1, x_2 & \geq & 0 \end{array}$$

Multiply each constraint by -1 and add slack variables. The resulting initial full tableau follows.

x_0	x_1	x_2	x_3	x_4	x_5	RHS
1	-40	-50	0	0	0	0
0	-1	-2	1	0	0	-60
0	3	-5	0	1	0	-15
0	-4	-3	0	0	1	-90

(6.1)

The reduced costs equal 40 and 50, which, being nonnegative, satisfy the primal optimality (dual feasibility) requirement. However, the primal feasibility requirement is violated by all three slack variables. These primal basic variables correspond to reduced costs of nonbasic variables in the dual. From the dual point of view, the solution is feasible but fails the optimality requirements. The maximum infeasibility in the primal is $x_5 = -90$, which corresponds to the most favorable reduced cost in the dual. Therefore, we choose x_5 to be the exiting variable, so row 3 is the pivot row.

Next, we choose the entering variable. This is the opposite of the primal simplex method, which chooses the entering variable before choosing the exiting variable. The pivot term must be strictly negative, so that when the row is divided by the pivot term, the -90 RHS value becomes positive. In tableau (6.1) both x_1 and x_2 are eligible to enter the basis, because they would provide pivot terms of -4 and -3, respectively. Which one do we choose? We need to retain the primal optimality (i.e., dual feasibility) property. If x_1 entered the basis, the value in row 0 of the x_2 column would become $-50 - \frac{-40}{-4}(-3) = -20$, which is negative. But if x_2 entered the basis, the value in row 0 of the x_1 column would become $-40 - \frac{-50}{-3}(-4) = 26\frac{2}{3}$, which is positive. Therefore, we choose x_1 to enter the basis. With hindsight, you can see that we were selecting the smaller of the two ratios $\frac{40}{4}$ and $\frac{50}{3}$.

After performing elementary row operations, the new tableau follows. The objective function value of the primal has gotten worse. That makes sense because the initial solution was too good to be true. It was infeasible. In the dual, the solution remains feasible and has improved.

x_0	x_1	x_2	x_3	x_4	x_5	RHS
1	0	-20	0	0	-10	900
0	0	-1.25	1	0	$-.25$	-37.5
0	0	-7.25	0	1	$.75$	-82.5
0	1	$.75$	0	0	$-.25$	22.5

(6.2)

In tableau (6.2) we would ordinarily choose x_4 to exit the basis, since it has the most negative value. However, since I've looked ahead, I'm choosing x_3 instead. This is perfectly legal. The correctness and finite termination of the simplex method do not depend on choosing the most favorable reduced cost. All that is necessary is to choose a variable with favorable reduced cost, if any exists. Variable x_2 enters the basis because $\frac{20}{1.25} < \frac{10}{.25}$. The next tableau follows.

x_0	x_1	x_2	x_3	x_4	x_5	RHS
1	0	0	-16	0	-6	1500
0	0	1	$-.8$	0	$.2$	30
0	0	0	-5.8	1	2.2	135
0	1	0	0.6	0	$-.4$	0

(6.3)

Tableau (6.3) contains the optimal solution because all primal variables are feasible. The optimal primal solution is degenerate because $x_1 = 0$. The meaning of primal degeneracy in the dual is that there are multiple dual optimal solutions. You can see this by imagining doing another dual simplex iteration that chooses x_1 to exit the basis. The RHS values of the other primal basic variables would not change, because a multiple of 0 would be added to them. The primal solution would be the same feasible solution it is now. The reduced costs of the primal in row 0 would change, corresponding to a different optimal dual solution.

To summarize the dual simplex method on tableaux,

- Begin with a dual feasible but primal infeasible basic solution.
- Choose a negative primal variable to exit the basis. This choice determines the pivot row.
- Choose the primal variable to enter the basis with a minimum ratio test that compares the ratios of reduced costs to pivot row values, for all negative pivot row values.
- Perform elementary row operations to update the tableau to the new choice of basis.

There is one case that was not encountered in the above example, and is omitted from the preceding summary. What happens if there is no negative value in the pivot row (excluding the RHS) to serve as the pivot term? The pivot row would be an equality constraint with nonnegative coefficients on all variables, and a strictly negative RHS. Since all variables are restricted to be nonnegative, this constraint could not be satisfied. The LP would be infeasible. Elementary row operations on the original system of equalities would have yielded a contradiction. The algorithm would terminate. In the dual, the pivot row corresponds to a ray of unboundedness, which implies primal infeasibility by the weak duality theorem.

Summary: The dual simplex method makes decisions in reverse order. It first selects the leaving basic primal variable, and afterwards determines the entering variable. It is perfect for re-solving a solved LP to which a new constraint violated by the optimal solution has been added.

6.2.1 An Example of Both the Primal and Dual Simplex Algorithms, with Geometry

---

Preview: This section presents a detailed example to ensure that you have a firm understanding of the algorithms and their geometry.

The primal LP is

$$\begin{array}{rcl} \max & 3x_1 + 5x_2 & \text{subject to} \\ x_1 + 2x_2 & \leq & 32 \\ x_1 + x_2 & \leq & 20 \\ x_1, x_2 & \geq & 0 \end{array}$$

We call this the *original primal* to distinguish it from the LP with slack variables added. We will visualize the geometry in the space of the original primal. *Question: Why?*[4] Figure 6.2 depicts the constraints, feasible region, and basic feasible solutions of the original primal.

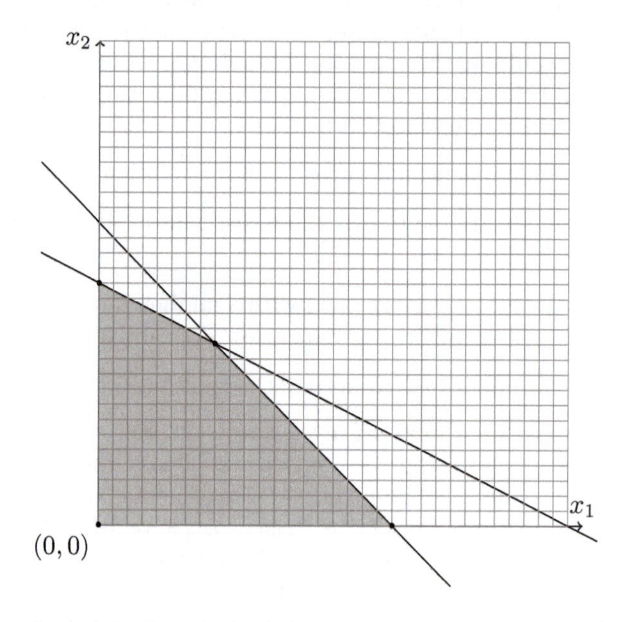

FIGURE 6.2: In the original primal, the primal simplex algorithm starts at $(x_1, x_2) = (0, 0)$.

Add the slack variables to get an equivalent primal.

$$\begin{array}{rcl} \max 3x_1 + 5x_2 & \text{subject to} & \\ x_1 + 2x_2 + x_3 & = & 32 \\ x_1 + x_2 + x_4 & = & 20 \\ x_1, x_2, x_3, x_4 & \geq & 0 \end{array}$$

We call this LP the primal.

Introduce a variable x_0 that equals the objective function value.

$$\begin{array}{rcl} \max x_0 & \text{subject to} & \\ x_0 - 3x_1 - 5x_2 & = & 0 \\ x_1 + 2x_2 + x_3 & = & 32 \\ x_1 + x_2 \quad + x_4 & = & 20 \\ x_1, x_2, x_3, x_4 & \geq & 0 \end{array}$$

[4]It is easier to visualize in two dimensions than in four.

Call this LP the *extended primal.*

Write the extended primal in full tableau format.

1	-3	-5	0	0	0
0	1	2	1	0	32
0	1	1	0	1	20

The tableau represents the basic solution in the extended primal $\mathbf{x}_{\mathbb{B}} = \{x_0, x_3, x_4\} = (0, 32, 20)$ (and $x_1 = x_2 = 0$). In the primal, the corresponding basic solution is at $\mathbf{x}_{\mathbb{B}} = \{x_3, x_4\} = (32, 20)$ and $x_1 = x_2 = 0$. In the original primal, the solution is at the origin, $(x_1, x_2) = (0, 0)$ (see Figure 6.2).

The dual to the original primal, which we call the original dual, is

$$\min 32\pi_1 + 20\pi_2 \quad \text{subject to} \tag{6.4}$$
$$\pi_1 + \pi_2 \geq 3 \tag{6.5}$$
$$2\pi_1 + \pi_2 \geq 5 \tag{6.6}$$
$$\pi_1, \pi_2 \geq 0 \tag{6.7}$$

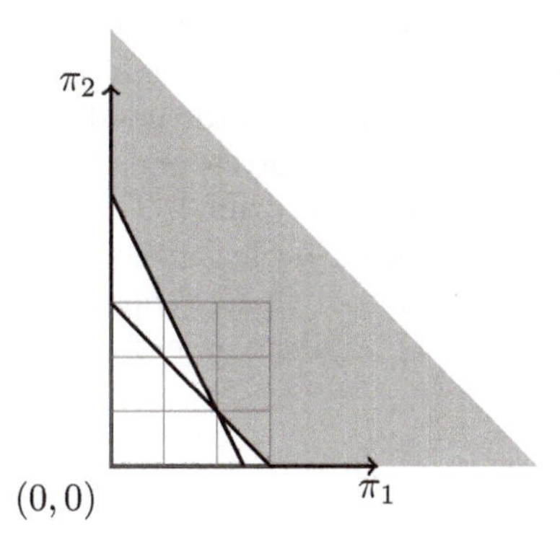

FIGURE 6.3: In the original dual, the dual simplex algorithm starts at $(\pi_1, \pi_2) = (0, 0)$, which violates constraints (6.5),(6.6).

Figure 6.3 depicts the original dual geometrically in the space of the variables (π_1, π_2). Do not try to see geometric similarities between the feasible region of the original dual and the feasible region of the original primal. They do not usually have the same dimension. Adding slacks to the original dual gives what we will call the dual. As you would expect, it is the dual of what we are calling the primal.

$$\begin{array}{lcl} \min 32\pi_1 + 20\pi_2 & \text{subject to} & \\ \pi_1 + \pi_2 - \pi_3 & = & 3 \\ 2\pi_1 + \pi_2 \quad - \pi_4 & = & 5 \\ \pi_i & \geq & 0 \ i = 1, \ldots, 4 \end{array}$$

Let's rewrite the dual's constraints to make their meaning more clear.

$$\pi_1 \begin{pmatrix} 1 \\ 2 \end{pmatrix} + \pi_2 \begin{pmatrix} 1 \\ 1 \end{pmatrix} + \pi_3 \begin{pmatrix} -1 \\ 0 \end{pmatrix} + \pi_4 \begin{pmatrix} 0 \\ -1 \end{pmatrix} = \begin{pmatrix} 3 \\ 5 \end{pmatrix}; \quad \pi_i \geq 0 \ \forall i$$

The geometric meaning of the dual in the original primal is to express the gradient of the objective function, $(3,5)$, as a nonnegative combination of the vectors $(1,2),(1,1),(-1,0),(0,-1)$. The dual variables are the nonnegative weights placed on those vectors. The vectors are orthogonal to the original primal constraints $x_1+2x \leq 32, x_1+x_2 \leq 20, -x_1 \leq 0, -x_2 \leq 0$, and point away from feasibility.

Every basic solution in the primal corresponds by complementary slackness to a basic solution in the dual. The complementary slackness conditions in this example are as follows:

$$
\begin{aligned}
\pi_1(32 - x_1 - 2x_2) &= 0 \\
\pi_2(20 - x_1 - x_2) &= 0 \\
x_1(3 - \pi_1 - \pi_2) &= 0 \\
x_2(5 - 2\pi_1 - \pi_2) &= 0.
\end{aligned}
$$

In terms of the slack variables, the complementary slackness conditions have the following simple form:

$$\pi_1 x_3 = \pi_2 x_4 = x_1 \pi_3 = x_2 \pi_4 = 0. \tag{6.8}$$

Never make the mistake of associating x_1 in a primal with π_1 in its dual. The correspondence is always between the slack variables of one with the original variables of the other (and, by symmetry, vice versa).

The basic dual solution that corresponds to $\mathbf{x} = (0,0,32,20)$ in the primal has $\pi_1 = \pi_2 = 0$, and basic variables π_3 and π_4 equal to -3 and -5, respectively. The dual is therefore infeasible. Its objective function value is 0, which is too good to be feasible. Such a solution is called *superoptimal.* Infeasibility in the dual corresponds to suboptimality in the primal. The dual variables are the reduced costs of the primal. The only original primal objective functions for which this dual solution would be feasible are of form $\max c_1x_1 + c_2x_2$ for $c_1 \leq 0$ and $c_2 \leq 0$. Figure 6.4 depicts the polyhedral cone at $(0,0)$ in the original primal formed by the vectors $(-1,0)$ and $(0,-1)$, and the objective function vector $(3,5)$ which is not in the cone.

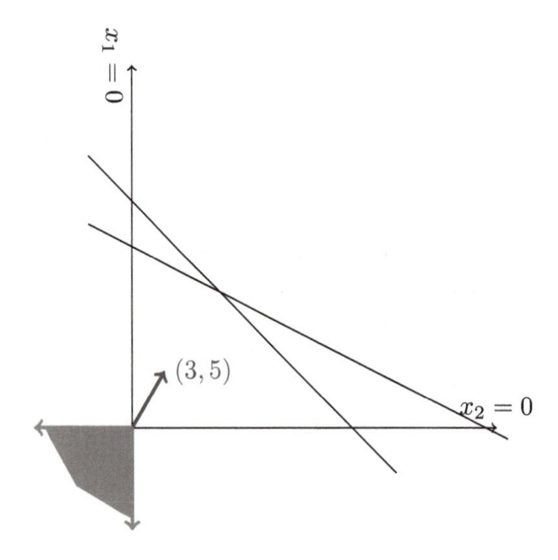

FIGURE 6.4: In the original primal, $\mathbf{c} = (3,5)$ is not in the cone generated by $(0,-1)$ and $(-1,0)$ at $(0,0)$.

Returning now to the initial simplex tableau of the extended primal, perform the first iteration of the simplex method. If x_1 is increased by θ, the objective will increase by 3θ. If x_2 is increased by θ the objective will increase by 5θ. Choose x_2 to enter the basis at value θ. To keep $A\mathbf{x} = \mathbf{b}$ true, x_3 must decrease by $\theta B^{-1}_{1,:}A_{:,2} = 2\theta$, and x_4 must decrease by $\theta B^{-1}_{2,:}A_{:,2} = \theta$. Basic variable x_3 hits 0 at $\theta = 32/2$, a smaller value than the value of $\theta = 20/1$ at which x_4 would hit 0. Therefore, x_3 leaves the basis.

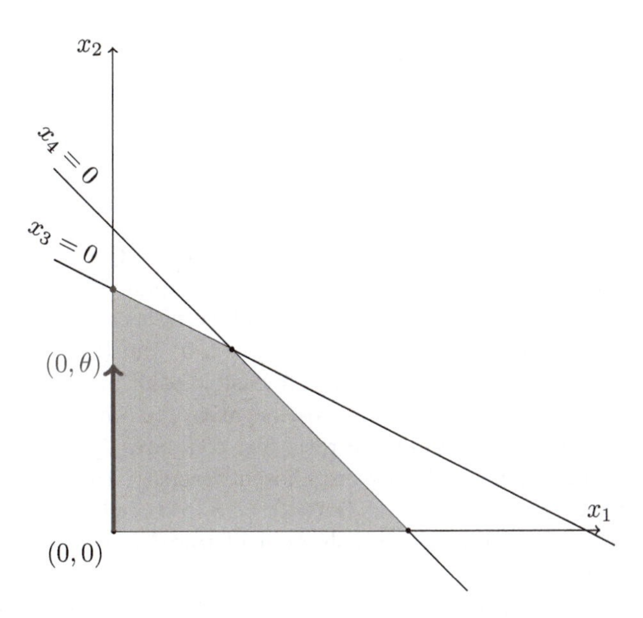

FIGURE 6.5: As nonbasic variable x_2 increases, both x_3 and x_4 decrease. The former hits 0 first and exits the basis.

Figure 6.5 shows the geometry of the leaving basic variable x_3. As $x_2 = \theta$ increases, we move upwards from the origin to the point $(0, \theta)$. We are moving towards both of the original primal constraints which is the same as saying that x_3 and x_4 are both decreasing. We hit the $x_1 + 2x_2$ constraint first, which is the same as saying that x_3 hits 0 before x_4 does. The constraint $x_1 \geq 0$ remains binding as we move because x_1 is a nonbasic variable. Only the values of the basic variables change to keep $A\mathbf{x} = \mathbf{b}$ true.

The number in the row corresponding to x_3 (often called the pivot row) and the column corresponding to x_2 (often called the pivot column) is displayed in boldface in Figure 6.6. It is often called the pivot value. Perform elementary row operations to establish an identity matrix in the columns of the new basis $\mathbf{x}_{\mathbb{B}} = (x_0, x_2, x_4)$. The row operations are to divide the pivot row by the pivot value, and then to add multiples of the pivot row to the other rows to make the rest of the pivot column values equal to 0.

In the second tableau, the matrix B^{-1} of the primal and the extended primal are, respectively,

$$\begin{bmatrix} 1/2 & 0 \\ -1/2 & 1 \end{bmatrix} \text{ and } \begin{bmatrix} 1 & 2.5 & 0 \\ 0 & 1/2 & 0 \\ 0 & -1/2 & 1 \end{bmatrix}.$$

The values of the dual variables in the first tableau are in the top row. Remember that by Equation (6.8) the original dual variables π_1, π_2 are associated with the primal variables x_3, x_4 and are found in the columns of those variables. The values of the dual slack variables are found in the columns of the original primal variables x_1, x_2.

In the second tableau, the dual solution is therefore $\pi_1 = 2.5, \pi_2 = 0, \pi_3 = -\frac{1}{2}, \pi_4 = 0$. Let's do an iteration of the dual simplex method to find the same values directly. The

1	-3	-5	0	0	0
0	1	**2**	1	0	32
0	1	1	0	1	20
1	$-1/2$	0	2.5	0	80
0	$1/2$	1	$1/2$	0	16
0	$1/2$	0	$-1/2$	1	4

FIGURE 6.6: In the first iteration of the primal simplex algorithm, row operations convert the pivot column, which contains the value 2 in boldface, into a column of the identity matrix.

first and second tableaux are shown in Figure 6.7. The basic solution depicted in the first tableau is $\pi_3 = -3, \pi_4 = -5$ and nonbasic variables π_1, π_2 equal to zero. The solution is infeasible. In the original primal geometry, the infeasibility is because $(3,5)$ is not in the cone of nonnegative linear combinations of the vectors $(-1,0)$ and $(0,1)$. Instead, the dual constraints $(3,5) = \pi_3(-1,0) + \pi_4(0,-1)$ have the infeasible solution $\pi_3 = -3, \pi_4 = -5$.

In terms of the dual simplex algorithm, the most negative (i.e., most infeasible) variable x_4 is selected to leave the basis. The minimum ratio test compares $\frac{-32}{-2}$ with $\frac{-20}{-1}$ to choose x_1 as the entering variable. If instead x_2 were chosen to enter, the reduced cost of x_1 in the next tableau would equal $-32 - (20/1)(-2) = 8 > 0$, which would violate the optimality condition. The pivot term -2 in column 1 of row 3 is printed in boldface. As usual, to calculate the values of the next tableau, perform elementary row operations to make the coefficients of $\pi_{\mathbb{B}} = (\pi_0, \pi_3, \pi_1)$ form an identity matrix. Be sure you understand why $\boldsymbol{\pi}_{\mathbb{B}} \neq (\pi_0, \pi_1, \pi_3)$. The new basic solution is $\pi_0 = 80, \pi_3 = -1/2, \pi_1 = 5/2$.

In the original dual geometry, this basic solution is not feasible because it violates constraint 6.5. In the original primal geometry, this basic solution is not feasible because, as shown in Figure 6.8, the vector $(3,5)$ is not in the cone of nonnegative linear combinations of the vectors $(-1,0)$ and $(1,2)$. Instead $(3,5) = (-1/2)(-1,0)+(5/2)(1,2)$. *Question: Why does the value of* π_3 *get multiplied by the vector* $(-1,0)$ *from the primal point of view?*[5]

π_0	π_1	π_2	π_3	π_4	RHS
1	-32	-20	0	0	0
0	-1	-1	1	0	-3
0	$\mathbf{-2}$	-1	0	1	-5
1	0	-4	0	-16	80
0	0	$\mathbf{-1/2}$	1	$-1/2$	$-1/2$
0	1	$1/2$	0	$-1/2$	$5/2$
1	0	0	-8	-12	84
0	0	1	-2	1	1
0	1	0	1	-1	2

FIGURE 6.7: In the 1st iteration of the dual simplex algorithm, π_4 leaves the basis and π_1 enters. In the 2nd iteration, π_3 leaves the basis and π_2 enters.

Continuing with the dual simplex algorithm, $\pi_3 = -1/2 < 0$ is the only variable eligible

[5]By Equation (6.8) the variables x_1, π_3 are complementary. Therefore, π_3 is the shadow price for the constraint $x_1 \geq 0 \Leftrightarrow (-1,0) \cdot (x_1, x_2) \leq 0$.

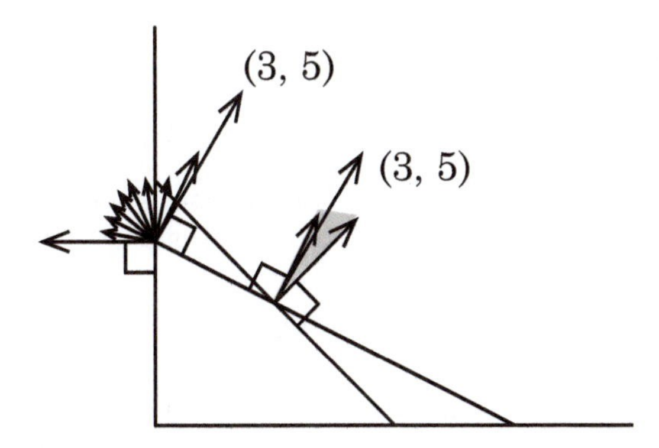

FIGURE 6.8: The second basic solution encountered by the simplex algorithm is not dual feasible because $(3,5)$ is not in the cone formed by $(-1,0)$ and $(1,2)$. The third basic solution encountered is dual feasible because $(3,5)$ is in the cone formed by $(1,2)$ and $(1,1)$.

to leave the basis; π_2 enters the basis rather than π_4 because $4/(1/2) = 8 < 32 = 16/(1/2)$. The resulting solution $\pi_0 = 84, \pi_2 = 1, \pi_1 = 2, \pi_3 = \pi_4 = 0$ is feasible and hence optimal. In the original primal geometry this solution is feasible because $(3,5) = 1(1,1)+2(1,2)$ is in the cone of nonnegative combinations of $(1,1)$ and $(1,2)$. The values of the corresponding primal optimal solution may be obtained from the reduced costs, in the dual, of the complementary variables. From Equation (6.8) the reduced cost of π_3, which is $= -8$, tells us the value of x_1. The reduced cost of π_4, which is -12, tells us the value of x_2. The values themselves are multiplied by -1 because the dual is a minimization problem.

1	$-1/2$	0	2.5	0	80
0	$1/2$	1	$1/2$	0	16
0	**1/2**	0	$-1/2$	1	4
1	0	0	2	1	84
0	0	1	1	-1	12
0	1	0	-1	2	8

FIGURE 6.9: The second iteration of the primal simplex method on the primal brings x_1 into the basis and removes x_4.

Return to the primal simplex algorithm in Figure 6.9. The two tableaux show the second iteration. In the upper tableau, if nonbasic variable x_1 is set to θ, the objective function will improve (increase) by $\theta/2$.

Here are the details of why the reduced cost of x_1 is $1/2$. At present, $\mathbf{x}_{\mathbb{B}} = (x_0, x_2, x_4)$ and the nonbasic variable $x_1 = 0$. If x_1 is increased to θ, the LHS of the equation $A\mathbf{x} = \mathbf{b}$ in the primal is increased by $\theta A_{:,1} = \theta(1,1)$ (while the LHS of $A\mathbf{x} = \mathbf{b}$ in the extended primal is increased by $\theta(3,1,1)$). To counteract this change in the LHS of $A\mathbf{x} = \mathbf{b}$, construct a synthetic $A_{:,1}$ column comprising a linear combination of the basic variable columns $A_{:,2}, A_{:,4}$. (In the extended primal the synthetic $A_{:,1}$ column $(3,1,1)$ is a linear combination of the basic variable columns $A_{:,0}, A_{:,2}, A_{:,4}$.) The linear combination that works for the primal is

$$\begin{aligned} A_{:,1} &= \begin{pmatrix} 1 \\ 1 \end{pmatrix} = \begin{pmatrix} 2 \\ 1 \end{pmatrix} 1/2 + \begin{pmatrix} 0 \\ 1 \end{pmatrix} 1/2 = B \begin{pmatrix} 1/2 \\ 1/2 \end{pmatrix} \\ &= B(B^{-1}A_{:,1}) = (BB^{-1})A_{:,1} = A_{:,1}. \end{aligned}$$

The direct impact of $x_1 = \theta$ on the objective function is an increase of $c_1\theta = 3\theta$. The indirect impact is a decrease of θ times the cost of the synthetic $A_{:,1}$ column,

$$\theta c_{\mathbb{B}}^T(B^{-1}A_{:,1}) = \theta(5,0)\begin{pmatrix} \frac{1}{2} \\ \frac{1}{2} \end{pmatrix} = \frac{5}{2}.$$

The net impact on the objective function in the primal is thus an increase of $\theta(3-2.5) = \theta/2$.
Similarly, the linear combination that works for the extended primal is

$$\begin{aligned} A_{:,1} &= \begin{pmatrix} 1 \\ 1 \end{pmatrix} = \begin{pmatrix} 2 \\ 1 \end{pmatrix} \frac{1}{2} + \begin{pmatrix} 0 \\ 1 \end{pmatrix} \frac{1}{2} = B \begin{pmatrix} \frac{1}{2} \\ \frac{1}{2} \end{pmatrix} \\ &= B(B^{-1}A_{:,1}) = (BB^{-1})A_{:,1} = A_{:,1}. \end{aligned}$$

In the extended primal, to keep $A\mathbf{x} = \mathbf{b}$ true, the basic variables must decrease by $\theta(-\frac{1}{2}, \frac{1}{2}, \frac{1}{2})$. The number $-\frac{1}{2}$ in the x_1 column of the top row now has the following interpretation: when x_1 is increased to θ, the basic variable x_0 must decrease by $-\frac{1}{2}\theta$ to keep the equation $A\mathbf{x} = \mathbf{b}$ of the extended primal true. From both points of view, primal and extended primal, the reduced cost of x_1 is $\frac{1}{2}$.

The tableau simplex method is designed to provide all needed numerical values. The vector $B^{-1}A_{:,1}$ is the column of x_1 in the current tableau, since the current tableau always depicts $B^{-1}A\mathbf{x} = B^{-1}\mathbf{b}$. Since the columns of x_0, x_3, x_4 in the initial tableau form an identity matrix, those columns are B^{-1} in the current tableau. That is,

$$B^{-1} = \begin{bmatrix} 1 & 2.5 & 0 \\ 0 & .5 & 0 \\ 0 & -.5 & 1 \end{bmatrix}$$

with respect to the extended primal and

$$B^{-1} = \begin{bmatrix} .5 & 0 \\ -.5 & 1 \end{bmatrix}$$

with respect to the primal. Let's verify that the rightmost column of the current tableau is $B^{-1}\mathbf{b}$ for both primal and the extended primal. Multiply the 2×2 inverse matrix by $\mathbf{b} = (32, 20)$ to get $B^{-1}\mathbf{b} = (16, 4)$. Multiply the 3×3 inverse matrix by $\mathbf{b} = (0, 32, 20)$ to get $B^{-1}\mathbf{b} = (80, 16, 4)$.

Figure 6.10 shows the geometry of the second iteration in the original primal. Starting at the point $(0, 16)$, move horizontally to the right and vertically slightly downwards as x_1 is brought into the basis. Geometrically, we move along an edge of the polytope because all constraints that were binding at $(0, 16)$ remain binding, except $x_1 \geq 0$. This is the same as saying that the nonbasic variable x_3 remains at 0. As we move along the edge, we get closer to the constraint $x_1 + x_2 \leq 20$ and the constraint $x_2 \geq 0$. We hit the former constraint well before we would hit the latter. This is the same as saying that in the minimum ratio test, $4/(\frac{1}{2}) = 8$ is much smaller than $16/(\frac{1}{2}) = 32$. Therefore, the slack variable x_3 for the constraint $x_1 + x_2 \leq 20$ reaches 0 and leaves the basis when the basic variable x_2 is still strictly positive. We arrive at the new basic feasible solution where $x_1 + x_2 \leq 20$ and $x_1 + 2x_2 \leq 32$ are both binding, and the basic variables are $x_1 = 8, x_2 = 12$. The objective

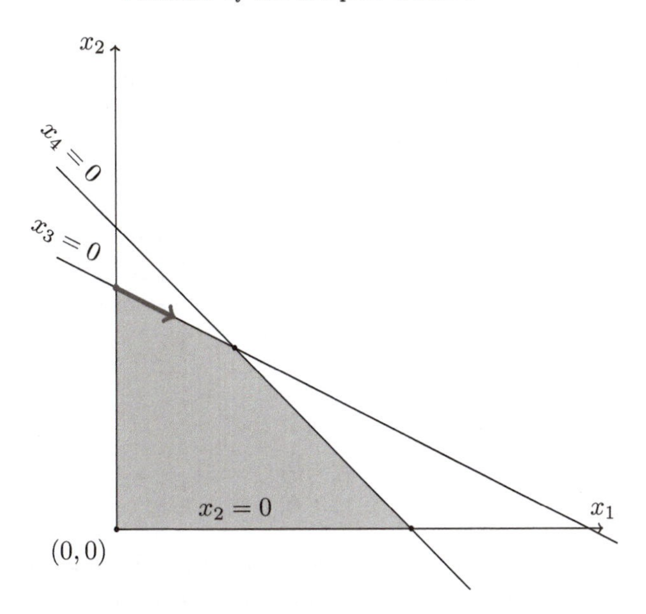

FIGURE 6.10: In the 2nd iteration of the primal simplex algorithm, increasing the nonbasic variable x_1 moves us towards constraints $x_1 + x_2 \leq 20 \Leftrightarrow x_4 \geq 0$ and $x_2 \geq 0$ and keeps constraint $x_1 + 2x_2 \leq 32 \Leftrightarrow x_3 \geq 0$ binding.

value is 84. This solution is optimal. There are two geometric ways to see the optimality. The first way is shown in Figure 6.10. The hyperplane $\mathbf{c} \cdot \mathbf{x} = 3x_1 + 5x_2 = 84$ passes through the point $(8, 12)$. Every point in the feasible region is in the closed halfspace $\{\mathbf{x} | \mathbf{c} \cdot \mathbf{x} \leq 84\}$. Therefore, $(8, 12)$ is optimal. The second way is shown in Figure 6.8 at the point $(8, 12)$. The optimality criterion in the primal, which is precisely the feasibility criterion in the dual, is geometrically satisfied because $(3, 5)$ is in the cone of nonnegative linear combinations of the vectors $(1, 2)$ and $(1, 1)$.

Summary: If you can work out an example on your own at the same level of detail as given here, you understand both the algebra and geometry of the simplex method.

6.3 Primal-Dual Algorithm

Preview: The primal-dual algorithm solves a given LP by solving a sequence of easier LPs. Its power is especially evident when it is applied to combinatorial problems.

The primal-dual method was grown by Kuhn from his Hungarian algorithm for the assignment problem, a problem which we will see in both this chapter and in Chapter 11. In their classic book on combinatorial optimization Papadimitriou and Steiglitz [241] showed that well-known algorithms such as Dijkstra's shortest path algorithm can be viewed as primal-dual algorithms, and that the primal-dual method is a general method for solving weighted optimization problems as a series of simpler unweighted feasibility problems. When the field of approximation algorithms developed, the primal-dual algorithm emerged as one of the fundamental ways to build approximation algorithms [125, 287]. Thus the primal-dual method is one of the oldest and yet one of the most vibrant versions of the simplex method.

The primal-dual algorithm operates on LPs of form

$$\min \mathbf{c}\cdot\mathbf{x} \quad \text{subject to} \tag{6.9}$$
$$A\mathbf{x} = \mathbf{b}; \tag{6.10}$$
$$\mathbf{x} \geq \mathbf{0}; \tag{6.11}$$

where without loss of generality $\mathbf{b} \geq \mathbf{0}$. This form has the advantage that part of the complementary slackness conditions is satisfied merely by primal feasibility, namely the condition $(A_{j,:}\mathbf{x} - b_j)\pi_j = 0\ \forall j$.

The dual is

$$\max \mathbf{b}\cdot\boldsymbol{\pi} \quad \text{subject to} \tag{6.12}$$
$$A^T\boldsymbol{\pi} \leq \mathbf{c} \tag{6.13}$$

The algorithm starts with an initial basic feasible solution $\boldsymbol{\pi}$ to the dual. Let I denote the set of indices i of the rows of (6.13) which are binding at $\boldsymbol{\pi}$, i.e., for which $A^T_{i,:}\boldsymbol{\pi} = c_i$. The set I defines the *restricted primal*, the feasibility LP

$$A\mathbf{x} = \mathbf{b}; \tag{6.14}$$
$$\mathbf{x} \geq \mathbf{0}; \tag{6.15}$$
$$x_i = 0 \quad \forall i \notin I. \tag{6.16}$$

The restricted primal enforces the complementary slackness conditions with $\boldsymbol{\pi}$, namely $(A^T_{i,:}\boldsymbol{\pi} - c_i)x_i = 0\ \forall i$. If the restricted primal has a feasible solution, then it is optimal to the original primal because it is feasible in the original primal and it satisfies complementary slackness with the feasible solution $\boldsymbol{\pi}$ to the dual. Of course we'd have to be lucky for our initial dual feasible solution $\boldsymbol{\pi}$ to be optimal. So most likely the restricted primal is not feasible.

Rewrite the restricted primal as $A_{:,I}x_I = \mathbf{b}; x_I \geq 0$. If it is infeasible, then by a theorem of the alternative (part 2 of Theorem 8.1.1) there exists λ such that $\lambda\cdot\mathbf{b} > 0$ and $\lambda^T A_{:,I} \leq \mathbf{0}$. (This vector λ can be found explicitly by running phase I on the restricted primal, which is explained in detail later.) Now we can take a step in the direction λ, in the dual. For small enough $\theta > 0$, $\boldsymbol{\pi} + \theta\lambda$ must be feasible in the dual because for all binding constraints $i \in I$ in the dual, $A^T_{i,:}\lambda \leq 0$. As usual with the simplex method, once we've found a good direction to travel in, we go as far as we can while maintaining feasibility. The maximum feasible value of θ will be the minimum of

$$\theta^* = \frac{c_i - A^T_{i,:}\boldsymbol{\pi}}{A^T_{i,:}\lambda} : A^T_{i,:}\lambda > 0.$$

If there is no minimum, the dual is unbounded. Otherwise, we update $\boldsymbol{\pi} = \boldsymbol{\pi} + \theta^*\lambda$ and iterate. Since $\theta^* > 0$ the objective strictly increases at each iteration, so no dual basis can be repeated. Therefore, the algorithm must terminate in a finite number of iterations. Each iteration takes finite time because phase I of the simplex method takes finite time. This algorithm may seem ridiculous because it solves an LP by solving an LP at each iteration, but as explained in the beginning of the section it is remarkably powerful in combinatorial optimization. It also is faster than it looks because the feasibility linear programs are closely related to each other.

Now let's see how to find λ explicitly. Set up a phase I objective in the restricted primal to get the LP

$$\min \textstyle\sum_i \bar{x}_i \quad \text{subject to} \tag{6.17}$$
$$A_{:,I}\mathbf{x}_I + \bar{\mathbf{x}} = \mathbf{b} \tag{6.18}$$
$$\mathbf{x}_I, \bar{\mathbf{x}} \geq \mathbf{0}. \tag{6.19}$$

The phase I problem is feasible because it admits the solution $\bar{\mathbf{x}} = \mathbf{b}; \mathbf{x}_I = \mathbf{0}$. (That is why we required $\mathbf{b} \geq \mathbf{0}$ initially). It is not unbounded because the optimal objective can't be less than zero. If at optimality the objective value is strictly positive, then the corresponding optimal dual solution λ^* satisfies $\lambda^* \cdot \mathbf{b} > 0$ by the strong duality theorem. Also it is dual feasible so $A_{:,I}\lambda^* \leq \mathbf{0}$. The other part of dual feasibility, that $\lambda^* \leq \mathbf{1}$, doesn't matter since we use λ^* as a direction to travel, not as an absolute step size.

6.3.1 The Primal-Dual Algorithm Applied to Assignment Problems

Let's apply the primal-dual algorithm to a special class of LPs called the *assignment problem*. There are n processors indexed $i = 1, 2, \ldots, n$ and n jobs indexed $j = 1, 2, \ldots, n$ to be performed. Each processor can do any one of the jobs, and each job must be completed. The cost for processor i to do job j is c_{ij}. The problem is to assign processors to jobs so as to minimize the sum of the costs.

To formulate the assignment problem as an LP, define variables $x_{ij} : 1 \leq i \leq n; 1 \leq j \leq n$ to represent the fraction of job j assigned to processor i. In reality we want each x_{ij} variable to equal 1 or 0, but an LP is not allowed to constrain variables to take integer values. Fortunately, as we will see in Chapter 11, the basic feasible solutions to our LP will have only integer values. The constraints are nonnegativity, $x_{ij} \geq 0$ for all i and j, and:

$$\text{For all } i = 1, \ldots, n: \quad \sum_{j=1}^{n} x_{ij} = 1; \quad \text{every processor does one job.}$$

$$\text{For all } j = 1, \ldots, n: \quad \sum_{i=1}^{n} x_{ij} = 1; \quad \text{every job is done once.}$$

The objective is to minimize $\sum_{i=1}^{n} \sum_{j=1}^{n} c_{ij} x_{ij}$.

The dual problem is

$$\max \sum_{i=1}^{n} \alpha_i + \sum_{j=1}^{n} \beta_j \qquad \text{subject to}$$
$$\alpha_i + \beta_j \leq c_{ij} \quad i = 1, \ldots, n, j = 1, \ldots, n$$

The dual can be interpreted as assigning a value α_i to each processor i and a value β_j to each job j to maximize total value, subject to not exceeding c_{ij} for the value of the pair i, j. The dual variables are unrestricted in sign.

The first step of the primal-dual algorithm requires an initial dual feasible solution. It is always easy to find a feasible solution to the dual of an assignment problem. For example, set $\beta_j = 0$ for all j and set $\alpha_i = \min_{1 \le j \le n} c_{ij}$.

Let α, β be the current dual feasible solution. Let $K = \{i, j | \alpha_i + \beta_j = c_{ij}\}$ be the set of index pairs of the dual's binding constraints at α, β. The restricted primal has variables $x_{ij} : (i, j) \in K$ and constraints

$$\text{For all } i = 1, \ldots, n: \quad \sum_{j:(i,j) \in K} x_{ij} = 1;$$
$$\text{For all } j = 1, \ldots, n: \quad \sum_{i:(i,j) \in K} x_{ij} = 1;$$
$$\text{For all } (i, j) \in K: \quad x_{ij} \ge 0.$$

As pointed out by Papadimitriou and Steiglitz [241], the restricted primal has been stripped of the costs c_{ij}. Instead, it is the relatively simple feasibility problem of matching processors to jobs when only pairs in K may be used. The dual of the restricted primal is feasible because its constraints are a subset of the constraints of the dual of the unrestricted primal, which has feasible solution α, β. Therefore, if the restricted primal is feasible, it is optimal because it satisfies complementary slackness with the feasible dual solution α, β (by Corollary 2.1.4). If the restricted primal is infeasible, then its dual must have unbounded optimum. The ray of unboundedness a, b must satisfy the constraints:

$$\text{For all } (i, j) \in K: \quad a_i + b_j \le 0; \tag{6.20}$$
$$\textstyle\sum_{i=1}^n a_i + \sum_{j=1}^n b_j > 0. \tag{6.21}$$

As usual with the simplex method, we move from the current feasible solution α, β along the ray $\alpha + \theta a, \beta + \theta b : \theta \ge 0$ as far as possible without violating feasibility. The maximum possible value of θ is strictly positive because $\alpha_i + \beta_j < c_{ij}$ for all $(i, j) \notin K$. That is, constraints for $(i, j) \notin K$ are not binding at α, β. For the binding constraints, (6.20) assures us that increasing θ maintains feasibility, because the left-hand-sides can only decrease or not change. To be specific, the maximum value of θ is

$$\theta^* = \min_{\substack{(i,j) \notin K \\ a_i + b_j > 0}} \frac{c_{ij} - \alpha_i - \beta_j}{a_i + b_j}.$$

From $\theta^* > 0$ and (6.21) the new dual solution $(\alpha, \beta) + \theta^*(a, b)$ has strictly larger objective value than the old dual solution (α, β). Therefore, the set K cannot occur twice and the algorithm will terminate finitely.

Summary: The primal-dual algorithm is actually a general solution method for LPs. To use it effectively, tailor it to your class of LPs in order to solve the subproblems very quickly.

6.4 Self-Dual Parametric Algorithm

Preview: The self-dual parametric simplex algorithm can start from any basic solution, with no requirement of primal or dual feasibility. It employs both primal and dual simplex pivots.

Here is a refreshingly different variant of the simplex method. Unfortunately, it requires both "primal" and "dual" pivots, which makes it unsuitable for fast implementation. (Most implementations don't like to switch back and forth between primal and dual operations.) From a theoretical point of view, it is a good introductory example of a *homotopy* method, and it is similar to variants of the simplex method whose average-case performance has been proved to be very fast. I'll begin with an informal description of the algorithm, continue with a few examples that point up potential difficulties, and end with a precise statement of the algorithm.

Start from a basic solution which may be neither primal nor dual feasible. Alter the problem by adding a parameter λ to every variable, whether primal or dual, that is infeasible. For large enough values of λ, the altered problem is at optimality since it is both primal and dual feasible. As λ decreases from ∞ it eventually reaches a value at which one or more variables reach zero and would become negative if λ were reduced further. For the present assume only one variable threatens to go negative. The degenerate case can be handled by the usual perturbation technique. Pivot the variable out of the basis using a primal or dual simplex pivot depending on whether the variable is a dual or primal variable, respectively. *Question: Why use a dual pivot on a primal variable?*[6] The new basis will be optimal for some range of λ below which some other variable will go negative. Repeat the process until λ reaches zero. At that point you will have an optimal basic solution to the unperturbed original problem. In the absence of degeneracy, λ strictly decreases at each step. Therefore, a choice of basis cannot repeat and the method must terminate in a finite number of steps. As the examples will demonstrate, infeasibility or unboundedness in the primal will require minor modifications to this informal algorithm description.

Example 6.2 For this example, the LP is

$$\begin{array}{rcl} \min 10x_1 - 20x_2 + 30x_3 & & \text{subject to:} \\ x_1 + 2x_2 - x_3 & \leq & 80 \\ 2x_1 - 3x_2 - x_3 & \geq & 20 \\ x_1 - x_2 + 2x_3 & \geq & 40 \\ x_i \geq 0, & i = 1, 2, 3 & \end{array}$$

Adding slack variables and variable x_0 to equal the objective function we get the equivalent LP:

[6] Because the dual simplex method eliminates primal infeasibility.

$$\begin{array}{rcl} \min x_0 & & \text{subject to:} \\ x_0 - 10x_1 + 20x_2 - 30x_3 & = & 0 \\ 0x_0 + x_1 + 2x_2 - x_3 + x_4 & = & 80 \\ 0x_0 - 2x_1 + 3x_2 + x_3 \qquad + x_5 & = & -20 \\ 0x_0 - x_1 + x_2 - 2x_3 \qquad + x_6 & = & -40 \\ x_i \geq 0, \quad i = 1, \ldots, 6 & & \end{array}$$

The initial tableau contains negative terms in the RHS column and a positive term in in the x_0 (negative reduced costs) row. Hence it is both primal and dual infeasible. But it does represent a basic solution. The algorithm will move from basic solution to adjacent basic solution, guided by λ, until it reaches an optimal basic feasible solution. For the initial tableau, add λ to all the negative terms in the RHS column, and subtract λ from all the positive terms in the x_0 row.

x_0	x_1	x_2	x_3	x_4	x_5	x_6	RHS
1	-10	$20-\lambda$	-30	0	0	0	0
0	1	2	-1	1	0	0	80
0	-2	3	1	0	1	0	$\lambda - 20$
0	-1	1	-2	0	0	1	$\lambda - 40$

For all $\lambda \geq 40$ the tableau is optimal. Once λ drops below 40, variable x_6 becomes negative and primal feasibility is lost. To find an optimal solution for $40 - \epsilon < \lambda < 40$ for small $\epsilon > 0$, imagine the tableau at $\lambda = 39.9999$. It is dual feasible but primal infeasible. Apply the dual simplex algorithm to remove x_6 from the basis. The two candidates to enter the basis are x_1 with ratio 10/1 and x_3 with ratio 30/2. Therefore, x_1 enters the basis. The new tableau is shown below. The elementary row operations are similar to those in the dual simplex method, except that the RHS terms are affine functions of λ rather than constants.

x_0	x_1	x_2	x_3	x_4	x_5	x_6	RHS
1	0	$10-\lambda$	-10	0	0	-10	$400-10\lambda$
0	0	**3**	-3	1	0	1	$\lambda + 40$
0	0	1	5	0	1	-2	$60-\lambda$
0	1	-1	2	0	0	-1	$40-\lambda$

The tableau above is optimal for $10 \leq \lambda \leq 40$. At $\lambda < 10$ the reduced cost of x_2 becomes negative, so x_2 must enter the basis. The minimum ratio test to determine the leaving variable compares, at $\lambda = 10$,

$$(\lambda + 40)/(3) = (10 + 40)/3 = 50/3$$

for x_4 with

$$(60 - \lambda)/(1) = (60 - 10)/1 = 50$$

for x_5. Therefore, x_4 leaves the basis. The elementary row operations are similar to those in the primal simplex method, except that $\frac{\lambda - 10}{3}$ times the second row must be added to the top row to make the reduced cost of x_2 equal to zero. Since the RHS terms are already affine functions of λ, the objective function value in the RHS column of the top row becomes a quadratic function of λ. I won't bother to calculate it, though if you do, it will give correct values. There is no need to do so since at optimality you can compute the objective value directly. The next tableau is shown below.

x_0	x_1	x_2	x_3	x_4	x_5	x_6	RHS
1	0	0	$-\lambda$	$\frac{\lambda-10}{3}$	0	$\frac{\lambda-40}{3}$	
0	0	1	-1	$1/3$	0	$1/3$	$\frac{\lambda+40}{3}$
0	0	0	6	$-1/3$	1	$-7/3$	$\frac{140-4\lambda}{3}$
0	1	0	1	$1/3$	0	$-2/3$	$\frac{160-2\lambda}{3}$

The tableau above is primal and dual feasible for $0 \leq \lambda \leq 10$. Since the problem we actually want to solve occurs at the value $\lambda = 0$, the optimal solution is $x_1 = 160/3$, $x_2 = 40/3$, $x_3 = x_4 = x_6 = 0$, $x_5 = 140/3$. The objective function value is $10x_1 - 20x_2 + 30x_3 = 1600/3 - 800/3 = 800/3$.

In our second example, the primal has unbounded optimum. It turns out that it is more complicated to determine unboundedness with the self-dual parametric algorithm than it is with the standard simplex algorithm.

Example 6.3 The LP is

$$\begin{aligned} \max 30x_1 + 40x_2 \quad & \text{subject to} \\ -x_1 + 2x_2 & \leq -5 \\ 3x_1 - x_2 & \geq 2 \\ x_i & \geq 0 \quad i = 1, 2 \end{aligned}$$

Adding slack variables to get an initial basis gives the equivalent LP:

$$\begin{aligned} \max\ 30x_1 + 40x_2 \quad & \text{subject to} \\ -x_1 + 2x_2 + x_3 \qquad & = -5 \\ -3x_1 + x_2 \quad + x_4 & = -22 \\ x_i & \geq 0 \quad i = 1, 2, 3, 4 \end{aligned}$$

The initial tableau is

x_0	x_1	x_2	x_3	x_4	RHS
1	-30	-40	0	0	0
0	-1	2	1	0	-5
0	-3	1	0	1	-2

Add λ to every negative term in the top row and in the right-hand-side (RHS).

x_0	x_1	x_2	x_3	x_4	RHS
1	$-30+\lambda$	$-40+\lambda$	0	0	0
0	-1	**2**	1	0	$-5+\lambda$
0	-3	1	0	1	$-2+\lambda$

The tableau above is optimal for the range $40 \le \lambda < \infty$. At $\lambda = 40$, primal variable x_2 enters the basis. The minimum ratio test compares $\frac{-5+\lambda}{2} = \frac{-5+40}{2} = 17.5$ with $\frac{-2+\lambda}{1} = \frac{-2+40}{1} = 38$. Therefore, x_3 leaves the basis. The pivot term is in boldface. The new tableau is:

x_0	x_1	x_2	x_3	x_4	RHS
1	$-50 + \frac{3\lambda}{2}$	0	$20 - \frac{\lambda}{2}$	0	
0	$-1/2$	1	$1/2$	0	$\frac{\lambda-5}{2}$
0	$-\frac{5}{2}$	0	$-1/2$	1	$\frac{\lambda+1}{2}$

This tableau is optimal for the range $33\frac{1}{3} \le \lambda \le 40$. As λ decreases below $33\frac{1}{3}$, the LP appears to have unbounded optimum as x_1 increases from 0. In particular, at $\lambda = 0$, add $\theta(1, \frac{1}{2}, 0, \frac{5}{2})$ to (x_1, x_2, x_3, x_4) to increase the objective function by 50θ, maintain $A\mathbf{x} = \mathbf{b}$, and not drive any variables negative as θ increases. The vector $d = (1, \frac{1}{2}, 0, \frac{5}{2})$ is a direction to create a ray $\mathbf{x} + \theta d$ of unboundedness.

Reading Tip: *Before you read further, try to figure out why we may not jump to the conclusion that the primal is unbounded.*

The answer is that we don't have a starting point $\mathbf{x}$ for the ray! At $\lambda = 0$ the current tableau gives the basic solution $x_1 = 0, x_2 = -\frac{5}{2}, x_3 = 0, x_4 = \frac{1}{2}$ which is not feasible. All we need is one feasible solution $\mathbf{x}$ at which to start the ray. If a feasible $\mathbf{x}$ exists, the LP is unbounded. Otherwise, the LP is infeasible. From the dual point of view, the vector $\mathbf{d}$ shows that the dual is infeasible. So by the strong duality theorem, the primal is infeasible or unbounded. Algorithmically, we switch from the self-dual parametric algorithm to the dual simplex algorithm, because all we want is a feasible primal solution. We could instead switch to a primal Phase I algorithm — that would be valid as well. To use the dual simplex method, set $\lambda = 0$ and replace the primal objective vector with an arbitrary one that makes the dual feasible. I've used values $c_j = 0 : j \in \mathbb{B}$ and $c_j = 1 : j \notin \mathbb{B}$, but any values that make the reduced costs bad would be OK.

x_0	x_1	x_2	x_3	x_4	RHS
1	-1	0	-1	0	0
0	$\mathbf{-1/2}$	1	$1/2$	0	$-\frac{5}{2}$
0	$-\frac{5}{2}$	0	$-1/2$	1	$1/2$

Variable x_2 exits the basis and variable x_1 enters. The pivot term is in boldface. The new tableau is

x_0	x_1	x_2	x_3	x_4	RHS
1	0	-2	-2	0	5
0	1	-2	-1	0	5
0	0	-5	-3	1	13

This tableau represents the primal feasible solution $\mathbf{x} = (5, 0, 0, 13)$. Plug in the original objective vector $\mathbf{c} = (30, 40, 0, 0)$ to compute its objective value, $\mathbf{c} \cdot \mathbf{x} = 150$. The ray of unboundedness is

$$\mathbf{x} + \theta(1, \frac{1}{2}, 0, \frac{5}{2}) = (5 + \theta, \frac{\theta}{2}, 0, 13 + \frac{5\theta}{2}) : \ \theta \ge 0$$

with objective value $150 + 50\theta$. If instead the dual simplex method had revealed primal infeasibility, we would have concluded that the LP was infeasible.

In one sense, the self-dual parametric algorithm is a variation of the primal and dual simplex algorithms. In another sense, it introduces a completely new idea for solving problems. We want to solve problem P. Create an infinite set of problems $P(\theta)$, parameterized by θ, such that $P(0) = P$ and $P(t)$ is easy to solve for some easily determined value t. ($P(\theta)$ should be a continuous function of θ in the sense that both the objective and constraint functions are continuous functions of θ.) Starting from $\theta = t$, reduce $|\theta|$ to 0 continuously, tracking the solution to $P(\theta)$ as θ changes. When θ reaches 0, problem P is solved. Is P too hard to solve directly? Change it to an easy problem, solve that, then change the problem back to P. That may seem like a crazy idea, but it can work. This kind of algorithm is called a homotopy method. Homotopy methods have been used successfully to find equilibria [83],[84],[85],[259] and to solve systems of polynomial equations [190],[254] and various other nonlinear problems [148].

Summary: The self-dual parametric algorithm introduces the concept of homotopy algorithms, and has the advantage of requiring neither primal nor dual feasibility to start. To date, however, its use of both primal and dual simplex pivots prevents it from being computationally competitive with the dual or primal simplex algorithms.

6.5 Dantzig-Wolfe Decomposition

This method was invented by Dantzig and Wolfe [72] to solve problems with a special structure, shown pictorially in Figure 6.11. Only a small portion of the constraints employ all of the variables. The rest of the constraints form a block-diagonal matrix. Decomposition has the advantage of requiring less memory than the standard method because each block is solved separately. It is reminiscent of the primal-dual method in that it solves a single LP by solving a sequence of simpler LPs, which sounds roundabout but often turns out to be efficient.

However, memory is very rarely a limitation on modern computers, and if the entire constraint matrix can be stored (which essentially depends on the total number of nonzeros), the decomposition method is no faster than standard methods. Therefore, this method is obsolete for linear programming, and I have relegated its derivation to the exercises. On the other hand, the decomposition concept leads to the Lagrangian relaxation method for integer programming [145, 118, 117, 146], which is anything but obsolete (see, *e.g.*, [96, 55, 35, 256]). Dantzig-Wolfe decomposition also has a lovely interpretation as a decentralized planning method that uses prices to coordinate among the planning components. See Chapter 23 of Dantzig's original book on linear programming for details [73]. The dual version of Dantzig-Wolfe decomposition is called *Benders decomposition* [22].

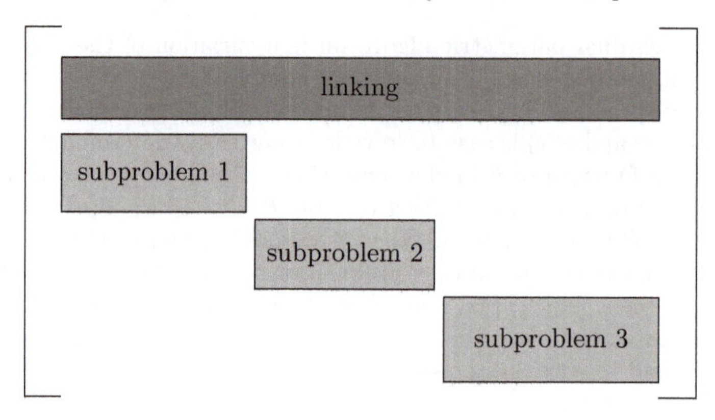

FIGURE 6.11: Coefficients are zero in unshaded areas of the constraint matrix. The variables of the large LP are partitioned to form several smaller *subproblem* LPs that are independent of each other except for the master *linking* constraints in the top part of the constraint matrix.

6.6 High Level Computational Issues

Preview: This section presents several "big-picture" computational principles and applies them to the simplex method.

When you implement an algorithm, you should always consider several principles. They are:

- Compute only what you need.
- Data structures usually make or break your algorithm. [185].
- Changes in values very often can be computed in less time than it takes to compute the values from scratch. This principle ties in closely with the choice of data structures.
- Mathematical correctness does not ensure numerical accuracy. Nonzero values that are close to zero are especially problematic.
- Never make assumptions about the input data that users are going to supply.
- The best theoretical guarantees of algorithm performance are often based on conservative choices of parameters or strategies. Don't permit actual errors in output, but do try for bolder choices.
- Whenever one portion of your code accounts for more than half of your CPU time, concentrate on that portion.
- Most real data sets are sparse, that is, only a small fraction of the data are nonzero.
- Is the algorithm being controlled by irrelevant data?

With these principles in mind, let's examine the simplex method.

Compute changes in values: When the basis changes, only one column of B has changed. Computing the change in B^{-1} is an order of magnitude quicker than computing the new values of B^{-1} from scratch. This principle is employed even in the full tableau method. It should be employed in the algorithm of Table 5.1.

Compute only what you need: The full tableau method computes $B^{-1}A_{:,N}$, but of these columns only the one corresponding to the entering variable is needed. This makes the algorithm in Table 5.1 superior. Historically that was called "the revised simplex method."

Numerical accuracy and values near 0*:* The tableau version of the simplex method is not numerically stable. It starts to fail at about 25 to 30 rows. There are ways to avoid the cumulative round-off errors of row operations. You can recalculate B^{-1} from scratch every, say, 50 or 100 iterations. This is called "reinverting the basis." You also get better stability if you remember, in effect, the set of row operations represented by B^{-1} rather than maintaining the matrix $B^{-1}A$. This is called the product form of the inverse.

If a square submatrix of A is singular, it cannot represent a basis, which by definition must consist of linearly independent columns. Therefore, we don't have to worry about cases where B is singular, i.e., B^{-1} does not exist. Those cases will not arise. The cases to worry about are bases for which B is *almost* singular. If the determinant of B is nonzero, but close to zero, the algorithm can become numerically unstable. It is helpful to avoid bringing a column into B that brings B's determinant close to 0, or that has values much smaller (but nonzero) or much larger than the values in the other columns. Therefore, it is a good idea to scale the rows and columns of the LP to have less unequal values prior to solving it.

Check the input data: The LP constraints can fail to be full rank. If so, add variables to the LP or remove redundant constraints. The LP can be infeasible without the user knowing it. If the LP is infeasible, the output should provide information as to the cause of infeasibility.

Sparseness and data structures: All of the matrix A data should be stored using sparse data structures. Methods to store and update B^{-1} when B is sparse have been developed by researchers in linear algebra.

Compute only what you need, again: Do we really need B^{-1}? No, we just have to be able to solve certain equations involving B, namely, $Bx_{\mathbb{B}} = \mathbf{b}$, $\boldsymbol{\pi}^T B = c_{\mathbb{B}}^T$ and $B\mathbf{h} = A_{:,k}$ where x_k is the entering variable. It turns out to be less expensive computationally to maintain the ability to solve these equations, as B changes, than to keep the matrix B^{-1} explicitly. In good implementations of the simplex method, computing the dual variables, calculating the minimum ratio test, and updating the ability to solve equations involving B all require a significant portion of computation time, but no one of these three accounts for the great majority of computation time.

Do we really need temporary (artificial) variables? The answer is usually no. Examine the system of constraints (5.30),...,(5.35) Variables y_2 and y_3, respectively, have exactly the same coefficients as do the slack variables x_4 and x_6, times a factor of -1. This remains true no matter what elementary row operations have been performed on the system of constraints. Also, it is impossible for a matching pair of variables such as y_2 and x_4 to be in the basis at the same time. *Question: Why?*[7] Therefore, a temporary variable inserted into a constraint that has a slack variable is not necessary from a data storage and computational point of view. The variables inserted into equality constraints are even simpler to eliminate from the computations, theoretically speaking. Use the constraint to eliminate one variable from the LP. That is, if the equality constraint is $\mathbf{a} \cdot \mathbf{x} = a_0$, and $a_t \neq 0$, replace x_t by $(a_0 - \sum_{j \neq t} a_j x_j)/a_t$ in other constraints where x_t appears, and replace the equality constraint by $\mathbf{l}_t \leq (a_0 - \sum_{j \neq t} a_j x_j)/a_t \leq \mathbf{u}_t$. However, if the equality constraint has many

[7]Their columns are linearly dependent.

nonzero coefficients, and x_t appears in many other constraints, this replacement may greatly increase the number of nonzero coefficients in the LP. This phenomenon is an example of what is called "fill-in". To avoid fill-in, choose a variable x_t that appears in few other constraints, or employ the temporary variable.

Irrelevant data: Imagine that you multiply the coefficients of a variable x_j by a scalar $\alpha > 0$. The LP does not change in any essential way, as long as you remember to multiply the jth objective function coefficient, too. But the value of the reduced cost (the dual variable corresponding to x_j) changes by a factor of α. This change could cause a different choice of the entering variable in the primal simplex algorithm. Similarly, if you multiply a constraint by a positive scalar, the dual simplex method could make a different choice of the exiting variable. Therefore, the rule we have been using to choose the entering (respectively exiting) variable in the primal (respectively dual) simplex method is influenced by irrelevant data. It would make more sense to scale, for each $j \in N$, the reduced cost of x_j by a value like $||B^{-1}A_{:,j}||$ before selecting the entering variable for the primal simplex method. Scaling the reduced costs significantly reduces the number of simplex iterations [26, 130, 238]. Of course, the rule itself must not require too heavy a computational cost. Consequently, approximate scaling such as Paula Harris's *devex* pricing [143] and its successors [101] are used in practice.

Summary: The dual and bounded variable simplex algorithms are conceptually very similar to the primal algorithm. The former selects entering and leaving basic variables in reverse order, because it begins with dual feasibility and works towards primal feasibility. The latter permits nonbasic variables to be at either their lower or upper bounds, rather than at a fixed lower bound of zero. The full tableau method aids in understanding, but is numerically unstable and inefficient in both time and space usage.

6.7 Notes and References

The dual simplex method is due to Carlton Lemke [205], whose other great algorithmic invention was the Lemke-Howson pivoting algorithm [206, 207] for the linear complementarity problem, still widely used for Nash equilibria. Harold Kuhn invented the primal-dual method to solve assignment problems [197]. He called his algorithm the "Hungarian Method" to honor prior work of the Hungarian mathematicians Dénes König and Jenö Egerváry. Variations [198] and extensions to general linear programming [75] followed. Papadimitriou and Stieglitz's exposition of the primal-dual algorithm as a general method to reduce optimization problems with numerical objectives to a sequence of non-numerical feasibility problems [241] deepened the appreciation of the method. Interest in primal/dual methods blossomed again with Goemans and Williamson's approximation algorithms, which they pointed out could be seen as primal/dual methods [124, 125]. For additional history of the method see [200, 202], and [104]. For additional history of linear and combinatorial optimization, enjoy the unmatched scholarship of [264].

6.8 Problems

E Exercises

1. For the linear program

$$
\begin{aligned}
\text{Minimize } 2x_1 - 10x_2 + 13x_3 & \qquad \text{subject to:} \\
2x_1 + 3x_2 + x_3 + x_4 &= 10 \\
3x_1 - 2x_2 - 4x_3 \qquad + x_5 &= -2 \\
-20 \le x_1 \le 5; 2 \le x_2 \le 4; -10 \le x_3 \le 16 & \\
0 \le x_4; 0 \le x_5. &
\end{aligned}
$$

   (a) Find the basic solution when x_1, x_2, x_3 are all at their lower bounds and x_4, x_5 are basic. Is the solution feasible? Is it degenerate?

   (b) Find the basic solution when x_1, x_2, x_3 are all at their upper bounds and x_4, x_5 are basic. Is the solution feasible? Is it degenerate?

   (c) Find the basic solution when x_1 is at its lower bound, x_2, x_3 are at their upper bounds, and x_4, x_5 are basic. Is the solution feasible? Is it degenerate?

   (d) Find the basic solution when x_2 and x_4 are at their lower bounds, x_3 is at its upper bound, and x_1, x_5 are basic. Is the solution feasible? Is it degenerate?

   (e) Set up an initial full tableau from the basic solution where x_1, x_2 are at their lower bounds and x_3 is at its upper bounds. For which nonbasic variable or variables is it good for the objective function to make the variable enter the basis, and what variable would leave the basis?

2. Solve the LP of Figure 6.1 starting at $(x_1, x_2) = (4, 2)$. Verify that you follow the path shown in the figure.

3. Solve the following LP graphically, and then solve with the bounded simplex method starting with x_1, x_2 at their upper bounds. Track the steps of the bounded simplex method on your graph of the LP.

$$
\begin{aligned}
\text{Minimize } 40x_1 + 21x_2 & \qquad \text{subject to:} \\
2 \le x_1 \le 10; 5 \le x_2 \le 15 & \\
x_1 + x_2 &\ge 9 \\
2x_1 - 3x_2 &\le 1 \\
-x_1 + 2x_3 &\le 25
\end{aligned}
$$

4. Set up the following problem for solution by the dual simplex method. Find the initial basic solution and the corresponding dual solution. Show that the former is infeasible and the latter is feasible. Find the leaving and entering variables for the first iteration.

$$
\begin{aligned}
\text{Minimize } 20x_1 + 32x_2 + 45x_3 & \qquad \text{subject to:} \\
x_1 + x_2 + 2x_3 &\ge 10 \\
3x_1 + x_2 + x_3 &\ge 12 \\
x_1, x_2, x_3 \ge 0 &
\end{aligned}
$$

5. Apply the dual simplex method to the following infeasible LP. Identify the contradiction in the primal constraints that the dual simplex method finds. Identify the direction π of unboundedness in the dual that this corresponds to. Graph the dual's feasible region and verify geometrically that π is a direction.

$$\begin{aligned} \text{Minimize } x_1 + x_2 & \qquad \text{subject to:} \\ x_1 + x_2 &\geq 20 \\ -3x_1 - 2x_2 &\geq 5 \\ x_1, x_2 &\geq 0 \end{aligned}$$

6. Imitate the example in Section 6.2.1 for the problem, Maximize $5x_1$ subject to the constraints $x_1 \leq 12$ and x_1 nonnegative.

7. Repeat Exercise 6 with the additional constraint $2x_1 \leq 18$.

8. Construct a small example of a dual simplex tableau that indicates primal infeasibility, and show explicitly the ray of unboundedness in the dual.

9. Solve the LP to minimize $7x_1$ subject to $x_1 \geq 0, x_1 \geq 3, x_1 \geq 5$ by the dual simplex method. Solve it again by the primal-dual method starting with a dual feasible solution in which the dual variable for $x_1 \geq 3$ is nonzero and the dual variable for $x_1 \geq 5$ is zero.

10. For the 3×3 assignment problem with $c_{ii} = 10\ \forall i$, $c_{12} = 3, c_{23} = 5, c_{31} = 4$, $c_{21} = 1, c_{32} = 50, c_{13} = 2$:

    (a) Find the optimal solution by inspection.

    (b) Write the primal LP. Use variables $x_{ij} \geq 1$ to denote the fraction of job j done by worker i.

    (c) Write the dual LP.

    (d) By inspection of the dual, show that the primal optimal solution value is greater than or equal to 12.

    (e) Find an optimal dual solution in which the variable corresponding to constraint $\sum_i x_{i3} = 1$ is zero.

    (f) For dual variables π_j corresponding to $\sum_i x_{ij} = 1$ and μ_i corresponding to $\sum_j x_{ij} = 1$, verify that the solution $\mu_1 = 2, \mu_3 = 3, \pi_1 = 1$ and other variables 0 is feasible. Initiate the primal-dual algorithm from that solution by setting up the restricted primal, verifying that the restricted primal is infeasible and writing the phase I LP for the restricted primal.

11. Set up the following problem for solution by the self-dual parametric algorithm. Identify the optimal solution for very large values of θ. Identify the value of θ where the first change of basis occurs. Identify the entering variable, the leaving variable, and compute the resulting new tableau and optimal solution.

$$\begin{aligned} \text{Minimize } 20x_1 - 30x_2 + 40x_3 - 50x_4 & \qquad \text{subject to:} \\ x_1 + x_2 + x_3 + x_4 &\geq 15 \\ 3x_1 - 6x_2 \qquad - 8x_4 &\geq 25 \\ 4x_1 + 5x_2 + 2x_3 + x_4 &\leq 65 \\ x_i &\geq 0\ \forall i \end{aligned}$$

The $\leq$ symbol in the third constraint is not a typographical error.

M Problems

12. Imitate the example in Section 6.2.1 using your own 2×2 matrix A and 2-vectors b, c.

13. Can you find a point $\mathbf{x}$ such that $A\mathbf{x} \leq \mathbf{b}; x \geq 0$ where $b \ngeq 0$ using the dual simplex method instead of temporary variables? Start from the basic infeasible solution $\mathbf{x} = 0$.

14. Solve the assignment problem of Exercise 10 with the primal-dual algorithm, beginning with the feasible dual solution given. You may solve the restricted primal Phase I LPs by inspection and their duals by complementary slackness and inspection.

15. Consider the problem to minimize x_0 subject to $x_0 \geq 0$ and $x_0 \geq \beta_i$ for $i = 1, \ldots m$. Let the m corresponding dual variables be $\boldsymbol{\pi}_1, \ldots, \boldsymbol{\pi}_m$. The β_i are strictly positive scalar data. Suppose $\beta_i < \beta_{i+1}$ for all $i = 1, 2, \ldots, m-1$. Analyze what the primal-dual algorithm will do when applied to this problem starting from the dual feasible solution $\boldsymbol{\pi}_1 = 1; \boldsymbol{\pi}_i = 0 \forall i \geq 2$. What is the least number of iterations the algorithm could take to reach optimality? What is the largest number of iterations it could take to reach optimality?

16. Solve the LP in Problem 11.

17. Change the RHS vector in Problem 11 from $(15, 25, 65)$ to $(100, -40, 60)$ and solve the resulting infeasible LP with the self-dual parametric algorithm.

18. Derive a version of the perturbation method that applies to the bounded simplex algorithm, and prove that it averts cycling.

19. Prove that if the initial dual solution is chosen in a particular simple way, the primal-dual method terminates in at most n iterations when applied to an $n \times n$ assignment problem. Hint: If in the restricted primal all $x_{ij} = 0$, one iteration of the primal-dual algorithm will allow $sum x_{ij}$ to increase from 0 to at least 1.

20. Derive the primal-dual method for linear programs with variables x_{ij} and y_j of form

$$
\begin{array}{lcl}
\min & \sum_{j=1}^{m} d_j y_j + \sum_{i=1}^{n} \sum_{j=1}^{m} c_{ij} x_{ij} & \text{subject to} \\
\forall i, j & x_{ij} - y_j \leq 0; & \\
\forall i & \sum_{j=1}^{m} x_{ij} = 2; & \\
\forall j & 0 \leq y_j \leq 1; & \\
\forall i, j & 0 \leq x_{ij} &
\end{array}
$$

21. For the dual simplex algorithm, develop a perturbation method that guarantees a nonzero change in the objective function at each step (making the objective value worse at each step) thereby ensuring finite termination.

22. For the self-dual parametric simplex variant, develop a perturbation method that guarantees strict decrease in θ at each step, thereby ensuring finite termination.

23. Construct a small example of a dual simplex tableau and perform an iteration. Convert your LP to its dual, and perform an iteration of the primal simplex method. Explain why the calculations performed are not identical, even though the choices of bases match.

24. Prove or disprove: In the primal and dual simplex methods, the entering (exiting) variable cannot be the exiting (entering) variable in the next iteration. (This is four problems in one, two for the primal and two for the dual simplex method.)

25. For the LP $\max 333x_1$ subject to $0 \leq x_1 \leq 10; x_1 \leq 12$, guess the values of the dual variables at the primal's optimum. Do not discard the constraint $x_1 \leq 12$. Solve the LP with the simplex method, starting at the BFS $x_1 = 0$. Confirm that your guess is correct. What happens to the optimal objective value if the lower bound on x_1 is increased from 0 to 0.5? If the upper bound is increased from 10 to 10.5? If the upper bound is increased from 10 to 20? Compare the shadow price predictions with the actual values.

26. Using the primal simplex algorithm, solve the problem $\max 100x_1 + 120x_2 + 150x_3$ subject to $x_i \geq 0$, $x_1 + x_2 + x_3 \leq 12$; $3x_1 + x_2 + 2x_3 \leq 25$. Start from the basic feasible solution where $x_1 = x_2 = x_3 = 0$. Find the values of $\mathbf{x}_{\mathbb{B}}, B$, and $\mathbf{c}_{\mathbb{B}}$ at the optimum solution. Find the values of the dual variables. Use the dual variables to predict the impact on the optimum objective function value of each of the following changes (made separately, one at a time.)

    (a) Change the right-hand-side of the first constraint from 12 to 11.

    (b) Change the right-hand-side of the second constraint from 25 to 26.

    (c) Change the lower bound on x_3 from 0 to 0.1.

    (d) Change the lower bound on x_1 from 0 to 0.1.

27. Suppose the dual simplex method has reached an optimal solution, and that the jth RHS term $B^{-1}_{j,:}b = 0$.

    (a) Give an example to show that the optimal dual solution may be unique.

    (b) Give a simple set of conditions under which the optimal dual solution is not unique.

    (c) Give a simple set of conditions under which there exists an infinite ray of optimal solutions.

28. The finite termination of the simplex method proves that if both primal and dual are feasible, they have the same optimal objective value. The weak duality theorem tells us that if the primal has unbounded optimum, then the dual must be infeasible. Use these two facts (and no others) to prove Theorem 4.3.2. (Hint: Consider an LP with a zero cost vector.)

29. Suppose a ray of unboundedness is found in the dual during the dual simplex method. (i) (easy) What is the contradiction that is apparent in the primal constraints? (ii) (not tricky, but less easy) How can you exhibit an explicit contradiction in the primal constraints that were present at the beginning of the simplex method?

30. Let A be a matrix with m columns and n rows. Consider the problem

$$\begin{array}{lcl} \min & \sum_{i=1}^{n} f_i(x_i) & \text{subject to} \\ A\mathbf{x} & = & \mathbf{b} \\ \mathbf{x} & \geq & \mathbf{0} \end{array}$$

where for each $i = 1 \ldots n$ the function $f_i(x_i)$ is piecewise linear, defined in the range $[0, \infty)$.

(a) Derive a variation of the simplex method similar to the usual primal method which operates directly on this problem, not on a reformulation. Begin by defining basic solution and basic feasible solution. Assume that you will be given an initial basic feasible solution according to your definition of BFS.

(b) Give a simple example for which your algorithm fails to find an optimum solution, even though one exists. Your example must be feasible. You may choose the starting BFS.

(c) Under what natural condition is your algorithm guaranteed to find an optimal solution? Prove that you are correct. You may assume that cycling does not occur.

(d) Does complementary slackness have a meaning in your algorithm?

31. Show how to initialize the Phase I restricted primal (6.17),(6.18),(6.19) from the optimal solution to the Phase I restricted primal in the previous iteration of the primal-dual algorithm. Hint: Suppose x_i was a basic variable in the optimum solution that was found in the previous iteration. What must be true about its reduced cost in the current Phase I restricted primal?

32. Apply the self-dual parametric algorithm to the following problem.

$$
\begin{array}{rcl}
\min 10x_1 - 20x_2 - 30x_3 & & \text{subject to:} \\
x_1 + 2x_2 - x_3 & \leq & 8 \\
2x_1 - x_2 - x_3 & \geq & 20 \\
6x_1 - x_2 - 2x_3 & \geq & 40 \\
x_i \geq 0, & i = 1, 2, 3 &
\end{array}
$$

(i) Show that for a strictly positive λ the primal has unbounded optimum, but the endpoint of the ray of unboundedness is infeasible at $\lambda = 0$. (ii) Apply the modification of the algorithm given in the chapter so that it terminates correctly.

33. Consider a *minimization* LP that has an optimal solution. Let $z(\theta)$ be the objective function value of the feasible basic solution during the course of the self-dual parametric algorithm, as θ ranges from ∞ down to 0. Prove or disprove: $f(\theta)$ is

(a) nondecreasing.

(b) piecewise linear.

(c) piecewise quadratic.

(d) convex.

(e) concave.

(f) continuous.

34. Determine whether or not Bland's rule works for the bounded simplex method.

D Problems

35. Prove or disprove finite convergence of the primal-dual algorithm.

36. The decomposition method was invented to solve large LPs of the form

$$\begin{array}{rcl} \min \ \sum_{k=1}^{K} \mathbf{c}^k \cdot \mathbf{x}^k & & \text{subject to} \\ \sum_{k=1}^{K} A^k \mathbf{x}^k & \geq & \mathbf{b} \\ D^k \mathbf{x}^k & \geq & \mathbf{d}^k \quad k = 1, \ldots, K \\ \mathbf{x}^k \geq \mathbf{0} \quad k = 1, \ldots, K & & \end{array}$$

The $\sum_{k=1}^{K} A^k x^k \geq b$ constraints and the objective function form the *master* problem. For each k, the $D^k x^k \geq d^k$ and $x^k \geq 0$ constraints form a *slave* problem. The key property of the slave problems is that no two slave problems share any variables.

(a) Prove that there exist integers n^k, m^k and finite sets of vectors $v^{k,i}, w^{k,j} : i = 1, \ldots, n^k; j = 1, \ldots, m^k;\ k = 1, \ldots, K$ such that the above LP is equivalent to

$$\min \sum_{k=1}^{K} \left(c^k \cdot \left(\sum_{i=1}^{n^k} \alpha^{k,i} v^{k,i} + \sum_{j=1}^{m^k} \beta^{k,j} w^{k,j} \right) \right) \qquad \text{subject to} \tag{6.22}$$

$$\sum_{k=1}^{K} \left(\sum_{i=1}^{n^k} A^k v^{k,i} \alpha^{k,i} + \sum_{j=1}^{m^k} A^k w^{k,j} \beta^{k,j} \right) \geq b \tag{6.23}$$

$$\alpha^{k,i} \geq 0; \quad \beta^{k,j} \geq 0 \ \ \forall i, j, k \tag{6.24}$$

$$\sum_{i=1}^{n^k} \alpha^{k,i} = 1; \quad k = 1, \ldots, K \tag{6.25}$$

(b) Where does your proof in part (a) use the $x^k \geq 0$ constraints? Why would you probably not want to write the LP in part (a) explicitly?

(c) Let $\boldsymbol{\pi}$ and γ be the vectors of dual variables corresponding, respectively, to constraints (6.23) and (6.25) at a basic feasible solution to the LP in part (a). Fix k. Write a "slave" LP that uses D^k, d^k rather than A^k data and which if solved finds an i such that variable $\alpha^{k,i}$ has strictly negative reduced cost , or finds a j such that variable $\beta^{k,j}$ has strictly negative reduced cost, or determines that neither such an i nor j exists.

(d) The general form of the decomposition method should now be clear to you. Starting from a feasible solution that uses a small number of the variables from each slave problem, solve the master problem to get the dual variables $\boldsymbol{\pi}, \gamma$. Solve the slave problems to find a variable to enter the basis of the master problem. Add this variable to the master problem, solve it for new dual variable values, and iterate until optimal. Why is it not very computationally expensive to solve the master problem repeatedly? Why is it not very computationally expensive to solve the slave problems repeatedly? If more than one slave problem finds a variable with favorable reduced cost, would you add just the one with best reduced cost to the master problem, or would you add all of the ones with favorable costs?

(e) What would happen in the algorithm if the problem had an unbounded optimum?

I Problems

37. Specify how to integrate the dual simplex algorithm with the simplex method for bounded variables.

38. (a) Show that the following is a valid formulation of the assignment problem: minimize
$$\sum_{i=1}^{n}\sum_{j=1}^{n} c_{ij}x_{ij}$$
subject to
$$\sum_{i=1}^{n} x_{ij} \geq 1 \ \forall j; \ \sum_{j=1}^{n} x_{ij} \leq 1 \ \forall i; \ x_{ij} \geq 0 \ \forall i, j.$$
    (b) Determine whether or not the polyhedron of feasible solutions has degeneracy.
    (c) Write the dual.
    (d) Find a vector **c** such that the dual has the most possible amount of degeneracy. (Choose your own measure of degeneracy.)
    (e) Find a vector **c** such that the dual has the least possible amount of degeneracy. (Choose your own measure of degeneracy.)
    (f) Characterize degeneracy in the dual.

39. (a) Suppose that, prior to solving the LP (2.1) with the self-dual parametric algorithm, the cost vector **c** is multiplied by a large constant so that the c_i magnitudes dwarf the b_j magnitudes. Analyze how the algorithm will behave.
    (b) Suppose instead that the RHS vector b is multiplied by a large constant prior to solving. Analyze how the algorithm will behave.

40. Suppose that a problem of the form given at the beginning of Problem 36 (the kind of problem for which the decomposition method was invented) is to be solved with software that uses the usual simplex algorithm and does not store matrix data in sparse data format.
    (a) How much space would it take to store: a basis matrix B? A full tableau?
    (b) How much space would be needed for B and for a full tableau for the master problem of the decomposition method? How much space would be needed for a slave problem?
    (c) If the problem were sparse, and the software used data structures for sparse matrices, how much space would be needed for B if you were using the usual simplex algorithm? How much space would be needed for the master problem and for a slave problem if you were using the decomposition method?

41. (*) Explain why the dual simplex method tends to be faster than the primal simplex method in practice more than one would expect based on the ratio of the number of columns to the number of rows (the aspect ratio).

Chapter 7

Shadow Prices, Sensitivity Analysis, and Column Generation

Preview: "Change is the only constant," as an old saying goes. After you solve an LP model, it is very likely that you will want to modify it and solve again, because the data have changed or to perform "what-if" assessments. This chapter is about predicting the approximate impact of a change in an LP on the optimal solution, and computing the new optimal solution precisely.

One reason LP is so useful in practice is that, on solving a primal LP, one gets the dual solution as well. The dual variables predict how the objective value would change if the constraint right-hand sides were changed. Dual variables are called *shadow prices* when they are used for these predictions. If we have a basic optimal primal solution, its choice of basis can be used to predict how the solution and objective value would change if the problem were changed.

The predictions are all based on one principle: what would the current choice of basis do with the altered data?

If there is no degeneracy, the predictions are correct for sufficiently small changes, and may be correct for larger changes. If there is degeneracy, the predictions may fail to be correct even for tiny changes.

Memorize the following rule:

When a prediction about a change in constraints fails to be accurate, it always fails by being optimistic.

We will rigorously justify this rule with Theorem 7.1.1. For now, the intuitive explanation is that the prediction only uses local information about constraints. That is, the prediction is based only on the constraints that are binding at the optimum solution. As one moves away from the current optimum, additional constraints can come into play, and make things worse.

Example 7.1 Consider

$$\max 5x_1 \quad \text{subject to}$$
$$x_1 + x_2 \leq 20;$$
$$x_1, x_2 \geq 0$$

The optimum primal solution is $20, 0$ with dual variable value 5. The primal slack variable has value 0 and the dual slack variables have values $0, 5$. The predicted effect of increasing the RHS value 20 is to increase the objective at a rate of 5. *Question: Which 5, the dual variable or the dual slack variable?*[1] This is accurate for any increase. However, if the LP also had the constraint $x_1 \leq 100$, then at a right-hand-side value of 100.1 the prediction would be optimistic. If we decrease the RHS, the prediction is a reduction in value at a rate of 5. This is accurate until the RHS drops to 0. After that, the prediction is optimistic. Instead of suffering a decrease in value of 5, the problem becomes infeasible (which is like a value of $-\infty$).

Question: What is the dual of this rule?[2]

7.1 The Mathematical Justification of Shadow Prices

Preview: Section 3.6 introduced the concept of a dual variable as a shadow price, the value of an additional unit of a resource, or the cost of changing the RHS of a constraint by one unit. This section puts shadow prices on a firm mathematical footing.

Let A be an $m \times n$ matrix. Suppose basis $\mathbb{B}$ provides an optimal nondegenerate solution $\mathbf{x}_{\mathbb{B}} = B^{-1}\mathbf{b} > \mathbf{0}$ to the LP $\min \mathbf{c} \cdot \mathbf{x}$ subject to $A\mathbf{x} = \mathbf{b}; x \geq \mathbf{0}$. Change the value of b_i by $\pm\epsilon$. For sufficiently small $\epsilon > 0$, $B^{-1}\mathbf{b} \pm \epsilon B^{-1} e_i \geq \mathbf{0}$. *Question: How big can ϵ be?*[3] Changing $\mathbf{b}$ does not affect the reduced costs. Hence basis $\mathbb{B}$ provides an optimal solution to the changed LP. The resulting change in objective function value is

$$\pm\epsilon \mathbf{c}_{\mathbb{B}}^T B^{-1} e_i = \pm\epsilon \boldsymbol{\pi}_i \tag{7.1}$$

where $\boldsymbol{\pi} = \mathbf{c}_{\mathbb{B}}^T B^{-1}$ is the optimal dual solution. This *sensitivity* information is very useful, because many real-world constraints are somewhat flexible. Such constraints are called "soft," as opposed to "hard constraints" that cannot be altered. If you know how much it is worth to you to change the RHS of a constraint, you are in a good position to negotiate for a change. If the constraint represents an upper bound on the availability of a particular resource, you know how much it is worth to you to buy (or sell) some of that resource.

Example 7.2 In the manufacturer and resource purchaser LP of Section 3.1, if constraint 1 is an upper bound on gallons of gas used in the primal, then $\boldsymbol{\pi}_1$ would tell you how much you'd be willing to pay to have one more gallon of gas, or be paid to have

---

[1]The dual variable.

[2]When a prediction about a change in objective function coefficients fails to be accurate, it always fails by being pessimistic.

[3]$\min_{1 \leq j \leq m} B^{-1}_{j,:} b / B^{-1}_{ji}$.

one less. We say that π_1 is the *shadow price* of a gallon of gas. If the market price is lower than the shadow price, it is to your advantage to buy more; if the market price is higher, it is to your advantage to sell some of what you have.

Example 7.3 In the diet and vitamin salesman LP of Section 3.1, if constraint 2 is a lower bound on vitamin D intake, then π_2 is the increase in cost you will incur if the lower bound is increased by one unit, which equals the decrease in cost you will enjoy if the lower bound is decreased by one unit. Not at all by coincidence, this is the price the vitamin salesman from the dual LP will offer you for a pill containing one unit of vitamin D.

If there is a slack variable x_k in the constraint $A_{i,:}x = b_i$, then the reduced cost of x_k equals the negative of the dual variable π_i for constraint i if the constraint had been a $\leq$ constraint prior to the introduction of x_k into the model. *Question: Why the negative?*[4] If the constraint had been a $\geq$ constraint, the value is not multiplied by -1 because in that case $A_{:,k} = -e_i$. If you are ever confused about the meaning of a dual variable, remember that relaxing a constraint can only be harmless or beneficial, and tightening a constraint can only be harmless or harmful. If the constraint was originally an equality constraint, use Equation (7.1) to determine the meaning.

Many shadow prices will be zero, because all nonbasic dual variables equal zero. Mathematically, this is simply complementary slackness. If the upper bound on a resource in the manufacturer LP is not binding at optimality, additional units of that resource are worth nothing to you, because you aren't even using what you have available. If the lower bound on a nutrient in the diet problem is not binding at optimality, your cost does not increase if the lower bound is increased. Your diet already contains more than the minimum required amount, so as to satisfy other constraints at minimum cost.

It may be startling to think of a resource as being worth zero, or a constraint as costing you zero to comply with. Keep in mind that the shadow price is the *marginal* value of the resource, or the *marginal* cost of enforcing a constraint. The rule I asked you to memorize at the beginning of this chapter says that the prediction, if inaccurate, is optimistic.

Example 7.4 Continuing Example 7.2, reduce the available amount of gasoline available from b_1 to 0 gallons. If the gasoline constraint is not binding when b_1 gallons are available, then initially the gasoline reduction will cost you nothing. Eventually, however, the constraint will become binding with respect to the current optimal solution. When it does, the cost will increase (it could stay zero for a while longer if there is dual degeneracy), and could eventually rise to ∞, meaning that the LP becomes infeasible because there isn't enough gasoline available to meet all the constraints.

The rule of optimism also applies when a constraint is relaxed. A positive shadow price of $\pi_i > 0$ means that initially, each unit of change in the constraint right-hand-side improves your objective function by π_i. But eventually, as the constraint is relaxed further, the marginal benefit to you of more relaxation decreases, and typically reaches zero.

So far, I have explained the shadow price meaning of dual variables mainly with respect to so-called "functional" constraints of form $A\mathbf{x} \leq \mathbf{b}$ or $A\mathbf{x} \geq \mathbf{b}$. But once an LP is

[4]Since $A_{:,k} = e_k$, the kth basis vector, the reduced cost is $\mathbf{c}_k - c_{\mathbb{B}}^T B^{-1} A_{:,k} = -c_{\mathbb{B}}^T B_i^{-1} = -\pi_i$.

converted to the standard form for the simplex method, with constraints $A\mathbf{x} = \mathbf{b}; x \geq 0$, the mathematics does not distinguish between the original variables and the slack variables of the LP. All are simply nonnegative variables. Increasing the lower bound on a slack variable from 0 to ϵ is equivalent to tightening the associated inequality constraint by ϵ. Decreasing the lower bound on a slack variable from 0 to $-\epsilon$ is equivalent to relaxing the associated inequality constraint by ϵ. From this point of view, it is clear that the reduced cost is a shadow price, because the reduced cost is precisely the predicted impact on the objective function of increasing a nonbasic slack variable from 0 to ϵ.

Continuing with this point of view, we can see the meaning of a reduced cost of the original (non-slack) variables in an LP. Suppose nonbasic variable x_i represents the amount performed of activity i. We perform zero of that activity. Increase the lower bound of x_i from 0 to $\epsilon > 0$. The reduced cost of x_i equals the impact on the objective function, just as though we started to bring x_i into the basis at level ϵ. At optimality, if there is no dual degeneracy, all activities not being performed have nonzero reduced costs. If we were forced to perform a positive amount of one of these activities, our objective value would worsen at the rate given by the reduced cost.

Example 7.5 In the manufacturer LP of Section 3.1, suppose x_3 denotes the number of dinner plates to be produced, and x_3 is not basic in the optimal solution. The CEO decides that at least 1000 dinner plates must be manufactured because they are the flagship product of the company, even though that decision will hurt the bottom line. The estimated impact of enforcing the lower bound $x_3 \geq 1000$ equals $1000z_3$, where z_3 equals the reduced cost of x_3 (a negative number because we would like to maximize revenue). The actual impact could be worse, because shadow price predictions can be optimistic but not pessimistic.

Example 7.6
In the diet problem of Section 3.1, suppose x_5 denotes the number of ounces of orange juice to include in the meal, and x_5 is not basic in the optimal solution. Some of your customers insist on drinking at least 6 ounces of orange juice. Those customers will each pay at least $6z_5$ more for their meals, where z_5 is the reduced cost of orange juice.

Let's prove our rule of optimism. It turns out that there is a simple proof if we look at the dual.

Theorem 7.1.1 *Let z^* denote the optimal objective function value for the LP* (2.1) *or* (2.2) *Let α denote the absolute value of the reduced cost (shadow price) for a particular constraint that is binding at an optimal solution. Let $z^{\oplus}(\epsilon)$ denote the optimal objective function value of the LP if that constraint is relaxed by $\epsilon > 0$. Then $|z^{\oplus}(\epsilon) - z^*| \leq \alpha\epsilon$. Let $z^{\ominus}(\epsilon)$ denote the optimal objective function value of the LP if that constraint is tightened by $\epsilon > 0$. Then $|z^{\oplus}(\epsilon) - z^*| \geq \alpha\epsilon$. Therefore, predictions based on shadow prices are either accurate or optimistic.*

Proof: Consider the case of LP (2.1). Let $\boldsymbol{\pi}$ denote the dual variables, let b denote the dual objective function coefficients, and let $\boldsymbol{\pi} = \boldsymbol{\pi}^* \geq \mathbf{0}$ denote an optimal dual solution. Then $\boldsymbol{\pi}_j^* = \alpha$. When the primal constraint corresponding to $\boldsymbol{\pi}_j$ is relaxed by $\epsilon > 0$, b_j is decreased by ϵ. Since the dual feasible region is unchanged, the solution $\boldsymbol{\pi}^*$ remains feasible, while its objective value decreases by $\boldsymbol{\pi}_j^*\epsilon$. Since $\boldsymbol{\pi}^*$ might no longer be optimal, the optimal objective value of the dual might decrease by less than that amount. But it cannot decrease by more. By the weak duality theorem (2.1.1) the optimal objective function value of the primal cannot decrease by more than $\pi_j^* \epsilon = \alpha\epsilon$. When the primal constraint is tightened by $\epsilon > 0$, b_j is increased ϵ. Again, the dual's feasible region is unchanged, and the solution π^* remains feasible, while its objective value increases by $\alpha\epsilon$. Hence the optimal dual solution value has increased by at least $\alpha\epsilon$. The case of LP (2.2) is the same with "increase" and "decrease" exchanging roles. ■

Summary: At optimality, the dual variable π_j that is the complement of a nonbasic primal variable x_i predicts the rate of change in the objective function if x_i changes from 0. If you are uncertain about whether the change is positive or negative, remember that expanding the feasible region cannot worsen the optimal solution, and shrinking the feasible region cannot improve it. Memorize the rule of Theorem 7.1.1, that these predictions are overly optimistic when they are not accurate.

7.2 Sensitivity Analysis

Preview: You've solved your LP. Then the data change slightly. How do you re-solve quickly?

In sensitivity analysis scenarios, we have an optimal solution to a problem, and then some part of the problem changes. The cost vector or resource vector changes, a new constraint is added, a new variable is added, etc. The method of sensitivity analysis is to start from the choice of basis that was optimal and proceed from there to the solution to the altered problem. The idea is that this previously optimal choice of basis is apt to be optimal or fairly close to optimal, if the problem has not altered drastically. We expect therefore that it will take considerably less computational effort to start from that basis than to solve the altered problem from scratch. This is called doing a warm start as opposed to a cold start. In practice it typically reduces computational effort by an order of magnitude.

The specifics of sensitivity analysis for typical problem changes are given below.

Changed cost vector. If any elements of $\mathbf{c}_{\mathbb{B}}$ have changed, all of the dual variables $\boldsymbol{\pi} = c_{\mathbb{B}}^T B^{-1}$ and reduced costs $\mathbf{c}_j - \boldsymbol{\pi} A_{:,j}$ must be recomputed. If not, the reduced costs of nonbasic variables simply change by the amounts by which the cost coefficients change. If any of the recomputed reduced costs are favorable, the (primal) simplex method must be used.

Changed resource (RHS) vector. Compute the new value of $B^{-1}\mathbf{b}$. If it is nonnegative, the choice of basis remains optimal. Otherwise, one or more RHS terms are negative. But fortunately, all of the reduced costs are unchanged, hence not favorable. Use the dual simplex method.

Added constraint. Add the constraint to the model, making the slack variable for the new constraint the additional basic variable. If the slack variable is nonnegative, you are optimal. Otherwise, use the dual simplex method.

Added variable. Compute the reduced cost of the new variable. If it is unfavorable you are optimal. Otherwise, pivot it into the basis to start up the primal simplex method.

Change of column. If a column of the A matrix is changed, there are two cases. Case 1: the column is not in the current basis. Procedure: remove the column, then follow the procedure for an added variable. Case 2: the column is basic. This can be troublesome because the new column values might make the current basis fail to have full rank, or to be so nearly singular that you lose numerical stability. One easy way to deal with this situation is to arbitrarily pivot out that column and then treat it as a change to a nonbasic column (Case 1).

I have made this section brief because modern software will re-solve most of these cases without special algorithmic intervention by the user.

Summary: If data of your solved LP changes, start from the basis that was optimal.

7.3 Degeneracy and Shadow Prices

Preview: This section explores the pitfalls of using shadow prices when there is degeneracy.

Equation (7.1) can fail to be valid for any $\epsilon > 0$ if the LP optimal solution is degenerate. The proof of Theorem 7.1.1 does not assume nondegeneracy. Hence we may expect that sometimes the shadow price prediction will be overly optimistic for any $\epsilon > 0$. The general rule of thumb is not to put too much faith in shadow price predictions when the primal optimal solution is degenerate.

Consider the two-dimensional example in Figure 7.1. The original variables are x_1, x_2 and the slack variables are y_1, y_2, y_3. The constraints are

$$\begin{array}{rcl} x_1 + x + 2 + y_1 & = & 2 \\ 2x_1 + x_2 \quad + y_2 & = & 3 \\ x_1 + 2x_2 \qquad + y_3 & = & 3 \\ x_1, x_2, y_1, y_2, y_3 \geq 0 & & \end{array}$$

The objective is to maximize $x_1 + 5x_2$. Suppose the current degenerate basis consists of x_1, x_2 and y_3 at the point $x_1 = x_2 = 1$. Consider bringing y_2 into the basis. The other nonbasic variable y_1 will remain fixed at 0. It appears that we will move along the edge defined by $y_1 = 0$, in the direction of the arrow in the figure. That movement has the best reduced cost of $-1 + 5 = 4$. Geometrically, in the space of x_1, x_2, the movement is in the direction $(-1, 1)$, increasing the objective by 4 for each unit of change in the variable x_2. But $y_3 = 0$ and decreases with movement in the direction of the arrow. Therefore, the constraint $y_3 \geq 0$ blocks us as soon as we try to move. The y_3 variable exits the basis, and the variable y_2 enters the basis at value 0.

In the next iteration, the nonbasic variables are y_1 and y_3. Since y_3 just exited the basis it has an unfavorable reduced cost. But y_1 has a favorable reduced cost of $-2(1) + 5 = 3$. As y_1 enters the basis, y_3 remains nonbasic at 0. We move along the edge defined by $y_3 = 0$ and reach the optimum vertex, denoted in Figure 7.1 with an asterisk. Geometrically, in the space of x_1, x_2, we move in the direction $(-2, 1)$, increasing the objective by 3 for each unit of change in the variable x_2. The prediction of an improvement in the objective at rate 4 was optimistic for any strictly positive amount of movement.

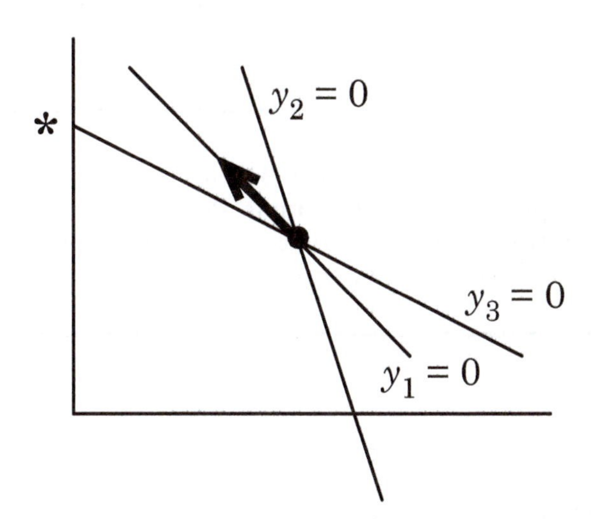

FIGURE 7.1: When the slack variable y_2 is nonbasic, all movement in the direction of the arrow is blocked.

For a three-dimensional example, imagine a pyramid with a square base and four triangular faces A,B,C,D, labeled so that A shares an edge with B and an edge with D. Let x_A, x_B, x_C and x_D be the slack variables associated with the sides. At the degenerate pyramid peak, let x_A be a degenerate basic variable (i.e., equal to 0) and let x_B, x_C, x_D be nonbasic. Choose x_C to enter the basis. The direction of movement is defined by the intersection of the two planes defined by B and D, a horizontal line. That direction of movement is blocked immediately by A. Therefore, the predicted rate of change in the objective based on the reduced costs of x_C is useless. The nonbasic variables at the next iteration will be x_A, x_B, x_D. Variable x_C will be basic at value 0. If instead we had chosen x_B to enter the basis, the direction of movement would have been along the edge shared by C and D, because x_C, x_D would have remained nonbasic.

The example of the pyramid also shows that predictions can differ from one iteration to the next. The reduced cost of a nonbasic variable can change from unfavorable to favorable, or vice versa, from iteration to iteration, all while remaining at the same vertex. Consider the reduced cost of a nonbasic variable x_B. If x_A is basic, then as we've seen the direction of movement as x_B enters the basis is down the edge shared by C and D. If instead x_C were basic, the direction of movement would be down the edge shared by A and D. The reduced cost of x_B could be quite different. For instance, if the objective vector were horizontal and perpendicular to the edge shared by C and the pyramid base, the signs of the reduced costs would differ.

Summary: Shadow price predictions can be inaccurate for any nonzero change when the optimum solution is primal degenerate. Typically the dual then has multiple optimal solutions, some of which may provide a more accurate shadow price than others.

7.4 Parametric Programming

Preview: This section provides a finer understanding of how the optimal solution and objective value change in sensitivity analysis.

The standard parametric programming setup is either to

$$\min \quad (\mathbf{c}^T + \theta \mathbf{d}^T)\mathbf{x} \quad \text{subject to} \tag{7.2}$$

$$A\mathbf{x} \quad \geq \quad \mathbf{b}, \mathbf{x} \geq \mathbf{0} \tag{7.3}$$

or to

$$\max \quad \mathbf{c}^T\mathbf{x} \quad \text{subject to} \tag{7.4}$$
$$A\mathbf{x} \geq \mathbf{b} + \theta\mathbf{f} \tag{7.5}$$
$$\mathbf{x} \geq \mathbf{0} \tag{7.6}$$

The goal is to find the optimal solution values over a range of θ, usually $0 \leq \theta < \infty$. As $\theta \to \infty$ the $\mathbf{d}$ or $\mathbf{f}$ term dominates the $\mathbf{c}$ or $\mathbf{b}$ term. So, ignoring scaling, we are finding the optimal solution as the cost vector (respectively RHS vector) changes continuously from $\mathbf{c}$ to d (respectively from $\mathbf{b}$ to $\mathbf{f}$). Unlike what occurs in the sensitivity analysis scenarios, the vector never reaches $\mathbf{d}$ (respectively $\mathbf{f}$), because θ never reaches ∞.

The main theorem regarding parametric programming is that the value function is a piecewise linear function of θ and is either convex or concave.

Theorem 7.4.1 *Let $z(\theta)$ (respectively $Z(\theta)$) denote the optimal objective function value of problem* (7.2) *(respectively* (7.4)*). Then $z(\theta)$ (respectively $Z(\theta)$) is either:*

1. *$+\infty$ $\forall\theta$ meaning that the problem is infeasible (respectively has unbounded optimum) $\forall\theta$;*
2. *$-\infty$ $\forall\theta$, meaning that the problem has unbounded optimum (respectively is infeasible) $\forall$ θ;*
3. *a piecewise linear concave function of θ on a closed interval, and $-\infty$ for other values of θ.*

If (7.2) *were a maximization problem (respectively* (7.4) *were a minimization problem), the three cases would instead be (i) $-\infty$ $\forall\theta$; (ii) ∞ $\forall\theta$; (iii) a piecewise linear convex function of θ on a closed interval, and ∞ for other values of θ.*

Proof: We prove the theorem for $z(\theta)$ and get the result for $Z(\theta)$ by duality. If the feasible polyhedron is empty, then $z(\theta) = \infty$ $\forall\theta$. For the other two cases, the feasible region is a nonempty polyhedron P. The nice thing about P is that it is the same for all θ.
Claim: The set $\{\theta : z(\theta) \text{ is finite }\}$ is empty or a closed interval. *Proof of Claim*: By the representation theorem, P has a finite set of directions $\{v^j\}$ such that $z(\theta) = -\infty$ iff for at least one j, $v^j \cdot (c + \theta d) < 0$. For each j, the set $\theta : \theta < \frac{v^j \cdot c}{v^j \cdot \mathbf{d}}$ is an open interval of form $(-\infty, t)$ or (t, ∞). Hence its complement is a closed interval. The intersection over all j of these closed intervals must be empty or a closed interval. This proves the claim. *Question: Actually, we didn't have to use the representation theorem to prove the Claim. Why not?*[5]
The second case occurs when the intersection is empty.

In the third case, the set $\{\theta : z(\theta) \text{ is finite }\}$, or equivalently, the values of θ for which (7.2) has an optimal solution, is a closed interval. By the fundamental theorem of LP, for each such θ there is an optimal solution that is an extreme point of P. Hence $z(\theta)$ equals the minimum over all extreme points $\mathbf{y}^i$ of $\mathbf{y}^i \cdot (c^T + \theta d^T)$. This is the minimum of a finite set of linear functions, so it is concave and piecewise linear as illustrated in Figure 7.2.

For $Z(\theta)$, observe that the dual of Problem (7.4) has the same form as Problem (7.2). By the strong duality theorem the objective function values are the same. Hence $Z(\theta)$ is piecewise linear and concave.

[5]The intersection of infinitely many closed intervals is empty or a closed interval. So we didn't need to assume that the set of directions d^j was finite, or even countable.

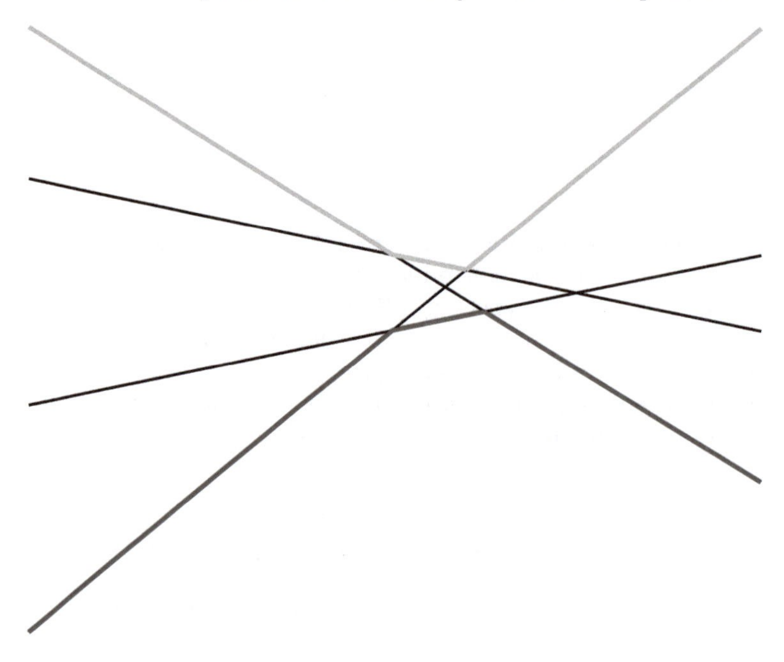

FIGURE 7.2: The minimum (respectively maximum) of a finite set of linear functions is concave (respectively convex) and piecewise linear.

If the min and max are switched, the logic remains the same for cases (i) and (ii). Infeasibility (respectively unboundedness) in a max problem corresponds to $-\infty$ rather than ∞ (respectively ∞ rather than $-\infty$). In case (iii), $z(\theta)$ equals the maximum of a finite set of linear functions, and is therefore convex and piecewise linear. ■

Theorem 7.4.1 adds specific structure to the rule of optimism. A prediction based on shadow prices corresponds to the scenario (7.4). The shadow price prediction is a linear function of θ. The rule of optimism tells us only that the true optimal value will be less than or equal to the predicted value, i.e., that the function $Z(\theta)$ lies on or below the linear prediction function. Theorem 7.4.1 tells us that $Z(\theta)$ coincides with the prediction in a neighborhood of $\theta = 0$ (assuming primal nondegeneracy) and is concave and piecewise linear. Hence the rate at which the prediction differs from the true value is nondecreasing.

Summary: Parametric programming tracks the optimum solution of an LP as the RHS or objective vector is continuously varied from one vector towards another vector. The optimum solution traverses a sequence of choices of basis, resulting in a piecewise linear movement of the objective function.

7.5 Column Generation

Preview: Column generation and its dual, constraint generation, are powerful techniques to solve problems that would otherwise be out of the range of linear programming because the LP would be too large. These techniques do not store or even generate the entire LP model explicitly. Instead they work with a subset of of the columns or constraints, and generate additional ones as needed. The specifics of these techniques depend on the form of the LP model.

The general mathematics of column generation is simple; the specifics needed to solve a particular class of problems within a reasonable amount of time can be challenging to develop.

Suppose A has many more columns than rows and we wish to solve the LP $\min \mathbf{c} \cdot \mathbf{x}$ subject to $A\mathbf{x} \geq \mathbf{b}; \mathbf{x} \geq \mathbf{0}$. LPs like this occur frequently in practice. For example, in real cutting stock and logistics problems, the number of rows often ranges from 10 to 1000, while the number of columns exceeds 10^9. Column generation takes the following form:

Algorithm 7.5.1 (Column generation: general form) *Input: A description of matrix A and cost vector $\mathbf{c}$ adequate to create them explicitly if desired; constraint vector $\mathbf{b}$.*
Step 1: Assemble a subset of the columns of A with index set S such that the polyhedron $A_{:,S}\mathbf{x}_S \geq \mathbf{b}; x_S \geq \mathbf{0}$ is not empty.
Step 2: Solve LP $\min \mathbf{c}_S \cdot \mathbf{x}_S$ subject to $A_{:,S}\mathbf{x}_S \geq \mathbf{b}; \mathbf{x}_S \geq 0$. Set $\boldsymbol{\pi} = \mathbf{c}_{\mathbb{B}}B^{-1}$, the vector of dual variables.
Step 3: Search the columns $A_{:,i} : i \notin S$ for one or more columns such that $\mathbf{c}_i - \boldsymbol{\pi}^T A_{:,i} < 0$. The search may be explicit or implicit. If no such columns exist, the current solution is optimal and terminate. Otherwise, augment S with the indices of some of these columns and go to Step 2.

Every time the algorithm returns from Step 3 to Step 2, it solves the new LP starting from the optimal basic solution of the previous LP. This can be viewed as sensitivity analysis when columns are added, or it can viewed as equivalent to the simplex method with a peculiar way of selecting the entering basic variable. The latter view proves that the column generation algorithm is valid, provided that the search procedure finds at least one favorable column if one exists. Here are a few guidelines for using column generation:

- Re-starting from the previously optimal basis is essential for you to have a chance at computational practicality.
- Column generation is theoretically valid if each iteration of Step 3 augments S with one column. In practice, the algorithm often runs much faster if multiple columns are added each iteration. One popular way to operate column generation is to add the best K columns found to the LP and solve for the new optimum before adding another set of K columns. Here K usually is a constant between a few dozen and a few hundred, depending on problem size.

- If a nonbasic column at the end of step 2 has very poor reduced cost and S is large enough to start causing memory problems, remove the column. Step 3 will find it again if it is needed later.

- It is very common for column generation to encounter degeneracy. Steps 2 and 3 may iterate hundreds of times (or more) without the optimal objective value in Step 2 improving at all. Then, when enough columns have been generated to work together to replace a set of basic columns, the objective value will move. This is one reason to augment S with multiple columns.

- If the columns of A are composed only of 0's and 1's and the constraints have the form $A\mathbf{x} \geq \mathbf{b}$, the LP is apt to be fairly easy to solve, and even the associated integer program (in which the $\mathbf{x}_i$ must take integer values) is apt to be relatively easy to solve. When $\mathbf{x}_i \in \{0, 1\}$ these problems are called *set covering* problems. If instead the constraints are $A\mathbf{x} \leq \mathbf{b}$ or $A\mathbf{x} = \mathbf{b}$ the problem is called *set packing* or *set partitioning*, respectively. Set packing tends to be computationally more difficult than set covering; set partitioning is harder yet.

- Many problems are hard to solve because they involve many yes/no (so-called discrete) decisions. Column generation can make some of these problems tractable by pushing the nasty discrete part of the problem down to the level of creating columns, while the higher level problem is just an LP or a relatively easy integer program.

- The art of column generation lies in the problem-specific techniques needed to perform Step 3 correctly, yet quickly enough. Look for formulations where, given $\boldsymbol{\pi}$, the search for a good column can be performed faster than by solving a general LP. Can you search by solving a network problem (see Chapter 11), by dynamic programming (see Chapter 14), or by rapid enumeration of columns (see below)?

Rapid enumeration of columns. Networks and dynamic programming are important topics in their own right and are covered elsewhere in this text. Rapid column enumeration is an idea special to column generation, so I explain it here. If the matrix A has m rows, it would seem that generating a possible column $A_{:,j}$ and calculating its reduced cost must take time at least proportional to m. Very often in real LP models, the columns of A are sparse, meaning that a small fraction of their entries are nonzero. When the columns are sparse, it is likely to be easy to generate them and compute their reduced costs in time proportional to the number of their nonzero entries. The idea of rapid enumeration of columns (which can also apply to non-sparse columns) is to generate and test columns even faster than that, by using the computational principle (see Section 6.6) that it can be cheaper to compute a change in value than to compute the value from scratch. The general idea is this: suppose you have computed the reduced cost of column $A_{:,j}$. Suppose further there is another column $A_{:,k}$ that differs from $A_{:,j}$ in only a few entries. Then the reduced cost of $A_{:,k}$ differs by only a few $\boldsymbol{\pi}_i(A_{:,k} - A_{:,j})$ terms and $\mathbf{c}_k - \mathbf{c}_j$. Computing the difference could take far less time than time proportional to the number of nonzero entries. If you can find a long sequence of columns to check such that each column in the sequence differs in only a few terms from the previous column, you can in effect enumerate the columns in the sequence with a few calculations each, on average, even if the computation for the first column in the sequence is slow. The use of this idea is, of course, highly problem-specific. I'll give a simple example to make the idea clearer. Suppose each column of A corresponds to a selection of t delivery locations from a set of size T, together with a permutation of the selected locations. (The permutation is the order in which the deliveries would be made.) For each selection of t sites, there are $t!$ columns, one for each permutation. The set of all permutations on t elements can be enumerated recursively by passing the largest element through the permutations of the other $t-1$ elements. (For example, for $t=5$ the permutation 54321 would be succeeded by 45321 and then by 43521. The permutation 43215 would be succeeded by 53421, then 35421, and then 34521.) In the resulting sequence of $t!$ columns, most columns differ from their predecessor by a swap between two consecutive delivery locations. In many applications, the difference in reduced cost could be computed with just a few computations. When those $t!$ columns have been checked, change the selection by removing one of the t locations and adding a new one. That change is apt to be inexpensive, as well. The computation time per column, on average, would be proportional to 1 rather than proportional to t^2.

Constraint generation. The dual of column generation is, of course, constraint generation. It may seem strange that there are useful LP models containing huge numbers of constraints, but it is true. In constraint generation algorithms, an LP containing a small subset of the constraints is solved. Let $\mathbf{x}^*$ be the solution obtained. The algorithm must then determine whether or not $\mathbf{x}^*$ violates any of the constraints that heretofore have been omitted from the LP. If $\mathbf{x}^*$ does, the algorithm must find a violated constraint, which can then be added to the LP to start the dual simplex method. This computational problem should sound familiar. It is the *separation* problem defined in Chapter 4. We will see more about separation when we study the ellipsoid algorithm in Section 10.2.

7.5.1 The Cutting Stock Problem

Gilmore and Gomory [122, 123] invented column generation to solve a paper cutting stock problem in the 1960s. Surprisingly, these problems are alive and well to this day. While I was writing the previous chapter of this book, I was asked to solve several cutting stock problems for local factories that cut rolls of sheet metal to start their manufacturing processes. One of those problems was mathematically identical to the one Gilmore and Gomory solved. Their problem was as follows: the manufacturer has a set of items to produce from some material such as paper or sheet metal. Each item requires a specific length of contiguous material to be produced. In particular, $i \in I$ indexes the set of lengths, and n_i items require length L_i. Unprocessed material is stocked in strips or rolls of length L each, called *blanks*, where for feasibility $L \geq L_i \ \forall i \in I$. When a blank is used to produce items, it is cut into pieces of the necessary length. The unused material is discarded as scrap. The problem is to decide how to cut the blanks to minimize total scrap while producing all the items. Equivalently, the problem is to cut the minimum number of blanks needed to produce all items.

Example 7.7 For example, blanks have length 32 feet, and of the 105 items to produce, 15 need length 5, 35 need length 6, 5 need length 10, 40 need length 12, and 10 need length 15.

Reading Tip: *Try to solve the example by hand to get a feel for the problem. After you've found a good solution or an optimal solution, you should understand the equivalence between minimizing scrap and minimizing blanks.*

The straightforward formulation has variables x_{ik} = the number of pieces of length L_i to cut from the kth blank. But it is a computational nightmare. Can you see why?

The main trouble is that this model depends heavily on requiring variables to be integers. The LP derived from an IP by ignoring integrality constraints is called the *LP relaxation*. As you will see in Chapter 15, integer programming algorithms solve a series of LP relaxations.

When the gap between an integer program and its LP relaxation is large, these algorithms tend to be less effective. The cutting stock problem has a large gap.

Reading Tip: *Try to figure out why by constructing a small example. If you couldn't do this yourself on your first reading, try again on your second reading.*

Example 7.8 From blanks of length 100, cut 20 pieces length 40 each, or 49 pieces length 51 each. In the first case, $x_{1,k}$ = the number of pieces length 40 to cut from the kth blank. Then $x_{1,k} = 2.5$ for $k = 1, \ldots, 8$ is an optimal LP solution. The trim is zero and the number of blanks is 8. But the optimal integer solution has trim 200 and uses 10 blanks. In the second case, $x_{1,k}$ = the number of pieces length 51 to cut from the kth blank. Then $x_{1,k} = \frac{100}{51}$ for $k = 1, \ldots, 24$ and $x_{1,25} = \frac{99}{51}$ is an optimal LP solution with trim 1, using 25 blanks. But the optimal integer solution has trim 49^2 and uses 49 blanks.

The model also has a huge amount of symmetry because any blank's cutting plan could be swapped with any other blank's plan. If there were 100 blanks and 5 different plans used, a solution could be one of as many as

$$\frac{100!}{(20!)^5} \approx 10^{67}$$

different but symmetrically equivalent solutions. Symmetry in an LP is not important, but it can be disastrous in an integer program.

Gilmore and Gomory instead proposed the following model: Let $j \in J$ index the set of possible cutting plans for a single blank. In particular, let p_{ij} = the number of pieces of length L_i to cut from the blank according to plan j. To be feasible, plan j must satisfy

$$\sum_{i \in I} L_i p_{ij} \leq L.$$

The decision variables are $\forall j \in J,\ \mathbf{x}_j$ = the number of blanks to cut according to plan j. The optimization model is

$$\begin{array}{llll} \min & \sum_{j \in J} \mathbf{x}_j & \text{subject to:} & (7.7) \\ \forall i \in I: & \sum_{j \in J} p_{ij} \mathbf{x}_j & \geq n_i & (7.8) \\ \forall j \in J: & \mathbf{x}_j \geq 0; \mathbf{x}_j \text{ integer} & & (7.9) \end{array}$$

This model also requires the variables to be integer, but it does not depend as heavily on integrality as the previous set of variables would. Also, it is easy to generate a feasible solution from a fractional solution to this model by rounding up fractional variables to integers. And it is easy to prove upper bounds on how much worse the rounded solution can be than the optimum solution. The challenge in this model is to solve the LP version, that is, the model above without the $\mathbf{x}_j$ integrality constraints. The number of possible cutting plans, and therefore the number of variables, is typically too large to be computationally practical. For the baby example given above, the number is about three dozen, but in general the number easily ranges from thousands to millions.

You may have wondered why the constraints (7.8) are in $\geq$ rather than $=$ form. There are two reasons. The first is because, as I mentioned earlier, set covering tends to be computationally easier to solve than set partitioning. The second is to reduce the number of possible columns. By changing the constraints from equalities to inequalities, we can restrict our set J of columns to the set of *maximal* columns, that is, cutting plans that leave a piece of scrap smaller than all L_i, hence too small to use. In practice, the manufacturer can omit some of the cuts in some of the cutting plans if the overall solution calls for a few extra items.

The key is to solve Step 3 of the column generation procedure. Let $\pi_i : i \in I$ be the dual variables at the optimum from Step 2, where the LP with a subset of columns is solved. The reduced cost of cutting pattern j equals

$$1 - \sum_{i \in I} \pi_i p_{ij}.$$

This formula should make sense to you. Variable π_i is the shadow price of the constraint that requires n_i pieces of length L_i. It is the cost of producing the n_ith piece of length L_i, with respect to the LP solved in Step 2. If cutting plan j is used, it will cost 1 blank in direct cost, and it will indirectly save $\pi_i p_{ij}$ as the basic variables change to produce p_{ij} fewer pieces of length L_i. Therefore, the cutting plan with best reduced cost is the j that maximizes $\sum_{i \in I} \pi_i p_{ij}$ over all $j \in J$, i.e., subject to the feasibility constraint $\sum_{i \in I} L_i p_{ij} \leq L$. Hence, Step 3 defines integer variables $p_i \geq 0 : i \in I$ and solves

$$\max_{i \in I} \pi_i p_i \text{ subject to } \sum_{i \in I} L_i p_i \leq L.$$

This optimization problem is a so-called "knapsack" problem, characterized by nonnegative integer variables, a linear objective function, and a single linear inequality constraint. In Chapter 14 you will see that knapsack problems, though theoretically difficult to solve in the worst case, are easy to solve in practice if L is not too large. (The name of the problem derives from the story of a traveler deciding how many of each type of item $i \in I$ to pack in a knapsack that holds at most L pounds. Each item i has value π_i but weighs L_i pounds.)

In practice, cutting a blank into usable pieces can be more complicated than in the problem just discussed. For example, cutting could be 2-dimensional rather than 1-dimensional. In these cases, a simple knapsack problem will not suffice for Step 3. But the column generation algorithm, together with a good heuristic for 2-dimensional cutting plans, can still be an excellent way to solve the problem.

Summary: Columnn generation takes to an extreme the guideline to compute only what you need. Its implementation is problem-specific, but if you can figure out how to generate good columns quickly given the dual variable values, you can solve much bigger cases than you could otherwise.

7.6 Problems

E Exercises

1. Below is an optimum tableau.

$$\begin{aligned} z + 3x_3 + 2x_4 &= 13 \\ x_2 + x_3 &= 3 \\ x_1 + x_3 + x_4 &= 5 \\ -2x_3 - x_4 + x_5 &= 2 \end{aligned}$$

The original problem was

$$\begin{aligned} \max z = 2x_1 + x_2 & \\ x_2 &\leq 3 \\ x_1 - x_2 &\leq 2 \\ x_1 + x_2 &\leq 10 \\ x_i &\geq 0 \end{aligned}$$

- Write the dual problem, and its optimal solution including values of dual slack variables and the objective function.
- Estimate the effect on the objective function if the first constraint is changed to $x_2 \leq 1.5$.

  Perform sensitivity analysis for the following cases. This includes solving for the new optimal solution.
- b changes to $b' = (1, 2, 3)$.
- c changes to $c' = (5, -3)$.

2. Below is an optimum solution to an LP.

$$\begin{aligned} z + 3x_3 + 2x_4 &= 10 \\ x_2 + x_3 &= 2 \\ x_1 + x_3 + x_4 &= 4 \\ -2x_3 - x_4 + x_5 &= 4 \end{aligned}$$

The original problem was

$$\begin{aligned} \max z = 2x_1 + x_2 & \\ x_2 &\leq 2 \\ x_1 - x_2 &\leq 2 \\ x_1 + x_2 &\leq 10 \\ x_i &\geq 0 \end{aligned}$$

- Write the dual problem, and its optimal solution including values of dual slack variables and the objective function.
- Estimate the effect on the objective function if the first constraint is changed to $x_2 \leq 1.5$.
- Estimate the effect on the objective function if the third constraint is changed to $x_1 + x_2 \leq 9.5$.

  Perform sensitivity analysis for the following cases. This includes solving for the new optimal solution.
- b changes to $b' = (10, 12, 18)$.
- b changes to $b' = (3, -4, 5)$.
- c changes to $c' = (15, -8)$.

3. Consider a parametric LP with maximization objective and RHS $b - \theta\hat{b}$. Prove that the range R of θ for which the LP is feasible is convex.

4. Verify the correctness of Theorem 7.1.1 in the case of LP (2.2) by switching the roles of "increase" and "decrease."

5. Construct a one-variable parametric LP that is infeasible for $\theta = 0$, feasible for $\theta = 1$, and infeasible for $\theta = 2$.

6. Assume that blanks (the stock) in a cutting stock problem are all length L and that there are orders for lengths 25,45,60,and 130 in the amounts $b_1 \ldots b_4$ respectively. Suppose at some iteration in the procedure the LP solution produces $b_2 + 0.5$ of length 45 and $b_3 + 1.5$ of length 60. How will this be reflected in the next column generation subproblem?

7. Solve the cutting stock problem for making 30 pieces length 40 each from blanks length 150 by hand using column generation, but without the knapsack subproblem. Instead, use the three cutting patterns: one piece length 40 from a blank; two pieces length 40 each from a blank; three pieces length 40 each from a blank. (The former two are not maximal but use them anyway.) Start with the feasible solution of 30 of the first cutting pattern. Solve the LPs by inspection but calculate the dual values exactly. Interpret the values given by the dual variables to all three patterns.

M Problems

8. Consider the problem

$$\min 20x_1 + 30x_2 + 40x_3$$

subject to

$$\begin{aligned} x_1 + 2x_2 + 3x_3 &\leq 1000 \\ 500x_1 + 600x_2 + 700x_3 &\leq -1000 \\ x_i &\geq 0 \end{aligned}$$

   (a) Convert the problem to standard $A\mathbf{x} = \mathbf{b}, x \geq 0$ form by adding slacks. Starting from a basis of slack variables, solve the problem by the dual simplex method.

   (b) Provide explicitly the ray of unboundedness in the dual.

   (c) Provide a row vector $\boldsymbol{\pi}^T$ which shows $A\mathbf{x} = \mathbf{b}, x \geq 0$ is infeasible, and explain why.

9. Construct an example of an LP that contains a constraint whose shadow price remains strictly positive no matter how much the constraint is relaxed. Interpret the meaning of this phenomenon in the dual.

10. Prove or disprove that the following can occur in a RHS parametric analysis:

    (a) The linear program starts out feasible and becomes infeasible.

    (b) The linear program starts out with a finite optimal solution but for some θ has an unbounded optimum.

    (c) The LP goes from infeasible to feasible to infeasible.

    (d) The LP goes from feasible to infeasible to feasible.

    (e) The LP goes from infeasible to feasible to unbounded.

11. Suppose at an optimal solution to an LP with constraints $Ax = b, x \geq 0$, the optimum basic feasible solution is nondegenerate. Prove that if one of the b_j values is changed by a small enough amount $\epsilon > 0$, then the corresponding dual variable (shadow price) correctly predicts the effect on the objective function value. Illustrate for the case:

$$\begin{array}{rl} \max 100x_1 + 100x_2 & \text{subject to} \\ x_1 + x_2 \leq & \beta \\ 2x_1 + x_2 \leq & 12 \\ x_1 + 2x_2 \leq & 12 \\ x_i \geq 0 & \end{array}$$

    by comparing $\beta = 9$ with $\beta = 8$. At the value $\beta = 8$ how do the different choices of optimal basis give accurate or inaccurate predictions? (Remember β could increase or decrease slightly.)

12. You know that dual variables predict the effects of changes to the constraint right-hand-side data, but that these predictions are not always accurate for large changes in value.

    (a) Construct a small numerical example of a maximization LP with constraints of form $Ax \leq b$ in which there is a nonzero dual variable at the optimal solution that **correctly** predicts the effect on the objective value for **increasing** its associated constraint's right-hand-side, b_j for some j, no matter how much b_j is increased.

    (b) Construct a small numerical example of a maximization LP with constraints of form $Ax \leq b$ in which there is a nonzero dual variable at the optimal solution that **correctly** predicts the effect on the objective value for **decreasing** its associated constraint's right-hand-side, b_j for some j, no matter how much b_j is decreased.

    (c) Can you construct a numerical example of a maximization LP with constraints of form $Ax \leq b$ in which there is a nonzero dual variable at the optimal solution that **correctly** predicts the effect on the objective value for changing its associated constraint's right-hand-side, b_j for some j, no matter how much b_j is **increased or decreased**? Construct such an example or prove that none exists.

    (d) Construct a small numerical example of a maximization LP with constraints of form $Ax \leq b$ in which there is a nonzero dual variable at the optimal solution that **incorrectly** predicts the effect on the objective value for changing its associated constraint's right-hand-side, b_j for some j, no matter how little b_j is **increased or decreased**.

13. State precisely the dual rule of thumb from the question at the beginning of the chapter. Prove it.

14. State the dual of the cutting stock LP. Suppose you want to solve this dual with constraint generation. Derive the constraint generation subproblem. (Of course, this should be mathematically equivalent to the column generation subproblem in the Gilmore-Gomory cutting stock column generation procedure. This problem asks you to derive it from the point of view of the dual.) Explain why constraint generation ultimately finds an optimal solution to your dual LP even though most of its constraints may never be generated. In addition, give an intuitive geometric interpretation of constraint generation and why it is valid.

15. Suppose a basic column in an LP tableau changes in value and its variable must be removed from the basis. Is there a way to choose an entering variable such that the resulting basic solution is either primal feasible or dual feasible, so that the primal or dual simplex method could be used? Prove or disprove.

16. Prove that the following cannot occur: the optimal solution value of $\max \mathbf{c} \cdot \mathbf{x}$ subject to $\mathbf{x} \geq \mathbf{0}$; $A\mathbf{x} = \mathbf{b} + \theta\mathbf{b}'$ has values 10 at $\theta = 0$, 20 at $\theta = 5$, and 41 at $\theta = 10$. First, prove it using Theorem 7.4.1. Second, prove it directly using only elementary algebra.

17. Prove Theorem 7.1.1 as a corollary to Theorem 7.4.1.

18. Construct an instance of the cutting stock problem for which there exists an optimal basic feasible solution of the LP relaxation none of whose basic columns are used in any optimal integer solution.

D Problems

19. Suppose the LP relaxation of a cutting stock problem is solved by the simplex method, and the LP solution is rounded up to give all integer values. Let v^* denote the objective function value so attained. Let v^{opt} denote the optimal objective function value. Construct an instance for which the ratio $\frac{v^*}{v^{opt}}$ is as large as you can achieve. Prove the tightest upper bound you can for the ratio $\frac{v^*}{v^{opt}}$.

20. A feasible solution $\mathbf{x}$ to a cutting stock problem is *minimal* if there does not exist a feasible solution $\check{\mathbf{x}} \lneq \mathbf{x}$. Any feasible solution $\mathbf{x}$ can easily be decreased to a minimal feasible solution by iteratively decrementing one variable at a time. Repeat Problem 19 if the LP solution is rounded up and then decreased to be minimal.

21. Repeat Problem 19 under the assumption that the LP solution is optimal, but was not necessarily found by the simplex method (i.e., is not necessarily a BFS).

22. Repeat Problem 20 under the assumption that the LP solution is optimal, but was not necessarily found by the simplex method.

Chapter 8

Advanced Topics on Polyhedra

Preview: This chapter explores relationships among the duality theorems that you've seen in previous chapters. It also gives several new results involving separation, convexity, complementary slackness, and two new concepts, facets and polarity.

8.1 The Geography of Mount Duality

The law of conservation of difficulty is a rule of thumb. It says that if a mathematical result is deep or difficult, it can't be proved easily unless another deep result is employed. There are a bunch of theorems in linear programming which all take some work to prove from scratch. But given one of these theorems, several others are usually fairly easy to prove. These include various theorems of the alternative such as Theorem 4.3.2 and the strong duality theorem of linear programming for various forms of LPs. I think of proving these theorems in terms of mountain climbing. The theorems compose a mountain range. It takes some hard work to climb up to the range, but once there it is not so difficult to get from one peak to another. The LP duality, alternatives, separating hyperplane, and representation theorems seem to be neighboring mountain peaks. Navigating to and from them is not always trivial, but is much less work than when starting at the mountain base.

In this text we've proved four of these theorems from scratch. Each proof used some machinery. The proofs of the theorem of the alternative (Theorem 4.3.2) and of projection (Theorem 4.2.2) employed Fourier-Motzkin elimination; strong duality needed the correctness of the simplex method, which required a resolution of degeneracy by Bland's rule or by perturbation; the separating hyperplane theorem required the deep results on compactness in $\Re^n$ from real analysis. In this section we discuss some of the relatively easy paths between pairs of the theorems.

8.1.1 Variants of Theorems of the Alternative and Strong Duality

We call a pair of systems of equalities and inequalities an *alternative pair* if exactly one of them must have a feasible solution. Theorem 4.3.2 states that $A\mathbf{x} \leq \mathbf{b}$ (system I); and $\boldsymbol{\pi} \geq \mathbf{0}$, $\boldsymbol{\pi}^T A = \mathbf{0}, \boldsymbol{\pi}^T \mathbf{b} < 0$ (system II) are an alternative pair.

Theorem 8.1.1 *The following systems are alternative pairs. For any data* $A, \mathbf{b}$ *exactly one of the pair is feasible.*

1. $A\mathbf{x} \leq \mathbf{b}$ *(system I);* $\boldsymbol{\pi} \geq \mathbf{0},\ \boldsymbol{\pi}^T A = \mathbf{0}, \boldsymbol{\pi}^T\mathbf{b} < 0$ *(system II).*
2. $A\mathbf{x} = \mathbf{b}, \mathbf{x} \geq \mathbf{0}$ *(system I);* $\boldsymbol{\pi}^T\mathbf{b} < 0, \boldsymbol{\pi}^T A \geq \mathbf{0}$ *(system II).*
3. $A\mathbf{x} > 0$ *(system I);* $\boldsymbol{\pi}^T A = \mathbf{0}, \boldsymbol{\pi} \geq \mathbf{0}, \boldsymbol{\pi} \neq \mathbf{0}$ *(system II).*
4. $A\mathbf{x} \leq \mathbf{b}, \mathbf{x} \geq \mathbf{0}$ *(system I);* $\boldsymbol{\pi} \geq 0;\ A^T\boldsymbol{\pi} \geq \mathbf{0};\ \boldsymbol{\pi} \cdot \mathbf{b} < 0$ *(system II).*

Proof: The first and fourth were proved in Chapter 4. To prove the third, named Gordan's Theorem for Paul Gordan who proved it in 1873 [133], replace system I with the equivalent $A\mathbf{x} \geq \mathbf{1}$ (where $\mathbf{1}$ is a vector of 1's.) *Question: Why are the systems equivalent?*[1] The alternative system is therefore $\boldsymbol{\pi} \geq \mathbf{0}, \boldsymbol{\pi}^T A = \mathbf{0}, \boldsymbol{\pi}^T\mathbf{1} > 0$ which is clearly equivalent to system II.

For the second, write system I in the form of the first system I:

$$\begin{bmatrix} A \\ -A \\ -I \end{bmatrix} \mathbf{x} \leq \begin{bmatrix} \mathbf{b} \\ -\mathbf{b} \\ \mathbf{0} \end{bmatrix}.$$

Obtain the alternative system

$$({\boldsymbol{\pi}^+}^T, {\boldsymbol{\pi}^-}^T, \phi^T) \begin{bmatrix} A \\ -A \\ -I \end{bmatrix} = \mathbf{0};\ ({\boldsymbol{\pi}^+}^T, {\boldsymbol{\pi}^-}^T, \phi^T) \begin{bmatrix} \mathbf{b} \\ -\mathbf{b} \\ \mathbf{0} \end{bmatrix} < 0;\ ({\boldsymbol{\pi}^+}^T, {\boldsymbol{\pi}^-}^T, \phi^T) \geq \mathbf{0}.$$

Let $\boldsymbol{\pi} = \boldsymbol{\pi}^+ - \boldsymbol{\pi}^-$. The system simplifies to $\boldsymbol{\pi}^T\mathbf{b} < 0,\ \boldsymbol{\pi}^T A = \phi^T \geq \mathbf{0}$ which simplifies to $\boldsymbol{\pi}^T\mathbf{b} < 0, \boldsymbol{\pi}^T A \geq \mathbf{0}$.

Just as the weak duality theorem holds for different primal/dual pairs, so do the fundamental and strong duality theorems.

[1] Any $\mathbf{x}$ satisfying $A\mathbf{x} > \mathbf{0}$ can be scaled up to satisfy $A\mathbf{x} \geq \mathbf{1}$.

Theorem 8.1.2 *For each of the following primal/dual pairs of LPs,*

(1) $\min \mathbf{c}\cdot\mathbf{x}$ *subject to* $A\mathbf{x} \geq \mathbf{b}; \mathbf{x} \geq \mathbf{0}$ *(primal)*
$\max \mathbf{b}\cdot\boldsymbol{\pi}$ *subject to* $\boldsymbol{\pi}^T A \leq \mathbf{c}^T, \boldsymbol{\pi}^T \geq \mathbf{0}^T$ *(dual).*

(2) $\min \mathbf{c}\cdot\mathbf{x}$ *subject to* $A\mathbf{x} = \mathbf{b}; \mathbf{x} \geq \mathbf{0}$ *(primal)*
$\max \mathbf{b}\cdot\boldsymbol{\pi}$ *subject to* $\boldsymbol{\pi}^T A \leq \mathbf{c}^T$ *(dual).*

(3) $\max \mathbf{c}\cdot\mathbf{x}$ *subject to* $A\mathbf{x} = \mathbf{b},\ \mathbf{x} \geq \mathbf{0}$ *(primal)*
$\min \mathbf{b}\cdot\boldsymbol{\pi}$ *subject to* $\boldsymbol{\pi}^T A \geq \mathbf{c}^T$ *(dual).*

Exactly one of the following is true:

1. *Both are infeasible.*
2. *One has an unbounded optimum and the other is infeasible.*
3. *Both are feasible.*

Moreover, in case 3, both primal and dual have optimal solutions, $\mathbf{x}^*$ *and* $\boldsymbol{\pi}^*$ *respectively, that:*

1. *are basic feasible solutions,*
2. *have equal objective function value* $\mathbf{c}\cdot\mathbf{x}^* = \mathbf{b}\cdot\boldsymbol{\pi}^*$*, and*
3. *satisfy complementary slackness.*

Proof: We've proved strong duality for the pair (2) via the correctness of the simplex method. Consider the primal of pair (1). Its range of attainable objective function values is the same as that of

$$\begin{array}{ccc} \min & \mathbf{c}\cdot\mathbf{x} + \mathbf{0}\cdot\mathbf{y} & \text{subject to} \\ & [A| - I](\mathbf{x},\mathbf{y}) = \mathbf{b}; & \\ & (\mathbf{x},\mathbf{y}) \geq \mathbf{0} & \end{array}$$

Next, consider the dual of pair (1). Its range of attainable objective function values is the same as that of

$$\begin{array}{ccc} \max & \mathbf{b}\cdot\boldsymbol{\pi} & \text{subject to} \\ & \boldsymbol{\pi}^T[A| - I] \leq (\mathbf{c},\mathbf{0})^T. & \end{array}$$

The equivalent LPs just defined have the form of the pair (2), with primal constraint matrix $[A| - I]$ and primal objective vector $(\mathbf{c},\mathbf{0})$. The strong duality property of (2) therefore implies the strong duality property of (1).

Strong duality of the pair (3) follows from strong duality of (2) by changing $\max \mathbf{c}\cdot\mathbf{x}$ to $\min(-\mathbf{c})\cdot\mathbf{x}$ and changing $\boldsymbol{\pi}$ to $-\boldsymbol{\pi}$, since $\boldsymbol{\pi}^T A \geq (-\mathbf{c})^T \Leftrightarrow -\boldsymbol{\pi}^T A \leq \mathbf{c}^T$.

The complementary slackness property follows from the proof of the weak duality theorem.

For all LPs except the duals of (2) and (3), the nonnegativity constraints make it impossible for the feasible region to contain a line. Therefore, they have optimal basic feasible solutions by Theorem 4.4.8. In the duals of (2) and (3), the rows of A must be linearly independent unless the constraints $A\mathbf{x} = \mathbf{b}$ are contradictory, in which case the primal is infeasible, or $A\mathbf{x} = \mathbf{b}$ contains redundant constraints, in which case the redundancies can be removed. Assuming now that the rows of A are linearly independent, suppose that the dual's feasible region contains a direction $\mathbf{d} \neq \mathbf{0}$. This means that $A^T\mathbf{d} = \mathbf{0}$ but $\mathbf{d} \neq \mathbf{0}$. Then the columns of A^T are linearly dependent, a contradiction. ■

8.1.2 Degeneracy and Complementary Slackness

If basic optimal solutions $\mathbf{x}^* \in \Re^n, \boldsymbol{\pi}^* \in \Re^m$ to the primal-dual pair (1) are both not degenerate, they satisfy strict complementary slackness. That is, following Definition 2.3, for each $1 \leq i \leq n$, exactly one of the values

$$\mathbf{x}_i^*, \quad \mathbf{c}_i - (A^T)_{i,:}\boldsymbol{\pi}^*$$

equals zero, and for each $1 \leq j \leq m$ exactly one of the values

$$\boldsymbol{\pi}_j^*, \quad A_{j,:}x^* - b_j$$

equals zero. What happens to strict complementary slackness if there is degeneracy?

Reading Tip: *The paragraph you just read makes an assertion without proof. In mathematical books, the absence of proof should signal that the assertion either has been proven earlier, or follows directly from facts already proved. (The latter is the case here.) In either case, make sure you know precisely why the assertion is true. If you didn't see it immediately, figure out why. Here, the stumbling block is that it connects two hitherto unrelated concepts, degeneracy and complementary slackness (Exercise 5).*

Most textbooks would skip the next example and jump directly to the theorem. I chose to illustrate the empirical side of mathematics. Play with specific examples, see a pattern, formulate a precise conjecture, and prove it.

The simplex algorithm helps us see how degeneracy affects complementary slackness.

Example 8.1

$$\max \quad 3x_1 + 3x_2 \quad \text{subject to} \tag{8.1}$$
$$x_1 \leq 2 \tag{8.2}$$
$$x_1 + x_2 \leq 10 \tag{8.3}$$
$$x_1, x_2 \geq 0 \tag{8.4}$$

Add slack variables x_3, x_4 and bring x_2 into the basis to get an optimal tableau

x_0	x_1	x_2	x_3	x_4	RHS
1	−3	−3	0	0	0
0	1	0	1	0	2
0	1	**1**	0	1	10

x_0	x_1	x_2	x_3	x_4	RHS
1	0	0	0	3	30
0	**1**	0	1	0	2
0	1	1	0	1	10

Variable x_1 has zero reduced cost. Pivot it into the basis to arrive at a different optimal tableau.

x_0	x_1	x_2	x_3	x_4	RHS
1	0	0	0	3	30
0	1	0	1	0	2
0	0	1	−1	1	8

In the primal geometry, we've moved from the optimal solution $(0, 10)$ along the heavy edge in Figure 8.1 to the optimal solution $(2, 8)$.

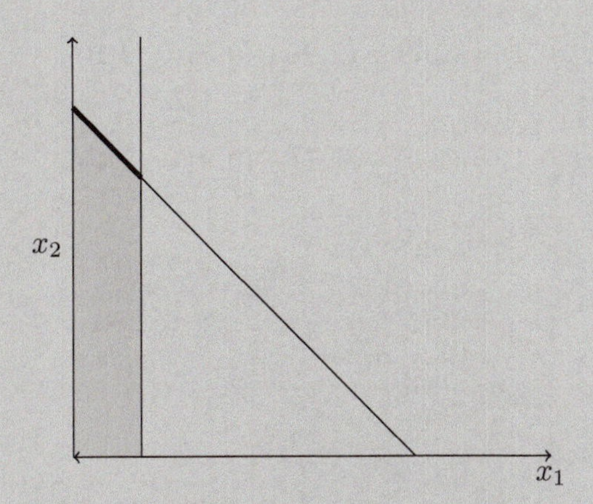

FIGURE 8.1: The segment in bold between $(0, 10)$ and $(2, 8)$ is the set of optimal solutions to the LP (8.1–8.4).

Neither of the optimal basic solutions $(0, 10, 2, 0)$ and $(2, 8, 0, 0)$ satisfies strict complementary slackness with the optimal dual solution $\boldsymbol{\pi} = (0, 3, 0, 0)$. The first fails at $x_1 = \pi_3 = 0$; the second fails at $x_3 = \pi_1 = 0$. But their midpoint $x = (1, 9, 1, 0)$ (or any other strict convex combination) succeeds: $x_1 > 0, x_2 > 0, x_3 > 0, \pi_2 > 0$. In terms of the simplex algorithm, the failure of strict complementary slackness at $(0, 10, 2, 0)$ is that the reduced cost of x_1 is $\pi_3 = 0$. That failure is precisely what gives us multiple optimal primal solutions.

Before we leave this example, let's see what is happening in the dual LP

$$\begin{aligned} \min \quad & 2\pi_1 + 10\pi_2 \quad \text{subject to} \\ & \pi_1 + \pi_2 \geq 3 \\ & \pi_2 \geq 3 \\ & \pi_1, \pi_2 \geq 0 \end{aligned}$$

Adding slack variables π_3, π_4, the dual becomes

$$\begin{aligned} \min \quad & 2\pi_1 + 10\pi_2 \quad \text{subject to} \\ & \pi_1 + \pi_2 - \pi_3 \quad = 3 \\ & \pi_2 \quad - \pi_4 = 3 \\ & \pi_1, \pi_2, \pi_3, \pi_4 \geq 0 \end{aligned}$$

Figure 8.2 shows that three constraints, $\pi_1 \geq 0; \pi_1 + \pi_2 \geq 3; \pi_2 \geq 3$ are all binding at the unique optimum $(0, 3)$. They correspond to primal variables x_3, x_1, x_2, respectively. Their orthogonal vectors (shown dashed) are $(1, 0), (1, 1)$, and $(0, 1)$. The three optimal primal solutions we examined are sets of multipliers to express the dual objective vector as nonnegative linear combinations of the binding constraints' orthogonal vectors. Thus, the dual objective vector $(2, 10)$ (shown solid) can be expressed as:

- Primal solution $x^* = (0, 10, 2, 0)$
  $x_3^*(1,0) + x_1^*(1,1) + x_2^*(0,1) = (2,0) + (0,0) + (0,10) = (2,10)$.

- Primal solution $x^* = (2, 8, 0, 0)$
  $x_3^*(1,0) + x_1^*(1,1) + x_2^*(0,1) = (0,0) + (2,2) + (0,8) = (2,10)$.

- Primal solution $x^* = (1, 9, 1, 0)$
  $x_3^*(1,0) + x_1^*(1,1) + x_2^*(0,1) = (1,0) + (1,1) + (0,9) = (2,10)$.

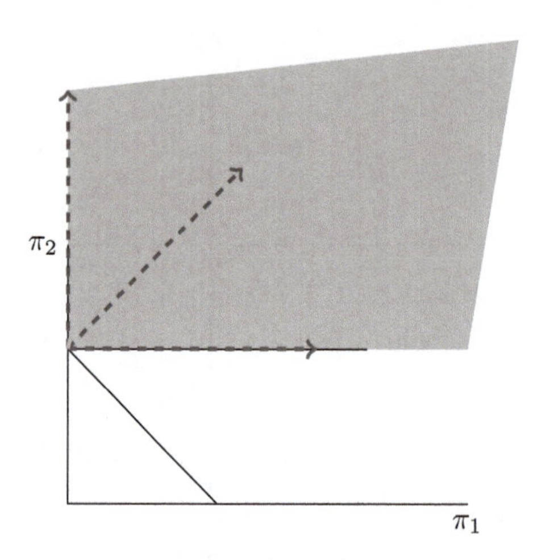

FIGURE 8.2: There are multiple ways to express the objective vector as a convex combination of the binding constraints's orthogonal vectors.

Question: What geometric fact in Figure 8.2 tells you that $x_2^* > 0$ in all optimal primal solutions?[2]

This example suggests that strict complementary pairs exist even when there is degeneracy, because degeneracy in an LP gives its dual multiple optima. If $\pi_j > 0$ at one or more of those optima, $\pi_j > 0$ at any strict convex combination of those optima. Geometrically, define the polyhedra P^*, D^* to be the sets of optimal solutions to the primal and dual, respectively. Choose an $\mathbf{x}^*$ in the interior of P^* and a $\boldsymbol{\pi}^*$ in the interior of D^* (Section 8.2 defines the proper technical term, "relative interior".) If for every complementary pair of indices i, j either $x_i^* > 0$ or $\pi_j^* > 0$, those two points are strictly complementary slack.

The following theorem assures that the interior points of P^* and D^* always have strict complementary slackness. This geometric interpretation will come in handy when we study interior point LP algorithms. Essentially, by working simultaneously with the primal and dual, *interior point algorithms always evade degeneracy.* This can be a tremendous computational advantage over simplex algorithms.

Theorem 8.1.3 *If both primal* **(P)** *and dual* **(D)**

$$\begin{array}{cccc} & \min \mathbf{c}\cdot\mathbf{x} & \max \mathbf{b}\cdot\boldsymbol{\pi} & \\ \textbf{(P)} & A\mathbf{x} \geq \mathbf{b} & A^T\boldsymbol{\pi} \leq \mathbf{c} & \textbf{(D)} \\ & \mathbf{x} \geq \mathbf{0} & \boldsymbol{\pi} \geq \mathbf{0} & \end{array}$$

[2]Its orthogonal vector $(0,3) + \theta(0,1)$ is on one side of the dual objective vector $(0,3) + \theta(2,10)$. The other two orthogonal vectors, $(0,3) + \theta(1,0)$ and $(0,3) + \theta(1,1)$ are both on the other side.

are feasible, there exist optimal solutions, $\mathbf{x}^*$ *and* $\boldsymbol{\pi}^*$*, respectively, that are mutually strictly complementary slack:*

$$\begin{aligned} &\textit{For all } i, \textit{ exactly one of} \quad \mathbf{x}_i^*, (\mathbf{c}_i - (A^T)_{i,:}\pi^*) \quad \textit{equals zero;} && (8.5)\\ &\textit{For all } j, \textit{ exactly one of} \quad \boldsymbol{\pi}_j^*, (A_{j,:}\mathbf{x}^* - b_j) \quad \textit{equals zero.} && (8.6)\\ & && (8.7)\end{aligned}$$

Proof: *By weak duality's Theorem 2.1.2 at most one of the two values in (8.5) and (8.6) is nonzero for all optimal solution pairs* $\mathbf{x}^*, \boldsymbol{\pi}^*$ *Question: How is this similar to other duality proofs?*[3]*. It suffices to prove that for each* i *there exists a pair of optimal solutions* x^{i*}, π^{i*} *such that at least one of* x_i^{i*}, $(\mathbf{c}_i - (A^T)_{i,:}\pi^{i*})$ *is not zero. For if so, by primal-dual symmetry, for each* j *there exists a pair of optimal solutions* x^{j*}, π^{j*} *such that at least one of* π_j^{j*}, $(A_{j,:}x^{j*} - \mathbf{b}_j)$ *is not zero. Then any strict convex combination over all* i *and* j *of the optimal solution pairs must be optimal and be mutually strictly complementary slack.*

Without loss of generality, let $i = 1$*. Let* z^* *denote the optimal solution value to both* **(P)** *and* **(D)***, as assured by Theorem 2.1.3. If no optimal solution to* **(P)** *with* $x_1 = 0$ *exists, the optimal value of the following feasible LP must be zero.*

$$\textbf{(P1)} \qquad \begin{aligned} &\max \mathbf{x}_1 \quad \textit{subject to}\\ &A\mathbf{x} \geq \mathbf{b}\\ &\mathbf{c} \cdot \mathbf{x} \leq z^*\\ &\mathbf{x} \geq \mathbf{0}\end{aligned}$$

The dual of **(P1)** *with variables* $\boldsymbol{\pi}, t$ *is*

$$\textbf{(D1)} \qquad \begin{aligned} &\min -\mathbf{b} \cdot \boldsymbol{\pi} + z^* t \qquad \textit{subject to}\\ &-A^T\boldsymbol{\pi} + \mathbf{c}t \geq \begin{pmatrix} 1\\0\\0\\ \vdots \\ 0\end{pmatrix}\\ &\boldsymbol{\pi} \geq \mathbf{0}; t \geq 0\end{aligned}$$

By Theorem 2.1.3, **(D1)** *has an optimal solution* $\boldsymbol{\pi}^*, t^*$ *with objective value* $z^*t^* - \mathbf{b}\cdot\boldsymbol{\pi}^* = 0$.
Case 1: If $t^* > 0$*, then*

$$\frac{\boldsymbol{\pi}^*}{t^*} \geq \mathbf{0}; \mathbf{b} \cdot \frac{\boldsymbol{\pi}^*}{t^*} = z^*; A^T\frac{\boldsymbol{\pi}^*}{t^*} \leq \mathbf{c} - \begin{pmatrix} \frac{1}{t}\\0\\0\\ \vdots \\ 0\end{pmatrix}$$

whence $\frac{\boldsymbol{\pi}^*}{t^*}$ *is optimal in* **(D)** *and the first constraint,* $A_{1,:}^T\boldsymbol{\pi} \leq \mathbf{c}_1$*, is slack at* $\frac{\boldsymbol{\pi}^*}{t^*}$ *by at least* $\frac{1}{t} > 0$.
Case 2: If $t^* = 0$*, then*

$$\boldsymbol{\pi}^* \geq \mathbf{0}; \boldsymbol{\pi}^* \cdot \mathbf{b} = 0; A^T\boldsymbol{\pi}^* \leq \begin{pmatrix} -1\\0\\0\\ \vdots \\ 0\end{pmatrix}.$$

[3]It is easy to show both things cannot happen; it is harder to show at least one must happen.

Geometrically, $\boldsymbol{\pi}^*$ is a direction of the feasible region of (**D**), *orthogonal to the objective vector* **b**. *Its first coordinate is* ≤ -1. *Let* $\hat{\boldsymbol{\pi}}$ *be any optimal solution for* (**D**). *Then the vector* $\hat{\boldsymbol{\pi}}+\boldsymbol{\pi}^*$ *is also optimal for* (**D**), *and the first constraint,* $A_{1,:}^T \pi \leq \mathbf{c}_1$, *is slack at* $\hat{\boldsymbol{\pi}}+\boldsymbol{\pi}^*$ *by at least* 1.
In both cases, a suitable optimal solution to (**D**) *exists.* ■

8.1.3 From a Theorem of the Alternative to Strong Duality and Back Again

Suppose $\mathbf{x}$ is a feasible solution to the primal LP min $\mathbf{c} \cdot \mathbf{x}$ subject to $A\mathbf{x} \geq \mathbf{b}$. Suppose also that $\mathbf{x}$ is *locally optimal*, that is, for some $\epsilon > 0$, $||\mathbf{y}-\mathbf{x}|| < \epsilon$ implies $\mathbf{c} \cdot \mathbf{y} \geq \mathbf{c} \cdot \mathbf{x}$ or $A\mathbf{y} \not\geq \mathbf{b}$. Let I be the set of constraints binding at $\mathbf{x}$. By local optimality there can't be a vector $\mathbf{d}$ such that $\mathbf{c} \cdot \mathbf{d} < 0$ and $A(\mathbf{x}+\mathbf{d}) \geq \mathbf{b}$. By scaling $\mathbf{d}$ we can make $A_{j,:}(\mathbf{x}+\mathbf{d}) \leq \mathbf{b}_j$ for all $j \notin I$. Therefore, local optimality is equivalent to the nonexistence of $\mathbf{d}$ such that $\mathbf{c} \cdot \mathbf{d} < 0$ and $A_{I,:}\mathbf{d} \leq \mathbf{0}$. By a theorem of the alternative (version 2 in Theorem 8.1.1) this is equivalent to the existence of $\boldsymbol{\pi} \geq \mathbf{0}$ such that $\boldsymbol{\pi}^T A_{I,:} = \mathbf{c}$. Geometrically this says that the primal objective vector $\mathbf{c}$ is a nonnegative combination of the rows of A that are binding at $\mathbf{x}$.

Now the dual is $\max \mathbf{y}^T\mathbf{b}$ subject to $\mathbf{y}^T A = \mathbf{c}^T; \mathbf{y} \geq \mathbf{0}$. By definition of I, $A_{I,:}\mathbf{x} = \mathbf{b}_I$. Let $\mathbf{y}_I^* = \boldsymbol{\pi}$ and let the other components of $\mathbf{y}^*$ equal zero. So $\mathbf{y}^* \geq \mathbf{0}$ and $\mathbf{y}^{*T} A = \boldsymbol{\pi}^T A_{I,:} + \mathbf{0} = \mathbf{c}$. Thus $\mathbf{y}^*$ is feasible in the dual. Moreover $\mathbf{y}^*$ and $\mathbf{x}$ are complementary slack. By the weak duality theorem, $\mathbf{y}^*$ is optimal and $\mathbf{b} \cdot \mathbf{y}^* = \mathbf{c} \cdot \mathbf{x}$. Using a theorem of the alternative, we have proved the strong duality theorem for LPs in the form min $\mathbf{c} \cdot \mathbf{x}$ subject to $A\mathbf{x} \geq \mathbf{b}$.

It is even easier to prove a theorem of the alternative from the strong duality theorem. For example, consider the primal LP

$$\min \mathbf{0} \cdot \mathbf{x} \text{ subject to: } A\mathbf{x} = \mathbf{b}, \mathbf{x} \geq \mathbf{0}.$$

The dual is

$$\max \mathbf{b} \cdot \boldsymbol{\pi} \text{ subject to: } \boldsymbol{\pi}^T A \leq \mathbf{0}.$$

The dual has feasible solution $\boldsymbol{\pi} = \mathbf{0}$ with objective value 0. If there exists $\mathbf{x} \geq \mathbf{0} : A\mathbf{x} = \mathbf{b}$ the primal has a feasible solution with objective value 0. Then by the weak duality theorem, there could not exist $\boldsymbol{\pi} : \boldsymbol{\pi}^T A \leq \mathbf{0}, \boldsymbol{\pi}^T\mathbf{b} > 0$. On the other hand, if there doesn't exist $\boldsymbol{\pi} : \boldsymbol{\pi}^T A \leq \mathbf{0}, \boldsymbol{\pi}^T\mathbf{b} > 0$, then the dual has an optimal solution with value 0. By strong duality, the primal must have an optimal solution, which of course must be feasible. Hence there must exist $\mathbf{x} \geq \mathbf{0} : A\mathbf{x} = \mathbf{b}$.

We've proved an alternative theorem using LP duality. *Question: Why is what we proved equivalent to the 2nd pair in Theorem 8.1.1?*[4] The easy part of a theorem of the alternative is that both systems can't have feasible solutions. The proof of the easy part came from the weak duality theorem. The hard part of a theorem of the alternative is that at least one system has a solution. The proof of the hard part came from the strong duality theorem. This is an example of the law of conservation of difficulty.

8.1.4 Helly's Theorem – A Hiking Trail on Mount Duality

Try drawing a set of line segments in $\Re$ such that every pair of line segments intersects. You won't be able to do it without there being at least one point common to all the line segments. This is the Helly Theorem in one dimension. In two dimensions, the three

[4]The primal is the same. In the dual, $\boldsymbol{\pi}$ is unrestricted. Replace $\boldsymbol{\pi}$ by $-\boldsymbol{\pi}$ to get the equivalence.

sides of a triangle have no point in common, though every two sides intersect. However, in two dimensions, any collection of convex sets such that every triple of sets has nonempty intersection must have at least one point common to all the sets. *Question: Guess the general statement of the theorem.*[5] Starting from LP duality, we prove the general theorem in a sequence of steps.

We begin with a Helly theorem for half-spaces.

Theorem 8.1.4 *For $A \in \Re^{m \times n}$ and $\mathbf{b} \in \Re^m$, suppose $\nexists \mathbf{x} : A\mathbf{x} \leq \mathbf{b}$. Then there exists a subset $\mathbb{N} \subset \{1 \ldots m\}$ of the row indices with $|\mathbb{N}| \leq n+1$ such that $\nexists \mathbf{x} : A_{\mathbb{N},:}\mathbf{x} \leq \mathbf{b}_{\mathbb{N}}$.*

Proof: Consider the primal LP max $\mathbf{0} \cdot \mathbf{x}$ subject to $A\mathbf{x} \leq \mathbf{b}$. Its dual min $\mathbf{b} \cdot \boldsymbol{\pi}$ subject to $\boldsymbol{\pi} \geq \mathbf{0}, A^T\boldsymbol{\pi} = \mathbf{0}^T$ has feasible solution $\boldsymbol{\pi} = 0$. By hypothesis the primal has no feasible solution. By strong LP duality, the dual has an unbounded optimum. Since the dual has constraints $\boldsymbol{\pi} \geq 0$, its feasible region has no line so it must contain a basic feasible solution.

Apply the simplex method to the dual. The algorithm must terminate with an extreme ray $\boldsymbol{\pi} + \theta\boldsymbol{\pi}^* : \theta \geq 0$ where $\boldsymbol{\pi}$ is a basic feasible solution, $\boldsymbol{\pi}^*$ has one more nonzero component than $\boldsymbol{\pi}$, $A^T\boldsymbol{\pi}^* = \mathbf{0}^T$, and $\boldsymbol{\pi}^* \cdot \mathbf{b} < 0$. Since $\boldsymbol{\pi} \in \Re^m$ has at least $m - n$ binding nonnegativity constraints, it has at most n nonzero components. Hence the vector $\boldsymbol{\pi}^*$ has at most $n+1$ nonzero components. Let $\mathbb{N}$ denote the indices of these nonzero components. Then $|\mathbb{N}| \leq n+1$ and $\boldsymbol{\pi}^* \cdot \mathbf{b} = \boldsymbol{\pi}^*_{\mathbb{N}} \cdot \mathbf{b}_{\mathbb{N}} < 0; \boldsymbol{\pi}^*_{\mathbb{N}} \geq 0; (A_{\mathbb{N},:})^T\boldsymbol{\pi}^*_{Nn} = 0$. Multiply the system $A_{\mathbb{N},:}x \leq \mathbf{b}_{Nn}$ by $\boldsymbol{\pi}^*_{Nn}$ to get the contradiction $0 = \mathbf{0} \cdot \mathbf{x} = (\boldsymbol{\pi}^*_{Nn})^T A_{Nn,:}\mathbf{x} \leq \boldsymbol{\pi}^*_{Nn} \cdot \mathbf{b}_{Nn} < 0$. Therefore, $\nexists \mathbf{x} : A_{Nn,:}\mathbf{x} \leq \mathbf{b}_{Nn}$. ■

Helly theorems are traditionally stated affirmatively, that if all $n+1$-tuples of a collection have nonempty intersection then the entire collection has nonempty intersection. Following tradition, we give the following corollary.

Corollary 8.1.5 *Let $\mathcal{H}$ be a finite set of closed halfspaces in $\Re^n$ such that for every subset $S \subset \mathcal{H}$ of $n+1$ halfspaces, $|S| = n+1$, the intersection is not empty, $\bigcap_{H \in S} H \neq \phi$. Then $\bigcap_{H \in \mathcal{H}} H \neq \phi$.*

Proof: By the contrapositive, suppose $\bigcap_{H \in \mathcal{H}} H = \phi$. Let $A\mathbf{x} \leq \mathbf{b}$ denote the inequalities corresponding to the halfspaces $\mathcal{H}$. Then $\nexists \mathbf{x} : A\mathbf{x} \leq \mathbf{b}$. By the theorem, there exists a subset of at most $n+1$ of the inequalities $A\mathbf{x} \leq \mathbf{b}$ that has no solution. The corresponding set of closed halfspaces has empty intersection. This is the negation of the hypothesis of the corollary. ■

Application to Minimal Contradictions:
Your software tells you that your LP is infeasible. What do you do? How do you find out what is causing the infeasibility? Modern software will give the option of generating a minimal set of contradictory constraints for you. A minimal contradictory constraint set is infeasible, but would become feasible if any one of its members were removed. Theorem 8.1.4 gives an upper bound on how large a minimal set can be, namely, $n+1$.

Next, extend the corollary from halfspaces to polyhedra.

[5] In n dimensions, for any collection of convex sets every $n+1$-tuple of which has nonempty intersection, there exists a point common to all the sets.

Theorem 8.1.6 *Let $\{P_1, P_2, \ldots P_N\} = \mathcal{P}$ be a finite set of polyhedra in $\Re^n$ such that for every subset $S \subset \mathcal{P}$ of $n+1$ polyhedra, $|S| = n+1$, the intersection is not empty, $\bigcap_{P \in S} P \neq \phi$. Then $\bigcap_{P \in \mathcal{P}} P \neq \phi$.*

Proof: Let $\mathcal{H}$ denote the set of all closed halfspaces that define the polyhedra in $\mathcal{P}$. Let $\{H_1, H_2, \ldots, H_{n+1}\} \in \mathcal{H}$. For each $i : 1 \leq i \leq n+1$ there exists $j_i : 1 \leq j_i \leq N$ such that H_i is one of the halfspaces that defines P_{j_i}. Therefore, $P_{j_i} \subset H_i$. By hypothesis,

$$\phi \neq \bigcap_{j_i : 1 \leq i \leq n+1} P_{j_i} \subset \bigcap_{1 \leq i \leq n+1} H_i \neq \phi.$$

Apply Theorem 8.1.4 to $\mathcal{H}$ to conclude

$$\bigcap_{P \in \mathcal{P}} P = \bigcap_{H \in \mathcal{H}} H \neq \phi.$$

■

It is a remarkable fact that the Helly Theorem for polyhedra implies the Helly Theorem for convex sets.

Corollary 8.1.7 *Helly's Theorem: Let $\mathcal{Q}$ be a finite set of convex sets in $\Re^n$ such that for every subset $S \subset \mathcal{Q}$ of $n+1$ convex sets, $|S| = n+1$, the intersection is not empty, $\bigcap_{Q \in S} Q \neq \phi$. Then $\bigcap_{Q \in \mathcal{Q}} Q \neq \phi$.*

This beautiful proof is due to Chvátal [51]. For every subset $S \subset \mathcal{Q}$, $|S| = n+1$ select a point q_S contained in the intersection of the members of S. For every $Q \in \mathcal{Q}$ construct polytope P_Q to be the convex hull of all $q_S \in Q$. Since P_Q is the convex hull of points in Q, and Q is convex, then $P_Q \subset Q$. By construction, for every $S \subset \mathcal{Q}, |S| = n+1$ the intersection $\bigcap_{P \in S} P$ is nonempty, for it contains at least the point q_S. Therefore, by Theorem 8.1.6 there exists a point q common to all the polytopes P_Q. Since $P_Q \subset Q$, q is common to all the convex sets Q. ■

Question: Would it suffice to let P_Q be the convex hull of the q_S of all S containing Q?[6]

Theorem 8.1.4 says that if a polyhedron $P = \{\mathbf{x} : A\mathbf{x} \leq \mathbf{b}\} \subset \Re^n$ is empty, there exists a subset of at most $n+1$ of the constraints $A\mathbf{x} \leq \mathbf{b}$ that forms a contradiction. A theorem of Carathéodory's makes a curiously similar-sounding statement if the polyhedron P is *not* empty. If P is bounded and $\mathbf{x} \in P$, there exists a subset of at most $n+1$ extreme points of P that can be used to form $\mathbf{x}$.

Theorem 8.1.8 *Let $P \subset \Re^n$ be a polytope. Let $\mathbf{x} \in P$. Then there exists a set of at most $n+1$ extreme points of P whose convex hull contains $\mathbf{x}$.*

Proof: *By hypothesis P is not empty. Let $\mathbf{v}^1, \ldots, \mathbf{v}^N$ be the extreme points of P, and let $V = [\mathbf{v}^1\mathbf{v}^2 \ldots \mathbf{v}^N]$ be the $n \times N$ matrix they compose. By Theorem 4.5.2 P is the convex hull of its extreme points. Therefore, there exists a feasible solution to the system $V\pi = x, \mathbf{1} \cdot \pi = 1, \pi \geq 0$. Equivalently, the polyhedron $\Pi = \{\pi : V\pi = x, \mathbf{1} \cdot \pi = 1, \pi \geq 0\}$ is not empty. By Theorem 4.4.5 Π has an extreme point $\hat{\pi}$ which by Proposition 4.4.2 is a BFS. By definition, at least N of the constraints that define Π must be binding at $\hat{\pi}$.*

[6] Yes.

There are only $n+1$ equality constraints, so at least $N-(n+1)$ of the nonnegativity constraints $\boldsymbol{\pi} \geq 0$ must be binding at $\hat{\boldsymbol{\pi}}$. Hence $\hat{\boldsymbol{\pi}}$ contains at most $n+1$ nonzero terms. The corresponding columns of V are the desired extreme points. ■

I hope you have wondered if the common term $n+1$ and the similarity of the statements of Theorems 8.1.8 and 8.1.4 are coincidences. They are not. If you apply a theorem of the alternative to the former, you will arrive at the latter.

8.1.5 Separation Characterization of Convexity

The main result of this section is a pleasing converse to the separating hyperplane theorem. Before we get to that, let's state a trivial separating hyperplane property for polyhedra.

Proposition 8.1.9 *Let P be a polyhedron and let $\mathbf{w} \notin P$. Then there exist $\boldsymbol{\pi}, \pi_0$ defining a hyperplane $\{\mathbf{x} : \boldsymbol{\pi} \cdot \mathbf{x} = \pi_0\}$ that strictly separates P from $\mathbf{w}$ in the sense that $\boldsymbol{\pi} \cdot \mathbf{w} > \pi_0$ and $\forall \mathbf{x} \in P, \boldsymbol{\pi} \cdot \mathbf{x} < \pi_0$.*

Proof: Without loss of generality let $P = \{\mathbf{x} | A\mathbf{x} \leq \mathbf{b}\}$. By assumption $\exists i : \; A_{i,:}\mathbf{w} > \mathbf{b}_i$. Then $\mathbf{b}_i < \pi_0 \equiv .5b_i + .5A_{i,:}\mathbf{w} < A_{i,:}\mathbf{w}$ and $\boldsymbol{\pi} \equiv A_{i,:}$ define the desired hyperplane. ■

Therefore, the separation property of polyhedra is elementary. Its easy proof does not employ a theorem of the alternative. The more general Theorem 4.1.1 applies to all closed convex sets in $\Re^n$, not only to polyhedra. One might wonder whether the theorem can be generalized beyond closed convex sets. The following result shows that the answer is "no."

Theorem 8.1.10 *Let $S \subset \Re^n$. Suppose that for every $p \notin S$ there exists a separating hyperplane $\boldsymbol{\pi}, \boldsymbol{\pi}_0$ between S and p, such that $\boldsymbol{\pi} \cdot x \geq \boldsymbol{\pi}_0 \; \forall x \in S$ and $\boldsymbol{\pi} \cdot p < \boldsymbol{\pi}_0$. Then S is closed and convex.*

Proof: Suppose to the contrary S is not convex. Then there exist $\mathbf{x} \in S, \mathbf{y} \in S, 0 < \lambda < 1$ such that $\mathbf{p} = \lambda\mathbf{x} + (1-\lambda)\mathbf{y} \notin S$. By hypothesis let $\boldsymbol{\pi}, \pi_0$ separate S from $\mathbf{p}$. Then $0 < \lambda < 1$ and $\boldsymbol{\pi} \cdot \mathbf{x} \geq \pi_0$ and $\boldsymbol{\pi} \cdot \mathbf{y} \geq \pi_0$ imply $\boldsymbol{\pi} \cdot \lambda\mathbf{x} \geq \lambda\pi_0$ and $\boldsymbol{\pi} \cdot (1-\lambda)\mathbf{y} \geq (1-\lambda)\pi_0$. Hence $\boldsymbol{\pi} \cdot \mathbf{p} \geq \pi_0$, a contradiction with $\boldsymbol{\pi} \cdot \mathbf{p} < \pi_0$. Therefore, S is convex.

Suppose now to the contrary S is not closed. Then its complement is not open. Hence there exists a point $\mathbf{p} \notin S$ such that $\forall \epsilon > 0 \; \exists \mathbf{x} \in S$ such that $||\mathbf{x} - \mathbf{p}|| < \epsilon$. By hypothesis let $\boldsymbol{\pi}, \pi_0$ separate S from $\mathbf{p}$. Without loss of generality let $||\boldsymbol{\pi}|| = 1$. Let $\epsilon = \frac{\pi_0 - \boldsymbol{\pi} \cdot p}{2} > 0$ and let $\mathbf{x} \in S$ with $||\mathbf{x} - \mathbf{p}|| < \epsilon$. Then $\boldsymbol{\pi} \cdot \mathbf{x} \leq \boldsymbol{\pi} \cdot \mathbf{p} + (||\mathbf{x} - \mathbf{p}||)(||\boldsymbol{\pi}||) \leq \boldsymbol{\pi} \cdot \mathbf{p} + \epsilon < \pi_0$, contradicting $\boldsymbol{\pi} \cdot \mathbf{x} \geq \pi_0$. Therefore, S is closed. ■

Theorem 8.1.10 is a little stronger than an exact converse to the separating hyperplane theorem. A strict converse would make the stronger assumption that $\boldsymbol{\pi} \cdot \mathbf{x} > \pi_0 \; \forall \mathbf{x} \in S$. I used the weaker assumption because it made the proof no harder.

What happens if we try to use the existence of a weakly separating hyperplane $\boldsymbol{\pi}, \pi_0$: $\boldsymbol{\pi} \neq \mathbf{0}$ such that $\boldsymbol{\pi} \cdot \mathbf{x} \geq \pi_0 \; \forall \mathbf{x} \in S$ and $\boldsymbol{\pi} \cdot \mathbf{p} \leq \pi_0$ as the hypothesis of Theorem 8.1.10? Consider the set $S \in \Re^2$ consisting of the two points $(1,0)$ and $(2,0)$. The hyperplane defined by $x_2 = 0$ can be used to weakly separate any point in $\Re^2$ from S. *Question: What is not precisely correct about this statement?*[7] But S is not convex. The hypothesis would be too weak to reach the desired consequence.

[7] You have to use the hyperplane as $-x_2 = 0$ for points above the x_1 axis.

Let's conclude this section with a theorem about closed convex sets that is an easy walk to and from the separating hyperplane theorem.

Theorem 8.1.11 *Let $C \subset \Re^n$ be closed and convex. Let $\mathcal{H}$ be the set of closed halfspaces that contain C. Define*

$$\mathcal{C} \equiv \bigcap_{H \in \mathcal{H}} H.$$

Then $C = \mathcal{C}$.

Proof: By definition $C \subseteq \mathcal{C}$. We must prove $C \supseteq \mathcal{C}$, or equivalently, $\mathbf{p} \in \mathcal{C} \Rightarrow \mathbf{p} \in C$. By the contrapositive, we will do this by proving that $\mathbf{p} \notin C \Rightarrow \mathbf{p} \notin \mathcal{C}$. If $\mathbf{p} \notin C$, then by the separating hyperplane theorem (Theorem 4.1.1) $\exists \boldsymbol{\pi}, \pi_0$ such that $\boldsymbol{\pi} \cdot \mathbf{p} < \pi_0$ and $\boldsymbol{\pi} \cdot \mathbf{c} > \pi_0$ $\forall \mathbf{c} \in C$. Hence $\boldsymbol{\pi}, \pi_0$ yields a closed halfspace containing C but not $\mathbf{p}$. Then by definition of $\mathcal{H}$,

$$\mathbf{p} \notin \{\mathbf{x} | \boldsymbol{\pi} \cdot \mathbf{x} \geq \pi_0\} \in \mathcal{H}.$$

Since $\mathbf{p}$ is not in at least one member of $\mathcal{H}$, $\mathbf{p}$ is not in $\mathcal{C}$, the intersection of all members. ∎

Thus it is an easy walk from the separating hyperplane theorem to Theorem 8.1.11. To walk in the other direction, let C be closed and convex, let $\mathbf{p} \notin C$, and suppose Theorem 8.1.11 is true, but we don't know whether the separating hyperplane theorem is true. That $C = \mathcal{C}$ implies $\mathbf{p} \notin \mathcal{C}$, which implies $\exists \hat{H} \in \mathcal{H}$ such that $\mathbf{p} \notin \hat{H}$. Let $\boldsymbol{\pi}, \pi_0$ be such that

$$\hat{H} = \{x | \boldsymbol{\pi} \cdot \mathbf{x} \geq \boldsymbol{\pi}_0\}.$$

We have $\boldsymbol{\pi} \cdot \mathbf{c} \geq \pi_0$ $\forall \mathbf{c} \in C$ and $\boldsymbol{\pi} \cdot \mathbf{p} < \pi_0$. Let $\epsilon = \pi_0 - \boldsymbol{\pi} \cdot \mathbf{p} > 0$. Then the hyperplane

$$\{\mathbf{x} | \boldsymbol{\pi} \cdot \mathbf{x} = \pi_0 - \frac{\epsilon}{2}\}$$

strictly separates $\mathbf{p}$ from C. *Question: Why did I use $\frac{\epsilon}{2}$ instead of ϵ to define the hyperplane?*[8]

8.1.6 Projection and Representation

If P is the sum of the set of convex combinations of $\mathbf{y}^1 \cdots \mathbf{y}^N$ and the cone of nonnegative combinations of $\mathbf{d}^1 \cdots \mathbf{d}^M$,

$$P = \left\{ \sum_{i=1}^{N} \lambda_i \mathbf{y}^i + \sum_{j=1}^{M} \mu_j \mathbf{d}^j \mid \lambda_i \geq 0 \ \forall i, \mu_j \geq 0 \ \forall j, \sum_{i=1}^{N} \lambda_i = 1 \right\}, \tag{8.8}$$

then P is a polyhedron. This is called a representation theorem because P can be represented in the form $\{\mathbf{x} | A\mathbf{x} \leq \mathbf{b}\}$. We proved it in Chapter 4 using the projection theorem. The proof was to project the polyhedron

$$\left\{ (\mathbf{x}, \lambda, \mu) \mid \sum_{i=1}^{N} \lambda_i \mathbf{y}^i + \sum_{j=1}^{M} \mu_j \mathbf{d}^j - \mathbf{x} = 0, \lambda \geq 0, \mu \geq 0, \sum_{i=1}^{N} \lambda_i = 1 \right\}$$

onto the $\mathbf{x}$ variables. By Theorem 4.2.2 P is a polyhedron. *Question: What question should you be asking yourself right now?*[9]

[8] At a RHS of $\pi_0 - \epsilon$ the hyperplane would contain $\mathbf{p}$. To strictly separate, we need a value in the open interval $(0, \epsilon)$.

[9] Is there an easy way to prove the projection theorem from the converse of the representation theorem?

Let's prove the projection theorem for pointed polyhedra. We will use both the representation theorem and its converse. Suppose $P \subset \Re^n$ is a polyhedron. By Theorem 4.5.3 there exist $\mathbf{y}^1 \cdots \mathbf{y}^N$ and $\mathbf{d}^1 \cdots \mathbf{d}^M$ such that (8.8) holds. Let S denote a subset of the indices $1, \ldots, n$ and for any vector $\mathbf{w}$ let w_S denote the projection of $\mathbf{w}$ onto S. Then the projection of P onto S is

$$P_S = \left\{ \sum_{i=1}^{N} \lambda_i \mathbf{y}^i_S + \sum_{j=1}^{M} \mu_j \mathbf{d}^j_S \mid \lambda_i \geq 0 \ \forall i, \mu_j \geq 0 \ \forall j, \sum_{i=1}^{N} \lambda_i = 1 \right\}.$$

By Theorem 4.5.1, P_S is a polyhedron.

If the polyhedron P were not pointed, we would need a generalization of the representation theorem that applied to all polyhedra in order to prove that P_S is a polyhedron. The generalization is valid, except that the points $\mathbf{y}^i$ cannot be taken to be the extreme points of P. Exercise 32 shows you one way to prove the generalization.

Summary: In general, theorems of the alternative and strong LP duality theorems are closely related in that having a proof of one permits fairly easy proofs of others. Some other theorems, such as the projection and representation theorems, are related to each other as well.

8.2 Facets of Polyhedra

Preview: Every polyhedron has the form $P = \{x | A\mathbf{x} \leq \mathbf{b}\}$, but for every polyhedron P there are infinitely many valid choices of A and b that yield P. Is there a "best" or canonical choice of A, b?

In two dimensions, a facet of a polyhedron is a line segment, ray, or line that forms part of the boundary of the polyhedron. In three dimensions, a facet of a polyhedron is a 2-dimensional (flat) region that forms part of the surface of the polyhedron. The main result of this section is that the constraints that define the facets of a polyhedron are necessary and sufficient to describe the polyhedron. For example, in Figure 8.3, the constraints $x_1 + x_2 \leq 10$, $x_1 \geq 0$, and $x_2 \geq 0$ are necessary and sufficient to describe the shaded triangle. The constraints $x_1 + x_2 \geq 0$, $x_1 + 2x_2 \leq 20$, and $x_1 \leq 10$ are valid but superfluous. As another example, the description of a cube must include a constraint for each of the 6 sides of the cube, and need not contain any other constraints.

At this point you may be wondering why such an obvious statement requires proof. I would answer with an example of another obvious statement. A *simplex* in $\Re^n$ is a polytope with nonzero n-dimensional volume and $n + 1$ extreme points. A 2-dimensional simplex is a triangle; a 3-dimensional simplex is a tetrahedron. Consider the intersection of a line

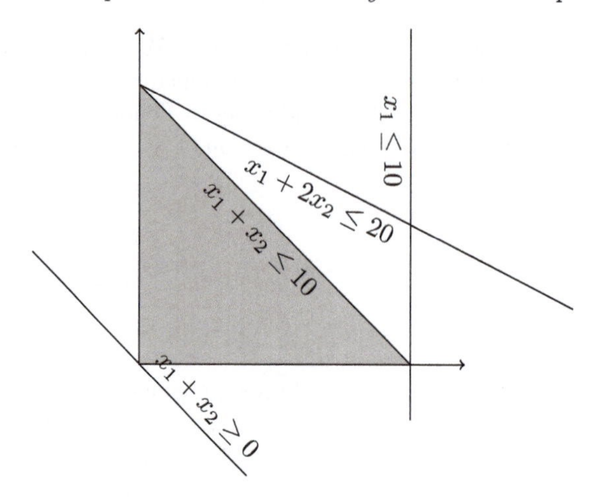

FIGURE 8.3: Three constraints are necessary and sufficient to describe the shaded region.

with a triangle. Its length cannot be greater than the length of the longest side (i.e., facet). Consider the intersection of a plane with a tetrahedron. Its 2-dimensional area cannot be greater than the area of the largest side (i.e., facet). It may seem obvious that, in general, the intersection of a hyperplane with a simplex in $\Re^n$ cannot have $n-1$-dimensional volume greater than that of the largest facet. However, this statement is false in $\Re^5$.

8.2.1 Affine Independence and Dimension

A linear combination of points in $\Re^n$ is a weighted sum of the points. A set of points is *linearly independent* iff the only linear combination that equals zero uses weights that are all zeroes. That is, the points $\mathbf{x}^j : j \in J$ are linearly independent iff

$$\sum_{j \in J} \alpha_j \mathbf{x}^j = \mathbf{0} \Rightarrow \alpha_j = 0 \ \forall j \in J.$$

Definition 8.1 (Affine Independence) *A set* $\mathbf{x}^0, \mathbf{x}^1, \ldots, \mathbf{x}^m$ *of* $m+1$ *points in* $\Re^n$ *is* affinely independent *iff the* m *points* $\mathbf{x}^1 - \mathbf{x}^0, \mathbf{x}^2 - \mathbf{x}^0, \ldots, \mathbf{x}^m - \mathbf{x}^0$ *are linearly independent.*

Geometrically this means that if you place yourself at one of the points, and treat your location as the origin, do the rest of the points look linearly independent? Linear independence implies affine independence, but the converse is false. For example, any three noncolinear points in $\Re^2$ are affinely independent, but no three points in $\Re^2$ are linearly independent. *Question: How can two points in* $\Re^2$ *fail to be affinely independent?*[10]

An equivalent definition of affine independence is that a set of points $\mathbf{x}^j : j \in J$ in $\Re^n$ is *affinely independent* iff the points $(\mathbf{x}^j, 1)$ in $\Re^{n+1}$ are linearly independent.

When $|J| = n+1$ it is easy to see the equivalence in terms of matrices. The latter statement is equivalent to the following matrix being nonsingular:

$$\begin{bmatrix} \mathbf{x}^0 & \mathbf{x}^1 & \mathbf{x}^2 & \ldots & \mathbf{x}^n \\ 1 & 1 & 1 & \ldots & 1 \end{bmatrix} \tag{8.9}$$

[10]If and only if they are the same point.

Subtract the first column from each of the other columns. This does not affect the determinant of the matrix and it yields

$$\begin{bmatrix} \mathbf{x}^0 & \mathbf{x}^1 - \mathbf{x}^0 & \mathbf{x}^2 - \mathbf{x}^0 & \dots & \mathbf{x}^n - \mathbf{x}^0 \\ 1 & 0 & 0 & \dots & 0 \end{bmatrix} \tag{8.10}$$

which has the same determinant (up to a factor of -1) as its submatrix

$$\begin{bmatrix} \mathbf{x}^1 - \mathbf{x}^0 & \mathbf{x}^2 - \mathbf{x}^0 & \dots & \mathbf{x}^n - \mathbf{x}^0 \end{bmatrix} \tag{8.11}$$

which is nonsingular iff its columns are linearly independent. In fact, the volume of the n-dimensional simplex whose extreme points are $\mathbf{x}^0, \mathbf{x}^1, \dots \mathbf{x}^n$ is $\frac{1}{n!}$ times the determinant of (8.11). What does it mean if the determinant is zero? It means that the simplex has zero n-dimensional volume. (The convex combination of the points might even fail to be a simplex in lower dimension.) All of this motivates our definition of dimension.

Definition 8.2 (Dimension of a set) *A set S has dimension m, written $dim(S) = m$, if S contains $m+1$ but not $m+2$ affinely independent points.*

From the definition it is immediate that $Q \subset S \Rightarrow dim(Q) \leq dim(S)$. Here is a useful lemma for establishing the dimension of a set:

Lemma 8.2.1 *If the set $\{x \in \Re^n : A\mathbf{x} = \mathbf{b}\}$ is nonempty, it has dimension n minus the rank of A, and rank(A) = rank(A,b).*

Proof: Let $A\mathbf{x}^0 = \mathbf{b}$. Then the number of affinely independent solutions to $A\mathbf{x} = \mathbf{b}$ equals one plus the number of linearly independent solutions to $A\mathbf{x} = 0$. This is because if $\mathbf{x}^0, \mathbf{x}^1 \dots \mathbf{x}^m$ are affinely independent solutions to $A\mathbf{x} = \mathbf{b}$, then $\mathbf{x}^1 - \mathbf{x}^0, \mathbf{x}^2 - \mathbf{x}^0, \dots \mathbf{x}^m - \mathbf{x}^0$ are linearly independent solutions to $A\mathbf{x} = 0$. Conversely if $\mathbf{y}^1 \dots \mathbf{y}^m$ are linearly independent solutions to $A\mathbf{x} = 0$, then $\mathbf{x}^0 + \mathbf{y}^1, \dots \mathbf{x}^0 + \mathbf{y}^m$ and $\mathbf{x}^0$ are affinely independent solutions to $A\mathbf{x} = \mathbf{b}$. Finally, the number of linearly independent solutions to $A\mathbf{x} = 0$ equals n minus the rank of A.

For the second statement, $rank(A, \mathbf{b}) - 1 \leq rank(A) \leq rank(A, \mathbf{b})$. Suppose without loss of generality that the first m rows of $(A, \mathbf{b})$ are l.i., and the first $m-1$ rows of A are l.i., but $A_{m,:} = \sum_{1 \leq i \leq m-1} \alpha_i A_{i,:}$. Then $\mathbf{b}_m = A_{m,:}\mathbf{x}^0 = \sum 1 \leq i \leq m - 1\alpha_i A_{i,:}\mathbf{x}^0 = \sum 1 \leq i \leq m - 1\alpha_i \mathbf{b}_i$, whence $(A, \mathbf{b})_{m,:} = \sum_{1 \leq i \leq m-1} \alpha_i (A, \mathbf{b})_{i,:}$ contradicting the linear independence of the first m rows of $(A, \mathbf{b})$. ■

Example 8.2 Let

$$A = \begin{bmatrix} 1 & 3 \\ 2 & 6 \end{bmatrix}; \mathbf{b} = \begin{pmatrix} 5 \\ 10 \end{pmatrix}; \mathbf{b}' = \begin{pmatrix} 5 \\ 11 \end{pmatrix}.$$

Then $\exists \mathbf{x} : A\mathbf{x} = \mathbf{b}$ and $rank(A) = rank(A, b) = 1$ but $\nexists \mathbf{x} : A\mathbf{x} = \mathbf{b}'$ and $rank(A, \mathbf{b}') = 2$.

8.2.2 Faces and Facets

Definition 8.3 (Supporting Hyperplane, Face, Facet) *A hyperplane H* supports *a polyhedron P if $P \bigcap H \neq \phi$ and P lies entirely in one of the two closed halfspaces defined by the hyperplane. The intersection of a polyhedron and a supporting hyperplane is a* face *of the polyhedron. A* facet *of a polyhedron P is a face that has dimension one less than the dimension of P itself. A linear inequality $\mathbf{c} \cdot \mathbf{x} \leq c_0$* induces *the face F of polyhedron P if $F = P \bigcap \{\mathbf{x} : \mathbf{c} \cdot \mathbf{x} = c_0\}$ and $\mathbf{c} \cdot \mathbf{x} \leq c_0 \ \forall \mathbf{x} \in P$.*

Reading Tip: *Invent your own examples of these definitions before reading further.*

Example 8.3 Hyperplane H_1 in Figure 8.4 intersects the polytope at a vertex, which is a 0-dimensional face – not a facet. The linear inequality that is satisfied by all points in the polytope, and is binding at points in H_1, induces that face. Rotate H_1 slightly about that face to get infinitely many non-equivalent linear inequalities that also induce it.
Hyperplane H_2 intersects the polytope at an edge, which is a 1-dimensional face and hence a facet. The linear inequality that is satisfied by all points in the polytope, and is binding at points in H_2, induces that facet. It is the unique inequality that induces that facet, not counting equivalent inequalities formed by positive scaling.

Unlike H_1 and H_2, hyperplanes H_3 and H_4 do not support the polytope. *Question: Why not?*[11]

It follows immediately from the definitions that a face is a polyhedron, because if $H = \{\mathbf{x} : \mathbf{c} \cdot \mathbf{x} = c_0\}$, then $H \bigcap P$ is defined by a finite set of linear inequalities, namely, those that describe P and $\mathbf{c} \cdot \mathbf{x} \leq c_0$ and $\mathbf{c} \cdot \mathbf{x} \geq c_0$. However, that description of the face is not very useful, because it depends on $\mathbf{c}, c_0$. Intuitively, there should be a description of a face that is independent of the particular choice of the linear inequality $\mathbf{c}, c_0$ which induces it. The next theorem states that a face can always be described by setting some of the linear inequalities to linear equalities, among the linear inequalities that define the polyhedron. The inequalities that we set to equalities are the ones that are binding constraints when one maximizes $\mathbf{c} \cdot \mathbf{x}$ over the polyhedron.

Theorem 8.2.2 *Let $A \in \Re^{m \times n}$, let $P = \{\mathbf{x} | A\mathbf{x} \leq \mathbf{b}\}$ and let F be a face of P. Then there exists a subset $I \subseteq \{1 \ldots m\}$ of the rows of A such that*

$$F = \{\mathbf{x} | A_{i,:}\mathbf{x} = \mathbf{b}_i \ \forall i \in I; A_{i,:}\mathbf{x} \leq \mathbf{b}_i \ \forall i \notin I\}. \tag{8.12}$$

Corollary 8.2.3 *The number of distinct faces of a polyhedron is finite.*

Proof: By definition, there exists a supporting hyperplane $H = \{\mathbf{x} : \mathbf{c} \cdot \mathbf{x} = c_0\}$ whose intersection with P is F. Without loss of generality, let the face F be the set of optimal solutions to the primal LP $\max \mathbf{c} \cdot \mathbf{x}$ subject to $A\mathbf{x} \leq \mathbf{b}$. Its dual is $\min \boldsymbol{\pi} \cdot \mathbf{b}$ subject to $\boldsymbol{\pi}^T A = c^T$. The primal has optimum solution value c_0. Hence the dual has an optimal solution. Let $\boldsymbol{\pi}^*$ be an optimal solution to the dual, and let $I \subset \{1 \ldots m\}$ index the nonzero

[11] H_3 doesn't intersect it; H_4 goes through its interior.

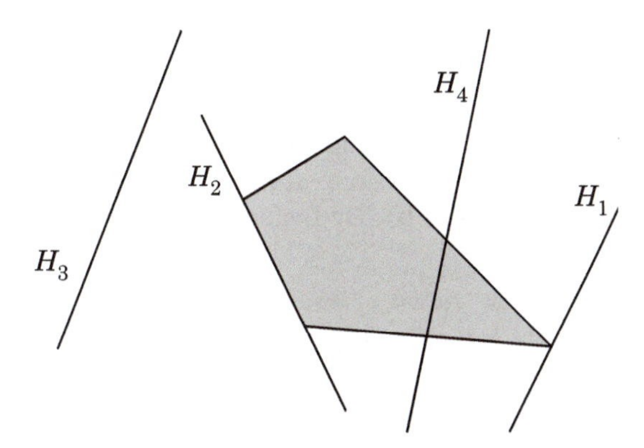

FIGURE 8.4: H_1 and H_2 support the polytope, yielding 0 and 1-dimensional faces, respectively.

elements of $\boldsymbol{\pi}^*$. We claim that this choice of I satisfies Equation (8.12). This is true because if $\mathbf{x}$ satisfies the right-hand side of (8.12), it precisely satisfies the complementary slackness conditions with respect to $\boldsymbol{\pi}^*$ and the primal feasibility conditions. Therefore, $\mathbf{x}$ is optimal in the primal and so $\mathbf{x} \in F$. Conversely, if $\mathbf{x}$ does not satisfy the right hand side of (8.12), then either it violates primal feasibility, in which case $\mathbf{x} \notin P$ and hence $\mathbf{x} \notin F$, or it violates complementary slackness, and hence for some $i \in I$, $\boldsymbol{\pi}_i^* > 0$ but $A_{i,:}\mathbf{x} < \mathbf{b}_i$. But then $\mathbf{c} \cdot \mathbf{x} = (\boldsymbol{\pi}^*)^T A\mathbf{x} < \boldsymbol{\pi}^* \cdot \mathbf{b} = c_0$, and $\mathbf{x} \notin F$. For the corollary, observe that the total number of possible facets is bounded by 2^m, the number of possible choices of the set I. ■

Reading Tip: *Why must H be supporting for F to be the set of optimal solutions to the LP? What theorem is implicitly invoked to conclude that the dual has an optimal solution? For a given face, is the subset I unique?*

We need another preliminary result before coming to our main theorems. This is another seemingly obvious statement which requires proof. It says, in the case that P is a full-dimensional polyhedron, that P has a point in its interior. Let's first define our terminology.

Definition 8.4 (Interior, Relative Interior) *Let P be the polyhedron $\{\mathbf{x}|A\mathbf{x} \leq \mathbf{b}\}$. A point $\mathbf{y}$ is in the* interior *of P if $A\mathbf{y} < \mathbf{b}$. Now let $I = \{i|A_{i,:}\mathbf{x} = \mathbf{b}_i \ \forall \mathbf{x} \in P\}$ be the index set (possibly empty) of inequalities that hold with equality for all points in P. A point $\mathbf{y} \in P$ is in the* relative interior *of P if $A_{i,:}\mathbf{y} < \mathbf{b}_i \ \forall i \notin I$.*

Lemma 8.2.4 *Every nonempty polyhedron has a relative interior point.*

Proof: By definition of I, for every $i \notin I$ there exists $\mathbf{y}^i \in P$ such that $A_{i,:}\mathbf{y}^i < \mathbf{b}_i$. Let $\mathbf{y}$ be any strict convex combination of the $\mathbf{y}^i : i \notin I$. Then $A_{i,:}\mathbf{y} < \mathbf{b}_i \ \forall i \notin I$. By convexity $\mathbf{y} \in P$, so $\mathbf{y}$ is our desired point. ■

Theorem 8.2.5 *Let polyhedron $P = \{\mathbf{x} \in \Re^n | A\mathbf{x} \leq \mathbf{b}\}$. Let I be, as in Definition 8.4, the index set of inequalities that are binding at all points in P. Then $dim(P)$ + the rank of $(A_{I,:}, \mathbf{b}_{I,:}) = n$.*

Corollary 8.2.6 *If F is a facet of polyhedron $P = \{\mathbf{x}|A\mathbf{x} \leq \mathbf{b}\}$, then there exists $i \notin I$ such that $A_{i,:}\mathbf{x} \leq \mathbf{b}_i$ induces F.*

Proof: Let r denote the rank of $(A_{I,:}, b_I)$. Let $\mathbf{y}$ be a relative interior point of P, which exists by Lemma 8.2.4. By Lemma 8.2.1 there exist $n-r+1$ affinely independent points $\mathbf{x}^i$ (but not $n-r+2$) satisfying $A_{I,:}\mathbf{x}^i = 0$. For sufficiently small $\epsilon > 0$ the points $\mathbf{y} + \epsilon\mathbf{x}^i$ are also in the relative interior of P. Therefore, $dim(P) \geq n-r$. On the other hand, let Q denote the subspace defined by $A_{I,:}\mathbf{x} = \mathbf{b}_I$. By Lemma 8.2.1 $dim(Q) = n-r$. Since $P \subset Q$, $dim(P) \leq n-r$. ■

Reading Tip: *Why does adding $\epsilon\mathbf{x}^i$ to $\mathbf{y}$ keep you in the relative interior? Is the conclusion of the corollary true when F is a face but not a facet?*

We now come to the principal result of this section. Roughly speaking, it says that the facet-inducing inequalities are the necessary and sufficient set of linear inequalities needed to describe a polyhedron. The precise statement is complicated by the possible presence of linear equalities satisfied by the polyhedron.

Theorem 8.2.7 *Let $P = \{\mathbf{x}|A\mathbf{x} \leq \mathbf{b}\}$ and let I, as in Definition 8.4, index the inequalities that are binding at all points of P. Let r denote the rank of $(A_{I,:}, b_I)$. Let Θ denote a minimal set of equalities and inequalities that represent P. Then Θ consists of one facet-inducing inequality for each facet of P, together with r equality constraints that are equivalent to $A_{I,:}\mathbf{x} = \mathbf{b}_I$.*

Proof: First we must show that Θ must contain a facet-inducing inequality for each facet of P. Let F be a facet of P. Without loss of generality, let $A_{t,:}\mathbf{x} \leq \mathbf{b}_t$ uniquely induce F among the inequalities of Θ defining P. (If t is not unique, then Θ is not minimal). Define $Q = \{\mathbf{x}|A_{i,:}\mathbf{x} \leq \mathbf{b}_i \; \forall i \neq t; A_{i,:}\mathbf{x} \geq \mathbf{b}_i \; \forall i \in I\}$. We must show that $Q - P$ is nonempty. F is a polyhedron defined by $A\mathbf{x} \leq \mathbf{b}$ and $A_{t,:}\mathbf{x} \geq \mathbf{b}_t$. Let $\mathbf{p}$ be a relative interior point of F. Then for all $i \notin I$ and $i \neq t$, $A_{j,:}\mathbf{p} < \mathbf{b}_i$. Let B be a ball of sufficiently small radius $\epsilon > 0$ around $\mathbf{p}$ intersected with $\{\mathbf{x}|A_{I,:}\mathbf{x} = \mathbf{b}_I\}$. Then $A_{i,:}\mathbf{q} < \mathbf{b}_i$ for all $i \notin I; i \neq t$ for all $\mathbf{q} \in B$. Since $t \notin I$ there does not exist $\boldsymbol{\pi}$ such that $\boldsymbol{\pi}^T A_{I,:} = A_{t,:}$. By Theorem 4.3.1 there must exist $\mathbf{r} \in \Re^n$ such that $A_{t,:}\mathbf{r} > 0$ and $A_{I,:}\mathbf{r} = 0$. Set $\mathbf{q} = \mathbf{p} + \epsilon\mathbf{r}$. See Figure 8.5. Since $A_{t,:}\mathbf{p} = \mathbf{b}_t$, then $A_{t,:}\mathbf{q} > \mathbf{b}_t$. So $\mathbf{q} \in Q - P$ as desired.

Second, we must show that Θ does not contain any face-inducing inequalities for faces of dimension less than $n-r-1$. Let F be such a face, induced by inequality $A_{t,:}\mathbf{x} \leq \mathbf{b}_t$, where $t \notin I$. We need to show that this inequality can be removed from Θ without allowing any new points to satisfy the description Θ of P. By contradiction, suppose such a point $\mathbf{y}$ exists. By assumption $A_{t,:}\mathbf{y} > \mathbf{b}_t$, and $\mathbf{y}$ satisfies all other constraints in Θ. In particular, $A_{I,:}\mathbf{y} = \mathbf{b}_I$. Now let $\mathbf{w}$ be a point in the relative interior of P, so $A_{t,:}\mathbf{w} < \mathbf{b}_t$. Since $\mathbf{w} \in P$, $A_{I,:}\mathbf{w} = \mathbf{b}_I$. Let $\mathbf{z}$ be the convex combination of $\mathbf{y}$ and $\mathbf{w}$ such that $A_{t,:}\mathbf{z} = \mathbf{b}_t$, so $\mathbf{z} \in F$. Then for all $i \notin I, i \neq t$, $A_{i,:}\mathbf{z} = \lambda A_{i,:}\mathbf{z} + (1-\lambda)A_{i,:}\mathbf{y}$ for some $1 > \lambda > 0$. Therefore, $A_{i,:}\mathbf{z} < \mathbf{b}_i \; \forall i \notin I, i \neq t$. Therefore, the only equalities that could satisfied by all points of F are $A_{I,:}\mathbf{x} = \mathbf{b}_I$ and $A_{t,:}\mathbf{x} = \mathbf{b}_t$. The rank of these equalities is at most $r+1$. Therefore, F has dimension at least $n-r-1$, a contradiction. ■

Theorem 8.2.7 tells us that every facet is necessary to the description of a polyhedron. This still leaves open the possibility that some facets may be more important than others. For example, in Figure 8.6, there are three facets that account for most of the "shape" of the polyhedron, and two that seem barely significant. Is this contrived example typical or atypical of what happens in higher dimensions? There is a modest amount of empirical evidence that polyhedra tend to have a small fraction of facets that are "important". The

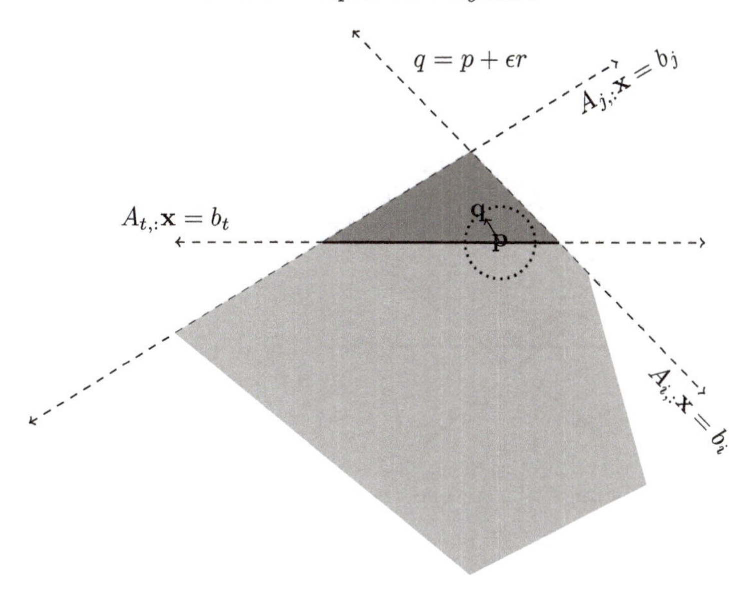

FIGURE 8.5: Move in the direction $\mathbf{r}$ from the relative interior point $\mathbf{p}$ to violate the facet-defining constraint $A_{t,:}\mathbf{x} \leq \mathbf{b}_t$ while remaining feasible with respect to the other constraints.

evidence is mainly from what are called *shooting experiments*. In a shooting experiment one shoots imaginary bullets in random directions from a point in the relative interior of the polyhedron P. If P is a polytope, of course all bullets exit P. Whether or not P is a polytope, with probability 1 each bullet that exits P exits through the relative interior of a facet (i.e., the chance of a random bullet hitting a lower dimensional face is zero). One counts, for each facet, the number of times it is hit by a bullet. The idea is that the frequency of being hit is a measure of the importance of the facet. The first shooting experiments were reported by Kuhn in 1963 [199, 203]. Kuhn shot from the centroid of the asymmetric traveling salesman problem (defined in Chapter 15) and found that most of the shots hit nonnegativity constraint induced facets. Kuhn repeated the experiment on a larger scale in 1991, shooting more than 100,000 bullets [201]. He found that the 20 nonnegativity facets accounted for 80% of the bullets, even though they account for only about 5% of the facets. Gomory, Johnson, and Evans [131] performed similar shooting experiments on master cyclic group polyhedra, though they shot from outside the polyhedron. They too found that a relatively small number of facets accounted for the majority of the bullet hits. The complexity of performing shooting experiments is analyzed in [157].

Summary: A facet of an m-dimensional polyhedron is an $m-1$-dimensional intersection of a valid inequality (i.e., a supporting hyperplane) with the polyhedron. If a polyhedron is full-dimensional, a set of such inequalities, one per facet, is necessary and sufficient to define it. If a polyhedron in $\Re^n$ is $n-k$-dimensional, k linearly independent equality constraints and one inequality per facet are necessary and sufficient to define it. There is some evidence that in many polyhedra the facets are not of equal importance.

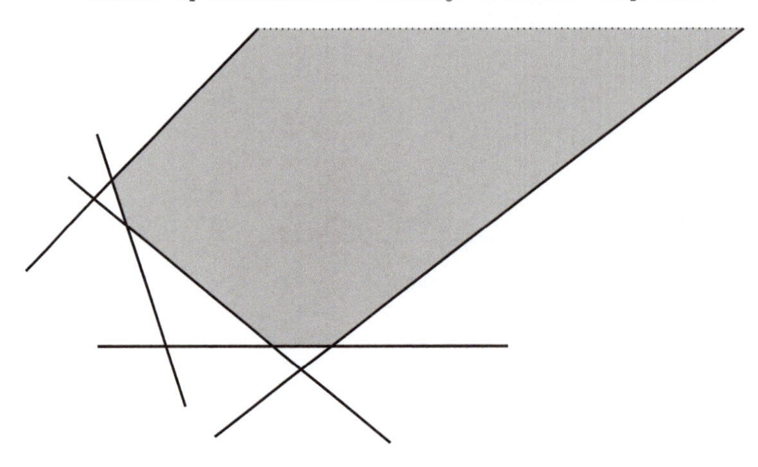

FIGURE 8.6: The two unbounded facets and the third long one account for most of the shape of the polyhedron.

8.3 Column Geometry

Preview: This section explains the column geometry, a way to visualize the simplex method that explains its name and provides intuition as to why it is efficient computationally.

When George Dantzig first thought of the simplex algorithm, he did not name it as such, nor did he expect it to be computationally efficient. Why not? It was because he visualized it geometrically as shown in Section 2.2. In that geometry, the algorithm has nothing to do with simplices. Nor does it seem likely to be efficient, because it traverses all the way around a polyhedron's exterior instead of traveling directly through the interior to reach the optimum. After several algorithms invented by other people turned out to perform poorly in tests, Dantzig reconsidered his own algorithm, which he had not yet tested because it had appeared less promising. This time, he visualized his algorithm in a different geometry, called the *column space*. In that geometry, simplices play a pivotal role, and the algorithm seems likely to be efficient. On that geometrically intuitive basis he tested his algorithm. It was very efficient, and became a workhorse of operations research from then to the present day.

We can derive the general form of the column space geometry by applying the definition of duality from Section 1.2. For constraints $A\mathbf{x} = \mathbf{b}; x \geq 0$ where A is an $m \times n$ matrix, the usual geometric visualization takes place in $\Re^n$. Each row of A is an n-vector that together with the RHS term defines a hyperplane in $\Re^n$. The matrix A is being viewed as a set of rows. The dual view of A is as a set of columns. The dual visualization therefore takes place in $\Re^m$. Each column of A is a point in $\Re^m$, as is the RHS vector b.

The specifics of the column space geometry are as follows: Consider this variation of the LP form:

$$\min \mathbf{c}\cdot\mathbf{x} \ \text{ subject to } \ \begin{array}{l} A\mathbf{x}=\mathbf{b}; \\ \mathbf{1}\cdot\mathbf{x}=1; \\ \mathbf{x}\geq\mathbf{0}. \end{array} \qquad (8.13)$$

The constraint $\mathbf{1}\cdot\mathbf{x}=1$ is called a *convexity constraint* because together with $\mathbf{x}\geq 0$ it forces $\mathbf{x}$ to be a vector of weights that makes $A\mathbf{x}$ a convex combination of the columns of A. It entails no loss of generality because it is possible to place a finite upper bound on each variable. (Details are given in Section 10.2.) *Question: Why do finite upper bounds permit us to enforce the convexity constraint?*[12] Let $\mathbb{B}$ be the index set of a basic solution, and let $B=A_{:,\mathbb{B}}$. *Question: Why is* $|\mathbb{B}|\neq m$?[13] The constraints of (8.13) say that $Bx_{\mathbb{B}}=\mathbf{b}$, $\mathbf{x}_{\mathbb{B}}\geq\mathbf{0}$, and $\sum_{i\in\mathbb{B}}x_i=1$. Geometrically these constraints have a precise meaning: $\mathbf{b}$ is a convex combination of the columns of B. For example, in Figure 8.7, choices of basis indices $(2,4,5)$ and $(1,2,4)$ give feasible basic solutions, because $\mathbf{b}$ is in the convex hull of the basic points $A_{:,i}: i\in\mathbb{B}$. The basis index set $(1,2,3)$ is infeasible because $\mathbf{b}$ is not in the convex hull of $A_{:,1},A_{:,2},A_{:,3}$. The convex hull of $m+1$ points in $\Re^m$ is called a *simplex*, from which the name of the simplex algorithm derives.

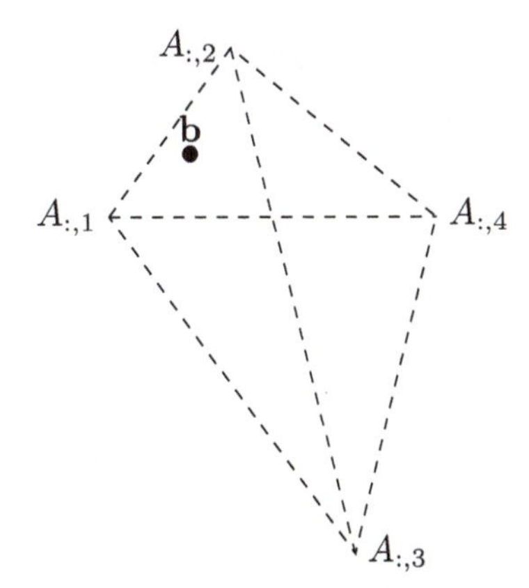

FIGURE 8.7: The LP constraints require b to be a convex combination of the basic columns $A_{:,i}: i\in\mathbb{B}$, as with $\mathbb{B}=\{1,2,3\}$ or $\{1,2,4\}$. If $\mathbf{b}$ is not in the simplex formed by the $m+1$ basic columns, the basic solution is not feasible, as with $\mathbb{B}=\{2,3,4\}$ or $\{1,3,4\}$. When $m=2$ a simplex is a triangle.

The weights forming the basic solution are the solution to $Bx_{\mathbb{B}}=\mathbf{b}; \mathbf{1}\cdot\mathbf{x}=1$. Define

$$\bar{B}=\left[\begin{array}{c} B \\ 1\,1\,\ldots\,1 \end{array}\right]; \bar{\mathbf{b}}=\left(\begin{array}{c}\mathbf{b}\\ 1\end{array}\right).$$

Then $\mathbf{x}_{\mathbb{B}}=\bar{B}^{-1}\bar{\mathbf{b}}$. Geometrically, the weights $\mathbf{x}_{\mathbb{B}}$ express $\mathbf{b}$ as an *affine* combination of the columns of B. When $\mathbf{b}$ is contained in the convex hull of B's columns, these weights are nonnegative, i.e., they express a *convex* combination of the columns of B.

To visualize the objective function value, add a dimension to each point $A_{:,i}$ by inserting

[12]Scale the variables by the sum of the bounds to get $\mathbf{1}\cdot\mathbf{x}\leq 1$ and then add a slack variable to convert to an equality constraint.

[13]The $m+1$st constraint $\mathbf{1}\cdot\mathbf{x}=1$ makes $|\mathbb{B}|=m+1$.

the objective function coefficient c_i to get the vectors

$$\begin{pmatrix} c_i \\ A_{:,i} \end{pmatrix} \forall i.$$

Call these vectors the *basic points* and call the hyperplane they define (i.e., their affine hull) the *basic hyperplane*. Think of the c_i dimension as the vertical dimension and think of $\Re^m$ as being a 2D plane as in Figure 8.8. The larger the height of a point, the larger is its objective function coefficient. The objective value of the basic solution $\mathbf{x}_{\mathbb{B}}$, algebraically, is $\mathbf{c}_{\mathbb{B}} \cdot \mathbf{x}_{\mathbb{B}}$. Since $\mathbf{x}_{\mathbb{B}}$ is the set of weights on the columns of B that produce $\mathbf{b}$ as a weighted average, $\mathbf{c}_{\mathbb{B}} \cdot \mathbf{x}_{\mathbb{B}}$ is the corresponding weighted average of the heights of the basic points. Geometrically, the objective value is the height of the point where the line $(\theta, \mathbf{b}) : \theta \in \Re$ intersects the basic hyperplane.

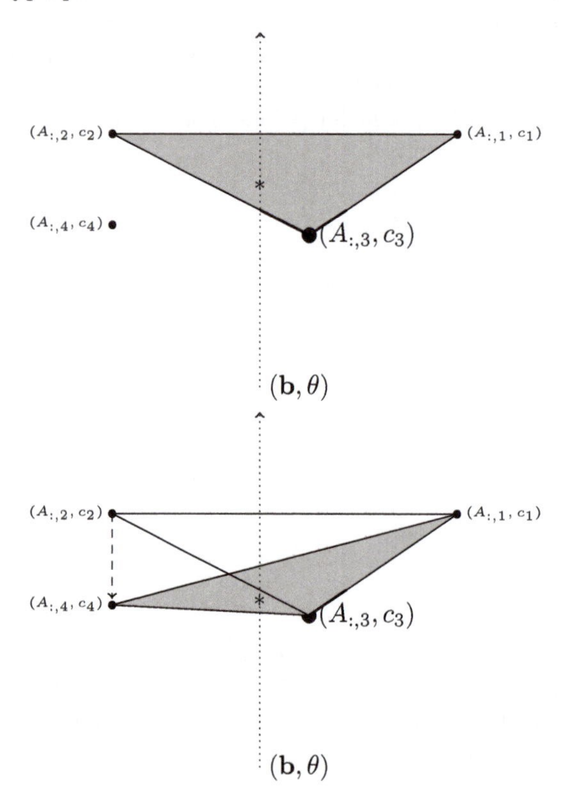

FIGURE 8.8: The "height" of a basic point is the objective function coefficient $\mathbf{c}_i$. The objective function value of a basic solution is the height of the point where the convex hull — a simplex — of the basic points intersects the line $(\theta, \mathbf{b}) : \theta \in \Re$.

The reduced cost of a nonbasic variable $\mathbf{x}_j$ as always is defined as $\mathbf{c}_j - \mathbf{c}_{\mathbb{B}}^T B^{-1} A_{:,j}$. In column space the reduced cost has a direct geometric meaning. The value $\mathbf{c}_j$ is the height of the point $(\mathbf{c}_j, A_{:,j}$. The value $\mathbf{c}_{\mathbb{B}}^T B^{-1} A_{:,j}$ is the height of the intersection of the basic hyperplane with the line $(\theta, A_{:,j}) : \theta \in \Re$. Therefore, *the reduced cost tells how far the nonbasic point is below or above the basic hyperplane.* Since we are minimizing, nonbasic points below the basic hyperplane have good reduced costs.

Now we are ready to describe the simplex algorithm in the column geometry. The linear programming problem is to find a set of $m+1$ basic columns of A whose convex hull —

which is an m-dimensional simplex — intersects the RHS line $(\theta, \mathbf{b})$ at as low a point as possible. The least reduced cost rule selects the nonbasic variable that is farthest below the basic hyperplane to put into the basis. Absent degeneracy, there is only one choice for the leaving variable that will retain an intersection between the RHS line and the convex hull of the new basic points. Imagine that you grasp the m-dimensional simplex between your fingers at the point of the leaving basic variable. Keep the simplex attached at the other m basic points, and pull your fingers down to the point of the entering basic variable. The simplex swings down, or *pivots*, on the m basic points to give a new simplex corresponding to the new choice of basis.

In this geometry, the simplex method does not appear to be stupidly going around the outside of a polyhedron. Instead, it is seeking the lowest simplex that intersects the RHS line by iteratively picking the point that is lowest with respect to the current simplex. It seems plausible that it will skip over most of the points in favor of the lowest points, and hence will take about $m+1$ iterations to reach optimality. Now you should see why the algorithm is named the simplex method, and why it appeared to be efficient in the column space geometry. For a more detailed exposition, see [272]. You can see a video of my class demonstrating the column geometry on YouTube here:

`https://www.youtube.com/watch?v=Ci1vBGn9yRc`

What sense can we make of the fact that two equally valid geometric visualizations lead to two completely different intuitive predictions about how fast the simplex method performs? Dantzig used to say that shows that we have no idea what is going on in high dimensions.

Summary: In the column geometry, columns of the $m \times n$ constraint matrix are points in $\Re^m$. A basis corresponds to $m+1$ of these points whose convex hull is a simplex which must contain the RHS vector $\mathbf{b}$ for the basic solution to be feasible. To each point add an $m+1$st dimension consisting of its objective function coefficient. The basic points now are the vertices of an m-dimensional simplex lying on a hyperplane in $\Re^{m+1}$. Points below the hyperplane have good reduced costs (when minimizing). The simplex algorithm selects the point farthest below the hyperplane to enter the basis. The simplex drops one of its vertices and "pivots" on its m other vertices to form a new simplex with the selected point.

8.4 Polarity

Preview: This section introduces polarity, a fascinating duality between pairs of polyhedra wherein the vertices and facets exchange roles.

If you have ever studied the five Platonic solids, you may have encountered an interesting relationship between the cube and the octahedron, and between the dodecahedron and the icosahedron. The cube has 6 facets and 8 extreme points; the octahedron has 8 facets and 6 extreme points. Both have 12 edges. The dodecahedron has 12 (pentagonal) facets and 20 extreme points; the icosahedron has 20 (triangular) facets and 12 extreme points. Both have 30 edges. Moreover, if you chop off the vertices of a cube, creating a facet where each vertex had been, you can get an octahedron, and vice versa. Similarly, chopping off the vertices of a dodecahedron can produce an icosahedron, and vice versa. The simplest of the Platonic solids, the tetrahedron, has 4 facets and 4 extreme points. It enjoys the same kind of relationship with itself.

In this section we study the algebraic meaning of this geometric relationship, which is called *polarity*. Polarity is a duality via the incidence function

$$\mathfrak{I} : \mathcal{F} \times \mathcal{V} \mapsto \{0, 1\}$$

where $\mathcal{F}$ is the set of facets, $\mathcal{V}$ is the set of extreme points, and $\mathfrak{I}(f, v) = 1$ if facet f and vertex v are incident. Polarity gives us a hiking trail between the representation theorem (Theorem 4.5.2) and the facet-characterization theorem (Theorem 8.2.7) for polytopes.

For simplicity, we will assume that our polyhedron is a full dimensional polytope. We also assume that the origin is an interior point of the polytope. This second assumption entails no further loss of generality with respect to extreme points and facets because a full-dimensional polyhedron can always be translated so that the origin is in its interior.

Definition 8.5 (Polar of a polytope) *Let $P \subset \Re^n$ be a polytope containing the origin as an interior point. The polar of P is the set $P^\perp \equiv \{\boldsymbol{\pi} | \boldsymbol{\pi} \cdot \mathbf{x} \leq 1 \ \forall \mathbf{x} \in P\}$.*

In words, the polar of P is the set of all valid inequalities of P, scaled to have RHS 1.

Theorem 8.4.1 *If P is a polytope containing the origin as an interior point, then its polar $P^\perp$ is a polytope containing the origin as an interior point, and the polar of the polar $P^{\perp^\perp} = P$.*

Proof: let $\mathbf{x}^1 \dots \mathbf{x}^N$ be the extreme points of P. We claim that $P^\perp = \{\boldsymbol{\pi} | \boldsymbol{\pi} \cdot \mathbf{x}^i \leq 1 \ \forall 1 \leq i \leq N\} \equiv Q$. On the one hand, $P^\perp \subseteq Q$ because $P^\perp$ is defined by a set of inequalities that includes all the inequalities defining Q. On the other hand, let $\mathbf{x} \in P$ so $\boldsymbol{\pi} \cdot \mathbf{x} \leq 1$ is one of the (infinitely many) constraints defining $P^\perp$. Since P is a polytope, there exist $\lambda_i \geq 0; 1 \leq i \leq N$ such that $\mathbf{x} = \sum_{i=1}^N \lambda_i \mathbf{x}^i$ and $\sum_{i=1}^N \lambda_i = 1$. Then the constraint $\boldsymbol{\pi} \cdot \mathbf{x} \leq 1$ is a nonnegative linear combination of the constraints $\boldsymbol{\pi} \cdot \mathbf{x}^i \leq 1$, which are the constraints defining Q. Hence $P^\perp = Q$, and therefore $P^\perp$ is a polyhedron.

If $P^\perp$ has a direction $\mathbf{d} \neq 0$, then $\mathbf{d} \cdot \mathbf{x}^i \leq 0 \ \forall 1 \leq i \leq N$. This implies $\mathbf{d} \cdot \mathbf{x} \leq 0 \ \forall \mathbf{x} \in P$, which is not consistent with the origin being an interior point of P (the point $-\epsilon \mathbf{d}$ would not be in P for any $\epsilon > 0$). Therefore, $P^\perp$ is a polytope.

Let $K = \max_{1 \leq i \leq N} ||\mathbf{x}^i||$. Then for all $\mathbf{y}$ such that $||\mathbf{y}|| < 1/K$ we have $\mathbf{y} \cdot \mathbf{x}^i \leq 1 \ \forall i$ whence $\mathbf{y} \in P^\perp$. Therefore, the origin is an interior point of $P^\perp$ and $P^\perp$ is full-dimensional.

Let $\mathbf{x} \in P$. By definition $\boldsymbol{\pi} \cdot \mathbf{x} \leq 1 \ \forall \boldsymbol{\pi} \in P^\perp$. That is, $\boldsymbol{\pi} \cdot \mathbf{x} \leq 1$ is a valid inequality of $P^\perp$. By definition, $\mathbf{x} \in P^{\perp^\perp}$. Therefore, $P \subseteq P^{\perp^\perp}$.

Conversely, suppose $\mathbf{w} \notin P$. Then $\mathbf{w}$ can't be expressed as a convex combination of

$\mathbf{x}^1 \ldots \mathbf{x}^N$. By the separating hyperplane Theorem 4.1.1 there exists $\boldsymbol{\pi}^0, \pi_0^0$ such that $\mathbf{x}^i \cdot \boldsymbol{\pi}^0 < \pi_0^0 \; \forall i$ and $\mathbf{w} \cdot \boldsymbol{\pi}^0 > \pi_0^0$. P contains the origin, so $\pi_0^0 \neq 0$. Without loss of generality let $\pi_0^0 = 1$. Then $\boldsymbol{\pi}^0 \in P^\perp$. Since $\mathbf{w} \cdot \boldsymbol{\pi}^0 > 1$, $\mathbf{w} \cdot \boldsymbol{\pi} \leq 1$ can't be a valid inequality of $P^\perp$. Therefore, $\mathbf{w} \notin P^{\perp\perp}$. By the contrapositive, $P^{\perp\perp} \subseteq P$. *Question: What is the contrapositive argument here?*[14] We conclude that $P = P^{\perp\perp}$. ■

Corollary 8.4.2 *Let P be a polytope containing the origin as an interior point. Then $P^\perp = \{\boldsymbol{\pi} | \boldsymbol{\pi} \cdot \mathbf{x}^i \leq 1 \, \forall \, 1 \leq i \leq N\}$, where $\mathbf{x}^1 \ldots \mathbf{x}^N$ are the extreme points of P, and $P = \{\mathbf{x} | \mathbf{x} \cdot \boldsymbol{\pi}^j \leq 1 \; \forall 1 \leq j \leq M\}$, where $\boldsymbol{\pi}^1 \ldots \boldsymbol{\pi}^M$ are the extreme points of $P^\perp$. Moreover, these sets of inequalities are minimal representations of $P^\perp$ and P, respectively.*

When we describe a polytope as being the set of convex combinations of its extreme points, every extreme point is necessary to the description, and every point in the polytope that is not extreme is superfluous. This follows from the definition of an extreme point. By Theorem 8.2.7, when we describe a full-dimensional polytope as being the set of points satisfying valid inequalities, the facet-defining valid inequalities are all necessary to the description, and every valid inequality that does not define a facet is superfluous. When we construct the polar of a polytope, we turn extreme points into facets, and facets into extreme points. The polar of a regular n-gon centered at the origin is simply another regular n-gon, so in two dimensions you can't easily appreciate the way polarity interchanges the roles of vertices and facets.

Corollary 8.4.3 *The inequalities in Corollary 8.4.2 all define facets.*

Proof: As indicated above, this can be argued to follow from Theorem 8.2.7. It is best seen directly as follows. Let $\mathbf{x}^i$ be an extreme point of P. By Proposition 4.4.2 there are n linearly independent binding constraints at $\mathbf{x}^i$, where $n = dim(P)$. Each of these binding constraints is a valid inequality of P. Since linear independence implies affine independence, there are n affinely independent vectors $\boldsymbol{\pi}^j \in P^\perp$ such that $\boldsymbol{\pi}^j \cdot \mathbf{x}^i = 1$. Therefore, the inequality $\boldsymbol{\pi} \cdot \mathbf{x}^i \leq 1$ defines a facet of $P^\perp$. The statement about the inequalities defining P follows from the symmetry of the relationships between P and $P^\perp$. ■

Summary: The different versions of the strong duality theorem and the theorems of the alternative can each be used to prove the others. The fundamental theorem of LP has a close relationship to Helly's Theorem. In $\Re^n$, the defining inequalities of an n-dimensional polyhedron that are binding at $n-1$ dimensional subsets of the polyhedron form facets, which are necessary and sufficient to describe the polyhedron. Polarity forms pairs of polyhedra each of whose extreme points are the facets of the other.

---

[14] The statement $\mathbf{w} \notin P \Rightarrow \mathbf{w} \notin P^{\perp\perp}$ is the contrapositive of $\mathbf{w} \in P^{\perp\perp} \Rightarrow \mathbf{w} \in P$. The latter means $P^{\perp\perp} \subset P$.

8.5 Problems

E Exercises

1. For each of the following sets of constraints,

   - Write the alternative system of constraints using Theorem 8.1.2.
   - Find a feasible solution to the alternative system.
   - Use your feasible solution to explicitly prove that given set of constraints has no feasible solution.

   (a) $x_1 = -5, x_1 \geq 0$.

   (b) $x_1 + 2x_2 \leq 10, -3x_1 - x_2 \leq -35, x \geq 0$.

   (c) $x_1 + x_2 \leq 11; x_1 + x_2 \geq 12$.

2. Define infinitely many choices of A, b that yield, as $\{x|A\mathbf{x} \leq \mathbf{b}\}$, the three-dimensional cube with side-length 5 and extreme points at $(2,2,2)$ and $(7,7,7)$. What is a minimal choice of A, b?

3. Define $\mathcal{C} = \{A\pi|\pi \geq 0\}$. Prove that any rescaling of the columns of A by strictly positive values does not change $\mathcal{C}$.

4. Assuming that strong duality holds for the pair (1) in Theorem 8.1.2, prove that strong duality holds for pair (2).

5. Show that if a pair of primal and dual optimal solutions are not degenerate, they are the unique optima and satisfy strict complementary slackness.

6. Determine whether or not the following sets of points are affinely independent.

   (a) 1 (a point in $\Re^1$)

   (b) 0

   (c) $0; 1$ (two points in $\Re^1$)

   (d) $3; -3$

   (e) $0; 1; 2$

   (f) $1; 2; 5$

   (g) $(1,1); (4,5); (-4,-5)$

   (h) $(2,3); (4,2); (60,50)$

   (i) $(1,2); (-3,-4); (10,15)$

   (j) $(1,2,3); (4,8,-2); (-2,-4,8)$

   (k) $(1,2,3); (4,8,-2); (-2,-4,8), (8,0,2)$

   (l) $(0,-1,1); (1,1,1); (-1,-1,-1), (1,-1,1)$

7. State theorems of the alternative for the following systems:

   (a) $A\mathbf{x} \geq \mathbf{b}$

   (b) $A\mathbf{x} = \mathbf{b}; x \leq 0$

   (c) $A\mathbf{x} \geq \mathbf{b}; x \geq 0$
   (d) $A\mathbf{x} < 0$

8. Prove that the tetrahedron defined by $x_i \geq 0 : i = 1, 2, 3; x_1 + 2x_2 + 3x_3 \leq 12$ is 3-dimensional.

9. Prove that the polygon defined by $x_i \geq 0 : i = 1, 2, 3; x_1 + 2x_2 + 3x_3 = 12$ is 2-dimensional.

10. Define $P \subset \Re^2$ as $\{(x_1, x_2 | x_1 \geq 0, x_2 \geq 0, x_1 + x_2 \leq 6, x_1 + x_2 \geq 0, x_2 \leq 6\}$.

    (a) Prove that $x_1 \geq 0$ induces a facet of P. Remember that you must first establish the dimension of P.
    (b) Prove that $x_1 + x_2 \geq 0$ induces a face but not a facet of P.
    (c) Prove that $x_1 + x_2 \leq 6$ induces a facet of P.
    (d) What subset of the given constraints provides a minimal description of P?

11. Define $P \subset \Re^3$ as $\{(x_1, x_2, x_3) | x_i \geq 0 : i = 1, 2, 3; x_3 = 4; x_1 + x_2 + x_3 \leq 8$.

    (a) Use the definitions of affine independence and dimension to prove rigorously that the dimension of P is 2.
    (b) Which of $x_i \geq 0 : i = 1, 2, 3$ induce facets of P?
    (c) Does $x_3 = 4$ define a facet of P?

12. Suppose all terms in the lower and upper bound vectors $\mathbf{l}$ and $\mathbf{u}$ are finite. Convert

$$\min \mathbf{c} \cdot \mathbf{x} \quad \text{subject to } A\mathbf{x} = \mathbf{b}, \mathbf{l} \leq \mathbf{x} \leq \mathbf{u}$$

    to the form

$$\min \mathbf{c} \cdot \mathbf{x} \quad \text{subject to } A\mathbf{x} = \mathbf{b}, \mathbf{x} \geq \mathbf{0}, \sum_i x_i = 1.$$

13. Let H be the hyperplane in $\Re^2$ passing through the points $(2, 8)$ and $(6, 6)$.

    (a) Where does $(4, 0)$ project to on H?
    (b) Where does $(0, 0)$ project to on H?
    (c) Where does $(x, 0)$ project to on H?

14. Let H be the hyperplane in $\Re^3$ passing through the points $(1, 0, 10)$, $(0, 1, 8)$, and $(2, 2, 40)$.

    (a) Where does $(1, 1, 0)$ project to on H?
    (b) Where does $(-1, -4, 0)$ project to on H?
    (c) Where does $(x_1, x_2, 0)$ project to on H?

15. Find a 3-dimensional polyhedron such that every 3-tuple of its facets have nonempty intersection, but the intersection of all the facets is empty.

16. Construct 3 convex sets in $\Re^2$ such that each pairwise intersection has strictly positive 2-dimensional area, but the 3-way intersection is empty.

17. Find the paragraph in Section 8.3 that contains two awful puns.

M Problems

18. Prove Theorem 8.1.1 for the 1st, 3rd, and 4th pairs using strong duality theorems.

19. (a) Find and prove the correctness of an alternative system to $A\mathbf{x} < b$ using the strong duality theorem. (b) Prove the validity of your alternative system by mimicking the proof of the theorem of the alternative via Fourier-Motzkin elimination. (c) Prove the validity of your alternative system directly from a theorem of the alternative. (d) Read ahead in Chapter 10 about tiny perturbations to strict inequality systems. Can you prove the validity of your alternative system using those results?

20. Prove that a set of points $\mathbf{x}^j : j \in J$ in $\Re^n$ is *affinely independent* iff the points $(\mathbf{x}^j, 1)$ in $\Re^{n+1}$ are linearly independent.

21. Let $P = \{x | A\mathbf{x} \leq \mathbf{b}\}$ be a nonempty polyhedron. A constraint $\boldsymbol{\pi} \cdot \mathbf{x} \leq \boldsymbol{\pi}_0$ is valid for P if it is satisfied by all $\mathbf{x} \in P$. Prove or disprove: $\boldsymbol{\pi} \cdot \mathbf{x} \leq \boldsymbol{\pi}_0$ is valid for P iff $\exists \mathbf{y} \geq 0$ such that $\mathbf{y}^T A = \boldsymbol{\pi}^T$ and $\mathbf{y} \cdot \mathbf{b} \leq \boldsymbol{\pi}_0$.

22. Prove Theorem 8.1.3 using a theorem of the alternative rather than LP strong duality.

23. Prove Theorem 8.1.3 by applying strong duality to the LP

$$\begin{gathered} \max v \\ \mathbf{1}v \leq \mathbf{x} + \mathbf{c} - A^T\boldsymbol{\pi} \\ \mathbf{1}v \leq \boldsymbol{\pi} + A\mathbf{x} - \mathbf{b} \\ A\mathbf{x} \geq \mathbf{b} \\ A^T\boldsymbol{\pi} \leq \mathbf{c} \\ \mathbf{c} \cdot \mathbf{x} - \mathbf{b} \cdot \boldsymbol{\pi} \leq 0 \\ \mathbf{x} \geq \mathbf{0}; \boldsymbol{\pi} \geq \mathbf{0} \end{gathered}$$

24. Construct an optimal simplex tableau in which the primal variable x_1 is nonbasic, has reduced cost $\pi_{n+1} = 0$, and such that: (i) If x_1 is pivoted into the basis with the (primal) simplex algorithm, it enters at value zero; (ii) If a degenerate primal basic variable x_j (i.e., $x_j = 0$) is removed from the basis by a dual simplex pivot, π_{n+1} remains zero. What does this imply about proving Theorem 2.1.5?

25. Prove Theorem 8.1.11 with the definition of $\mathcal{H}$ changed to

$$\{\mathbf{x} | \boldsymbol{\pi} \cdot \mathbf{x} \geq \boldsymbol{\pi}_0\} = H \in \mathcal{H} \Leftrightarrow C \subset H \text{ and } C \cap \{\mathbf{x} | \boldsymbol{\pi} \cdot \mathbf{x} = \boldsymbol{\pi}_0\} \neq \phi.$$

A hyperplane $\{\mathbf{x} | \boldsymbol{\pi} \cdot \mathbf{x} = \boldsymbol{\pi}_0\}$ satisfying the above is called a *supporting hyperplane* of C. Hint: You must use the fact that C is closed, since otherwise C might fail to have supporting hyperplanes.

26. Hiking Trail: Assuming Theorem 8.1.8 is true, prove Theorem 8.1.4.

27. Hiking Trail: assuming Theorem 8.1.6 is true, prove Theorem 8.1.8.

28. (a) Draw the polar of a square.
    (b) Determine the polar of the unit cube in $\Re^3$.
    (c) Show that the polar of any finitely generated cone $F = \{\sum_i \theta_i w^i | \theta_i \geq 0 \forall i\}$ is $F^\perp = \{\pi | \pi^T w^i \leq 0 \forall i\}$.
    (d) Show that the polar of $F^\perp$ is F.

(e) Show that a finitely generated pointed cone F is a polyhedron.

29. Suppose the origin is in the interior of polytope P, and the shooting experiment is performed from the origin. It is found that a few faces account for the great majority of the bullet hits. What does this imply about the polar of P?

30. Use the projection theorem and the separation proposition to prove a theorem of the alternative.

31. Without using Theorem 8.2.7 prove that if a polyhedron described by $A\mathbf{x} \leq \mathbf{b}$ has full dimension, and a constraint from $A\mathbf{x} \leq \mathbf{b}$ is redundant (removing it does not change the solution set), then that constraint does not form a facet of the polyhedron. Assume the system $A\mathbf{x} \leq \mathbf{b}$ has no row that is a positive multiple of another row.

D Problems

32. Let $P \subset \Re^n$ be a polyhedron containing a line. Prove the representation theorem for P as follows: intersect P with each of the 2^n closed orthants. P equals the union of these intersections. Prove that each intersection is a pointed polyhedron. Apply the representation theorem to each intersection. Prove that every direction of P is a direction of at least one intersection and that every direction of an intersection is a direction of P itself. Finally, prove that the extreme points of the intersections, taken together with the directions, generate all of P.

33. Let $P = \{\mathbf{x}|A\mathbf{x} \geq \mathbf{0}\}$ be a polyhedral cone. You may assume that P is full-dimensional. Prove that every facet of P contains $\mathbf{0}$. Then use Theorem 8.2.7 and elementary reasoning to prove a theorem of the alternative.

I Problems

34. A regular polyhedron in $\Re^3$, by definition must have facets that are all congruent to each other and are regular (equilateral and equiangular) polygons, and must be symmetric with respect to both facets and vertices. Prove that the regular tetrahedron, cube, octahedron, dodecahedron, and icosahedron are the only regular convex polyhedra in $\Re^3$. Hint: Consider the possible numbers of facets that can share a vertex, and use Euler's formula $v - e + f = 2$ where v, e, f equal the number of vertices, edges, and facets of a convex polyhedron, respectively.

35. Construct a regular cube out of clay or some other firm material that can be cut. (I use milk chocolate; cheese would work well, too.) Gradually shave each vertex with cuts whose orthogonal vectors point from the vertex to the cube's center. Stop when you have a polyhedron whose facets are a mix of squares and equilateral triangles. Verify Euler's formula for that polyhedron. Continue shaving until you have an octahedron. Verify Euler's formula for your octahedron, and verify that vertices and facets have exchanged roles. If you are using an edible material, shave the corners of the octahedron until you get a cube, shave the cube's corners to get an octahedron, etc., until you are both intellectually and gustatorially satiated.

36. Can you use polarity to teleport between Theorems 8.1.8 and 8.1.6, under the restriction that all of the polyhedra are polytopes?

37. Find a hiking trail from the polyhedral representation theorem (that a pointed polyhedron is the set of convex combinations of its extreme points plus nonnegative multiples

of its extreme directions) to the theorem that the facets of a polyhedron are necessary and sufficient to define it. Hint: Use polarity.

38. Find a hiking trail between the two theorems in Problem 37 but going in the opposite direction.

39. What happens if you apply polarity to the projection theorem (that the projection of a polyhedron is a polyhedron)?

40. An n-simplex is a full-dimensional polytope in $\Re^n$ with $n+1$ extreme points. A triangulation of the n-cube is a covering by n-simplices such that every non-empty intersection of two of the simplices is a face (not necessarily a facet) of each. The 2-cube can be triangulated with two simplices. Prove that the minimum possible triangulation of the 3-cube has 5 simplices. (Hint: The volumes of the simplices must sum to 1.) (*) For $n = 4, 5, 6$ and 7 the minimum numbers are $16, 67, 308$, and 1493 [58, 155, 156, 220].

Chapter 9

Polynomial Time Algorithms

Preview: Optimization algorithms rely on fast lower-level algorithms from computer science. Section 9.1 explains the theoretical and practical meanings of the term "fast". Sections 9.2 and 9.3 describe fundamental data structures and fast algorithms that are often used by optimization algorithms.

Elsewhere in this text I use the word "problem" loosely, sometimes meaning a category such as LP, and sometimes referring to a particular numerical case. In this chapter, a *problem* is always a category. A specific member of the category is called an *instance*. For reasons that will become clear later, any problem we study contains infinitely many instances. Since instances are always encoded in binary or some other finite alphabet, there is no upper bound on the size of the instances in the problems we study.

Definition 9.1 (Problem, Instance) *A* problem *is a category of questions following a specific structure. The structure calls for a set of numerical values to be input, and for output that satisfies specified properties with respect to the input. A specific set of input data comprises an* instance *of the problem.*

Example 9.1 LP is a problem. Minimize $3x_1 + 4x_2$ subject to $5x_1 + 6x_2 \geq 4, x_1 \geq 0, x_2 \geq 0$ is an instance of LP.

Example 9.2 Factoring is the problem that calls for a positive integer as input, and requires as output a multiset of prime numbers whose product equals the input. To factor $2^{32} + 1$ is an instance of factoring.

A problem may be similar to but more general than another problem. A problem may require yes/no output instead of or as well as numerical output.

Example 9.3 A k-coloring of $G = (V, E)$ is a function $f : V \mapsto \{1, 2, \ldots, k\}$ such that $f(v) \neq f(w)$ for all edges $(v, w) \in E$.
Graph k-Coloring takes a positive integer k and a graph $G = (V, E)$ as input. As output, it requires a k-coloring of G if one exists, and "No" if not.

Graph 3-Coloring takes a graph $G = (V, E)$ as input. As output, it requires a 3-coloring of G if one exists, and "No" if not. It is a problem in its own right, despite its being in essence a special case of Graph k-coloring. That is, its set of instances corresponds precisely to the instances of Graph k-coloring for which $k = 3$.

Summary: An *instance* of a problem is a specific single case, such as determining whether or not 1,111,111 is prime, or finding the optimal objective value of the LP example (2.19). A *problem* is a category of instances. A problem must contain infinitely many instances, or equivalently, for all N must contain an instance of size at least N.

9.1 Running Time of Algorithms

"Sooner or later any problem will get big enough that you can't solve it. But it's nicer if you make it later. " — Adam Rosenberg

---

Preview: This section explains how to measure the amount of time an algorithm takes, both in theory and in practice.

How much time does a computer take to find the maximum of a set of numbers? A numerical answer, such as "0.018 seconds," is meaningless, because the length of time depends on the hardware, the computer program, and how many numbers are in the set. A meaningful answer should meet these criteria:

- It measures the number of basic steps — arithmetic operations, comparisons, storage and retrieval of data — required, rather than time units.
- It only applies to a particular algorithm, not to the problem to be solved.
- It is a function of the size of the particular instance to be solved. *The function tells how the algorithm run time grows as the input size increases.*

In this case, suppose the numbers are $A[i] : i = 1, \ldots, n$. An simple algorithm sets $k = 1$ and for $i = 2, 3, \ldots, n$ compares $A[k]$ with $A[i]$. Whenever the latter is larger, the algorithm sets $k = i$. At termination, $A[k]$ is the maximum value. (If there are ties, $A[k]$ is the first maximum value.) This algorithm finds the maximum in a number of steps proportional to n. We say the algorithm *runs in linear time*, or *takes $O(n)$ time*, or for short, *is $O(n)$*. If instead we wish to sort the numbers in ascending order, the fastest algorithms require time proportional to $n \log n$. We say that these sorting algorithms run in $O(n \log n)$ time.

The precise definition of $O()$ time bounds is:

Definition 9.2 **($O(\ )$ time bound)** *An algorithm has time bound $O(f(n))$ if there exist constants N and K such that for every input of size $n \geq N$ the algorithm will not take more than $Kf(n)$ steps.*

What exactly do we mean by size of an input? The precise meaning is the actual length, the number of digits or bits, of the input data. This is the mathematical definition, but it is nonintuitive and awkward to work with. Instead, often one or two natural parameters describe the size of the particular instance to be solved, and we state an algorithm's time requirements as functions of these parameters. You can usually get away with using these natural parameters instead of input length, because they usually give the right answer.

Example 9.4 Consider more carefully the time required to find the maximum of n numbers. The natural parameter to use is n. However, the bound $O(n)$ is not technically correct, because comparing two numbers takes more time if the numbers are very large (many digits long). The correct analysis must be performed in terms of the combined input length of the numbers, L, which takes into account the number of digits. There might be a few very large numbers or many small numbers of total length L. We can assume each number on the list is itself stored as a list, from least significant digit to most significant digit. *Question: Why can we assume this?*[1] Once the numbers are stored in this way, any two numbers from the list can then be compared in time proportional to the length of the shorter. Let c be a constant such that two numbers of lengths α and β can be compared in time $c\min\{\alpha, \beta\}$. Let k_j be the length of the jth number in the list, so $\sum_{j=1}^{n} k_j \leq L = O(L)$. *Question: Why don't we say that the sum of the k_j equals L?*[2] To compare k_{j+1} with the maximum of $k_1, \ldots, k_j$ therefore takes time at most ck_{j+1}. Hence, it takes time $O(L)$ to find the maximum of a list of numbers L bits long, regardless of whether the list contains many small numbers or a few large numbers.

So for finding the maximum of a set of n numbers, $O(L)$ is technically correct while $O(n)$ is not. On the other hand, since numbers are usually of limited precision in practice, the more natural $O(n)$ bound is a good surrogate. Moreover, we do find the maximum of a set of numbers in linear time. The length of the instance is L, which can be bigger than n because the numbers can be large. The time required is $O(L)$, which can be bigger than n because the numbers can be large. If we incorrectly use n instead of L, the two technical errors cancel out!

Example 9.5 For a more subtle example, consider how much time is required to verify that a 3-coloring of a graph $G = (V, E)$ is valid. Naively, the edge set E occupies space proportional to $|E|$. However, if there are 10^6 vertices, their labels require about $\log_2 10^6 \approx \log_2 2^{20} = 20$ bits of space. This increases the space for E by a factor of $O(\log |V|)$. The time required to process an edge also increases by a $O(\log |V|)$ factor. Figure 9.1 shows the natural and exact measures of input length and time. As in Example 9.4, the inaccuracies cancel out, and the time required is linear.

---

[1]The input can be converted into that form in time $O(L)$. Since $O(L) + O(L) = O(L)$ we can afford to do the conversion and get a $O(L)$ time bound as long as the rest of the calculations take $O(L)$.

[2]Some bits of input are needed to separate the numbers from each other.

	Naive Natural	Precisely Correct
Encoding Length per Vertex	$O(1)$	$O(\log \lvert V \rvert)$
Encoding Length of 3-coloring	$O(\lvert V \rvert)$	$O(\lvert V \rvert)$ or $O(\lvert V \rvert \log \lvert V \rvert)$
Encoding Length per Edge	$O(1)$	$O(\log \lvert V \rvert)$
Encoding Length of E	$O(\lvert E \rvert)$	$O(\lvert E \rvert \log \lvert V \rvert)$
Encoding Length of Instance	$\mathbf{O(\lvert E \rvert)}$	$\mathbf{O(\lvert E \rvert \log \lvert V \rvert)}$
Time to Validate One Edge	$O(1)$	$O(\log \lvert V \rvert)$
Time to Validate All Edges	$\mathbf{O(\lvert E \rvert)}$	$\mathbf{O(\lvert E \rvert \log \lvert V \rvert)}$

FIGURE 9.1: It takes linear time to check the validity of a 3-coloring if we naively measure both instance size and computation time by the natural parameters. The precise size and time are both larger by a factor of $\log |V|$, cancelling out the naive errors.

The cancelation of errors still leaves us with a conflict. In Example 9.5, the technically correct run time is $O(|E| \log |E|)$ in terms of the natural parameter $|E|$. *Question: Why is it OK to use* $\log |E|$ *instead of* $\log |V|$*?*[3] But we would like to say the run time is linear.

The conflict between what is technically correct and what is natural to use is almost always resolved by the following conventions. We pretend that basic arithmetic operations such as addition, multiplication, and division take one unit of time. We also pretend that accessing an item in computer memory takes unit time. These conventions are usually well justified. As long as fewer than 64 bits of precision are required, the arithmetic operations do take unit time. As long as there are no more than 2^{64} items stored, a 64-bit computer can access an item in unit time. Under these conventions, we analyze the running times of algorithms in terms of natural parameters such as matrix dimensions, the number of vertices or edges in a graph, the number of jobs to be scheduled, etc.

As a rule of thumb, use the natural parameters without fear, unless the individual numbers in the data (for example, cost coefficients) are very long or very short. If they are very long, your problem may be harder than it looks; if they are very short, your problem may be easier. This will be discussed further in Section 14.9.1.

Another useful rule of thumb is that when comparing algorithms, *small multiplicative factors are not important*, and indeed, often cannot be measured precisely. The $O()$ definition ignores these factors via the constant K.

Example 9.6 Let's compare two algorithms for finding the maximum of the numbers $A[i] : i = 1, 2, \ldots, n$. Algorithm I is the simple one stated above. Algorithm II is as follows:

Step 1 Set $best := A[1]$.

Step 2 Set $i = 1$.

Step 3 If $A[i] > best$ set $best := A[i]$ and go to Step 2; otherwise proceed to Step 4.

Step 4 If $i = n$ stop. If $i < n$ set $i := i + 1$ and go to Step 3.

Algorithm II wastes time by starting over at $i = 1$ whenever it finds a larger value than it has found so far. If $A[i] = -i \; \forall i$ Algorithm II will take time proportional to n, but

[3] $2 \log |V| = \log |V|^2 \geq \log |E|$. Hence unless the graph is absurdly sparse, $\log |E| = \theta \log |V|$.

if $A[i] = i \; \forall i$ it will take time proportional to n^2. Since the $O()$ time bound definition must apply to all inputs, Algorithm II is $O(n^2)$
Algorithm II does one thing better than Algorithm I. It stores the value *best* instead of storing the integer k. This is better because Algorithm I has to find the value $A[k]$ many times. The computer must do an extra calculation to get the value $A[k]$ because first it must get the value k, and then it uses that value to get $A[k]$. It is quicker to just get the value *best*. If we were implementing Algorithm I carefully, we would pay attention to this detail. But to choose between Algorithm I and Algorithm II, we should only compare $O(n)$ with $O(n^2)$ and choose the former.
In practice, there might be no difference in speed between getting the values $A[k]$ and *best*. A good compiler will store the value $A[k]$ in a special location, separate from the set of $A[]$ values, so that it can be retrieved extremely quickly. A multi-core processor can pre-compute $A[k]$ in case it will be needed. This example illustrates why small multiplicative factors are often not important, and might not even be possible to assess.

When comparing *implementations* of the *same* algorithm, a small multiplicative factor can be very important. It would be quite an achievement to speed up a commercial implementation of the simplex algorithm by a factor of 3.

Definition 9.2 is a worst-case definition, since the bound applies to all sufficiently large inputs. Some algorithms that are slow in only rare cases will have misleading time bounds. The simplex algorithm is a famous example of this phenomenon. To get around this difficulty, we might say that the simplex method is $O(m^3)$ on average or in practice even though the best theoretical bound is exponentially large.

Definition 9.3 *An algorithm runs in polynomial time iff for some k it has a time bound of $O(L^k)$.*

The major distinction we make between algorithms is whether or not they take polynomial time. Most algorithms that are not polynomial are exponential, requiring more than c^{dL} time in the worst case, for some constants $c > 1$ and $d > 0$. Any exponential function will eventually exceed any polynomial function as L increases. From a theoretical point of view, that is why polynomial time algorithms are preferred to exponential time algorithms. In practice, polynomial time algorithms are usually faster than exponential time algorithms on real cases.

There aren't many polynomial time algorithms that require $10^9 L^{10}$ time, and there aren't many exponential time algorithms that require $10^{-9}(1 + 10^{-10})^{L/10^9}$ time. So in practice, when L is a realistic size, a polynomial time algorithm is usually faster than an exponential time algorithm.

Lower Bounds on Time

Occasionally we want to measure how slowly an algorithm runs, as opposed to how fast it runs.

Reading Tip: *Guess why we don't use the definition of $O()$ for algorithms as 9.2 with "$\leq$" replaced by "$\geq$".*

The standard symbol for a lower bound is $\Omega()$. Although this symbol is the capitalized Greek letter "omega," it is traditionally pronounced "omega" rather than "big omega". For a function, the definition of Ω is analogous to the definition of $O()$.

Definition 9.4 ($\Omega(\ )$ and $\theta(\ )$ for functions) *Function $f(n) = \Omega(g(n))$ iff there exist constants $c, N > 0$ such that $f(n) \geq cg(n)\ \forall n \geq N$. If $f(n) = \Omega(g(n))$ and $f(n) = O(g(n))$ we say $f(n) = \theta(g(n))$.*

However, for algorithm run times, we want the $\Omega()$ notation to mean that an algorithm *might* take a long time, not to mean that it *always* takes a long time. For example, consider the following stupid algorithm to sort n numbers $x_1, x_2, \ldots, x_n$ from smallest to largest: For $i = 1$ to $n-1$ check whether or not $x_i \leq x_{i+1}$. If all checks are successful, terminate. If not, exchange x_t with x_{t+1}, where t is the least index i such that $x_i > x_{i+1}$, and iterate. If the input data are in reverse order, there will be $1 + \binom{n}{2}$ iterations, for a total number of operations proportional to n^3. But if the input data are in increasing order, the total number of operations will be $O(n)$. We want to say that the algorithm can take cn^3 time, not that the algorithm is sure to take at least cn time. Therefore, we define $\Omega()$ time bounds by the worst-case time required by the algorithm.

Definition 9.5 ($\Omega(\ )$ and $\theta(\ )$ for algorithms) *An algorithm runs in $\Omega(f(n))$ time if there exists a constant $c > 0$ such that for infinitely many values of n there exists an instance of size n on which the algorithm takes time at least $cf(n)$. An algorithm takes $\theta(f(n))$ time iff it takes both $O(f(n))$ and $\Omega(f(n))$ time.*

Question: What is an equivalent definition that uses the words "for all n" rather than "for infinitely many n"?[4] Some texts employ the notation $\overset{\infty}{\Omega}()$ instead of $\Omega()$, but the latter has become standard. It should be clear to you that if an algorithm takes both $O(f(n))$ and $\Omega(f(n))$ time, the upper bound $O(f(n))$ cannot be improved. We say that such an upper bound is *sharp* or *tight*.

Summary: The formal definition of time bounds for an algorithm requires that we bound the run time as a function of its actual size, which is its length — the number of bits to encode it. In practice, we usually measure the size of a problem input in terms of natural parameters such as the number of variables, constraints, vertices, etc. We usually assess the speed of an algorithm by an upper bound on the time it could take, as a function of that measure of input size. Usually, polynomial time algorithms are better than exponential time algorithms. However, it will turn out in the next chapter that linear programming is an exception to the latter two of these rules of thumb!

---

[4] $\exists c > 0$ such that $\forall n$ there exists an instance of size $N \geq n$ on which the algorithm takes time at least $cf(N)$.

9.2 Some Core Polynomial Time Algorithms

Preview: This section introduces commonly used algorithms for basic operations such as sorting and solving simultaneous linear equations. These are operations that we generally take for granted when we devise and analyze optimization algorithms.

9.2.1 The Euclidean Algorithm for Greatest Common Divisor

Given two integers a and b, how can we find their greatest common divisor, or *gcd*, the largest integer c such that a/c and b/c are both integers? We will need to do this in polynomial time later in the chapter when we solve simultaneous equations. In principle, one could factor a and b into primes and extract their common prime factors. However, no polynomial time algorithm for factorization is known at the time of this writing. Indeed, our most widely used cryptographic systems such as RSA [249], on which privacy and security of communication via the internet depend, rely on the computational difficulty of factoring. Fortunately, a version of Euclid's algorithm solves the gcd problem in polynomial time.

Algorithm 9.2.1 (Division Version of Euclid's Algorithm) *Input: Integers* a, b *with* $a > b$.
Output: $gcd(a, b)$*, the greatest common divisor of* a *and* b.
Step 1: If $b == 0$ *return* a.
Step 2: Set $r = a \mod b$ *(the remainder when* a *is divided by* b*)*
Step 3: Call $c = gcd(b, r)$ *and return* c.

The original version of Euclid's algorithm sets $r = a - b$ and exchanges r with b if $r > b$. That version makes clear the correctness of the algorithm, since if c divides both b and a it divides $a - b$. However, it could take an exponential amount of time. *Question: Give a simple example that takes exponential time.*[5] Let's prove that the division version takes polynomial time.

Theorem 9.2.2 *Algorithm 9.2.1 runs in polynomial time.*

Proof: If $b > a/2$, then $r = a - b < a/2$. If $b \leq a/2$, then $r < b \leq a/2$. Thus $r < a/2$ always in Step 2. After two iterations, the values of both a and b are less than half their previous values. Therefore, the number of iterations is bounded by $2 \log a = O(L)$ where L is the input length. Each iteration takes time $O(L)$ because the lengths of a and b are less than L. Therefore, the algorithm runs in time $O(L^2)$. ■

9.2.2 Sorting

In the world of algorithms, *sorting* means to arrange items according to a well-defined ordering, such as ascending numerical or descending alphabetical. It is standardly assumed

[5] $a = 2^m, b = 2$ would take 2^m iterations.

that two items can be compared to determine whether they are equal, and if not, which item precedes the other, in time $O(1)$. (One could perform a tedious analysis to confirm that this is OK, like the one we did for finding the maximum of a set of numbers in Section 9.1. I won't subject you to that, even as an exercise.) *Searching* means to acquire a particular item from a set of data. Sorting and searching is a field unto itself. It is the sole topic of volume III of Knuth's classic "Art of Computer Programming" [184]. This section summarizes key results and constructs of the field that pertain to optimization.

Of course, n items can be sorted in $O(n^2)$ time by repeatedly selecting the item that comes first from amongst the unselected items. *Question: Why is this $O(n^2)$?*[6] Several algorithms such as *mergesort* and *heapsort* sort in $O(n \log n)$ time, which is significantly faster. Mergesort operates on a list of n numbers by splitting the list into two sublists of length $\lfloor \frac{n}{2} \rfloor$ and $\lceil \frac{n}{2} \rceil$, recursively sorting each sublist (say from least to largest), and then merging the two sorted sublists in time $O(n)$ by repeatedly selecting and removing the least remaining number in either sublist, which takes $O(1)$ time per selection because the only two possibilities are the first remaining element in each sublist. You will see how heapsort operates later in this chapter.

No search algorithm based on pairwise comparisons can be faster than $O(n \log n)$. The elegant proof of this lower bound is information-theoretic. The first comparison branches into two possibilities, the second into 4 possibilities, and in general the mth into (at most) 2^m possibilities. Since there are $n!$ possible outcomes, m must be large enough that $2^m \geq n!$. Hence $m \geq \log_2 n! \geq n \log_2 n - n \log_2 e$. The order $n \log n$ term dominates the order n term, giving the result.

Quicksort, another sorting algorithm, runs in $O(n^2)$ in the worst case, but if it uses random numbers to make decisions, it runs in $O(n \log n)$ time on average. Quicksort picks a random element x in the list and in $O(n)$ time splits the list into two sublists consisting of those elements $\leq x$ and those $> x$. Quicksort then recursively sorts each sublist. Because it uses random numbers its performance can vary on the same data set. Its $O(n \log n)$ average performance bound is valid for every possible data set. This type of guarantee is preferable to an average with respect to a probability distribution taken over all possible data sets. The latter depends on the very strong assumption that real life data sets are independent samples from a known probability distribution; the former depends only on the quality of the random number generator. Interestingly, however, one version of the most popular pseudorandom number generator fails to achieve a $O(n \log n)$ bound for quicksort [171] (see also [211]). There are also deterministic versions of quicksort that use a $O(n)$ subroutine to select an x with quantile close enough to $\frac{1}{2}$ [29],[158] to guarantee $O(n \log n)$ time. However, these versions are not known to be computationally competitive.

When implemented, quicksort outperforms both heapsort and mergesort for most values of n.

9.2.3 Searching

Searching depends on the data structure used to hold the data set. One can often search an ordered set of n values in $O(\log n)$ time with a *binary search* algorithm. Binary search maintains two locations, v^+ and v^-, which are strict upper and lower bounds, respectively, on the location of the value being searched for. Check the value at location $v = \frac{1}{2}(v^+ + v^-)$. If the desired value is not at location v, then v is either too high or too low. Update the upper or lower bound and repeat. This algorithm requires that the set be ordered but not

[6] The total work is order $n + (n-1) + (n-2) + \ldots + 1 = \binom{n+1}{2} \approx n^2/2 = O(n^2)$.

that the value being searched for be present. *Question: Why not?*[7] If the value is not present, binary search tells you where it would fit into the ordered set.

In practice, the data structure usually must support not only search operations but also build, insert, and delete operations. Sometimes more complex operations such as union are also required. A linked list allows insertions and deletions in $O(1)$ time, but forces searching to take $O(n)$ time. At the other extreme, an array allows searching in $O(1)$ or $O(\log n)$ time, depending on whether the array index is itself the item's identifier, but forces insert and delete to be $O(n)$ time.

The classic compromise between these two unattractive extremes is some kind of *balanced tree*, such as red-black trees, heaps, B-trees, or treaps. In a balanced tree containing n items, the maximum depth (distance to the root) is $O(\log n)$. *Question: What is the maximum depth of an unbalanced tree?*[8] Operations such as insert, delete, and search generally take time proportional to the tree depth. Balanced trees therefore usually allow search, insert, and delete operations in $O(\log n)$ time. They are not as simple to code as arrays or linked lists, but they are worth it. *Question: Why is it a really bad idea to build a tree one insertion at a time in the hope that the tree turns out to be balanced?*[9] The heap data structure is less complex than the others. It does not allow search in time $O(\log n)$, but it allows the following operations to be performed in $O(\log n)$ time: insert, delete, and search for the minimum item. *Question: How would you use a heap to sort?*[10] The treap is a powerful randomized data structure that supports both basic and several complex operations such as union.

Hash tables are another popular structure for data storage, searching, insertion, and deletion. A hash function $h : W \mapsto \{1, 2, \ldots M\}$ maps a huge set of possible identifiers W to a relatively small finite range of integers $1 \ldots M$ such that similar identifiers tend to map to very different integers. For example, W could be the set of all strings of alphabetic characters with length 10 or less, corresponding to a set of last names truncated to length 10. A hash function should be very likely to map similar names such as "Thompson" and "Thompsen" to dissimilar integers. The idea is that the function will map even a highly patterned data set into a random-looking and therefore scattered set of hash values. Essentially, a hash function behaves like a random number generator that uses the identifier as a starting seed value. When analyzing the performance of hashing, we always assume that $h(w)$ is equally likely to be any of the values $1 \ldots M$, independent of the values $h(v)$ for all $v \neq w$.

The hash table is an array of storage locations indexed by the integer range (e.g. $1 \ldots M$). Ideally, the data corresponding to an identifier $w \in W$ is stored at the location $h(w)$, called the "hash value" of w. This would make storage, retrieval, and deletion of data extremely fast, 0(1). A "collision" occurs when $h(w) = h(w')$ for two distinct identifiers w and w'. When the number of items to be stored in the hash table is less than approximately $\frac{2}{3}M$, there won't be too many collisions. The multiple entries can be stored in adjacent table locations without excessive additional cost. Storage, retrieval, and deletion will still take 0(1) time on average.

If the number of items to be stored is larger, one must either choose a larger value of M and build a new table (this is called "rehashing") or store multiple entries external to the table itself. The simplest way to store externally is to make each table entry the head of a linked list consisting of all data with the same hash value. One should not let the number of items exceed a modest multiple of M or the linked lists will become too long to use

[7] Binary search proves that the item is not present when $v^+ \leq v^- + 1$.

[8] $n - 1$.

[9] The data is likely to be sorted or partially sorted.

[10] Insert all the items, then repeatedly search for and remove the minimum item, all in time $O(n \log n)$.

efficiently. *Question: what is the expected length of the linked list at location $h(w)$ if $5M$ items are stored in the hash table and w is a new data item to be stored?*[11]

9.2.4 Linear Equations and Matrices

Theorem 9.2.3 *The following can be performed in polynomial time on integer input data.*

- *Matrix addition*
- *Matrix multiplication*
- *Calculating the determinant of a matrix*
- *Matrix inversion*
- *Solving a set of linear equations*

Proof: First we will prove polynomial time bounds on the number of arithmetic operations in terms of the natural parameters, which are the matrix dimensions. The length, L, of the input data is an upper bound on each matrix dimension. Therefore, we will have polynomial time bounds on the number of arithmetic operations in terms of L. Second, we will prove that each arithmetic operation requires time polynomial in the input length L. Since the product of two polynomials is a polynomial, these two steps will prove the theorem.

Let the two matrices be $m \times n$. Matrix addition requires mn addition operations.

Let the two matrices be $m \times n$ and $n \times p$, respectively. Matrix multiplication requires mp computations, each of which requires n multiplication and $n-1$ addition operations. The total number of operations is $O(mnp)$. If $m = n = p$ multiplication requires n^3 operations. This would seem to be as fast as possible. In 1969 Volker Strassen surprised the computer science community by showing that this multiplication can be performed with slightly less than $O(n^{2.81})$ operations. Strassen's algorithm, like others that followed it, is recursive and only makes computational sense to use when n is very large.

Let the matrix A be $n \times n$. The determinant of a matrix is invariant under the elementary row operation of adding a multiple of one row to another. Select j such that $A_{j1} \neq 0$. If no such j exists A has determinant 0. Move row j to the top of the matrix, and record the value 1 if j is odd and -1 if j is even. For each $2 \leq i \leq n$ subtract $(A_{i1}/A_{11})A_{1,:}$ from row i of A. The resulting matrix has zeros in column 1 except in the first row. Recursively apply the same procedure to the $(n-1) \times (n-1)$ submatrix consisting of rows $2 \ldots n$ and columns $2 \ldots n$, until a 1×1 submatrix is reached. The resulting matrix A' will satisfy $A_{ij} = 0 \; \forall i > j$. That is, all of the terms below the main diagonal equal zero. Such a matrix is called *upper triangular*. Its determinant is the product of its diagonal terms, $\prod_{i=1}^{n} A'_{ii}$. *Question: Why is this true?*[12] The determinant of A is the same value, times the product of the recorded ± 1 values. This procedure requires $n-1$ iterations, each requiring at most $n^2 - n$ subtractions, $n-1$ divisions, and $n^2 - n$ multiplications. Therefore, it requires $O(n^3)$

[11] Treat the hash values as independent random numbers in the range $1 \ldots M$. Since $h(w)$ is independent of the hash values of the stored data, the expected number of data values stored there equals $5M/M = 5$.

[12] Recursively expand by minors along the first column, or use elementary row operations to eliminate the off-diagonal terms.

operations. (*Question: Give a more accurate estimate of the number of operations.*[13]) If the matrix A is sparse, the number of operations required can be much less.

To invert an $n \times n$ matrix A, create the $n \times 2n$ matrix $[A|I]$ and perform elementary row operations to convert A to I. Each elementary row operation is equivalent to multiplication on the left by an $n \times n$ matrix. Let B denote the product of these matrices. Then $BA = I$. Hence $B = A^{-1}$ so the right half of the $n \times 2n$ matrix now equals A^{-1}. Much like the transformation to upper triangular form, the elementary row operations require $O(n^3)$ operations.

To solve an $n \times n$ system of linear equations $A\mathbf{x} = \mathbf{b}$, use elementary row operations to transform the first n columns of $[A|b]$ into an upper triangular matrix. This is like the procedure used in the calculation of the determinant, except that 0 terms on the diagonal are permitted if no nonzero alternative is available. If A is nonsingular, calculate the values of $x_n, x_{n-1}, \ldots x_1$ using rows $n, n-1, \ldots 1$ respectively, together with the values already calculated. If A is singular, a contradiction of form $0 = \alpha$ for $\alpha \neq 0$ may arise, in which case no solution exists.

Now we prove that each arithmetic operation takes polynomial time in L, the length of the input measured in number of bits. If all operations are performed on integers, it suffices to prove that every number involved in the operation has length bounded by a polynomial in L. Let M equal the largest absolute value of any entry in the input data. We assume that the data are encoded in binary, so $M \leq 2^L$. For matrix addition, the largest term has absolute value $\leq 2M$. For matrix multiplication, the largest term has absolute value $\leq nM^2 \leq 2^{3L}$ and therefore has length $\leq \log 2^{3L} = 3L = O(L)$.

For the other three problems, the calculations can involve rational numbers, which we will express as the ratios of integer pairs. Gaussian elimination is the key to all three problems, so we analyze it first. A straightforward implementation of Gaussian elimination can produce numbers that have exponential length. That is the bad news. The good news is that by Simpson's formula or equivalently the formula for the inverse of a matrix in terms of determinants, we know that each term in the solution is a ratio of two determinants of submatrices of A (times the RHS $\mathbf{b}$). If an $n \times n$ integer matrix A is encoded in length L, then $\max_{i,j} |A_{ij}| < 2^L$. Then $|det(B)|$ for any submatrix of A (including A itself) is an integer bounded by $n!(2^L)^n < (n2^L)^n$, which has length $O(n(\log n + L) = O(L^2)$. Therefore, the solution has polynomial length. Thus we can hope that every term in the intermediate calculations has polynomial length, too. We now prove that this is true.

Lemma 9.2.4 *Every term encountered during Gaussian elimination of the matrix A is the ratio of two integers, each of polynomially bounded length.*

Proof of Lemma: Initially we prove that the diagonal terms generated are bounded. Afterwards we will perform a a thought experiment to extend the bound to all terms. It will be convenient to define some special submatrices of A.

Definition 9.6 *Let A be an $n \times n$ matrix. The jth* leading principal submatrix *of A, denoted $A[j]$, is $A_{1\ldots j,1\ldots j}$, the matrix consisting of the first j rows and columns of A.*

The leading principal submatrices are handy because Gaussian elimination on them is not affected by other terms in A. Let A^j be the result of $j-1$ iterations of Gaussian elimination on A, so $A^1 = A$ and A^n is triangular. The determinant of a triangular matrix is the product of its diagonal terms. After $j-1$ iterations, the jth diagonal term does not change. Call its

[13] $n^3/3$ subtractions, $n^2/2$ divisions, and $n^3/3$ multiplications.

value "final." Therefore,

$$det(A[j]) = \prod_{i=1}^{j} A^j_{i,i} = \prod_{i=1}^{j} A^i \Rightarrow A^j_{j,j} = \frac{det(A[j])}{det(A[j-1])}.$$

Since we have already proved that the determinants of all submatrices are integers of length $O(L^2)$, this proves that all final diagonal terms during Gaussian elimination are of polynomial length $O(L^2)$. Figure 9.2 illustrates the thought experiment for off-diagonal terms. In the figure, the matrix in the upper left corner is the initial matrix A. The matrices below it show the successive iterations of Gaussian elimination on A. The value -6 is an off-diagonal value that appears after two iterations. Imagine that we exchanged columns 3 and 4 of A, as shown in the matrix to the right of A, in the top middle of the figure. The matrices below it show what would be the resulting successive iterations of Gaussian elimination. In these calculations, the value -6 is a diagonal term. Returning to the calculations based on A, the value -7 appears after two iterations as an intermediate values in the calculations. It is not present in the final triangular matrix in the lower left corner. Imagine that we exchanged columns 3 and 4 of A, and also exchanged rows 3 and 4 of A, as shown in the matrix in the upper right corner of the figure. Then the -7 value would have shown up as a diagonal term.

In general, any value that gets computed in a Gaussian elimination (excepting the zero terms below the diagonal and the terms in the first row, which are not computed) could be a diagonal term with the right reordering of the rows and columns. To be precise, let $x = A^j_{i,m} : i \geq j; m \geq j$. If we exchanged row i with row j and column m with column j in the original matrix A, the leading principal submatrix A^{j-1} would not be affected, and the jth diagonal term after $j-1$ iterations would be x. Therefore, x has length $O(L^2)$. ■

To express the terms in polynomial length, we need only reduce each term to a fraction in lowest terms. We can do that in polynomial time by applying Algorithm 9.2.1 to each new value in every iteration. By Theorem 9.2.2 each reduction takes polynomial time $O(L^2)$, for a total amount of work $O(n^3L^2) = O(L^5)$, a polynomial. The calculations are all of the form $A^j_{i,m} - A^j_{j,m}A^jA^j_{i,j}/A^j_{j,,j}$ which also takes total time $O(L^5)$. This proves that Gaussian elimination takes polynomial time. This implies polynomial bounds on the time to solve a set of linear equations. Once a matrix has been converted to triangular form, its determinant (which is not changed by adding multiples of a row to another row) equals the product of its diagonal terms. Therefore, determinants can be calculated in polynomial time. This in turn lets us calculate matrix inverses in polynomial time, since the (i,j) term is $(-1)^{i+j}$ times the ratio of two determinants. (Given what precedes, this is the quickest way to prove that matrix inversion can be done in polynomial time. But this is not the quickest way to invert a matrix.) ■

$$
\begin{array}{ccccc}
\begin{bmatrix} 2 & 1 & -1 & -2 \\ 4 & 3 & -1 & 1 \\ -2 & 0 & 3 & 1 \\ 6 & 4 & 1 & -8 \end{bmatrix} & \longrightarrow &
\begin{bmatrix} 2 & 1 & -2 & -1 \\ 4 & 3 & 1 & -1 \\ -2 & 0 & 1 & 3 \\ 6 & 4 & -8 & 1 \end{bmatrix} & \longrightarrow &
\begin{bmatrix} 2 & 1 & -2 & -1 \\ 4 & 3 & 1 & -1 \\ 6 & 4 & -8 & 1 \\ -2 & 0 & 1 & 3 \end{bmatrix} \\
\downarrow & & \downarrow & & \downarrow \\
\begin{bmatrix} 2 & 1 & -1 & -2 \\ 0 & 1 & 1 & 5 \\ 0 & 1 & 2 & -1 \\ 0 & 1 & 4 & -2 \end{bmatrix} & &
\begin{bmatrix} 2 & 1 & -2 & -1 \\ 0 & 1 & 5 & 1 \\ 0 & 1 & -1 & 2 \\ 0 & 1 & -2 & 4 \end{bmatrix} & &
\begin{bmatrix} 2 & 1 & -2 & -1 \\ 0 & 1 & 5 & 1 \\ 0 & 1 & -2 & 4 \\ 0 & 1 & -1 & 2 \end{bmatrix} \\
\downarrow & & \downarrow & & \downarrow \\
\begin{bmatrix} 2 & 1 & -1 & -2 \\ 0 & 1 & 1 & 5 \\ 0 & 0 & 1 & \boxed{-6} \\ 0 & 0 & 3 & \boxed{-7} \end{bmatrix} & &
\begin{bmatrix} 2 & 1 & -2 & -1 \\ 0 & 1 & 5 & 1 \\ 0 & 0 & \boxed{-6} & 1 \\ 0 & 0 & -7 & 3 \end{bmatrix} & &
\begin{bmatrix} 2 & 1 & -2 & -1 \\ 0 & 1 & 5 & 1 \\ 0 & 0 & \boxed{-7} & 3 \\ 0 & 0 & -6 & 1 \end{bmatrix} \\
\downarrow & & \downarrow & & \downarrow \\
\begin{bmatrix} 2 & 1 & -1 & -2 \\ 0 & 1 & 1 & 5 \\ 0 & 0 & 1 & -6 \\ 0 & 0 & 0 & 11 \end{bmatrix} & &
\begin{bmatrix} 2 & 1 & -2 & -1 \\ 0 & 1 & 5 & 1 \\ 0 & 0 & -6 & 1 \\ 0 & 0 & 0 & \frac{11}{6} \end{bmatrix} & &
\begin{bmatrix} 2 & 1 & -2 & -1 \\ 0 & 1 & 5 & 1 \\ 0 & 0 & -7 & 3 \\ 0 & 0 & 0 & \frac{-11}{7} \end{bmatrix}
\end{array}
$$

FIGURE 9.2: The value -6 in the third matrix on the left would be a diagonal value if we exchanged columns 3 and 4 as shown in the middle. The value -7 in the third matrix on the left would be a diagonal value if we exchanged columns 3 and 4 and also exchanged rows 3 and 4, as shown on the right.

Geometric Interpretation of Gaussian Elimination

The absolute value of the determinant of a matrix equals the volume of the parallelogram whose edges are defined by the rows of the matrix. The volume of a parallelogram does not change if one if its sides (facets) is moved parallel to itself. Gaussian elimination moves the sides so that the jth diagonal term of the matrix gives the height along the x_j axis. Figure 9.3 illustrates Gaussian elimination in two dimensions for the matrix

$$\begin{bmatrix} 4 & -1 \\ 12 & 12 \end{bmatrix} \rightsquigarrow \begin{bmatrix} 4 & -1 \\ 0 & 15 \end{bmatrix}.$$

The initial parallelogram has vertices at the origin and at $(4, -1)$, $(12, 12)$, and $(16, 11)$. The vertices are the rows and sums of rows of the initial matrix. Subtracting 4 times the first row from the second row moves the side with endpoints $(12, 12); (16, 11)$ parallel to itself along the dotted line in the figure to the endpoints $(0, 15); (4, 14)$. The new parallelogram has the same area (2-dimensional volume), which is $4 \times 15 = 60$. *Question: Can you see why the area is* 4×15*?*[14] The values 4 and 15 are the diagonal elements of the matrix after Gaussian elimination has been performed.

9.2.5 Square Roots

In a later chapter we will need to compute square roots to a high degree of accuracy. Gauss invented extraordinarily fast methods to compute various irrational values to many

[14]Figure 9.4 shows why.

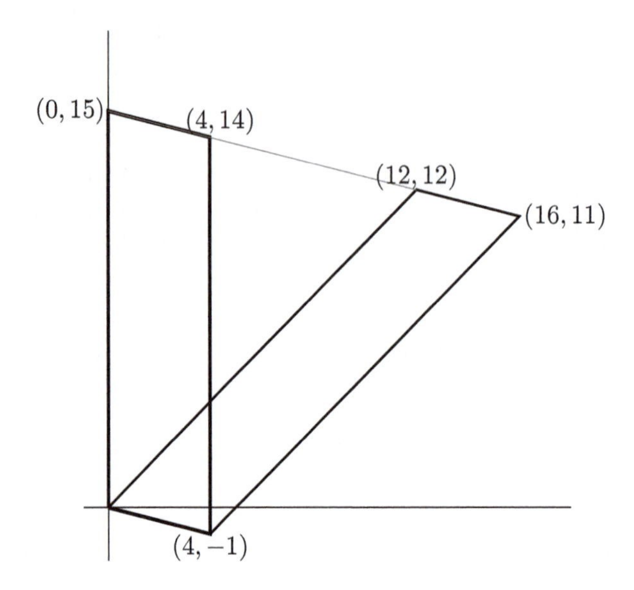

FIGURE 9.3: Gaussian elimination moves the top side of the parallelogram parallel to itself to make the upper left corner hit the vertical axis while preserving the two-dimensional volume.

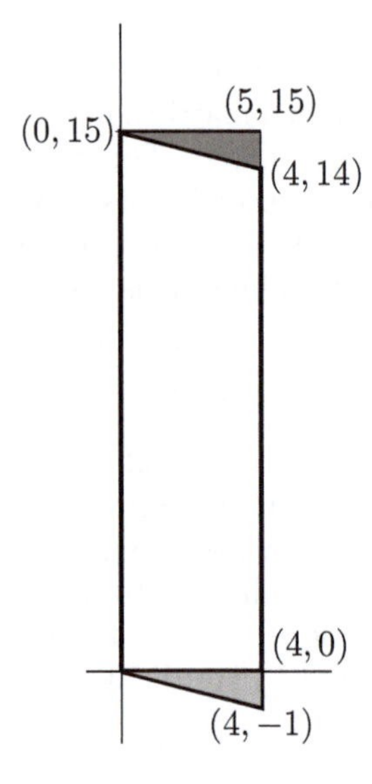

FIGURE 9.4: The dark gray and light gray triangles are congruent. Hence the parallelogram has the same two-dimensional volume as a rectangle with vertices at $(4, 0)$ and $(0, 15)$.

digits of accuracy, but we will do well enough with Newton's method. For differentiable functions $f : \Re \mapsto \Re$ Newton invented an algorithm for finding x such that $f(x) = 0$. Starting with an initial value x^0, his algorithm computes a series of values $x^j : j = 1, 2, \ldots$ as follows:

$$x^{j+1} = x^j - \frac{f(x^j)}{f'(x^j)}.$$

As illustrated in Figure 9.5, x^{j+1} is where the line tangent to $f(x)$ at $x = x^j$ hits 0. If $f(x)$ is an affine function, Newton's method will solve $f(x) = 0$ in one step. More generally, the smaller the magnitude of the second derivative of $f(x)$, the quicker Newton's method tends to converge.

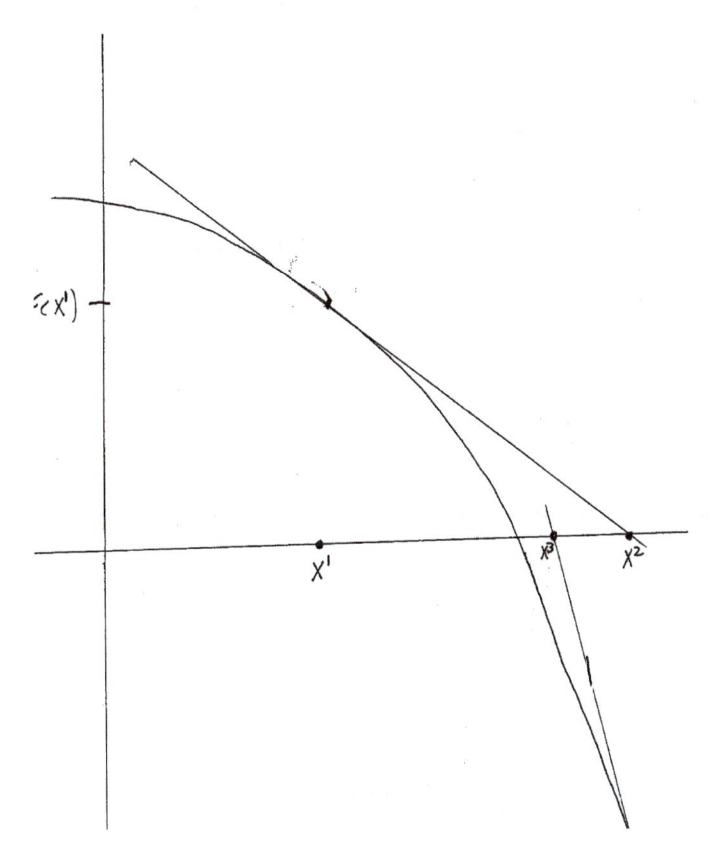

FIGURE 9.5: Starting from the previous estimate x^j, Newton's method estimates x^{j+1} to be the point where the line tangent to $f(x)$ at x^j hits zero.

In the case of the function $f(x) = x^2 - k$, Newton's method has a very intuitive meaning that does not depend on knowledge of calculus. Let x^j be our current estimate of $\sqrt{k}$. If $x^j = \sqrt{k}$, then $k/x^j = \sqrt{k}$. Otherwise, if $x^j < \sqrt{k}$, then $k/x^j > \sqrt{k}$; if $x^j > \sqrt{k}$, then $k/x^j < \sqrt{k}$. So if x^j is too big, k/x^j will be too small, and if x^j is too small, k/x^j will be too big. The geometric mean of these two values is precisely $\sqrt{x^j k/x^j} = \sqrt{k}$, but computing the geometric mean requires a square root calculation, which begs the question of how to compute square roots. Instead, approximate the geometric mean by the arithmetic mean to get the new estimate $x^{j+1} = \frac{x^j + k/x^j}{2}$. Since the arithmetic mean always exceeds the geometric mean (of different positive values), each estimate is greater than $\sqrt{k}$. The error of the estimate decreases rapidly. Define $\delta_j \equiv \frac{(x^j)^2-k}{k}$, a natural measure of the error of estimate x^j. As observed, $\delta_j \geq 0$. Then

$$\delta_{j+1} = \left(\frac{x^j}{2} + \frac{k/x^j}{2k}\right)^2 - 1 = \frac{1}{4}\delta_j \frac{\delta_j}{k+\delta_j} < (\delta_j)^2/4.$$

Therefore, the error of the estimate is halved and then squared each iteration. The squaring of the error means that the number of digits of accuracy in the estimate doubles each iteration.

Summary: Algorithms for sorting and searching require ordinal comparisons and are easily proved to run in low order polynomial time. Standard algorithms for solving nonlinear and systems of linear equations are also efficient, but require more careful analysis of the time required to achieve adequate precision.

9.3 Heaps

Preview: This section explains the details of how to perform operations on binary heaps. These details are not needed to establish upper bounds on the run time of algorithms that employ binary heaps. However, these details are needed to establish lower bounds on the worst-case run time, which is required to prove that an upper bound is best possible.

To define binary heaps, we first need to define binary trees.

Definition 9.7 (Binary Tree)

1. *An empty graph is a binary tree with empty root.*
2. *Let S and T be disjoint binary trees with roots s and t, respectively. One or both of S and T may be empty. Let r be a node disjoint from both S and T. Assign s to be the left child of r, and t to be the right child of r. If s (respectively t) is not empty, assign r to be its parent and create new arc (r, s) (respectively (r, t)). Then r, S, T, and the new arcs (if any) form a binary tree with root r. S (respectively T) is called the left (respectively right) subtree of r.*

If you've programmed in a language with dynamically generated addresses (pointers), you should see how well suited null addresses are to the binary tree definition.

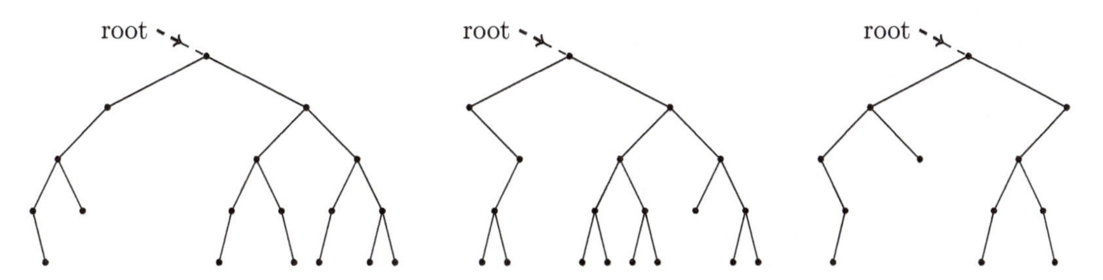

FIGURE 9.6: Each node of a binary tree may have a left child, a right child, both, or neither.

All binary trees can be constructed by following this definition. The simplest nonempty binary tree is a single node which is its tree's root. Figure 9.6 shows several binary trees. Every nonempty binary tree has one root, and every node, except the root, has one parent. If a subtree is empty, we do not count its root as being a child. Hence each node may have $0, 1$ or 2 children. A node with no children is called a **leaf**. Exercise 9 asks you to show that a binary tree with n nodes can have at most $(n+1)/2$ leaves.

The *depth* of a node in a binary tree is the distance, counted in number of edges, between it and the root. Hence the root has depth 0. The depth of the tree is the maximum depth of its nodes. *Question: What is the relationship between the depths of a node and its children's?*[15] At one extreme, a binary tree may have no nodes with two children, in which case the tree looks like a path and has only one leaf. Such a tree is very unbalanced in the sense that the right and left subtrees of all nodes, except the leaf, contain very different numbers of nodes. Figure 9.7 depicts the other extreme, a binary tree so perfectly balanced that every node has an equal number of nodes in its two subtrees. Balance is usually desirable because many operations on binary trees take time proportional to the distance (number of edges) between its nodes and the root. We think of unbalanced trees as long and stringy; we think of balanced trees as bushy and leafy. Exercise 19 asks you to show that complete binary trees have a maximum proportion of leaves.

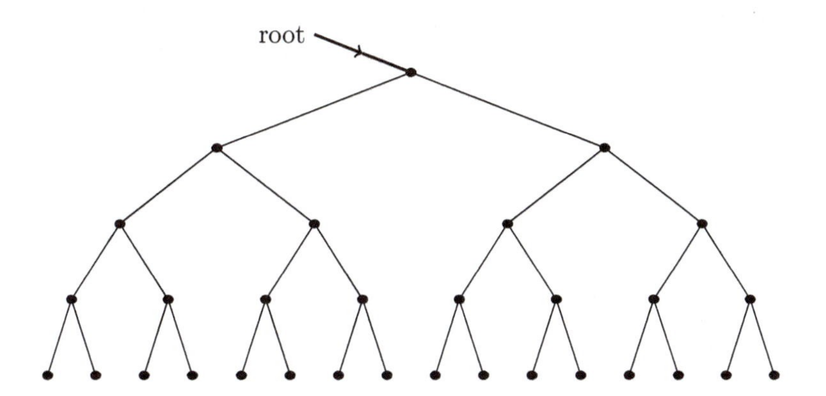

FIGURE 9.7: A complete binary tree has $2^n - 1$ nodes, each of which is on a path of length at most $n-1$ edges from the root.

[15]A child has depth one more than its parent's.

Definition 9.8 (Binary Heap) *A binary heap of* n *numbers is a binary tree with the following properties:*

1. *It has* $1 + \lfloor \log_2 n \rfloor$ *levels.*
2. *The first* $\lfloor \log_2 n \rfloor$ *levels form a complete binary tree.*
3. *The last level is filled in from left to right.*
4. *Every parent is at least as good as its children. If we are minimizing, every node is greater than or equal to its parent. If we are maximizing, every node is greater than or equal to its children. This is called the* **heap property**.

Heaps are usually defined either as minimizing heaps or maximizing heaps. I hate having to memorize whether minimizing or maximizing goes with $\geq$ or $\leq$. It is simpler to think of heaps as embodying the complaint heard in every generation, that the children are lazy, disrespectful, spoiled, and in all other ways inferior to their parents. Figure 9.8 shows several minimizing binary heaps, where "better" means "smaller." Figure 9.9 shows several binary trees that are not binary heaps.

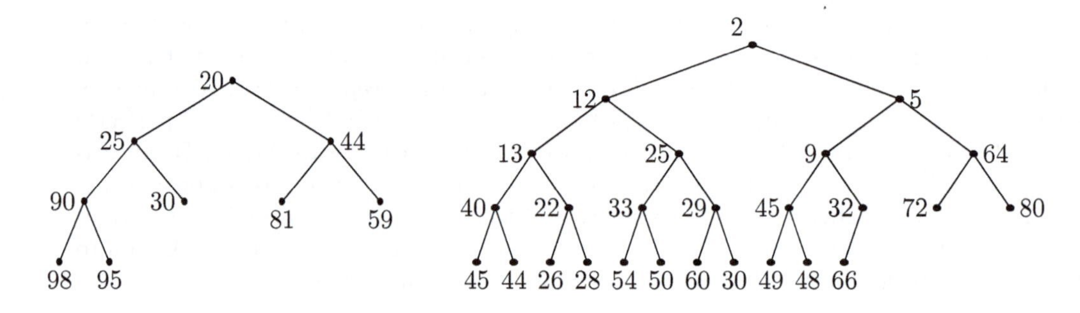

FIGURE 9.8: A binary heap of n elements has a unique prescribed shape. In the minimizing heaps shown, parents have better, smaller values than their children's.

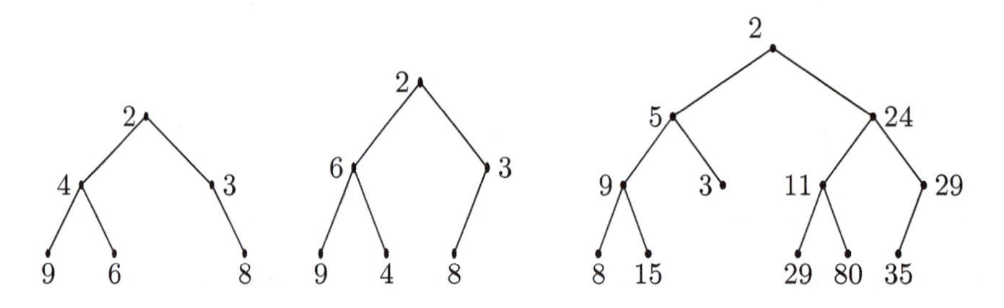

FIGURE 9.9: A binary tree can fail to be a binary heap because it has the wrong shape, has a child better than a parent, or both.

Theorem 9.3.1 *The following operations can be performed on an n-element binary heap in time $O(\log n)$ so as to retain the properties of a binary heap.*

1. *Improve-value. Change a number to a better value.*
2. *Devalue. Change a number to a worse value.*
3. *Insert. Add a new element, resulting in an $n+1$-element binary heap.*
4. *Remove-best. Remove a best number, resulting in an $n-1$-element binary heap.*

Proof:

1. Let x be the number whose value has improved. While x is better than its parent, swap it with its parent. By property 1 at most $\log_2 n$ swaps and $\log_2 n$ comparisons will occur.
2. Let x be the number whose value has worsened. While x has a better child, swap it with its best child. At most $\log_2 n$ swaps and $2\log_2 n$ comparisons will occur.
3. Let x be the value being added. Place x in the required leaf location to achieve the correct shape of an $n+1$-element binary heap. Apply Improve-value to x.
4. Remove the root. Let x be the rightmost element in the last level. Place x at the root. Apply Devalue to x.

By their definition, Improve-value and Devalue preserve the defining properties of a binary heap. If it is not clear to you that Insert results in a binary heap, perform the following *gedanken* (thought-experiment): temporarily set x to an infinitely bad value. When it is placed in its leaf location, it is worse than its parent. Hence the tree is a binary heap. Now improve x to its true value. If you believe that Improve-value works correctly, you must believe that applying Improve-value to x results in a binary heap. *Question: Think of the gedanken for Remove-best. What temporary value do you give to x?*[16] Insert and Remove-best add $O(1)$ time to Improve-value and Devalue, respectively. Therefore, all four operations require $O(\log n)$ time. ■

Example 9.7 Insert the element 22 into the 9-element heap of Figure 9.8. The new element is initially the left-child of the element 30, to maintain the mandatory heap shape. It swaps locations with its parent 30, and again swaps with its parent 25. The operation terminates because $22 \geq 20$. Figure 9.10 depicts the sequence of steps.

Theorem 9.3.1 begs the question of how to create the heap to begin with.

Reading Tip: *If you have not already tried the four operations of Theorem 9.3.1 on a small example of your own, do so. Then, before reading further, try to figure out a way to create a heap efficiently.*

Starting from an empty heap, insert n elements one at a time. The mth insert will take $O(\log m + 1) = O(\log n)$ time, for a total of $O(n \log n)$ time. If we are using the heap to sort n elements, the remove-best operations will take $O(n \log n)$ time. The $O(n \log n)$ construction time would not increase the order of the runtime, but it might, say, double the runtime. Instead, here is a way to construct the heap in linear time.

[16] Infinitely good.

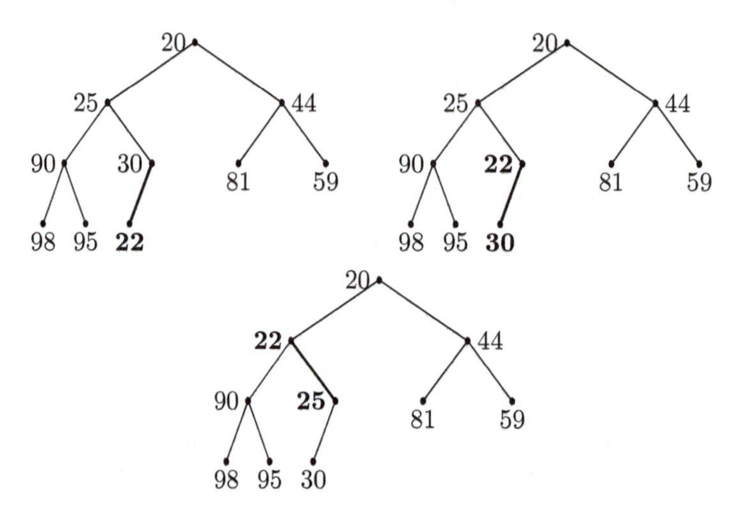

FIGURE 9.10: Insert the new element with value 22 in the next open space in the bottom row. Then repeatedly swap it with its parent until it is not better than its parent.

Theorem 9.3.2 *A binary heap of n elements can be created in time $O(n)$.*

Proof: Let $K = 1 + \lfloor \log_2 n \rfloor$ be the number of levels. Let r be an element in a binary tree that has the shape of a binary heap but may not have the heap property. Let the subtrees rooted at r's children have the heap property. Apply Devalue to r. The resulting subtree has the heap property.

Therefore, if we apply Devalue to every location in a binary tree, starting with level $K-1$, and processing all locations in level k before processing locations in level $k-1$, the resulting tree will have the heap property.

Given any n elements, place them in arbitrary order in the shape of a binary heap. Apply Devalue as just described. The result will be a binary heap. The number of swaps required is at most $(n+1)/2$ at level $K-1$, $2(n+1)/4$ at level $K-2$, $3(n+1)/8$ at level $K-3$, and in general $j(n+1)/2^j$ at level $K-j$. The total number of swaps is at most

$$\sum_{j=1}^{K-1} j(n+1)/2^j \leq (n+1)\sum_{j=1}^{\infty} j/2^j \tag{9.1}$$

$$= (n+1)\sum_{j=1}^{\infty}\sum_{i=1}^{j} 1/2^j \tag{9.2}$$

$$= (n+1)\sum_{i=1}^{\infty}\sum_{j=i}^{\infty} 2^{-j} = (n+1)\sum_{i=1}^{\infty} 2^{1-i} \tag{9.3}$$

$$= = (n+1)2 = O(n). \tag{9.4}$$

■

Pay attention to the interchange of summations from the end of (9.2) to the beginning of (9.3). It is the same duality that is at the heart of linear programming duality! Omitting the factor of $n+1$, the sum being evaluated can be visualized in two dimensions as the following array.

$$\begin{array}{cccc}
\frac{1}{2} & & & \\
\frac{1}{4} & \frac{1}{4} & & \\
\frac{1}{8} & \frac{1}{8} & \frac{1}{8} & \\
\frac{1}{16} & \frac{1}{16} & \frac{1}{16} & \frac{1}{16} \\
\vdots & & &
\end{array}$$

The double summation at the end of (9.2) treats this array one row at a time, since j is the outer summation index. The double summation at the beginning of (9.3) treats this array one column at a time.

Summary: An algorithm runs in time $O(g(n))$ if its runtime is at most $O(g(n))$ on all inputs of size n. An algorithm runs in polynomial time iff for some k its runtime is $O(n^k)$. As rules of thumb, usually n can be taken to be a natural measure of size such as the number of variables, polynomial time algorithms are fast in practice, and non-polynomial time algorithms are slow in practice. Basic operations on matrices such as multiplication, triangulation, and inversion can be performed in polynomial time. Fundamental operations for storing and retrieving data can typically be done in time $O(\log n)$ where n is the number of data elements. The logarithmic time is often achieved by storing n elements in a tree with maximum depth $h = O(\log n)$ such that each operation takes time $O(h)$.

9.4 Notes and References

There are significant gaps between the linear algebra algorithms as described in this chapter and their modern implementations. Achieving numerical accuracy and stability is a field unto itself. For references on solving systems of equations and computing eigenvalues see [295, 296, 297]. Factorization and handling large sparse matrices is another field of study which has had major impact on the speed of LP software. See for example [119, 126].

9.5 Problems

E Exercises

1. Explain why the category of LPs with two variables contains infinitely many instances.

2. Explain why the category of LPs with two variables and two constraints contains infinitely many instances.

3. Let $x_1, \ldots, x_n$ and $y_1, \ldots, y_n$ be integers.

   (a) If you treat the numbers naively as having $O(1)$ length, how much time does it take to compute the n^2 products $x_i y_j : \forall i, j$?

   (b) If L equals the total number of base 10 digits to represent the data, and the numbers all have the same length, how much time does it take to compute the n^2 products $x_i y_j$ in terms of L and n? In terms of L?

   (c) Show how to compute the n products $x_1 y_j : j = 1, \ldots, n$ in a single multiplication.

4. Let $\mathbf{x}, \mathbf{y}$ be n-vectors of integers.

   (a) Show that $\mathbf{x} \cdot \mathbf{y}$ takes $O(n)$ time to compute if you treat the numbers naively as having $O(1)$ length.

   (b) How much time does it take you to compute $\mathbf{x} \cdot \mathbf{y}$ if the two vectors together have length L when encoded?

   (c) Repeat 4b if moreover each entry of each vector has the same length.

5. Arrange the following functions in increasing $O()$ order. That is, if you place g after f, then g should at least as "big" in the sense that $f(n) = O(g(n))$.
   $16n + 3; 2 + n + \frac{1}{2}n^2 + \frac{1}{6}n^3; \log n + n^{0.02}; e^n; 20n^2; \sqrt{5n}; 7(\log n)^2; \frac{1}{2}n \log n$.

6. Let $A_{ij} = (-1)^{ij}(2i + j)$ for $i, j = 1, 2, 3$ and $b = (6, 39, 120)$. Solve $A\mathbf{x} = \mathbf{b}$ with Gaussian elimination. Also calculate the determinant of A.

7. Apply four iterations of Newton's method to estimate the square root of 1, starting from the initial estimate $\frac{1}{2}$. Observe the rate at which the error decreases.

8. Verify that both parallelograms in Figure 9.3 have area 60. *Hint: Project to get the altitude. Geometrically, move in direction* $(1, 4)$ *from* $(0, 0)$ *until reaching the set of affine combinations of* $(12, 12)$ *and* $(16, 11)$.

9. Prove by induction that a binary tree with n nodes can have at most $(n+1)/2$ leaves.[17]

10. Use operation Improve-value to change the value 81 in the smaller heap of Figure 9.8 to 40.

11. Use operation Improve-value to change the value 81 in the smaller heap of Figure 9.8 to 1.

[17] The induction is on n. At $n = 1$ the bound is correct. For $n > 1$ let r be the root and let l_r be the number of leaves. Suppose r has two children s and t. For $i = s, t$ let n_i and l_i be the number of nodes and the number of leaves in the tree rooted at i. By construction $n = n_s + n_t + 1$ and $l_r = l_s + l_t$. By induction $l_s + l_t \leq (n_s + n_t + 2)/2 = (n+1)/2$. If r has one child s, by induction $l_r = l_s \leq (n_s + 1)/2 = n/2 \leq (n+1)/2$.

12. Use operation Devalue to change the value 20 in the smaller heap of Figure 9.8 to 96.

13. Insert the value 4 into the larger heap of Figure 9.8.

14. Perform operation Remove-best on the larger heap of Figure 9.8.

15. Insert the value 9 into the larger heap of Figure 9.8. Handle tied values without unnecessary steps.

16. Let the values in a heap, starting from the root, then the root's children from left to right, then the root's grandchildren from left to right, etc. be 5,20,10,22,40,15,50,23,24,41,66,85,17,59,56. Draw the heap and verify that it satisfies the heap property.

    (a) Is the heap a minimization or maximization heap?

    (b) Improve the value 41 to 21.

    (c) Devalue the value 5 to 44.

    (d) Insert a new element with value 18.

    (e) Remove the best element (the root).

17. For the values in Exercise 16:

    (a) Starting from an empty heap, create a minimization heap by repeatedly inserting the values in the given order, i.e., add 5, add 20, etc.

    (b) Starting from an empty heap, create a maximization heap by repeatedly inserting the values in the given order, i.e., add 5, add 20, etc.

    (c) Convert the heap given in 16 to a maximization heap using the algorithm from the proof of Theorem 9.3.2.

    (d) Compare the two maximization heaps you've created, and the amount of work expended.

18. True or False: If each node in a binary tree has either zero or two children, the tree is complete.

19. Recall that the depth of a node in a rooted tree is the number of edges in the path from the node to the root. The depth of the tree is the maximum of its nodes' depths.

    (a) For a binary tree with n vertices, what is the maximum possible depth? What is the minimum possible depth?

    (b) Prove that for $2^n - 1$ nodes the complete binary tree uniquely minimizes the average node depth.

    (c) Prove that for $2^n - 1$ nodes the complete binary tree uniquely minimizes the number of leaves.

    (d) Prove that for m nodes a heap minimizes the average node depth.

    (e) Prove that a heap of m nodes has the maximum possible number of leaves of any binary tree of m nodes.

20. How many different heaps can you make from 2 distinct elements? From 3? From 4? From 5? Don't double-count by making both maximizing and minimizing heaps.

21. How many different binary trees can you make with $2, 3, 4$ and 5 vertices?

22. Mergesort can be defined as a recursive algorithm based on the observation that two sorted lists of length m can be combined into a single sorted list of length $2m$ in time $O(m)$. Work out the details of the algorithm and verify the $O(n \log n)$ time bound.

23. Quicksort can be defined as a recursive algorithm that sorts n items by selecting a random item r, separating the other items into those that precede r and those that succeed r, and calling itself recursively on those two subsets. Prove that the worst case time requirement is $O(n^2)$.

24. (a) By enumeration, calculate the expected number of items stored at location $h(w)$ in a hash table size 3 containing $1, 2$, and 3 items, when w is being retrieved from the table.
    (b) By enumeration, calculate the expected number of items stored at location $h(w)$ in a hash table size 3 containing $1, 2$, and 3 items, when w is not in the table.
    (c) Explain why your answers from (a) and (b) differ, and why the answer that is larger is larger.

M Problems

25. Let $x_1, \ldots, x_n$ and $y_1, \ldots, y_n$ be integers. Show how the n^2 products $x_i y_j : \forall i, j$ can be computed by a single multiplication.

26. Let $f(x) = \sum_{j=0}^{n} a_j x^j$. Show how to compute the coefficients of the polynomial $(f(x))^2$ with $\frac{3n^2}{4}$ multiplications and a modest number of additions. (Use the natural parameter n). Use your method recursively to do the computation with approximately $n^{1.6}$ multiplications.

27. Arrange the following functions in increasing $O()$ order. That is, place g after f only if $f(n) = O(g(n))$.

    $4n + n^{\log n}; \log n + n^{0.0.2}; n^{2.1}; n!;\; e^n; n^n; 2^{2n/\log n}; (\log n)^{\log n}; n^2 \log\log n;\; n^2 \log n/(\log\log n); n^2\sqrt{\log n};\; n^{\log\log n}; n^{\sqrt{\log n}}$.

28. Suppose you have a special computer whose hardware permits 11-way comparisons in a single step. That is, given 11 elements, it can branch to any of the 11! different ordinal rankings of the 11 elements, in a single computational step. Prove that the worst-case running time of any comparison-based sorting algorithm on your special computer is not less than $O(n \log n)$.

29. Apply four iterations of Newton's method to estimate the square root of 4, starting from the initial estimate 1. Compare your answer with the answer to Exercise 7. Generalize the result of your comparison.

30. Prove that a straightforward implementation of Gaussian elimination, which does not simplify rational numbers by removing common factors, can produce numbers that have exponential length.

31. What permutation of the integers $1 \ldots n$ will cause the algorithm of Theorem 9.3.2 to take the least amount of work? What permutation will take the most work?

32. Suppose a bad implementation of quicksort always chooses the $\lfloor \frac{m}{2} \rfloor$ element from a sublist of m elements to split the sublist. Find two permutations of the integers $1 \ldots n$ for which the implementation will run as fast as possible. Find a permutation for which it will run as slowly as possible.

33. Repeat Problem 32 for an implementation that always chooses the first element from the sublist.

34. Determine how quicksort will perform on an input of n values that are all the same. Why could this make quicksort perform poorly in practice in some applications? Propose two ways to alter quicksort to handle this issue, one randomized and the other deterministic.

35. If $5M$ items are stored in the hash table, and w is an item being retrieved, the expected length of the linked list at location $h(w)$ is greater than 5. Why? Hint: Study the simple cases of 1 or 2 items stored in a table of size $M = 2$.

36. (a) Find the probability that k items are stored at location j in a hash table size n that contains m items.

    (b) Find the expected number of items stored at location j in a hash table size n that contains m items.

    (c) Let w be a random item stored in a hash table (all stored items are equally likely to be selected). Find the expected number of items stored at w's location (including w).

37. Let $h(n)$ be the number of different heaps that can be made from n distinct elements. Find a recursive formula for $h(n)$. Look up the Catalan numbers and show that they satisfy your recursion.

D Problems

38. Prove that Gaussian elimination in fixed precision arithmetic can be very inaccurate. More precisely, for $k = 3$ and some N prove that for all $n \geq N$ there exists an , $n \times n$ matrix M and m-vector b, all of whose elements are have k decimal digits of precision, such that Gaussian elimination on $Mx = b$, always rounding to k digits of precision, yields solutions arbitrarily inaccurate as n increases.

39. Prove that the average case time of Quicksort is $O(n \log n)$.

40. Speed up the average case time to extract the best node in a heap.

41. Consider the following procedure for adding an element a from an ordered set to a rooted binary tree T, each of whose vertices is an element of that set. If T is empty, make a its root. Otherwise, let t be the root of T and let T_L and T_R denote t's (possibly empty) left and right subtree, respectively. If $a \leq t$, recursively add a to T_L; if $a > t$ recursively add a to T_R.

    Suppose now this procedure is employed to build a binary tree T from a randomly ordered list of distinct numbers $a_1, \ldots, a_n$. Prove that the expected depth of T is $\theta(\log n)$. *Hint: For $i < j$ let p_{ij} be the probability that the jth number inserted is a descendant of the ith number inserted.* What does your result suggest about how balanced a binary tree is apt to be? Why is that suggestion misleading for use in practice?

I Problems

42. Under what conditions does Newton's method work to find square roots of complex numbers? Hint: Apply Newton's method to find the square root of -1 from an initial value 1 and from an initial value $.5 + i$. Apply Newton's method to find the square root of i from an initial value 1.

43. A straightforward implementation of the "duplicate row" presolve rule in Section 2.3 takes $O(m^2 n)$ time, where m is the number of rows and n is the number of variables, since each row must be compared to each other row. Show how to use hashing to greatly reduce the expected amount of time required.

44. (a) Describe a conceptually simple way to enumerate all possible heaps containing n distinct elements, such as the integers $1, 2, \ldots, n$.

    (b) Write a computer program that enumerates the heaps. Use a standard language such as Java, Python, C, $C++$, or $C\sharp$.

    (c) Analyze the speed of your program.

    (d) (*) [244] Can you make your program run in $O(nM(n)\log^2 n)$ time, where $M(n) \equiv$ the number of possible heaps? Identify the functionality missing from the language that makes it non-trivial to make your program run efficiently.

Chapter 10

Speed of the Simplex Method and Complexity of Linear Programming

Preview: This chapter shows that the simplex algorithm does not run in polynomial time, but a different one, the ellipsoid algorithm, does solve linear programs in polynomial time.

10.1 Worst-Case Behavior of the Simplex Method

Preview: The construction of an example for which the simplex method performs poorly calls for an enjoyable blend of geometric and algebraic reasoning.

However fast is the average-case behavior of the simplex method, its worst-case behavior is not polynomially bounded. This was proved by Klee and Minty [181]in a famous paper titled, "How Good is the Simplex Method?". They constructed an n-dimensional instance with $2n$ constraints for which the simplex method takes $2^n - 1$ iterations to reach optimality, if the nonbasic variable with best reduced cost is selected as the entering variable. Their instance is built inductively. In one dimension, the instance is a line segment which requires one iteration to achieve optimality. In $n-1$ dimensions, the instance is shaped roughly like an $n-1$ dimensional cube for which the simplex method travels through every vertex to reach the optimum. In n dimensions, the instance consists of two $n-1$ dimensional instances. The simplex method travels through every vertex of the first $n-1$ dimensional instance. When it reaches the last vertex, it travels to the corresponding vertex of the second $n-1$ dimensional instance, and then travels through every vertex of the second instance in the reverse order that it traversed the first instance. Figure 10.3 illustrates this inductive idea. The starting point is a single simplex iteration from the optimum, but the shape of the Klee-Minty polytope is perturbed from the shape of a perfect cube so as to fool the simplex algorithm into going the long way around.

One could modify the traditional simplex method so as to solve the Klee-Minty example in a single iteration. Instead of selecting the non-basic variable with greatest reduced cost, select the non-basic variable which would yield the largest change in the objective function value. However, Jeroslow subsequently proved that for that selection rule, and for many other rules that depend only on values in the current tableau, there exist instances for which the simplex method takes exponentially many steps.

Norman Zadeh invented the following rule to circumvent the Klee-Minty example and Jeroslow's generalization: Keep track of how often a variable has been selected to enter the basis during all previous iterations. From among those with beneficial reduced cost, select as the entering variable one which has been selected least often in the past. This and other so-called "affirmative action" selection rules require only polynomial time on the Klee-Minty and related examples, because they seek out dimensions that have not been explored previously. This tactic defeats the inductive construction method because after linearly many iterations the algorithm jumps from the first copy of the $n-1$ dimensional example to the second copy [92]. However, Oliver Friedmann proved that Zadeh's rule can take more than polynomially many steps [106, 107]. Similarly, so can random selection [108] as well as two other history-based rules [275].

Instead of asking whether for each selection rule there exists an instance for which the simplex method requires many iterations, we could ask whether for each polyhedron there exists a short path from the starting point to the optimum. Hirsch conjectured in 1957 [64] (see also [182]) that the answer to this question is in the affirmative, in a very strong sense. He conjectured that for any polyhedron defined by d constraints in n dimensions, there exists a path between any two extreme points of length at most $d-n$ edges. Klee and Walkup [183] constructed a counterexample, an unbounded 4-dimensional polyhedron defined by 8 constraints. The conjecture for polytopes remained open for 45 more years, until Francisco Santos [258] constructed a 43-dimensional counterexample defined by 86 constraints, which he parlëyed into counterexamples for fixed n with shortest paths length $(1+\epsilon)d$ for some $\epsilon > 0$. The weaker so-called "Polynomial Hirsch Conjecture," that there always exists a path of length less than some polynomial $p(d)$, is an open problem. If it too is false, all hope for a polynomial-time version of the primal simplex method would be lost.

10.1.1 Derivation of the Klee-Minty Example

Klee and Minty's construction utilizes an illuminating combination of geometric and algebraic reasoning. The construction comprises three parts. First, specify the sequence of basic solutions to be traversed; second, ensure that the sequence improves the objective function value at each step; third, trick the best-reduced-cost rule into following the desired sequence.

Let's specify the sequence of vertices that we want the simplex method to traverse. The polytope will be a perturbed n-dimensional cube. We will refer to the polytope's vertices as though it were a perfect cube. Thus the vertices are specified by binary n-vectors $x_1, x_2, \ldots, x_n$. Without loss of generality, denote the starting vertex $0, 0, \ldots, 0, 0$ and the optimum vertex $0, 0, \ldots, 0, 1$. Let S_n denote the sequence in n dimensions. In two dimensions the sequence $S_2 = 00, 10, 11, 01$. In three dimensions, S_3 is $000, 100, 110, 010$ followed by $011, 111, 101, 001$. In four dimensions, S_4 is $0000, 1000, 1100, 0100, 0110, 1110, 1010, 0010$ followed by $0011, 1011, 1111, 0111, 0101, 1101, 1001, 0001$. In n dimensions, S_n is S_{n-1} with $x_n = 0$, followed by S_{n-1} in reverse order, with $x_n = 1$.

Geometrically, in two dimensions the sequence is given in Figure 10.1. In three (respectively four) dimensions the sequence is given in Figure 10.2 (respectively 10.3). The three-dimensional (in general, n-dimensional) instance comprises two copies of the two-dimensional (in general, $n-1$-dimensional) instance, one copy placed above the other (above with respect to the third (in general, nth) dimension). The sequence in the lower copy is identical to the sequence in two (in general, $n-1$) dimensions. The sequence then jumps to the upper copy and retraces the sequence in reverse order, to end directly above the starting vertex.

The sequence S_n always gives priority to lower index terms in the binary n-vector. It flips the value of the earliest component in the binary n-vector that would not repeat a vertex that has already been traversed. We will utilize this prioritization in part 3 of the construction.

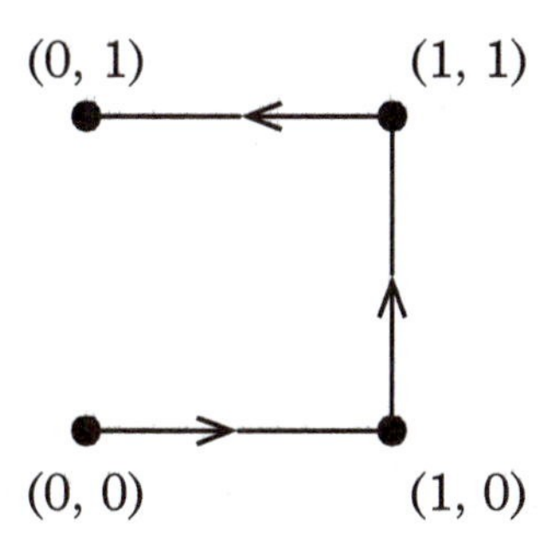

FIGURE 10.1: The path followed in two dimensions.

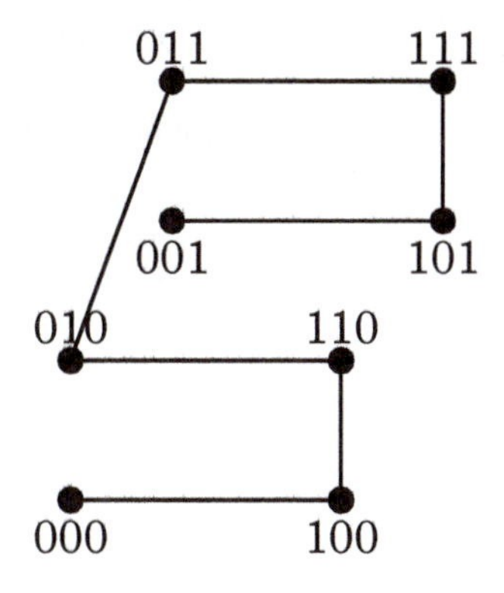

FIGURE 10.2: The path followed in three dimensions.

For the second step, we switch to a purely geometric view. In two dimensions, we maximize x_2 and perturb the square slightly to make the objective improve along the sequence S_2 (see Figure 10.4). The corresponding LP is

$$\max \quad x_2 \quad \text{subject to} \tag{10.1}$$

$$0 \le x_1 \le 1 \tag{10.2}$$

$$-\epsilon x_1 + x_2 \ge 0 \tag{10.3}$$

$$\epsilon x_1 + x_2 \le 1 \tag{10.4}$$

In three dimensions we maximize x_3 on a perturbed cube composed of two copies of the perturbed square. The lower copy is tilted to make $S_2 = 00, 10, 11, 01$ increase in x_3

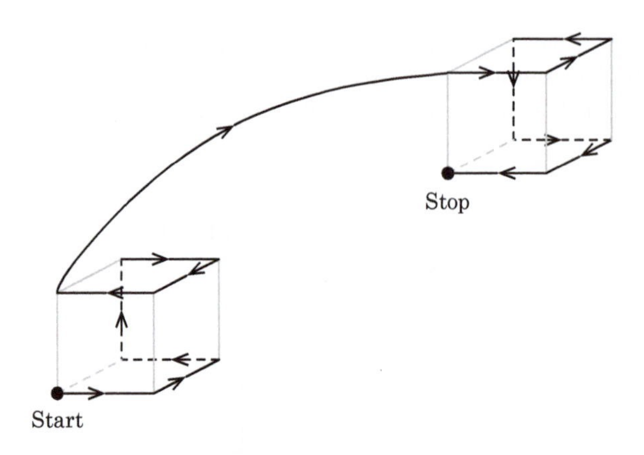

FIGURE 10.3: The path followed in four dimensions.

value; the upper copy is tilted the other way to make the sequence $10, 11, 10, 00$ increase in x_3 value. Exactly how do we tilt the copies? The perturbed square was built so that the sequence S_2 increases in the value of x_2. Therefore, *we must tilt the lower copy such that x_3 increases when x_2 increases.* If the lower copy were flat, not tilted, it would lie on the hyperplane $x_3 = 0$. Referring to Figure 10.5, the orthogonal vector must be changed from $0, 0, 1$ to $0, -\epsilon, 1$ to give it the correct tilt. The upper copy must be tilted the opposite way so that x_3 increases when x_2 decreases. The orthogonal vector changes from $0, 0, 1$ to $0, \epsilon, 1$.

The LP is defined as

$$\max \quad x_3 \quad \text{subject to} \tag{10.5}$$
$$0 \leq \quad x_1 \leq \quad 1 \tag{10.6}$$
$$-\epsilon x_1 + x_2 \quad \geq \quad 0 \tag{10.7}$$
$$\epsilon x_1 + x_2 \quad \leq \quad 1 \tag{10.8}$$
$$-\epsilon x_2 + x_3 \quad \geq \quad 0 \tag{10.9}$$
$$\epsilon x_2 + x_3 \quad \leq \quad 1 \tag{10.10}$$

Thinking algebraically, the pattern is now clear. The n-dimensional instance is defined as

$$\max x_n \text{ subject to} \tag{10.11}$$
$$0 \leq x_1 \leq 1 \tag{10.12}$$
$$-\epsilon x_{i-1} + x_i \geq 0 : i = 2 \ldots n \tag{10.13}$$
$$\epsilon x_{i-1} + x_i \leq 1 : i + 2 \ldots n \tag{10.14}$$

We prove inductively that the sequence S_n increases in objective value in the linear program 10.11. Constraints $x_1 \geq 0$ and $\epsilon x_1 + x_2 \leq 1$ imply $x_2 \leq 1$. Constraints $x_1 \geq 0$ and $-\epsilon x_1 + x_2 \geq 0$ imply $x_2 \geq 0$. By induction the same reasoning gives $0 \leq x_i \leq 1 : 1 \leq i \leq n$ in any feasible solution.

In the first half of S_n we are in the lower copy of S_{n-1}, where the constraint $-\epsilon x_{n-1} + x_n \geq 0$ is binding. Hence $x_n = \epsilon x_{n-1} \leq \epsilon$. Similarly, in the second half of S_n, the constraint

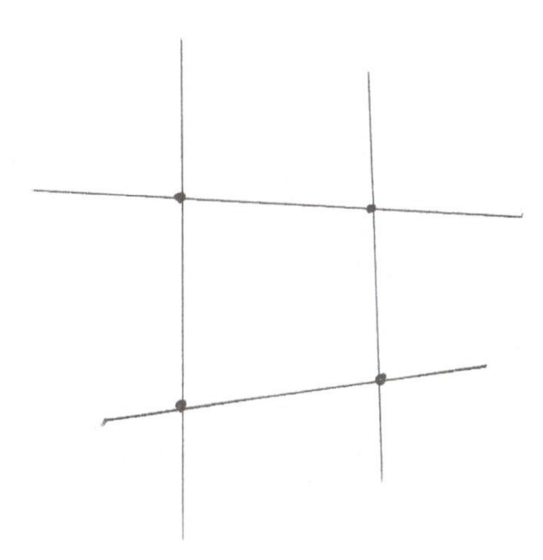

FIGURE 10.4: The two-dimensional Klee-Minty instance is a perturbed square.

$\epsilon x_{n-1} + x_n \leq 1$ is binding, whence $x_n \geq 1 - \epsilon$. As long as $0 < \epsilon < \frac{1}{2}$ all vertices in the second half of S_n have larger objective value than any of those in the first half.

Appeal to induction to see what happens within each copy of S_{n-1}. In the lower copy $-\epsilon x_{n-1} + x_n = 0$. Thus $x_n = \epsilon x_{n-1}$. Maximizing x_n is therefore equivalent to maximizing x_{n-1}, which by induction increases within the sequence S_{n-1}. In the upper copy $\epsilon x_{n-1} + x_n = 1$, so $x_n = 1 - \epsilon x_{n-1}$. Maximizing x_n is equivalent to minimizing x_{n-1}. By induction, x_{n-1} decreases as we traverse S_{n-1} in reverse order. Therefore, for all $n \geq 2$ the sequence S_n increases in the value x_n.

For the third part of the construction, it is helpful to think algebraically. We must trick the greatest-reduced-cost rule into following our desired sequence S_n. As we have remarked, this means we must always give higher priority to lower-indexed variables. The reduced cost $c_1 - \mathbf{c}_{\mathbb{B}}^T B^{-1} A_{:,1}$ must exceed all other reduced costs in magnitude; the reduced cost $c_2 - \mathbf{c}_{\mathbb{B}}^T B^{-1} A_{:,2}$ must exceed all reduced costs but that of x_1 in magnitude, and so on.

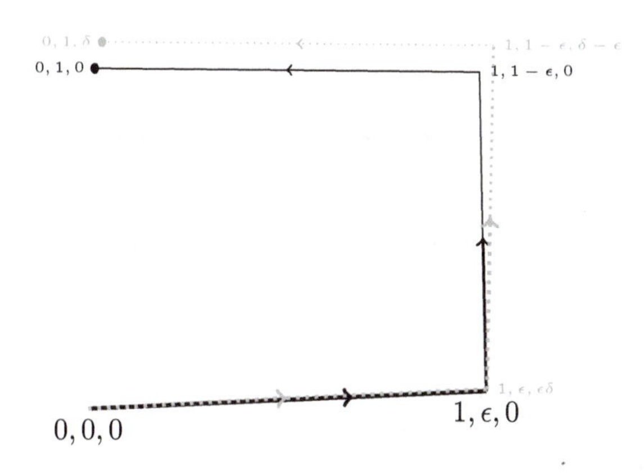

FIGURE 10.5: Tilt the $x_3 = 0$ hyperplane so that x_3 increases as x_2 increases.

Reading Tip: *Try to figure out how to do this before reading further.*

One can multiply any constraint or the objective function in an LP by a positive constant without changing the essentials of the LP. Also, one can multiply the coefficients of any variable (both in the constraints and in the objective function) in an LP by a positive constant without changing the essentials of the LP. The multiplication merely changes the scale in which the variable is defined. *Question: To change the units of x_i from meters to kilometers, what multiplier should be applied to the coefficients of x_i?*[1] Consider $c_j - \mathbf{c}_{\mathbb{B}}^T B^{-1} A_{:,j}$ when x_j is nonbasic and is rescaled. The values of the column $A_{:,j}$ and the coefficient c_j change by the same multiplier, while $\mathbf{c}_{\mathbb{B}}$ and B are unchanged. Therefore, the reduced cost of x_j changes by the multiplicative factor that was applied to the coefficients of x_j.

Letting M be some suitably large constant, multiply the coefficients of x_i by M^{n-i}. Also, multiply the coefficients of the slack variables associated with x_i by M^{n-i}. This completes the Klee-Minty construction.

Theorem 10.1.1 *The simplex method with the greatest-reduced-cost rule for the entering variable requires $2^n - 1$ iterations to solve the Klee-Minty n-dimensional instance.*

Proof: Part two of the construction ensures that the sequence S_n strictly increases in objective value. Therefore, at each step, the non-basic variable needed to enter the basis so as to follow the sequence S_n has strictly positive reduced cost. According to part one of the construction, S_n is defined such that that nonbasic variable is the lowest indexed one with strictly positive reduced cost. By part three of the construction, for sufficiently large M that nonbasic variable's reduced cost exceeds all other eligible reduced costs. Therefore, the simplex method will select that nonbasic variable to enter the basis. Hence the simplex method will traverse the vertices of the polytope in order of the sequence S_n. By part one of the construction, this traversal entails $2^n - 1$ steps. ■

Theorem 10.1.1 tells us that the simplex algorithm does not run in polynomial time, but only if we use the rule of thumb that the size of an input can be measured by the natural parameter n. A rigorous statement that the simplex method does not take polynomial time must employ the formal definition of time bounds in terms of the length of the input as given in Definition 9.2. *Question: Why?*[2] We state the result here. Problem 7 guides you through the proof.

Corollary 10.1.2 *The simplex method with the greatest-reduced-cost rule for the entering variable is not a polynomial-time algorithm.*

Summary: In the worst case, the simplex method does not run in polynomial time because it can traverse exponentially many extreme points. On the one hand, this reveals a weakness in the theoretical criterion, that a fast algorithm has polynomially bounded worst-case run time. Apparently, if its non-polynomial-time cases are rare enough, an algorithm can be fast in practice. On the other hand, the Klee-Minty and subsequent worst-case analyses of other variable selection rules

[1] $\frac{1}{1000}$.

[2] We've assumed the size of the instance to be polynomial in n. If M or ϵ requires a superpolynomial (in n) number of digits to encode, our assumption would be false.

spurred researchers to find polynomial time LP algorithms. The next section presents the first breakthrough, the ellipsoid algorithm. You will learn about interior point algorithms in Chapter 18.

10.2 Complexity of Linear Programming: The Ellipsoid Algorithm

Summary: We've seen that the simplex method with the greatest-reduced-cost rule is not a polynomial-time algorithm, even though it is very fast in practice. Here we describe the first polynomial time algorithm that was discovered for LP, and state an important consequence of that algorithm.

Naum Schôr [266], David Judin, and Arkadi Nemirovski [166] developed the *ellipsoid algorithm* to solve convex optimization problems. Subsequently, Leonid Khachian [179, 178] proved that when the ellipsoid algorithm is applied to LP with integer data, it requires only polynomial time (see [13] for a nice exposition in English, and [28] for a thorough survey). The algorithm has not been found in general to be competitive computationally with the best known LP algorithms. However, it possesses both historical and theoretical importance.

The ellipsoid algorithm solves the feasibility problem by constructing a sequence $E^1, E^2, \ldots$ of ellipsoids that each contain the feasible region. Each ellipsoid has slightly less volume than the preceding one. Eventually, the algorithm either finds a feasible solution, or it reaches an ellipsoid with such small volume that it can conclude no feasible solution exists. As illustrated in Figure 10.6, the center of E^k is tested for feasibility. If the center is feasible, the algorithm terminates successfully. If not, at least one constraint is violated by the center. That constraint restricts the feasible region to half (or less) of the ellipsoid E^k. The ellipsoid E^{k+1} is constructed so as to contain that half-ellipsoid, and hence to contain the feasible region.

Khachian's algorithm begins by converting the set of inequalities $A\mathbf{x} \leq \mathbf{b}$ to an equivalent set $A'\mathbf{x} < \mathbf{b}'$. The resulting feasible region must either be empty or have strictly positive full-dimensional volume. Perhaps Khachian's key insight was that in the latter case, the volume is bounded away from zero by a value $\epsilon > 0$. Therefore, if for some k the ellipsoid E^k has volume less than ϵ, then the feasible region must be empty. If the feasible region is not empty, the algorithm must terminate with a feasible ellipsoid center before the ellipsoid volume drops below ϵ. The rate at which the volumes of successive ellipsoids decrease, though slow, is quick enough to guarantee termination in a polynomial number of iterations. The value of ϵ decreases as the number of variables or the precision of the data increases. Thus the value of ϵ is very small only if the length of the input is large.

To minimize a linear objective $\mathbf{c} \cdot \mathbf{x}$, rather than merely determine the feasibility of the system $A\mathbf{x} \leq \mathbf{b}$, solve the feasibility problem of the following system:

$$A\mathbf{x} \leq \mathbf{b}; A^T\boldsymbol{\pi} = \mathbf{c}; \mathbf{c} \cdot \mathbf{x} - \mathbf{b} \cdot \boldsymbol{\pi} = 0; \boldsymbol{\pi} \geq \mathbf{0}. \tag{10.15}$$

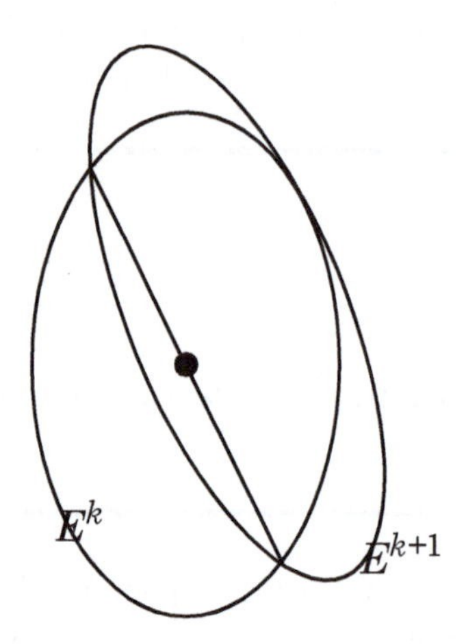

FIGURE 10.6: If the center of ellipsoid E^k is not feasible, the feasible region must be contained in one half of E^k. Ellipsoid E^{k+1} contains that half.

By the strong duality Theorem 2.1.3, the set of optimal primal and dual solutions to the optimization problem is precisely the set of feasible solutions to the system (10.15). If no feasible solutions exist, determine feasibility of $A\mathbf{x} \leq \mathbf{b}$ with the ellipsoid algorithm. If that system has a feasible solution, the optimum must be unbounded. Any feasible solution to

$$A\mathbf{x} \leq \mathbf{0}; \mathbf{c} \cdot \mathbf{x} \leq -1$$

will provide a direction for the ray of unboundedness. Therefore, the LP optimization problem can always be solved with at most three uses of an LP feasibility algorithm.

Later we will see that in some cases, the ellipsoid algorithm can solve the LP feasibility problem in time polynomial in the dimension n even when the number of constraints m is greater than any polynomial in n. In such cases the LP feasibility problem (10.15), which has $n + m$ variables, would be too big for the ellipsoid method to solve in time polynomial in n. The reduction from optimization to feasibility just described would fail.

Instead, repeatedly solve the feasibility problem with a bound on the objective value. Binary search on the bound will determine the optimal objective value to sufficient accuracy that the exact solution is found. Again, the transition to the exact solution is assured by the limited precision of the original data.

To calculate ϵ we need a sequence of lemmas. We assume without loss of generality that integers are encoded as binary strings. Using a different finite alphabet would affect string length by only a constant factor.

Lemma 10.2.1 *Let L denote the total number of digits used to encode an integer n by n matrix A. Then the determinant of A is an integer that satisfies $|det(A)| < n!2^L$.*

Corollary 10.2.2 *Under the conditions of Lemma 10.2.1, if A^{-1} exists, it can be expressed as the product of a scalar that is the reciprocal of a positive integer less than $n!2^L$, and a matrix each of whose coefficients is an integer less than $n!2^L$ in absolute value.*

Proof: By definition, the determinant is the sum of $n!$ products each of n distinct terms of A (half with their signs changed). The absolute value of each of the products is less than 2^L since the total number of digits in the n terms is less than L. *Question: Why is the number of digits strictly less than L?*[3] Hence $|det(A)| < n!2^L$. The corollary then follows from the standard formula for the inverse of a matrix. *Question: Why can the scalar be assumed to be positive?*[4] ■

Lemma 10.2.3 *Let A, b be integer $m \times n$ matrix and m-vector, respectively. Suppose that there exists L such that every subset of $n+1$ rows of $[Ab]$ can be encoded by a string of length at most L. Let $\delta = \frac{1}{(n+2)!n!2^{2L}}$. Then $P^0 \equiv \{\mathbf{x} : A\mathbf{x} \le \mathbf{b}\}$ is nonempty iff $P^\delta \equiv \{\mathbf{x} : A\mathbf{x} \le \mathbf{b} + 1\delta\}$ is nonempty.*

Proof: All we must prove is that if P^0 is empty, so is P^δ. The converse, that if P^0 is not empty, then P^δ is not empty, is so evident that it needs no proof. If P^0 is empty, then by Helly's Theorem there must exist a subset of $n+1$ rows of $[Ab]$, indexed by, say, $S \subset \{1 \dots m\}$, such that $A_{S,:}\mathbf{x} \le \mathbf{b}_S$ has no solution. By Theorem 4.3.2 there exists $\boldsymbol{\pi} \ge 0$ such that $\boldsymbol{\pi}^T A_{S,:} = 0$ and $\boldsymbol{\pi}^T b_S < 0$. Scaling $\boldsymbol{\pi}$ by some positive constant, there exists $1 \ge \boldsymbol{\pi} \ge 0$ such that $\boldsymbol{\pi}^T A_{S,:} = 0$ and $\boldsymbol{\pi}^T b_S < 0$. Consider the LP $\min \boldsymbol{\pi}^T b_S$ subject to $1 \ge \boldsymbol{\pi} \ge 0; \boldsymbol{\pi}^T A_{S,:} = 0$. The feasible region is a nonempty polytope. By Theorem 4.4.8, there exists a basic feasible solution $\boldsymbol{\pi}^*$ to this LP with $\boldsymbol{\pi}^* \cdot \mathbf{b}_S < 0$.

Since $\boldsymbol{\pi}^*$ is feasible, $\boldsymbol{\pi}^* \ge 0$. Since $\boldsymbol{\pi}^*$ is a basic solution, $\boldsymbol{\pi}^* = B^{-1}\mathbf{b}'$ for some square $n+1$ by $n+1$ matrix B and binary vector $\mathbf{b}'$. The matrix B consists of rows of A, I and $-I$. Therefore, it can be encoded by a string of length at most L. The vector b is integer. By Corollary 10.2.2 there exists an integer vector w such that $|w_j| < (n+1)!2^L \; \forall j$ and a positive integer $\tau < n!2^L$ such that $\boldsymbol{\pi}^* = \frac{1}{\tau} w$. *Question: Where does the upper bound on w_j come from?*[5]

Since b_S is integer, $\boldsymbol{\pi}^* \cdot \mathbf{b}_S < 0 \Rightarrow \boldsymbol{\pi}^* \cdot \mathbf{b}_S \le \frac{-1}{\tau} < \frac{-1}{n!2^L}$.

From the upper bound on each w_j, $\boldsymbol{\pi}^* \cdot 1\delta < (n+1)!2^L(n+1)\delta = \frac{(n+1)!2^L(n+1)}{2^{2L}(n+2)!n!} < \frac{1}{n!2^L}$.
Putting the last two inequalities together gives

$$\boldsymbol{\pi}^* \cdot (b_S + 1\delta) < \frac{-1}{n!2^L} + \frac{1}{n!2^L} = 0.$$

By Theorem 4.3.2, $A_{S,:}\mathbf{x} \le \mathbf{b}_S + 1\delta$ has no solution. Therefore, P^δ is empty. ■

Lemma 10.2.4 *Under the conditions of Lemma 10.2.3, if P^0 is empty, then the volume of P^δ is zero. If P^0 is nonempty, then the volume of P^δ is at least*

$$\left((n+2)!n!2^{3L}\right)^{-n}.$$

Proof: The first statement is Lemma 10.2.3. For the second statement, let $\mathbf{y} \in P^0$. Let $\mathbf{x}$ be such that $|\mathbf{x}_j - y_j| < \delta 2^{-L} \forall j$. Then $|(A\mathbf{x} - Ay)_j| = |A_{j,:}(\mathbf{x} - y)| < 2^L \delta 2^{-L} = \delta$. Therefore, $\mathbf{x} \in P^\delta$. The volume of P^δ must be at least the volume of an n-dimensional cube of side length $2\delta 2^{-L}$, which equals $\delta^n 2^{n-Ln}$. Ignoring the unimportant 2^n term and substituting the value of δ gives the desired lower bound. *Question: Why do we ignore the 2^n term?*[6] ■

Lemma 10.2.5 *Under the conditions of Lemma 10.2.3, and if P^0 is nonempty and contains no line, then $P^0 \bigcap \{\mathbf{x} \in \Re^n : |\mathbf{x}_j| \le 2^{2L} n! n \; \forall j\} \ne \phi$.*

---

[3] Some digits have to be used as delimiters between terms of A.

[4] If not, it and the matrix can be multiplied by -1.

[5] Since $\mathbf{b}'$ is binary, each term of w is at most the sum of $n+1$ terms each less than $n!2^L$ in absolute value.

[6] It would only increase the lower bound by 2^n, a tiny amount compared with 2^{-3Ln}.

Proof: By Theorem 4.4.8 P^0 contains a basic feasible solution $B^{-1}b$ for some matrix B whose terms, by Corollary 10.2.2, are all bounded by $2^L n!$ and some vector b the sum of whose terms is bounded by 2^L. ■

Judin and Nemirovski proved that the ellipsoids can be constructed such that the volume of E^{k+1} is at most $e^{-\frac{1}{2(n+1)}}$ times the volume of E^k [166].

Theorem 10.2.6 *Let $E^k \subset \Re^n$ be an ellipsoid, and let $h = \boldsymbol{\pi} \cdot \mathbf{x} > \boldsymbol{\pi}_0$ be an open halfspace not containing the center of E^k. Then there exists an ellipsoid E^{k+1} such that*

1. $E^{k+1} \supseteq h \bigcap E^k$;
2. *Volume*$(E^{k+1}) < e^{-\frac{1}{2n+2}}$ *Volume*(E^k).

Proof: We prove the theorem for the case E^k is a ball, and then prove this case entails no loss of generality. Assume that E^k is the unit ball $\mathbf{x} : \mathbf{x}\cdot\mathbf{x} \leq 1$. Observe $\boldsymbol{\pi}_0 \geq 0$ since $0 \notin h$. If $\boldsymbol{\pi}_0 > 0$ set $\boldsymbol{\pi}_0 = 0$. This change in $\boldsymbol{\pi}_0$ adds to the set h, making requirement 1 more stringent. Hence it is sufficient to consider only this case. Rotate $\boldsymbol{\pi}$ so that $\boldsymbol{\pi} = e_1 = (1, 0, \ldots 0)$. This rotation involves no loss of generality because E^k is spherically symmetric. It is evident geometrically that E^{k+1} should have its center on the positive side of the $\mathbf{x}_1$ axis, at, say, αe_1, that its axes should be parallel to the n coordinate axes, and that by symmetry all the radii except for the one corresponding to $\mathbf{x}_1$ should have equal length, say, β. In algebraic terms, define

$$E^{k+1} = \left\{ \mathbf{x} : \frac{(\mathbf{x}_1 - \alpha)^2}{(1-\alpha)^2} + \sum_{j=2}^{n} \frac{\mathbf{x}_j^2}{\beta^2} \leq 1 \right\}. \tag{10.16}$$

This defines an ellipsoid with center αe_1, radius parallel to the $\mathbf{x}_1$ axis $1 - \alpha$, and all other radii β parallel to axes $\mathbf{x}_2 \ldots \mathbf{x}_n$. We want the defining inequality in (10.16) to be tight at the points $e_1, e_2, \ldots e_n$. I've already chosen the denominator $(1 - \alpha^2)$ of the $\mathbf{x}_1$ term to ensure equality at e_1. To ensure equality at e_2, β must satisfy $\frac{\alpha^2}{(1-\alpha)^2} + \frac{1}{\beta^2} = 1$, from which $\beta = \frac{1-\alpha}{\sqrt{1-2\alpha}}$.

In one dimension, E^k is the interval $[0, 1]$, $E^k \bigcap h = [0, 1]$, and the smallest ellipsoid E^{k+1} meeting requirement 1 is the interval $[0, 1]$. Ellipsoid E^{k+1} therefore has center $\alpha = \frac{1}{2}$. In two dimensions, it is not too hard to see that $\alpha = \frac{1}{3}$ and $\beta = \sqrt{4/3}$ works well, containing the half circle and possessing area $4\pi\sqrt{3}/9 \approx 0.77\pi$ which is nicely less than π, the area of the circle E^k. These two examples suggest for n dimensions, we set

$$\alpha = \frac{1}{n+1}.$$

This value gives

$$\beta = \frac{n/(n+1)}{\sqrt{(n-1)/(n+1)}} = \frac{n}{\sqrt{n^2-1}}.$$

We now must show that requirement 2 is met. Let k_n denote the volume of the unit ball. Then $k_n \prod_{j=1}^{n} r_j$ is the volume of an ellipsoid with radii r_j. The ratio of volumes of requirement 2 is therefore $(1 - \alpha)\beta^{n-1} = \frac{n}{n+1}(1 + \frac{1}{n^2-1})^{\frac{n-1}{2}}$. Recall that $(1 + 1/k)^k$ approaches e from below as $k \to \infty$. Set $k = n^2 - 1$. Then

$$\left(1 + \frac{1}{n^2-1}\right)^{\frac{n-1}{2}} = (1 + 1/k)^{k/(2n+2)} < e^{1/(2n+2)}. \tag{10.17}$$

On the other hand, from the Taylor series of e at 0, $e^{\frac{1}{n+1}} = 1 + \frac{1}{n+1} + \frac{1}{2(n+1)^2} + \frac{1}{6(n+1)^3} + \ldots < \frac{1}{1-\frac{1}{n+1}} = \frac{n+1}{n}$. Hence

$$\frac{n}{n+1} < e^{-\frac{1}{n+1}}. \tag{10.18}$$

Combining inequalities (10.17),(10.18) gives requirement 2.

Last, we prove that no loss of generality was entailed by assuming E^k is a unit ball. By translation, there is no loss of generality in assuming that 0 is the center of E^k. Such an ellipsoid (with positive volume) is always defined as $E^k = \{\mathbf{x} : \mathbf{x}^T A\mathbf{x} \leq 1\}$, where A is a symmetric positive definite matrix. Any symmetric positive definite matrix A can be expressed as $M^T M = A$ for some nonsingular matrix M. Consider the transformation $T : \mathbf{x} \mapsto M\mathbf{x}$. The inner product of $M\mathbf{x}$ with $M\mathbf{y}$ equals $(M\mathbf{x})^T M\mathbf{y} = \mathbf{x}^T M^T M\mathbf{y} = \mathbf{x}^T A\mathbf{y}$. Any point $\mathbf{x}$ such that $\mathbf{x}^T A\mathbf{x} \leq 1$ is such that $M\mathbf{x} \cdot M\mathbf{x} \leq 1$, and vice versa. Therefore, $\mathbf{x} \in E^k$ iff $M\mathbf{x}$ is in the unit ball. The linear transformation T turns the ellipsoid E^k into the unit ball. Equivalently, the linear transformation T^{-1} turns the unit ball into the ellipsoid E^k. If $P \subset Q$, then $T(P) \subset T(Q)$ and $T^{-1}(P) \subset T^{-1}(Q)$ for any linear transformation T. Also, since M is nonsingular, T and T^{-1} preserve ratios of volumes. Therefore, we can map any ellipsoid E^k to the unit ball $T(E^k)$, employ Equation (10.16) to compute the ellipsoid $\hat{E}^{k+1}$ that contains half of the unit ball, and then map back to $E^{k+1} = T^{-1}(\hat{E}^{k+1})$, while maintaining requirements (1,2). ∎

Theorem 10.2.7 *Under the conditions of Lemma 10.2.3, the feasibility problem $A\mathbf{x} \leq \mathbf{b}$ can be solved by the ellipsoid algorithm in at most $6n^3 \log n + 12Ln^2$ iterations.*

Proof: Without loss of generality assume the polyhedron P^0 defined by $P^0 = \{\mathbf{x} : A\mathbf{x} \leq \mathbf{b}\}$ has no lines. If the $\mathbf{x}$ variables are unrestricted, convert the feasibility problem into an equivalent form with nonnegativity constraints. This conversion does not affect the conditions of Lemma 10.2.3.

Begin with the ellipsoid $E^0 = \{\mathbf{x} : ||\mathbf{x}|| \leq 2^{2L}(n+1)!\sqrt{n}\}$. By Lemma 10.2.5 $E^0 \bigcap P^0 \neq \phi$ iff $P^0 \neq \phi$. *Question: Where does the $\sqrt{n}$ term come from?*[7] Moreover, the extra $\frac{n+1}{n}$ factor in the radius of E^0 compared with the bound in 10.2.5 ensures that the n-dimensional cube within P^δ from the proof of Lemma 10.2.4 is also contained in E^0.

Run the ellipsoid algorithm until it finds a feasible solution, or the volume of E^k is less than $\left((n+2)!n!2^{3L}\right)^{-n}$. In the latter case P^0 is empty, by Lemma 10.2.4, and the algorithm terminates. For the next Theorem, we will need the algorithm to terminate if the ellipsoid volume shrinks slightly more slowly, by a factor of $e^{-\frac{1}{2(n+2)}}$ instead of $e^{-\frac{1}{2(n+1)}}$. For our later convenience we use the former here. The volume of E^k is at most

$$\frac{\pi^{n/2}}{\lfloor(n/2)\rfloor!}(2^{2L}(n+1)!\sqrt{n})^n \times e^{-\frac{k}{2(n+2)}} < e^{2n(L+n\log n)}e^{-\frac{k}{2(n+2)}}.$$

Let $k = 12n^3 \log n + 12Ln^2$). The following straightforward computation shows that this value of k forces the volume of E^k below the termination value:

$$\begin{array}{rcl} \log((n+2)!n!2^{3L})^n + 2nL + 2n^2\log n & < n\log(n^{2n}2^{3L}) + 2nL + 2n^2\log n \\ \leq \qquad 5n^2\log n + 5nL & < \dfrac{6n^3\log n + 6Ln^2}{n+2}. \end{array}$$

∎

[7]If $|\mathbf{x}_j| \leq K\ \forall j$, then $||\mathbf{x}|| \leq K\sqrt{n}$.

Theorem 10.2.7 does not rigorously prove that the ellipsoid algorithm solves the linear programming feasibility problem in polynomial time, because each iteration requires the calculation of square roots, which theoretically requires infinite precision. We must prove that it is sufficient to calculate the square root approximately, within a polynomially bounded amount of time. The idea is to compute an ellipsoid $\tilde{E}^{k+1}$ that contains the ellipsoid E^{k+1} which would be computed had we infinite precision, and is only a tiny bit larger than E^{k+1}. As long as the decrease in volume compared to $\tilde{E}^k$ is enough, the number of iterations will still be bounded by a polynomial.

The square root terms arise when computing the symmetric positive definite matrix M such that $M^TM = A$. Matrix M is then used to define the linear transformation T that maps E^k to the unit ball. Since we are aiming for a purely theoretical result, not a computationally practical result, we will find a simple loose bound rather than a more complicated but tighter bound. Compute all square roots to have error less than 2^{-2L^4} so all computed values $\tilde{v}$ of true values v satisfy $|\tilde{v}/v - 1| < 2^{-L^4}$. These computations can be done in polynomial time using Newton's method, as proved in Section 9.2.5. Define $\tilde{E}^{k+1}$ as in (10.16) with a tiny increase in size, as follows:

$$\tilde{E}^{k+1} = \left\{ \mathbf{x} : \frac{(\mathbf{x}_1 - \alpha)^2}{(1-\alpha)^2} + \sum_{j=2}^{n} \frac{\mathbf{x}_j^2}{\beta^2} \leq 1 + 2^{-L^2} \right\}. \tag{10.19}$$

The cumulative effect of n^2 errors of order 2^{-L^4} in the computation of M^-1 at most order 2^{-L^3} since $n^2 < L$. Therefore, the inexactly computed transformation of the larger ellipsoid $\tilde{E}^{k+1}$ contains the exact (but not computed) ellipsoid $T^{-1}E^k$. It remains to show that the larger volume of $\tilde{E}^{k+1}$ is still sufficiently less than the volume of the ellipsoid of the previous iteration that the number of iterations remains bounded by a polynomial. The ratio of volumes stated in Requirement 2 was $\text{Volume}(E^{k+1}) < e^{-\frac{1}{2n+2}}\text{Volume}(E^k)$. The enlarged ellipsoid $\tilde{E}^{k+1}$ satisfies the weaker inequality

$$\text{Volume}(\tilde{E}^{k+1}) < (1 + 2^{-L^3})^n e^{-\frac{1}{2n+2}}\text{Volume}(\tilde{E}^k).$$

We have

$$(1 + 2^{-L^3})^n < (1 + 2^{-nL^2})^n \leq e^{1/L^2} < e^{1/2(n+1)^2}$$

and then

$$e^{1/2(n+1)^2 - \frac{1}{2n+2}} = e^{-\frac{n}{2(n+1)^2}} < e^{-\frac{1}{2n+4}}.$$

Hence

$$\text{Volume}(\tilde{E}^{k+1}) < e^{-\frac{1}{2(n+2)}}\text{Volume}(\tilde{E}^k).$$

We anticipated the value $-1/2(n+2)$ as opposed to $-1/2(n+1)$ in the proof of Theorem 10.2.7. We sum up the preceding discussion with a statement that the ellipsoid algorithm solves LPs in polynomial time.

Theorem 10.2.8 *Under the conditions of Lemma 10.2.3, the feasibility problem* $A\mathbf{x} \leq \mathbf{b}$ *can be solved by the ellipsoid algorithm (with Newton's method for square root estimation) in polynomial time.*

Corollary 10.2.9 *Under the same conditions as Theorem 10.2.8, optimization LPs can be solved in polynomial time.*

Proof: As discussed already, LP optimization can be solved by at most three calls to a feasibility algorithm, by testing for primal feasibility, combining primal and dual feasibility, and if necessary finding a ray of unboundedness. ■

Another way to solve LP optimization using LP feasibility is to perform binary search on the optimal objective function value. A problem at the end of this chapter guides you to a proof that the number of iterations required is bounded by a polynomial, thus giving an alternative proof of Corollary 10.2.9.

10.2.1 Polynomial Equivalence of Separation and Optimization

The bound of Theorem 10.2.7 on the number of iterations of the ellipsoid algorithm does not depend on the number of constraints in the LP, but only on the number of columns n and the precision required to express any n constraints of the LP. Moreover, the computations required to update E^k to E^{k+1} also depend only on n and the precision required to express coefficients. Therefore, the ellipsoid algorithm runs in polynomial time, even if the number of constraints is exponential, as long as the following *separation problem* can be solved in polynomial time:

Definition 10.1 (Separation Problem) *For a set $A\mathbf{x} \leq \mathbf{b}$ of linear inequalities with integer coefficients, given $\mathbf{y}$, either determine that $Ay \leq \mathbf{b}$, or exhibit a linear inequality $A_{j,:}\mathbf{x} \leq \mathbf{b}_j$ that is violated by $\mathbf{y}$.*

We say that the separation problem can be solved in polynomial time if it can be solved in time polynomial in L, where L is an upper bound on the encoding length of all subsets of n rows of $A\mathbf{x} \leq \mathbf{b}$ and $\mathbf{y}$.

Martin Grötschel, Lászlo Lovász and Alexander Schrijver [135, 136] brought attention to this feature of the ellipsoid algorithm. Moreover, they proved an important *equivalence* between separation and optimization. Polarity is the idea behind this equivalence. Suppose one can optimize a linear function over a polyhedron P in polynomial time. Optimization over P is mathematically identical to separation over the polar P^* of P. Apply the ellipsoid algorithm to the polar $P*$ to yield a polynomial time algorithm for optimization over the polar P^*. Optimization over P^* is mathematically identical to separation over P. Therefore, separation over P and optimization over P are *polynomially equivalent*: either both are solvable in polynomial time, or neither one is. A rigorous proof of polynomial equivalence is complicated by the necessary restrictions on the encoding length of the polar P^*, and is well beyond the scope of this book. The reader is referred to the book by Grötschel et al. [137]. They extend the equivalence between separation and optimization from polyhedra to convex sets. A precise statement of the equivalence is greatly complicated by how a representation of a convex set is defined and the consequent degree of inexactness permitted in the definitions of optimization and membership. The proof itself is even more intricate. The end result can be employed to yield remarkably short proofs that certain problems can be solved by an algorithm in polynomially bounded time.

A separation algorithm can be used within the simplex method. Suppose we have an LP with many constraints. Solve a partial model that contains only by a subset of the constraints. Run a separation algorithm to determine whether or not the solution satisfies the other constraints. If the answer is yes, the solution is optimal for the full LP. If the answer is no, add the violated constraint to the partial model and re-solve. The method may seem familiar to you. It should. It is the dual of column generation. There is no guarantee that this method will terminate in polynomial time, but it has been useful in practice. We will see an example in Chapter 11.

Summary: The ellipsoid algorithm, the first to solve LP in polynomial time, encloses the feasible region in a sequence of shrinking ellipsoids. It either finds a feasible solution at an ellipsoid center or reaches an ellipsoid whose volume is so small that the feasible region must be empty. The ellipsoid algorithm is not fast enough in practice to be used commercially, but it is a powerful theoretical tool because it only requires that any infeasible point can be separated in polynomial time from the feasible region with a valid constraint, regardless of the number of valid constraints. By polarity this establishes the polynomial equivalence of optimization and separation, even for convex sets that are not polyhedra.

10.3 Problems

E Exercises

1. Write out the Klee-Minty LP for the case $n = 4$.

2. Let
$$M = \begin{bmatrix} \frac{1}{2} & 0 \\ 0 & \frac{1}{3} \end{bmatrix}.$$
Compute $A = M^T M$ and draw the ellipse $E = \{\mathbf{x} : \mathbf{x}^T A \mathbf{x} \leq 1\}$. How does the area of E compare with the area of a unit circle? How could you know how they compare by examining M or A?

3. Repeat Exercise 2 with
$$M = \begin{bmatrix} \frac{1}{2} & \frac{1}{2} \\ 0 & \frac{1}{2} \end{bmatrix}.$$

4. Consider the LP instance $\max 15\mathbf{x}_1 + 10\mathbf{x}_2$ subject to $\mathbf{x}_1 + \mathbf{x}_2 \leq 300; \mathbf{x} \geq 0$.
   (a) Verify that, starting from $\mathbf{0}$, the simplex algorithm will require one iteration if it selects the entering variable with best reduced cost.
   (b) Convert the LP to an equivalent one that will trick the simplex algorithm into taking two iterations.

5. Prove $A = M^T M$ is symmetric positive definite if M is nonsingular.

M Problems

6. Calculate specific values for ϵ and M for the Klee-Minty LP in the case $n = 4$.

7. This problem guides you to prove Corollary 10.1.2.

(a) Let $\epsilon = \frac{1}{10}$ in the Klee-Minty construction. Calculate a value of M that ensures x_i receives higher priority than x_{i+1}. Hint: Use formulas from the ellipsoid algorithm analysis.

(b) Prove that the simplex method with greatest-reduced-cost rule for the entering variable is not a polynomial time algorithm, according to the precise definition of polynomial time as a function of the length of the input. Hint: Use part (i) to prove that the length of the Klee-Minty construction can be bounded by a polynomial in the dimension n; then apply Theorem 10.1.1 together with the property that a polynomial of a polynomial is a polynomial.

8. Prove that Zadeh's affirmative action rule for selecting the entering variable solves the Klee-Minty instance in $O(n^2)$ iterations.

9. Prove that if the simplex method selects the entering variable at random from among the nonbasic variables with strictly positive reduced costs, then the expected number of iterations to solve the Klee-Minty example is $O(n^2)$ [177].

10. The Least-Recently-Used (LRU) affirmative action rule for the simplex method is to select as the entering variable, among all nonbasic variables with good reduced cost, that which least recently has been basic. For example, suppose the algorithm has traversed the following sequence of basis index sets on 8 variables: $1, 2, 3, 4;\ \ 2, 3, 4, 6;\ \ 2, 4, 6, 8;\ \ 2, 4, 5, 8;\ \ 1, 2, 4, 5;$. Then the index with highest LRU priority is 7 (since it has never been basic). The next least recently used is 3, followed by 6, followed by 8. Prove that the LRU rule solves the n-dimensional Klee-Minty example in $O(n^2)$ iterations.

11. Adrian, Ronnie, and Chen are given the same $A\mathbf{x} \leq \mathbf{b}$ feasibility LP to solve via the ellipsoid algorithm. Adrian works with the data exactly as given. Ronnie divides the coefficients of the first row, the constraint $A_{1,:}\mathbf{x} \leq b_1$, by 10 before applying the ellipsoid algorithm. Chen multiplies the first column $A_{:,1}$ data by 100 before applying the algorithm. How do Ronnie's computations differ from Adrian's? How do Chen's computations differ?

12. To begin the ellipsoid algorithm, one needs an initial ellipsoid E^1 which contains the feasible region. Assuming that the feasible region is a polytope, derive a valid initial ellipsoid. Prove that the volume of your ellipsoid is sufficiently small to ensure that the algorithm terminates in polynomial time.

13. Show how to initialize the ellipsoid algorithm if the feasible region is unbounded.

14. Prove algebraically that the ellipsoid E^{k+1} contains $E^k \bigcap h$ in Theorem 10.2.6.

Let A be an integer $m \times n$ matrix, $\mathbf{b}$ an integer m-vector, and $\mathbf{c}$ and $\mathbf{w}$ integer n-vectors. Consider the problem

$$\begin{aligned} \max \mathbf{c}^T\mathbf{x} \ \ &\text{subject to} \\ A\mathbf{x} &\leq \mathbf{b} \\ \sum_{i=1}^{n} |w_i x_i| &\leq w_0 \end{aligned}$$

(a) Apply a projection modeling technique from Chapter 3 and Corollary 10.2.9 to prove that this problem can be solved in polynomial time.

(b) Replace the nonlinear constraint by the 2^n linear constraints

$$\sum_{i=1}^{n} e_i w_i x_i \leq w_0 \quad \forall e \in \{-1, 1\}^n. \tag{10.20}$$

Devise a polynomial time algorithm to separate a vector $\mathbf{x}$ from the feasible region defined by $A\mathbf{x} \leq \mathbf{b}$ and the constraints (10.20). Conclude that the problem can be solved in polynomial time without the reformulation of part 14a.

(c) In this problem, the separation algorithm is very easy to invent. Had you not been able to invent it, how could you have known it must exist, based on part 14a?

D Problems

15. Here is a different idea for accomplishing part two of the Klee-Minty construction. Let w_j denote the jth vertex in the sequence S_n. Define $y_j = w_j + \frac{j}{2^n} e_n$. (The vector e_n is the nth unit basis vector.) Let the Klee-Minty polytope be the convex hull of $\mathbf{y}_1, \ldots, y_{2^n}$. By definition, the objective function would increase by $\frac{1}{2^n}$ each step. Therefore, the goal of part two would be achieved. Nonetheless, there is a flaw in this construction. What is it?

16. (Alternate proof of Corollary 10.2.9)

    (a) Use Lemmas in this chapter to find a lower bound on how close the objective function values of two basic feasible solutions can be, but still differ from each other.

    (b) Prove that the crashing algorithm (Algorithm 4.4.6) runs in polynomial time.

    (c) Use the fundamental theorem of LP to prove that for sufficiently small $\epsilon > 0$, if no feasible solution is better than some basic feasible solution $\mathbf{x}^*$ by more than ϵ, then $\mathbf{x}^*$ is optimal.

    (d) Combine these steps to prove that in the ellipsoid algorithm, binary search for the optimal objective function value can be terminated in a polynomial number of steps, where each step consists of a feasibility check.

17. (*) Disprove the linear Hirsch conjecture, that there exists α such that for all polyhedra with d facets, for all pairs of extreme points, there exists a path of length at most αd.

I Problems

18. Use the ellipsoid algorithm to prove that finding optimal multipliers in a Lagrangian relaxation can be done in polynomial time iff the Lagrangian subproblem can be solved in polynomial time [23, 38].

19. (*) Read [289] or [62], which give smoothed analysis upper bounds on the number of iterations of the simplex method that are polynomial in the number of variables (d) and sublinear (in fact poly-logarithmic) in the number of constraints (n). Determine which of the following is true:

    - The sublinear term implies that smoothed analysis of the simplex algorithm is not an analysis of the actual simplex algorithm, which takes a number of iterations on average linear in the number of constraints.

- The sublinear bound correctly describes the average performance of the simplex algorithm on the class of instances in which the dual is infeasible (this could be sublinear because all that is needed is a single column row to prove primal unboundedness).
- The sublinear bound is correct with respect to the simplex algorithm because n represents only the number of constraints that are not upper or lower bounds on variables.

Chapter 11

Network Models and the Network Simplex Algorithm

Preview: Networks are mathematical models in which material flows along *edges* between *nodes*. Problems on networks with linear costs and a single type of material compose a special form of LP. This form has several advantages, principally, integer optimal solutions and shorter computation times. A specialized version of the simplex algorithm, network simplex, achieves the computational advantage.

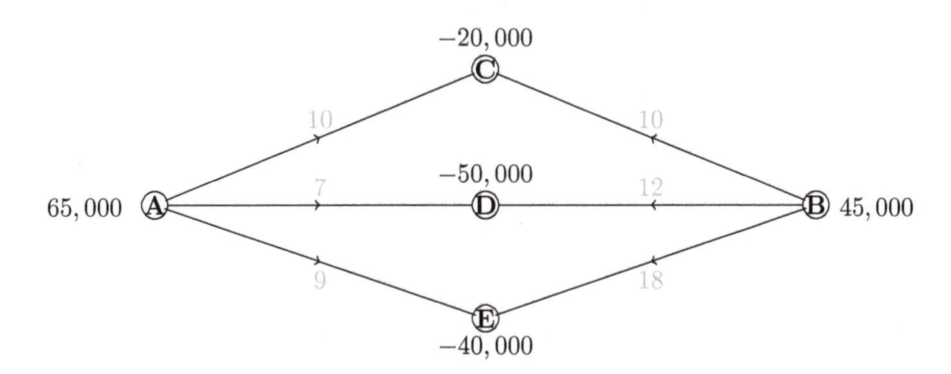

FIGURE 11.1: Minimize transportation costs from ports A and B to refineries C, D, and E.

Figure 11.1 shows a small example of a minimum cost network flow problem. Crude oil from two ports, A and B, must be transported to three refineries, C, D, and E. The supply (demand) at each port (refinery) is given as a positive (negative) number. Shipping costs per barrel from each port to each refinery are given on the arcs between them. The objective is to minimize total transportation costs.

On the one hand, even this simple example cannot be solved by sorting the arc costs from lowest to highest and sending as much as possible on the arcs in that order. On the other hand, it is straightforward to model the problem as a linear program with decision variables x_{ij} = the number of barrels to send from port $i \in \{A, B\}$ to refinery $j \in \{C, D, E\}$. The resulting LP's constraints are sparse, and their few nonzero coefficients are all 1's and −1's. *Question: What are the constraints?*[1] Thus, minimum cost network flow problems can

[1] Nonnegativity: $x_{ij} \geq 0$. Demand: $\sum_i x_{iC} = 20,000$, $\sum_i x_{iD} = 50,000$, $\sum_i x_{iE} = 40,000$. Supply: $\sum_j x_{Aj} = 65,000, \sum_j x_{Bj} = 45,000$.

be modeled as LPs with a special structure. As we will see, this structure can be exploited to solve very large instances quickly, and it also yields integer optimal basic solutions.

Proceeding from the specific to the general, we first need the mathematical definition of a graph.

Definition 11.1 (Graph) *A Graph $G = (V, E)$ is a finite set V of points, called vertices or nodes, and a set $E \subset V \times V$ of node pairs, called edges or arcs.*

Several graphs are shown in Figures 11.2, 11.3, and 11.4. A graph is an abstract model that depicts which pairs of elements in a set are related in some way. For example, Figure 11.3 is the graph of a subway system, where each node represents a station, and an arc represents a track that directly connects a pair of stations without passing through any intermediate stations. In a graph of friendships such as Figure 11.2, each node represents a person, and an arc between two people's nodes means they are friends. The ordering of the nodes in the arc is irrelevant in this example. If i and j are friends, j and i are friends.

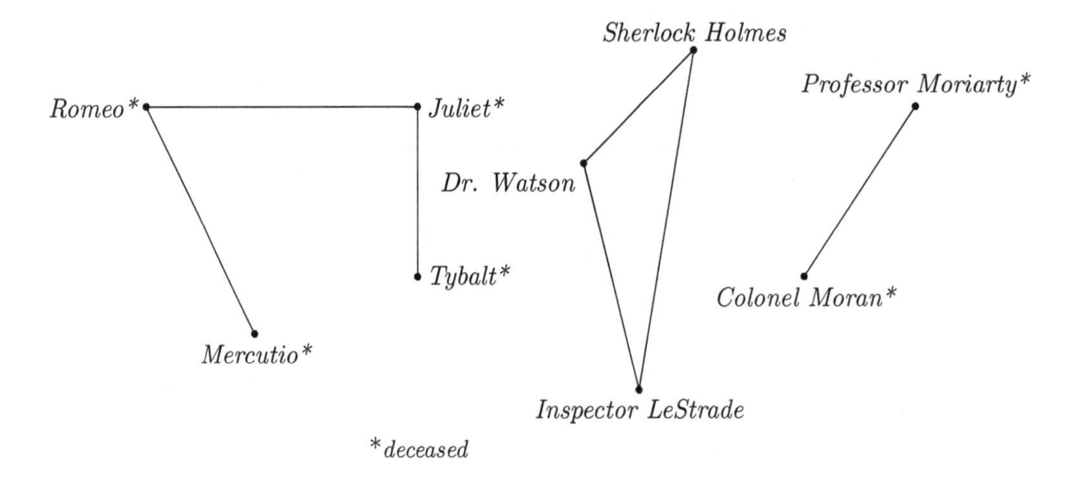

FIGURE 11.2: Undirected arcs representing friendships in a graph connect pairs of nodes.

In many other cases, the order of nodes matters. An arc for which node order matters (respectively does not matter) is a *directed* (respectively *undirected*) arc or edge. For example, a graph of web site connections such as Figure 11.4 represents each web page as a node, and a directed arc (i, j) denotes a link from page i to page j, not the reverse. In a graph representing a road system, directed arcs may be necessary if there are one-way streets. In the great majority of cases, an arc (i, j) that is not directed can be represented adequately by the two directed arcs (i, j) and (j, i). In this book, therefore, arcs are by default assumed to be directed. If an arc is directed, I will usually call it an arc or edge; if it is undirected, I will always call it an undirected arc or edge. *Question: If $M_{ij} = 1$ denotes an arc from i to j, what kind of matrix represents a graph with undirected arcs?*[2]

A *network* is an abstract model that usually depicts the capability of movement or flow of something (fuel, data packets, soldiers, etc.) from node to node in a graph. Nodes can represent actual physical locations such as cities and seaports, virtual locations such as customers, points in time, and corporations, or combinations of physical and virtual locations such as airports at specific times. The data required to describe a network typically

[2]Symmetric.

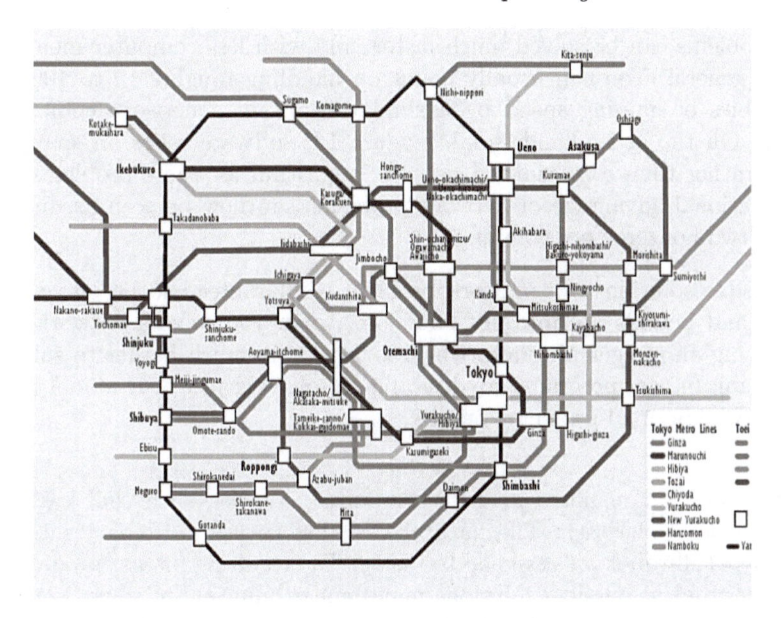

FIGURE 11.3: A graph depicts the stations and rail lines of the Tokyo subway system[53].

are: a supply amount at each node, a cost per unit of flow for each arc, and lower and upper bounds on flow along each arc, in addition to the description of the underlying graph. *Question: How is a demand amount at a node represented?*[3] A typical network is shown in Figure 11.5. If no lower bound on an arc is specified, the bound is assumed to be zero. If no upper bound is specified, the bound is assumed to be infinite. Standard network models always follow two rules of flow conservation:

Conservation of flow in an arc The amount of flow into an arc equals the amount of flow out of the arc. In other words, flow is neither lost nor gained within an arc.

Conservation of flow at nodes The supply at a node, plus the total flow into the node, equals the total flow out of the node.

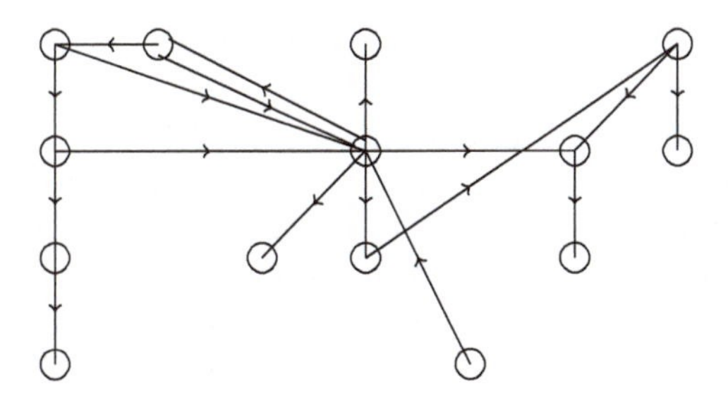

FIGURE 11.4: In this graph, nodes represent web pages, and directed arcs represent hyperlinks. A graph of the web in 2012 has more than 3.5 billion nodes and 125 billion arcs[228].

It is straightforward to model a network flow problem as an LP. Since LPs can be solved readily, you should be wondering why we study networks as a separate topic. There are two main reasons:

[3] It is represented by a negative supply amount.

- Network problems can be solved much faster, and with less computer memory, than can LPs in general. You can usually count on handling roughly 10 to 100 times as many variables or gaining speed by a similar factor in a network compared to a general LP. On the other hand, some modern LP software relies on sparse matrix techniques rather than a specialized network algorithm for these problems. Most of the computational advantage of network problems can now be seen as due to their sparseness (two nonzeros per column).

- A basic feasible solution to a network problem is guaranteed to be integer, as long as supplies and bounds are integer. Therefore, when you solve a network problem, you are solving an integer program, which is ordinarily much harder to solve than a linear program. Integer programs are desirable to solve because they model the yes/no decisions that many real problems require.

Example 11.1 You must assign n robots to n tasks, such that each robot performs exactly one task. The natural variables to use are binary integer x_{ij} where $x_{ij} = 1$ iff robot i is assigned to task j. The resulting integer program with n^2 variables and $n!$ possible solutions may appear daunting to solve, but in fact it is very tractable computationally because it is a network problem.

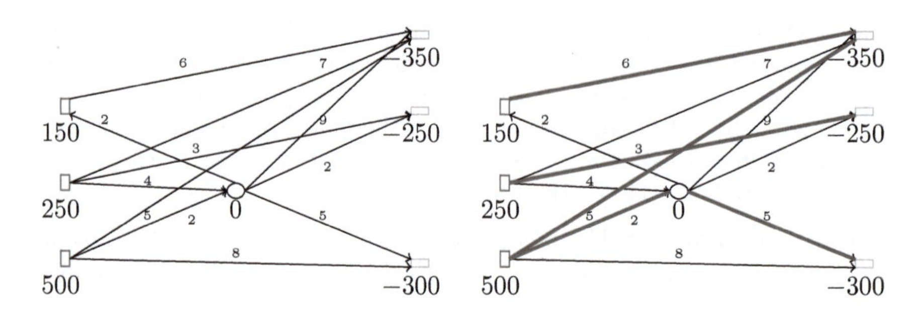

FIGURE 11.5: A network model usually includes arc costs and node supplies (positive values) and demands (negative values). Arcs that have nonzero flow in the optimal solution are shaded in the right copy. Deduce the amount of flow on each arc.

Clients usually like to understand an optimization model before they agree to use it. Hence, a good visualization helps convince a client to buy into your optimization model. Since network models are much easier to depict visually than are LPs or IPs in general, they are often easier to sell. Even if your optimization model is not a standard network problem, it may be a network problem with some additional constraints. If that is the case, draw the network part of the model for your client first, and explain the other parts of the model later.

11.1 Network Models

Preview: Our core network model is to minimize the total cost of flow on arcs, subject to flow balance constraints at the nodes, and subject to bounds on individual arc flows.

The standard network problem asks to satisfy supply and demand requirements at minimum total cost, without violating upper and lower bounds on arc flows. This is called the *Minimum Cost Flow* problem. As an LP the minimum cost flow problem on an underlying graph $G = (V, E)$ has index set V, with data s_j denoting the supply at node $j \in V$. A value $s_j < 0$ denotes demand at node j. Total supply and demand must be equal, that is, $\sum_{j \in V} s_j = 0$. The variables are $x_{ij} : (i, j) \in E$ denoting the amount of flow on the arc from node i to node j. Arc data consist of c_{ij}, l_{ij}, and u_{ij}, denoting costs per unit, lower bounds, and upper bounds on flow, respectively, for arcs $(i, j) \in E$.

Often we will focus on the simple case of the minimum cost flow model where all lower bounds are zero and all upper bounds are infinite. The primal LP for this case follows.

$$\min \sum_{(i,j) \in E} c_{ij} x_{ij} \qquad \text{subject to} \tag{11.1}$$

$$\sum_j x_{ij} - \sum_j x_{ji} = s_i \qquad \forall i \in V \tag{11.2}$$

$$x_{ij} \geq 0 \quad \forall (i, j) \in E \tag{11.3}$$

Reading Tip: *Understand the primal LP before reading the explanation that follows. Formulate the dual LP yourself.*

The LP's functional constraints, (11.2) enforce *flow balance* at each node. The LHS (left-hand-side) is the net flow out of node i, which must equal the RHS s_i, the supply at i. Each variable x_{ij} appears in exactly two functional constraints, once with a coefficient of 1 in the constraint for node i, and once with a coefficient of -1 in the constraint for node j. Therefore, the dual LP has $|V|$ variables $\boldsymbol{\pi}_v : v \in V$, and $|E|$ constraints, one for each arc $(i, j) \in E$, of form $\boldsymbol{\pi}_i - \boldsymbol{\pi}_j \leq c_{ij}$. The dual LP follows.

$$\max \sum_{i \in V} s_i \boldsymbol{\pi}_i \qquad \text{subject to} \tag{11.4}$$

$$\boldsymbol{\pi}_i - \boldsymbol{\pi}_j \leq c_{ij} \quad \forall (i, j) \in E \tag{11.5}$$

The primal LP for the general minimum cost network flow problem replaces the non-negativity constraints with upper and lower bounds as follows.

$$\min \textstyle\sum_{(i,j) \in E} c_{ij} x_{ij} \qquad \text{subject to} \tag{11.6}$$

$$\textstyle\sum_j x_{ij} - \sum_j x_{ji} = s_i \quad \forall i \in V \tag{11.7}$$

$$l_{ij} \leq x_{ij} \leq u_{ij} \qquad \forall (i, j) \in E \tag{11.8}$$

The dual LP has unrestricted variables $\boldsymbol{\pi}_i$ corresponding to the flow conservation constraints at the nodes $i \in V$, and nonnegative variables μ_{ij} and ι_{ij} corresponding to the upper bounds $-x_{ij} \geq -u_{ij}$ and lower bounds $x_{ij} \geq l_{ij}$, respectively, for $(i,j) \in E$. (I rewrote the upper bound constraints as $\geq$ constraints because it is easier for me to think about nonnegative variables than about nonpositive variables.) The dual LP is therefore

$$\max \quad \textstyle\sum_{i\in V} s_i\boldsymbol{\pi}_i + \sum_{(i,j)\in E} l_{ij}\iota_{ij} - u_{ij}\mu_{ij} \quad \text{subject to} \tag{11.9}$$

$$\boldsymbol{\pi}_i - \boldsymbol{\pi}_j + \iota_{ij} - \mu_{ij} \quad = c_{ij} \ \forall (i,j) \in E \tag{11.10}$$

$$\boldsymbol{\pi}_i \text{ unrestricted }; \quad \iota_{ij}, \mu_{ij} \geq 0. \tag{11.11}$$

If arc (i,j) in the primal has no upper bound on flow, the corresponding variable μ_{ij} does not exist in the dual.

We need to get three interrelated technicalities out of the way. Sum the Equations (11.2), giving the equation $0 = \sum_{v\in V} s_v$. We always assume that the data satisfy this equation, for otherwise no feasible solution can exist. Under this assumption there is a redundancy, since any $|V|-1$ of these constraints imply the other one. The primal constraint matrix is not full rank. A basis will consist of $|V|-1$ variables, rather than $|V|$. Operationally, the redundant constraint can be ignored or used to check the correctness of the other computations. In the dual, there is an extra variable. Since $\sum_{v\in V} s_v = 0$, any feasible dual solution is essentially unchanged if a constant α is added to each variable. Only the *differences* between $\boldsymbol{\pi}_v$ values matter. Operationally, this means that we can choose any $\boldsymbol{\pi}_v$ at the outset and set its value arbitrarily. Usually we set its value to 0.

11.1.1 Examples of Minimum Cost Flow Models

This section gives several examples of minimum cost flow models, beginning with a concrete case and continuing with more abstract ones.

Transport Through a Network

An oil company contracts with a trucking company to transport gasoline from its refineries to its customers. Each refinery has a known supply; each customer has a known demand. Supplies and demands are measured in truckloads. For each refinery and each customer, it is known how much the transportation company will charge per truckload for delivery from that refinery to that customer.

The model: Let M denote the set of refineries and C denote the set of customers, with indices m and c respectively. Let s_m be the supply at refinery m in truckloads and let d_c be the number of truckloads demanded by customer c. Construct a graph $G = (M \bigcup C, E)$ where the supply at $m \in M$ equals s_m, and the supply at $c \in C$ equals $-d_c$. For each $m \in M$ and $c \in C$, arc (m,c) has cost equal to the price of sending one truckload from m to c. All arcs have lower bounds $l_{mc} = 0$ and no upper bounds, i.e., $u_{mc} = \infty$.

What to do if supply does not equal demand.
If in reality total demand exceeds total supply, then necessarily not all real demand will be met. Model this situation with an extra "dummy" refinery m_0 that has fictitious supply equal to the amount of excess demand $\sum_{c\in C} d_c - \sum_{m\in M} s_m$. For each customer node c create an arc (m_0, c). Choose an upper bound u_{m_0c} on how much of c's demand will not be met in reality, and/or choose a cost to penalize each unit of unmet real demand at c. *Question: Why should one choose these values for each $c \in C$?*[4] Model excess real supply similarly, with a dummy customer node c_0 that has fictitious supply $\sum_{c\in C} d_c - \sum_{m\in M} s_m < 0$. The cost of sending flow from m to c_0 might reflect the cost of storage at m.

[4]Some customers may be more valuable or at risk to lose if their demands are not met.

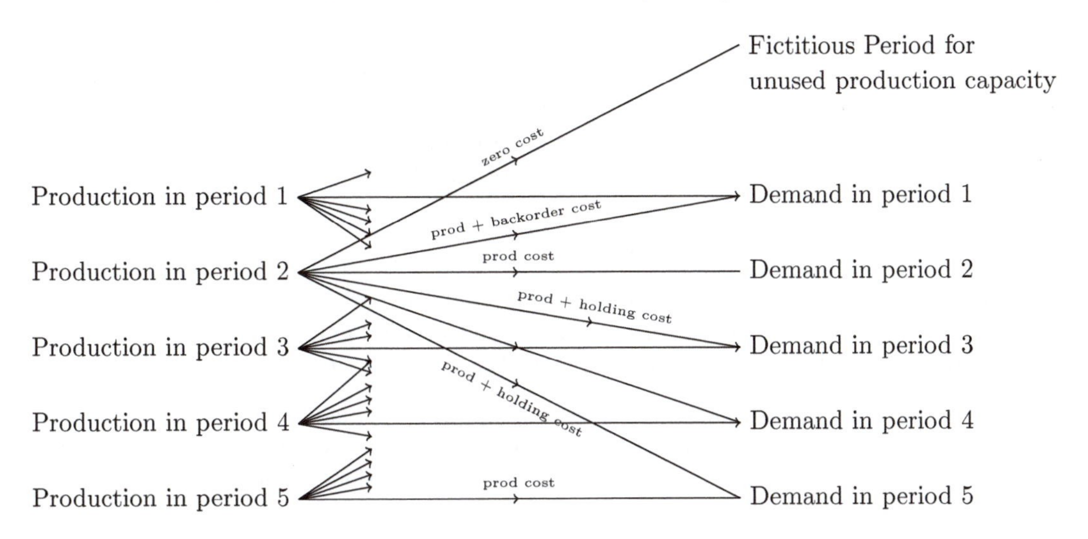

FIGURE 11.6: Nodes on the left represent production; nodes on the right represent demand. Demand filled by production from a later time period is called a *backorder*. The cost on the corresponding arc equals the production cost plus the backorder penalty.

This first example is concrete in the sense that each node represents a physical location. In the next example, some of the nodes represent locations in time rather than in space.

Production with Backorders and Storage

A firm must plan the production levels of its product for each of the next 12 months. For each month, the following data are known: customer demand, unit production cost and capacity, minimum production level, selling price, storage cost per unit of product stored from that month to the next, penalty cost for late delivery of product in future months to meet demand this month.

The model: Let $t = 1, \ldots, |T| \in T$ index time periods. Let the data be as follows.

- d_t = # units that are demanded in period t;
- u_t = max # units that can be produced in period t;
- l_t = min # units that may be produced in period t;
- c_t = cost per unit of production in period t;
- b_{tz} = backorder penalty of delivering one unit demanded in period z in period t, for $z = 1, \ldots, t-1$.
- h_t = cost per unit to hold from period t to period $t+1$.

Create a graph $G = (V, E)$ with node set $V = \{W_t : t \in T\} \bigcup \{D_t : t \in T\} \bigcup D_0$. Nodes W_t represent production capability in period t; nodes D_t represent demand in period $t \in T$; node D_0 is a dummy node representing unused production capability. See Figure 11.6. For all $t \in T, z \in T$, arc (W_t, D_z) represents the use of production in period t to meet demand in period z. For $z = 0$ the arc represents unused production capability.

At node $W_t : t \in T$ set supply to u_t. At node $D_t : t \in T$ set supply to $-d_t$. At dummy node D_0 set supply to $\sum_{t \in T} u_t - d_t$, the excess production capacity. Set the lower bound on flow of each arc to 0. Enforce the lower bounds on production l_t by setting an upper bound $u_t - l_t$ on the flow on arc (W_t, D_0). All other arcs have no upper bound. For arcs $(W_t, D_t) : t \in T$ set the unit cost of flow to the production cost c_t. For arcs (W_t, D_z) where $z > t$, set the unit cost of flow to the production cost plus the holding cost, $c_t + \sum_{j=t}^{z-1} s_j$. For arcs (W_t, D_z) where $z < t$, set the unit cost of flow to the production cost plus the backorder penalty, $c_t + b_{tz}$.

Complications:

- Product aging: To apply a different cost to the $i+1$st period of an item's storage than to the ith, replace the s_t data with s_{tz} data akin to the b_{tz} data. If an item cannot be stored for more than k periods, exclude arcs W_t, D_j for $j > t + k$.
- Upper bound on holding capacity: If the number of units held from period t to $t+1$ may not exceed H_t, model holding as flow from W_t to W_{t+1} and put an upper bound of H_t on the flow on arc (W_t, W_{t+1}). This model alteration is not compatible with the preceding one. Were we modeling the problem as an LP, we could bound the sum of flows across several arcs to enforce the upper bound on storage capacity. This illustrates a limitation of network models. Upper bounds on flow may only be imposed on single arcs, not on sets of arcs.
- Overtime production: If in period t the upper bound u_t may be exceeded by paying workers overtime or outsourcing, create an additional node W_t' to represent the alternate means of production. As long as the production costs exceed those at W_t, the optimal solution will not send flow from W_t' to any $D_z : z \in T$ unless the flow on arc (W_t, D_0) equals 0.
- Not meeting demand: If it is permitted to not fulfill all demand in one or more periods t create a dummy node W_0 that represents non-delivery. Arcs from W_0 to D_t can have costs reflecting a per-unit penalty for not meeting demand, and can have upper bounds to limit the degree to which demand is not met, in period t.
- Other complications are taken up in the Problems.

Airplane Maintenance

During the next T days, an airline needs to have a_t engines ready for use at its hub on day $t = 1, 2, \ldots T$. After a day of use, an engine must undergo maintenance. The standard maintenance procedure costs c_3 dollars and takes 48 hours, so an assembly used on day t is ready for use on day $t + 3$. By paying overtime, the airline can have an assembly ready in 24 hours at cost $c_2 > c_3$ dollars. After use on a Sunday, Monday or Tuesday, the airline can pay $c_4 < c_3$ dollars to get the assembly ready in 72 hours because no weekend labor time is needed. The airline will have K_t assemblies ready to be used on days $t = 1, 2, 3$ based on prior schedules. What schedule of maintenance minimizes the total cost while meeting the requirements?

Note: The original published version in 1954 [160] was a fictitious problem about a caterer needing a daily supply of clean napkins. The purpose of the fiction was apparently to keep the solution of the real problem secret from competing airlines. However, the network model had already been formulated by some other researchers, and the pretense was fully exposed by 1956 [247].

The model: Create nodes r_t and $n_t : t = 1, 2, \ldots, T$ to represent assemblies ready at the beginning of day t and not ready at the end of day t, respectively. Set the supply of node r_t to K_t for $t = 1, 2, 3$. All other supplies equal 0. For each $t = 1, \ldots, T$ create the following arcs, except for those that would involve a non-existent node:

- (r_t, n_t) with lower bound a_t and cost 0. (It is) OK but not necessary to place an upper bound of a_t on the arc.) This arc represents use of assemblies on day t.
- (n_t, r_{t+3}) with cost c_3 and lower bound 0 to represent standard maintenance of assemblies that were used on day t.
- (n_t, r_{t+2}) with cost c_2 and lower bound 0 to represent overtime maintenance of assemblies that were used on day t.
- (r_t, r_{t+1}) with cost 0 and lower bound 0 to represent keeping an assembly that was ready for use on day t at the hub to be ready for use on day $t + 1$.
- (n_t, n_{t+1}) with cost 0 and lower bound 0 to represent keeping an assembly that needs maintenance as of the end of day t at the hub to still need maintenance at the end of day $t + 1$. This arc is needed because it might be cheaper, for example, to hold an assembly at the end of a Saturday to take advantage of the less expensive c_4 maintenance cost available later.
- For each $t = 1, 2, \ldots, T$ that corresponds to a Sunday, Monday or Tuesday create arc (n_t, r_{t+4}) with cost c_4 and lower bound 0 to represent less expensive maintenance of assemblies that were used on day t.

Instead of the arcs (r_t, r_{t+1}) and (n_t, n_{t+1}), I could alternatively have defined additional maintenance arcs $(n_t, r_{t+i}) : i > 3$ with cost c_3, arcs $(n_t, r_{t+i}) : i > 2$ with cost c_2, and similar additional arcs for the inexpensive maintenance. Why did I not choose this alternative? My primary reason is that it would greatly increase the number of arcs in the model. The model as given has $O(T)$ arcs. The alternative model would have a number of arcs proportional to T^2. There would have to be a very compelling reason to square the size of the model. The model I chose also lets us limit the numbers of ready and not-ready assemblies stored from one day to the next by putting upper bounds on the corresponding arcs. *Question: What is wrong with the model so far?*[5]

The model as stated is incomplete. The sum of the supplies is $\sum_{t=1}^{3} K_t$ rather than 0. This flaw reflects an incompleteness in the statement of the problem. In what condition must the assemblies be at the end of the T days? If they need not be ready for use, set the supply at node n_T equal to $-\sum_{t=1}^{3} K_t$. If K_{T+i} assemblies must be ready on days $T + i : i = 1, 2, 3$, create dummy nodes $r_{T+i} : i = 1, 2, 3$ with supplies $-K_{T+i} : i = 1, 2, 3$, respectively. Add the arcs as defined above of form (n_t, r_{T+i}). Balance out the remaining assemblies by setting supply at n_T equal to $\sum_{i=1}^{3} K_{T+i} - K_i$. However, if that value is strictly positive, more assemblies are needed after day T than the airline has, and the model is not valid.

Some complications are given in the Problems.

Mathematically Related Models

The *minimum cost circulation problem* is the minimum cost flow problem restricted to cases in which $s_i = 0 \; \forall i \in V$. This restriction is inconsequential. Any minimum cost flow instance can be converted to an equivalent circulation instance as follows: create a new node t; for all $v \in V$ such that $s_v > 0$, create an arc (t, v) with $l_{tv} = u_{tv} = s_v$ and zero cost $c_{tv} = 0$; for all $v \in V$ such that $s_v < 0$, create an arc (v, t) with $l_{vt} = u_{vt} = -s_v$ and zero cost $c_{vt} = 0$; set $s_v = 0 \; \forall v \in V \bigcup \{t\}$. Figure 11.7 illustrates the transformation.

Node capacities are upper bounds on the total flow through a node. Such bounds are realistic in some applications, but they are not included in our definition of the minimum

[5]Read on.

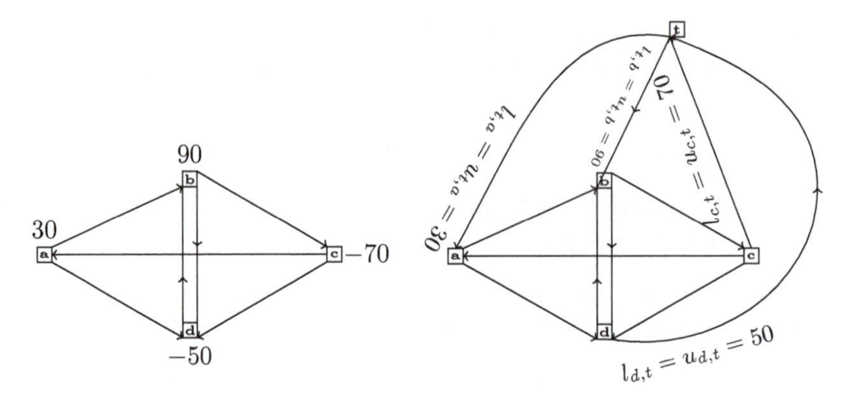

FIGURE 11.7: Replace node supplies and demands with forced flows from and to t to convert a flow instance into a circulation instance.

cost network flow problem. Model a capacity of K at node v as follows: replace v with two nodes, v^{in} and v^{out}. Replace every arc $(w, v) \in E$ going into v with an identical arc (w, v^{in}). Replace every arc $(v, w) \in E$ going out of v with an identical arc (v^{out}, w). Create a new arc (v^{in}, v^{out}) with cost 0, lower bound 0, and upper bound K. Figure 11.8 illustrates the transformation.

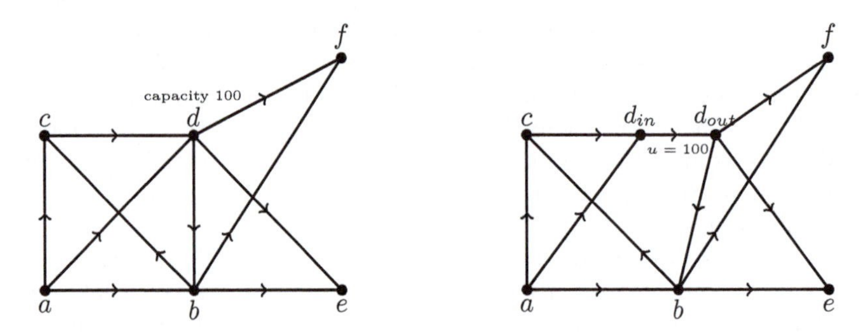

FIGURE 11.8: Node d has a capacity of 100. Replace it with an "in" and an "out" copy. Link the two copies with an arc that has flow upper bound 100.

The *transportation problem* is the minimum cost flow problem restricted to cases in which the only arcs are from supply nodes to demand nodes. In mathematical notation, the graph $G = (V, E)$ is such that $(i, j) \in E \Rightarrow s_i > 0$ and $s_j < 0$. Such a graph is called *bipartite* because V can be partitioned into two sets U and W such that all arcs go between U and W. The transportation problem has only historical interest because its specialized algorithms do not perform significantly better than do general minimum cost flow algorithms.

The *transshipment problem* generalizes the bipartite graphs of the transportation problem to graphs whose nodes are partitioned into K sets, or layers, $V^1, V^2, \ldots, V^K$ such that $(i, j) \in E \Rightarrow i \in V^k, j \in V^{k+1}$ for some k. That is, all arcs go between consecutive layers. Like the transportation problem, the transshipment problem possesses mainly historic interest. It might be of some theoretical interest that any minimum cost flow instance can be converted to an equivalent transshipment problem, and that any transshipment problem can be converted to an equivalent transportation problem [20].

Summary: The minimum cost network problem with bounds on arc flows readily models other network variants such as minimum cost circulation. It also readily accommodates complications such as node capacities, excess supply, and excess demand.

11.2 Integrality and Duality

Preview: All network problems have optimal primal and dual solutions that are integer, as long as the data are integer. This phenomenon is an example of the *total unimodular* property, or *TU* for short, whereby all basic solutions are integer.

Definition 11.2 (Totally Unimodular (TU)) *A matrix A is totally unimodular (TU) iff every square submatrix of A has determinant $0, 1$ or -1.*

Proposition 11.2.1 *Let A be TU. Then*

1. *A^T is TU.*
2. *The block matrices $[A|I]$ and $[A|-I]$ are TU.*
3. *The inverse of every nonsingular square submatrix of A has each of its entries equal to $1, -1$ or 0.*

Proof: The first statement is true because the determinant of a matrix equals the determinant of its transpose. The second statement is true because for every submatrix M of the block matrix that contains one or more columns not from A, repeated expansion by minors along those columns reduces the determinant of M to $0, 1$ or -1 times the determinant of a submatrix of A. To prove the third statement, let M be a $k \times k$ nonsingular square submatrix of A. Then the determinant of M equals ± 1. Each term of M^{-1} is therefore a product, ± 1 times the determinant of a $(k-1) \times (k-1)$ submatrix of M, which latter must equal $0, 1$ or -1. Therefore, the product equals $0, 1$, or -1. ■

The flow conservation constraints are so common and so important that we define a special matrix called the *arc-node incidence matrix* to represent them.

Definition 11.3 (Arc-Node Incidence Matrix) *The arc-node incidence matrix A^G of graph $G = (V, E)$ has $|V|$ rows and $|E|$ columns. It is defined with respect to an ordering of the nodes $v \in V$ and an ordering of the arcs $e \in E$. Its values are defined as follows:*

$$A^G_{v,e} = \begin{cases} 1 & \text{if } e = (v, j) \text{ for some } j \in V \\ -1 & \text{if } e = (j, v) \text{ for some } j \in V \\ 0 & \text{otherwise.} \end{cases}$$

In words, the column for arc (i, j) has a positive 1 in row i, a negative 1 in row j, and is 0 everywhere else. Dually, the row for node v has a positive 1 for each arc leaving v and a negative 1 for each arc entering v. *When it is clear what G is from context, we drop the superscript on A^G and write A.*

Example 11.2 Let $V = \{1,2,3,4\}$ and let $E = \{(1,2),(2,3),(3,4),(4,1),(1,3),(3,2),(2,1)\}$ as in Figure 11.9. For graph $G = (V, E)$ the arc-node incidence matrix A^G is:

$$\begin{bmatrix} 1 & 0 & 0 & -1 & 1 & 0 & -1 \\ -1 & 1 & 0 & 0 & 0 & -1 & 1 \\ 0 & -1 & 1 & 0 & -1 & 1 & 0 \\ 0 & 0 & -1 & 1 & 0 & 0 & 0 \end{bmatrix}$$

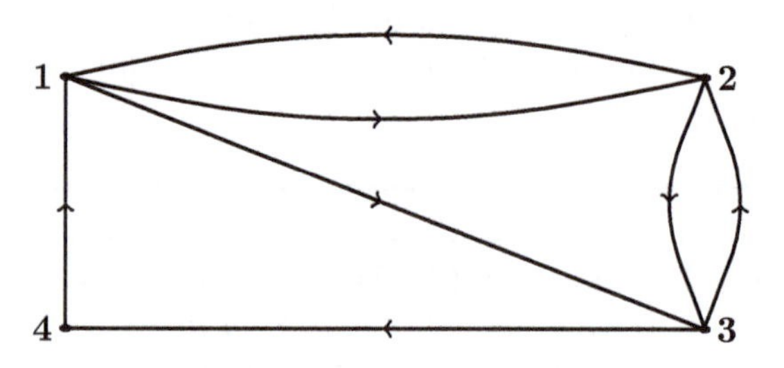

FIGURE 11.9: The arc-node incidence matrix for this graph has 4 rows and 7 columns.

In terms of A^G the minimum cost flow problem has the following LP formulation:

$$\min \sum_{e \in E} c_e x_e$$

subject to

$$A^G \mathbf{x} = \mathbf{s}$$
$$\mathbf{l} \le \mathbf{x} \le \mathbf{u}$$

For all graphs $G = (V, E)$ with $|E| \ge 1$ the arc-node matrix A^G is singular because $1^T A^G = 0$. That is, each column of A^G sums to 0. Each of the n equations $A^G \mathbf{x} = \mathbf{b}$ is implied by the other $n - 1$ equations. Therefore, these n equations only contribute $n - 1$ linearly independent binding constraints to basic solutions to LPs with constraints $A^G \mathbf{x} = \mathbf{b}; \mathbf{l} \le \mathbf{x} \le \mathbf{u}$. The upper and lower bounds $\mathbf{l} \le \mathbf{x} \le \mathbf{u}$ must contribute $|E| - |V| + 1$ binding constraints. Hence there will be only $|V| - 1$ basic variables, variables permitted to be unequal to their upper and lower bounds, in the LP. This may be confusing, but don't worry about it. Keep in mind that it is not important either theoretically or computationally. It causes awkwardness in some older expositions of LP that define basic solutions of $A\mathbf{x} = \mathbf{b}, \mathbf{x} \ge \mathbf{0}$ as having $n - m$ variables "fixed" at 0 and m "free" variables (where A is an $m \times n$ matrix).

Theorem 11.2.2 *Let A be the arc-node incidence matrix of graph G. Then A is TU.*

Proof: By contradiction, let $B = A_{K', J'}$ be a minimal square submatrix of A whose determinant is not $0, 1$ or -1. Let k' be the number of rows of B. Since every entry of A is $0, 1$

or -1, $k' \geq 2$. By minimality, every proper square submatrix of B (that is, every square submatrix of B except B itself) has determinant $0, 1$ or -1.

The matrix B cannot contain a column that has fewer than 2 nonzero elements. If it did, since $k' \geq 2$, expansion by minors along that column would compute the determinant of B to be either 0, or ± 1 times the determinant of a proper square submatrix of B, which would equal ± 1 times $0, 1$ or -1, a contradiction.

Since every column of A has only two nonzero elements, one 1 and one -1, every column of B contains one 1 and one -1. Hence the sum of the terms in each column equals 0, B is singular, and B has determinant 0, a contradiction. ■

Reading Tip: *Make sure you understand the use of minimality, which is essential to the proof.*

Theorem 11.2.3 *Let A^G be the arc-node incidence matrix of a graph $G = (V, E)$, let $\mathbf{b}$ be an integer vector of dimension $|V|$, and let $\mathbf{l}, \mathbf{u}$, be integer vectors of dimension $|E|$. Then every basic solution to the primal system $A^G\mathbf{x} = \mathbf{b}, \mathbf{l} \leq \mathbf{x} \leq \mathbf{u}$ is integer. If $\mathbf{c}$ is an $|E|$-dimensional integer vector, then every basic solution to the dual system $A^T\boldsymbol{\pi} + I\iota - I\mu = \mathbf{c}, \iota \geq \mathbf{0}, \mu \geq \mathbf{0}$ is integer, even if $\mathbf{b}, \mathbf{l}, \mathbf{u}$ are not integer.*

Corollary 11.2.4 *Let $A, \mathbf{b}, \mathbf{l}, \mathbf{u}$ be as in the theorem and let $\mathbf{c} \in \Re^n$ be any cost vector. Suppose the LP $\min \mathbf{c}^T\mathbf{x}$ subject to $A\mathbf{x} = \mathbf{b}, \mathbf{l} \leq \mathbf{x} \leq \mathbf{u}$ has an optimal solution. Then it has an optimal integer solution. If instead of $\mathbf{b}, \mathbf{l}, \mathbf{u}$ being integer, the vector $\mathbf{c}$ is integer, then the dual LP has an optimal integer solution.*

Proof: By parts (i) and (ii) of Proposition 11.2.1 and Theorem 11.2.2 the constraint matrix

$$\begin{bmatrix} A \\ I \\ -I \end{bmatrix}$$

is TU. By definition of TU, the constraint matrix remains TU after the redundant constraint is removed. By part (iii) of Proposition 11.2.1, any choice of basic variables yields solution $B^{-1}(b, l, u)$ where B^{-1} has all integer entries. Since the RHS vector (b, l, u) is integer by assumption, the basic solution is integer. The dual's constraint matrix is the transpose of the primal's constraint matrix. Hence by part (i) of Proposition 11.2.1 the former is TU. By the same reasoning as for the primal, the integrality of the vector $\mathbf{c}$ implies that the basic solutions are integer. The corollary then follows from Theorem 4.4.8. ■

Question: In the proof of the theorem, exactly why are the primal constraints TU?[6]

Later we will combine network integrality and strong LP duality to prove several duality theorems.

Summary: The constraint matrix of a min-cost flow problem is an arc-node incidence matrix. Such matrices are totally unimodular. Consequently, all basic solutions are integer if node supplies and arc flow bounds are integer.

[6]Use part (i) on A to prove A^T TU; then use part (ii) on A^T to prove $[A^T|I]$ TU; then use part (ii) again on $[A^T|I]$ to prove $[A^T|I|-I]$ TU; then use part (i) to get the result.

11.3 Network Simplex Algorithm

Preview: The simplex algorithm will solve minimum cost network problems correctly. This section derives a specialized version called the network simplex algorithm which will perform the same steps as the general version, but much faster. Many software implementations for LP include this algorithm. As a result, network problems that are a few orders of magnitude larger than ordinary LPs can still be solved in a reasonable amount of time. The derivation is a good review of the ideas of the simplex algorithm.

11.3.1 Preliminary Graph Concepts

This section defines basic concepts about graphs and states some of their well-known properties.

Definition 11.4 (Incident, Degree, Outdegree, Indegree) *In a graph $G = (V, E)$, the arc $(v, w) \in E$ is said to point or go from v to w. The arc (v, w) is said to be* incident *on nodes v and w, and nodes v and w are said to be* incident *on arc (v, w). The* degree *of a node is the number of arcs incident on it. The outdegree of $v \in V$ is the number of arcs $(v, w) \in E$ that point from v; the indegree of v is the number of arcs $(w, v) \in E$ that point to v.*

Proposition 11.3.1 *In the graph $G = (V, E)$ let $d(v)$ = the degree of node $v \in V$. Then $\sum_{v \in V} d(v) = 2|E|$.*

Proof: This is easy to see, because each arc in E contributes twice to the sum of the node degrees. However, it is instructive to prove this property with duality. Let $f : V \times E \mapsto \{0, 1\}$ be the incidence function defined as $f(v, e) = 1$ if node v and arc e are incident, and 0 otherwise. Then

$$\sum_{v \in V} d(v) = \sum_{v \in V} \sum_{e \in E} f(v, e) = \sum_{e \in E} \sum v \in V f(v, e) = \sum_{e \in E} 2 = 2|E|.$$

■

Definition 11.5 (Path, Simple Path, Cycle) *A sequence of k arcs $e_1, e_2, \ldots, e_k$ in a graph $G = (V, E)$ is a* path of length k from v to w *if there exist nodes $v = v_1, v_2, \ldots, v_{k+1} = w$ such that $e_i = (v_i, v_{i+1})$ for $i = 1, \ldots, k$. If all v_i are distinct, the path is* simple. *If all v_i are distinct except $v_1 = v_{k+1}$, the path is a cycle.*

In some texts, what is defined here as a path (respectively cycle) is called a *directed path* (respectively *directed cycle*).

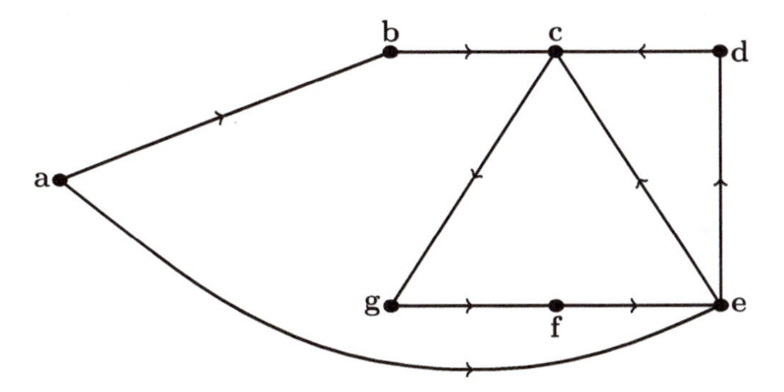

FIGURE 11.10: The sequence of nodes a, b, c, g, f, e corresponds to a simple path; sequence a, b, c, d, e, f, g corresponds to a simple orientable path; sequence d, c, e, d corresponds to an orientable cycle. When you describe a cycle such as that corresponding to the node sequence c, g, f, e, c it doesn't matter where you start.

Often we will use sequences of arcs that could be directed paths if we reversed the way some of the arcs point. We call these sequences *orientable paths.*

Definition 11.6 (Orientable Path and Cycle) *A sequence of k arcs $e_1, e_2, \ldots, e_k$ is an* orientable path *from v to w iff e_1 is incident on v, e_k is incident on w, and for all $i = 1, \ldots, k-1$ there exists a node incident on both arcs e_i and e_{i+1}. If no three arcs are incident on the same node and $v \neq w$ the orientable path is* simple. *If no three arcs are incident on the same node and $v = w$ the sequence is an* orientable cycle.

Figure 11.10 illustrates a path, an orientable path, and a cycle in a graph.

Definition 11.7 (Connected, Strongly Connected) *A graph $G = (V, E)$ is* connected *iff for every distinct $v \in V, w \in V$ it contains an orientable path from v to w. A graph G is* strongly connected *if for all distinct nodes $v \in V, w \in V$ it contains a path from v to w.*

Definition 11.8 (Tree, Spanning Tree) *A tree is a graph that is connected and has no orientable cycles. The arcs of a spanning tree of a graph $G = (V, E)$ compose a subset $T \subset E$ of the arcs such that the graph (V, T) is a tree.*

Example: Figure 11.12 shows several examples of trees. They would be trees regardless of the orientations of their arcs.

Proposition 11.3.2 *A graph $G = (V, E)$ is a tree iff for every pair of distinct vertices $v, w \in V$ there is a unique orientable simple path between v and w.*

Reading Tip: *Try to prove Proposition 11.3.2 yourself before reading further. It is so easy I almost made it an exercise, but I hate it when other authors do that.*

Proof: *($\Rightarrow$:) Suppose G is a tree. By definition, G is connected and therefore for all $v \in V, w \in V, v \neq w$ there exists an orientable simple path between v and w. If for some*

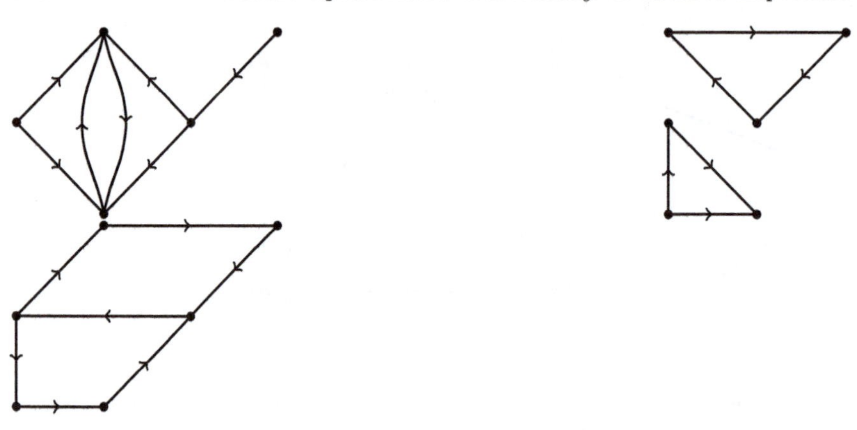

FIGURE 11.11: The graph on the left is connected but not strongly so; the graph in the middle is not connected; the graph on the right is strongly connected.

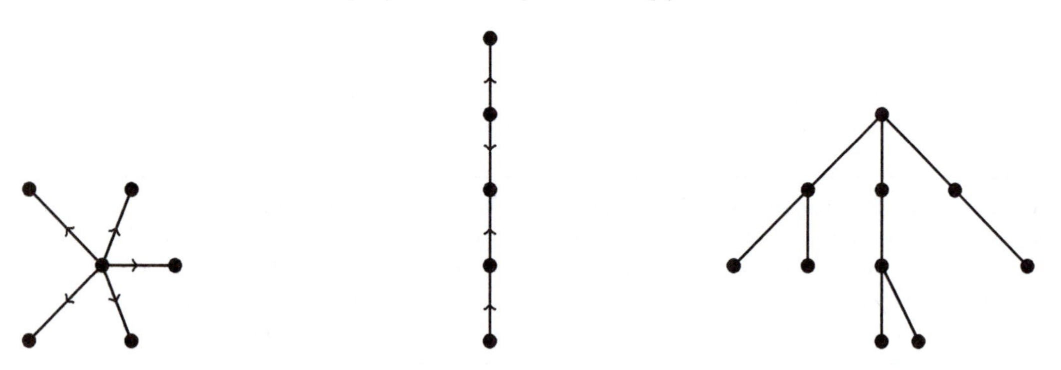

FIGURE 11.12: A tree always has one less arc than it has nodes. It is always connected and never contains an orientable cycle. The rightmost tree's arcs are undirected to emphasize that orientation does not affect the tree property.

pair v, w there exist two such paths, their concatenation must contain an orientable cycle, a contradiction. Hence the path is unique.
($\Leftarrow$:) If for every vertex pair $v \neq w$ there is a unique orientable path between them, then by definition G is connected. Suppose by the contrapositive that G contains an orientable cycle C. Denote the vertices that C traverses as $v_1, v_2, \ldots, v_k, v_1$. Then the edge between v_1 and v_2 is an orientable simple path between v_1 and v_2, as is the path that traverses the vertices $v_2, \ldots, v_k, v_1$. That contradicts the uniqueness of the path between v_1 and v_2. ■

Proposition 11.3.3 *Let $G = (V, E)$ be a tree. Then $|E| = |V| - 1$.*

Proof: *Let (u, v) be any arc in E. Remove it from E. In the resulting graph $\hat{G}$ there is no orientable simple path from u to v, because the removed arc is the unique such path in G. Also, $\hat{G}$ has no orientable cycle because G has none. On the other hand, for every $w \in V, w \neq u, w \neq v$, w must be connected to either u or v in $\hat{G}$. For if not, both the path from w to u and the path from w to v in G contain the arc (u, v), contradicting path uniqueness. Therefore, $\hat{G}$ comprises two trees of at least one vertex each. By induction on $|V|$, $\hat{G}$ has $|V| - 2$ edges, whence G has $|V| - 1$.* ■

Proposition 11.3.4 *A tree with at least two nodes contains at least two nodes of degree* 1. *Such nodes are called* leaves.

Proof: *Let P be a maximal orientable path in the tree. By definition P cannot contain a cycle. Hence it must be a simple path. Its end points must be distinct leaves.* ■

Here is a useful theorem about flows in a network.

Theorem 11.3.5 *Flow Decomposition. Let* $\mathbf{x}$ *be a feasible solution to a network problem* (11.6) *on a graph* $G = (V, E)$. *Then the flow* $\mathbf{x}$ *can be decomposed into at most* $|E| + |V|$ *flows, each of which is a cycle, or a path beginning at a node with positive supply and terminating at a node with negative supply.*

Proof: Let $V^\circ \subset V$ be the nodes with nonzero supply. The statement of the theorem is true when $|E|+|V^\circ| = 0$. By induction on $|E|+|V^\circ|$, $\mathbf{x}$ contains $|E|$ strictly positive components, for any arc with zero flow can be removed from the graph. Find a path or cycle in G with the following procedure.

Step 1 Select any $(i, j) \in E$ for which i has maximum supply and set path $P = (i, j)$.

Step 2 If no arc leaves the last node k in P, stop. Otherwise, select any arc (k, v) and append it to P.

Step 3 If v has already appeared once in P, stop. Otherwise, go to Step 2.

Case I: If the procedure terminates at Step 2, P is a path. The last node of P must have strictly negative supply, for otherwise by flow balance there would be an arc leaving it. The first node of P must have strictly positive supply since the sum of supplies in the graph equals 0. Hence the path has the properties as claimed in the statement of the Theorem. Case II: If the procedure terminates at Step 3, the part of P from the first occurrence of v to the second occurrence (at the end of P) is a cycle. Discard the portion of P from its first arc until the first occurrence of v.

Both cases: Now that P is found, let Δ equal the minimum flow on any arc in the path or cycle P. If P is a path, let Θ equal the minimum of the absolute values of the supplies at the first and last nodes of P, and set $\Delta = \min\{\Delta, \Theta\}$.

Subtract Δ from the flow on each arc in P. If P is a path subtract Δ from the supply at the start node of P and add Δ to the supply at the end node of P. Remove all arcs whose flow now equals 0. The remaining flow is valid on the remaining graph, which has either fewer arcs or fewer nodes with nonzero supply. This completes the induction. ■

11.3.2 Every Basis is a Spanning Tree

The first thing to understand about networks as LPs is that a basis is always a tree, in fact, a spanning tree.

Let A be an arc-node incidence matrix as per Definition 11.3 for a connected graph $G = (V, E)$ whose nodes are labeled $1 \dots n$. *Question: Why may we assume that G is connected?*[7]

[7] If not, any minimum cost network problem on G decomposes into separate problems, independent of each other, one for each connected component of G. Another answer: There will be more than one redundant constraint (one for each connected component) if G is not connected.

Let $\mathbb{B}$ be the indices of a subset of the columns of A. If the arcs in G corresponding to $\mathbb{B}$ contain a cycle, the columns $A_{:,\mathbb{B}}$ are not linearly independent.

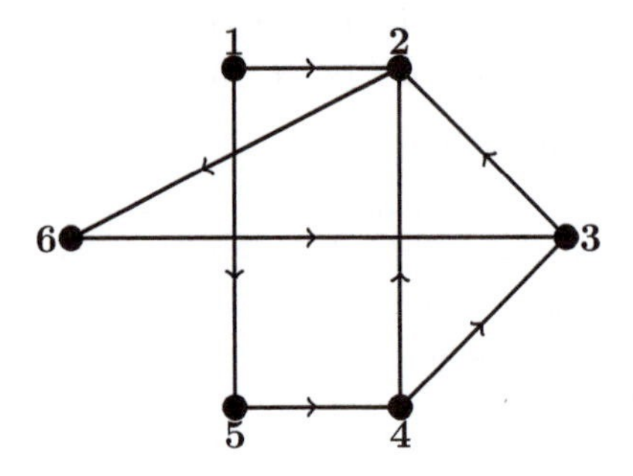

FIGURE 11.13: The orientable cycle 26342 causes linear dependence even though its arcs have inconsistent orientations.

Example 11.3 The directed cycle $2, 6, 3, 2$ in Figure 11.13 corresponds to the three columns

$$A_{\mathbb{B}} = \left[\begin{array}{c|ccc} & \mathbf{(2,6)} & \mathbf{(6,3)} & \mathbf{(3,2)} \\ \mathbf{1} & 0 & 0 & 0 \\ \mathbf{2} & 1 & 0 & -1 \\ \mathbf{3} & 0 & -1 & 1 \\ \mathbf{4} & 0 & 0 & 0 \\ \mathbf{5} & 0 & 0 & 0 \\ \mathbf{6} & -1 & 1 & 0 \end{array}\right].$$

The sum of the three columns is 0 since each 1 is canceled out by a -1. The directed cycle leaves each node once (giving a 1) and enters each node once (giving a -1). Hence the columns of $A_{\mathbb{B}}$ are linearly dependent.

The cycle $2, 6, 3, 4, 2$ corresponds to the four columns

$$A_{\mathbb{B}} = \left[\begin{array}{c|cccc} & \mathbf{(2,6)} & \mathbf{(6,3)} & \mathbf{(4,3)} & \mathbf{(4,2)} \\ \mathbf{1} & 0 & 0 & 0 & 0 \\ \mathbf{2} & 1 & 0 & 0 & -1 \\ \mathbf{3} & 0 & -1 & -1 & 0 \\ \mathbf{4} & 0 & 0 & 1 & 1 \\ \mathbf{5} & 0 & 0 & 0 & 0 \\ \mathbf{6} & -1 & 1 & 0 & 0 \end{array}\right].$$

Traverse the cycle, taking note of the arcs that point in reverse. Arcs $(2, 6)$ and $(6, 3)$ point forwards, arc $(4, 3)$ points in reverse, and arc $(4, 2)$ points forwards. Multiply by -1 the columns of $A_{\mathbb{B}}$ that correspond to the reverse arcs, giving

$$\begin{bmatrix} 0 & 0 & 0 & 0 \\ 1 & 0 & 0 & -1 \\ 0 & -1 & 1 & 0 \\ 0 & 0 & -1 & 1 \\ 0 & 0 & 0 & 0 \\ -1 & 1 & 0 & 0 \end{bmatrix}.$$

Now the columns sum to 0. This proves the columns of $A_{\mathbb{B}}$ are linearly dependent.

Proposition 11.3.6 *Let A be the arc-node incidence matrix of a connected graph $G = (V, E)$. Every set of $|V| - 1$ linearly independent columns of A corresponds to a spanning tree of G, and vice versa.*

Proof: Let B denote a set of $|V| - 1$ arcs in G, represented by corresponding linearly independent columns of A. Suppose B contained an orientable cycle. Traverse the orientable cycle, labeling each arc "plus" or "minus" depending on whether the arc points forward or backward when traversed. The sum of the "plus" columns minus the sum of the "minus" columns must equal the 0 vector. Therefore, B's columns would not be linearly independent, a contradiction. Hence B does not contain an orientable cycle. By Proposition 11.3.2, the $|V| - 1$ columns of B correspond to a spanning tree.

For the converse, suppose T is a spanning tree on G. By Proposition 11.3.2 it has $|V|-1$ arcs, corresponding to $|V| - 1$ columns of A. Since the rank of A equals $|V| - 1$, these columns form a basis if they are linearly independent. By contradiction, suppose they are linearly dependent. For $|V| = 2$ the singleton column would not be a linearly dependent set. By induction, suppose $|V| \geq 3$ is the smallest possible value for which some spanning tree's columns are linearly dependent. We will construct a smaller graph and spanning tree whose columns are not linearly independent to get a contradiction. Let

$$\sum_{e \in T} \pi_e A_{:,e} = 0$$

be a nonzero weighted sum of the columns of T that equals 0. By Proposition 11.3.4 T has a leaf $v \in V$. Since only one arc of T is incident on v, row v of $A_{:,T}$ contains exactly one nonzero term, which is in the column corresponding to the arc incident on v. Then $\pi_v = 0$ since no other term in the weighted sum can cancel out the nonzero term in row v. Let f be the arc of T incident on leaf v. Then

$$\sum_{e \in T \setminus \{e\}} \pi_e A_{:,e} = 0.$$

Therefore, the smaller graph $G = (V \setminus \{v\}, E \setminus \{f\})$ has spanning tree $(V \setminus \{v\}, T \setminus \{f\})$ whose columns are linearly dependent. ∎

11.3.3 Computing Basic Solutions

In any LP computing a basic solution is computationally straightforward because it requires merely the solution of simultaneous linear equations. In network problems, computing a basic solution is especially easy and quick because a basis is a tree (Proposition 11.3.6) and a tree always has a leaf (Proposition 11.3.4). Let's look first at the common simple case where all lower bounds are 0 and there are no upper bounds. The idea is to start at a leaf and work inwards into the tree. Why? Because if only one arc is incident on a node v, the supply at v dictates the flow on that arc.

Algorithm 11.3.7 (Network Basic Solutions with $L = 0$ and no upper bounds)
Step 1. Set initial supply values $t_v := s_v$ for all $v \in V$.

Step 2. Choose an arc (i, j) incident on a leaf node.
Step 3. If i is a leaf set $x_{ij} = t_i$ and adjust $t_j := t_j + t_i$. Remove i and (i, j) and go to
Step 4. Otherwise j is a leaf: set $x_{ij} := -t_j$, adjust $t_i := t_i - t_j$; remove j and (i, j); go to

Step 4.

Step 4. If no arcs remain, stop. Otherwise return to Step 2.

If the basic variables all take nonnegative values, the basic solution is feasible.

Example 11.4 Figure 11.14 shows the spanning tree forming the choice of basis. Numbers at the nodes are the supply values, initially equal to the actual supply values s_v. At each iteration one leaf is removed and the flow on its incident arc is fixed. In Figure 11.15 the graph on the left shows the choice of basis and supply values, and the graph on the right shows the basic solution values.

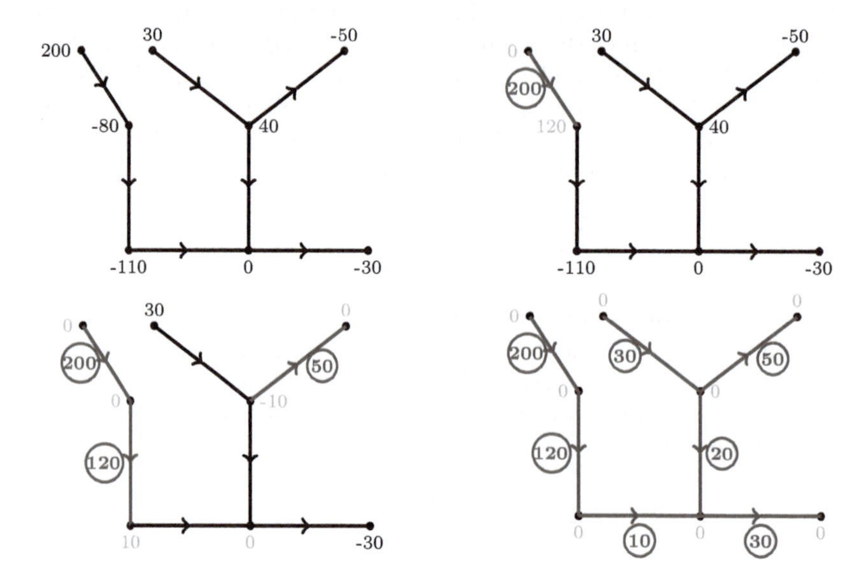

FIGURE 11.14: Apply the flow balance constraints to the supply values to compute basic solutions when $L = \mathbf{0}$ and there are no upper bounds. The first graph (upper left) shows the initial spanning tree and supplies; the second graph (upper right) shows the tree after one basic variable has been calculated. The lower left (respectively lower right) graph shows the tree after three (respectively seven) basic variables have been calculated.

Now let's look at the general case of lower bounds L and upper bounds U on arc flow. As in the simplex algorithm for LPs with general upper and lower bounds, each nonbasic variable x_{ij} is at its lower bound l_{ij} or upper bound u_{ij}. The basic variables must take values that will then satisfy the equations $A\mathbf{x} = s$. Consider the small example in Figure 11.15. The basic variables are x_{ab} and x_{bc}. The supply vector is $(100, -400, 300)$. The nonbasic variable is x_{ca}. If x_{ca} were at a lower bound of 0, the basic variable values would have to be $x_{ab} = 100$ and $x_{bc} = -300$. But if x_{ca} were at an upper bound of 444, the flow of 444 from c to a would force the flow x_{ab} to increase by 444, which then would force the flow x_{bc} to increase by 444. The basic variable values would be $x_{ab} = 544$ and $x_{bc} = 144$. *Question: What would the basic variable values be if x_{ca} were at a lower bound of 444?*[8] We would have arrived at the same values had we increased the supply at a by 444 and decreased the

[8]The same.

supply at c by 444. In general, we adjust the supply vector s to incorporate the values of the nonbasic variables, and then compute the basic variable values with Algorithm 11.3.7.

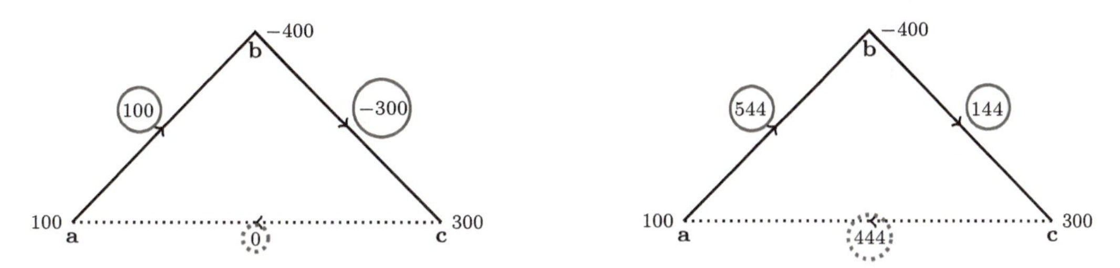

FIGURE 11.15: If nonbasic variable x_{ca} equals 444, supply at node a in effect increases by 444, while supply at node c in effect decreases by 444.

Algorithm 11.3.8 (Network Basic Solutions) *Execute Algorithm 11.3.7, with the following modification: After Step 1, execute the following step.*

Step 1.5: For each nonbasic arc (i, j)*, adjust* $t_j := t_j + \Delta$ *and* $t_i := t_i - \Delta$ *where* Δ *is the value at which* x_{ij} *is fixed (either* L_{ij} *or* U_{ij}*).*

If the basic variables all take values within their bounds, the basic solution is feasible.

Example 11.5 Figure 11.16 shows an example of Algorithm 11.3.8. In this example, nonbasic variable x_{ce} is at its lower bound $L_{ce} = 3$, and nonbasic variables x_{ea} and x_{bc} are at their upper bounds $U_{ea} = 10$ and $U_{bc} = 8$. The supply values t_a, t_b, t_c, t_d, t_e of $2, -5, -12, 22, -7$ respectively are adjusted to the values $12, -13, -7, 22, -14$, respectively. *Question: How can you know in advance that the value of* t_d *doesn't change?*[9] Once the t_j values have been adjusted, work inwards from the tree's leaves to find the basic variable values. Node a has supply 12 and is incident on only one basic arc, (a, b). Hence $x_{ab} = 12$. Similarly, $x_{de} = 14$ and $x_{dc} = 7$. There are two ways to compute the last basic variable x_{db}, because there is always a redundancy in the flow balance constraints. At node d, we get $x_{db} = 22 - 7 - 14 = 1$. At node b, we get $x_{db} + x_{ab} = 13 \Rightarrow x_{db} = 1$. I like to do both computations as a check for arithmetic errors.

By Proposition 11.3.6, every basic solution to a network problem can be computed with Algorithm 11.3.8. That algorithm employs only addition and subtraction of components of the vectors s, L, and U. Therefore, if s, L, U are all integers, every basic solution is integer. Thus Algorithm 11.3.8 provides an alternative proof of the first part of Corollary 11.2.4.

Algorithm 11.3.7 runs in time $O(|V|)$, a much smaller time than what it usually takes to solve $|V|$ equations in $|V|$ unknowns. In matrix terms, the basic columns form a triangular matrix because trees always have leaves. The first step of Gaussian elimination is already done for us. Algorithm 11.3.8 runs in time $O(|E|)$ at first glance because all of the nonbasic arc flows must be taken into account. However, if basic solutions are computed repeatedly during the course of the simplex algorithm, only the first computation will require $O(|E|)$ time. Subsequent iterations can run in time $O(|V|)$. Retain the modified supply values t_v at the end of Step 1.5. An iteration of the simplex algorithm will change at most 4 of these values, 2 for the entering variable and 2 for the exiting variable.

[9] It is not incident on any nonbasic arcs.

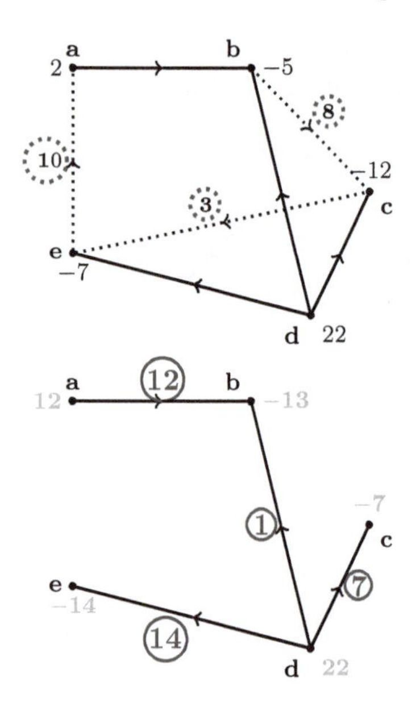

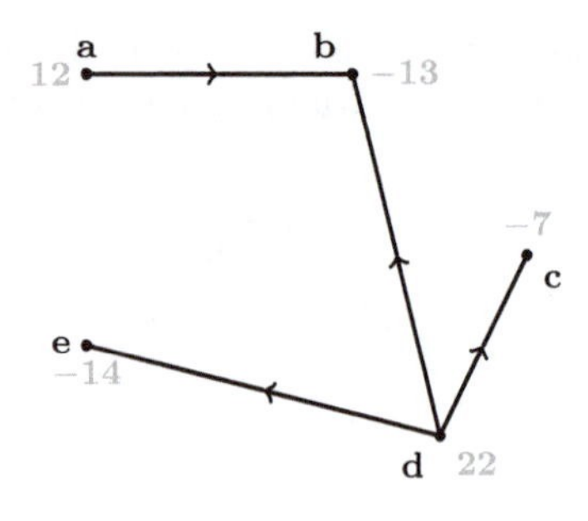

FIGURE 11.16: Flows on nonbasic arcs, shown within dotted circles, force the node supplies to be adjusted prior to computing the circled basic variable values.

Until roughly the mid-1990s, Algorithm 11.3.8 enjoyed an additional speed advantage over general algorithms to solve simultaneous linear equations. Addition and subtraction of integers used to be several times faster than real number computations, especially division. Circuitry to support real arithmetic operations has now been so optimized that this is no longer true.

You can see mathematically that basic solutions to the network problem are integers because the constraint matrix is totally unimodular. You can see the integrality computationally because a basis is a tree, the tree of equations can be solved one variable at a time with addition and subtraction, and since the coefficients are all in $\{0, 1, -1\}$ the addition and subtraction maintain integrality.

11.3.4 Computing Reduced Costs

There are two main ways to compute reduced costs in a network problem, using *cycle completion* and using *dual constraints.* Cycle completion is visually intuitive and mathematically correct, but not computationally efficient. Using dual constraints is correct and efficient, but not as immediately intuitive. I will start with cycle completion. When you understand cycle completion, you will see how beautifully the dual variables capture the cost of enforcing the supply constraints. Then you will be ready to understand intuitively how to use dual constraints to compute reduced costs.

Let's begin with a simple example with lower bounds $L = \mathbf{0}$. Consider the nonbasic arc (a, g) in Figure 11.17.

To get from a to g via basic arcs, we would use the orientable path $(a, b), (b, c), (g, c)$. Just as with the simplex method for LPs, consider increasing x_{ag} from 0 to θ. In the network, this means sending θ units of flow from a to g on arc (a, g). To maintain the flow balance constraints, we must decrease flow on arc (a, b) by θ. This change forces us to decrease flow on arc (b, c) by θ. Since arc (g, c) points in the other direction, we then must *increase* the flow on that arc by θ. Therefore, the reduced cost of x_{ag} equals $c_{ag} - c_{ab} - c_{bc} + c_{gc}$.

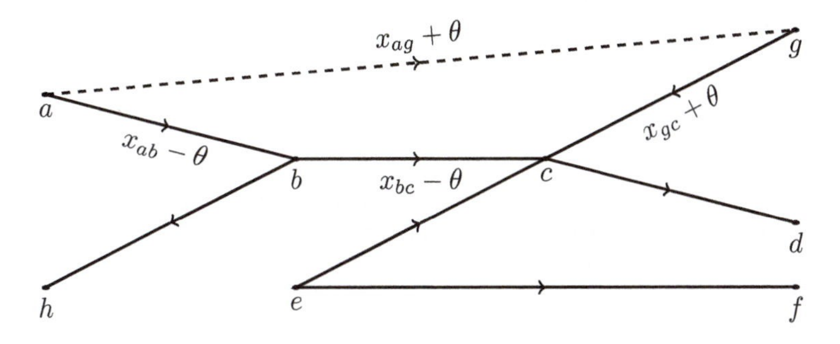

FIGURE 11.17: If flow on nonbasic arc (a, g) — shown as a dashed line — is increased, flow from a to g along the orientable path $(a, b), (b, c), (g, c)$ must be decreased to maintain flow balance.

When we send a unit of flow on arc x_{ag}, we substitute that flow for a unit of flow along basic arcs from a to g. The linear combination of basic variables $x_{ab} + x_{bc} - x_{gc}$ is the "synthetic carrot," in this case a synthetic unit of flow from a to g composed of unit flows on basic arcs. The cost of a synthetic unit of flow is $c_{ab} + c_{bc} - c_{gc}$. The reduced cost of x_{ag} is c_{ag} minus the cost of the synthetic flow. *Question: Write the reduced cost of nonbasic variables x_{eb} and x_{fa}.*[10]

The computation of reduced costs is the same in the general case of lower bounds L and upper bounds U. In the example just given, the values of the bounds never entered into our calculations. That is correct in general. In all LPs, the reduced cost vector $\mathbf{c} - \mathbf{c}_{\mathbb{B}}^T B^{-1} A$ is independent of the RHS vector. Bounds will only matter during the simplex algorithm, when the interpretation of a reduced cost as good or bad will depend on the sign of the reduced cost, whether we are minimizing or maximizing, and whether the nonbasic variable is at its lower or upper bound. We can now state the cycle completion algorithm for computing reduced costs.

Algorithm 11.3.9 (Cycle completion for reduced costs) *To compute the reduced cost of nonbasic arc variable x_{ij}:*
Step 1. Set $\pi_{ij} = c_{ij}$.

Step 2. Find the unique path P of basic arc variables from i to j.

Step 3. For each arc $(v, w) \in P$, adjust $\pi_{ij} := \pi_{ij} - c_{vw}$ if (v, w) points from i to j in P; adjust $\pi_{ij} := \pi_{ij} + c_{vw}$ if (v, w) points from j to i in P.

Justification of Algorithm: By Proposition 11.3.6 the basic arcs form a tree. Since a tree is connected, it contains a path P from i to j. Moreover, P is unique. If two distinct paths P, Q from i to j existed in the tree, the path P concatenated with the path Q in reverse would contain a cycle. Since a tree contains no cycles, this is impossible. The path P together with the arc (i, j) forms a cycle. The algorithm computes the cost of sending a unit of flow around that cycle. Increasing flow on every arc in a cycle, respecting orientation of the arcs, maintains all flow balance constraints. ■

Reduced Costs Via Dual constraints

It is more efficient to calculate reduced costs from the dual constraints than by cycle completion. This computation turns out to be very much like computing the primal variable values.

[10] $c_{eb} + c_{bc} + c_{ce}$ and $c_{fa} + c_{ab} + c_{bc} + c_{ce} - c_{fe}$.

Recall from (11.9) that the constraints in the dual LP for minimum cost flow are:

$$\pi_i - \pi_j + \iota_{ij} - \mu_{ij} = c_{ij} \ \forall (i,j) \in E \quad (11.12)$$

$$\pi_i \text{ unrestricted} ; \quad \iota_{ij}, \mu_{ij} \geq 0. \quad (11.13)$$

For the basic solution with basic variables from a spanning tree T, complementary slackness implies that in the corresponding dual basic solution $\iota_{vw} = \mu_{vw} = 0$ for all $(v,w) \in T$. Therefore, in the dual basic solution corresponding to basic variables T,

$$\pi_v - \pi_w = c_{vw} \ \forall (v,w) \in T. \quad (11.14)$$

Equation (11.14) has the following intuitive meaning:

> π_v is the cost of having an additional unit of supply at v when the arcs being used for additional transport are the basic arcs T.

To deal with an extra unit of supply at v we could transport it on arc (v,w) to w and then deal with the extra supply at w. That would cost c_{vw} for transport, plus whatever the cost of dealing with an extra unit of supply at w, which is π_w. Since the π values must be consistent with each other, $\pi_v = c_{vw} + \pi_w$. The same intuition can be stated in terms of demand. At node w, $-\pi_w$ is the cost of increasing demand by one unit. We could handle the increased demand at w by transporting a unit from v to w and then handling the resulting increased demand at v. The net cost would be $c_{vw} + (-\pi_v)$, since $-\pi_v$ is the cost of increasing demand at v by one unit. Hence $-\pi_w = c_{vw} - \pi_v$. I've called this an intuitive meaning, but you should see that it has a precise mathematical underpinning. The dual variable for the flow balance constraint at v is the instantaneous impact on the objective function of altering the RHS of that constraint, namely, the supply s_v.

Suppose we knew the value π_v for some node $v \in V$. By Equation (11.14) we could easily compute the value of π_w for every node w connected to v by an arc in T. Since T is connected, repeated use of Equation (11.14) would give us the values of all $\pi_j : j \in V$. Now the redundancy of the primal flow balance equations helps us. We may choose any one $j \in V$ and set π_j equal to any value we wish. This reasoning gives us our algorithm for computing the dual π values.

Algorithm 11.3.10 (Network Dual Basic Node Variables) :
Step 1. Select any $u \in V$ and set $\pi_u = 0$.

Step 2. Perform depth-first search on T from u, treating arcs as undirected to guide the search. When the search traverses node w, set π_w according to Equation (11.14).

In Step 2, depth-first search ensures that w is a leaf in the subtree of untraversed nodes of T. Since Algorithm 11.3.10 only performs addition and subtraction on the cost vector components $c_{ij} : (i,j) \in E$ it gives us an alternative proof of the second part of Corollary 11.2.4.

Example 11.6 For the spanning tree in Figure 11.18, set $\pi_1 = 0$. The resulting π values are shown on the left in Figure 11.19. If instead we select node 3 in Step 1, the resulting π values are shown on the right in Figure 11.19. The differences $\pi_i - \pi_j$ for all $i \in V, j \in V$ are the same in either case.

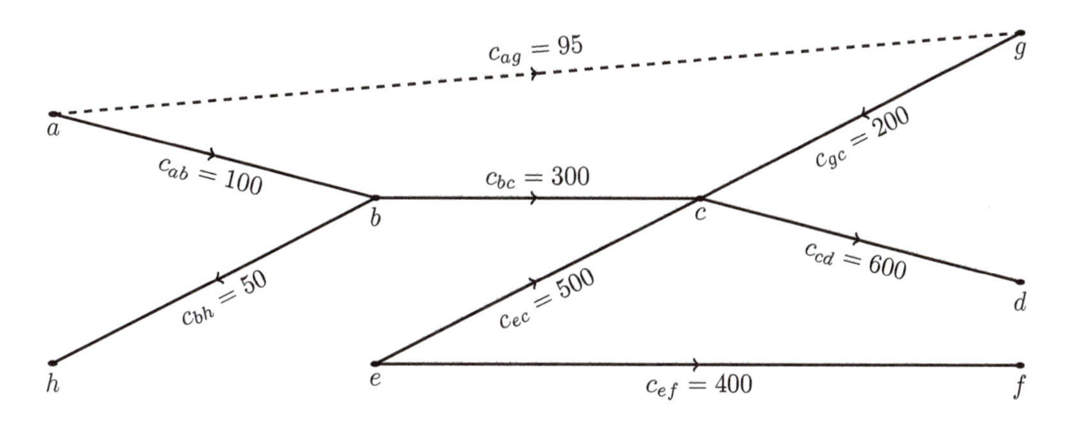

FIGURE 11.18: The cost coefficients are shown for each arc in the basis T, and for nonbasic arc (a, g). The reduced cost of x_{ag} is -105.

The reduced cost of any primal variable x_{ij} equals $c_{ij} - \pi^T A_{:,ij}$. By the definition of the arc-node incidence matrix (12.24),

$$\pi^T A_{:,ij} = \pi_i - \pi_j.$$

Therefore,

$$\text{The reduced cost of } x_{ij} = c_{ij} - \pi_i + \pi_j. \tag{11.15}$$

When x_{vw} is basic, i.e., $(v, w) \in T$, its reduced cost by (11.14) equals $(\pi_i - \pi_j) - \pi_i + \pi_j = 0$, as it should. We summarize the algorithm to compute reduced costs.

Algorithm 11.3.11 (Reduced Costs in a Network) *:*
Step 1. Use Algorithm 11.3.10 to compute the dual variable values π*.*

Step 2. For each nonbasic $(i, j) \in E$*, compute its reduced cost by Equation* (11.15) *as* $c_{ij} - \pi_i + \pi_j$*.*

Equation (11.15) has the following intuitive meaning, based on the intuitive meaning of Equation (11.14). The nonbasic arc (i, j) provides the opportunity to send flow directly from i to j at unit cost c_{ij}. If we send flow from i to j, in effect we increase the supply at j, and increase the demand at i. As I explained previously, the unit cost of increasing the supply at j is π_j; the cost of increasing the demand at i is $-\pi_i$. The net cost per unit of using nonbasic arc (i, j) is therefore $c_{ij} + \pi_j - \pi_i$, which is precisely what Equation (11.15) says. If you understand the meaning of the dual variables, you don't have to memorize Equations (11.14) and (11.15) because they make so much sense.

Example 11.7 Figure 11.19 depicts two sets of dual node variable values for the costs in Figure 11.18. The resulting π values, if π_a is set to zero, are on the left side of the figure. The right side of the figure shows the π values if instead we set $\pi_c = 0$ in Step 1. The differences $\pi_i - \pi_j$ for all $i \in V, j \in V$ are the same in both cases.

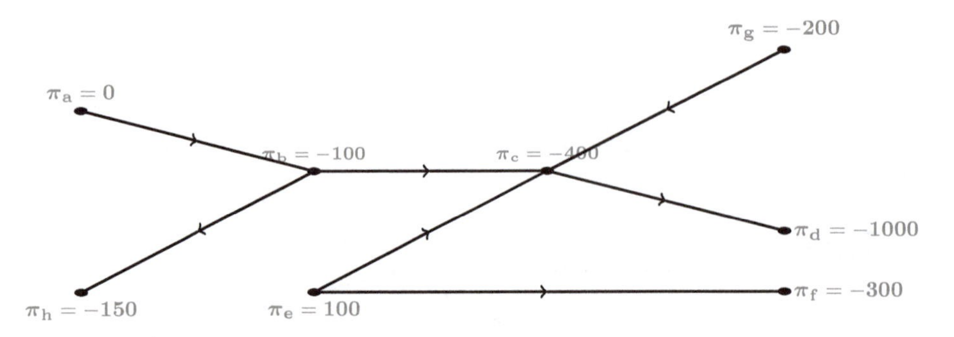

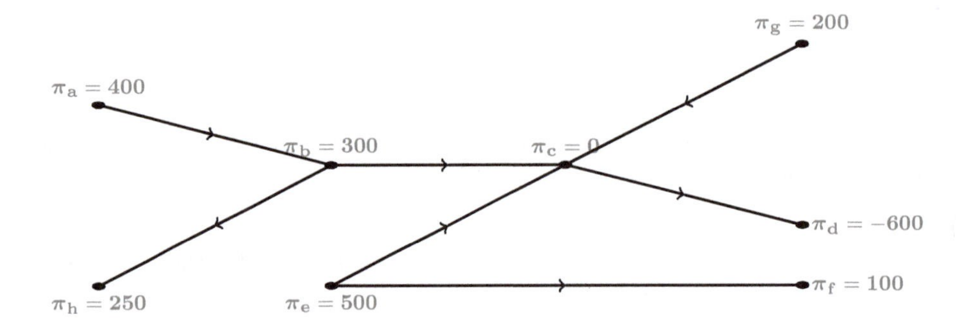

FIGURE 11.19: The differences $\pi_i - \pi_j$ are the same whether $\pi_a = 0$ (left) or $\pi_c = 0$ (right). The reduced cost of x_{ag} is $c_{ag} - \pi_a + \pi_g = 95 - 0 - 200 = 95 - 400 + 200 = -105$. The reduced cost of x_{ha} is $c_{ha} + 150$. The reduced cost of x_{gf} is $c_{gf} - \pi_g + \pi_f = c_{gf} - 100$.

The reduced cost of x_{ag} is $c_{ag} - \pi_a + \pi_g = -105$. This value would result from the cycle completion algorithm by the computation $c_{ag} + c_{gc} - c_{ab} - c_{bc} = -105$. *Question: What is the reduced cost of* x_{ah}*?*[11]

11.3.5 Putting the Pieces Together — The Network Simplex Algorithm

Given an initial basic feasible solution, the only piece missing from our specialized simplex algorithm for networks is the minimum ratio test that determines the leaving variable. The cycle completion method for computing reduced costs shows how to identify the leaving basic variable. When we change the value of the nonbasic variable x_{ij} by θ, we also must change the flow on every arc on the path P from i to j by $\pm\theta$. As $|\theta|$ increases, the flows on the arcs in P and the flow on the arc (i, j) move towards either their upper or lower bounds. The first arc whose flow hits its upper or lower bound leaves the basis. As in the bounded simplex method, the entering arc could be the leaving arc. The minimum ratio test has this particularly simple meaning because the denominator of each ratio is ± 1. We can now describe a single iteration of the network simplex algorithm.

[11] $c_{ah} - 150$.

Algorithm 11.3.12 (Iteration of Network Simplex Algorithm) *:*
Input: Graph $G = (V, E)$, lower and upper bounds on arc flow L, U, unit flow costs $c_{ij} : (i,j) \in E$, and a feasible choice of basis variables corresponding to spanning tree $T \subset E$ and a choice of lower or upper bound for each nonbasic variable.

Output: A new basic feasible solution with strictly smaller objective value if the input solution is not degenerate, and not larger objective value otherwise.

Step 1. If there was no previous iteration, compute the basic feasible solution $x_{ij} : (i,j) \in T$ with Algorithm 11.3.8. Otherwise, Step 6 from the previous iteration provides the x_{ij} values.

Step 2. Compute the reduced costs with Algorithm 11.3.11.

Step 3. If nonbasic variable $x_{ij} = L_{ij}$ (respectively $x_{ij} = U_{ij}$), its reduced cost is favorable if it is < 0 (respectively > 0). If no reduced cost is favorable, stop: the BFS is optimal. Otherwise, select a variable x_{rs} with favorable reduced cost to enter the basis.

Step 4. Mark x_{rs} increasing if $x_{rs} = L_{rs}$; otherwise $x_{rs} = U_{rs}$ and mark it decreasing. Determine the unique path $P \subset T$ from r to s. For each $(i,j) \in P$, mark (i,j)

- *Increasing if x_{rs} is increasing and (i,j) points from s to r in the path P, or x_{rs} is decreasing and (i,j) points from r to s in the path P.*
- *Decreasing otherwise, i.e., if x_{rs} is increasing and (i,j) points from r to s in the path P, or x_{rs} is decreasing and (i,j) points from s to r in the path P.*

Step 5. Find

$$(v,w) = \arg \min_{(i,j) \in P \bigcup \{(r,s)\}} \begin{cases} U_{ij} - x_{ij} \text{ if } x_{ij} \text{ is increasing;} \\ x_{ij} - L_{ij} \text{ if } x_{ij} \text{ is decreasing.} \end{cases}$$

Step 6. Set $\theta^* := U_{vw} - x_{vw}$ if x_{vw} is increasing and $x_{ij} := x_{vw} - L_{vw}$ if x_{vw} is decreasing. If $\theta = \infty$ the optimum is unbounded and the remainder of this step will generate a ray of unboundedness. For each $(i,j) \in P \bigcup \{(r,s)\}$ set $x_{ij} := x_{ij} + \theta^*$ if x_{ij} is increasing and $x_{ij} := x_{ij} - \theta^*$ if x_{ij} is decreasing. Add x_{rs} to the basis. Remove x_{vw} from the basis.

I've separated Steps 2 and 3, and Steps 4 and 5 for clarity. In an implementation, each pair of steps could be merged. The overall speed of the algorithm would not differ significantly if Steps 4 and 5 were merged. Steps 2 and 3 are the bottleneck steps, each requiring $O(|E|)$ time to check the nonbasic arcs. Merging these two steps should reduce the runtime by a slight constant factor.
Finding an initial basic feasible solution. There are several simple ways to create temporary variables for a Phase I of the simplex algorithm to find an initial BFS. The general idea is to choose a spanning tree basis $T \subset E$, set each nonbasic variable $x_{ij} : (i,j) \notin T$ to its lower

or upper bound, and compute the resulting basic solution. If that solution is not feasible, create for each $(i,j) \in T$ a temporary arc with lower bound 0 and upper bound equal to the amount of infeasibility, set the costs on temporary arcs to 1 and set other costs to 0. To be specific, if basic variable $x_{ij} < L_{ij}$, create temporary arc (j,i) with lower bound 0, upper bound $L_{ij} - x_{ij}$, and cost 1. If basic variable $x_{ij} > U_{ij}$, create temporary arc (i,j) with lower bound 0, upper bound $x_{ij} - U_{ij}$ and cost 1. In both cases the temporary variable will be nonbasic at its upper bound. The algorithm below follows this general idea. It includes a couple of tweaks to help Phase I find an initial BFS that is fairly good with respect to the Phase II objective function.

Algorithm 11.3.13 (Network Simplex Phase I Initialization) *:*
Step 0. Preprocess the graph $G = (V, E)$ with lower bounds L, upper bounds U, and costs c to ensure that all arcs have at least one finite bound. For all $(i,j) \in E$ with $L_{ij} = -\infty$ and $U_{ij} = \infty$, set $L_{ij} = 0$ and create new arc (j,i) with $L_{ji} = 0, U_{ji} = \infty$, and $c_{ji} = -c_{ij}$.

Step 1. Choose an initial spanning tree T arbitrarily or by finding a minimum cost spanning tree of $G = (V, E)$ with respect to costs $c_e : e \in E$.

Step 2. Compute the reduced costs with respect to cost vector c of all nonbasic arcs $e \in E \setminus T$ with Algorithm 11.3.11.

Step 3. Set each $x_e : e \in E \setminus T$ to L_e if its reduced cost is ≥ 0 and to U_e otherwise.

Step 4. Compute the basic solution $x_e : e \in T$ for basis T and nonbasic variables as set by Step 3.

Step 5. For each $e \in E$ with $x_{ij} < L_{ij}$ create temporary arc (j,i) with lower bound $L_{ji} = 0$, upper bound $U_{ji} = L_{ij} - x_{ij}$, and cost 1. For each $e \in E$ with $x_{ij} > U_{ij}$ create temporary arc (i,j) with lower bound 0, upper bound $x_{ij} - U_{ij}$ and cost 1.

Summary: Network models are easy to visualize and explain. The minimum cost flow model can incorporate most other models, yet solves quickly and yields integer optimal basic solutions. Its primal and dual basic solutions can be calculated rapidly because the tree structure of a basis makes the system of equations triangular. Its dual variables have a clear interpretation as the cost of disposing one additional unit of supply at the nodes.

11.4 Problems

E Exercises

1. In the network of Figure 11.16 let nonbasic variables $x_{ea} = U_{ea} = 10$ and $x_{bc} = U_{bc} = 8$ as in the example, but let $x_{ce} = U_{ce} = 8$. For each basic variable let the lower bound be 0 and the upper bound ∞.

(a) Find the basic solution. Is it feasible? Is it degenerate?

(b) Repeat with $U_{ce} = 10$.

(c) Repeat with $U_{ce} = 12$.

2. Find the basic variable values in the example of Figure 11.16 if the nonbasic variable values were $x_{bc} = -10$, $x_{ce} = 19$, and $x_{ea} = 30$.

3. Explain why Step 1.5 of algorithm 11.3.8 does not change $\sum_j t_j$.

4. Suppose you calculate the flow on a basic arc two ways but get different values. (The last flow can be calculated in two ways.) If you've made no arithmetic errors, what does this mean?

5. Define the arcs and their parameter values that would be needed for the inexpensive (cost c_4) maintenance in the alternative airplane maintenance model.

6. 16 tons of coal must be transported from Newcastle to London. Transportation costs per ton are:

Newcastle to Brighton	200 £
Newcastle to Dover	350 £
Brighton to Dover	120 £
Dover to Brighton	80 £
Brighton to London	450 £
Dover to London	250 £

(a) Model this as a minimum cost flow problem. Depict it visually.

(b) Write the primal LP.

(c) Why do the arcs $(B, L), (D, L), (B, D)$ not form a basis? Explain in terms of the graph and in terms of the submatrix.

(d) Write the dual LP.

(e) Find the optimum solution by inspection.

(f) Find the primal basic solution with arcs $(N, B), (N, D), (B, L)$ in the basis. Write the matrix comprising the columns of the basic variables.

(g) For the preceding question, find the dual basic solution using cycle completion, and determine its feasibility.

(h) Repeat the preceding question using the dual's equations.

(i) Find the primal basic solution with arcs $(N, D), (D, B), (B, L)$ in the basis.

(j) For the preceding question, find the dual basic solution using cycle completion, and determine its feasibility.

(k) Repeat the preceding question using the dual's equations.

7. *Continuation of 6.* Solve the problem with the network simplex algorithm from initial basis (ND, DB, DL).

8. *Continuation of 6.* Suppose all arc flow lower bounds are zero and all upper bounds are infinite except for the constraints $x_{N,B} \leq 10$ and $x_{D,L} \leq 9$.

   - Find the optimum solution by inspection.

- Find the primal basic solution with arcs $(N, B), (N, D), (B, L)$ in the basis, and nonbasic variable $x_{D,L}$ at its upper bound. Would the basic solution be feasible if $x_{D,L}$ were at its lower bound?
- Find the primal basic solution with arcs $(N, D), (D, B), (B, L)$ in the basis, $x_{N,B}$ at its upper bound, and $x_{D,L}$ at its lower bound. Would the basic solution be feasible if instead $x_{D,L}$ were at its upper bound?

9. *Continuation of 6.* To the original problem append upper bounds $x_{N,B} \leq 10; x_{D,L} \leq 9; x_{B,L} \leq 14$. Solve the problem with the network simplex algorithm from an initial basic solution with $x_{B,L}$ at its upper bound, $x_{N,B}$ at its lower bound, and $x_{D,L}$ basic.

10. Which of the following matrices are TU?

$$\begin{bmatrix} 1 & -1 \\ 1 & -1 \end{bmatrix} \begin{bmatrix} \frac{1}{2} & 0 \\ 0 & 2 \end{bmatrix} \begin{bmatrix} -1 & -1 \\ 1 & -1 \end{bmatrix} \begin{bmatrix} 1 & 0 & 1 \\ -1 & 1 & -1 \\ 0 & 1 & -1 \end{bmatrix} \begin{bmatrix} 0 & 0 & 1 \\ -1 & -1 & 0 \\ -1 & -1 & -1 \end{bmatrix}$$

11. Prove or Disprove: if a nonsingular square matrix B is TU, then B^{-1} is TU.

12. A *circulation* problem is a minimum cost network flow problem in which there is zero supply at every node. Show how to transform any min cost network flow problem into a min cost circulation problem. Hint: Use nonzero lower bounds on arc flows.

M Problems

13. Modify the airplane landing gear assembly maintenance model to account for the following complications:

    (a) Upper bounds on the number of ready assemblies that may be stored from day t to day $t + 1$, and upper bounds on the number of assemblies that are scheduled to begin overtime maintenance at the end of day t.

    (b) The airline can purchase additional landing gear assemblies to arrive ready to be used for the first time on day t at cost $c_0(1.5T - t)$ dollars each, for $t = 1, \ldots T$. (An assembly bought later costs less in the model because it will have a longer average lifetime after day T than will an assembly bought earlier).

    (c) Upper bounds on the total number of assemblies that may be stored from day t to day $t + 1$. This number includes assemblies ready but not used on day t, and assemblies that need maintenance but do not begin maintenance on day t.

14. Of the 81 2×2 matrices with each entry $\in \{-1, 0, 1\}$, how many are TU?

15. Prove Proposition 11.3.4 as follows: First, show that the average degree of the nodes in a tree is strictly less than 2. Hence there must be at least one leaf. Second, show that removing a leaf from a tree leaves a tree. Apply induction.

16. Prove Proposition 11.3.4 as follows: First, show that the average degree of the nodes in a tree is strictly less than 2. Hence there must be at least one leaf. Second, start at the leaf and move along edges of the tree until you cannot move further.

17. For graph $G = (V, T)$ prove that any two of the following properties imply the third:

    (a) $|T| = |V| - 1$.

    (b) The graph G has no orientable cycles.

(c) The graph G is connected.

18. For an undirected graph $G = (V, E)$ with edge costs $c_e : e \in E$, the Minimum Spanning Tree problem (MST) is to find a subset $T \subset E$ containing $|T| = |V| - 1$ edges that is a tree and has minimum possible cost $\sum_{e \in T} c_e$. Although the MST can be solved readily in polynomial time by a greedy algorithm, there is no known practical LP formulation for it. Prove that the following is an incorrect minimum cost network flow formulation: Distinguish one vertex $s \in V$ as the source. Set the supply at s to $|V| - 1$ and set the supply at all $v \neq s$ to -1 (i.e., a demand of 1).

19. Suppose that a minimum cost network problem has unbounded optimum. Consider the iteration of the network simplex method that reveals the unboundedness. Prove that the graph must contain an oriented cycle composed of arcs with no upper bounds, such that the sum of the arc costs is strictly negative.

20. Modify the network simplex algorithm for graphs with upper and lower bounds on edge flows to avoid cycling using Bland's rule. Be sure to consider the case when the entering variable ties with others to be the leaving variable.

21. Prove Proposition 11.3.4 algorithmically. *Hint: Start at any node and move along arcs. Prove that you must stop.*

22. Assuming that Proposition 11.3.4 is true, prove Propositions 11.3.2 and 11.3.3 by induction.

23. Modify the min cost flow network model of production with storage and backorders to enforce an upper bound on the total number of items stored between period t and period $t + 1$. For example, any item produced in period 1 or period 2 and delivered in any period $i \geq 3$, would count towards the total number of items stored between periods 2 and 3.

24. Show that a naive implementation of Algorithm 11.3.7 might take $\Omega(|V|^2)$ time. Hint: Make it take a long time to find a leaf.

25. You calculate the flow on basic arc (d, z) two ways. Evaluating flows at d yields $x_{dz} = 200$ but evaluating flows at z yields $x_{dz} = 260$. Find the sum (taken over all nodes in the graph) of the supplies.

26. Let $G = (V, E)$ be an undirected graph. A *matching* is a subset $M \subset E$ such that $e \in M, e' \in M \Rightarrow e \cap e' = \phi$. In words, two edges in M are incident on the same node. Find a smallest example that proves the LP $\max_{e \in E} x_e$ subject to $x_e \geq 0 \; \forall e \in E$, $\sum_{e: v \in e} x_e \leq 1$ can fail to have an optimal solution that is integer.

27. You have an unlimited supply of n-dimensional blocks with dimensions $b_1, \ldots, b_n$ and a larger box with dimensions $a_1, \ldots, a_n$. What is the maximum number of blocks that will fit into the box if all blocks must have the same orientation? Model this problem as a weighted bipartite matching problem.

28. Write and interpret the dual of the LP of Problem 26.

29. This question explores how far from TU can a $0, \pm 1$ matrix be. Let $\kappa(n)$ be the largest possible determinant of an $n \times n$ matrix each of whose entries equals $0, 1$ or -1. For example $\kappa(1) = 1$ and $\kappa(2) = 2$.

    (a) Find $\kappa(3)$.

(b) Find $\kappa(4)$.

(c) Prove that $\kappa(n) \geq \varphi(n)$ where $\varphi(n)$ is the nth Fibonacci number. (The Fibonacci sequence is defined by $\varphi(1) = \varphi(2) = 1$, $\varphi(n) = \varphi(n-2) + \varphi(n-1)$ for $n \geq 3$.) Conclude that $\kappa(n)$ grows at least exponentially quickly but $\kappa(n) = o((n/e)^{n+1})$.

30. Calculate an upper bound (e.g., $O(|V|), O(|E|)$) for each step of Algorithm 11.3.12. Verify that the bottleneck steps are as stated in the text.

31. Show how to convert any transshipment problem with three layers $V = V^1, V^2, V^3$ into an equivalent transportation problem. (Hint: Make two copies of the middle layer and force the flow out of one copy to equal the flow out of the other copy.) Argue inductively that all transshipment problems can be converted into transportation problems.

D Problems

32. An assignment, or complete bipartite matching, may be defined as an integer vector in the polytope

$$P = \left\{ \mathbf{x} \in \Re^{n \times n} : \mathbf{x} \geq \mathbf{0};\ \sum_{j=1}^{n} x_{ij} = 1\ \forall i; \sum_{i=1}^{n} x_{ij} = 1\ \forall j \right\}.$$

(a) Prove that the assignments correspond to the extreme points of P.

(b) Suppose that a subset $F \subset \{1, 2, \ldots, n\} \times \{1, 2, \ldots, n\}$ of the x_{ij} variables are constrained to equal zero, $x_f = 0\ \forall f \in F$. Prove or disprove: The feasible assignments correspond to the extreme points of $P \cap \{\mathbf{x} : x_f = 0 \forall f \in F\}$.

(c) Devise a polynomial time algorithm to maximize $\max_{i,j} c_{ij} x_{ij}$ over all assignments $\mathbf{x}$ such that $x_f = 0\ \forall f \in F$.

(d) Given costs $c_{ij} \geq 0$, devise a polynomial time algorithm to minimize $\max_{i,j} c_{ij} x_{ij}$ over all assignments $\mathbf{x}$ such that $x_f = 0\ \forall f \in F$. .

33. Prove that the following statements are equivalent:

- Matrix A is totally unimodular.
- The block matrix $[A|I]$ is totally unimodular.
- Matrix A^T is totally unimodular.
- For all integer vectors $\mathbf{b}$, all extreme points of the polyhedron $\{\mathbf{x} \geq \mathbf{0} : A\mathbf{x} = \mathbf{b}\}$ are integer.

34. Show how to implement Algorithm 11.3.7 to take $O(|V| \log |V|)$ time.

35. Modify the network simplex method with upper and lower bounds to avoid cycling using the perturbation/lexicographic method. Be sure to consider cases with negative lower bounds.

36. (i) Develop a version of the network simplex method based on the dual simplex method for graphs with lower bounds 0 and no upper bounds. Your algorithm should start at an infeasible basic solution to the primal whose complementary dual solution is feasible, and work towards primal feasibility while maintaining dual feasibility. (ii) Extend your algorithm to graphs with lower bounds L and upper bounds U.

37. Prove or disprove that the LP of Problem 26 has the half-integral property (there exists an optimal solution in which each variable has value $0, 1$, or $\frac{1}{2}$).

I Problems

38. Suppose $G = (V, E)$ is an undirected graph. Compare the time and space requirements of representing members of E as undirected edges versus a pair of directed edges.

39. Find a data structure that lets you implement Algorithm 11.3.7 in time $o(|V| \log |V|)$.

40. Show how to implement Algorithm 11.3.7 in time $O(|V|)$ for each iteration except the first.

41. On small instances, it is easy to execute Step 4 of Algorithm 11.3.12 (or the similar Step 2 of Algorithm 11.3.9) visually. However, it is not obvious how to execute Step 4 more efficiently than in time $O(|V|)$ on a computer.

    (a) Devise a data structure and algorithm that finds the path P in time $O(|P|)$.

    (b) Show how to initialize your data structure in time $O(|V|)$.

    (c) Show how to update your data structure from one iteration to the next in time $O(|V|)$. Can you reduce the time to $O(|P|)$?

42. Choose a random spanning tree T on a complete graph of n nodes as follows: set T to consist of one random arc. Until $|T| = n - 1$, add to T an arc chosen randomly from the set of arcs that are incident on exactly one node among those incident on arcs in T. Prove a $O(\log^2 n)$ bound on the expected length of the path in T between two random nodes. Conclude that the expected fraction of basic variables whose values change in an iteration of the simplex method is $o(n)$, if the present basis is a random spanning tree as defined here. Explain why this bound does not necessarily apply to subsequent iterations of the algorithm.

Chapter 12

Shortest Path Models and Algorithms

Preview: Finding a least-cost path from one node to another in a graph with arc costs is a special kind of network problem, traditionally called a *shortest path* problem. Many real-world problems contain shortest-path subproblems. Also, some real-world problems can be modeled as shortest path problems. There are various specialized shortest path algorithms, tailored to different types of arc costs.

12.1 Modeling Shortest Path Problems

Recall from Definition 11.5 that a path in a graph arises from a sequence of nodes such that each consecutive pair of nodes is an arc. The sequence of nodes $v_0, v_1, \ldots, v_k$ forms a path in graph $G = (V, E)$ if $(v_i, v_{i+1}) \in E$ for all $0 \leq i < k$. The path itself is defined to be the sequence of arcs $(v_0, v_1), \ldots, (v_{k-1}, v_k)$, but we often think of the path in terms of its nodes instead. The path formed by $v_0, v_1, \ldots v_k$ is said to start at v_0 and end at v_k, or to be a v_0, v_k-path. Nodes $v_1, \ldots, v_{k-1}$ are called *interior* nodes. The path is *simple* if all its nodes are distinct. When all nodes are distinct except $v_0 = v_k$ the path is called a *cycle.*

In shortest path problems, each arc $e \in E$ has a numerical value c_e, referred to as its *length* or *cost.* The length or cost of path $v_0, v_1, \ldots, v_k$ is the sum of its arc values $\sum_{i=0}^{k-1} c_{v_{i,i+1}}$.

The classical shortest path problem is, given graph $G = (V, E)$, arc values $c_e : e \in E$, and two specified nodes $s \in V; t \in V$, find a simple s, t-path in G of minimum possible length. *Question: What plausible condition on the arc values guarantees that the shortest path be simple?*[1] It is so often taken for granted that the path must be simple that the problem is not called "the shortest simple path problem." *Question: Devise a graph on two nodes, s and t, such that no shortest (non-simple) s,t path exists, even though the arc (s,t) exists.*[2] There are some real-world shortest path problems for which the paths may not repeat arcs but need not be simple, and even some for which the paths may contain repeated arcs.

Some real problems can be readily modeled as shortest path problems. A great many real problems can be modeled in principle as shortest path problems, but only on graphs that are too large to be computed or stored explicitly. Shortest path problems also frequently occur as subproblems of a more complex problem. For the latter reason alone, there is a

[1] $w_e > 0$ for all $e \in E$.
[2] $w_{st} = w_{ts} = -1$.

fairly good chance that you will have to solve some of these problems, and even write your own code to do so, if you make a career in optimization.

The main practical things to remember about shortest paths are the following:

- In a shortest path model, a node does not have to represent a mere physical location. It may instead represent
  - a time;
  - a physical location at a time;
  - a stage of completion of a task;
  - a state or status of the world. An arc then represents an activity or event that causes a transition from one state to another.
- Instances without cycles or without negative arc weights can be solved very quickly.
- Instances with cycles and negative weights require different and slower solution methods.
- Cycles whose total arc costs are negative make the shortest path problem either meaningless or hard.

For understanding optimization, shortest path problems are the simplest ones that are still complex enough to illustrate many concepts, including:

- Special properties of the data can render a problem easier to solve.
- Data structures are crucial to algorithm performance.
- Optimal primal and dual solutions have complementary slackness.
- The primal and dual LPs are two different ways of looking at the same thing.

Formulation: Shortest and Quickest Paths

When you request directions from a website, your GPS device, or your cellphone app, the underlying software solves shortest path problems to select the routes to offer. If the weights on arcs are the street lengths between nodes, the shortest path on the graph is literally the shortest (legal) path to your destination. If the weights on arcs are travel times from one node to another, the shortest path on the graph is the quickest path. These formulations are so simple that they hardly seem worth mentioning. Be aware, however, that the ubiquity of these applications relies on our ability to solve these shortest path problems extremely quickly.

Formulation: Most Reliable Path

Suppose that the graph $G = (V, E)$ represents a communications network whose links can fail. For each $e \in E$, the link represented by e has probability p_e of failing, independent of whether other links fail. What is the most reliable communications path from $s \in V$ to $t \in V$? The probability of successful transmission on a path P equals

$$\prod_{e \in P} (1 - p_e).$$

To model this problem as a shortest path problem, set the weight on arc e to $w_e = -\log(1 - p_e)$. Then

$$\sum_{e \in E} w_e = -\log \prod_{e \in P} (1 - p_e).$$

Since the log function is an increasing function, minimizing $\sum_{e \in E} w_e$ maximizes the probability of successful transmission.

Formulation: Minimum cost replacement and maintenance

A truck is needed throughout the next T years. Each year, the truck must undergo maintenance or be sold and replaced by a new truck. Maintenance for a truck starting its kth year of use costs c_k for $k \geq 2$; a new truck in year k costs d_k; a truck after k years of use may be sold for s_k. What policy minimizes total cost?

Define nodes $i = 1, \ldots, T+1$ to represent the beginning of year i. Each node represents a different time. Node i represents the beginning of year i, or in the case $i = T+1$, the end of year $i-1$. For all $i < j$ create an arc (i, j) with cost[3] equal to the price of a new truck in year i, plus the maintenance costs of keeping the truck for $j - i$ years, minus the selling price of a $j - i$ year-old truck in year j. This arc represents the decision to buy a new truck at the beginning of year i and use it until the beginning of year j. That is,

$$c_{i,j} = c_i + \sum_{k=2}^{j-i} d_k \quad - s_{j-i}.$$

In the case $j = i + 1$ the middle term is vacuous.

Another way to model this as a shortest path problem is to create a node for each $i = 0, \ldots, T$ and each $m = 1, \ldots, i$, a starting node $[0, 0]$, and a terminating node t. Node $[i, m]$ denotes the state of the world at the *end* of year i in which the truck on hand is m years old. In this model, two arcs emanate from node $[i, m]$. The arc $([i, m], [i+1, m+1])$ represents the decision to use the truck that was on hand at the end of year i during the year $i + 1$. Its cost is d_m. The other arc $([i, m], [i+1, 1])$ represents the decision to sell the old truck and buy a new one at the beginning of year $i + 1$. Its cost is $c_{i+1} - s_m$. For $i = T$ a single arc $([T, m], t)$ with cost $-s_m$ represents the sale of the used truck at the end of year T. This second model may seem more natural because it mimics the actual annual decision-making. The first model compresses all decisions about a truck, as if the truck's lifetime is decided at the time of purchase. However, the second model has order T^2 nodes compared with the first model's order T nodes. *Question: How about the number of edges?*[4]

Formulation: Selecting Stock Inventory to Minimize Waste

In the 1980s I solved this problem for a paper company that makes corrugated cardboard from rolls of paper. The solution reduced their paper usage by millions of pounds and saved them about $12 million dollars a year.

Paper rolls are approximately one yard in diameter and come in every possible width from 18 to 78 inches. Orders for cardboard are for various lengths, and come in every possible width from 18 to 78 inches. The rolls are so bulky, heavy, and expensive that the company only keeps 5 roll widths in stock. Even if there were no inventory holding cost for the paper rolls, and no shortage of space to store them, it would still not be good to use all possible widths, because changing out rolls causes significant downtime. When an order is processed, the roll with width closest but at least equal to the order width is used. Any excess paper is called "trim" or "waste" and is discarded or sold for mere pennies per pound. Given projected total lengths of orders for each possible width, which widths should the company stock to minimize trim? (For clarity, I've altered the problem slightly. Actual widths were measured by the quarter inch, and there was a nonzero minimum trim to accommodate irregularities and machine vibration.)

[3] A realistic model would discount future costs to convert all numbers to present value.

[4] Both models have order T^2 edges.

Solution: Let $i = 1, \ldots, N$ index the order widths in *decreasing* order from 72 to 18. Let L_i denote the total length, in inches, of orders for width i. Imagine that we process all orders of width index i before processing orders of the smaller width of index $i+1$. Let node $(i,k) : i = 1, \ldots, N+1; k = 1, \ldots, 5$ represent the state of the world where all orders of width index $\geq i$ must be processed, and $k-1$ different roll widths were previously used to process the orders of width index $< i$. The start node is $(1,1)$. The destination node is $(N+1, 5)$ where $N+1$ is a dummy index value representing a width less than 18. The cost of the arc from (i,k) to $(m, k+1)$ is the amount of paper (measured in square inches) wasted to process the orders of width indices $i, i+1, \ldots, m-1$ with a roll of width index i, which equals

$$\sum_{j=i}^{m-1} L_j(j-i).$$

Figure 12.1 illustrates part of the network with $L_1 = 800, L_2 = 600$, $L_3 = 1000$, and $L_4 = 220$.

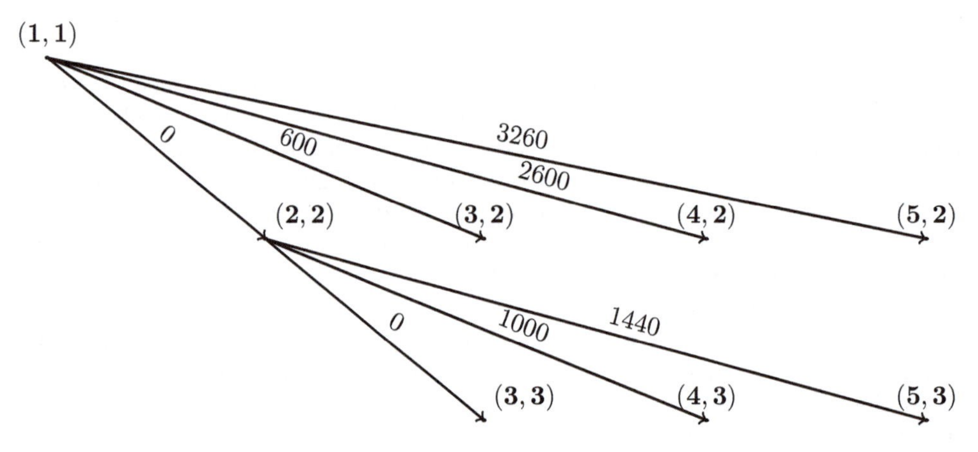

FIGURE 12.1: A node in row k, column i represents $k-1$ widths having been used to process the orders of the $i-1$ greatest widths. An arc represents using width i to process orders of a set of widths.

More of the story: I immediately told the company that this was not the problem they wanted me to solve. Instead, they should ask me which roll widths ought to be stocked if K widths were used and what the resulting trim amount would be, for a range of K. I was thinking that, since they didn't already know how to minimize trim, there was no way they could know that 5 was the "right" value for K. I also thought that it would be difficult to quantify the combination of inventory, material handling, and setup costs, and the awkwardness of fitting rolls into the limited space, if K were changed. If I could tell them how much waste they would get with, say, $K = 3, 4, \ldots, 10$ different widths, they could choose a tradeoff that they would be happy with. As it turned out, they increased K from 5 to 6.

Modified Solution: Increase the range of the index k from $1, \ldots, 5$ to $1, \ldots, 10$. As we will see later in this chapter, when an algorithm finds a shortest path from s to t, it also finds shortest paths from s to many nodes other than t. In fact, a slight change to the termination criterion makes the algorithm find shortest paths from s to all other nodes. The important shortest paths are those from $(1,1)$ to $(N+1, k)$ for $k = 3, \ldots 10$.

12.2 Acyclic Graphs

If the graph has no cycles, a shortest $s - t$ path can be found quickly with a simple iterative algorithm. The following algorithm works even if there are negative arc weights.

Algorithm 12.2.1 Shortest Path Algorithm I, for Acyclic Graphs

Step 1 *(Topological sort) Relabel the nodes with the integers* $1, 2, \ldots, n$ *so that* $s = 1$ *and* $i < j$ *for all* $(i, j) \in E$. *Nodes* v *for which there exists a path from* v *to* s *may be ignored or discarded.*

Step 2 *Set* $i = 1$. *Set* $d_j = \infty$ *and* $pred(j) = null$ $\forall j > 1$. *Set* $d_1 = 0$.

Step 3 *Scan the arcs emanating from node* i. *For each* $(i, j) \in E$
$\{$
$d' = d_i + w_{ij}$
If $d' < d_j$ $\{d_j = d';\ pred(j) = i\}$
$\}$

Step 4 $i = i + 1$; *If* $i == t$ *terminate ELSE go to Step 3*$\}$

When algorithm 12.2.1 terminates d_t equals the length of a shortest $s - t$ path. The path itself, going backwards from t, is $t, pred(t), pred(pred(t)), \ldots, s$. If $d_t = \infty$ no $s - t$ path exists. If shortest paths from s to all other nodes are desired, set $t = n$ at the end of Step 1.

Algorithm 12.2.1 is valid because it maintains the following property at the end of Step 3:

$$\forall j \geq 1 \;\; d_j = \text{ length of a shortest } s - j \text{ path whose interior nodes are } \leq i. \qquad (12.1)$$

We call such a property an *invariant.* At $i = 1 = s$ property (12.1) holds because the only possible path from s is the arc (s, j). *Question: Why can't a path from s to j have any other arc?*[5] For $i > 1$, inductively assume (12.1) holds at $i - 1$. If the shortest $s - j$ path with no interior nodes $> i$ does not traverse i, property (12.1) holds at i because it holds at $i - 1$. Otherwise, the last interior node must be i, because all arcs from i go to higher-numbered nodes. Every $s - j$ path that ends with arc (i, j) consists of an $s - i$ path concatenated with arc (i, j). All interior nodes of an $s - i$ path must be $\leq i - 1$. Since (12.1) holds at $i - 1$, the shortest $s - j$ path that contains arc (i, j) has length $d_i + w_{ij}$. This proves that the invariant is correct.

As usual, the algorithm runtime depends heavily on the choice of data structure. If G is represented as a (non-sparse) matrix W where $W_{ij} = w_{ij}$ for $(i, j) \in E$ and $W_{ij} = \infty$ otherwise, each iteration of Step 3 will take $O(|V|)$ time, and the algorithm will require time (and memory) $O(|V|^2)$. Unless G is extremely dense, it is better to represent it with arc lists (or equivalently to represent W as a sparse matrix). For each node $i \in V$, list the pairs j, w_{ij} for which $(i, j) \in E$. Step 3 then takes time proportional to the out-degree of (the number of arcs leaving) i. The total time required is then $O(|E|)$. Faster runtime is not possible in general because every arc must be considered. *Question: Why?*[6]

Step 1 of the algorithm, the topological sort, deals with a well-solved low-level computer science problem. For a graph stored in terms of outgoing arcs from nodes, first store the

[5] Arcs (i, k) are forbidden for $k \geq 2$, as k would be interior. Arcs (i, k) don't exist for $k < 1$.
[6] A single arc with large negative weight could affect the optimal solution.

lists of incoming arcs, which takes time $O(|E|)$. Then the topological sort can be done in time $O(|E|)$ (technically speaking, $O(|E|+|V|)$) by repeatedly selecting a node that has no incoming arcs to be next in the order, and removing its outgoing arcs from the graph. If no such node exists the graph has a cycle. The topological sort can also be accomplished in linear time by a depth-first search [56].

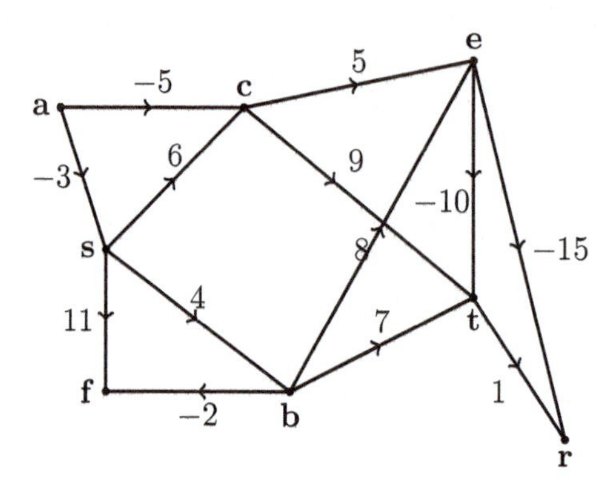

FIGURE 12.2: A topological sort of this graph could yield *ascbfetr* or *asbtcerf* or *asbctefr*, etc.

We summarize the discussion of Algorithm 12.2.1 with a theorem stating its correctness and runtime.

Theorem 12.2.2 *Given a specified node $s \in V$, Algorithm 12.2.1 finds a shortest $s-t$ path, or shortest $s-v$ paths for all $v \in V$, in time $O(|E|)$ on graphs $G=(V,E)$ containing no (directed) cycles.*

Example 12.1 The graph $G=(V,E)$ is depicted in Figure 12.2. The data are as follows. $V=\{s,a,b,c,e,f,r,t\}$.
$w_{as}=-3, w_{ac}=-5, w_{sb}=4, w_{sc}=6, w_{sf}=11,\ w_{bf}=-2, w_{be}=8, w_{bt}=7, w_{ce}=5,$
$w_{er}=-15, w_{ct}=9, w_{et}=-10, w_{tr}=1.$
Step 1: Remove node a. Remove arcs as, ac. Remove node s. Remove arcs sb, sc, sf. Remove node c. Remove arcs cb, cr, ct. Remove node b. Remove arcs bf, be, bt. Remove node e (you could remove node f now instead). Remove arc et. Remove node f (you could remove node t now instead). Remove node t (you could terminate Step 1 now if you are checking for t at each node removal). Remove arc tr. Remove node r. Discard node a because it precedes the starting node s. The resulting ordering is s, c, b, e, f, t, r.

Steps $2-4$: The table shows the status of d_j and $pred(j)$ after each node i is scanned. The algorithm stores only its most recent values of d_j and $pred(j)$ and outputs only the final values rather than the entire table. *Question: Why?*[7]

j	s	c	b	e	f	t	r	i
d_j	0	6	4	∞	11	∞	∞	s
pred(j)	null	s	s	null	s	null	null	
d_j	0	6	4	∞	11	15	-9	c
pred(j)	null	s	s	null	s	c	c	
d_j	0	6	4	12	6	11	-9	b
pred(j)	null	s	s	b	b	b	c	
d_j	0	6	4	12	6	2	-9	e
pred(j)	null	s	s	b	b	e	c	
d_j	0	6	4	12	6	2	-9	f
pred(j)	null	s	s	b	b	e	c	

The shortest path from s to t, tracing backwards from t, is *tebs* with length 2.

12.3 The Principle of Optimality

It is curious that when we seek a shortest $s-t$ path we find shortest paths from s to many other nodes as well. This phenomenon is partly explained by the *principle of optimality*, which is the following.

Definition 12.1 (Principle of Optimality for Shortest Paths) *A shortest path problem satisfies the principle of optimality if every shortest $s-t$ path P that has interior node v must consist of a shortest $s-v$ path joined to a shortest $v-t$ path. If so, every optimal $s-t$ path contains optimal paths from s to all its interior nodes, and optimal paths to t from all its interior nodes.*

This principle may seem obvious. The intuitive argument is that if there were a shorter $s-v$ (or $v-t$) path, it could replace the $s-v$ $(v-t)$ portion of P. That replacement would shorten P, contradicting P's optimality. However, the graph in Figure 12.3 violates the principle of optimality. In that graph, arc weights between a and b equal -1, and all other arc weights equal 2. The shortest $s-b$ path, *sab*, has length 1, as does the shortest $b-t$ path *bat*. But the optimal $s-t$ paths *sabt* and *sbat* fail to use *bat* and *sab*, respectively.

Next we show that on a large class of graphs, the principle of optimality is not violated. It will be handy to have a precise definition and notation for a path contained in another path.

Definition 12.2 (Subpath) *In a graph $G=(V,E)$ let P be a simple path specified by the sequence of m distinct nodes $P[1], P[2], \ldots, P[m]$. By definition of a path $(P[k], P[k+1]) \in E$ or $P([k+1], P[k]) \in E$ $\forall 1 \leq k \leq m-1$. For any $1 \leq j \leq k \leq m$, the* subpath of P from $P[j]$ to $P[k]$ *is the path $P[j], P[j+1], \ldots, P[k]$ in G and it is denoted $P_{P[j]P[k]}$. Thus if $a=P[j]$ and $b=P[k]$ the subpath is denoted P_{ab}. If P is not simple, the concept of a subpath is the same, but the starting and ending nodes must be unambiguously specified by their position in the sequence P.*

[7]The whole table may be size $\Omega(|V|^2)$.

Example 12.2 Let G be a graph on vertices $V = \{1, 2, 3, 4, 5\}$ containing all possible edges $E = \{(i, j)|i \in V, j \in V, i \neq j\}$. Let P be the path $1, 2, 3, 4, 5, 3, 2$. The subpaths P_{15} and P_{32} are well-defined, but there are two subpaths from 1 to 2 and three subpaths from 2 to 2 and from 3 to 3.

Theorem 12.3.1 *All graphs that contain no negative cycles satisfy the principle of optimality.*

Proof: Let P be a shortest $s - t$ path in G. Suppose v is interior to P. Express P as the concatenation of $s - v$ path P_{sv} and $v - t$ path P_{vt}, written as $P = P_{sv} \circ P_{vt}$. Suppose to the contrary there exists an $s - v$ path P' which is shorter than P_{sv}. Then $P' \circ P_{vt}$ is shorter than P. If $P' \circ P_{vt}$ is simple, we have a contradiction to the optimality of P. If it is not simple, then it must traverse some node x twice (or more). The subpath of $P' \circ P_{vt}$ from the first traversal of x to the last traversal x constitutes a cycle, which by hypothesis is not negative. Replacement of that cycle by the single node x can only leave the path's length unchanged, or shorten it. Remove all cycles to obtain a simple path whose length is no more than the length of $P' \circ P_{vt}$, which is less than the length of P, and contradicting the optimality of P. ■

Corollary 12.3.2 *For all graphs that contain no negative cycles, if there exists an $s - t$ path then there exists a shortest $s - t$ path that is simple.*

Proof: Let P be a shortest path that is not simple. As in the proof of the theorem, remove all cycles from P to obtain a simple path that has the same or shorter length as P. ■

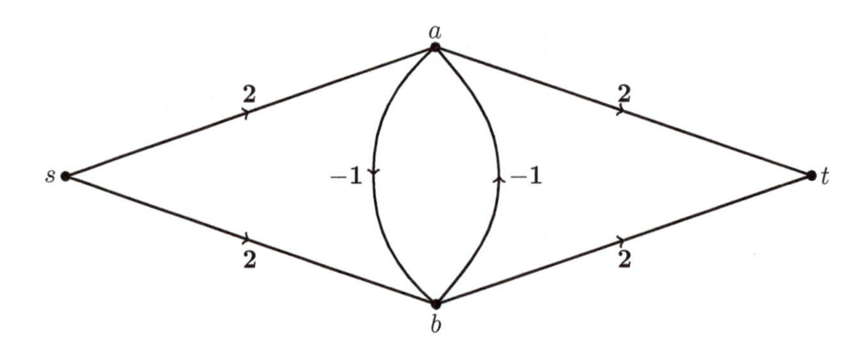

FIGURE 12.3: The shortest paths from s to a and from a to t each have length 1, contradicting the principle of optimality.

Another reason that shortest $s - t$ path algorithms often find shortest paths from s to other nodes is that as an algorithm branches out from s, it does not "know" in advance which branch will lead to t. You might be tempted to work backwards from t while working forwards from s, but it is not obvious how to do so correctly. The example in Figure 12.4 shows why. The path sat might be discovered before the path $sbct$ is discovered.

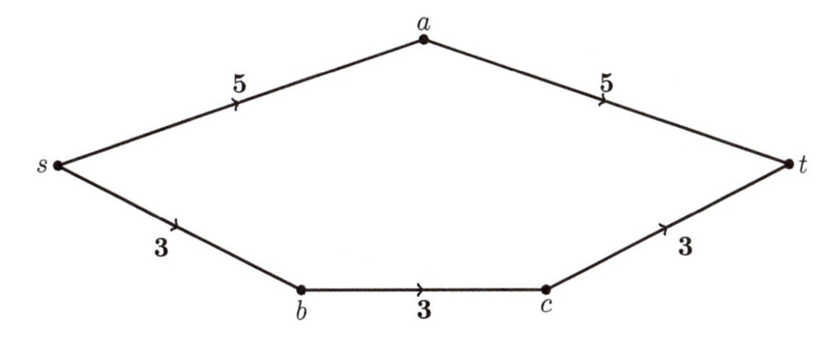

FIGURE 12.4: Looking forward from s and backward from t might make *sat* appear optimal, though *sbct* is superior.

12.4 Primal and Dual LPs for the Shortest Path Problem

The shortest path problem can be modeled as a minimum cost flow problem as follows. Set the supply at the start node s to 1. Set the supply at the end node t to -1. Set all other supply values to 0. Set the cost of arc e to w_e. Set all arc flow lower bounds to 0 and all arc flow upper bounds to 1. We shall see later that this model will fail if the graph contains a cycle of negative length. Ignoring that failure for now, what is the model's dual when considered as an LP?

When you become expert at LP, you can answer a question like this by purely general algebraic reasoning. Until then, it is usually better to start with a small explicit example. Consider the 4-node instance in Figure 12.5.

The minimum cost LP is:

$$\min 3x_{sa} + 10x_{sb} + 30x_{st} + 9x_{ab} + 13x_{at} + 5x_{bt} \quad \text{subject to} \tag{12.2}$$

$$x_{sa} + x_{sb} + x_{st} = 1 \tag{12.3}$$

$$-x_{sa} + x_{ab} + x_{at} = 0 \tag{12.4}$$

$$-x_{sb} - x_{ab} + x_{bt} = 0 \tag{12.5}$$

$$-x_{st} - x_{at} - x_{bt} = -1 \tag{12.6}$$

$$x_{ij} \geq 0 \ \ \forall i, j \tag{12.7}$$

Reading Tip: *Make sure you understand the given primal LP. Without looking ahead, write the dual LP, and figure out its meaning.*

Flow upper bounds are not necessary because all arc values are nonnegative. The constraint (12.6) looks awkward because all coefficients are negative. I've written it that way to be consistent with the other flow constraints, which are all in the form "flow out minus flow in equals supply." We could discard the constraint because as usual the flow conservation constraints contain one redundancy. *Question: Find a redundancy.*[8] I'm keeping it because it will make the dual easier to understand.

[8]The sum of constraints ((12.3)) and ((12.6)) equals the negative of the sum of the other two.

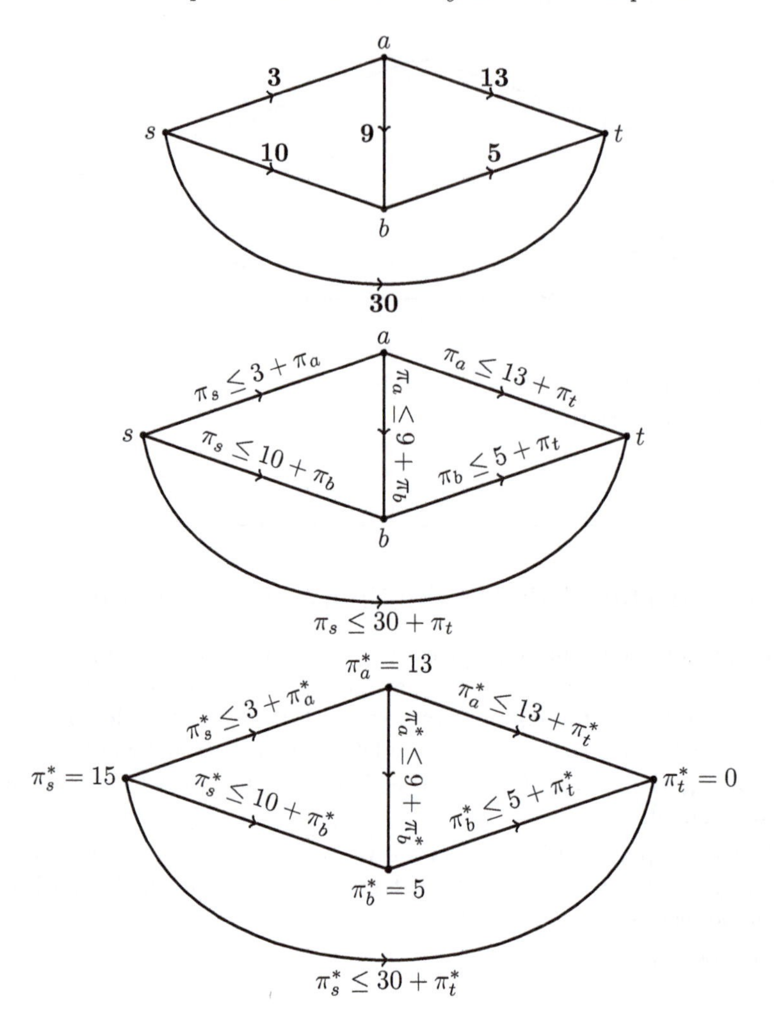

FIGURE 12.5: The shortest path from s to t traverses b and has length 15. The dual constraints do not permit $\boldsymbol{\pi}_s$ to exceed any of the values $30+\boldsymbol{\pi}_t = 30$, $3+\boldsymbol{\pi}_a = 16$,$10+\boldsymbol{\pi}_b = 15$.

Apply the standard rules to obtain the dual:

$$\max \boldsymbol{\pi}_s - \boldsymbol{\pi}_t \qquad \text{subject to} \tag{12.8}$$

$$\boldsymbol{\pi}_s - \boldsymbol{\pi}_a \leq 3 \tag{12.9}$$

$$\boldsymbol{\pi}_s - \boldsymbol{\pi}_b \leq 10 \tag{12.10}$$

$$\boldsymbol{\pi}_s - \boldsymbol{\pi}_t \leq 30 \tag{12.11}$$

$$\boldsymbol{\pi}_a - \boldsymbol{\pi}_b \leq 9 \tag{12.12}$$

$$\boldsymbol{\pi}_a - \boldsymbol{\pi}_t \leq 13 \tag{12.13}$$

$$\boldsymbol{\pi}_b - \boldsymbol{\pi}_t \leq 5 \tag{12.14}$$

$$\boldsymbol{\pi}_i \text{ unrestricted} \tag{12.15}$$

To understand this dual, first set $\boldsymbol{\pi}_t = 0$. On general principle we know we may do this, because the primal has a redundant constraint, which implies a redundant variable in the dual. In particular, adding any constant θ to every variable keeps the objective function and all constraint left-hand-sides unchanged. The reduced LP dual is

$$\max \pi_s \quad \text{subject to} \tag{12.16}$$
$$\pi_s - \pi_a \le 3 \tag{12.17}$$
$$\pi_s - \pi_b \le 10 \tag{12.18}$$
$$\pi_s \le 30 \tag{12.19}$$
$$\pi_a - \pi_b \le 9 \tag{12.20}$$
$$\pi_a \le 13 \tag{12.21}$$
$$\pi_b \le 5 \tag{12.22}$$
$$\pi_i \text{ unrestricted} \tag{12.23}$$

This dual assigns a value to each node in the graph, including a value of 0 to t. The objective is to maximize the value π_s assigned to s. The value of each node is bounded by the value of the arc that connects it to t. More generally, an arc from i to j restricts the value on i to be at most w_{ij} larger than the value on j.

How large can we make π_s? The arc st constrains $\pi_s \le 30$. But π_s is also bounded above by $\pi_a + 3$ and $\pi_b + 10$. Thus, to make π_s large we want π_a and π_b to be large. Consider π_b first. Its only constraint is $\pi_b \le 5$ because of the arc (b,t). Hence, in an optimal solution, we should set $\pi_b = 5$. Next, consider π_a. It is bounded by $\pi_a \le \pi_b + w_{ab} = 5 + 9 = 14$ because of the arc (a,b); it is also bounded by $\pi_a \le 13$ because of arc (a,t). The latter is more constraining. Hence we should set $\pi_a = 13$. Finally, we consider π_s. It has upper bounds of $w_{st} = 30, \pi_b + w_{sb} = 15$, and $\pi_a + w_{sa} = 16$. Hence we set $\pi_s = 15$. Look at the optimal dual solution we've built. For every node v the value π_v is the length of the shortest path from v to t. The binding constraints in the dual are those corresponding to arcs $(s,b), (b,t)$ and (a,t). This is duality at work. Those arcs are the ones in the shortest paths from s to t and from a to t.

This model admits of a charming physical interpretation which is attributed to Minty. (The interpretation does not account for arc directions.) Make a physical model of the graph in which nodes are small sticky balls, and arcs are strings of length equal to their value in inches, stuck at their ends to the nodes they connect. Hold the node t fixed at one end of a yardstick. Pull the node s along the yardstick away from t. The farthest distance you can pull s is the length of a shortest path from s to t. The strings that are taut correspond to the arcs that are part of the shortest path.

Let's generalize from our specific 4-node example. We are given a graph $G = (V, E)$ with arc weights $w_e : e \in E$, and two distinguished nodes $s, t \in V$. By Definition 11.3 the arc-node incidence matrix $A \in \Re^{|V| \times |E|}$ is

$$A_{ie} = \left\{ \begin{array}{l} 1 \ \text{ if } e = (i,j) \text{ for some } j \in V \\ -1 \ \text{ if } e = (j,i) \text{ for some } j \in V \\ 0 \text{ otherwise} \end{array} \right\} \tag{12.24}$$

If G has no negative cycles, the following LP models the shortest (simple) $s - t$ path problem as a minimum cost flow problem.

$$\min \ \textstyle\sum_{(i,j)\in E} w_{ij} x_{ij} \quad \text{subject to} \tag{12.25}$$
$$A_{s,:}\mathbf{x} = 1 \tag{12.26}$$
$$A_{t,:}\mathbf{x} = -1 \tag{12.27}$$
$$A_{i,:}\mathbf{x} = 0 : \ s \ne i \ne t \tag{12.28}$$
$$\mathbf{x} \ge \mathbf{0} \tag{12.29}$$

Constraint (12.27) is redundant and may be removed, or its corresponding dual variable may be set to 0. The dual is then equivalent to

$$\max \pi_s \quad \text{subject to} \tag{12.30}$$
$$\pi_j \leq \pi_i + w_{ij} \quad \forall (i,j) \in E \tag{12.31}$$
$$\pi_t = 0 \tag{12.32}$$

For any feasible dual solution π, π_i is at most the length of a shortest $i-t$ path. This must be true for $i=t$ because G has no negative cycles and it is feasible to get from t to t at zero cost. The upper bounds on the other π values are ensured by constraints (12.31). To see why, let π_i^* denote the length of a shortest $i-t$ path. If $\pi*_i < \pi_i$, then let j denote the 2nd node on an optimal $i-t$ path. If $\pi_j \leq \pi_j^*$, then $\pi \leq \pi_j + w_{ij} = \pi^*$ is a contradiction. Otherwise, $\pi^* j < \pi_j$, and repetition of the same argument starting at j will lead to a contradiction. The inequality $\pi_i \leq \pi_i^*$ illustrates that the feasibility criterion in the dual of an LP is an optimality criterion in the primal.

What then does optimality in the dual mean? Imagine a graph with strictly positive w_{ij}. The solution $\pi_i = 0 \ \forall i \in V$ is feasible in the dual. But from the primal point of view, that solution is too good to be true. There is no zero-length path from $i \neq t$ to t. Only when π_s takes its maximum possible value does it equal the length of an actual $s-t$ path in G. Optimality in the dual enforces feasibility in the primal.

At any optimum dual solution π^*, π_s^* is the length of a shortest path from s to t. This follows from the strong duality theorem. Moreover, there exists an $s-t$ path P such that for every node i in P, π_i^* is the length of a shortest $i-t$ path. However, it is not necessarily true that for all $i \in V$, π_i^* is the length of a shortest path from i to t, even if π^* is basic. *Question: Construct a three-node example that verifies this statement.*[9]

Given an optimal dual solution π^*, complementary slackness leads to an optimal primal solution. To visualize why, think of the physical interpretation of the model. Taut strings are the arcs in the shortest $s-t$ path. Mathematically, a taut string is a constraint of (12.31) that is binding. Therefore, to find the first arc of a shortest $s-t$ path, find an arc (s,j) such that $\pi_s^* = \pi_j^* + w_{ij}$. Repeat the same reasoning to find the rest of the arcs.

12.5 Shortest Paths in Graphs That Might Have Negative Cycles

If a graph possesses a negative cycle C, a path from s to some node of C, and a path from some node of C to t, then there is no lower bound on the length of a shortest non-simple $s-t$ path. A path can cycle through C an arbitrary number of times, driving the total length below any proposed finite lower bound. In financial engineering, such a cycle could represent an arbitrage possibility to be exploited vigorously until a market correction sets in. In most other applications, an unbounded optimum signals an error in the model or data. In either case, an algorithm that detects the presence of a negative cycle would be desirable. The main negative-cycle detection algorithms are all so closely related to shortest path algorithms that we study the following problem:

Problem 12.1 Shortest Paths Problem: *Given a graph $G=(V,E)$ with arc weights $w_e : e \in E$ and distinguished node $s \in V$, find a shortest simple $s-j$ path for each $j \in V$, or find a negative cycle.*

[9] $V=\{s,a,t\}, w_{st}=w_{sa}=w_{at}=1$; dual solution $\pi_t^*=0, \pi_s^*=1, \pi_a^*=0$.

There are three reasons why we don't study the following problem instead: Given a graph with arc weights and distinguished nodes s and t, find a shortest simple $s-t$ path. I've just explained the first reason, which is that negative cycle detection is desirable. The second reason, as discussed in Section 12.3, is that finding a shortest path to t usually entails finding shortest paths to many other nodes, so we might as well try to find paths to all of them. The third reason looks ahead to Chapter 14. Suppose there exists a polynomial-time algorithm $\mathcal{A}$ that solves that problem. Given a graph $G=(V,E)$ without arc weights, assign weights $w_e=-1\ \forall e\in E$. Apply $\mathcal{A}$ to find a shortest path P^*. If P^* has length $1-|V|$, it is a simple $s-t$ path that traverses every node of G. Such a path is called a *Hamiltonian* $s-t$ *path*. If the length of P^* exceeds $1-|V|$, then G does not admit a Hamiltonian $s-t$ path. Therefore, $\mathcal{A}$ allows the existence or nonexistence of a Hamiltonian $s-t$ path to be determined in polynomial time. However, it is extremely unlikely that this determination can be accomplished in polynomial time, as will be explained in Chapter 14. It is highly impractical to try to invent algorithm $\mathcal{A}$.

One of the fastest algorithms for Problem 12.1 is a classic one by Bellman and Ford. The algorithm determines the distance π_j, which is the length of a shortest (simple) $s-j$ path, for each node $j\in V$. It starts by assigning pessimistically large π_j values initially to all nodes except s, which gets a value of $\pi_s=0$. It then repeatedly updates the π_j values by applying the inequality (12.31) $\pi_j\le\pi_i+w_{ij}$. If a negative cycle containing j exists, the updating would never terminate because each decrease in π_j would trigger a further decrease in π_j. Therefore, if the algorithm runs for too many iterations, we can terminate it with the sure knowledge that the graph has a negative cycle. This is an important algorithmic idea. Prove that if your algorithm will find a solution, it will do so within some number of steps known *a priori*. Terminate the algorithm if it exceeds that number of steps.

Algorithm 12.5.1 Shortest Path Algorithm II (Bellman-Ford): For Graphs Without Negative Cycles

INPUT: Graph $G=(V,E)$, *arc weights* $w_{ij}\ \forall(i,j)\in E$.

Step 1: Initialize *Set* $\pi_s=0$. *Set* $pred(j)=null;\pi_j=\infty\ \forall j\in V, j\neq s$. *Set* $m=0$. *Set* $change=True$.

Step 2: Main Loop *WHILE* ($m\le|V|-1$ *AND* $change$) *DO*

{ $change=False$;
FOR each $(i,j)\in E$
{$d'=\pi_i+w_{ij}$;
IF $d'<\pi_j$ {$change=True;\pi_j=d';pred(j)=i$}
}
$m=m+1$;
}

Step 3 *IF* ($m<|V|$ *OR* $change==False$) *terminate with output* $\pi_j, pred(j)\ \forall j\in V$. *ELSE terminate with output "G contains a negative cycle."*

To identify a negative cycle explicitly, begin with any arc (i,j) such that $\pi_i+w_{ij}<\pi_j$. Backtrack to pred(i), pred(pred(i)), etc. until reaching s or backtracking to a node v for the second time. In the latter case, the path from the second occurence to the first occurence of v is a negative cycle. In the former case, iterate the process, starting from a node that has never been reached, until reaching a node that had been reached in a previous iteration, or backtracking to a node v for the second time. Continue iterating until a negative cycle is found.

Example 12.3
Let $V = \{s, a, b, c, d\}$, let E consist of all 20 possible directed arcs, let $w_{ba} = wdc = -5$ and let $w_e = 6$ for all other $e \in E$. Suppose that in Step 2 arcs (i, j) are processed in order $i = s, a, b, c, d$. That is, arcs leaving s are processed first, arcs leaving a are processed second, etc., and arcs leaving d are processed last. Figure 12.6 shows the values of $\pi_j, pred(j)$ at different stages of the algorithm.

outer loop	inner loop	s	a	b	c	d
0	(s, j)	0	$6, s$	$6, s$	$6, s$	$6, s$
0	(a, j)	0	$6, s$	$6, s$	$6, s$	$6, s$
0	(b, j)	0	$1, b$	$6, s$	$6, s$	$6, s$
0	(c, j)	0	$1, b$	$6, s$	$6, s$	$6, s$
0	(d, j)	0	$1, b$	$6, s$	$1, d$	$6, s$
1	(d, j)	0	$1, b$	$6, s$	$1, d$	$6, s$

FIGURE 12.6: Bellman-Ford Algorithm on a complete graph with 5 nodes and no negative cycles.

Now set $w_{cb} = -4$, leaving the other weights unchanged. Figure 12.7 shows the values of $\pi_j, pred(j)$ at different stages of Step 2 of the algorithm. Step 2 could be safely terminated after stage $3(a, j)$ because $\pi_s < 0$. I've continued to stage $4(d, j)$ for the sake of illustration.

outer loop	inner loop	s	a	b	c	d
0	(s, j)	0	$6, s$	$6, s$	$6, s$	$6, s$
0	(a, j)	0	$6, s$	$6, s$	$6, s$	$6, s$
0	(b, j)	0	$1, b$	$6, s$	$6, s$	$6, s$
0	(c, j)	0	$1, b$	$2, c$	$6, s$	$6, s$
0	(d, j)	0	$1, b$	$2, c$	$1, d$	$6, s$
1	(c, j)	0	$1, b$	$-3, c$	$1, d$	$6, s$
2	(b, j)	0	$-8, b$	$-3, c$	$1, d$	$3, b$
2	(d, j)	0	$-8, b$	$-3, c$	$-2, d$	$3, b$
3	(a, j)	$-2, a$	$-8, b$	$-3, c$	$-2, d$	$-2, a$
3	(c, j)	$-2, a$	$-8, b$	$-6, c$	$-2, d$	$-2, a$
3	(d, j)	$-2, a$	$-8, b$	$-6, c$	$-7, d$	$-2, a$
4	(b, j)	$-2, a$	$-12, b$	$-6, c$	$-7, d$	$-2, a$
4	(c, j)	$-2, a$	$-12, b$	$-11, c$	$-7, d$	$-2, a$
4	(d, j)	$-2, a$	$-12, b$	$-11, c$	$-7, d$	$-2, a$

FIGURE 12.7: Step 2 of Bellman-Ford Algorithm on a complete graph with 5 nodes and negative cycles *dcbd*, *cbac*, *dcbad*, *cbasc*, etc.

Step 2 has detected the presence of a negative cycle.

Theorem 12.5.2 *Algorithm 12.5.1 solves Problem 12.1 in* $O(|V||E|)$ *time.*

Proof: The inner loop of Step 2 takes $O(1)$ time. Hence the outer loop takes $O(|E|)$ time, and the entire step takes $O(|V||E|)$ time. Steps 4 and 5 each take $O(|E|)$ time. This proves the stated time bound. An intuition as to why the algorithm is correct is that it decreases the variables in the dual LP (12.30)-(12.32) just enough to ensure feasibility. A rigorous proof of correctness based on that intuition follows. First, if the algorithm terminates at Step 3, it has found an optimal solution. The solution π must be feasible in the dual LP because $m < |V|$ or *change* $==$ *FALSE* implies that every dual constraint is satisfied. Every upper bound enforced during Step 2 is a logical consequence of constraints (12.31),(12.32). Therefore, every feasible solution to the dual LP satisfies those upper bounds, regardless of which π_i is being maximized. At Step 3 each π_i equals the least of its upper bounds. Hence each π_i is at its maximum possible value in the dual.

Second, at any point of Step 2 of the algorithm, if $\pi_j < \infty$ there exists an $s - j$ path of length π_j. At the moment π_j is set to $\pi_i + w_{ij}$, it must be that $\pi_i < \infty$. By induction on the steps of the inner loop of Step 2, there exists an $s - i$ path of length π_i. By the algorithm's definition, $(i, j) \in E$. Hence the $s - i$ path combined with the arc (i, j) is an $s - j$ path of length π_j.

Third, if there exists an $s - j$ path of $\leq k$ edges with length L, then $\pi_j \leq L$ after k iterations of the outer loop of Step 2. The zero-edge path s makes this statement trivially true for $k = 0$, the base case of the inductive proof. For $k \geq 1$, let (i, j) be the last arc in the $s - j$ path. The rest of the path must be an $s - i$ path of $\leq k - 1$ edges with length $L - w_{ij}$. Inductively, $\pi_i \leq L - w_{ij}$ at the end of the $k - 1$st iteration of the outer loop. Therefore, during the kth iteration the inequality $\pi_j \leq \pi_i + wij \leq L - w_{ij} + w_{ij} = L$ will be enforced. ■

By Theorem 12.3.1 and its corollary, for every $j \in V$ such that an $s - j$ path exists, there exists a shortest $s - j$ path which is also simple. Any simple path in G has at most V nodes. Combining the three statements proves the theorem. ■

Implementation issues: The number of times the outer loop of Step 2 is executed can depend heavily on the order in which the arcs in E are processed. As a simple illustration, process the arcs (i, j) in the first part of Example 12.3 in order $i = d, c, b, a, s$. You will need three iterations of the outer loop, rather than two. For a more extreme example, let $V = \{1, 2, \ldots, n\}$, let $E = \{(i, i+1) : 1 \leq i \leq n-1\}$, and let $w_e = 1\ \forall e \in E$. Processing the arcs (i, j) in increasing order of i requires two iterations; processing the arcs in the reverse order requires n iterations.

How should one order the arcs for processing? Clearly, the arcs (s, j) should be processed first during the first iteration of the outer loop. At different iterations of the outer loop, one may want to use different orderings. Outgoing arcs from node i should not be processed unless π_i has decreased since the previous scan. Beyond these guidelines, the decisions are murkier. In general, it seems that higher priority should be given to arcs (i, j) for which the value π_i has recently been decreased. However, a conflict emerges between the goals of finding shortest paths and finding a negative cycle. If the former are sought, it is better for many π_i to be updated during each outer loop iteration, so that fewer iterations are required. But if the latter is sought, it is better for few π_i to be be updated each iteration. This is because the number of iterations is necessarily n, but the iterations themselves will be faster if π_i values decrease infrequently.

Searching for an explicit negative cycle can be done with a kind of breadth-first search instead of the depth-first type search described above. Breadth-first search is more complicated to implement, but is apt to terminate more quickly if the negative cycle detected has few arcs. See [48] for an extensive discussion of how to improve the computational performance of Bellman-Ford and other shortest path and negative cycle detection algorithms.

Next we tackle a problem slightly different from Problem 12.1, namely to find shortest paths between all pairs of nodes in a graph, or to find a negative cycle. Of course, we could run Algorithm 12.5.1 $|V|$ times, at total time cost $O(|E||V|^2)$ (or once if there is a negative cycle). It should not surprise you that there is a more efficient way than that to find shortest paths between all node pairs. You've seen from the principle of optimality that a shortest path between two nodes might be part of many shortest paths between other pairs of nodes. But it should surprise you that, for very dense graphs, it takes the same order of time to find all these paths as to run Algorithm 12.5.1 once! (That is true in a theoretical worst-case sense. Algorithm 12.5.1 is faster in practice, even on dense graphs.) The standard algorithm is by Floyd and Warshall. It is similar to Algorithm 12.2.1 in that it iteratively maintains an invariant. Read the algorithm and try to guess the invariant. (Hint: It has to do with interior nodes.)

Algorithm 12.5.3 Shortest Path Algorithm III (Floyd-Warshall): For All Pairs on Graphs Without Negative Cycles

Input: *Graph* $G = (V, E)$, $V = \{1, 2, \ldots, n\}$, *weights* w_e $\forall e \in E$.

1. *Set* $Pred(ij) = null; d^0_{ij} = \infty\ \forall i \in V, j \in V, i \neq j$.
   Set $d^0_{ii} = 0\ \forall i \in V$. $\forall (i,j) \in E$ *set* $d^0_{ij} = w_{ij}; pred(ij) = i$.

2. *FOR* $k = 1$ *to* n
   { *FOR* $i = 1$ *to* n
   { *FOR* $j = 1$ *to* n
   $\{d' = d^{k-1}_{ik} + d^{k-1}_{kj};$
   IF $d' < d^{k-1}_{ij}$ $\{d^k_{ij} = d'; pred(ij) = pred(kj);\}$
   ELSE $d^k_{ij} = d^{k-1}_{ij};$ } *IF* $d^k_{ii} < 0$ *STOP* ** *negative cycle detected* ** } }

The invariant is that d^k_{ij} = the length of a shortest $i - j$ path with all interior nodes $\leq k$. Also, $pred(ij)$ is the penultimate (next-to-last) node on such a path. *Question: What is the antepenultimate (third-to-last) node on such a path?*[10] In formal terms, let $P^k(ij)$ denote the shortest path from i to j found after the kth iteration of the outer loop of the algorithm. Then $P^k(ij) = \{P^k(i\ pred(ij)), j\}$.

The d^0_{ij} values are initialized correctly because a single arc is the only possible path with no interior nodes. The invariant is maintained because if G contains no negative cycles, the shortest $i - j$ path with interior nodes $\leq k$ either traverses k or is identical to the shortest $i - j$ path with interior nodes $\leq k - 1$. See Figure 12.8.

A negative cycle will terminate the algorithm with some $d^k_{ii} < 0$ such that $d^{k-1}_{ii} \geq 0$. The negative cycle found will be the combination of the shortest path $P^{k-1}(ik)$ from i to k with the shortest path $P^{k-1}(ki)$ from k to i.

Theorem 12.5.4 *Algorithm 12.5.3 solves Problem 12.1 for all pairs of nodes in* $O(|V|^3)$ *time.*

Proof: Correctness of the algorithm follows from the invariant stated above. The time bound is correct, and tight, because there are three nested loops of $|V| = n$ iterations each, and innermost loop takes $O(1)$ time. ■

[10] $pred(i\ pred(ij))$.

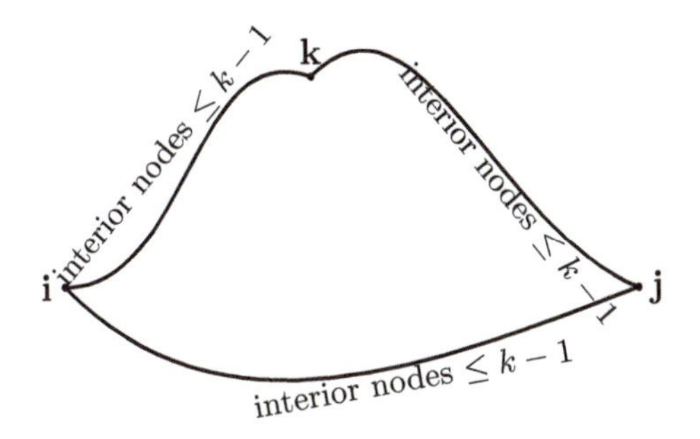

FIGURE 12.8: The shortest $i - j$ path with interior nodes $\leq k$ either combines the $i - k$ and $k - j$ paths with interior nodes $\leq k - 1$ or does not traverse k.

Example 12.4 Finding a negative cycle with Algorithm 12.5.3

Let $V = \{1, 2, 3, 4, 5\}$. Let $w_{12} = 10, w_{23} = 20, w_{34} = 30, w_{45} = 40, w_{51} = 50$, $w_{4,1} = -75, w_{13} = 100, w_{25} = 100$. The following table depicts the progress of the algorithm. Values that change from k to $k + 1$ are in boldface.

*from**to*	1	2	3	4	5
1	0	10, 1	100, 1	∞, *null*	∞, *null*
2	∞, *null*	0	20, 2	∞, *null*	100, 2
3	∞, *null*	∞, *null*	0	30, 3	∞, *null*
4	−75, 4	∞, *null*	∞, *null*	0	40, 4
5	50, 5	∞, *null*	∞, *null*	∞, *null*	0
$k = 1$					
1	0	10, 1	100, 1	∞, *null*	∞, *null*
2	∞, *null*	0	20, 2	∞, *null*	100, 2
3	∞, *null*	∞, *null*	0	30, 3	∞, *null*
4	−75, 4	**−65, 1**	**30, 1**	0	40, 4
5	50, 5	**60, 1**	**150, 1**	∞, *null*	0
$k = 2$					
1	0	10, 1	**30, 2**	∞, *null*	**110, 2**
2	∞, *null*	0	20, 2	∞, *null*	100, 2
3	∞, *null*	∞, *null*	0	30, 3	∞, *null*
4	−75, 4	−65, 1	**−45, 2**	0	**35, 4**
5	50, 5	60, 1	**80, 2**	∞, *null*	0
$k = 3$					
1	0	10, 1	30, 2	**60, 4**	110, 2
2	∞, *null*	0	20, 2	**50, 3**	100, 2
3	∞, *null*	∞, *null*	0	30, 3	∞, *null*
4	−75, 4	−65, 1	−45, 2	**−15, 3**	35, 4
5	50, 5	60, 1	80, 2	110, 3	0

When the algorithm terminates with $d^3_{44} = -15$, it has found the negative cycle that is the combination of $P^2(43)$ and $P^2(34)$. From the table at $k = 2$, $P^2(43) = \{pred(4\,pred(4\,pred(43))), pred(4\,pred(43)), pred(43), 3\} = \{4, 1, 2, 3\}$ with length -45.

Similarly, $P^2(34) = \{pred(34), 4\} = \{3, 4\}$ with length 30. The cycle is $\{4, 1, 2, 3, 4\}$ with length $-45 + 30 = 15$.

The algorithm requires $O(|V|^3)$ time. This is the same order of magnitude as the $O(|E||V|)$ time of Algorithm 12.5.1 for very dense graphs! For sparse graphs the $O(|V|^3)$ time bound is necessarily less good.

12.6 Dijkstra's Algorithm for Graphs with Nonnegative Costs

When all arc-lengths w_{ij} are nonnegative, a primal algorithm invented by Dijkstra is available instead. It is faster than Algorithm 12.5.1 both theoretically and in practice.

Algorithm 12.6.1 Shortest Path Algorithm IV (Dijkstra)
Input: Graph $G = (V, E)$*, distinguished node* $s \in V$*, arc lengths* $w_{ij} \geq 0 : \ (i, j) \in E$.
Step 1 *Set* $\pi_s = 0$*;* $pred(j) = null$ *and* $\pi_j = \infty \ \forall s \neq j \in V$*. Mark all* $v \in V$ *temporary.*
Step 2 *Select* $u \in V$ *such that* u *is marked temporary and* $\pi_u \leq \pi_v$ *for all* v *that are marked temporary. If no such* u *exists, or* $\pi_u = \infty$*, terminate.*
Step 3 *Scan* u*: FOR all arcs* $(u, j) \in E$ *such that* j *is marked temporary DO*
{
  Set $d' = \pi_u + w_{uj}$*;*
  If $d' < \pi_j$ $\{\pi_j = d';\ pred(j) = u\}$
}
Step 4 *Mark* u *permanent;*
Go to Step 2.

Figure 12.9 shows the progress of Algorithm 12.6.1 on a 4-node graph. The first image shows the graph at the beginning of the first iteration of Step 2. The succeeding images show the graph at the end of each iteration of Step 3. Each value π_v is shown next to its node v. Permanent node π_v values are in bold. The node that was just scanned is circled. Arcs from predecessors are doubled.

Theorem 12.6.2 *When Dijkstra's Algorithm 12.6.1 terminates on a graph* $G = (V, E)$ *with nonnegative arc lengths,* π_v *equals the length of a shortest* $s - v$ *path in* G *and* $pred(v)$ *is the penultimate node on such a path, for all* $v \in V$ *for which an* $s - v$ *path exists. For all other nodes* v*,* $\pi_v = \infty$.

Proof: The proof is by induction on the number of nodes marked *permanent*. The natural inductive hypothesis is that π_j and $pred(j)$ are correct for all permanent j.

Inductive hypothesis: At the beginning of Step 2, π_j equals the length of a shortest $s - j$ path for all permanent j, and $pred(j)$ is the penultimate node on such a path.

Let u be the node whose label becomes permanent in the current iteration of the algorithm and let (v, u) be the arc from permanently labeled node v from which the temporary label value $\pi_u = \pi_v + w_{uv}$ derives. By induction, there exists an $s - v$ path of cost π_v. Hence there exists an $s - u$ path, of length π_u whose penultimate node is v.

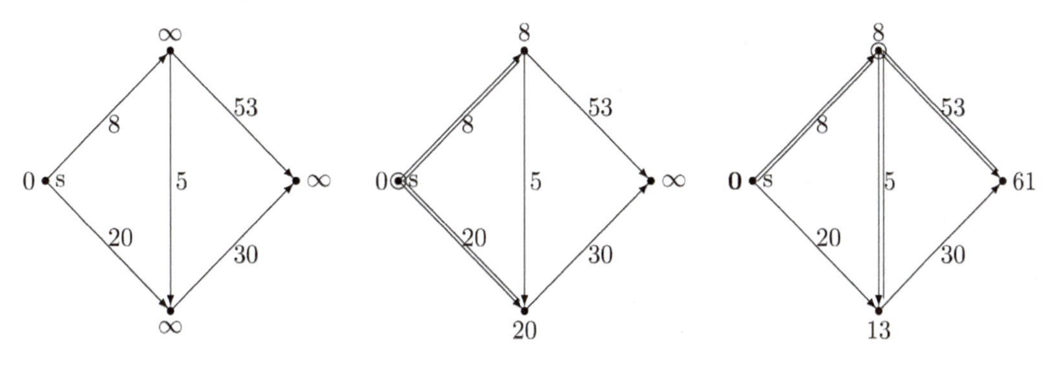

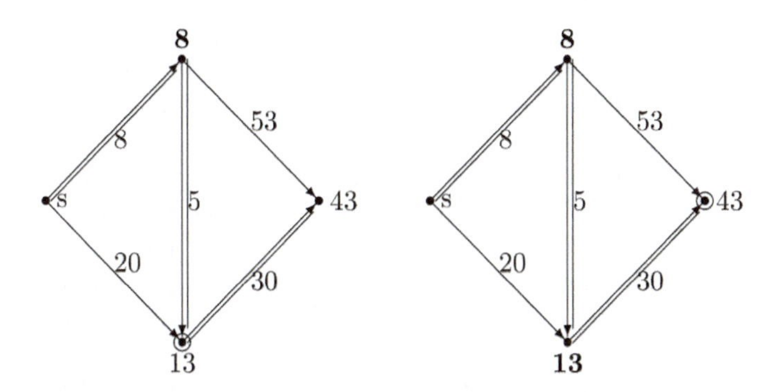

FIGURE 12.9: Dijkstra's algorithm chooses one node at a time to be scanned and made permanent.

To complete the proof, we must show that no shorter $s-u$ path exists in G. For any path P let $w(P)$ denote the length of P. Let Q be an arbitrary $s-u$ path in G. Since $\boldsymbol{\pi}_s$ is permanent and $\boldsymbol{\pi}_u$ is not, Q must contain at least one arc (i,j) such that $\boldsymbol{\pi}_i$ is permanent and $\boldsymbol{\pi}_j$ is not permanent (j could be u). Choose one such (i,j). Let Q_{si} (respectively Q_{ju}) denote the subpath of Q from s to i (respectively j to u). Then

$$\begin{aligned}
\boldsymbol{\pi}_u \leq & \quad \boldsymbol{\pi}_j && \text{by Step 2 since } \boldsymbol{\pi}_j \text{ is temporary} \\
\leq & \quad \boldsymbol{\pi}_i + w_{ij} && \text{by Step 3 since } \boldsymbol{\pi}_i \text{ is permanent} \\
\leq & \quad w(Q_{si} + w_{ij}) && \text{by the inductive hypothesis applied to node } i \\
\leq & \quad w(Q_{si} + w_{ij}) + w(Q_{ju}) && \text{by nonnegativity of edge lengths} \\
= & \quad w(Q)
\end{aligned}$$

Therefore, $\boldsymbol{\pi}_u$ is the length of an optimal $s-u$ path in G. By Step 3 the predecessor of u on the shortest $s-u$ path is correctly set to v, and the inductive step is complete. ∎

Question: What happens if Step 3 does not check that j is marked temporary?[11]

[11] The algorithm will run correctly but a tiny bit more slowly.

12.6.1 Running Time of Dijkstra's Algorithm

Dijkstra's algorithm exemplifies several important principles. This section will acquaint you with three.

Applications The density of your data can control which algorithm and data structures are best for you.

Algorithm Implementation Determine how many of each kind of operation your algorithm requires. Select your data structures accordingly.

Algorithm Analysis For upper bound analysis, all cases must be considered but inner loops may be analyzed separately from outer loops. For lower bound analysis, only one set of cases need be considered, but all loop levels must be analyzed simultaneously.

If $G = (V, E)$ is very dense, i.e., $|E| = \theta(|V|^2)$, store G and its arc weights in two-dimensional arrays of size $|V| \times |V|$, and store $\boldsymbol{\pi}$ and permanent/temporary information in one-dimensional arrays of size $|V|$. Each iteration of Step 2 takes $O(|V|)$ time. Each iteration of Step 3 takes $O(|V|)$ time. The total time is therefore $O(|V|^2)$. This upper bound is optimal (within some constant factor) for any algorithm because each arc must be accessed and there are $\Omega(|V|^2)$ arcs.

You should be asking yourself why each arc must be accessed. There is an elegant proof that uses the idea of an "adversary." Whenever the algorithm accesses data, the adversary may give any answer it wants, as long as it is consistent with the data that it has given previously. In the case of shortest paths, consider any shortest path algorithm. Let graph G have node set $1, 2, \ldots n$ and let E consist of all possible arcs (i, j) where $i < j$. Whenever the algorithm asks for an arc weight w_{ij}, the adversary answers $j - i$ until all but one arc weight has been queried. At the last query, the adversary can answer either $j - i$ or 0. Both answers are consistent with all previously given data, but they result in different optimal solution values, either $n - 1$ or $n - 1 - (j - i)$. Therefore, the algorithm cannot be sure to find a correct solution without querying all $|E|$ arc weights.

If G is not dense, the array-based implementation described above is not so good, even if the G and w data are stored as sparse matrices. *Question: Why?*[12]

To find a better implementation for sparse data, determine precisely what operations must be performed with the data, and how many of each operation are needed. A value $\boldsymbol{\pi}_v$ must be initialized for each node $v \in V$, and maintained for each temporary node. The temporary node with least $\boldsymbol{\pi}_v$ value must be identified and removed from the set of temporary nodes. Each time a node is made permanent its outgoing arcs must be identified, and each such arc may cause a temporary $\boldsymbol{\pi}_u$ value to be decreased. Therefore, the algorithm needs a data structure that will be initialized for $|V|$ values, remove its smallest value $|V|$ times, and reduce a stored value up to $|E|$ times. Data structures that support these operations are often called *priority queues.*

Fibonacci heaps, invented by Fredman and Tarjan [105], have the best theoretical performance known for these operations in these quantities. Initialization requires $O(|V|)$ time; m removals of the smallest value require total time $O(m \log |V|)$; m reductions in stored values require total time $O(|E|)$. The overall runtime of Dijkstra's algorithm with Fibonacci heaps is therefore $O(|E| + |V| \log |V|)$. The implementation and analysis are complicated, in part because sometimes a stored value reduction takes more than $O(1)$ time. That only occurs after more than $O(1)$ reductions that do take only $O(1)$ time, so that the total time for m such operations is $O(m)$. The very general idea of the data structure is to allow the

[12]Step 3 will take less total time, but Step 2 will still require $\theta(|V|^2)$ total time.

data to fragment into a lot of small heaps for a while instead of reorganizing after each operation.

The binary heap, which is a much simpler data structure than the Fibonacci heap, enables a runtime of $O(|E| \log |V|)$. Initialization requires $O(|V|)$ time; each removal of the smallest value requires $O(\log |V|)$ time; each reduction of a stored value requires $O(\log |V|)$ time. Chapter 9 explains how the binary heap is stored and how its operations are performed. Here we will treat the binary heap as a "black box" which we use without knowing what is inside it. We only need some high-level information about the black box, given in the following theorem.

Theorem 12.6.3 *A binary heap stores variables, each of which has a value. The values are subject to a transitive ordering relation $\leq$. The time required to build a heap of n elements is $\theta(n)$. The heap supports the following operations in $O(\log n)$ time:*

Remove-Minimum *Removes a variable with minimum value from the heap.*

Decrease-Value *Decreases the value of a variable in the heap.*

Increase-Value *Increases the value of a variable in the heap.*

Insert *Inserts a new variable into the heap.*

If the Decrease-Value operation is performed on a variable whose current value is the unique maximum in the heap and whose new value is the unique minimum in the heap, then the time required is $\theta(\log n)$.

The last statement of the theorem is not typically included in the description of the performance of a binary heap, but it will be needed for our analysis.

Theorem 12.6.4 *Dijsktra's algorithm requires $O(|E| \log |V|)$ time when supported by binary heaps. This bound is tight for all $|V| \leq |E| \leq \binom{|V|}{2}$.*

Proof: The algorithm creates the heap of $|V| - 1$ elements once, at cost $O(|V|)$. It removes the element with minimal $\boldsymbol{\pi}_v$ from the heap at most $|V| - 1$ times at cost $O(\log |V|)$ each, for a total cost $O(|V| \log |V|)$. The algorithm updates a $\boldsymbol{\pi}_v$ value at most $|E|$ times since each arc in the graph can trigger at most one update. The cost per update is $O(\log |V|)$ for a total cost $O(|E| \log |V|)$. Since s can't be connected to more than $|E|$ nodes in the graph, we can assume $|E| \geq |V|$. Therefore, the $O(|E| \log |V|)$ term dominates the other two terms, yielding the upper bound as claimed.

To prove that the upper bound is tight, it is necessary and sufficient to prove the existence of an infinite set of graphs, increasing in size $|E|, |V|$, that require $\Omega(|E| \log |V|)$ time. We must be careful about two things. First, the graphs must be allowed to vary depending on the arc density, since the Theorem statement claims so. Second, it does not suffice to prove that the number of $\boldsymbol{\pi}_v$ updates is $\Omega(|E|)$ and that there are updates that require $\Omega(\log |V|)$ time. There could be many updates, only a few of which require a lot of time. (Here "many" means $\geq c|E|$ for some positive constant c, and "a lot" means $\geq c' \log |V|$ for some positive constant c'.) We must prove that there are many updates that *each* require a lot of time.

Let $|V|$ be arbitrary and let M be any value in the range $2|V| \leq M \leq \binom{|V|}{2}$. Denote the nodes $1, 2, \ldots, |V|$. Create arc set $E = (1,2), (1,3), \ldots, (1, |V|)$; $(|V|, 2), (|V|, 3), \ldots, (|V|, |V|-1)$; $(|V|-1, 2), (|V|-1, 3), \ldots, (|V|-1, |V|-2)$; …, $(|V|-$

$k, 2), (|V| - k, 3), \ldots, (|V| - k, |V| - k - 1); \ldots$ until there are M arcs. We assume that the algorithm scans arcs emanating from u in lexicographic order. Play the adversary against the algorithm. Set $M = |E| + 1$. When the algorithm scans a node u and asks for the arc length w_{uv}, answer according to the following rule:

- If (u, v) is the last arc from u to be queried, respond with $w_{uv} = 0$.
- Otherwise, respond with $w_{uv} = M$. Decrement $M = M - 1$.

These arc values create a path of length 0 from s. Every permanent label has value 0. Because the M arcs are scanned in order of the node they go to, each time a temporary label is updated its value is the oldest – hence the unique largest – value in the heap, and it is changed to the unique least value in the heap. By Theorem 12.6.3, each of the first $M/2$ updates requires $\Omega((\log(|V - 1|/2|) - 1)$ time for a total amount of time $\Omega(|E| \log |V|)$. ■

Summary: For modeling, the most important idea is that a node need not represent a physical location. It may instead represent

- a time;
- a physical location at a time;
- a stage of completion of a task;
- a state or status of the world.

For understanding optimization, shortest path problems are the least complicated ones that are still complex enough to illustrate many concepts, including:

- Special properties of the data can render a problem easier to solve.
- Data structures are crucial to algorithm performance.
- Optimal primal and dual solutions have complementary slackness.
- The primal and dual LPs are two different ways of looking at the same thing.

12.7 Problems

E Exercises

1. Let graph G have vertex set $V = \{1, 2, 3, 4, 5, 6\}$ and directed arcs $\{j, j+1\}$ with cost $4j^2$ for $j = 1 \ldots, 5$ and directed arcs $\{j, j+2\}$ with cost $-(j+5)$ for $j = 1, \ldots, 4$. Find the length of a shortest path from 1 to 6.

2. For both models of the truck replacement and maintenance example in the beginning of the chapter, draw the graph, including all node labels and edge costs, for a 5-year problem with $c_k = k^2, d_k = 20$, and $s_k = 18 - 2k$.

3. Extend Figure 12.1 with the datum $L_5 = 900$.

4. (Continuation) Why is the datum L_1 irrelevant?

5. Let $V = \{1, 2, 3, 4, 5\}$ and let $c_{12} = 40 = c_{21}; c_{13} = 30 = c_{31}$; $c_{14} = 60 = c_{41}$; $c_{25} = 90$; $c_{32} = c_{23} = 8$; $c_{43} = 10 = c_{35} = c_{53}$, $c_{45} = 100$; $c_{34} = 20 = c_{43}$. Find a shortest path from 1 to 5 using Dijkstra's algorithm.

6. Repeat Problem 5 to find a shortest path from 4 to 2.

7. Let G be as in Problem 5. Apply the Floyd-Warshall algorithm to find all shortest paths, using vertex order $\{1, 2, 3, 4, 5\}$.

8. Repeat Problem 7 with vertex order $\{5, 4, 3, 2, 1\}$.

9. Modify Algorithm 12.2.1 for the case $w_{ij} \geq 0$.

10. Devise a graph with negative arcs but no negative cycles for which Dijkstra's algorithm does not find an optimal solution.

11. Devise a graph with a negative cycle for which Dijkstra's algorithm finds an optimal solution.

M Problems

12. Consider the truck replacement example from the beginning of the chapter. Modify the shortest path model to account for the following circumstances:
    - The cost of purchasing a new truck varies from year to year.
    - The cost of maintaining a truck depends on both its age and the year it was purchased.
    - The salvage value of a truck depends on both its age and the year it was purchased.
    - There are two truck models, each with its own purchase, maintenance, and salvage data.

13. Determine the number of shortest paths from $(0, 0)$ to (m, n) on a 2D grid.

14. Which shortest path would an implementation of Dijkstra's algorithm be apt to find in Problem 13?

15. Which shortest path would an implementation of the Bellman-Ford algorithm be apt to find in Problem 13?

16. Determine the number of shortest paths from $(0, 0, 0)$ to (m, n, t) on a 3D grid.

17. Let $G = (V, E), s \in V, t \in V, w_e : e \in E$ be a shortest path instance with no negative cycles. (i) Prove that for any optimal dual solution $\boldsymbol{\pi}^*$ to the shortest path LP, there exists an $s - t$ path P such that for every node i in P, π_i^* is the length of a shortest $i - t$ path. (ii) Prove that there exists an optimal dual solution $\boldsymbol{\pi}^*$ such that for every node $i \in V$, π_i^* is the length of a shortest $i - t$ path.

18. One could argue that no algorithm to find all-pairs shortest paths can run in time $o(|V|^2)$ in the worst case, because there are $|V|(|V|-1)$ pairs, and a positive fraction of the shortest paths could each contain more than $c|V|$ arcs for some positive constant c. What is the flaw in this argument?

19. Suppose the maximum degree of any node in $G = (V, E)$ is 5. One might hope to take advantage of the low degree of the nodes to speed up the Floyd-Warshall Algorithm 12.5.3. Prove that it is impossible to achieve $o(|V|^3)$ time for any algorithm that iteratively finds all-pairs shortest paths containing no interior node labeled $\geq i$ for $i = 1, 2, \dots |V|$.

20. Let $G = (V, E)$ be a directed acyclic graph with distinguished nodes s and t. Invent an algorithm to determine the number of $s - t$ paths in G. Make your algorithm run as quickly as you can.

21. Analyze the time required by the Bellman-Ford Algorithm 12.5.1 to find a negative cycle once it has detected the existence of one.

22. Show exactly how to extract a negative cycle from the Floyd-Warshall Algorithm 12.5.3.

23. Can you speed up the Floyd-Warshall algorithm to run on sparse graphs, e.g., graphs $G = (V, E)$ with $|E| = O(|V| \log |V|)$?

24. Can you speed up the Bellman-Ford algorithm to run on sparse graphs?

25. Devise a polynomial time algorithm to find a negative cycle with the least number of arcs.

26. Consider a topological sort in Step 1 of Algorithm 12.2.1 that repeatedly removes a node that has no incoming arcs, together with its outgoing arcs. Suppose that the graph $G = (V, E)$ is represented by the following data structure: an array listing the nodes of the graph; for each node $v = 1 \dots |V|$, an array $In_v[\,]$ with $In_v[0]$ = number of arcs entering v, $In_v[1 \dots In_v[0]]$ the nodes j for which $(j, v) \in E$, and a similar array $Out_v[\,]$ listing the arcs leaving v. (i) A stupid implementation would, at each iteration, search the nodes in V to find one with $In_v[0] = 0$. Show that this could take total search time $\Omega(|V|^2)$, and show how to reduce the time to $O(|V|)$. (ii) Show that the other steps of the algorithm could take time $\Omega(|E||V|)$. (iii) Show how to reduce that time to $O(|E|)$ by changing the data structure.

27. Negative cycles and linear programming: Let $G = (V, E)$ be a graph with arc weights $w_e : e \in E$. (i) Write a linear program that has unbounded optimum iff G has a negative cycle. (ii) Can you write a linear program that finds a negative cycle in G with the largest possible number of arcs? If not, what goes wrong? (iii) Use duality to find a system of linear inequalities that is feasible iff G has no negative cycles.

28. (a) Construct a dense graph without a negative cycle on which the Bellman-Ford Algorithm 12.5.1 can take $\Omega(|V||E|)$ time.
    (b) Repeat the previous question for the version of Algorithm 12.5.1 described in the text that only processes arcs outgoing from nodes whose π values have changed since their outgoing arcs were processed.
    (c) Prove or disprove: For every graph $G = (V, E)$ without a negative cycle there exists a permutation σ of E such that Algorithm 12.5.1 runs in time $O(|E|)$, if arcs are processed in the order given by σ.

D Problems

29. (a) Prove that the Bellman-Ford Algorithm 12.5.1 can take time $\Omega(|V||E|)$ regardless of the order in which vertices are scanned, even if the algorithm scans in a different order from one iteration to the next. *Hint: Construct a graph with a long negative cycle but no short negative cycles.*

    (b) Prove that if an edge-weighted graph $G = (V, E)$ with distinguished vertex $s \in V$ contains no negative cycles, and for each $v \in V$ contains a directed $s - v$ path, then there exists a spanning tree T with edge set $F \subset E$ such that for all $v \in V$ the $s - v$ path in T is a shortest $s - v$ path in G. *Hint: Use a basic theorem of LP and Proposition 11.3.6.*

    (c) Prove that for every graph G as in part 29b, it is possible for the Bellman-Ford algorithm 12.5.1 to terminate within two iterations. *Hint: Use part 29b.*

30. The *transitive closure* of graph $G = (V, E)$ is the graph $H = (V, F)$ defined by $(i, j) \in F \Leftrightarrow \exists$ path P in G from i to j. How can you construct H efficiently, given G?

31. Examine the LP for the shortest $s - t$ path, and its dual. Explain what happens in the dual, and what the strong duality theorem means, when (i) there are negative arc weights but no negative cycles; (ii) there are negative cycles.

32. The Minimum Spanning Tree problem (MST) for an undirected graph $G = (V, E)$ with arc weights $w_e : e \in E$ is to find a spanning tree $T \subset E$ of G such that $\sum_{e \in T} w_e$ is minimized.

    (a) Prove that the following greedy algorithm solves the MST: select any $v \in V$ and set $V_T = \{v\}$ and $T = \phi$. Until $|T| = |V| - 1$, select an $(i, j) \in E : i \in V_T; j \notin V_T$ with minimum weight w_{ij}, add j to the set V_T and add (i, j) to the set T.

    (b) Prove that if the greedy algorithm is implemented with binary heaps, it runs in time $O(|E| \log |V|)$.

    (c) Prove that the $O(|E| \log |V|)$ bound is tight. (Hint: The analysis is similar to the analysis of Dijkstra's algorithm.)

33. Prove Theorem 12.6.4 without assuming that edges are scanned in a pre-determined order. Hint: Show that every leaf moves to the root and use the property that $\lceil \frac{n}{2} \rceil$ of the elements in an n-element heap are leaves.

34. (a) Let α_i be the value of the label made permanent in the ith iteration of Dijkstra's algorithm. Prove $\alpha_i \geq \alpha_{i+1}$ $\forall i$.

    (b) Suppose all arc costs are nonnegative integers $\leq W$. Without using a complicated data structure, find an algorithm to find shortest paths from a start node s in time $O(|E| + |V|W)$.

35. (Ahuja *et al.* [4]) Suppose all arc costs are nonnegative integers $\leq W$. Without using Fibonacci heaps, find an algorithm to find shortest paths from a start node s in time $O(|E| + |V| \log W)$.

36. (Karp [174]) The mean cost of a (directed) cycle C is the average cost of its arcs, that is, $\frac{1}{|C|} \sum_{e \in C} c_e$. Find an $O(|V||E|)$ algorithm to find the minimum over all cycles in

the graph $G = (V, E)$ of the mean cost. Hint: Let $v \in V$. Show that if the minimum mean cost equals 0, this value equals

$$\min_{w \in V} \max_{0 \leq k < |V|} \frac{\pi_w^n - \pi_w^k}{n - k}$$

where $\pi_j^k \equiv$ the cost of a shortest path from v to j consisting of exactly k arcs.

37. (*) The Bellman-Ford Algorithm 12.5.1 is a dual algorithm from the perspective of the linear program (12.25) to (12.28), because it operates on the dual variable vector π. Dijkstra's Algorithm 12.6.1 is also a dual algorithm from this perspective. Find a primal algorithm that is competitive.

I Problems

38. Think of a real-world shortest path problem in which the paths may not repeat arcs but need not be simple. Think of a real-world shortest path problem in which the paths may contain repeated arcs.

39. Show how any problem involving a finite sequence of binary decisions with deterministic outcomes can be modeled as a shortest path problem. Explain why, in general, such a model is useless.

40. In a weighted *mixed* graph, every edge has a weight. However, some edges are directed and the other edges are undirected. Prove that finding a minimum weight simple path in a mixed graph is NP-hard, as is the detection of negative weight cycles [8].

41. The name "Edsger Dijkstra" is particularly appealing to formulators of linear and integer programs because it contains the oft-used string "ijk." Find another name that is more appealing.

42. The original Rubik's Cube puzzle is a $3 \times 3 \times 3$ cube with 9 1×1 cubes visible on each facet. The goal of the puzzle is to rotate facets to make each facet monochromatic. Consider the problem of minimizing the number of rotations to reach the goal configuration. Explain why in principle this is a shortest path problem. Hypothesize why it would be computationally impractical. Prove your hypothesis.

Chapter 13

Specialized Algorithms for Maximum Flow and Minimum Cut

Preview: The maximum flow problem is to send as much as possible through a network from a given node s to a given node t, while respecting upper bounds on the arcs. Its dual, the minimum cut problem, is to interdict all possible flow from s to t by removing arcs from the network at minimum possible cost, where cost equals the sum of the upper bounds of the removed arcs. This section derives the duality twice: directly from the classical algorithm by Ford and Fulkerson, and from linear programming duality. It also shows two ways devised by Edmonds and Karp to speed up the algorithm, and some further algorithmic improvements.

The maximum flow problem is about sending as much as possible through a network whose arcs have upper bounds on flow. Let $G = (V, E)$ be a graph with nodes V, two distinct special nodes $s \in V, t \in V$, directed arcs E, and upper bounds $u_{ij} \geq 0$ on each arc $(i, j) \in E$. The goal is to set flow values on the arcs of G to maximize the amount of flow that originates at s and arrives at t. Flow is not permitted to originate at any node other than s, which is called the *source*, nor to accumulate at any node other than t, which is called the *sink*. Figure 13.1 shows a small example for which the upper bounds are $u_{sa} = 5, u_{at} = 18$, $u_{sb} = 20$, $u_{ba} = 30$, $u_{bt} = 6$, and the maximum possible $s - t$ flow is 24.

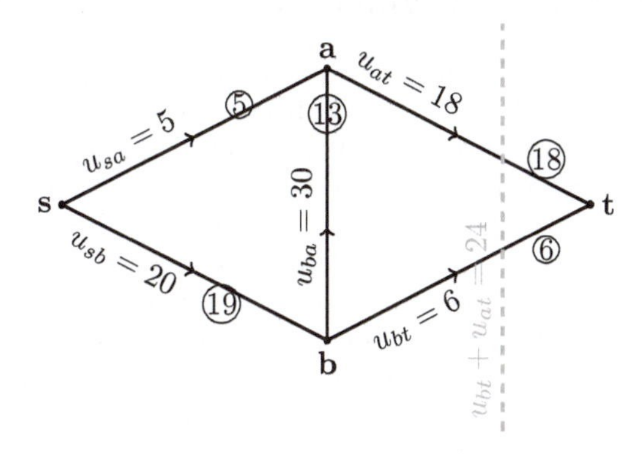

FIGURE 13.1: Send 5 on path s, a, t, send 13 on path s, b, a, t, and send 6 on path s, b, t to maximize $s - t$ flow. The circled values are the flow amounts on the arcs. Separate s from t by cutting arcs (a, t) and (b, t) to see that the network cannot send more than 24.

Definition 13.1 (Maximum Flow Problem) *Let $G = (V, E)$ be a directed graph with special nodes $s \in V, t \in V$ and arc upper bounds $u_{ij} \geq 0 : (i, j) \in E$. An $s-t$ flow on G is a set of values $f_e : e \in E$ such that $0 \leq f_e \leq u_e \; \forall e \in E$ and, for all $v \in V \setminus \{s, t\}$, $\sum_{uv \in E} f_{uv} = \sum_{vw \in E} f_{vw}$. The Maximum Flow Problem seeks an s–t flow with maximum possible value $\sum_{sv \in E} f_{sv} - \sum_{vs \in E} f_{vs} = \sum_{vt \in E} f_{vt} - \sum_{tv \in E} f_{tv}$.*

Since it is never beneficial to use a positive flow $f_{vs} > 0$ or $f_{tv} > 0$, it is usually assumed that no arcs (v, s) or (t, v) exist. The value of an $s-t$ flow then simplifies to $\sum_{sv \in E} f_{sv} = \sum_{vt \in E} f_{vt}$.

The problem originates from the Cold War era. The standard story is that U.S.A. operations researchers at RAND were asked to figure out how quickly the Russian army could convey troops to potential battlefronts in Western Europe. A joke that later circulated in the operations research community was that until Ford and Fulkerson completed their research, the Russians themselves didn't know how quickly they could transport their troops. However, Schrijver [263] has found that Russian researchers began to study maximum flow in their rail system for non-military purposes as early as 1930. In contrast, the U.S.A. researchers were secretly motivated by the dual problem of how to best interdict the Russian railway network by air strikes. And indeed, Ford and Fulkerson's [98] earliest technical report, dated 1954, proved the maximum flow/minimum cut duality. Their now-classic algorithm first appeared in a 1955 technical report [97].

Ford and Fulkerson [99] also studied the dynamic problem of maximizing total flow over time, starting with zero flow on each arc. They proved that after an initial startup phase, the network (if used optimally) would settle into a steady state that conveyed the same number of troops each day. Virtually all succeeding papers have focused on the static problem of determining the optimal steady state. The dynamic version of the problem has been nearly forgotten [263].

13.1 Ford and Fulkerson's Augmenting Path Algorithm

You may be wondering why an algorithm is even needed to solve the Max Flow problem. Why not repeatedly send as much flow as possible from s to t along a path until every $s-t$ path is blocked by an arc at its upper bound? The example in Figure 13.1 shows why not. Suppose we first send a flow of 18 on the path s, b, a, t. Next we could send a flow of 2 on the path s, b, t. But then we'd be stuck. Every path from s to t would now be blocked by either (s, b) at its upper bound 20 or (a, t) at its upper bound 18. Yet our total flow would be only 20. Ford and Fulkerson's insight was that we might need to *undo* a decision we've made previously. We are currently (pun intended) sending 18 from b to a. That flow uses up all of (a, t)'s capacity so that we can't use arc (s, a). Let's permit ourselves to send flow *in reverse* from a to b. Each unit of such flow isn't physically real; instead it is a *reduction* in the real flow on arc (a, b). To this end, define a graph of *residual capacity* as follows.

Definition 13.2 (Residual Capacity Graph) *Let $G = (V, E)$ be a directed graph with special nodes $s, t \in V$ and arc upper bounds $u_{ij} \geq 0 : (i,j) \in E$. Let $f_e : e \in E$ be an $s-t$ flow on G. The residual capacity graph with respect to f, denoted G^f, has node set V and two arcs (i,j) and (j,i) for each arc $(i,j) \in E$. The upper bounds on arcs in G^f depend on both the u_{ij} and f. For each arc $(i,j) \in E$, the corresponding arc $(i,j) \in G^f$ has upper bound $u_{ij} - f_{ij}$, and the corresponding arc (j,i) has upper bound f_{ij}. (If E contains both arcs (i,j) and (j,i), then G^f may contain two pairs of such arcs, or equivalently one arc (i,j) with upper bound $u_{ij} - f_{ij} + f_{ji}$ for each $(i,j) \in E$.)*

Figure 13.2 shows the flow of 20 described above and the associated residual capacity graph for the maximum flow instance of Figure 13.1.

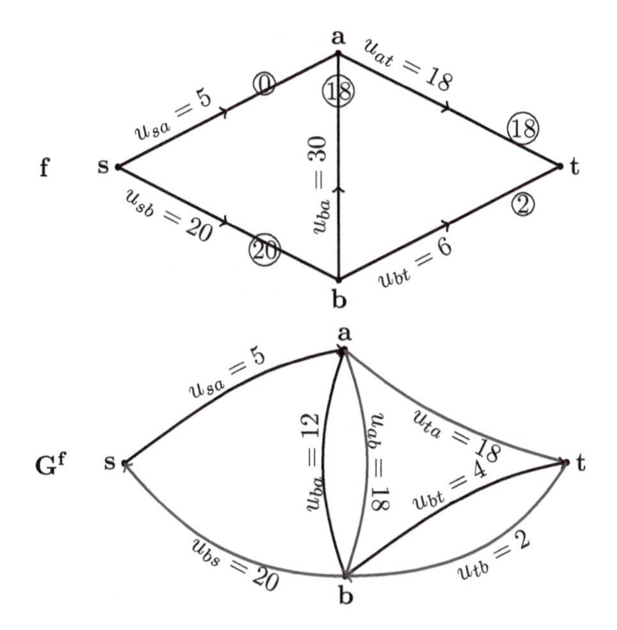

FIGURE 13.2: For $s-t$ flow $f_{sa} = 0, f_{sb} = 20, f_{ba} = f_{at} = 18, f_{bt} = 2$, the residual capacity graph G^f contains the augmenting path s, a, b, t.

The Ford-Fulkerson algorithm repeatedly increases the flow from s to t along paths in the residual capacity graph.

Algorithm 13.1.1 (Ford-Fulkerson Max Flow) *Input:* $G = (V, E), s \in V, t \in V, u_{ij} \geq 0 : (i,j) \in E$.

1. *Set $f_e = 0 \; \forall e \in E$.*

2. *Construct G^f as defined in Definition 13.2.*

3. *Find an $s-t$ path P in G^f such that every arc in P has a strictly positive upper bound. If no such P exists, terminate with optimal flow f.*

4. *Let $\Delta =$ the minimum upper bound in G^f of the arcs in P. For each $(i,j) \in P$ increase $f_{ij} - f_{ji}$ by Δ (by increasing f_{ij}, decreasing f_{ji}, or both.) Return to 2.*

Any path in G^f such that each of its arcs has strictly positive upper bound is called an

Augmenting Path, or *AP* for short. With this terminology, step 3 of the algorithm is to find an $s-t$ AP, and terminate if none exists.

Duality comes into play when we prove that the algorithm is correct. First let's define the dual min-cut problem precisely.

Definition 13.3 (Minimum Cut Problem) *Let $G=(V,E)$ be a directed graph with special nodes $s,t \in V$ and arc values $u_{ij} \geq 0 : (i,j) \in E$. An $s-t$ cut is a partition of V into sets S,T such that $s \in S, t \in T$ (and $S \bigcup T = V$ and $S \bigcap T = \phi$). The value of an $s-t$ cut is $\sum_{i \in S, j \in T} u_{ij}$. The minimum cut problem seeks an $s-t$ cut of minimum possible value.*

The minimum cut problem is a natural model for minimizing the cost of interdiction of a transportation network. Suppose we want to interdict all routes from s to t by bombing some of the arcs. The cost of destroying arc (i,j) is u_{ij}, measured in the expected number of planes required or lost. Then the minimum $s-t$ cut tells us which arcs to destroy to minimize our cost.

Weak duality between maximum flow and minimum cut is fairly easy to understand. For any $s-t$ cut S,T, every unit of flow leaving s must traverse an arc from a node in S to a node in T. Therefore, the sum of the upper bounds on all such arcs is an upper bound on the total flow from s to t. The Cold War origin of the problem inspires a nice visual image of the weak duality. Imagine the arcs of G as being roads on a map. Drop an iron curtain from the sky so that s and t are on opposite sides of the curtain. Every soldier who travels from s to t must pass through the curtain from the s side to the t side. Therefore, the total capacity (in the direction from the s side to the t side) of the roads hit by the iron curtain is an upper bound on the total possible flow from s to t.

You should guess that the strong duality theorem states that if you drop the iron curtain in the best possible way, the resulting upper bound equals the maximum possible $s-t$ flow. We prove the strong duality next.

Theorem 13.1.2 *If Algorithm 13.1.1 terminates, flow f is optimal for G. Moreover, the set $S = v \in V : \exists s - vAP$ defines a minimum cut in G whose value equals the $s-t$ flow of f. If all u_{ij} are rational, the algorithm terminates in a finite number of steps with rational flow f. If all u_{ij} are integer, the algorithm terminates with integer flow f.*

Proof: Let S be as defined in the statement of the theorem. By convention $s \in S$ since we consider there to be a path of no arcs from s to s. By step 3 $t \notin S$. Let $T = V \setminus S$. Then S,T is an $s-t$ cut. For every arc $(i,j) \in E$ such that $i \in S$ and $j \in T$ it must be that $f_{ij} = u_{ij}$. Otherwise, j would be in S. *Question: Why?*[1] For every arc $(j,i) \in E$ such that $i \in S$ and $j \in T$ it must be that $f_{ij} = 0$. Otherwise, G^f would contain an arc (i,j) with strictly positive upper bound f_{ij} and then j would be in S. Therefore, all of the flow between nodes of S and nodes of T is from S to T, and its total amount equals

$$\sum_{(i,j) \in E; i \in S, j \in T} u_{ij}$$

which is the value of the cut S,T. Since flow is conserved at all nodes except s and t, it must be that all of the flow originates at s and ends at t. This flow must be optimal because no flow can exceed the value of any $s-t$ cut. If all u_{ij} are integers, the total flow must increase

[1] Arc (i,j) in G^f would have strictly positive upper bound $u_{ij} - f_{ij}$.

by a strictly positive integer in every iteration of step 4; the total flow thus increases by at least 1 in every iteration, and in every iteration f is integer. If the u_{ij} are not integer but are rational, the total flow must increase by at least $1/M$ in every iteration, where M is the least common multiple of the denominators. In either case, since the u_{ij} are finite, the algorithm must terminate in a finite number of steps. ■

Corollary 13.1.3 *If all upper bounds are rational, the value of the maximum flow equals the value of a minimum cut.*

13.2 The Edmonds-Karp Modifications to the Ford-Fulkerson Algorithm

The Ford-Fulkerson algorithm can take a painfully long time to reach optimality. For example, let $G = (V, E)$ contain the four nodes s, A, B, t and the five arcs $(s, a), (s, b), (a, b), (a, t), (b, t)$. As shown in Figure 13.3, arc (a, b) has capacity $u_{ab} = 1$ and the other four arcs all have capacity 10^{12}. Obviously, the maximum flow could be found in two iterations by augmenting by 10^{12} twice, once each along paths s, a, t and s, b, t. But instead the algorithm could take 10^{12} times as many iterations as follows: augment by 1 along path s, a, b, t; augment by 1 along path s, b, a, t; repeat $10^{12} - 1$ times. Ford and Fulkerson also constructed a small example with irrational data for which the algorithm can fail even to converge towards the optimum solution [100].

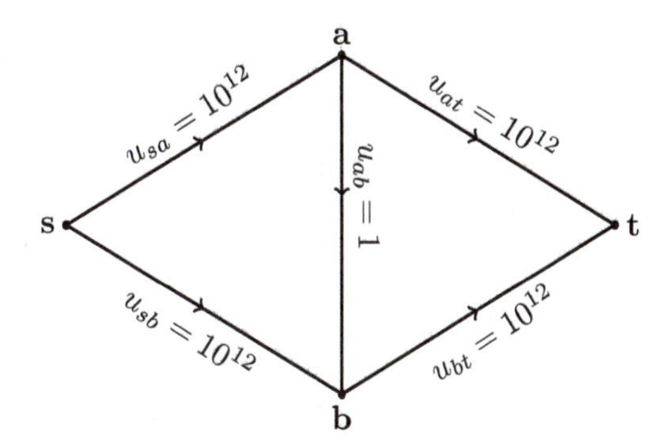

FIGURE 13.3: Repeated augmentations along paths *sabt*,*sbat* take 2×10^{12} iterations to reach optimality.

Edmonds and Karp [88] invented two different modifications to ensure that the algorithm runs quickly. The first requires almost no change to Ford and Fulkerson's algorithm, and has led to several generalizations. The second introduces a powerful scaling idea, which has been used subsequently for other problems. Before you read about these modifications, ask yourself how you could try to improve on the Ford-Fulkerson algorithm. Study the example of Figure 13.3. What is bad or stupid about the choice of s, a, b, t for the first augmenting path?

Two ideas should occur to you. The path s, a, b, t is less direct than the other paths. It uses three arcs instead of two. Once you see that you can get flow from s to a, why not go directly to t? Also, the path s, a, b, t has much smaller capacity than the other available paths s, a, t and s, b, t. It seems stupid to choose the path with tiny capacity. The first idea leads to the Edmonds-Karp *shortest augmenting path* modification, to always choose an augmenting path with the least possible number of arcs. The tricky part of this modification is the analysis to prove a good upper bound on the total number of iterations. The second idea leads to the following modification: always choose an augmenting path with greatest possible capacity. Edmonds and Karp proved fairly easily that the number of iterations is small, but the resulting algorithm is too slow in both theory and practice because it takes too long to find a maximum capacity AP. An additional idea such as scaling is needed to make this approach competitive.

Let's tackle the first idea first. Define a *shortest augmenting path*, or *SAP* for short, to be an augmenting path that has a minimum number of arcs.

Theorem 13.2.1 *[88] Algorithm 13.1.1 takes at most $|E||V|/2$ iterations if it always chooses SAPs.*

Corollary 13.2.2 *Algorithm 13.1.1 runs in time $O(|E|^2|V|)$ if it always chooses SAPs.*

Corollary 13.2.3 *Corollary 13.1.3 is valid for all nonnegative real upper bounds u_{ij}.*

Proof: With respect to a flow f, define a $v-w$ augmenting path to be a path from node v to node w in the residual graph G^f all of whose arcs have strictly positive capacity. For any path P define $|P|$, the length of P, to be the number of arcs in P. Define a $v-w$ SAP to be a $v-w$ AP of minimum length. Define

$$\sigma_v^f \equiv \text{ length of an } s-v \text{ SAP}; \tau_v^f \equiv \text{ length of a } v-t \text{ SAP}. \tag{13.1}$$

Lemma 13.2.4 *Let f be a flow and let f' be the flow that results from augmenting along a SAP with respect to f. Then $\sigma_v^{f'} \geq \sigma_v^f$ and $\tau_v^{f'} \geq \tau_v^f$.*

Proof of Lemma: By induction on σ_v^f assume that for all $w \in V$ such that $\sigma_w^f < \sigma_v^f$ the inequality $\sigma_w^{f'} \geq \sigma_w^f$ is correct. By contradiction suppose an $s-v$ SAP P' with respect to f' exists with fewer arcs than exist with respect to flow f, that is, $|P'| < \sigma_v^f$. Think of flow f existing in the present time and flow f' existing in a future "prime time" after we alter f with the SAP Q. If path P' were an AP now, by definition $|P'| \geq \sigma_v^f > |P'|$, a contradiction. Why is P' not an AP now? Path P' must contain at least one arc (i,j) that has 0 capacity now with respect to f but will have strictly positive capacity with respect to f'. Therefore, either $(i,j) \in E$ and $f_{ij} = u_{ij}$ and (j,i) is an arc in Q, or $(j,i) \in E$ and $f_{ji} = 0$ and (j,i) is an arc in Q. Either way, Q contains the arc (j,i). If more than one such arc exists, let (i,j) be the last such arc in P'. Therefore, the subpath P'_{jv} is a $j-v$ AP with respect to f. Also, node $j \neq v$ because otherwise the subpath of Q from s to j would be an AP shorter than σ_j^f. By induction, $\sigma_j^f \leq \sigma_j^{f'}$.

We now create an $s-v$ AP with respect to f that has length less than σ_v^f. See Figure 13.4. We have shown that P'_{jv} is a $j-v$ AP with respect to f. By definition of Q, Q_{sj} is an $s-j$ AP with respect to f. Therefore, $Q_{sj} \circ P'_{jv}$, the concatenation of the two paths, is an $s-v$ AP with respect to f. But by the inequality above and the principle of optimality 12.3.1

$$\sigma_v^f \leq |Q_{sj} \circ P'_{jv}| = \sigma_j^f + |P'_{jv}| \leq \sigma_j^{f'} + |P'_{jv}| = |P'| < \sigma_v^f,$$

a contradiction. The proof for τ_v is symmetric. ■

The Lemma tells us that $\sigma_t \equiv \tau_s$ is nondecreasing from iteration to iteration, as long as SAPs are used. The algorithm's progress therefore divides into $|V|-1$ phases. During

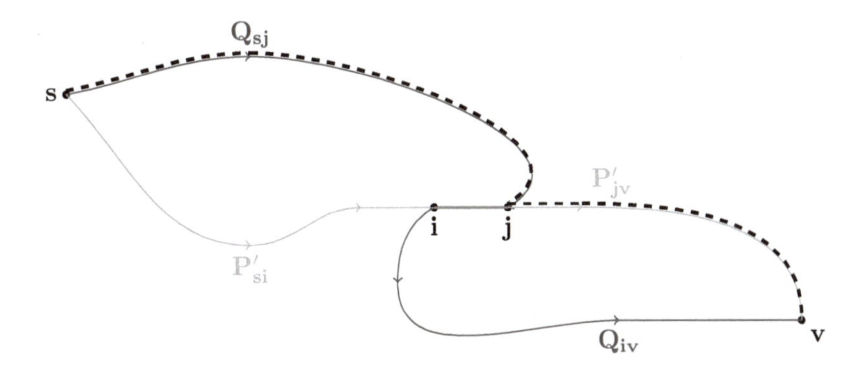

FIGURE 13.4: Combine Q_{sj}, the subpath of Q to j, with P'_{jv}, the subpath of P' from j to v, to get a shorter augmenting path from s to v. The shorter path is indicated by the dashed curve.

phase k, the SAPs have length k. That is, during phase k, $\sigma_t = k$. In practice, many of the phases are empty.

Next we bound the number of iterations that can occur within any two consecutive phases. Let P be the SAP used in an iteration of the algorithm now. Let (i, j) be an arc in P whose capacity in G^f is minimal in step 4. Hence $\sigma_j = \sigma_i + 1$ and $\tau_i = \tau_j + 1$. The next time the algorithm employs the arc (i, j) in an AP it must do so in the reverse direction (j, i) since no residual capacity from i to j remains. Let σ', τ' denote the σ, τ values at the future "prime time" when the SAP contains (j, i). Thus $\sigma'_i = \sigma'_j + 1$ and $\tau'_j = \tau'_i + 1$. Combining these equalities with Lemma 13.2.4 gives

$$\sigma'_i = \sigma'_j + 1 \geq \sigma_j + 1 = \sigma_i + 2.$$

Since by the Lemma, $\tau'_i \geq \tau_i$, at prime time the SAP containing (j, i) has length

$$\sigma'_i + \tau'_i \geq \sigma_i + \tau_i + 2.$$

Therefore, prime time must occur at least 2 phases after the present time. Hence, in any two consecutive phases, at most $|E|$ augmentations can occur. The $|V| - 1$ phases can be partitioned into $\lfloor |V|/2 \rfloor$ pairs (there will be one singleton if $|V|$ is even), each containing at most $|E|$ augmentations, for a total bound of $|E||V|/2$ as desired. ■

For the first corollary, an SAP can be found in $O(|E|)$ time with breadth-first-search. Indeed, a straightforward implementation of the unmodified algorithm is apt to use breadth-first-search and therefore to find SAPs fortuitously. For the second corollary, the theorem just proved does not rely on rationality of the data. Therefore, the Ford-Fulkerson algorithm with SAPs is sure to terminate, and the corollary follows from Theorem 13.1.2. ■

Zadeh [304] constructed a sequence of graphs for which the SAP algorithm requires $O(|E||V|)$ iterations. Therefore, that part of the analysis of the theorem is tight, to within a constant factor.

Next we derive the Edmonds-Karp bound on augmentations along maximum capacity APs. Let $u^* = \sum v \in V \setminus \{s\} u_{sv}$ or some other upper bound on the maximum $s - t$ flow.

Theorem 13.2.5 *Let $G = (V, E)$ have integer upper bounds $u_e : e \in E$. If the Ford-Fulkerson Algorithm 13.1.1 always chooses an augmenting path with maximum residual capacity, it will terminate in at most $|E| \lceil \log_2 u^* \rceil$ iterations, where u^* is value of a maximum flow.*

Proof: We will show that after $|E|$ iterations, the algorithm finds at least half the remaining possible flow. The key fact we use, that every flow can be described as the sum of at most m paths, is stated precisely in the following lemma.

Lemma 13.2.6 *Let f be a feasible $s-t$ flow on graph $G=(V,E)$. Then f can be decomposed into at most $|E|$ $s-t$ paths of flow. That is, there exist $|E|$ $s-t$ paths $P^1 \ldots P^{|E|}$ and nonnegative scalars $\beta_1 \ldots \beta^{|E|}$ such that $f=\sum_{i=1}^{|E|} \beta_i P^i$.*

Proof: The proof is the Ford-Fulkerson algorithm in reverse. Formally speaking, assume inductively that the lemma is true for all graphs with $|E|-1$ or fewer arcs. Select any $s-t$ path P in G such that every arc in P has strictly positive flow. Let $\Delta = \min_{e \in P} f_e$. Set $\beta^{|E|} = \Delta$ and $P^{|E|} = P$. Subtract Δ from f_e for every $e \in P$ and remove from G all arcs $e \in P$ for which $f_e = \Delta$. By definition at least one such arc exists. G now has strictly fewer arcs than it had initially. By induction on $|E|$ the lemma is proved. ■

A tantalizing consequence of the lemma is that the Ford-Fulkerson algorithm could solve any maximum flow problem in at most $|E|$ iterations if it were prescient. For our purposes, the lemma implies that if the value of $s-t$ flow f on G $\sum_{v \in V} f_{sv} = z$, then there exists an $s-t$ path along which every arc e has flow $f_e \geq z/|E|$. Now comes the subtle part of the proof. Let z^* be the value of a maximum flow in G. Let f be an $s-t$ flow in G with value $z < z^*$. Let G^f be the graph of residual capacities. By definition, there must exist a feasible $s-t$ flow in G^f, say $f^\diamond$, with value $z^* - z$. Apply the lemma to G^f and $f^\diamond$, *not* to G and f. The graph G^f must therefore contain an augmenting path with capacity at least $(z^*-z)/|E|$. This is true for every $s-t$ flow f that has value z. Suppose now that, starting with $s-t$ flow f, $|E|$ augmentations along maximum capacity APs are performed. If every one of those iterations increases the flow by at least $\frac{z^*-z}{2|E|}$, then altogether the flow has been increased by at least $\frac{z^*-z}{2}$. If one of those APs increases flow by less than $\frac{z^*-z}{2|E|}$, let y denote the total flow that has been achieved when that AP is about to be chosen. By the lemma, $(z^*-y)/|E| \leq \frac{z^*-z}{2|E|}$, which implies that $y \geq \frac{z^*+z}{2}$. Therefore, in either case, the flow has increased by at least $\frac{z^*-z}{2}$. In other words, the gap between the current and optimal flows is cut by at least half.

Since the data are integers, the algorithm will find an optimum flow within $|E| \log_2 z^*$ iterations.

13.3 Further Improvements of Maximum Flow Algorithms

Many improved algorithms for the maximum flow problem have been invented since the classical results described here. This subsection gives a brief overview of key improvements.

One line of attack has been to find SAPs more quickly than in $O(|E|)$ time. This idea is plausible because there is a lot of repeated computation from iteration to iteration. First, the residual capacity graph G^f needs not be rebuilt from scratch at each iteration. Instead, it should simply be updated along the augmenting path. Second, the search for an SAP in G^f also involves a lot of repeated computation. It seems that there ought to be a way to use less than $O(|E|)$ work on average to find SAPs. Dinits [80] was the first to do so. He observed that during phase k, the only arcs (i,j) that can be used are such that (i) $f_{ij} < u_{ij}$ or $f_{ji} > 0$,; (ii) $\sigma_i + 1 = \sigma_j$; and (iii) $\tau_i = \tau_j + 1$. At the beginning of phase k, his algorithm builds a "layered" residual capacity graph that only contains such arcs. Depth-first search on the layered graph finds all needed SAPs for phase k in $O(|E|)$ total time. There may be up to $|E|$ augmentations in the phase, each costing $O(|V|)$ to process, for a total computation time of $O(|V|(|E|+|E||V|)) = O(|V|^2|E|)$. Tardös *et al.* [274] generalized the Dinits layering method to other augmenting path algorithms. They found that additional arcs violating (i) may be necessary, and therefore the correct generalization enforces only (ii) and (iii). When flow is augmented along a path P, any straightforward data structure will require time $\Omega(|P|)$ to update the flows on the edges of P. Once augmenting paths are being found quickly, this updating takes more time than any other part of the algorithm. To speed up the updating, Sleator and Tarjan [267]invented a more sophisticated data structure that permits flow updates in $O(\log|P|)$ time. They employed this "dynamic tree" structure to get a $O(|V||E|\log|V|)$ algorithm.

Another line of attack has been to relax the requirement of Theorem 13.2.5 that the AP have maximum capacity in G^f. This is plausible because it is so time-consuming to find such an AP. Instead, quickly find an AP whose capacity is, say, within a constant factor of the maximum possible. *Scaling* is a popular way to accomplish this. Write the arc capacities in binary with leading zeros, and approximate the values by the first digit, then by the first two digits, and so on until the arc capacities are treated precisely. The intuitive idea is this: the first approximation stage treats each arc as "big" or "small" and seeks APs that comprise only "big" arcs. Finding these APs consists merely of finding paths from s to t in an auxiliary graph of only "big" arcs. Once the flow f has value close enough to the optimum, no more "big" augmentations are possible. Therefore, the capacities of any remaining "big" arcs can be rounded down to the power of 2 that had differentiated "big" from "small." Inductively, the succeeding stages behave similarly. At a cost of only a factor of $\log_2 u^*$, where $u^* \equiv \max_{e\in E} u_e$, the resulting algorithms are quite fast. Gabow improved on a straightforward implementation of this idea to get an algorithm with time bound $O(|E||V|\log u^*)$ [110]. The potential of augmenting path algorithms appears to have been exhausted. Further improvements call for some way to increase flow along more than one path at a time. Karzanov [176] introduced the important concept of a "preflow." Instead of requiring that f satisfy conservation of flow at all nodes except s and t, permit the inflow at nodes to exceed the outflow. The idea is that there may be a path from s to v that then branches into multiple paths to t. Do the work once to get a lot of flow to v, instead of once for each of the multiple paths. Goldberg and Tarjan [128] used preflows and the dynamic tree data structure of Sleator and Tarjan [267] to get a $O(|E||V|\log(|V|^2/|E|))$ algorithm. Ahuja and Orlin [5] then combined the ideas of preflows and scaling to get an $O(|E||V|+|V|^2\log u^*)$ algorithm. A series of speedups and improved analyses soon appeared in the literature, including important ones by Cheriyan and Maheshwari [47], Ahuja, Orlin and Tarjan [6], and Goldberg and Rao [127].

Two concepts have proved to be very important. One, due to Goldberg and Rao [127] is to alter the idea of shortest APs by giving preference to arcs with "large" residual capacity, as the scaling algorithms do. Redefine the length of an AP to be the number of its "small" residual capacity arcs. Lemma 13.2.4 no longer applies, making the algorithms and their analysis complicated. Another concept, called "push-relabel", is due to Goldberg and Tarjan [128]. Push-relabeling changes a preflow along just one arc, so it can be done very quickly as compared with any change made to a preflow or flow along a path. The change in flow is the "push"; the consequent changes in the node labels, which are proxies for σ_v and τ_v, constitute the "relabel."

More recently, the maximum flow problem has been divided into multiple cases, depending on the density of the graph (the relationship between $|E|$ and $|V|$). Different techniques have been employed for these cases and yielded remarkably fast algorithms, notably ones by King, Rao, and Tarjan [180], and Goldberg and Rao [127]. Orlin [237] assembled several of these algorithms, and combined multiple techniques to solve the remaining case, to produce the first $O(|E||V|)$ algorithm for maximum flow. This time bound seems to be the best possible one, because a maximum flow can consist of $\Omega(|E|)$ paths of length $\Omega(|V|)$. Therefore, any algorithm that expresses the optimum flow as a collection of flows on paths must take $\Omega(|E||V|)$ time. However, Orlin's algorithm runs even faster, in time $O(|E||V|/\log|V|)$, on very sparse graphs in which $|E| = O(|V|)$. Other researchers have previously found algorithms faster than $O(|E||V|)$ for the case of unit capacities. Therefore, the search for even faster algorithms continues. For an excellent survey of the entire history of maximum flow algorithms, see [129].

13.4 Maximum Flow as a Linear Program

The LP formulation of the maximum flow problem defines x_{ij} to be the amount of flow on arc $(i,j) \in E$. The objective function and constraints are as follows:

$$\max \sum_{(s,i)\in E} x_{si} \text{ subject to:} \tag{13.2}$$

$$\sum_{(i,j)\in E} x_{ij} - \sum_{(k,i)\in E} x_{ki} = 0 \ \forall i \in V \setminus \{s,t\} \tag{13.3}$$

$$0 \le x_{ij} \le u_{ij} \qquad \forall (i,j) \in E. \tag{13.4}$$

The dual of the max flow LP is

$$\min \sum_{e\in E} u_e \chi_e \equiv \textstyle\sum_{(i,j)\in E} u_{ij}\chi_{ij} \quad \text{subject to} \tag{13.5}$$

$$\chi_{ij} \ge 0 \quad \forall \quad (i,j) \in E \tag{13.6}$$

$$\pi_i - \pi_j + \chi_{ij} \ge 0 \quad \Leftrightarrow \quad \pi_j \le \pi_i + \chi_{ij} \ \forall (i,j) \in E \tag{13.7}$$

$$\pi_t - \pi_s \quad \ge \quad 1 \tag{13.8}$$

The intuitive meaning of the dual variables χ_e is the amount of money we are going to charge for each unit of flow on arc e. The meaning of the π_v dual variables is the amount we will charge for sending a unit of flow from s to v.

Let S,T be an $s-t$ cut of G. Create the dual solution

$$\pi_i = 0 \ \forall i \in S; \pi_j = 1 \ \forall j \in T; \chi_{ij} = \begin{cases} 1 \text{ if } i \in S, j \in T \\ 0 \text{ otherwise} \end{cases}.$$

Constraint (13.6) is clearly satisfied. Since $s \in S$ and $t \in T$, constraint (13.8) is satisfied. Check the four cases $i \in S$ or $i \in T$, $j \in S$ or $j \in T$ to see that constraint (13.7) is also satisfied. The value of this dual solution is $\sum_{i\in S, j\in T} u_{ij}$ which is by definition the value of the $s-t$ cut. By Corollary 13.2.3 the minimum value of this type of dual solution equals the value of a maximum flow. Therefore, the dual of the max flow LP always has an optimal solution that is integer, regardless of whether the upper bounds u_e are integer.

Corollary 13.4.1 *The dual of the max flow LP, defined by* (13.5) ... (13.8), *has an optimal solution in which every variable equals* 0 *or* 1.

Corollaries 13.4.1 and 13.2.3 imply that the LP strong duality theorem is true for LPs of form (13.2) ... (13.4). If we could prove 13.4.1 without using Theorem 13.2.1 or 13.1.2, then we ought to be able to prove Corollary 13.2.3 from the LP strong duality theorem. That will be the task of the next section. There we will consider network problems in general, since it is no harder to prove integrality for them than for maximum flows.

13.4.1 Max Flow Min Cut

We've already used the Edmonds-Karp SAP modification of the Ford-Fulkerson algorithm to prove Corollary 13.2.3, that the amount of maximum $s-t$ flow in a graph with upper bounds on arcs equals the value of a minimum $s-t$ cut. Here we derive the same strong duality result from LP duality. Recall that the maximum flow problem on graph $G = (V, E)$ with upper bounds $u_e > 0 : e \in E$ and special nodes $s \in V, t \in V$ may be defined as follows:

$$\begin{aligned} \max \sum_{v \in V} x_{vt} & \qquad \text{subject to} \\ \sum_{w \in V} x_{vw} - \sum_{u \in V} x_{uv} &= 0 \ \forall v \in V \setminus \{s, t\} \\ 0 \leq x_{vw} \leq u_{vw} & \qquad \forall (v, w) \in E \end{aligned}$$

This definition assumes that G contains no arcs out of t, that is, $\nexists (t, v) \in E$. Recall also that an $s-t$ cut in G is a partition of V into S and T such that $s \in S$ and $t \in T$, and that the value of an $s-t$ cut is the sum

$$\sum_{i \in S, j \in T} u_{ij}.$$

Theorem 13.4.2 *For any maximum $s-t$ flow problem, the maximum possible amount of $s-t$ flow equals the minimum possible value of an $s-t$ cut.*

Proof: Let the maximum flow problem be called the primal. Let A denote A^G, the arc-node incidence matrix of G. From Section 13.4 the primal is equivalent to

$$\max x_0 \quad \text{subject to} \tag{13.9}$$

$$\mathbf{Ax} = \begin{pmatrix} x_0 \\ 0 \\ \vdots \\ 0 \\ -x_0 \end{pmatrix} \tag{13.10}$$

$$0 \leq x \leq u. \tag{13.11}$$

Make four changes to the primal that will not change the optimal solution value and will make the dual easier to interpret.

- Remove the redundant first row of constraint (13.10), which corresponds to flow at node s.
- Remove the nonnegativity constraint $x_0 \geq 0$. This has no effect on the optimal solution value since x_0 is being maximized and $x = 0$ is always feasible.
- Relax the constraints (13.10) from $\mathbf{Ax} = (0, \ldots, 0, -x_0)$ to $\mathbf{Ax} \leq (0, \ldots, -x_0)$. This alteration allows the amount of inflow at nodes to exceed the amount of outflow. (A

feasible solution to the relaxed problem is called a *preflow*.) Since we are measuring the net $s-t$ flow by the amount of flow that reaches t, this change has no effect on the optimal solution value.

- Add a new set of arcs $F = \{(s,v)|v \in V \setminus \{s\}\}$ to G. Let $y_e : e \in F$ denote the flow on these arcs. Set upper bounds $z_e = M \;\; \forall e \in F$ where M is the loose upper bound $1 + \sum_{(s,v)\in E} u_{sv}$ on the maximum flow. Change the objective function from $\max x_0$ to $\max x_0 - \sum_{e\in F} y_e$. This alteration permits extra flow from s to other nodes in G, but since each unit of flow subtracts 1 from the objective function value, there is no effect on the optimal solution value.

The new primal LP is

$$\max x_0 - \sum_{e\in F} y_e \text{ subject to} \tag{13.12}$$

$$A\mathbf{x} - Iy \leq \begin{pmatrix} 0 \\ 0 \\ \vdots \\ 0 \\ -x_0 \end{pmatrix} \tag{13.13}$$

$$0 \leq x_e \leq u_e : e \in E; 0 \leq y_e \leq M : e \in F. \tag{13.14}$$

In constraint (13.13) there is no row for node s. Let the dual variables corresponding to constraints (13.13) be $\boldsymbol{\pi}_i : i \in V\backslash\{s\}$. Let the dual variables corresponding to the constraints $\mathbf{x} \leq u$ be $\chi_e : e \in E$ and those corresponding to the constraints $y_e \leq M : e \in F$ be $\chi_e : e \in F$. The dual LP is then

$$\min \sum_{e\in E\bigcup F} u_e\chi_e \quad \text{subject to} \tag{13.15}$$

$$A^T\boldsymbol{\pi} + \chi_e \geq 0 : \; e \in E \tag{13.16}$$

$$-\boldsymbol{\pi}_j + \chi_{sj} \geq -1 \; \forall (s,j) \in F \tag{13.17}$$

$$\boldsymbol{\pi}_t \geq 1 \tag{13.18}$$

$$\chi \geq 0; \quad \boldsymbol{\pi} \geq 0. \tag{13.19}$$

Question: To what primal variables do constraints (13.16)-(13.18) *correspond?*[2] It is now convenient to introduce the dual variable $\boldsymbol{\pi}_s$ corresponding to the constraint on flow through s that we omitted from the primal. Constrain $\boldsymbol{\pi}_s = 0$ so that it has no effect on dual feasibility or on the value of dual solutions. Plugging in the definition of A, the dual becomes:

$$\min \sum_{e\in E} u_e\chi_e + M\sum_{e\in F}\chi_e \quad \text{subject to} \tag{13.20}$$

$$\boldsymbol{\pi}_i - \boldsymbol{\pi}_j + \chi_{ij} \geq 0 \; \forall (i,j) \in E \tag{13.21}$$

$$\chi_{ij} \geq 0 \quad \forall \quad (i,j) \in E \tag{13.22}$$

$$0 \leq \boldsymbol{\pi}_i \leq 1 + \chi_{si} \forall (s,i) \in F \tag{13.23}$$

$$\boldsymbol{\pi}_t \geq 1; \boldsymbol{\pi}_s = 0 \tag{13.24}$$

The primal is feasible because $\mathbf{x} = 0$ is feasible. The dual is feasible because $\boldsymbol{\pi}_s =$

[2]The first two sets correspond to $x_e : e \in E$ and $y_e : e \in F$, respectively. Constraint (13.18) corresponds to the primal variable x_0.

$0, \pi_v = 1\ \forall v \in V \setminus \{s\}$, $\chi_{sv} = 1 \forall (s,v) \in E$, $\chi_e = 0\ \forall e \in F$ is feasible with objective value $M-1$. By Theorem 2.1.3 the primal and dual have optimal solutions with equal value. By definition the value of the primal is the maximum possible amount of $s-t$ flow in G. By Corollary 11.2.4 there is an integer optimal dual solution.

Let π^*, χ^* be an optimal basic feasible solution to the dual, which we know must exist from the above and from Theorem 4.4.8. By Corollary 11.2.4 pi^* and χ^* are integer. Since all variables and objective function coefficients in (13.20) are nonnegative, a single $\chi^*_e : e \in F$ equal to 1 would make the objective value exceed $M-1$. Hence $\chi^*_e = 0\ \forall e \in F$. From constraint (13.23) then $0 \le \pi^*_i \le 1\ \forall i$. By integrality each π^*_i is 0 or 1. Also, (13.24) implies that $\pi^*_t = 1$. Let $S = \{v \in V : \pi^*_v = 0$ and let $T = V \setminus S = \{v \in V : \pi^* = 1\}$. Constraint (13.24) implies that $s \in S$ and $t \in T$. Hence (S,T) is an $s-t$ cut in G. From $u_e > 0$, constraint (13.22) and the minimization objective, $\chi^*_{ij} = \max\{0, \pi^*_j - \pi^*_i\}\ \forall (i,j) \in E$. This is because variable χ^*_{ij} is constrained to be at least that large, and any larger value would strictly increase the objective value. Therefore, we have a complete characterization of π^*, χ^* in terms of the $s-t$ cut (S,T), namely,

$$\pi^*_v = 0\ \forall v \in S;$$
$$\pi^*_v = 1\ \forall v \in T;$$
$$\chi_{ij} = \begin{cases} 1\ \forall i \in S, j \in T, (i,j) \in E \\ 0 \text{ otherwise} \end{cases}.$$

The value of this solution is by definition the value of the cut (S,T). By the same characterization, every $s-t$ cut corresponds to a dual solution whose value equals the value of the cut. Therefore, the value of the dual equals the minimum possible value of an $s-t$ cut in G. ■

The heart of the proof just given consists of examining the dual LP to prove that an integer optimal solution must yield an $s-t$ cut. Read that part of the proof carefully. It relies on every constraint in the dual, on dual integrality, and on the nonnegativity of the upper bounds u_e. (In Chapter 14 you will see that the minimum cut problem with negative upper bounds, which is called the *max cut* problem, cannot be modeled as a network problem or even an LP by any known method. Therefore, the proof here *must* make use of the nonnegativity of u.)

The changes I made to the primal made the proof easier than usual. *Caution: Making an unproven assumption about the form of the dual solution is the most common error in proofs of combinatorial duality based on LP.* Do not assume that you already know the form of the integer dual solution. Instead, base your reasoning only on the integrality of the solution, the form of the dual LP, and (possibly) that the solution is a BFS.

13.4.2 Bipartite Matching

A bipartite graph $H = (U, W, F)$ has node set $U \bigcup W$ and arc set $F \subset U \times W$. The (unweighted) bipartite matching problem is to choose as many arcs from F as possible such that no node is incident on more than one chosen arc. This is a discrete problem in the sense that one's choices are either zero or one rather than any value between zero and one. Each arc is either chosen or not chosen. For example, U could be a set of workers, W could be a set of jobs, and the arc (u,w) would exist iff worker u is capable of performing job w. We can model bipartite matching as an *integer program*, a linear program with additional constraints that variables must be integer. The integer programming model is:

$$\max_{(i,j)\in F} x_{ij} \quad \text{subject to} \tag{13.25}$$

$$\sum_{(i,k)\in F} x_{ik} \leq 1 \quad \forall i \in U; \tag{13.26}$$

$$\sum_{(k,j)\in F} x_{kj} \leq 1 \quad \forall j \in W; \tag{13.27}$$

$$0 \leq x_{ij} \leq 1; x_{ij} \text{ integer } \forall (i,j) \in F. \tag{13.28}$$

However, we don't have to resort to integer programming to solve bipartite matching. Instead, create the following maximum flow problem: add a new node s with arcs from s to each $u \in U$; add a new node t with arcs from w to t for each $w \in W$; set all upper bounds to 1. In algebraic notation, let $G = (V, E)$; $V = \{s\} \bigcup U \bigcup W \bigcup \{t\}$; $E = F \bigcup \{(s,u) \forall u \in U\} \bigcup \{(w,t) \forall w \in W\}$; $u_e = 1 \ \forall e \in E$. Now every integer $s-t$ flow in G corresponds to a matching in H and vice versa.

What does the max-flow/min-cut duality theorem tell us about bipartite matching? It turns out to be difficult to interpret minimum cuts in the graph G as we have defined it. So, we use a trick. Alter the upper bounds in G to be $x_{su} \leq 1 \forall u \in U$; $x_{wt} \leq 1 \forall w \in W$; and $x_e < \infty \ \forall e \in F$. These upper bounds still enforce constraints (13.26),(13.27). They rule out the possibility that both $u \in S$ and $w \in T$ if $(u,v) \in F$, in any minimum $s-t$ cut.

A minimum $s-t$ cut S, T (indeed, any $s-t$ cut with finite value) must have the following form: Let $S_U = S \bigcap U; S_W = S \bigcap W$. Similarly let $T_U = T \bigcap U; T_W = T \bigcap W$. Then $S = \{s\} \bigcup S_U \bigcup S_W$, $T = \{t\} \bigcup T_U \bigcup T_W$, and $F \bigcap S_U \times T_W = \phi$. The value of the cut is $|T_U| + |S_W|$.

Suppose we know the set S_U. Then we know the set $T_U = U \setminus S_U$. We also know that for every arc $(u,w) \in F$ such that $u \in S_U$, node $w \in S_W$. Since the value of the cut is $|T_U| + |S_W|$, S_W may not contain any additional nodes from W if it is to have minimum value. Therefore, once S_U is known, the entire S, T cut is known. Its value is $|U| - |S_U| + |\{w : \exists u \in S_U, (u,w) \in F\}|$.

In the case $|U| = |W|$ we might hope for a *perfect matching*, a matching with $|U|$ arcs so that every node in U and W is incident on one arc of the matching. Since the $s-t$ cut $S = \{s\}, T = U \bigcup W \bigcup \{t\}$ has value $|U|$, the max-flow/min-cut strong duality implies that a perfect matching exists if and only if no $s-t$ cut has value $< |U|$. From the preceding discussion, this is true iff for all $S_U \subseteq U$ the set $S_W = \{w \in W : (u,w) \in F \text{ and } u \in S_U\}$ has cardinality $|S_W| \geq |S_U|$. What does this mean? Visualize the nodes U as clouds and the nodes W as children. If arc $(u,w) \in F$, child w is in cloud u's shadow. For any set S_U of clouds, S_W is the set of children who are shaded by that set of clouds. The condition for a perfect matching to exist is that every set of k clouds shades at least k children, for all $1 \leq k \leq |U|$. The weak duality is evident. We've proved the strong duality from the max-flow/min-cut theorem. The result is known as Hall's theorem. Hall stated the result in terms of systems of distinct representatives, but we state it in the equivalent graph theoretic terms. The *neighborhood* of a set S of nodes in a graph is the set of nodes that share at least one arc with a node in S but are not in S. The neighborhood of S, also called the *boundary* of S, is denoted ∂S, and a member of ∂S is called a *neighbor* of S.

Theorem 13.4.3 *[142] A bipartite graph $G = (U, W, F)$ with $|U| = |W|$ has a perfect matching iff every subset $S \subseteq U$ has at least $|S|$ neighbors, that is, $\forall S \subseteq U, |\partial S| \geq |S|$.*

13.4.3 Köenig-Egerváry Theorem

If you don't know the rules of chess, skip the next two sentences. Place some rooks on an $n \times m$ chessboard. If none of them can attack another, we say they are *independent.* In mathematical terms, place an X in some of the cells of an $n \times m$ rectangular grid. If no row or column contains more than one X, we say that the set of Xs is *independent.* Given an arbitrary placement of Xs in the grid, you might want to solve the following problem:
Maximum Independent Set: *Given a rectangular grid some of whose cells contain an X, find a largest independent set of X's.*

The dual problem is to remove all the Xs in the matrix by deleting as few rows and/or columns as possible. Equivalently, the problem is to contain all the Xs in as few rows and columns as possible.
Minimum Containment: *Given a rectangular grid some of whose cells contain an X, select r rows and c columns so that every X is contained in at least one of the selected rows or columns, and $r + c$ is as small as possible.*

Weak duality, as usual, is easy to see. If there is an independent set of size k, at least k rows and columns must be selected, because each selection can contain at most one member of the independent set. The strong duality is called the Köenig-Egerváry Theorem. We prove it using network duality.

Theorem 13.4.4 (Köenig-Egerváry [89],[192]) *For any rectangular grid some of whose cells contain an X, the cardinality of a maximum independent set of Xs equals the minimum number of rows and columns that contain all the Xs.*

Proof: Let the grid contain m rows and n columns. Create graph $G = (V, E)$ with $m+n+2$ nodes, labeled $R_1 \dots R_m, C_1 \dots C_n, s$ and t. For $i = 1 \dots m$ create arc (s, R_i) with upper bound 1. For $j = 1 \dots n$ create arc (C_j, t) with upper bound 1. For every i, j such that the cell in row i, column j of the grid contains an X, create arc (R_i, C_j) with upper bound ∞. (You should recognize the trick we used for bipartite matching.) Every integer $s - t$ flow in G corresponds to an independent set in the grid whose elements are the cells i, j such that arc R_i, C_j has unit flow. The corresponding cells are independent because two elements in the same row i would require a flow of 2 on arc (s, R_i), and symmetrically two elements in the same column j would require a flow of 2 on arc (C_j, t). Conversely, every independent set in the grid corresponds to an integer $s - t$ flow in G. Therefore, the value of a maximum $s - t$ flow equals the cardinality of a maximum independent set.

13.4.4 Dilworth's Theorem

A partially ordered set, or poset for short, is a set of elements S and a relation $\succeq$ on some pairs of S such that

- (Reflexivity) For all $s \in S$, $s \succeq s$.
- (Antisymmetry) For all $s \in S$ and $t \in S$, $s \succeq t$ and $t \succeq s$ imply $s = t$.
- (Transitivity) For all s, t, u in S, $s \succeq t$ and $t \succeq u$ imply $s \succeq u$.
- Any two elements s and t such that neither $s \succeq t$ nor $t \succeq s$ are called *unrelated.*

A *chain* of length n in a poset is a sequence of distinct elements $s_1 \succeq s_2 \succeq \dots \succeq s_n$. A set of elements all pairs of which are unrelated is called an *antichain.*

Example 13.1 Any nonempty subset of $\Re^n$ is a poset with respect to the relation $\mathbf{x} \succeq \mathbf{y} \Leftrightarrow \mathbf{x} \geq \mathbf{y}$.

Example 13.2 Figure 13.5 depicts a poset on 13 elements, $a, b, c, \ldots, k, l, m$. Relations such as $a \succeq c$ and $k \succeq j$ are not drawn. The sequences of elements $\{a, b, c, e\}$ and $\{a, k, l, m\}$ are both chains of length 4; $\{a, b, k, l, j\}$ is a chain of length 5. *Question: What is the length of the longest chain?*[3] The elements a and g are unrelated. The set $\{c, f, m\}$ is an antichain.

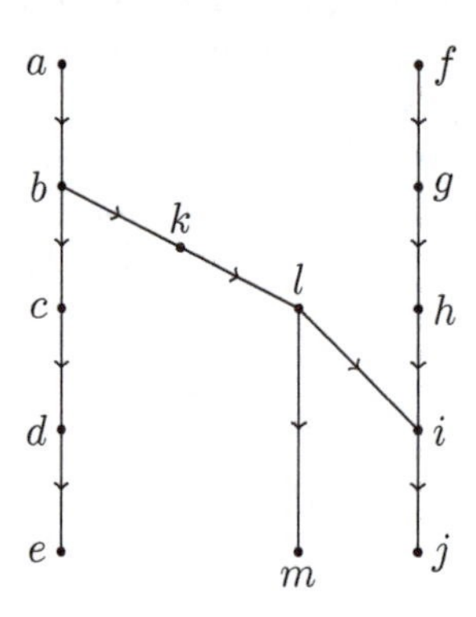

FIGURE 13.5: A partially ordered set on 13 elements. Relations that can be inferred from transitivity are not shown. $\{a, b, k, l, m\}$ is a maximal chain; $\{b, k, i, j\}$ is a non-maximal chain; $\{a, g\}, \{c, k, h\}$ and $\{d, m, i\}$ are all maximal antichains.

Example 13.3 Any nonempty subset of $\Re^n$ is a poset with respect to the relation $\mathbf{x} \succeq \mathbf{y} \Leftrightarrow \mathbf{x} \geq \mathbf{y}$.

Dilworth's Theorem states that the least number of chains into which the elements of a poset can be partitioned equals the cardinality of its largest antichain. The weak duality is obvious, as usual, because each chain can contain at most one member of an antichain.

Example 13.4 The elements of the poset in Figure 13.5 can be partitioned into the chains (a, b, k, l, m), (c, d, e), and (f, g, h, i, j). This partition, together with any of the many antichains of three elements, demonstrates the strong duality property for this poset. *Question: What do the complementary slackness conditions mean in this case?*[4]

To prove Dilworth's Theorem by LP duality, we will first derive a generalization of the max-flow min-cut duality theorem.

[3] 6.

[4] Every maximum antichain comprises one element from each of the three chains; every chain in a minimum partition contains an element from every maximum antichain.

Theorem 13.4.5 *Let $G = (V, E)$ be a graph whose directed edges $e \in E$ have lower and upper flow bounds $0 \le l_e \le u_e$, respectively. Let s, t be distinct distinguished nodes of G. Then the maximum total feasible flow on G from s to t equals the minimum over all partitions S, T of V where $s \in S$ and $t \in T$ of*

$$\sum_{i \in S, j \in T} u_{i,j} - \sum_{q \in T, r \in S} l_{q,r}.$$

Conversely, the minimum total feasible flow from s to t is the maximum over all such partitions S, T of

$$\sum_{i \in S, j \in T} l_{i,j} - \sum_{q \in T, r \in S} u_{q,r}.$$

Proof: *The LP formulation has variables $l_{i,j} \le f_{i,j} \le u_{i,j}$ defined to be the flow on edge (i, j). The objective is to maximize z; the constraints (other than the lower and upper bounds) are $A\mathbf{f} = (z, \mathbf{0}, -z)$. Here A is the usual node-arc incidence matrix defined by its nonzero terms $A_{i,(i,j)} = 1$ and $A_{j,(i,j)} = -1$. The dual is*

$$\begin{aligned} &\min \sum_{e \in E} u_e \delta_e - l_e \mu_e \text{ subject to} \\ &A^T \boldsymbol{\pi} + \delta - \mu = \mathbf{0} \Leftrightarrow \pi_i - \pi_j = \mu_{i,j} - \delta_{i,j} \ \forall (i, j) \in E \\ &\pi_t - \pi_s = 1 \\ &\delta_e \ge 0, \mu_e \ge 0 \text{ for all } e \in E \end{aligned}$$

As usual, adding the same constant to each π_i has no effect on the dual. Without loss of generality, set $\pi_t = 1$ and $\pi_s = 0$. Similarly, for any $e \in E$, adding the same constant to δ_e and μ_e has no effect on dual feasibility unless it makes one or both of the variables strictly negative. If $u_e > l_e$, at optimality it must be that $\min\{\delta_e, \mu_e\} = 0$. If $u_e = l_e$ any optimal solution can be reduced to make $\min\{\delta_e, \mu_e\} = 0$, which modification augments the set of binding constraints.

Therefore, if the dual has an optimal solution, it has an optimal basic feasible solution satisfying $\min\{\delta_e, \mu_e\} = 0$ for all $e \in E$. By total unimodularity this solution is all integer. Following any $s - t$ path or $t - s$ path shows that $\pi_t = 1$ and $\pi_s = 0$ imply $\delta_e \le 1$ and $\mu_e \le 1$. An optimal BFS can therefore be interpreted as $\delta_{i,j} = 1 \Leftrightarrow i \in S, j \in T$ and $\mu_{i,j} = 1 \Leftrightarrow i \in T, j \in S$. The objective function value therefore equals the sum of upper bounds of arcs from nodes in S to nodes in T minus the sum of lower bounds of arcs from nodes in T to nodes in S.

The converse follows from the fact that a maximum $t - s$ flow $\mathbf{f}$ must be a minimum $s - t$ flow (with total amount of flow multiplied by -1). ■

Figure 13.6 depicts the conversion of a poset to a network flow in the proof of Dilworth's Theorem. Represent each poset element by two nodes, an "in" node and an "out" node, and require a flow of 1 from the former to the latter. Add dummy nodes s and t to the network. Model a chain in the poset as a unit flow from s to t. Then a minimum feasible flow corresponds to a minimum covering by chains, which can be reduced to a minimum partition into chains. Set large upper bounds on the arcs. That forces the optimal dual cutset to correspond to an antichain.

Theorem 13.4.6 (Dilworth [79],[82]) *The cardinality of a largest antichain in a poset equals the least cardinality of a partition of the poset elements into chains.*

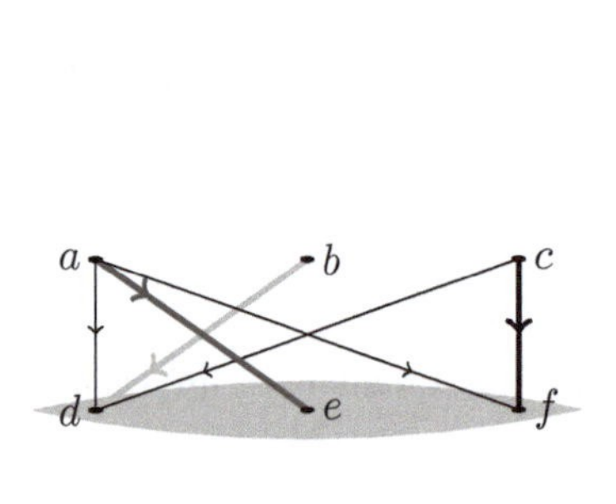

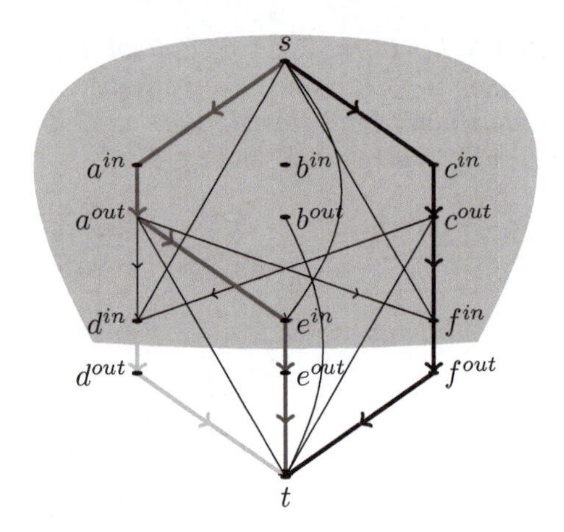

FIGURE 13.6: A partition into chains of the poset on the left corresponds to a feasible flow in the network on the right. The antichain $\{d, e, f\}$ on the left corresponds to the cutset on the right.

Proof: *Let the poset have elements S and relation $\succeq$, where $s \notin S$ and $t \notin S$. Create graph $G = (V, E)$ with*

- *For all $v \in S$, $v^{in} \in V$ and $v^{out} \in V$. Also, $s \in V$ and $t \in V$.*
- *For all $v \in S$, $(v^{in}, v^{out}) \in E$ with lower bound and upper bound $2|S| + 1$.*
- *For all $v \in S$, $(s, v^{in}) \in E$ and $(v^{out}, t) \in E$ each with lower bound 0 and upper bound 1.*
- *For all $v \neq w$ such that $v \succeq w$, $(v^{out}, w^{in}) \in E$ with lower bound 0 and upper bound $2|S| + 1$.*

First, every chain $v_1 \succeq v_2 \succeq \ldots \succeq v_m$ in the poset corresponds to a unit flow in G on the $s - t$ path

$$s, v_1^{in}, v_1^{out}, v_2^{in}, v_2^{out}, \ldots, v_m^{in}, v_m^{out}, t.$$

Second, by the definition of G, every $s - t$ path in G corresponds to a chain in the poset. Third, every partition of S into k chains corresponds to a feasible $s - t$ flow of k in G, because for every $v \in S$, the arc (v^{in}, v^{out}) has lower bound 1. Fourth, every integer feasible flow of size k in G corresponds to a partition of S into k chains.

By the integrality property of network flows, there exists a minimum size feasible flow that is integer. Therefore, the size of a minimum feasible flow in G equals the minimum number of chains needed to partition S.

Let $A \subset S$ be a maximal antichain in the poset. For all $v \in S \setminus A$, either $v \succeq a$ for some $a \in A$ or $a \succeq v$ for some $a \in A$. Otherwise, v would be unrelated to all members of A, contradicting maximality. Also, it is impossible for $v \succeq a$ and $a' \succeq v$ for $a \in A, a' \in A$. For if so, $a' \succeq v \succeq a$, implying either $a' = v = a$ which contradicts $v \notin A$, or $a' \succneqq a$ which contradicts A's being an antichain.

To see the correspondence between cutsets and antichains that we are aiming for, let A be a maximal antichain. Define the cutset $\mathcal{A}, \mathcal{A}^C$ of G as follows:

- *$s \in \mathcal{A}, t \notin \mathcal{A}$.*
- *For all $a \in A$, $a^{in} \in \mathcal{A}$ and $a^{out} \notin \mathcal{A}$.*

- *For all $v \in S \setminus A$ such that $v \succeq a$ for some $a \in A$, $v^{\in} \in A$ and $v^{out} \in A$.*
- *For all $v \in S \setminus A$ such that $a \succeq v$ for some $a \in A$, $v^{\in} \notin A$ and $v^{out} \notin A$.*

By construction, the lower bound on flow provided by cutset $\mathcal{A}, \mathcal{A}^C$ is

$$\sum_{i \in \mathcal{A}, j \notin \mathcal{A}} l_{i,j} - \sum_{q \notin \mathcal{A}, r \in \mathcal{A}} u_{q,r} = |A|.$$

That is the correspondence we seek.

By Theorem 13.4.5, there exists a cutset B, B^C of G with $s \in B, t \notin B$ which provides a lower bound equal to the size of a minimum feasible flow. The latter has been shown to equal the size of a minimum partition of S into chains, which is less than $|S|+1$. It remains to prove that the cutset yields an antichain with the same value.

The large upper bound $2|S|+1$ on arcs (v^{out}, w^{in}) for all $v \succeq w$ enforces the property

$$v \succeq w \text{ and } w^{in} \in Bv \Rightarrow v^{out} \in B.$$

Similarly,

$$v^{out} \in B \Rightarrow v^{in} \in B.$$

Let $A \subset S$ comprise the members a of S such that $a^{in} \in B$ and $a^{out} \notin B$. The properties just stated imply that A is an antichain, and the cutset B, B^C yields a lower bound on flow equal to $|A|$. ■

That Dilworth's Theorem can be derived from LP strong duality was first shown in [82].

Summary: Maximum flow algorithms that are fast enough to be useful in practice have been known for half a century. Decades of additional research have produced an extraordinary wealth of ideas, culminating in a $O(|V||E|)$ maximum flow algorithm, which employs different techniques for different ranges of edge density. The dual to maximum flow from s to t is the minimum cut between s and t, which is the minimum possible sum of arc capacities from S to T of a partition S, T of the nodes such that $s \in S$ and $t \in T$. This duality can be derived from LP duality or algorithmically, and applied to prove several combinatorial min-max theorems.

13.5 Problems

E Exercises

1. For the graph $G = (V, E)$ with $V = \{s = 1, 2, 3, 4, 5, 6 = t\}$ and arc upper bounds $u_{i,j} = (i-j)^2$ for $1 \leq i < j \leq i+3$ and $u_{i,i-1} = 3i$ for $i = 3, 4, 5$, find a maximum $s-t$ flow using Ford and Fulkerson's algorithm, and selecting augmenting paths by depth-first search. The first augmenting path will be $(1, 2, 3, 4, 5, 6)$; the second augmenting path will be $(1, 3, 4, 5, 6)$, etc.)

2. Repeat Problem 1 using the breadth-first search to implement the Edmonds-Karp shortest augmenting path algorithm.

3. Repeat Problem 1, selecting a *longest* augmenting path at each iteration. Don't forget the arcs $(3,2),(4,3)$, and $(5,4)$.

4. Repeat Problem 1, selecting a maximum capacity augmenting path at each iteration.

5. Repeat Problem 1, selecting a *minimum* capacity augmenting path at each iteration. Compare the number of augmentations required.

6. Repeat Problem 1, using scaling to approximate the choice of maximum capacity augmenting paths.

7. Decompose an optimal solution to Problem 1 into as few paths as possible. Compare the number of paths to $|E|$.

8. Apply the Köenig-Egerváry Theorem to prove that the symbol $\top$ cannot be covered by a single vertical or horizontal line, the symbol $\sqcap$ cannot be covered by two lines (each horizontal or vertical), and the symbol $\exists$ cannot be covered by three such lines.

M Problems

9. It is tempting to extend Corollary 13.1.3 to irrational data with the following argument: if u_{ij} are irrational, let f be an optimal $s-t$ flow. Construct the residual capacity graph G^f. Since f is optimal, G^f cannot contain an AP. Define the cut S, T as in the statement of Theorem 13.1.2. As in the proof of the theorem, the value of the cut equals the amount of the $s-t$ flow f. What is wrong with this argument?

10. Graph $G = (V, E)$ with upper bounds $u_e : e \in E$ and distinguished nodes s and t has a maximum $s-t$ flow value of 10. The $s-t$ cut S, T in G has value 10. Flow f uses all arcs in the cut S, T at full capacity, that is, $f_{ij} = u_{ij}$ $\forall i \in S, j \in T$. Yet f has value 8. How can this be?

11. State and prove a weak duality theorem that generalizes the Koenig-Egervary Theorem to three-dimensional grids. Construct a counterexample to strong duality.

12. Let $G = (V, E)$ have two distinguished nodes s and t, upper bounds $u_e \geq 0$ and lower bounds $l_e : 0 \leq l_e \leq u_e$ on the edges $e \in E$. Generalize Corollary 13.2.3 to apply to G. Your proof should require $u_e \geq 0$. Does it require $l_e \geq 0$?

13. On the ground set $\{1,2,3,4\}$, let S be the set of subsets of the ground set. Thus $|S| = 16$. Define the relation $A \succeq B \Leftrightarrow A \supseteq B$ for all $A \in S, B \in S$.
    - Verify that this relation defines a partially ordered set.
    - Find a largest antichain in the poset. Prove it is the largest by partitioning S into chains.
    - Prove that yours is the unique largest antichain.

D Problems

14. Prove or disprove: Theorem 13.4.5 remains true if $l_e < 0$ is permitted.

15. Prove Hall's Theorem as a corollary to Dilworth's Theorem.

16. (Hoffman and Schwartz [150]). Greene and Kleitman [134] proved the following generalization of Dilworth's theorem. Let $P = (V, \succeq)$ be a poset. A k-family of P is a subset $W \subseteq V$ that contains no chains of length $k+1$ (or more). For any chain C of P define $f^k(C) \equiv \min\{|C|, k\}$. It is a weak duality property that for every k-family W and every covering $\mathcal{C}$ of P by chains,

$$|W| \leq \sum_{C \in \mathcal{C}} f^k(C),$$

because W cannot contain more than $f^k(C)$ elements from each chain C. Let $\mathcal{W}$ denote the set of all k-families of P and let $\mathbb{C}$ denote the set of all coverings of P by chains. The strong duality theorem due to Greene and Kleitman is

$$\max_{W \in \mathcal{W}} = \min_{\mathcal{C} \in \mathbb{C}} \sum_{C \in \mathcal{C}} f^k(C).$$

Prove this theorem using LP duality and network integrality. Hint: It suffices to use a bipartite graph.

I Problems

17. Why doesn't the combination of Theorem 13.4.5 and Corollary 10.2.9 contradict the NP-hardness of MAXCUT?

18. How many maximum size antichains would there be in Problem 13 if the ground set were $\{1, 2, 3\}$?

19. Repeat Problem 13 for the ground set $\{1, 2, 3, 4, 5\}$, except for uniqueness.

20. ([269],[216]) Conjecture and prove a generalization of Problems 13 and 19 to the ground set $\{1, 2, \ldots, n\}$.

21. Prove Dilworth's Theorem directly by induction on the length of a longest chain.

22. ([90]) Use Dilworth's Theorem to prove that every sequence of $\alpha\beta$ numbers has an nondecreasing subsequence of length $\alpha + 1$ or has a nonincreasing subsequence of length $\beta + 1$. Hint: Define a partial order on the sequence $\{x_i\}$ on the elements $\{i, x_i\}$.

Chapter 14

Computational Complexity

14.1 Introduction

Preview: This section is a non-technical overview of computational complexity and the role it plays in solving problems.

Computational complexity is the assessment of how much effort is required to solve different problems. It provides a classification tool useful in tackling problems, especially discrete deterministic problems. Use it to tell, in advance, whether a problem is easy or hard. Knowing this won't solve your problem, but it will help you decide what kind of solution method is appropriate. If the problem is easy, you can probably solve it as a linear program or network model, or with other readily available software. If the problem is hard, you usually try solving it as an IP. If IP tools don't work, you will probably have to develop a specialized large-scale method, or seek an approximate solution obtained with heuristics.

Complexity theory can help you to understand and deal with hard problems. Complexity analysis can pinpoint the nasty parts of your problem, alert you to a possible special structure you can take advantage of, and help you model more effectively.

The following terminology from Chapter 9 is standard.

- For two nonnegative functions $f(n)$ and $g(n)$, $f(n)$ is bounded above by order $g(n)$, spoken "$f(n)$ is big-Oh of g(n)", written $f(n) = O(g(n))$, if for some positive constants c, N, $f(n) \leq cg(n)$ for all $n \geq N$.

- An algorithm, A, runs in polynomial time if there exist positive constants c, N, k such that for every input of length $n \geq N$ the algorithm terminates in at most cn^k steps.

- Equivalently, let $f_A(n)$ be the largest number of steps algorithm A takes on any input of length exactly n. Then A is a polynomial time algorithm iff $f_A(n) = O(n^k)$ for some k.

Any algorithm that does not run in polynomial time runs in "superpolynomial" time. Many superpolynomial time algorithms run in at least exponential time.

Definition 14.1 *An algorithm A requires at least exponential time if there exists a constant $c > 0$ such that for all N there exists an input of length $n \geq N$ for which A takes at least 2^{cn} steps. If also for some c' A runs in $O(2^{c'n})$, A is an exponential time algorithm.*

In Definition 14.1, 2 can be replaced by any constant > 1 without affecting the meaning. *Question: Think of a function that is superpolynomial but not as big as exponential.*[1]

14.1.1 An Historical View of NP-Hardness

In the heady years following the invention of linear programming (LP) and the simplex method, researchers explored many applications and extensions. One of these was integer programming (IP), which is the same as LP except for the seemingly minor additional requirement that certain variables take on only integer values. Two patterns soon emerged: one good, one bad.

The good pattern was that many problems could be modeled as IPs that (apparently) could not be modeled as LPs. Problem after problem, from fixed charge to traveling salesman, fell before the onslaught of IP modelers.

The bad pattern was that all the solution methods proposed turned out to be poor at solving IPs, except small cases. Gomory's cutting plane algorithm is a quintessential example. It was an extension of the simplex method, just as IP was an extension of LP. Like the simplex method, it could be proved to converge in a finite number of steps. By analogy, one would have expected Gomory's algorithm to solve IPs effectively. And it did, on small instances with, say, fewer than 20 constraints. But on moderate sized instances, the algorithm frequently bogged down, making many tiny ineffectual cuts, and sometimes failing to converge at all because of precision problems.

Researchers tried cutting planes, dynamic programming, branch and bound, and group theoretical methods, but all failed to solve the medium-sized cases.

The theory of computational complexity gives us a magic lens through which we can look at these two patterns, and see them, the good and the bad, as a whole.

Integer programming is an NP-hard problem. Like all NP-hard problems, it shares two properties:

- It models a broad variety of important problems;
- No known solution method consistently and exactly solves large instances efficiently.

14.1.2 Fast, Slow, Easy, and Hard

In complexity theory, we make the following generalizations: If an algorithm runs in polynomial time, it is *fast*; otherwise it is *slow*. Fast is good; slow is bad. A problem that we can solve by a fast algorithm is *easy*; a problem that we can't is *hard*.

Examples of easy problems include LP, minimum cost network flow, matching, minimum spanning tree, and sorting. Examples of hard problems include IP, traveling salesman, and precedence constrained scheduling. The class of easy problems is denoted P in Figure 14.1.

These generalizations work very well in practice, on the whole. Most polynomial time algorithms are fast and most exponential time algorithms are useless on large cases, in practice. Most easy problems can be solved quickly in practice (not necessarily by a theoretically fast algorithm – the simplex method is a good example); most hard problems cannot be solved quickly in practice if the cases are large.

Most uses of complexity in optimization involve a theoretically defined class of problems, the NP-complete problems. Theoreticians like me worry about whether the NP-complete problems really are hard. Indeed, the major unsolved question in theoretical computer science is to prove that no fast algorithm exists for an NP-complete problem. Practitioners

[1] $n^{\log n}$ or $2^{\sqrt{n}}$, for example.

like me don't worry about this question, because with the algorithmic technology available at present, the NP-complete problems are hard.

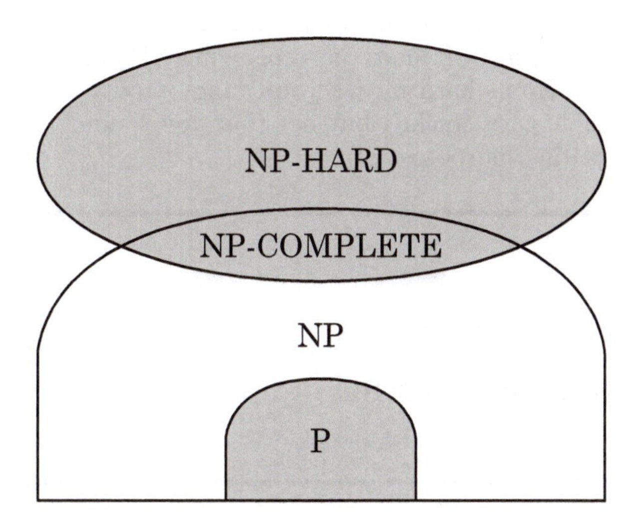

FIGURE 14.1: NP and related complexity classes.

Figure 14.1 depicts the class P of easy problems, the class of NP-complete problems, and the class of NP-hard problems, which contains the NP-complete class. You do not need to know the definitions to read and use this chapter, but I've included them in Section 14.10.

Since the main use of complexity theory is to distinguish easy from hard problems, it is usually just as helpful in practice to prove membership in the larger class, NP-hard, as in the smaller class, NP-complete. For almost all applications, the NP-hard class is simpler to work with. Therefore, I will treat NP-hardness almost exclusively in this chapter.

If your problem is hard, it is extremely likely to be "NP-hard". This is an empirical observation, and it permits this chapter to concentrate on the set of NP-hard problems. In terms of Figure 14.1, very few naturally occurring problems are found in the unshaded area between the set of easy problems P and the set of NP-hard problems.

Thousands of problems are already known to be NP-hard. If you want to be sure your problem is NP-hard, you must either determine that someone has already proved it, or use a known NP-hard problem to prove that your problem is hard as well.

If a problem is hard, you may be able to solve small instances readily, but any exact solution method will be in essence enumerative, and will require exorbitant time on large enough instances in the worst case.

This being said, we do not give up on a problem because it is hard. We are problem solvers. Our task does not end with an accurate diagnosis.

There are many reasons a hard problem can still be solved. NP-hard only means it takes a long time to solve exactly all cases of sufficiently large size. This leaves us with many loopholes. Most cases might be solvable quickly, or all could be solvable quickly to within a few percent of optimality. The cases you will have to solve might be small enough to be manageable. Identifying and focusing on the set of cases you have to solve is usually crucial to successful problem solving. I will call this set the "realistic cases".

Summary: As a general rule, polynomial time algorithms are fast and nonpolynomial algorithms are slow. Most problems encountered in practice, that cannot be solved by any known fast algorithm, are NP-hard. Hence this chapter will focus on recognizing NP-hard problems. Classifying a problem as hard means you cannot expect to solve exactly every possible case quickly, but not that you should despair of solving it for practical purposes.

14.2 Problems, Instances, and Real Cases

Preview: This brief section explains the absolutely essential concepts of problems, instances, and realistic cases.

In ordinary English usage, the term "problem" can refer to a specific numeric case, or to a class of cases having the same form. When studying computational complexity, we always call a specific case an *instance*. This may seem less natural than "case" but it is the standard term. We use the term *problem* to mean a class of instances taking the same form, as, "the traveling salesman problem." In Figure 14.1, a problem is represented by a single point. Maximize $\mathbf{c} \cdot \mathbf{x}$ subject to the constraints $A\mathbf{x} \leq \mathbf{b}; \mathbf{x} \geq \mathbf{0}$, is a problem. Maximize $4x_1 + 3x_2$ subject to the constraints $5x_1 + x_2 \leq 10; x_i \geq 0$, is an instance of that problem. *Question: Is Maximize $\sum_i x_i$ subject to the constraints $A\mathbf{x} \leq \mathbf{b}; x \geq \mathbf{0}$, a problem or instance?*[2] Using these terms, we rephrase an important statement from the previous section: a hard problem can have many easy-to-solve instances.

In practice, you will usually have to deal with something more general than a specific instance, but less general than a problem such as integer programming or line balancing. To refer to the set of instances of a problem that you expect to deal with, I will use the term "**realistic cases.**" The realistic cases usually are a family of related instances, often sharing structural or other properties. You don't know exactly what is in the family and what isn't, in advance, (and it almost always eventually turns out to include stuff you didn't think it would), but if you are a practitioner you know what I mean anyway. Identifying and taking advantage of features of the realistic cases is often crucial to solving your problem in practice. A simple example: suppose all the realistic cases are not more than a certain size. If this size is small enough, you may do perfectly well using, say, canned software for nonlinear or mixed integer programming. *Question: Why do you suppose we usually describe a problem rather than the realistic cases?*[3]

[2] A problem.

[3] The realistic cases take too long to explain and aren't fully known. Also, describing the realistic cases would cover much irrelevant detail. The *problem* is a model, an abstraction which includes the principal features.

Summary: An *instance* is a specific numerical case; a *problem* such as integer programming is a set of instances of the same form; the *realistic cases* are the instances of the problem that you expect might occur.

14.3 YES/NO Form

Preview: Most of the accumulated knowledge of problem complexity is stated in YES/NO form.

When we solve a problem, we almost always exhibit a schedule, route, part assignment, plan, or some other kind of solution. We laugh at the story of the mathematician who, seeing a fire in his wastebasket, points to the water faucet and a bucket, exclaims, "A solution exists!", and goes back to work, letting their house burn down. When I optimize the inventory policy for a corrugated paper plant to minimize trim, it doesn't occur to me to tell the manager his problem is solved, and average trim can be reduced to 1.82 inches, without providing the policy that achieves this value.

To do complexity analysis, we often must conquer our inclination to search out and provide solutions. Instead, we work with "Yes/No" versions of problems, where the output of an algorithm will be "Yes" or "No." *The first skill you must acquire is to take an optimization problem and convert it into "Yes/No" form.*

The optimization version of LP is to find $\mathbf{x}$ to maximize $\mathbf{c} \cdot \mathbf{x}$ subject to the constraints $A\mathbf{x} \leq \mathbf{b}; \mathbf{x} \geq \mathbf{0}$. To make a "Yes/No" version, we introduce a threshold value v, which is part of the input data. *Question: What do the rest of the input data consist of?*[4] We ask, is there a feasible $\mathbf{x}$ such that $\mathbf{c} \cdot \mathbf{x} \geq v$? Following Michael Garey and David Johnson's popular format [112], we define the problem in two parts. The first part describes the form of the input data. The second part is a "Yes/No" question. For threshold linear programming, we write:

Threshold LP
Instance: m by n matrix A; m-vector $\mathbf{b}$; n-vector $\mathbf{c}$; scalar v.
Question: Does there exist $\mathbf{x}$ such that $\mathbf{c} \cdot \mathbf{x} \geq v$ and $A\mathbf{x} \leq \mathbf{b}$?

In general, you can convert an optimization problem to "Yes/No" form by including a threshold value with the input data.

In some applications, problems naturally occur as feasibility or satisfice questions. These include many classroom and other facility scheduling problems. It should be clear how to convert these to "Yes/No" form. For example, if we are scheduling jobs with various processing times and deadlines on a machine, it may be appropriate to seek a schedule with no late jobs. In standard format the problem would be:

[4] Besides the threshold value v, the input contains the matrix A and the vectors $\mathbf{c}$ and $\mathbf{b}$.

1-Machine Scheduling with Deadlines
Instance: Processing times t_i and deadlines $d_i : i = 1, \ldots, n$.
Question: Is there a set of start times $s_i \geq 0 : i = 1, \ldots, n$ that is feasible, $s_j \notin [s_i, s_i + t_i)\ \forall i \neq j$, and that meets all deadlines, $s_i + t_i \leq d_i\ \forall i$?

> *When you convert to Yes/No form, hard problems should stay hard; easy problems should stay easy.*

We analyze the Yes/No version of a problem to classify its optimization version. Therefore, we want the two versions to have the same level of complexity. It is OK that they not have precisely the same complexity level, as long as the Yes/No version is easy if and only if the optimization version is easy.

When you convert to Yes/No form, the Yes/No version should never be more difficult to solve than the original version. That is, one should be able to use an algorithm that solves the optimization version, to answer the Yes/No question. *Question: What kind of error would occur if this property did not hold?*[5] In practice, unless you make a serious mistake, this property will hold.

The converse property usually holds, but proving it can take some effort. Fortunately, this is not necessary for NP-hardness proofs.

Often it is possible to restrict a Yes/No threshold problem by fixing the threshold value, without affecting the complexity of the problem. The 2-machine line-balancing problem provides a good example. The problem is to divide jobs with known processing times as evenly as possible between two machines. The threshold version is:

2-Machine Line Balancing
Instance: A set of processing times $t_1, \ldots, t_n$, and a number v.
Question: Can the indices $i = 1, \ldots n$ be partitioned into two sets, K and J, such that $|\sum_{i \in J} t_i - \sum_{i \in K} t_i| \leq v$?

A restricted case of the threshold version occurs when we require perfect balance between the machines, i.e., $v = 0$. In standard format we would write:

Partition
Instance: A set of processing times $t_1, \ldots, t_n$.
Question: Can the indices $i = 1, \ldots n$ be partitioned into two sets, K and J, such that $\sum_{i \in J} t_i = \sum_{i \in K} t_i$?

It turns out that this special case is just as hard as the threshold version. Since it is simpler to state and more restricted, it is better to work with.

On the other hand, an extreme restriction can be much easier than the threshold version. For example, one of the classic NP-hard problems on graphs, independent set, seeks a maximum size subset of nodes, no two of which are connected by an arc. What happens if we fix the threshold at an extreme value? We get the following problem:
Instance: A graph $G = (V, E)$.
Question: Does G contain an independent set of cardinality $|V|$, i.e., does there exist $S \subseteq V$; $|S| = |V|$; $(i, j) \notin E\ \forall i \in S, j \in S$?

In other words, does the given graph have no arcs? By fixing the threshold at an extreme value, we've substituted a very easy problem for a hard one. We could be misled into thinking that independent set was easy.

[5]The error would be that you might believe your optimization problem was hard, when in fact it was easy.

In my experience, you can usually tell when you are doing something silly as in the example just given. There are problems, such as Max 2-SAT, that may fool you (see Section 14.5), but all that will go wrong if you trust your judgment is that occasionally, you will spend a few hours thinking that your problem is easier than it actually is. And on the other hand, understanding the extreme case can often give you insight into solving the general case.

Sometimes the entire objective function can be removed without affecting the complexity of the problem. Integer programming feasibility is just as hard as threshold integer programming. *Question: State each of these problems in standard format, and explain why the former is just as hard as the latter?*[6]. LP optimization is just as easy as LP feasibility (combine the primal, dual, and strong duality constraints). *Question: Think of a problem that changes from hard to easy if the objective is removed.*[7]

Summary: For complexity analysis purposes, we state problems in Yes/No versions, in standard Instance – Question format. In general, to convert an optimization problem, include a threshold value in the input data format. Sometimes it suffices to choose a single, fixed threshold value to use for all instances. However, be careful not to choose an extreme threshold value that makes all instances easy to solve.

14.4 The Nuts and Bolts of NP-Hardness Proofs

Preview: This section introduces NP-hardness proofs. The main idea is to model a known NP-hard problem as your problem. Figure 14.3 shows the general form of a proof.

A basic way to solve problems in operations research is to have a toolbox of standard well-solved easy problems such as maximum flow and shortest path. You take your real problem and model it with the right choice of problem from the toolbox. That classifies your problem as easy, and lets you solve it by a standard method. This chapter teaches how

[6]The threshold problem is:
Instance: m by n Matrix A, m-vector $\mathbf{b}$ and n-vector $\mathbf{c}$, scalar v. **Question**: Is there an integer n-vector $\mathbf{x}$ such that $A\mathbf{x} \leq \mathbf{b}$ and $\mathbf{c} \cdot \mathbf{x} \geq v$?
The feasibility problem is:
Instance: m by n Matrix A, m-vector $\mathbf{b}$ and n-vector $\mathbf{c}$. **Question**: Is there an integer n-vector $\mathbf{x}$ such that $A\mathbf{x} \leq \mathbf{b}$?
Any threshold instance can be turned into an equivalent feasibility instance by including $\mathbf{c} \cdot \mathbf{x} \geq v$ in the feasibility constraints.

[7]Knapsack, Clique maximization, graph coloring, line balancing, and many others. See Section 14.5 for problem definitions.

to classify problems in the opposite way. You have a second toolbox, containing basic hard problems. You make the right choice of problem from the toolbox, and model it as yours. That classifies your problem as hard. This model in reverse is a *complexity proof*, and it sometimes pinpoints the difficulty in real problems. Pinpointing the difficulty can indicate where to aggregate, approximate, decompose, or simplify. If you know where the borders are between hard and easy, you will be better able to deal with your problem.

14.4.1 An Introductory Example

For our first example, I'll describe a situation that actually arose in the midst of a practical application. We were sequencing placements of electronic components on a circuit board. We wanted to minimize the production cycle time per board. The placement machine required a minimum of .25 seconds to place a component on the board. But if the previous component placed was far away on the board, or was a type of part located far away on the machine, the machine could take substantially more than .25 seconds to place. The machine had to begin and end at a fixed location (referred to as a fiducial), so our problem was a traveling salesman problem (TSP), a famous NP-hard problem.

In our TSP, the nodes represent placements, and the cost of arc from i to j is the time required to place j if the previous placement is i. There were so many arcs costing .25 that for practical purposes we could slightly simplify our problem: discard the arcs costing more than .25, and seek a Hamiltonian cycle in the simplified graph. (A cycle or path in a graph is Hamiltonian if it visits each node exactly once.) *Question: How could we arrive at this Hamiltonian cycle model as a yes/no version of our TSP?*[8]

The Hamiltonian cycle problem is another famous NP-hard problem, and we used heuristic methods to solve it. Then the problem changed. We found that some of the circuit boards had parts of varying sizes. The overall speed of the machine depended on the largest part it had picked up so far. We decided to sort the parts into size classes, and solve separate sequencing problems for each class in turn. However, these new sequencing problems were not Hamiltonian cycle problems! Consider the smallest size parts: the fiducial forces a fixed starting point, and we must visit each node (make each placement) once, but we no longer have to return to our starting point.

Would this extra bit of freedom make the problem any easier? Should we still have been using heuristics to solve it? We demonstrate that the new problem is NP-hard. This will serve as our first example of an NP-hardness proof.

We already know the following problem is hard.

UNDIRECTED HAM CYCLE
Instance: An undirected graph G.
Question: Does G contain a Hamiltonian cycle?

We would like to show our new problem is hard. In words, our problem is to find a Hamiltonian path which starts at a particular node. The starting point is referred to as "distinguished" because it is known in advance. We precisely define our new yes/no problem:

UNDIRECTED s-HAM PATH
Instance: An undirected graph $H = (V, E)$ with distinguished node $s \in V$.
Question: Does H contain a Hamiltonian path beginning at s?

Suppose s-HAM PATH were easy. Then we could have software to solve it quickly. We show we could use this imaginary software to solve HAM CYCLE quickly, giving a

[8] We find it as a threshold question with a particular fixed value. Does there exist a Hamiltonian cycle with total cost $\leq n(.25)$?

contradiction. (*Question: what will be the contradiction, and what will we be entitled to conclude?*[9] The way to use our imaginary software is to build a fast front end which has

INPUT: Undirected graph G with node set U and arc set E.

OUTPUT: Undirected graph H with distinguished node s.

PROPERTY: H has an s-Ham path iff G has a Ham cycle.

Question: What are the input and output instances of?[10]

Our imaginary software would work as follows: given input graph G, select any node u of U and make a copy of it, $\hat{u}$, connected to the same nodes u is connected to. Create a new node s connected only to u, and likewise a new node $\hat{s}$ connected only to $\hat{u}$. Call the resulting graph H. See Figure 14.2. The front end is fast, with the bulk of the time spent in copying G.

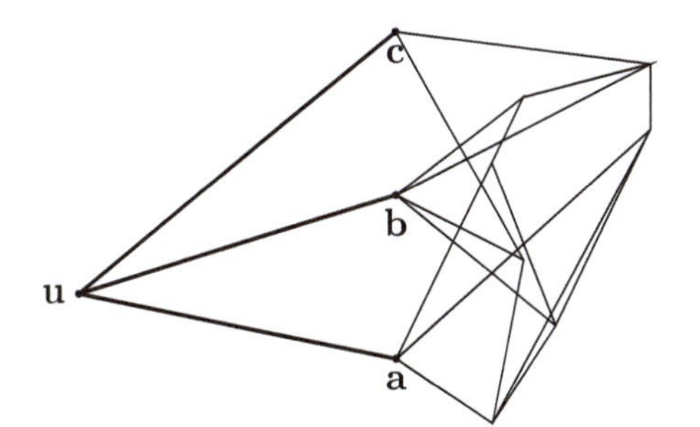

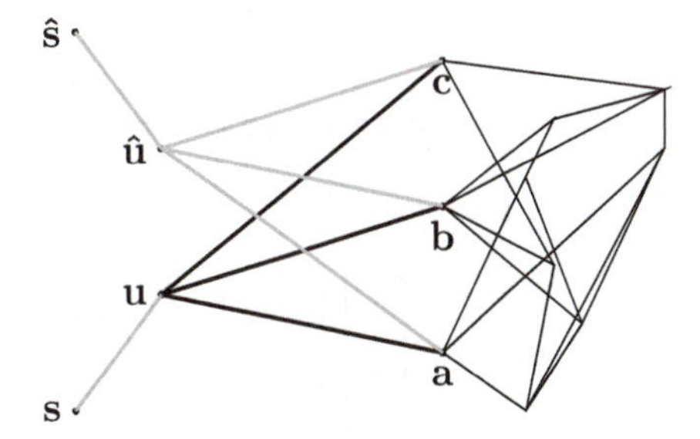

FIGURE 14.2: Connect a copy $\hat{u}$ of u to all of u's neighbors, then append degree-1 vertices $s, \hat{s}$ connected to $u, \hat{u}$, respectively, to transform HAM CYCLE to s-HAM PATH instance. Any s-HAM path must terminate at the degree-1 vertex $\hat{s}$.

If G has a Hamiltonian cycle, we can think of it as starting and ending at u. This cycle corresponds to a path in H starting at u, ending at $\hat{u}$, and visiting all the nodes in H except s and $\hat{s}$. Extend this path at the beginning and end to reach s and $\hat{s}$, respectively. This is a Hamiltonian path in H starting at s. Therefore, if we input a graph G that has a Hamiltonian cycle, then our front end will output a graph H that has a Hamiltonian path starting at s, and the imaginary software for s-HAM PATH, coupled with our front end, will correctly answer "YES".

If all we wanted was to answer "YES" whenever the input had answer "YES", we could use a very short program that always answered "YES". We also need to verify that if G has no Hamiltonian cycle, then our software will correctly answer "NO". We prove this by the contrapositive. *Question: State the contrapositive.*[11]

If our software answers "YES", then H contains a Hamiltonian path starting at s. Since the path is Hamiltonian, it must visit $\hat{s}$. By construction, only one arc is incident to $\hat{s}$. Therefore, the path has to visit $\hat{s}$ last. Truncating the first and last arcs in the path gives a path from u to $\hat{u}$, which corresponds to a Hamiltonian cycle in G.

We've shown that our software answers "YES" if and only if G contains a Hamiltonian cycle. The front end is fast, the imaginary software is fast, and thus we have fast software

[9]Being able to solve HAM CYCLE quickly, contradicts the fact that HAM CYCLE is hard. We are entitled to conclude that s-HAM PATH is hard.

[10]The input is an instance of HAM CYCLE; the output is an instance of s-HAM PATH.

[11]The contrapositive is as follows: if our software answers yes, then G has a Hamiltonian cycle.

to solve the HAM CYCLE problem. By contradiction, s-HAM PATH is NP-hard. We have completed our first NP-hardness proof.

Reading Tip: *Now you have your first chance to do your own. Close the book. Write the proof we've just done. You may peek at the definitions of HAM CYCLE and s-HAM PATH, and at Figure 14.2.*

After you've checked your proof, do Exercise 7.

14.4.2 The General Form of an NP-Hardness Proof

Figure 14.3 shows the basic outline of an NP-hardness proof. This is the structure you must follow to show your problem is NP-hard. You must select a known hard problem and *model it as a case of your problem.* This is counterintuitive because it is the opposite of what you might do to solve your problem, namely model your problem as an IP or as a case of some other more well-known problem.

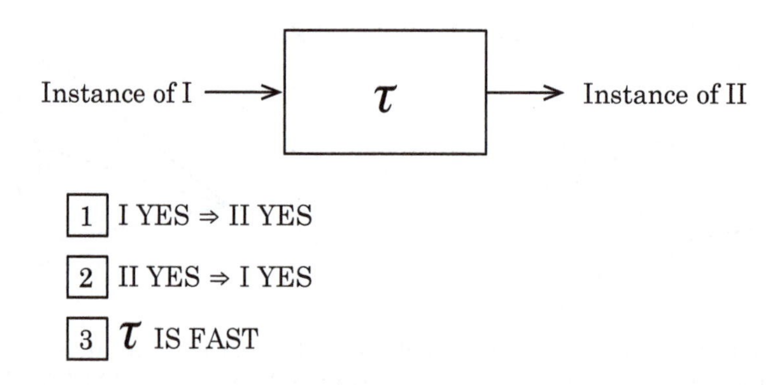

FIGURE 14.3: Basic structure of an NP-hardness proof.

You must concoct a transformation, $\mathcal{T}$, essentially a computer program like the front end in our s-HAM PATH proof, which has

INPUT : instance of I, a known NP-hard problem.

OUTPUT : instance of II, your problem.

Your transformation $\mathcal{T}$ must satisfy three properties:

1. $\mathcal{T}$ maps "yes" instances of I to "yes" instances of II.

2. $\mathcal{T}$ maps "no" instances of I to "no" instances of II.

3. $\mathcal{T}$ is fast.

Finding the right $\mathcal{T}$ often requires a flash of insight because it has to satisfy these properties simultaneously. NP-hardness proofs are more akin to integration than differentiation in calculus: they require creative pattern recognition rather than the following of a cookbook procedure.

Question: Exactly what does "fast" mean in property 3?[12]

One could state properties (1) and (2) more succinctly as, "$\mathcal{T}$ outputs a 'yes' instance of

[12] $\mathcal{T}$ runs in time $O(L^k)$ for some k, where L is the length of the input instance of I.

Π if and only if it is input a 'yes' instance of I." I have separated this into two properties because in my experience, *the most common mistake is not to satisfy property (2).* It is also common to satisfy property (2) but fail to prove it.

Occasionally a proof fails because it violates property (3). Property (3) means that $\mathcal{T}$ runs in polynomial time in the length of the input instance. *Question: Is this use of the word "fast" consistent with its definition in the previous section?*[13]

Question: Why does it help to know lots of NP-hard problems when trying to prove a new problem is NP-hard?[14]

14.4.3 Why this Proof Structure?

The most frequent point of confusion for beginners is the direction of the transformation $\mathcal{T}$. Many people want to input an instance of their problem, Π, and output an instance of some known hard problem I. That is the natural direction of a transformation if you are trying to *solve* your problem. You would convert your problem into some more well-known mathematical form such as IP or TSP. Complexity proofs require you to do the opposite. This first seems counterintuitive and takes some getting used to.

The explanation is as follows. Suppose I is IP, a known hard problem. If you had fast software to solve Π, you could put in $\mathcal{T}$ as a front end and have a fast IP solver! So it must be at least as difficult to solve Π as to solve IP.

Summary: You've seen your first NP-hardness proof, and you've written your first if you've followed the reading tip. Figure 14.3 shows the main elements. You must verify all three properties. The key is a transformation, or front end, which models a known NP-hard problem as a case of your problem. This is counterintuitive. It is the opposite of what you do to solve a problem, which is to model your problem as a case of a known problem.

14.5 Spotting Complexity

Preview: This section will help you get better at spotting complexity. After some general tips, it presents a collection of contrasting hard and easy problems. Studying these will develop your ability to recognize what is hard and what is easy. It will also give you practice understanding the concise style in [112] and other references.

---

[13] Yes.

[14] Knowing more hard problems gives you more choices of what to model as your problem. The more choices, the more likely you are to find one that you can do readily.

Here are some things to look for that tend to make problems hard:

- Dividing up work or resources perfectly evenly;
- Making sequencing decisions that depend not just on where you are, but also on where you've been;
- Splitting a bunch of objects into subsets, where the objects in each subset must satisfy some constraint;
- Finding a largest substructure which satisfies some property;
- Optimal packing or covering.

In Section 14.3 we sometimes restricted the threshold value. When a problem involves a bunch of interlocking decisions, and you are trying to minimize the number of bad events such as late jobs, mismatches, or items out-of-place, consider the restricted problem that tries to eliminate all bad events. Frequently the restricted problem will be hard if the original problem is hard. On the other hand, if you are trying to minimize a weighted sum of penalties, this restriction is more likely to alter the complexity of the problem. For example, in single machine scheduling, minimizing the total tardiness (sum of amounts by which late jobs are late) is hard, but determining whether all deadlines can be met is easy.

How do I evaluate the complexity of a problem? Of course, I don't follow a rigid method, but this is how I tend to proceed. The first thing I do is *get a clear idea of the situation.* I state a precise formulation of the problem, usually in words, sometimes with a picture. Second, I strip away the "story" part of the problem and *abstract it down to a math problem*: I turn people into nodes, processing times into numbers, retrieval orders into subsets, etc. Third, I *visualize several different cases of the problem.* I tinker with a few small instances, by hand. What is the problem like if all the weights are equal, or if all subsets are the same size, or there are only two subsets? If the problem has several complicating features such as deadlines and precedence constraints, I take turns eliminating them. Often the simpler case of the problem can immediately be recognized as NP-hard. And knowing *where* the complexity of your problem resides is important, too. Fourth, after I've visualized some cases, I usually get a feeling that the problem is easy, or that it is hard. If the former, I try to *solve it with standard methods* such as greed, and to model it as an LP, network, or other known easy problem. If the latter, I *focus on the simplest case that seems hard.* Why does it seem hard? What does it remind me of? If it is simple to state, I pore through lists in [112],[161], and other sources, trying to find it or something similar. If this doesn't work, I either take my simplest case back to the third step, or I roll up my sleeves and try a gadget proof.

It is extremely useful to spend a few hours reading and visualizing problems from a set of NP-hard problems. Contrasting easy problems are valuable, too. The more you do this, the more you will develop your own intuitive sense of complexity. Later, as you explore a new problem, you may find that it "seems" hard. What other problem does this remind you of? You may be well on your way to resolving the complexity question.

Reading Tip: *The following list of contrasting hard and easy problems is the core of this section. For each hard problem, think of a Yes instance and a No instance. For each easy problem, try to figure out how to solve it in polynomial time.*

Integer Programming (IP)
Instance: Integer $m \times n$ matrix A, integer m-vector $\mathbf{b}$, integer n-vector $\mathbf{c}$, integer z.
Question: Does there exist integer $\mathbf{x}$ such that $A\mathbf{x} \leq \mathbf{b}$ and $\mathbf{c}^T\mathbf{x} \geq z$?.

Contrast: Linear programming is easy, as is IP when the constraint matrix is totally unimodular, or has one of its dimensions fixed [208]. By "dimension" I mean either the number of variables or the total number of constraints, including nonnegativity.

The canonical NP-hard problem is next. It is similar to 0-1 IP with 0-1 coefficients, and only 3 nonzero coefficients per constraint.

3-SAT
Instance: A set of Boolean (True/False) variables, $X_1, \ldots, X_n$, and a set of clauses C_j, where each clause contains 3 distinct literals. (A *literal* is a variable X_i or its complement $\bar{X}_i$.)[15]
Question: Can each variable be set to True or False so that at least one literal in each clause is True?

Several important variations on 3-SAT are also hard. In **NAE-3-SAT (not-all-equal 3-SAT)**, the instance is defined the same as in 3-SAT, but the question is, can the variables be set so that each clause contains at least one true and one false literal? In Exact-3-SAT, the question is, can the variables be set so that each clause contains exactly one true literal? *Contrast*: 2-SAT, the same problem except that every clause contains 2 literals, is easy. The idea is that if one literal in a clause is false, it forces the other to be true. Instead of an exponentially branching set of possible consequences, 2-SAT has simple paths of unambiguous consequences. An instance turns out satisfiable if there are no contradictory cycles of implications. However, *Max 2-SAT*, which maximizes the number of clauses containing at least one true literal, is hard.

MAX-CUT (Maximum Cutset)
Instance: An undirected graph $G = (V, E)$ and threshold value k
Question: Does there exist a partition U, W of V such that $|\{(u, w) \in E : u \in U, w \in W\}| \geq k$?
This *unweighted* cutset problem was proved NP-complete by Garey, Johnson, and Stockmeyer [113].) *Question: How do you describe MAX-CUT in words?*[16]
Contrast: MIN-CUT is easy even if edges are weighted, since it is the dual of maximum flow. MAX BISECTION CUT is also hard. It is the same as MAX CUT, except we require $|U| = |W|$. MIN BISECTION CUT is hard, too. This seems surprising until we see that it is the same problem as MAX BISECTION CUT, on the complementary graph.

Maximum Clique (CLIQUE)
Instance: Graph $G = (V, E)$ and threshold value k.
Question: Does there exist a subset $C \subset V$ of vertices such that $(v, w) \in E$ for all $v \neq w, v \in V, w \in W$?
This is equivalent to seeking a maximum independent set, a subset $I \subset V$ of the vertices such that $(v, w) \notin E$ for all $v \in V, w \in V$, in the complementary graph.
Contrast: This problem is easy on planar graphs (graphs that can be drawn in the plane without any arcs crossing), since planar graphs cannot contain cliques of size 5. Finding a maximum independent set in a planar graph is hard.

3-Dimensional Matching (3DM), also called **3-matching** or **tripartite matching**

[15]In this book, the literals in a single clause must be from distinct variables. Thus the clause may not contain X_1 and $\bar{X}_1$, nor may it contain two occurrences of X_1 or $\bar{X}_1$. Some authors do not forbid multiple occurrences of a variable within a clause.

[16]Given an undirected graph, split the nodes into two sets to maximize the number of edges between them.

Instance: : Three disjoint base sets X, Y, and Z, with $n = |X| = |Y| = |Z|$, and a set $S \subseteq X \times Y \times Z$ of acceptable triples.
Question: Does there exist a collection $\mathcal{M} \subset S$ of disjoint acceptable triples such that $|\mathcal{M}| = n$?
Contrast: This is the problem of matchmaking if there are 3 sexes. When there are 2 sexes this is easy. *Question: Why?*[17]

Reading Tip: *In the standard reference on complexity by Garey and Johnson [112], as well as most other texts, problems are defined concisely. Get used to having to puzzle over a definition for a moment before you understand the problem. For example, 3DM asks if we can match up all* $3n$ *items into* n *triples. But the definition only asks if we can find* n *disjoint triples. It takes a moment to see that every item will end up in a triple, as there are only* $3n$ *altogether.*

Steiner Tree
Instance: Graph $G = (V, E)$ with edge lengths d_e, a subset $S \subset V$ of its nodes, and number k.
Question: Does there exist a tree $T \subset E$ in G that is incident on every node in S, of total length $\sum_{e \in T} d_e \leq k$?
Contrast: The minimum spanning tree problem is easy. The Steiner tree may have additional nodes not in the set S— try connecting the nodes of a 2×2 square in a grid.

Knapsack
Instance: Nonnegative vectors $\mathbf{c}$ and $\mathbf{w}$, and scalars v and K.
Question: Does there exist a 0-1 vector $\mathbf{x}$ such that $\mathbf{w} \cdot \mathbf{x} \leq K$ and $\mathbf{c} \cdot \mathbf{x} \geq v$?

Contrast: This is essentially IP optimization with a single functional constraint. This problem is theoretically hard only when the coefficients occur in high precision. Otherwise, it is easily solved with dynamic programming. See Section 14.9.1. **Partition**, the special case of Knapsack in which $\mathbf{c} = \mathbf{w}$ and $v = K = \frac{1}{2}(\mathbf{1} \cdot \mathbf{c})$, is also NP-complete.

3-Partition
Instance: A set of $3n$ numbers $1/4 < x_i < 1/2$, with total sum n.
Question: Can the numbers be partitioned into n subsets (of three numbers each) with subset sum 1? *Question: Why is "of three numbers each" in parentheses?*[18]

Contrast: Unlike Knapsack, this problem is hard even when the numbers do not occur in great precision. If the problem were to partition the numbers into 3 subsets such that their sums were equal, it would be like knapsack in that dynamic programming would quickly solve low-precision instances.

2-Machine Weighted Flowtime Minimization
Instance: A set of jobs with individual processing times and weights, and threshold number k.
Question: Does there exists a feasible schedule for two parallel machines with weighted sum of completion times less than or equal to k?
Contrast: If all weights are equal, or all times are equal, the problem is easy even for more than 2 processors.

Hamiltonian Problems: Ham Cycle, Ham Path, Longest Path: The Hamiltonian

[17] It is bipartite matching.
[18] The strict lower and upper bounds make it redundant.

problems, which require a route through a graph that visits every vertex once, have been defined in Section 14.4.1. Even for the cases where the graph is a grid-graph, (could be cut out of a square mesh), these problems are hard [159]. Anticipating Section 14.7 a bit, you should be able to see that if HAM CYCLE is hard, then the traveling salesman problem is hard even when all costs are 0 or 1. Also, since HAM PATH is hard, so is the problem of finding a longest path in a graph whose distances are all 0 or 1.
Contrast: Finding the shortest path in a graph with nonnegative costs is easy.

Minimum Feedback Arc Set
Instance: Directed graph $G = (V, E)$, threshold k
Question: Can the nodes of G be arranged in a line so that at most k arcs point backwards?
Question: Describe this problem in mathematical notation.[19]
Contrast: The case $k = 0$ asks the easy question of whether or not G is acyclic (see Chapter 9).

3-Coloring.
Instance: Undirected graph $G = (V, E)$.
Question: Can the nodes of G be partitioned into 3 sets such that no two nodes connected by an arc are in the same set?

This problem is hard even if the graph is planar and has no nodes of degree more than 4 [113].
Contrast: 2-coloring a graph is easy. 4-coloring a planar graph is easy, since the answer is always YES. But 4-coloring a graph in general is hard.

Shortest Path with Obstacles
Finding the shortest path between two points in 3 dimensions, which avoids a collection of polyhedral obstacles, is NP-hard.
Contrast: The 2-dimensional problem is easy.

Set Covering Instance: Collection $\mathcal{S}$ of subsets on a ground set H, threshold k.
Question: Does there exist a subcollection $\hat{\mathcal{S}} \subset \mathcal{S}$ of at most k subsets that covers H, i.e., such that $\cup_{S \in \hat{\mathcal{S}}} S = H$.
Set Packing, which asks for a maximum cardinality subcollection of disjoint subsets, is also hard.

Exact 3-Cover (X3C) is the special case of Set Covering in which $|S| = 3 \; \forall S \in \mathcal{S}$ and $|\hat{\mathcal{S}}|$ must equal $|H|/3$. Because it asks for a partition, X3C can equally well be thought of as a special case of Set Packing. *Question: Which of the previously listed problems is a special case of X3C?*[20]
Contrast: If instead $|S| = 2 \; \forall S \in \mathcal{S}$ and $|\hat{\mathcal{S}}|$ must equal $|G|/2$, this would be the problem of finding a perfect matching in a graph, famously proved to be solvable in polynomial time by Jack Edmonds [87], in the same paper that introduced the definition of polynomial time algorithms.

[19] Given directed graph $G = (V = \{1, 2, \ldots, n\} \equiv [n], E)$ and integer k, does there exist a permutation $\rho : [n] \mapsto [n]$ such that $|\{(i, j) \in E : \rho[j] < \rho[i]\}| \leq k$?
[20] 3DM

Vertex-Disjoint Paths
Instance: Undirected graph $G = (V, E)$, set of k pairs of nodes $s_i, t_i : i = 1, \ldots, k$ comprising $2k$ distinct nodes.
Question: Does G contain k vertex-disjoint paths, one each between s_i and t_i for $i = 1, \ldots, k$? This problem is NP-hard even if restricted to planar graphs, or if paths are required to be edge-disjoint rather than vertex-disjoint [261].
Contrast: Solvable in polynomial time if for any fixed k [250].

Bin Packing
Instance: A set of numbers $0 \leq x_i \leq 1 : i = 1, 2, \ldots, n$, and an integer m.
Question: Does there exist a partition $S_1, S_2, \ldots S_m$ of the indices $i = 1, \ldots, n$ such that $\sum_{i \in S_j} x_i \leq 1$ for all j?

The bin packing metaphor is to place the n numbers into as few bins as possible such that no bin's numbers sum to more than its capacity 1.
Contrast: Unlike some other NP-hard problems, it is easy to find a solution that is guaranteed to be close to optimal. If you pre-sort the numbers in increasing order and iteratively place them in the lowest indexed feasible bin, the solution will employ at most $\frac{11}{9}$ as many bins as the minimum possible [162].

In contrast to Bin Packing, for some problems it is hard to find solutions close in value to the optimum.

Approximate Solution of Maximum Clique
Given a graph G, it is hard to find a clique that is $1/100$ the size of the largest clique in G. The value $1/100$ could be set to any $1 > \epsilon > 0$ and the problem would be hard. *(Note: This particular problem cannot be phrased in Yes/No form, see [10]).*

Summary: Define in your own words each of the hard problems listed in this section. They compose your toolbox of basic hard problems.

14.6 Illustrations of Common Pitfalls

Preview: This section illustrates the most common reasoning errors in NP-hardness proofs. All the examples involve Hamiltonian paths and cycles. These rely on visual reasoning to aid comprehension.

The first error is to make $\mathcal{T}$ a simple but lengthy procedure, violating property (3). This error can often be detected by noticing that $\mathcal{T}$ produces instances of Π that are exponentially long. For example, here is a very easy problem:

FIND ZERO

Instance: A list L of nonnegative integers.

Question: Does L contain a 0?

Let us construct an invalid transformation from s-Ham Path to FIND ZERO. See Figure 14.4. For each permutation of the nodes $V - s$, start at s, traverse G, and record on a list the number of mistakes, or missing arcs. Then G contains a Hamiltonian path from s if and only if we record a 0. The error is that L is not polynomially long in the size of G. Precisely, if G has n nodes, then it has $O(n^2)$ arcs and the input to $\mathcal{T}$ is $O(n^2)$. But the list L contains $(n-1)!$ integers, and $(n-1)!$ grows faster than any polynomial function of n^2. If $\mathcal{T}$ produces more than a polynomial length output, it cannot possibly run in polynomial time, violating property (3).

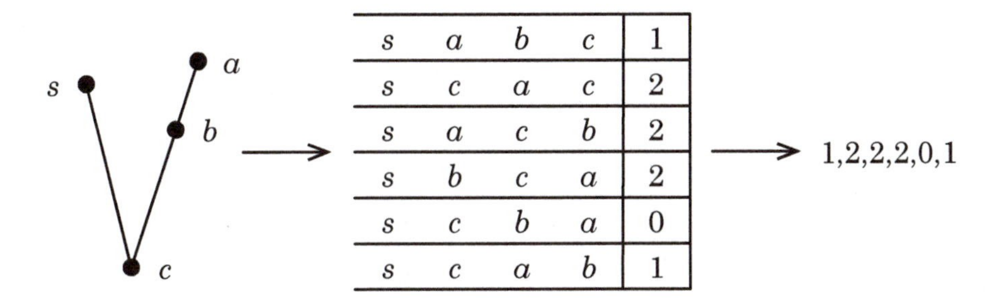

FIGURE 14.4: Invalid transformation of s-HAM PATH to FIND ZERO.

Although this mistake seems too simple to make, it occurs surprisingly often. You are most likely to make this mistake when transforming from a problem involving numbers. A number does not have to be very *long* (e.g., 30 binary characters) to be very *big* ($> 10^9$).

A related error is to make $\mathcal{T}$ solve the instance of problem I in some enumerative fashion. For example, one could solve LP by enumerating the extreme points of the polyhedron, sorting them by objective value, and selecting the maximum. The output of $\mathcal{T}$ would have polynomial length, and sorting n numbers can be done in $O(n \log n)$ time, yet $\mathcal{T}$ would not be a valid polynomial-time transformation because the LP could have exponentially many extreme points [34]. *Question: how could you construct a valid transformation from LP feasibility to FIND ZERO?*[21]

Suppose s-Ham Path is a known NP-hard problem, and we wish to use it to prove Ham cycle is NP-hard. This is the opposite of the first transformation we did in Section 14.4. Now we must take as input an instance of s-Ham Path, and transform it into an instance of Ham cycle. I will use this situation to illustrate two of the most common pitfalls.

Many beginners will take the graph G, and connect s by new arcs to every other node of G, to make H. They reason that a Hamiltonian path in G starting at s can be extended to form a cycle in H, simply by coming back to s. On the other hand, a Hamiltonian cycle

[21] Given an instance of LP feasibility, solve it with a polynomial time algorithm such as the ellipsoid. If the instance is feasible output the list $L = \{0\}$; otherwise, output the list $L = \{1\}$.

in H can be thought of as beginning at s; simply cut out the new arc and what remains is a Hamiltonian path from s in G. *Question: What is wrong with this reasoning?*[22]

Figure 14.5 illustrates the failure of this transformation. It will sometimes convert a No instance of s-Ham path to a Yes instance of Ham cycle (a definite no-no). The beginner here is guilty of wishful thinking when verifying Property (2). All that we can assume is that H has a Hamiltonian cycle. We cannot assume that this cycle uses exactly one of the new arcs. The beginner is dazzled by the vision of the Hamiltonian path turning into a cycle by means of the new arc, and perforce imagines that all Hamiltonian cycles consist of a path from G extended by a new arc. But as can be seen in Figure 14.5 the cycle might contain two new arcs and not correspond to a Hamiltonian path in G. *Question: Do cases where the cycle contains no new arcs cause an error?*[23] You might have noticed that if G is a 2-node graph with one arc, the transformation will convert this Yes instance into a No instance. This is not a serious error. It only occurs on graphs with fewer than 3 nodes, and transformations are permitted to fail on a finite number of instances.

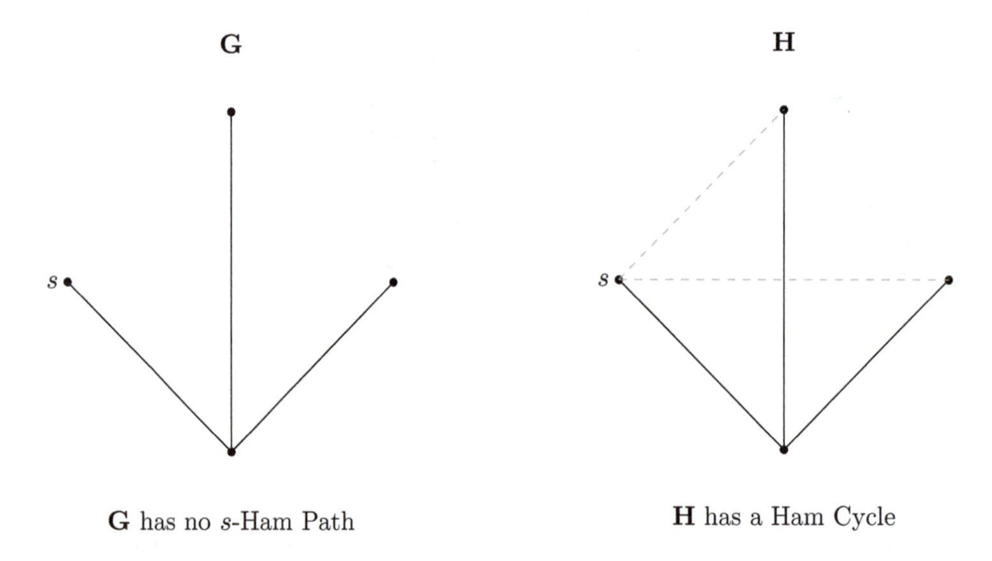

FIGURE 14.5: A No instance of s-Ham path could transform to a Yes instance of Ham cycle if edges from s to all other nodes are added.

This kind of wishful thinking is a very common source of error in verifications of Property (2). Don't get so caught up in your vision of how solutions to I transform to solutions to II, that you think all solutions to II have that form. To verify Property (2), you may only assume the answer is Yes, there is a solution; you may not make assumptions about the form of that solution.

Let's dig another pit for our beginner to fall into. The previous transformation didn't work because we connected s to too many nodes. Let us add only one new arc, so a cycle in H can't contain two new arcs. If G contains a Hamiltonian path starting at s, take the node v at the end of that path, and add a single new arc between v and s. The resulting graph H has a Hamiltonian cycle. The converse argument works this time to verify Property (2). A Hamiltonian cycle in H can't contain more than one new arc. Remove the new arc if it is present, otherwise remove either arc incident on s. The remaining arcs of the cycle form a Hamiltonian path from s in G.

[22] Property (3) is violated, as explained in the text following.

[23] No, those cases don't cause an error. The input instance has answer Yes.

The error is that we have violated property (3). How does the transformation $\mathcal{T}$ determine the node v? $\mathcal{T}$ is just dumb, fast front-end software. When it is input an instance, it doesn't know whether the answer is Yes or No, much less can it solve the instance. But when we "take" the node v at the end of the Hamiltonian path in G, we are implicitly assuming that $\mathcal{T}$ has been clever enough to find that Hamiltonian path.

When you verify property (1), you are permitted to assume the answer to the instance of I is Yes. When you verify property (2), you are permitted to assume the answer to the instance of II is Yes. But the transformation $\mathcal{T}$ is never permitted to make such an assumption. If a No instance is input to $\mathcal{T}$, it must produce a No instance. We often don't see No instances because we prove property (2) by the contrapositive, but these instances are crucial. *Question: The transformation $\mathcal{T}$ is not permitted to know whether the answer to the input instance is Yes or No. Why not?*[24]

If instead $\mathcal{T}$ selects an arbitrary node x in G to connect to s, Property (1) may fail, as shown in Figure 14.6.

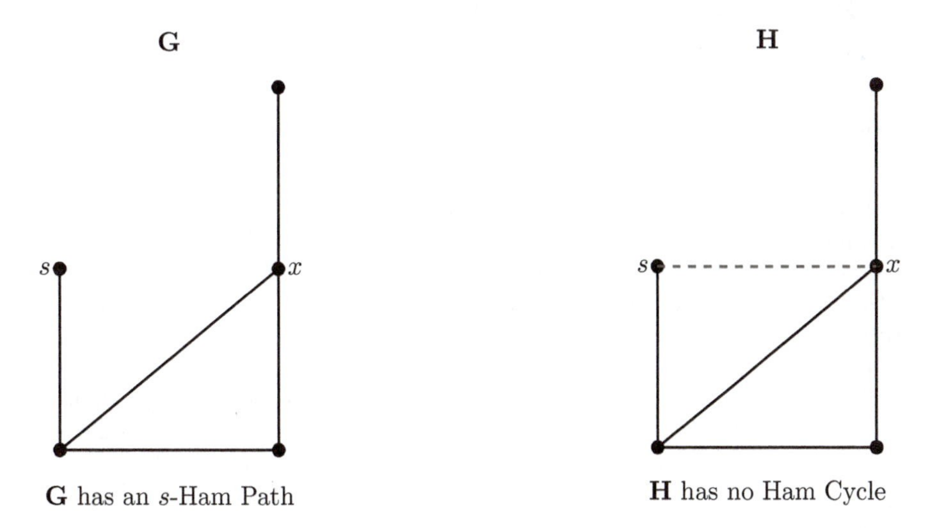

FIGURE 14.6: The transformation $\mathcal{T}$ cannot know which vertex to connect s to so as to turn G's s-Ham-path into a Ham cycle in H.

Finally, for a valid transformation, see Figure 14.7. Add two new nodes, t and u, to G. Connect t to every node except s in the new graph H, but connect u only to s and t. If there is a Hamiltonian path from s in G, it may be extended to a Hamiltonian cycle in G by appending t, then u. The required arcs from unknown node x to t, from t to u, and from u back to s, all are in H. Conversely, if there is a Hamiltonian cycle in G, it must use the arcs from t to u and from u to s. This is because H doesn't have any other arcs incident to u. Removing these arcs and the other new arc (the other one from t) leaves a Hamiltonian path from s in G.

Question: If we omitted u from the transformation, and connected s directly to t, which part of the proof would become invalid?[25]

Reading Tip: *Construct a small instance for which this invalid proof fails.*

The key to avoiding the "wishful thinking" error is to put an extreme substructure into

[24]If it knew, it could output 1 if Yes and 0 if No, and thereby transform any problem into the trivial problem: instance, a binary number x; question, is $x = 1$?

[25]The contrapositive proof of property (2) would be invalid. H might have a Hamiltonian cycle that doesn't use the arc (t, s).

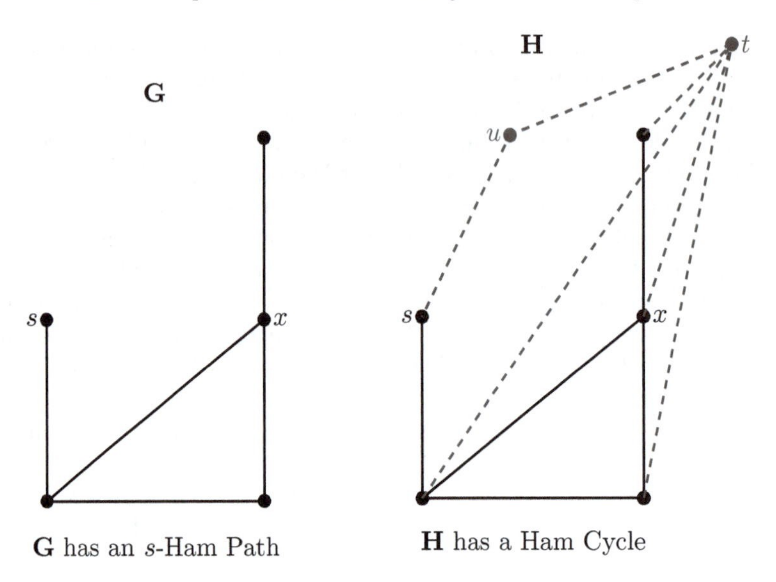

FIGURE 14.7: Valid transformation of s-HAM PATH to HAM CYCLE.

the instance which forces a solution to have the desired characteristic. In problems involving Hamiltonicity, nodes with only one or two incident arcs have strong forcing power. In other problems it may be helpful to use very large or very small costs or sizes, or other extreme conditions. *Question: In a scheduling problem with precedence constraints, what kind of job might be useful to force structure?*[26]

There is an interesting postscript to the s-Ham path to Ham cycle transformation. As described in Section 14.4.1, we had dealt with Ham cycle before encountering s-Ham path. So, we encoded the transformation as a front end to our existing Ham cycle module. Using the transformation for the opposite effect, solution instead of proving complexity, was the simplest way to solve s-Ham path. This approach had an additional advantage. Ham cycle is the cleaner, more widely studied problem of the two. We planned to acquire better software in the future for Ham cycle, and swap out our own module.

In the ordinary undirected graphs we have encountered so far, an arc (i, j) permits travel in either direction between i and j. In a directed graph an arc (i, j) permits travel only from i to j. For our last example in this section we transform the Hamiltonian cycle problem on directed graphs to Ham cycle.

Directed Ham cycle
Instance: A directed graph $G = (V, A)$.
Question: Does G contain a directed Hamiltonian cycle, i.e., a permutation $\pi(i)$ of the nodes V such that $(\pi(i), \pi(i+1)) \in A \ \forall i = 1, \ldots, |V| - 1$ and $(\pi(|V|), \pi(1)) \in A$?

Suppose we know Directed Ham cycle is hard and wish to prove Ham cycle is hard. See Figure 14.8. For each node v of G, make three copies, v_{in}, v_{middle}, and v_{out}, connected in a path as shown in the figure. For each arc (v, w) of G, make an arc from v_{out} to w_{in}. Call the new undirected graph H.

Property (1): Suppose G contains a Hamiltonian cycle $v^1, v^2, \ldots$. This means that for all i, G contains the arc (v^i, v^{i+1}). Therefore, for all i, H contains the arc $(v^i_{out}, v^{i+1}_{in})$. H also contains all the arcs between "in" and "middle" copies, and between "middle" and "out" copies of v^i. Therefore, H contains the Hamiltonian cycle $v^1_{in}, v^1_{middle}, v^1_{out}, v^2_{in}, v^2_{middle}, v^2_{out}, \ldots$

[26] A job which must be performed after all the other jobs, or before all the other jobs.

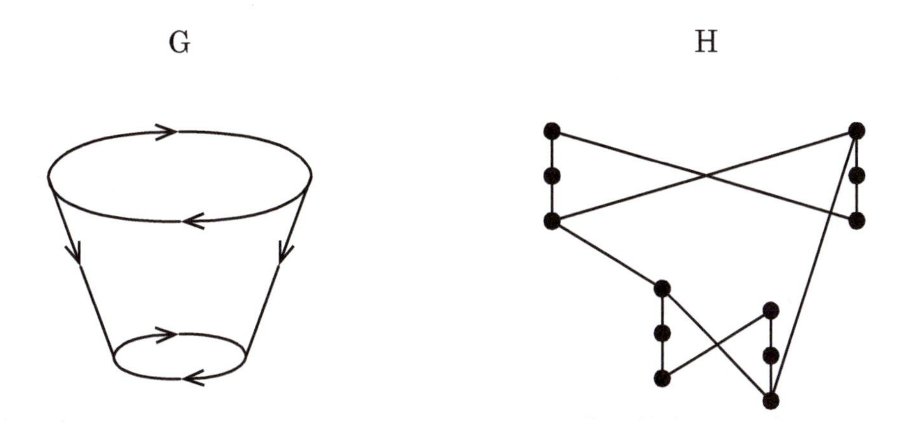

FIGURE 14.8: Transforming directed ham cycle to (undirected) HAM CYCLE.

Property (2): If H contains a Hamiltonian cycle C, we can think of C as starting at some node v_{in}^1 of H. The cycle C is Hamiltonian and must visit v_{middle}^1. But that node has only two arcs incident. So C must go from $v_{in}^1 \to v_{middle}^1 \to v_{out}^1$. From v_{out}^1 the cycle must go to an in copy of some other node, since all its arcs go to in copies (besides the arc to its own middle copy, which C has already used). From that in copy of, say, v^2, the cycle C must go to the middle and out copies of v^2, by the same reasoning as before (v_{middle}^2 has only two arcs). Thus C traverses the in, middle, and out copies of each node of G, returning to v_{in}^1. This corresponds to a Hamiltonian cycle in the directed graph G.

Property (3): The transformation requires time proportional to the size of the graph G, which is a polynomially bounded function of the size.

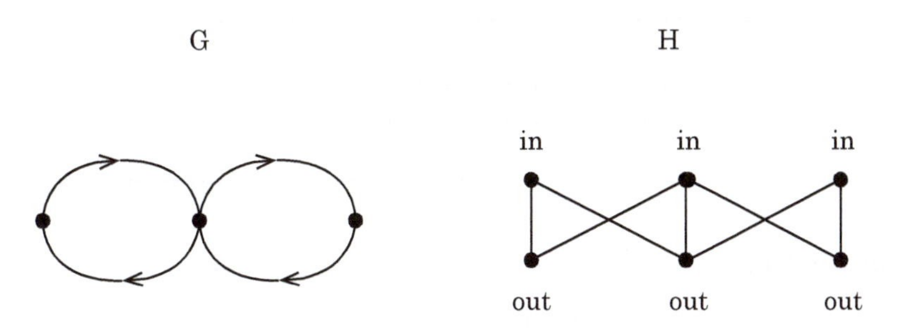

FIGURE 14.9: Omitting the "middle" nodes invalidates the transformation that was shown in Figure 14.8 from Directed to Undirected Ham Cycle. Without those nodes, H has a Ham Cycle though G does not.

Figure 14.9 shows the flaw in a transformation which omits the middle copies of the nodes. *Question: Which pit have we fallen into? Exactly where does the above proof become invalid?*[27] The middle copies enforced a specific structure on any Hamiltonian cycle in H because they each had degree two.

Reading Tip: *To acquire a firm understanding of correct NP-hardness proofs, I recommend that you to find transformations between several pairs of the following Hamilton variations.*

1. *Ham Cycle*
2. *s,t-Ham Path:* **Instance**: *Graph $G = (V, E)$ with two distinguished nodes s and t.* **Question**: *Does G contain a Hamiltonian path with endpoints s and t?*
3. *s-Ham Path*
4. *Ham Path* **Instance**: *Graph G.* **Question**: *Does G contain a Hamiltonian path?*

Summary: The most common pitfall in NP-hardness proofs is "wishful thinking," where you groundlessly assume a solution to II must have certain properties or structure. This makes your proof of property (2) invalid. You can often get around this difficulty by introducing extreme values or substructures in the instances of II, which *force* a solution to take a particular form. Other common pitfalls are to expect $\mathcal{T}$ to guess something which actually depends on knowing the solution, or do more than a polynomial amount of work.

14.7 Examples of NP-Hardness Proofs

Preview: This section contains a series of NP-hardness transformations, arranged in order of increasing difficulty.

14.7.1 The Easiest and Most Common Kind of NP-hardness Proof

I call this technique "parameter specialization". *Constrain some of the natural parameters of your problem, and arrive at a known NP-hard problem.* The more problems you know are NP-complete, the more powerful a weapon this will be in your arsenal. So many problems are known hard, and real problems are so complicated, that often this method will serve.

Perhaps the simplest examples involve fixing a general threshold parameter at a particular value. I've already hinted at this kind of transformation in Section 14.3, which describes Partition as a restricted case of 2-Machine Line Balancing.

[27] Wishful thinking. Property (2) fails because we can't be sure that C uses the arc (v^1_{in}, v^1_{out}).

Intuitively, when $V = \sum_i t_i/2$, the line balancing problem is the Partition problem. To be more precise, consider all the instances of line balancing in which $V = \sum_i t_i/2$. These instances compose the set of all instances of the Partition problem.

The transformation, which we would not ordinarily bother to write down, is as follows. Given an arbitrary instance $x_1, \ldots, x_n$ of Partition, set $t_i = x_i \ \forall\ i = 1, \ldots, n$ and set $V = \sum_i t_i/2$. The resulting instance of line balancing has answer Yes if and only if the given instance of Partition does.

Question: In a graph with nonnegative arc costs, the shortest path problem is a very well-known easy problem. How would you prove that the longest path problem is hard?[28]

Example: Scheduling with deadlines. A set of n jobs is to be sequenced on a single processor. The ith job requires time t_i, and incurs penalty p_i if it is not completed by its deadline d_i. The problem is to minimize the total penalty incurred, or equivalently to maximize the sum of the p_i values of jobs completed on time.

In the case where all jobs have the same deadline D, the problem is precisely the knapsack problem with knapsack capacity D, item sizes t_i and values p_i.

14.7.2 Specialization

This classification and terminology are due to [112]. In a formal sense, almost every NP-completeness proof is of this type, but this is what it means in practice: restrict the data of your problem in a simple way, and show that the resulting special class of your problem is the same as a known NP-hard problem.

For example, we transform HAM CYCLE to Directed HAM CYCLE. The latter problem is the same as HAM CYCLE, except the graph's arcs are directed (1-way) rather than undirected (2-way). The cases of Directed HAM CYCLE in which all arcs come in pairs (i, j) and (j, i) are equivalent to the undirected HAM CYCLE problem.

As another example, foreshadowed in Section 14.4.1, we transform HAM CYCLE to the threshold version of the Traveling Salesman Problem.

Instance: Complete graph $G = (V, E)$ with integer arc costs $c_e : e \in E$, and integer K.

Question: Does there exist a Hamiltonian cycle in G with total cost $\leq K$?

Restrict this problem to cases in which all arc costs c_e are 0 or 1, and $K = 0$. The question is then equivalent to whether G has a Hamiltonian cycle consisting only of 0-cost arcs. Thus it is the same as the Hamiltonian cycle problem. The transformation, which ordinarily we would not explicitly describe, is as follows. Take as input an instance of HAM CYCLE, a graph $H = (U, F)$. Set the cost of each arc in F to 0; then fill in the graph with the rest of the possible arcs, each costing 1. Call the resulting complete graph G. Finally, set $K = 0$.

Next we show 3DM (3D-matching) transforms to X3C (Exact 3-Cover). We are given a collection of acceptable triples on a base set. From Section 14.5, X3C asks if there is a partition of the base set composed of acceptable triples. To reduce it to 3DM, we first define X3C with some extra notation.

X3C – Exact-3-Cover

Instance: A collection S of subsets of $X = x_1, \ldots, x_{3n}$, each subset size 3.

Question: Is there a subcollection of n subsets in S whose union is X?

3DM (3-Dimensional Matching) is the special case of X3C where all $(x_i, x_j, x_k) \in S$ satisfy

[28]On an n-node graph with arc costs of 0 and 1, set the threshold value to $n - 1$ to specialize to the HAM PATH problem. This shows it is hard to find the longest path.

$1 \leq i \leq n < j \leq 2n < k \leq 3n$. In other words, there are n elements for each of three types. Only triples that comprise one of each type can possibly be acceptable. The $3n$ elements must be partitioned into n acceptable triples.

Most NP-hard problems arising in practical applications are fancy versions of one or more canonical NP-hard problems in the literature. If you strip away most of the embellishments and cluttering options, you often find that the much simpler problem underneath is still hard. Consider for example most vehicle fleet planning problems. There is a set of vehicles (e.g., trucks, ships) with various travel ranges, and capacities for conveying one or more commodities. There are one or more supply points, demands at various locations, and possibly restrictions on the timing of pickups, deliveries, etc. You seek a low cost or perhaps just a feasible schedule to meet demand.

Reading Tip: *Without reading ahead, where can you spot potential complexity in this problem?*

- If a low cost plan is sought, and travel distances affect cost, then even if there were only one supply point, and one vehicle with plenty of capacity and range, you would have a traveling salesman problem.

- It is NP-hard to split a set of numbers into two groups whose sums are equal. This is the Partition problem. If there are only two identical vehicles, a single supply point, and it is either infeasible or costly to split deliveries between vehicles, then the problem is hard. Why? In the case that total demand equals total fleet capacity, each vehicle has to go out full, hence Partition must be solved. This transformation works even if travel costs are relevant: just make them all equal.

- In Section 14.9.1 we will see that Partition is not as bad as some other hard problems because it is often susceptible to dynamic programming in practice. If there are m identical vehicles, however, splitting deliveries equally contains the 3-partition problem, which is hard and not susceptible to DP.

- If deliveries may not be split between vehicles, then the problem of minimizing the number of vehicles to meet demand is the hard Bin-packing problem. I'm assuming that fleet size affects costs.

- If there are multiple commodities, even the associated network flow problems are NP-hard.

If deliveries may not be shared among vehicles, it is hard merely to parcel out the deliveries to meet demand without exceeding capacity, or to minimize cost. If deliveries may be split among vehicles, then assigning deliveries among vehicles is like a transportation or network flow problem. However, even in this situation the delivery cost to one location depends on where else the vehicle goes. This is the traveling salesman-like part of the problem and it makes things difficult.

The point is that regardless of vehicle differences, customer peculiarities, multiple commodities, just the routing of vehicles or the partitioning of the orders is apt to be hard.

Many real applications are made even more difficult by uncertainty in future demand, salvage values of vehicle locations and commodity supplies, etc. There is relatively little applicable complexity theory here (but see Section 14.10.5). As a practical matter you can usually count on this making your problem *substantially* harder than otherwise.

14.7.3 Padding and Forcing

Padding and forcing transformations are one step up in difficulty from specialization. In general, these methods require you to do a little tinkering with small examples. They do *not* require great ingenuity or a big flash of insight. After you do the exercises in this chapter, you should be able to perform this kind of transformation. Most of the Hamiltonian variations fall into this category.

If you are willing to read through the literature for known NP-hard problems, these methods along with specialization will be enough to resolve complexity questions in the majority of practical situations.

Often we want to make the solution to an instance have a particular form or property, to get a transformation to work. One method, called *forcing*, is to use an "enforcer," a substructure which forces something to happen. We have already used enforcers in some transformations in Sections 14.4.1 and 14.6. For example, we stick a node incident to only one arc into an instance of Ham path, and force the solution (if there is one) to have that node as an endpoint. Similarly, when we put a node with only two arcs into an instance of Ham cycle, we force the solution (if there is one) to contain those two arcs. It is usually simple to contrive powerful enforcers for an NP-hard problem.

Another forcing example: the Subset Sum problem gives a set of integers $X = \{x_1, \ldots, x_n\}$ and integer K as an instance. The question is to find a subset of X whose elements sum to exactly K. *Question: Prove Subset Sum is hard, by specialization.*[29] To transform Subset Sum to Partition, include two additional elements in X. The first has value $y = M + K$; the second $z = M - K + \sum_{i=1}^{n} x_i$. Here M must be large enough to dwarf the sum of the original terms. The large value forces y and z to appear in different subsets in any solution to the Partition instance $\{X, y, z\}$. *Question: Find a simple valid value for* M.[30]

Almost any scheduling problem hard on K machines is also hard on $K+1$ machines. To transform the K machine problem, it usually works to add a single large job (or a set of highly interconnected jobs) to the instance. These added jobs completely tie up the $K+1$st machine, leaving a *de facto* K machine problem. I've picked this as the last example of forcing because it could just as well be thought of as padding, our next method.

Another useful method is to "pad" an instance, typically with trivial elements, to permit a property to be satisfied. This may seem like cheating, but it is perfectly legal. For example, some years ago, Jane Ammons, Chris Lofgren, Leon McGinnis and I encountered a machine configuration problem that looked almost like Partition (defined in Section 14.3). A manufacturing company had two identical flexible assembly machines, each of which could be configured with up to 24 tools. Associated with each of the 48 tools was a certain amount of work. The problem was to divide the tools among the machines to balance their workload. This is the same as Partition, except that the two subsets must have the same number of elements. We can define this *Equipartition* problem as:

Instance: Set of integers $X = \{x_1, \ldots, x_{2n}\}$.
Question: Is there a subset $S \subset X$ with $|S| = n$ such that $\sum_S x_i = \sum_X x_i/2$?

How could we show Equipartition is hard? Given an instance $x_1, \ldots, x_n$ of Partition, pad the instance with n additional zero elements $x_{n+1} = \ldots = x_{2n} = 0$.

Reading Tip: *Mentally fill in the details of the proof for the padding transformation just given.*

14.7.4 Gadgets: Transforming 3-SAT to 3-coloring

Now I'll do a more complicated transformation, from 3-SAT to 3-coloring. (The appearance of "3" in each problem name is just coincidence and doesn't help us.) The transformation requires us to invent one or two "gadgets," in this case little graphs to use as parts of

[29] The restriction of Subset Sum, in which $K = \sum_i x_i/2$, is Partition.
[30] $M = \sum_{i=1}^{n} x_i$.

the 3-coloring instance. I will show some of the blind alleys I went down while developing this proof, to show the kind of thinking one does when inventing gadgets.

Before trying to invent the transformation, I make sure I remember the problems' definitions. 3-coloring a graph takes a graph G as an instance. The question is, can G be legally colored using not more than 3 colors? That is, assign to each node of G a color A, B, or C, in such a way that no two nodes linked by an arc are assigned the same color. 3-SAT takes a set of clauses as an instance. Each clause is a set of 3 different terms, each of which is a complemented or uncomplemented True/False variable. For example, $\{X_1, \bar{X}_3, X_6\}$ is a possible clause. The question is, can the variables be assigned values of True or False so that at least one term in each clause is True?

First, I need a "gadget" or component to represent the Boolean variables X_i. What portion of a 3-coloring problem could correspond to a True-False, Yes/No decision?

A single node doesn't quite do it, since it could have any of *three* values. But if we rule out one color, the other two possible colors could correspond to T and F.

Call the colors A,B, and C. Add a triangle, which we may assume is colored as shown in Figure 14.10, since there is no *a priori* difference between colors. Connect X_1 to the node colored C. Now X_1 colored A can mean X_1 = True, and X_1 colored B can mean X_1 = False.

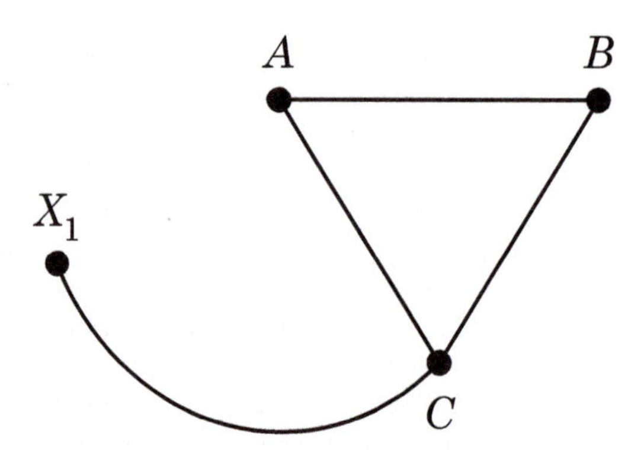

FIGURE 14.10: Connect node C to X_1 to force X_1 to have color A or B.

Conveniently we can get an $\bar{X}_1$,too. See Figure 14.11. Notice $\bar{X}_1$ must be A or B, and it is B iff X_1 is A.

The same construction gives us all of the variables X_1 through X_n, along with their complements. *Question: Why wouldn't it work for each X_i to have its own triangle marked ABC?*[31] I will call these $2n$ nodes the "variables nodes."

Now we need something to act as a clause. It will be a piece of a graph. I will call it GADGET. It must have the following properties:

1. GADGET connects to three variables nodes in some way. To help us think, let's work

[31] We couldn't be sure the triangles were colored consistently, unless we put edges from each A to each B and C, and each B to each C. That would be clumsy and make the graph unnecessarily large.

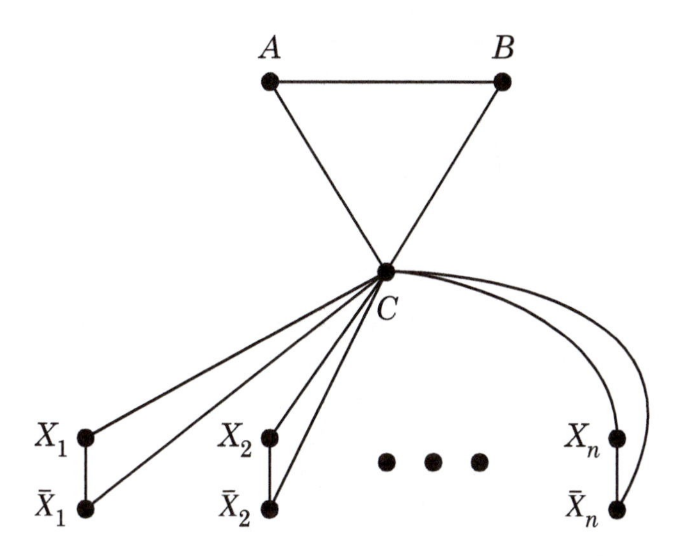

FIGURE 14.11: Force X_1 and $\bar{X}_1$ and the other variables and their complements to be colored A,B or B,A.

with $(X_1 \vee X_2 \vee \bar{X}_3)$ as a generic example. *Question: Is it an accident that my generic example contains both complemented and uncomplemented variables?*[32]

2. If one or more of the three variables nodes are colored A, then GADGET can be colored. *Question: What does this property mean?*[33]

3. But, if all three "literals" nodes are colored B, then GADGET can not be legally colored. *Question: What does this property mean?*[34]

My first idea for GADGET did not work. I tried a triangle (see Figure 14.12). The good: this structure satisfies property (3). If all three literals nodes were colored B, the triangle could not be legally colored since none of its nodes can be B. The bad: if all three nodes were colored A, then the triangle could not be legally colored. So property (2) fails.

Now notice, the first idea prohibited all three literals nodes from having the same color. The problem was that all A was the same as all B. To make A different from B we will have to actually know which is which. This means using the triangle labeled ABC, which is used to make the literals nodes work. This gave me an idea.

In Figure 14.13, the literals node X_1 has been given a companion node Y_1, which is connected by an edge to A. If X_1 is colored A (True), then Y_1 may be colored B or C. But if X_1 is colored B (False), then Y_1 must be colored C. This idea breaks the symmetry between colors A and B which foiled the idea in Figure 14.12. Give every literals node its own companion to get the working GADGET depicted in Figure 14.14. If all three literals nodes are B, then there is no freedom of choice. Their companions are all C and the inner triangle cannot be legally colored. This gives property (3). But, as long as at least one

[32]No. I'd be more likely to make a mistake if I were working with a special case. It would be like trying to prove a theorem about all triangles while visualizing an equilateral one.

[33]If a literal in the clause is true, the clause is satisfied.

[34]If all literals in the clause are false, the clause is not satisfied.

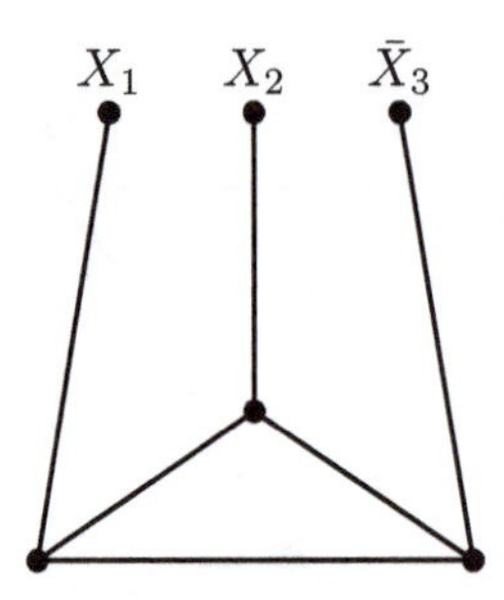

FIGURE 14.12: A triangular GADGET fails because it forbids all literals to be false (B), but it also forbids all literals to be true (A).

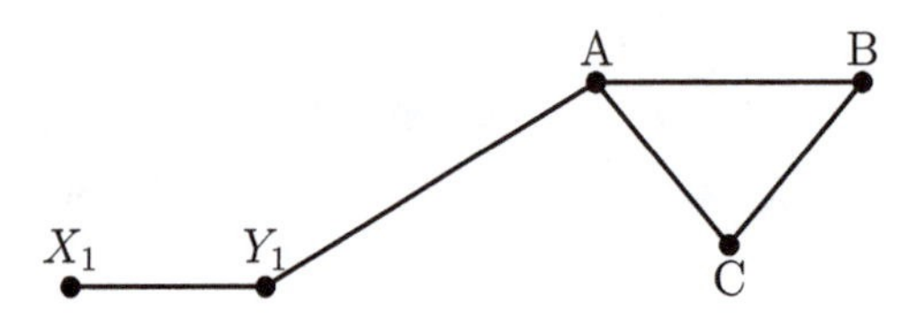

FIGURE 14.13: If X_1 is colored A (True) its companion Y_1 may be colored B or C; if X_1 is colored B (False) its companion Y_1 must be colored C.

literals node is A, then one of the companions can be B. The other companions can be C so the inner triangle can be legally colored.

You may understand the transformation from the figures. If you prefer symbolic logic, you may prefer algebra. For an arbitrary instance of 3-SAT with n variables X_i and m clauses C_j, create a graph $G = (V, E)$ as follows.
$V = \{a, b, c; x_i, \bar{x}_i : i = 1, \ldots, n; d_j^1, d_j^2, d_j^3, e_j^1, e_j^2, e_j^3 : j = 1, \ldots, m\}$. Nodes x_i and $\bar{x}_i$ correspond to the literals X_i and $\bar{X}_i$, respectively. (The d_j^l are the nodes of the inner triangles of the gadgets; the e_j^l are the companion nodes.)
The 6 categories of arcs are:

1. $(x_i, \bar{x}_i), (c, x_i)$, and $(c, \bar{x}_i) : i = 1, \ldots, n$ (literals)
2. $(a, b), (b, c), (c, a)$
3. $(d_j^1, d_j^2), (d_j^2, d_j^3), (d_j^3, d_j^1) : j = 1, \ldots, m$ (inner triangles)
4. $(a, e_j^l) : l = 1, 2, 3; j = 1, \ldots, m$ (connect companions to color A.)
5. $(e_j^l, d_j^l) : l = 1, 2, 3; j = 1, \ldots, m$ (connect companions to inner triangle.)
6. (L, e_j^l), where $\mathcal{L}$ is the lth literal in clause C_j, for $l = 1, 2, 3$ and $j = 1, \ldots, m$.

The accompanying proof would be:
(i): Suppose the instance of 3-SAT has a satisfying truth assignment. If $X_i = T$ in this assignment, color node x_i with A and $\bar{x}_i$ with B; otherwise, color node x_i B and $\bar{x}_i$ A. Color a with A, b with B, and c with C. Since the assignment is satisfying, clause j contains at least one literal that evaluates to True. Select one such literal $\mathcal{L}$; its corresponding node, either an x_i or $\bar{x}_i$, has color A. Color the companion e_j^l node that $\mathcal{L}$ is connected to with B, and color the d_j^l node that node is connected to with C. Color the other two e_j^l nodes

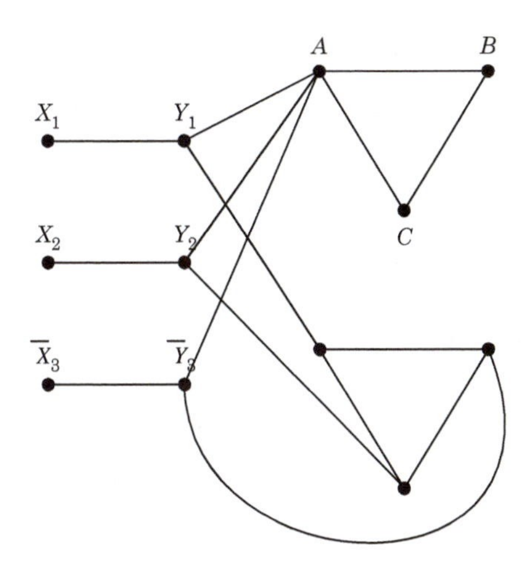

FIGURE 14.14: GADGET: final version.

with C, and the d_j^l nodes they connect to with A and B, respectively. It can be seen that the resulting 3-coloring is valid.

(ii): Suppose there exists a valid 3-coloring of G. The arcs in category (2) imply without loss of generality that a is colored with A, b with B, and c with C. By the category (1) arcs, for each i, the two nodes $x_i, \bar{x}_i$ are colored differently, one A and the other B. In the 3-SAT instance set $x_i =$ True iff x_i is colored A. For each clause C_j in the 3-SAT instance, at least one of the literals must be True, for if not then by (6) and (4) all three corresponding d_j^l nodes would be colored C, and by (5) and (3) the triangle formed by the three e_j^l nodes could not have been 3-colored.

In a typical NP-hardness proof, the explanatory comments in parentheses and references to arc categories would be omitted, as would the entire motivating discussion preceding the algebraic description. The gadget would probably not be illustrated since it is not complicated enough. All this makes NP-hardness proofs so hard to read that many experts rarely read other people's proofs! David Johnson, who used to receive dozens of NP-hardness results every month while he was writing his NP-completeness column [163] – [164], told me that he never read the proofs, except for occasionally looking to see what problem the reduction was from.

One of the awkward parts in writing down the proof is trying to describe correspondences in algebraic notation. It is relatively easy to say, "connect each literals node in the jth clause to an e node of the jth gadget." The notation can get clumsy, however.

You can often make your formal proof more readable by making a lemma about your gadget. I strongly recommend this style. To illustrate, I rewrite the reduction from 3-SAT to 3-coloring.

Proof: *Given an instance $L^1, \ldots, L^m$ of clauses on Boolean variables $X_1, \ldots X_n$ of 3-SAT, construct an instance $G = (V, E)$ of 3-coloring as follows. V contains three special nodes a, b, c; E contains the edges $(a, b), (a, c), (b, c)$. Then by symmetry on the colors A, B, C, a, b, c are colored A, B, C, respectively, in every 3-coloring of G, without loss of generality.*

Lemma 14.7.1 *There exists a gadget graph H containing three special nodes y_1, y_2, y_3 such that: (i) If each of the special nodes is connected by an edge to a node (not in H) colored B,*

H cannot be 3-colored; (ii) If at most two of the special nodes are connected to a node (not in H) colored B, and none are connected to a node colored C, then H can be 3-colored.

Proof: *The gadget is shown in Figure 14.14, together with three nodes, $X_1, X_2, \overline{X}_3$ connected to the three special nodes labeled $Y_1, Y_2, \bar{Y}_3$ in the figure. If all X nodes are colored B, all three special nodes must be colored C, preventing the 3-coloring to be extended to the lower triangle. This verifies property (i). For property (ii), suppose without loss of generality that Y_1 is permitted to be colored B. Color it B, and color its adjacent node in the lower triangle C. Color the other two special nodes C, and their adjacent nodes A and B to achieve a valid 3-coloring of H.* ∎

Returning to the construction, create nodes x_i and $\bar{x}_i$ to represent the literals X_i and $\bar{X}_i$ for $i = 1, \ldots, n$. Create edges $(x_i, \bar{x}_i), (c, x_i)$, and $(c, \bar{x}_i)$. These force that x_i be colored A and $\bar{x}_i$ be colored B, or vice versa, for each i, in any 3-coloring of G.

Let $L^j[1], L^j[2], L^j[3]$ be the three literals in clause L^j. For each clause L^j, create a copy of the gadget, denoted H^j, with special vertices y_1^j, y_2^j, y_3^j. Create edges $(L^j[k], y_k^j)$ for $k = 1, 2, 3$.

Suppose the instance of 3-SAT is satisfied by truth assignment $f(i) \in \{T, F\}$ for the variables X_i. For $i = 1, \ldots, n$, color node x_i A and node $\bar{x}_i$ B iff $f(i) = T$. Otherwise, color node $\bar{x}_i$ A and x_i B. Since $f()$ satisfies the instance, at least one of the literals in clause L^j is True. The node that represents the literal is colored A. By property (ii) of the Lemma, G has a valid 3-coloring.

Conversely, suppose G has a valid 3-coloring. By the contrapositive of property (i) of the Lemma, at least one of the representatives of $L^j[k] : k = 1, 2, 3$ is colored A. Set $x_i = T$ if x_i is colored A; set $\bar{x}_i = T$ if $\bar{x}_i$ is colored A. This is a valid truth assignment because, as observed, exactly one of the pair $x_i, \bar{x}_i$ is colored A and the other is colored B. It is a satisfying truth assignment because every clause L^j contains at least one true literal. Therefore, G has a 3-coloring iff the given instance of 3-SAT is satisfiable. ∎

In my experience, it usually is not too much trouble to cook up a gadget for an NP-hard problem. If you can't build enforcers and gadgets for a problem, it may well be an indication that the problem is not NP-hard.

Summary: This section has illustrated several kinds of NP-hardness transformations. One or more basic NP-hard problems are often embedded in a real problem. If so, specialization is the appropriate type of transformation. Padding and forcing arguments require only a little expertise and are worth learning. Gadget proofs take more work.

14.8 Complexity of Finding Approximate Solutions

If you cannot find an optimal solution to your problem in your desired time frame, the next best thing is usually an algorithm to find a feasible solution that has a "good" objective value. The standard performance measure for such algorithms (for minimization problems) is the smallest $\alpha > 1$ for which the approximate solution is sure to have objective value at most α times the optimal objective value. An algorithm with such a guarantee is called a "factor-α" approximation algorithm. The main techniques in approximation algorithms are linear programming (especially primal-dual reasoning), rounding (including the use of pseudo-polynomial-time algorithms), greed, and semi-definite programming.

Approximation algorithms have been studied for a long time. New techniques invented by Dorit Hochbaum, Michel Goemans, David Williamson, Yuri Nesterov, Arkadi Nemirovsky, and others brought the field forward considerably in the 1980s and 1990s [86]. But what do you do when, say, you find a factor-$\frac{3}{2}$ approximation algorithm for a problem, but you cannot find a smaller factor approximation? Computational complexity turns out to be very useful in those situations. It often tells you whether or not to keep trying, because it can be NP-hard to obtain approximate solutions better than some factor.

For some problems it is NP-hard to find a solution within any constant factor α of the optimum. The TSP is a good example. There is no polynomial-time factor-α approximation algorithm for the TSP, for any α, unless $P = NP$. Why? Because if there were a fast algorithm which always produced tours at most, say, 1000 times the length of an optimal tour, we could use it to solve Ham cycle quickly. *Question: What costs would the transformation assign to the arcs?*[35] Moreover, even if α is a function of n such as n^3 or 2^n, the same transformation proves that α cannot be attained unless $P = NP$. Likewise, for Set Covering no constant factor α is achievable if $P \neq NP$. Even worse, under a slightly stronger assumption than $P \neq NP$, the value $\alpha = (1 - \epsilon) \log n$ is not achievable for any $\epsilon > 0$.

On the other hand, many NP-hard problems admit factor-α guarantees for small constant values of α. Christofides [49] devises a factor-$\frac{3}{2}$ algorithm for the TSP when distances satisfy the triangle inequality. The Steiner Tree problem has a very simple factor-2 approximation: find the minimum spanning tree on the required vertices. Several simple greedy algorithms for Bin Packing have factor-2 approximation guarantees. For each $\alpha > 1$ there is a factor-α approximation algorithm for Knapsack, based on the pseudo-polynomial time algorithm (see Section 14.9.1) and limiting the precision of the data.

Approximation complexity is highly problem dependent, and more difficult to develop intuition about than exact solution. Two of the standard textbooks on approximation algorithms are written by Vijay Vazirani [287], and by Davids Williamson and Shmoys [300]. Another book by Gärtner and Matousek [114] focuses on approximation algorithms and bounds obtained from semi-definite programming.

A major breakthrough in complexity theory, the PCP theorem, enables proofs of many new, stronger results. ("PCP" is an acronym for "probabilistically checkable proofs." See Section 14.10 for a description.) One version of the PCP theorem by Håstad [144] concerns the 3-SAT Maximization problem. In this problem we are given a set of clauses, just as in 3-SAT, but the goal is to make as many of the clauses true as possible, rather than to make all of them true. Håstad proves that it is NP-hard to do better than a factor of 7/8 of optimal. Karloff and Zwick [170] devise an $\frac{8}{7}$-factor algorithm, which proves the factor of 7/8 to be tight.

14.9 Dealing with NP-Hard Problems

> Complexity theory is usually an excuse for laziness. Give me a problem: I'll solve it. — George Dantzig

[35] Costs of 0 for arcs in the original graph and 1 otherwise.

Preview: How do you solve a hard problem? The NP-hard classification leaves several important loopholes, described in this section.

Some NP-hard problems can be solved well with dynamic programming, and some can not. This is discussed in Section 14.9.1. The realistic cases might have special structure, permitting effective solution. The most important aspect of the realistic cases is their size. If they are not very large when modeled as integer programs, you should first try solving them with math programming software. One of the major accomplishments in optimization during the 1980s and 1990s was the practical solution of the medium-size IP [26], followed by the practical solution of medium-size MIPs in the following decade [2, 25]. At most half of this was due to increases in computer power; the rest was due to the monumental improvement of our algorithms [26, 27]. Section 14.9.2 discusses special structure less obvious than instance size.

If you can't readily solve your problem with off-the-shelf software, you may need to make a choice between heavy-duty mathematical programming methods, and heuristics. Tools for the former now accommodate some of the latter, so you may pursue both alternatives.

When is it appropriate to use mathematical programming? If the problem is *stable*, not apt to change within a year or two; and it is of *great economic value*, so getting that extra .1 or 1 or 4 percent is worth a great deal; and you are *confident of the accuracy of your model and data*, then it may well be worth the investment, in both time and money, to pursue a mathematical programming solution. Many contenders for the Franz Edelman Award fall into this category.

Approximation algorithms offer a compromise between math programming methods that seek optimal solutions, and general purpose heuristics that are relatively easy to implement but have no performance guarantees (see, *e.g.*, [280]). As discussed in Section 14.8, complexity theory can sometimes help you assess the prospects for an approximate solution.

When are heuristics appropriate? When you are unsure of the model, for example you can only approximately quantify the objective; when your data are not accurate, and you don't need an exact solution to a "guesstimated" problem; when the problem is transient or unstable, and robustness of the solution method becomes important; when the instances are really enormous or the problem very difficult in practice (e.g., jobshop scheduling, quadratic assignment); when solution speed is critical; when your client prefers a decent solution now to a great solution later; when the amount of money at stake doesn't justify using other techniques; when the expertise necessary for other methods is not available: in these situations you may choose to pursue heuristic solutions. If they are successful, you can always reassess the option of a more costly but more exact solution method.

Finally, you can often solve a hard problem by changing your model. Turn a constraint into an objective or vice versa; aggregate or extract what matters most; approximate on the model level rather than solution level. Here again computational complexity can help you as you choose among models, by assessing their difficulty.

14.9.1 Pseudo-Polynomial Algorithms, Dynamic Programming, and Unary NP-Hardness

Operations researchers don't think of the knapsack problem as hard, since it can easily be solved in practice by dynamic programming. The same is true for Partition, Equipartition, and Subset Sum. (See any standard introductory OR textbook.) Yet these problems are all

NP-hard. Complexity theory accommodates this phenomenon, and provides an additional classification tool to classify susceptibility to dynamic programming.

How big a table is required to solve the knapsack problem by dynamic programming? If the data are integers, and the knapsack capacity is K, the table is n by $K + 1$ large, requiring $O(nK)$ work. It suffices to store one row of the table at a time, but there is no way to avoid the factor of K (in the worst case) work requirement. Dynamic programming is fast only if K is not too large. This DP algorithm is called *pseudo-polynomial* because it is only fast when K is small. This ties back to our discussion in Section 9.1 about length. In that section we mentioned that the natural parameters of a problem are usually good surrogates for the input length. If the dynamic programming algorithm is polynomial in those natural parameters, we say the algorithm runs in pseudo-polynomial time.

However, theoretically speaking, K can be enormous. This is because it only takes $\log K$ digits to write the number K. To see why this matters, let's reduce Exact 3-Cover to Knapsack. I'll describe my thought process along the way.

Let the ground set be $1, \ldots, 3n$ and let $\mathcal{S} \equiv S_1, \ldots, S_m$ be the collection of 3-tuples of the ground set. Exact 3-Cover asks for a subset of $\mathcal{S}$ that partitions the ground set. I must convert this to a knapsack problem, which asks for a subset of a set of possible items that fits into the knapsack and has maximum value. My first insight is that it is natural to transform each member of $\mathcal{S}$ into a possible item, since both problems ask one to select a subset from a given set of possibilities. A solution to Exact 3-Cover should then correspond to an optimal knapsack solution.

The essence of Exact 3-cover is to pick triples that exactly split up the ground set. So, I think, an optimal knapsack solution should exactly fill up the knapsack. I need a knapsack problem that is optimized when the sizes of the items add up exactly right. My second insight is to set every item's size equal to its value. Then maximizing value subject to fitting in the knapsack is the same as finding a subset of items that exactly fill up the knapsack.

I need one more insight. How do I transform a partition of a ground set into a set of numbers that exactly add up to a particular value? I wish I could explain to you how to get the third insight – the image in Figure 14.15 just pops into my mind. Each S_j corresponds to an item that can be placed in the knapsack. Represent S_j as a binary string of length $3n$ with three 1's and the rest 0's. The ith digit in the string for S_j equals 1 if $i \in S_j$, and otherwise equals zero. Then a subset of $\mathcal{S}$ that partitions $\{1, \ldots, 3m\}$ is represented by a selection of the binary strings that sums to a binary string of all 1's. That sum is the knapsack capacity. *Question: Which of the three properties might fail?*[36]

Ground Set		1	2	3	4	5	6	7	8	9	10	11	12
$\{2, 3, 12\}$	--→	0	1	1	0	0	0	0	0	0	0	0	1
$\{2, 4, 6\}$	--→	0	1	0	1	0	1	0	0	0	0	0	0
$\{1, 5, 9\}$	--→	1	0	0	0	1	0	0	0	1	0	0	0
$\{1, 4, 11\}$	--→	1	0	0	1	0	0	0	0	0	0	1	0
$\{5, 6, 7\}$	--→	0	0	0	0	1	1	1	0	0	0	0	0
$\{8, 9, 10\}$	--→	0	0	0	0	0	0	0	1	1	1	0	0

FIGURE 14.15: Represent a 3-tuple as a binary string with 3 nonzero terms.

The transformation takes polynomial time (property (3)), and clearly satisfies property (1), that a partition of the ground set yields a set of items that exactly fit into the knapsack. But I run into trouble proving property (2). It might be wishful thinking to assume that

[36] (2), as explained next.

a set of numbers that sums to 111...111 comprises numbers that contribute exactly one of each power of 2. The trouble is, for example, that three numbers with last two digits 01 would give a sum with last two digits 11. (Exercise 18 guides you to an example that shows property (2) fails.)

I must prevent one power of 2 from spilling over to another power. Separate the powers from each other to avoid this trouble. Insert k 0's between the digits so that m 1's in one place can't spill over to the next. Figure 14.16 illustrates the padding. k must be large enough to prevent spillover, yet small enough to retain property (3), polynomial time transformation. Setting $k = m$ is adequate, although a moment's thought shows that $k = \lceil 1+\log m \rceil$ suffices.

Ground Set		1 2 3 4 5 6 7 8 9 10 11 12
$\{2,3,12\}$	--→	00000100001000000000000000000000000000000000000000000001
$\{2,4,6\}$	--→	00000100000000010000000001000000000000000000000000000000
$\{1,5,9\}$	--→	10000000000000000000100000000000000000001000000000000000
$\{1,4,11\}$	--→	10000000000000010000000000000000000000000000000000100000
$\{5,6,7\}$	--→	00000000000000000000100001000010000000000000000000000000
$\{8,9,10\}$	--→	00000000000000000000000000000000000100001000010000000000

FIGURE 14.16: Pad four zeroes between each digit to prevent any digit from spilling over onto another.

To complete this example, here is a precise statement of the transformation and a proof of its validity. The proof, though rigorous, is harder to understand than the preceding derivation.

Reading Tip: *Write your own proof before reading further.*

Proof: We are given an instance of Exact 3-Cover comprising ground set $0, \ldots, 3n-1$ and 3-tuples $S_1, \ldots, S_m$ of the ground set. Define m binary variables $x_1, \ldots, x_m$. For each $S_j = \{j_1, j_2, j_3\}$, let $a_j = v_j = 2^{mj_1} + 2^{mj_2} + 2^{mj_3}$. The knapsack instance is then

$$\max \sum_{j=1}^{m} v_j x_j \text{ subject to } \sum_{j=1}^{m} a_j x_j \leq U, \; x_j \text{ binary}$$

where $U \equiv \sum_{i=0}^{3n-1} 2^{mi}$. The question is, does there exist a feasible $\mathbf{x}$ with objective value at least U?

Property (3): We may assume $m \geq n$ because otherwise the input instance is trivially infeasible. The Knapsack instance has size $O(nm^2) = O(m^3)$ and is built by inserting the input data, which takes time polynomial in m.

Property (1): Let $S_{j_1}, \ldots, S_{j_n}$ partition the ground set. Then for each $i \in \{0, \ldots, 3n-1\}$ exactly one S_{j_t} contains i. By construction, for each i, $\sum_{t=1}^{n} a_{j_t}$ contains exactly one term equal to 2^{mi}. The values $x_t = 1 \forall t \in \{j_1, \ldots, j_n\}$, and $x_t = 0$ otherwise, are therefore feasible with objective value U.

Property (2): Let $x_1, \ldots, x_m$ be feasible with objective value $\geq U$. The constraint implies the objective value $\sum_{j=1}^{m} v_j x_j = U$, because $a_j = v_j \; \forall j$. Fix $i \in \{0, \ldots, 3n-1\}$. Consider the component 2^{mi} of the sum that defines U. All powers of 2 less than 2^{mi} are at most $2^{m(i-1)}$ by construction. Their sum cannot exceed $3m2^{m(i-1)} = 2^{mi}\frac{3m}{2^m} < 2^{mi}$. Therefore, the component 2^{mi} must arise from exactly one variable $x_j = 1$ where a_j is the sum of 2^{mi} and two other powers of 2. The set $j : x_j = 1$ therefore must contain exactly one j such that $i \in S_j$. Thus $\{S_j : j = 1\}$ is a partition of the ground set. ■

You should now see that my hard instances of Knapsack employ numbers $\Omega(n)$ long. No naturally occurring Knapsack instance contains numbers of that order of precision. There is another good thing about naturally occurring Knapsack cases. Even if K is large, dynamic programming will quickly find an approximate solution to Knapsack. Simply round the sizes to a few digits of precision, or truncate, if the capacity constraint isn't at all soft. (There are more effective heuristics, too.) This technique is also appropriate when the data are given to high precision, but you don't have confidence in the low order bits.

Most of the NP-hard problems I've encountered are not susceptible to dynamic programming. These problems are called *unary* NP-hard or *strongly* NP-hard. Problems that don't involve numbers at all, such as 3-SAT or HAM PATH, are unary NP-hard. Problems that involve numbers, but are obviously hard even when the numbers are small, are unary NP-hard, too. For instance, the TSP is hard when arc costs are 0 or 1 (HAM CYCLE), so TSP is unary NP-hard. Finally, some problems such as Bin Packing and 3-Partition are very number-oriented, yet are unary NP-hard.

Watch out for problems that are theoretically solvable in pseudo-polynomial time, but are like unary NP-hard problems in practice. Dynamic programming is great for 2-Machine Line Balancing; it is OK but slower by a factor of n for 3-Machine Line Balancing; it is impractical for 5-Machine Line Balancing. But theoretically, the 5-machine problem is solvable in pseudo-polynomial time. What is happening is this: if m is allowed to vary and get large, the m-machine problem is like 3-partition or bin packing, and is unary NP-hard. Theoretically, for any fixed value of m the problem is not unary NP-hard; but practically, $m = 5$ is already large.

Summary: A few NP-hard problems such as knapsack can often be solved in practice via dynamic programming. These problems are hard only in the sense of getting exact solutions when the numbers are many digits long. If your problem naturally occurs as a knapsack, partition, or other such problem, you are in luck if either: the data do not occur to great precision; or you do not require an absolutely precise solution. However, many NP-hard problems, including 3-partition, bin-packing, and all number-free problems, are "unary" or "strongly" NP-hard. These are not susceptible to quick solution by dynamic programming.

14.9.2 Special Structure

All good statisticians cheat by looking at the data. —H. Chernoff

A network flow problem is a kind of IP with special structure. That particular structure is very well known. Other helpful structures can be less apparent.

Here is a true story in which a problem that appeared difficult turned out to be quite easy.

I was having dinner with my friend Ivan Chase, a biologist. Perhaps because the soup contained a variety of fish, he remembered a question he had meant to ask me. He had a hundred or so fish in his lab, and wanted to run as many trials as possible of an experiment that required a small group of fish. No fish could participate in more than one trial, or the trials would not be independent. The difficulty was that some groupings among the fish were OK, and some were not. Ivan wanted to know what to do.

Immediately I visualized, quite incorrectly, that certain groups of fish somehow possessed a bad social dynamic, didn't get along properly, and would spoil the experiment. Obviously, Ivan had a 3DM problem (3-dimensional matching, see Section 14.5). He was trying to form, say, 33 groups of 3 out of 99 fish, when only certain groups of 3 were permitted. Then it occurred to me that what went wrong might be pairs of fish that didn't get along, rather than a complicated dynamic among three fish. In 3DM, abc, abd, and bcd might all be acceptable triples, while acd is not. But if the problems were due to pairwise interference this could not occur. Fortunately I remembered a problem proved NP-hard in Garey and Johnson's book [112], called Partition into Triangles. Given a graph $G = (V, E)$, the problem asks to partition V into $|V|/3$ triples, such that for each triple, G contains all three arcs between the members of that triple. So even if groups were unsuitable because of pairwise conflicts, Ivan's problem was still NP-hard.

I hesitated to tell Ivan that his problem was hard to solve, perhaps because I had a glimmer of the truth, but probably because I needed more time to think of a good heuristic for Partition into Triangles. I asked what made groups of fish unacceptable. Ivan explained that he couldn't put big fish together with little fish – differences of more than 15% in weight were not allowed. "I see," I responded, and discarding my complexity proof and heuristic ideas, gave him a simple procedure to maximize the number of groups. If the three heaviest fish weigh within 15% of each other, make them a group. Otherwise, eliminate the heaviest fish, which belongs to no acceptable group. Recurse.

The moral of the story is that a problem may look NP-hard, but be so highly structured as to be easy. The problem domain expert may tell you that certain jobs must be performed before others: if you visualize an arbitrary acyclic precedence constraint graph, you may be ignoring hidden structure. People close to the problem often think it is obvious what causes precedence constraints, mutual incompatibilities (forbidding two activities to be done at the same time), or preferences, and don't tell you about the natural structure of these constraints or objectives. In my experience, the operations researcher usually has to elicit information about special structure from the problem domain experts, to whom the structure is either too obvious or unimportant or both to bear mentioning.

In the case of Ivan's fish, there is an important intuitive explanation of why the special structure of the problem made it easy. 3DM is difficult because of the arbitrariness as to which triples are OK and which are not. If you decide to use the triple *abc* this might force d and e together, and simultaneously force $f, g,$ and h apart. Solving 3DM is like putting together a puzzle where what you assemble in one location affects what can fit together in many other locations. This looks difficult: making a bunch of interlocking yes/no decisions to simultaneously satisfy a collection of arbitrary-looking constraints. The constraint on weight difference so severely restricts the possible patterns of permissible triples that the problem becomes easy. *Question: What if the total mass of the fish in each group must equal 300 grams? Is this problem easy or hard?*[37]

Here is a method that sometimes works to elucidate special structure. Find a transformation $\mathcal{T}$ from 3-SAT, IP, knapsack, (or some other NP-hard problem you understand well) to your problem. Then apply $\mathcal{T}$ to a simple No instance of the hard problem whose LP relaxation is feasible. (For example, the constraints $2x \geq 1; 2x \leq 1$, are such a No instance of IP feasibility.) The resulting instance of your problem is apt to be a small No instance. However, if special structure is making your problem easy, this instance likely *cannot arise from the actual circumstances of your problem.* Show this small instance to your problem domain expert and ask whether it could happen, and if not, why not? Whatever answer you get will deepen your understanding of the real problem.

The method at work: Mike Carter and I [36] investigated classroom scheduling problems, where classes that meet at various times must be assigned to rooms, so that no classes use

[37] Hard. It is a 3-partition problem.

the same room at the same time. A simple transformation from graph coloring (see Section 14.7.4) shows that finding a feasible schedule for classes is hard. Each node in the graph is a class; each arc is a time period during which the two incident classes meet; each color is a room. Assigning colors to nodes so no two nodes sharing an arc have the same color, is the same as assigning rooms to classes so no two classes with overlapping time schedules have the same room.

To use the method, we began with the No instance of coloring a triangle with two colors (an odd cycle). The natural LP model would find a feasible fractional solution, in which each node was assigned half of each color. This instance transforms to an infeasible classroom scheduling instance, as shown in Figure 14.17. A practitioner criticized this infeasible instance, saying that one class met from 8–9 and 10–11, which didn't happen. This led us to realize that in many applications, each class uses a contiguous interval of time. These problems may be easier to solve by using properties of interval graphs.

In a further application of this method [IBID, pp. S31-32]: a practitioner criticized an infeasible instance we had generated because every possible pair of the four rooms was preferred by some teacher. This criticism made us realize that there could be a monotonicity structure among teacher preferences. This special structure permits scheduling to be done very easily in many applications, particularly in secondary schools.

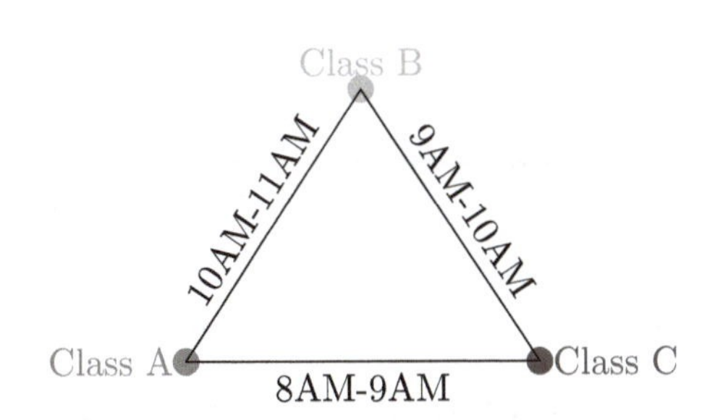

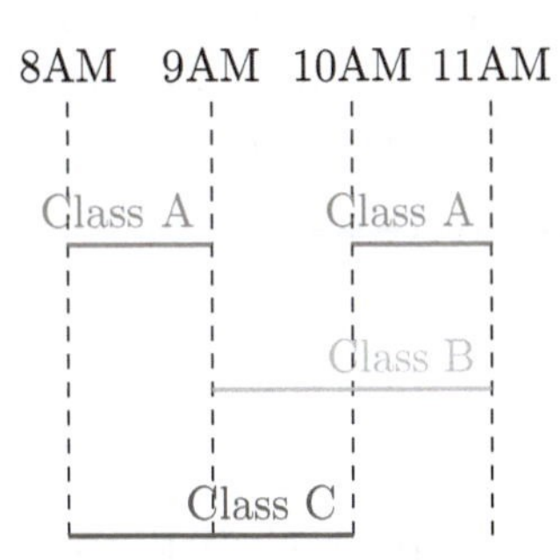

FIGURE 14.17: The infeasible 2-coloring instance has the feasible fractional solution where each class gets half of color (Room) I and half of color (Room) II. In the corresponding infeasible classroom assignment instance, Class A meets 8-9 and 10-11.

There are some important negative complexity results involving special structure. Section 14.5 states that Ham cycle (and therefore also TSP) is hard even for 2D grid-graphs, a very restrictively structured subset of the planar Euclidean graphs. This was proved by Itai, Papadimitriou, and Szwarcfiter [159]. Likewise, graph 3-coloring is hard on planar graphs. 3-SAT has a hard planar version as well: represent each variable and each clause by a node. Connect the nodes of the variables in a cycle. If a variable or its complement appears in a clause, place an arc between the corresponding nodes. Impose the special structure, that this graph must be planar, and 3-SAT is still hard. 3,4-SAT, in which no variable may appear in more than 4 clauses, is another useful hard restriction of 3-SAT [279].

Proofs of NP-hardness subject to special structure are often long and difficult. In general, these proofs either depend on gadgets, or on what Garey and Johnson call the method of *local replacement*, (see [112]). It often is best to impose the special structure step-by-step, generating a sequence of increasingly constrained NP-hard problems, finally arriving at the target problem. It can also help to start the transformation from one of the restricted NP-hard problems listed above.

While special structure may be a complexity theorist's headache, it may be just what you need to solve your problem. Even if a special structure doesn't make a problem change from hard to easy, it may give you an extra order of magnitude in instance size before your solution technique bogs down. Special structure often reduces the computational complexity of getting a good approximate solution. For example, in Section 14.8, you saw that no approximation guarantee for the TSP can be achieved in general, but if distances satisfy the triangle inequality, a factor-1.5 algorithm exists. Moreover, if distances are Euclidean, a factor-$(1+\epsilon)$ algorithm exists [9, 231] for any $\epsilon > 0$. We can think of this benefit of special structure from the point of view of the algorithm, rather than the problem. From this vantage, special structure frequently improves the performance of heuristics. For example, 2-opting for the TSP has much better performance guarantees if the distances are Euclidean, than if they are arbitrary or only satisfy the triangle inequality [37].

Summary: The realistic cases have special structure if the data cannot be arbitrary, but are restricted in some way. Special structure often derives from geometric or other physical considerations, common causes, or other factors obvious to the problem domain expert. One trick to elicit this information is to show the expert some pathological instances, and discover why they can't occur. Structure will sometimes make a problem easy, but many NP-hard problems remain hard when so restricted. Even so, special structure often permits solution of larger instances, or better quality heuristic solutions.

14.10 Formal Definition of NP and Other Complexity Classifications

---

Preview: This section defines terms such as NP-complete, and briefly surveys some other complexity classes. Two classes, $co-NP$ and P-space often have applications to optimization and mathematics; the others have relatively little bearing on practical considerations. Some other kinds of difficulty are not captured by the theory of computational complexity.

14.10.1 Unsolvable Problems

Some problems cannot be solved by any algorithm. While the halting problem is the canonical example, software testing is a more compelling example. No algorithm can infallibly detect bugs in software.

This complexity does give some insight into what is difficult and time consuming in our field. Generally speaking, it is an unsolvable problem to verify that a formal structure (e.g., a model, a computer program) does what we want it to do. That is one reason there has been relatively little formal work by researchers, in this very important area of modeling and model validation – it is a very hard problem. For a few formal attempts, see Dantzig's activity analysis [64], Geoffrion's structured modeling [116], and Hackman and Leachman's [140] continuous time framework for modeling production systems.

14.10.2 co-NP Completeness and Lack of Succinct Characterizations

Not all NP-hard problems are alike, although I've disguised that fact in this chapter. Here we need a more detailed picture of complexity classes. See Figure 14.18. Definitions are given in the next subsection. For here, we can make do with examples from IP.

This is an NP-complete problem:

IP Feasibility The set of all matrix-vector pairs $(A, \mathbf{b})$ such that for some integer vector $\mathbf{x}$, it is true that $A\mathbf{x} \leq \mathbf{b}$. Visualize this problem as a point in the region NPC in Figure 14.18.

The following is a co-NP-complete problem:

IP Infeasibility The set of all matrix-vector pairs $(A, \mathbf{b})$ such that for *no* integer vector $\mathbf{x}$, it is true that $A\mathbf{x} \leq \mathbf{b}$. Visualize this problem as a point in the region co-NPC in Figure 14.18.

The NP-complete problem comprises the YES instances to IP feasibility, while the co-NP-complete problem comprises the NO instances.

[htb]

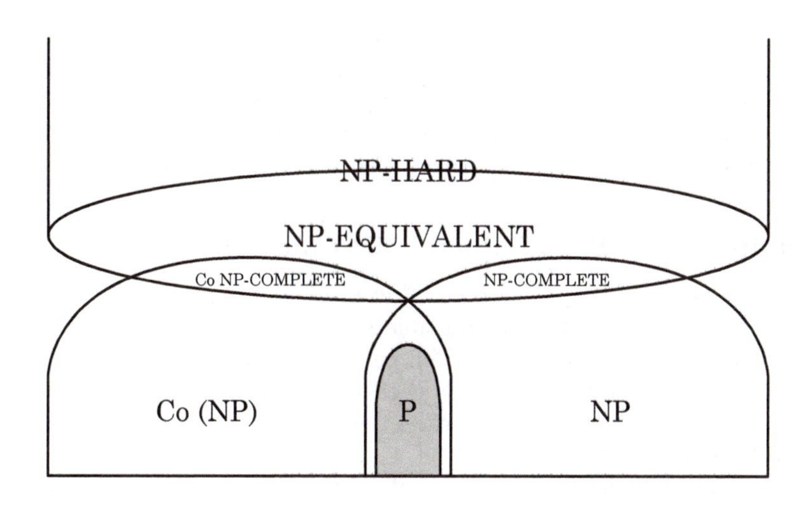

FIGURE 14.18: NP, co-NP, and related complexity classes.

NP-complete problems are not the same as co-NP-complete problems, though both are NP-hard. If you are lucky, you can solve an NP-complete problem quickly by guessing the answer, and verifying quickly that the answer is correct. For example, to solve the NP-complete IP feasibility problem just defined, you could luckily guess a feasible vector v, and quickly verify that $Av \leq \mathbf{b}$, and that v is integer.

However, co-NP-complete problems *cannot* be solved quickly by good guessing (to the best of our knowledge). Using formal terms we would say that co-NP-complete problems are not believed to be in NP. If an IP instance is infeasible, there isn't in general a way to prove infeasibility short of trying an exponential number of possibilities, and showing that each fails. Even a cutting plane proof is apt to be exponentially long. See [50] for a demonstration that finding a cutting plane proof can be identical to a complete solution by branch and bound.

The class co-NP-complete gives insight into how IP codes spend their time. It is commonplace for an IP solver code to find an optimal solution in a few minutes, then spend another couple of hours verifying optimality. What is happening? Let v^* be the optimal solution value. The IP solver is solving two problems. First, it solves the NP-complete problem of finding an x with objective value as good as v^*. Second, it solves the co-NP-complete problem of verifying that there isn't anything better. *Question: What is the No instance being solved?*[38] Complexity theory tells us that even if we hot-start our IP code with the optimal solution, it is still apt to take a long time to verify optimality, because any proof of optimality may have to be exponentially long.

At present, we are better at heuristically solving hard searching problems (NP-complete) than at heuristically solving hard verification problems (co-NP-complete). If you require proofs of optimality, you are likely to add significantly to your computing requirements.

Co-NP-completeness theory can also be applied to show the hopelessness of a quest for a simple characterization. The history of the so-called "3-Color Problem" furnishes a good example. In 1959, Grötzsch proved that every planar graph with no triangles can be 3-colored [138]. Since then, combinatorists have searched for the weakest possible conditions under which planar graphs can be 3-colored. Many partial results have been found (see, *e.g.*, [31, 270]), but the overall quest for "nice" necessary and sufficient conditions was not successful, because the set of 3-colorable planar graphs is NP-complete. Therefore, unless $NP = co(NP)$, they have no such succinct characterization. As another example, researchers spent a couple of decades seeking concise necessary and sufficient conditions for the existence of a core in spatial voting, but the co-NP-completeness of the problem implies that search was doomed from the start [17]. Attempts to find a concise test for P-matrices were similarly doomed, because recognizing them is co-NP-complete [60].

14.10.3 Definitions of NP, co-NP, and NP-Complete

A *string* is a finite sequence of 0's and 1's. Any instance of any of our problems can be represented as a string. The set of all possible strings is $\{0,1\}^*$. A Yes/No problem $L \subseteq \{0,1\}^*$ is a set of strings, consisting of all cases for which the answer is Yes. For example, the problem of whether an integer is even is the set of strings whose rightmost character is 0. IP feasibility is the set of all strings representing a matrix-vector pair (A, b) such that $A\mathbf{x} \leq \mathbf{b}$ for some integer vector x.

A computer program *recognizes* a problem L if, given a string $s \in \{0,1\}^*$ as input, it outputs "Yes" if $s \in L$ and outputs "No" if $s \notin L$. The class P of easy problems, introduced in Section 14.1.2 takes its name from "polynomial time." P is the set of all problems L for which there exists a computer program that recognizes L and runs in time polynomial in $|s|$, the length of the input string. Since a problem is itself a set of strings, we often refer to P as a *class* of problems rather than as a set of problems, to avoid confusion between sets and sets of sets.

The complement of a problem L is denoted $\bar{L} = \{0,1\}^* - L$. *Question: If $L \in P$ why must it be true that $\bar{L} \in P$?*[39] The complement of the class P would be denoted $co(P)$; it

[38] If we are minimizing, with integer coefficients, the instance is, given A, b, c, v^*, is there integer vector $\mathbf{x}$ such that $A\mathbf{x} \leq \mathbf{b}$ and $\mathbf{c} \cdot \mathbf{x} \leq v^* - 1$?

[39] Take the computer program that recognizes L and change "Yes" to "No" and vice versa in the output statements.

consists of the complements of all problems in P. By the question above, $co(P) = P$ so we don't use the term $co(P)$.

Now we define NP, which, contrary to popular belief, does not stand for "not polynomial." It means "non-deterministic polynomial" because if you allow lucky guesswork then instances of the problem for which the answer is "yes" can be solved in polynomial time. A certificate-checking computer program $\mathcal{C}$ to *nondeterministically recognize* a language L takes two distinct strings as input: s, the instance, and a "certificate" string c. The program $\mathcal{C}$ outputs "Yes" or "Maybe". If $s \notin L$, then no matter what c is, $\mathcal{C}$ outputs "Maybe". On the other hand, if $s \in L$, then for least one certificate string c the program $\mathcal{C}$ will output "Yes." NP is the set of all problems L for which there is certificate-checking computer program that nondeterministically recognizes L and runs in time polynomially bounded in $|s|$. Notice this implies $|c|$ is polynomially bounded in $|s|$ or the program would not have time to read all of c.

Determining that an integer program is feasible is a good example of a problem in NP. It might take you a long time to find a feasible solution $\mathbf{y}$ to $A\mathbf{x} \leq \mathbf{b}; \mathbf{x}$ integer. But once you have done so, $\mathbf{y}$ is a certificate for which it can easily be verified that $A\mathbf{y} \leq \mathbf{b}$ and that $\mathbf{y}$ is integer.

The class $co - NP$ is the set of problems that are complements of problems in NP. As far as is known, $co - NP$ is not the same as NP. Intuitively, this should make sense if you think about integer programs. No one has ever found a quick way to demonstrate that an IP is infeasible, even though there is a quick way to demonstrate that an IP is feasible.

A problem $L1$ is NP-complete if $L \in NP$ and every problem $L \in NP$, can be transformed to $L1$ in polynomial time, as defined in Section 14.4. The NP-complete problems are the hardest problems in NP in this sense: if you had a fast solution method for an NP-complete problem, you could solve any problem in NP quickly. *Question: How?*[40] A problem L is *unary NP-complete* if it is NP-complete and all numerical values necessary to specify an instance of L are encoded in unary (base 1) rather than binary. For example, the number 13 would have to be encoded as 1111111111111 rather than 1101.

The co-NP-complete problems are similarly defined as the hardest problems in co-NP. They turn out to be precisely the complements of the NP-complete problems. They are NP-hard: if you had a fast solution method for an NP-hard problem, you could solve any problem in NP quickly. They are also NP-equivalent: they are NP-hard and if you had a fast solution method for an NP-complete problem you could solve them quickly. There are NP-equivalent problems that don't seem to be in either NP or co-NP. Returning to Figure 14.18, this is an NP-hard problem not (thought to be) in NP or co-NP: the set of all A, b, c, k such that the optimal integer solution to $\min c \cdot \mathbf{x} : A\mathbf{x} \leq \mathbf{b}$ has objective value exactly k.

The PCP theorem characterizes NP in a different way that builds on the ideas of a certificate and certificate checker. For every "YES" instance there is a proof that it is a "YES". Think of the proof as a certificate together with a polynomial time computation that validates it. Instead of validating a certificate, check the proof for correctness. And instead of checking the entire proof, check a small, randomly chosen portion of the proof. If the proof is correct, our check will never find a mistake. But no "NO" instance has such a proof. Any purported proof must have at least one error. A probabilistic checker, if given a purported proof of a "NO" instance, is required to have probability at least $\frac{1}{2}$ of finding a mistake [12]. The PCP theorem states that NP is the class for which there exist probabilistic checkers that examine only $O(1)$ bits of the proof, and use only $O(logn)$ random bits to select which bits to examine [11]. See Chapters 11 and 22 of the textbook [10] for a full exposition.

[40] Attach the transformation to $L1$ as a front end.

14.10.4 Classes of Search Problems

At the time of this writing, there are several important problems for which no polynomial time algorithm is known, yet whose Yes/No versions are in P. Several complexity classes have been defined by Papadimitriou and by him with co-authors to categorize them. The classes all have the same flavor. They are search problems that require a valid solution to be exhibited. Solutions have the following two properties: first, there exists at least one valid solution; second, every valid solution can be verified to be valid in polynomial time.

Factoring composite integers is by far the most important problem of this flavor in practice. The privacy of communications over the web, such as credit card purchases and banking, depends on the computational difficulty of factoring large numbers. Determining whether or not a number is prime is in P. Therefore, determining that a number is composite is in P. Yet no known polynomial time algorithm can exhibit a factor of a composite number (other than 1 and itself). It's an intriguing situation. We can prove quickly such a factor exists, but we don't know how to find one quickly.

Other such intriguing problems of particular interest for optimization include:

- Find a locally optimal solution to a combinatorial optimization problem, with respect to a polynomial size neighborhood. At least one such solution exists, because there must be at least one global optimum. Local optimality can be verified in polynomial time because the neighborhood has polynomial size. The associated complexity class is PLS [165].

- Find a Nash equilibrium of a game. The Kakutani fixed point theorem guarantees that at least one equilibrium exists, but no polynomial time algorithm to exhibit one is known. The associated complexity class is PPAD [239]. The Nash equilibrium problem is complete for this class even for two player games [76].

14.10.5 P-space Completeness

There are other classes of problems that are widely believed to be harder than both the NP-complete and the co-NP-complete. The most important of these classes is the P-space complete problems. These are the hardest problems of those that can be solved using a polynomial amount of computer memory. Usually, if a problem is P-space complete, it will be harder to solve in practice than an NP-complete problem.

These include several natural questions about electronic circuits, such as the problem of finding a minimum size circuit to instantiate a given Boolean function. You can sometimes tell you are dealing with a problem like this, when it seems to involve an alternating sequence of difficult decisions between two opposing sides. Many games, including chess and checkers when generalized suitably to $n \times n$ boards, are in this category. Other things that seem to make problems P-space complete are the presence of feedback, periodicity, or stochasticity.

Several important problems are P-space complete. They include: stochastic scheduling, periodic scheduling, Markov decision problems, control of queueing networks, and solving systems of differential equations. Several motion planning problems involving jointed robot arms are P-space complete as well.

In general, adding stochasticity or uncertainty makes hard problems a whole level harder. The classic illustration of this phenomenon is the problem of project planning under uncertainty. This problem was Dantzig's motivation for inventing linear programming in the 1940s. Since that time, LP has been used to solve countless other problems, but many large instances of the original motivating problem are still computationally challenging.

14.10.6 Other Kinds of Difficulty

Computational complexity theory has relatively little to say about continuous nonlinear problems. In practice, convexity is usually the dividing line between easy and hard (global) optimization, though non-smoothness can make things difficult as well. The analog to determining whether the problem is easy or NP-hard, is to determine whether or not the problem has multiple local optima. The answer roughly dictates what kind of solution method to use or solution quality to expect. However, when NP-completeness theory is applied to this domain, problems thought of as simple in practice are theoretically difficult, e.g., determining whether a point is a local minimum of a quadratic function. Thus traditional NP-completeness theory has not been particularly useful.

It is not NP-hard for a human to run a 3-minute mile. I once visited a manufacturing site, and was told that they had a really hard problem assigning tools among the four machines in a production line. I thought the problem was hard because equalizing workload to avoid bottlenecks is a line balancing problem, with enough machines to make things moderately hard (see Section 14.9.1). I was completely wrong! It turned out that, due to differences among the machines, there was no choice as to which tools were assigned to each machine. The *decision* making part of the problem was *trivial.* The problem was hard because one of the machines did not have enough capacity for its tools.

Computational complexity theory presumes that the problem to be analyzed is well-posed. But in practice, it is often the fuzziness and open-endedness of a problem that makes it hard. If you can categorize or formalize all the possibilities, you may be able in principle to capture this hardness as computational hardness.

Summary: Computational complexity provides other classifications besides NP-hardness. Some problems, including program and model verification, are unsolvable. This partly explains the scarcity of research results and tools to aid in formal modeling. The P-space-hard problems are a step up in complexity from the NP-complete. These include several fundamental problems involving uncertainty, such as stochastic scheduling, queueing networks, and Markov decision processes. The class co-NP helps explain why optimization software spends so much time verifying optimality after it has found the optimal solution. Finally, a problem may be hard in a non-computational sense.

14.11 References and Conclusions

The deservedly classic reference on NP-completeness is the book by Garey and Johnson [112]. The treatment in [241] is excellent and is designed for the reader familiar with the basics of optimization, rather than computer science. For the latter, see [240]. For more advanced material such as the PCP theorem, see [10].

On the most fundamental level, understanding the basics of computational complexity should make you more conscious of how much time your solution methods require, and how large your problem instances are. Your choice of algorithm is often more significant than your choice of hardware, data structures, and other implementation decisions.

On the next level, classifying problems as easy or NP-hard will tell what kinds of solution methods may be available. Deciding on an appropriate method depends on many other considerations, including size, financial stake, model accuracy, and time constraints.

Computational complexity is an ineluctable phenomenon. Many problems are hard to solve. Yet, I would set against that statement the following observation:

The realistic cases are usually easy if you study them long enough.

Study the realistic cases carefully enough, and you will find enough special structure to enable their solution.

However, all is not rosy even if any problem can be solved, if studied enough. For one thing, it may not be worth it do so. For another thing, what often happens when you give control of a model and solution procedure to the user? The user runs it until it dies. There are always things the model does not know about that the user does. The user tries to push the model into taking care of these extras until the solution procedure fails.

What is happening is that the inherent computational complexity has resurged. You beat NP-hardness by knowing your problem really well, and finding a solution procedure which is effective within a small zone. The user pushes the model out of the zone to where the procedure is not effective.

It can be very tough to explain this phenomenon to a user who thinks, "I only made a slight change and now the model doesn't run." People who believe the computer is magic may not believe in its limitations. Sometimes you can convince your users that a problem is hard, by excitedly telling them what a fascinating research topic it makes.

One way or the other, we are brought back to the reality of the inherent computational difficulty of problems. Experience tells us that most real problems are hard. Computational complexity then tells us that we cannot build tools that have all of the following properties:

reality Solve realistic models.

ease Require little expertise to use.

optimality Find an exactly optimum solution.

speed Always fast.

generality Apply to a broad range of cases.

This is not bad news. On the contrary, it is the very best of news. It means that our optimization expertise will always be needed, and our work will always be challenging. In other words, we will get paid to have fun.

14.12 Problems

E Exercises

1. Explain why 2 could be changed to $\sqrt[3]{2}$ in Definition 14.1 without affecting the meaning of the definition. Explain why 2^{cn} could be replaced by $d(1.01)^{cn}$ without affecting the meaning, where $d > 0$ is a constant.

2. Write a version of the shortest path problem in standard "Yes/No" format. Write an instance of the problem you've defined.

3. Suppose you know IP Feasibility is NP-hard and you wish to prove 2-machine line balancing is hard. Describe precisely the form of the input and output of the transformation $\mathcal{T}$. Describe precisely the three things you would have to prove about $\mathcal{T}$.

4. Write a version of the minimum spanning tree problem in the standard "Yes/No" format.

5. State the threshold Yes/No version of the maximum independent set problem in standard format.

6. For each Yes/No problem defined in Section 14.3, write a specific instance for which the answer is Yes, and an instance for which the answer is No.

7. Modify the front end in Figure 14.2 as follows. Make the copy $\hat{u}$ of u as in the figure, and make a new node $\hat{s}$ connected only to $\hat{u}$ as in the figure, but do not make a new node s. Instead, change the label u to s. Prove s-HAM PATH is NP-hard using this modified front end. (I didn't use this front end in the text, though it is more concise, because I didn't want my first example of a proof to contain a confusing relabeling.)

8. For each problem defined in Section 14.5, think of an instance whose answer is YES, and an instance whose answer is NO.

9. Write a proof that Partition reduces to Equipartition, using the padding transformation in Section 14.7.3.

10. Reduce Clique to Independent-Set.

11. Reduce Vertex-Cover to Clique.

12. Reduce each of the following two problems to the other.
    Set Packing
    Instance: Integer n, a collection $\mathcal{S}$ of subsets of $\{1, 2, \ldots, n\}$, integers k and m.
    Question: Is there a subcollection $\mathcal{T} \subset \mathcal{S}$ such that $|\mathcal{T}| = k$ and $|\cap_{S \in \mathcal{T}} S| \leq m$?

    Max Set Union
    Instance: Integer n, a collection $\mathcal{S}$ of subsets of $\{1, 2, \ldots, n\}$, integers k and m.
    Question: Is there a subcollection $\mathcal{T} \subset \mathcal{S}$ such that $|\mathcal{T}| = k$ and $|\cup_{S \in \mathcal{T}} S| \geq m$?

13. Show that Minimum Feedback Arc Set is the same as finding a minimum size subset S of arcs, with the property that every directed cycle in G contains at least one arc in S. The related problem of finding a minimum size subset of nodes is also hard.

14. Find a counterexample similar to Figure 14.5, in which graph H has only three nodes.

15. Find 12 direct transformations, one in each direction, between each pair of the four variations on the Hamiltonian theme, with all graphs undirected.

16. Find two transformations, one in each direction, between directed $s - t$ Hamiltonian path and undirected Hamiltonian cycle.

17. Prove the following variation of Scheduling with Deadlines is NP-hard: there are two processors, and each job must be run on either one of them. Job i has processing time t_i and deadline d_i. The problem is to minimize the number of late jobs.

18. Consider the following numbers: $2^{35} + 2^{34} + 1$; $2^{33} + 2^{32} + 1; 2^{31} + 2^{30} + 1$;
$2^{25} + 2^{24} + 4$; $2^{23} + 2^{22} + 4; 2^{21} + 2^{20} + 4$;
$2^{15} + 2^{14} + 16$; $2^{13} + 2^{12} + 16; 2^{11} + 2^{10} + 16$. Supplement these numbers with other binary numbers containing three 1's each to get a set of numbers that sum to $2^{36} - 1$. Explain why this shows that property (2) fails in the middle of the proof that reduces Exact 3-Cover to Knapsack.

19. In Section 14.9.2, what is the complexity of Ivan's fish problem if the total mass of the fish in each group may not exceed 300 grams?

20. Prove Equipartition is hard even if all x_i must be strictly positive.

21. Use forcing to transform Equipartition to Partition. Hint: Given an instance $y_1, \ldots, y_n$ of Equipartition, add a "big M" term to each y_i, giving $x_i = y_i + M$.

22. Use padding to transform MAX CUT to BISECTION MAX CUT.

23. 4-SAT is defined the same as 3-SAT, except each clause contains exactly 4 literals. Transform 3-SAT to 4-SAT. (Hint: Force a dummy variable to be false; use it to pad the 3-clauses into 4-clauses.)

24. Prove that the following versions of counting problems are NP-hard. *They are actually of a higher category of complexity than the NP-complete.*
Instance: Integer matrix A, vector $\mathbf{b}$, and scalar k.
Question: Are there at least k different binary vectors $\mathbf{x}$ such that $A\mathbf{x} \leq \mathbf{b}$?
Instance: Integer matrix A, vector $\mathbf{b}$, and scalar k.
Question: Are there exactly k different binary vectors $\mathbf{x}$ such that $A\mathbf{x} \leq \mathbf{b}$?

M Problems

25. Prove that 2-coloring the vertices of a graph is easy.

26. Section 14.5 asserts that, in contrast to Max Cut, weighted minimum cut is easy because it is the dual of max flow. However, max flow is defined with two distinguished vertices s, t, but Max Cut is defined without distinguished vertices. Show a polynomial equivalence between weighted minimum cut and s,t-minimum cut for graphs with nonnegative weights.

27. Prove that Maximum Intersection (defined in Problem 12) is NP-complete.

28. The input to NEQ-3-SAT is the same as to 3-SAT, a Boolean expression in conjunctive normal form with exactly three literals per clause. The question is, does there exist a truth assignment to the variables such that each clause has at least one true literal and at least one false literal? Prove that NEQ-3-SAT is NP-hard.

29. Reduce 3-SAT to 3DM.

30. Reduce 3-SAT to Max Independent Set.

31. State a Yes/No version of the following weighted ordered set covering optimization problem, and prove it is NP-hard. Given: a collection $S_1, S_2, \ldots, S_m$ of subsets on the ground set $[n] \equiv \{1, 2, \ldots, n\}$. Find: an ordered subcollection $S_{\rho(1)}, S_{\rho(2)}, \ldots, S_{\rho(t)}$

that covers $[n]$, i.e. its union is $[n]$, and that minimizes (over all such ordered coverings) the weighted sum

$$\sum_{j=1}^{t} j|\{i \in [n] : i \in S_{\rho(j)}; i \notin S_{\rho(k)} \forall k < j\}|.$$

32. Prove that minimizing the number of late jobs on a single processor is NP-hard. An instance comprises processing times t_i and deadlines d_i for jobs $i = 1, 2, \ldots, n$. A schedule is a permutation π of the jobs. It results in completion times $c_{\pi(j)} = \sum_{k=1}^{j} t_{\pi(k)}$. Job $\pi(j)$ is late if $c_{\pi(j)} > d_{\pi(j)}$. *Hint: There is a simple proof. The difficulty is to select from which NP-hard problem listed in this chapter to reduce.*

33. Prove that precedence-constrained scheduling on identical parallel processors with unit processing times is NP-hard. An instance comprises an integer n, the number of jobs to be run, a directed graph $G = (V, E)$ with $|V| = n$, an integer m denoting the number of processors, and an integer deadline d. The question is whether there exists a feasible schedule of d or fewer time periods. A feasible schedule is an assignment $f : V \mapsto \{1, 2, \ldots, d\}$ such that $(i, j) \in E \rightarrow f(i) + 1 \leq f(j)$ (these are the precedence constraints) and such that for all $t \in \{1, 2, \ldots, d\}$, $|\{i : f(i) = t\}| \leq m$ (at most m jobs are run in each time period). *Hint:* $m = 3$ *suffices.*

34. Prove 2-SAT is solvable in polynomial time, but MAX 2-SAT is NP-hard.

35. Let $Z(\mathbf{x})$ denote the number of nonzero components of vector $\mathbf{x}$. Problem 29 in Chapter 3 shows that maximizing $Z(\mathbf{x})$ subject to $A\mathbf{x} = \mathbf{0}, \mathbf{x} \geq 0$ can be solved in polynomial time. Problem 36 shows that maximizing $Z(\mathbf{x})$ subject to $A\mathbf{x} = \mathbf{b}, \mathbf{x} \geq 0$ can also be solved in polynomial time.

    (a) Prove that minimizing $Z(\mathbf{x})$ subject to $A\mathbf{x} = \mathbf{0}$, $\mathbf{x} \neq \mathbf{0}$ is polynomially solvable.

    (b) Prove that it is NP-hard to find a solution to Bin Packing that has performance guarantee strictly less than 1.5. *Hint: If* $\alpha < 1.5$, *the approximation algorithm would never use 3 bins if 2 bins sufficed.*

    (c) Prove that minimizing $Z(\mathbf{x})$ subject to $A\mathbf{x} = \mathbf{b}$, $\mathbf{x} \neq \mathbf{0}$ is polynomially solvable.

    (d) Prove that minimizing $Z(\mathbf{x})$ subject to $A\mathbf{x} = \mathbf{0}$, $\mathbf{x} \geq \mathbf{0}; \mathbf{x} \neq \mathbf{0}$ is NP-hard.

36. An edge-coloring of a graph assigns colors to edges such that all edges incident on the same vertex have different colors. On your own, determine the complexity of minimizing the number of colors in an edge-coloring.

37. Prove that for any function $\alpha(n)$ that can be computed in polynomial time, there is no (polynomial-time) factor-α approximation algorithm for the TSP unless $P = NP$.

38. Suppose the edge lengths in an instance of Steiner Tree satisfy the triangle inequality. Show how to solve a minimum spanning tree problem that yields a factor-2 approximation. *Hint: Find a Hamiltonian path that is at most twice the length of the optimal Steiner tree. A path is a tree.*

39. Construct a Steiner Tree instance that satisfies the triangle inequality and asymptotically makes the factor-2 minimum spanning tree approximation bound tight. *Hints: Lengths are not Euclidean; the graph can have* $|S| + 1 = |V|$.

D Problems

40. Reduce $3 - SAT$ to Ham Path. You will need to invent gadgets.

41. Prove that LCP is NP-hard.

42. A string of characters $s[1, /ldots, m]$ is a *substring* of string $t[1, \ldots, n]$ iff there is a strictly increasing function $f : [1, \ldots, m] \mapsto [1, \ldots, n]$ such that $s[j] = t[f(j)]$ for all $j \in [1, \ldots, m]$. Prove that, given a set of strings $\mathcal{T}$, the problem of finding a longest string s that s is a substring of t for all $t \in \mathcal{T}$ is NP-hard.

43. The previous problem was to show it is NP-hard to maximize the length of a common substring for a given collection $\mathcal{T}$ of strings. Invent a related minimization problem and prove it is NP-hard.

I Problems

44. An $n \times n$ matrix M is a $Q-matrix$ if for all $q \in \Re^n$ there exist vectors $\mathbf{x} \in \Re^n, \mathbf{y} \in \Re^n$ such that:

    - $\mathbf{x} \geq \mathbf{0}; \mathbf{y} \geq \mathbf{0}$ (nonnegativity)
    - $\mathbf{x} \cdot \mathbf{y} = 0$ (complementarity)
    - $q = Mx - \mathbf{y}$

    In words, q is a nonnegative combination of columns of M and $-I$ that for all i does not employ the ith columns of both M and $-I$. Prove that the set of $Q - matrices$ is in Σ^2. *Open*: Prove that the set of $Q - matrices$ is $\Sigma^2 - complete$. Conclude that there is no succinct characterization of the $Q - matrices$ unless the polynomial hierarchy collapses.

Chapter 15

Formulating and Solving Integer Programs

Preview: Integer programs (IPs) are LPs with the additional constraint that some or all of the variables must be integer. In practice, IPs are the most useful form of hard problem in optimization, for two main reasons:

- Conversion of problems into IPs is usually much simpler than conversion into other forms.
- Software for solving IPs is far more advanced than for any other NP-hard problem except the TSP.

This chapter is a brief introduction to IP. It focuses on two topics: first, modeling hard problems as IPs; second, applying LP theory and algorithms to help solve IPs.

The integer programming problem (IP) is, given integer $m \times n$ matrix A, integer n-vector $\mathbf{c}$, integer m-vector b, and subset $I \subseteq \{1, 2, \ldots, n\}$, to

$$\text{Minimize } \mathbf{c} \cdot \mathbf{x} \quad \text{subject to} \tag{15.1}$$

$$A\mathbf{x} \geq \mathbf{b} \tag{15.2}$$

$$x_i \text{ integer } \forall i \in I \tag{15.3}$$

Just as with LP, constraint (15.2) may be replaced by $A\mathbf{x} \leq \mathbf{b}$ or $A\mathbf{x} = \mathbf{b}, \mathbf{x} \geq \mathbf{0}$ or $A\mathbf{x} = \mathbf{b}, \mathbf{0} \leq \mathbf{x} \leq \mathbf{u}$ or some other form of linear inequalities. Also, as in LP, the objective could equally well be to maximize or be omitted without losing the essence of the problem. Constraint (15.3) is what distinguishes LP from IP. When $I \subsetneq \{1, 2, \ldots, n\}$ – that is, when not all variables are restricted to integer values – the problem is called a MIP, which is short for "Mixed Integer Program." Sometimes the IP is called a "Pure IP" or PIP, when $I = \{1, 2, \ldots, n\}$.

Integer programming is very different from linear programming. LPs can be solved in polynomial time, whereas solving an IP is NP-hard. Every NP-complete problem can be posed as an integer program (of size polynomially bounded in the size of the original problem). What is it about constraint (15.3) that so drastically alters LPs? Imagine a production planning LP with variable x_i representing the number of units of product i to manufacture. In the LP, the value $x_i = 612.61$ is permitted. If we changed the model to an IP with $i \in I$, the values $x_i = 612$ and $x_i = 613$ would be permitted, but no value between those two would be. If that doesn't seem important enough to you to illustrate the great difference between LP and IP, good! The small values 0 and 1 most readily show the difference, rather than the large values 612 and 613.

An integer variable with lower bound 0 and upper bound 1 must equal either 0 or 1 and therefore is called a *binary* variable. *General* integer variables are those with ranges other than $\{0, 1\}$. A *general integer program* is an IP with one or more general integer variables.

15.1 Using Binary Variables to Model Problems Beyond the Scope of LP

Preview: Binary variables let IPs model yes/no decisions and logical relationships such as "or" and "not." They greatly expand the range of problems that can be modeled as LPs.

Just as in computer programming, when x is a binary variable we often think of $x = 1$ as meaning "yes" or "true" and $x = 0$ as meaning "no" or "false". All of the usual logical Boolean operations can be expressed easily with binary variables. "Not x" is $1 - x$. To set z equal to x AND y write $z \geq x + y - 1; z \leq x; z \leq y$. *Question: How do you set z equal to x OR y?*[1]

Simple IP Models

Here are some classic NP-hard problems that are modeled very simply with IP.

- Knapsack is a special case of IP.
- To partition the numbers $a_1, a_2, \ldots, a_n$ into two sets with equal sum, no objective function is needed. Let binary $x_i = 1$ if a_i is in the first set. Constrain

$$\sum_{i=1}^{n} a_i x_i = \frac{1}{2} \sum_{i=1}^{n} a_i$$

- For 3-SAT, no objective function is needed. Represent each Boolean variable X_i by a binary variable x_i. Represent the literal $\neg X_i$ by $1 - x_i$. Convert each clause into the constraint that the sum of the literals is greater than or equal to 1.
- For NAE-3-SAT, employ the additional constraint that the sum of the literals is less than or equal to 2.
- To find a maximum cardinality clique in graph $G = (V, E)$, index vertices by u and v, let binary $x_u = 1$ iff u is in the clique, and maximize $\sum_{u \in V} x_u$. To assure that the variables define a clique, put constraint $x_u + x_v \leq 1$ for all pairs $(u, v) \notin E$.
- To k-color graph $G = (V, E)$, let binary $x_{v,k} = 1$ iff vertex v is colored k. Require each vertex to be colored by $\sum_k x_{v,k} = 1$. Force the coloring to be valid with constraints $x_{v,k} + x_{u,k} \leq 1$ for all edges $(v, u) \in E$ and all colors k. *Question: What is unhelpful about this model if the graph might not be k-colorable*[2]*?*

[1] $z \geq x, z \geq y, z \leq x + y$.

[2] It doesn't find a partial coloring or minimally invalid coloring.

Many other graph-theoretic and other combinatorial problems require some thought to be modeled as IPs. In general, permuting or sequencing is harder to model than partitioning. Here is such an example.

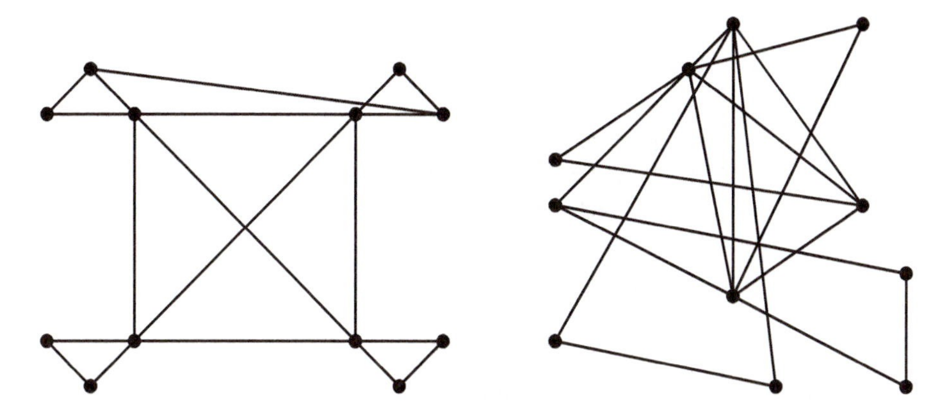

FIGURE 15.1: Are these two graphs isomorphic? Could you rearrange the nodes and preserve which nodes are connected to which?

Example 15.1 [Graph isomorphism]
An isomorphism between two directed graphs $G = (N, A)$ and $H = (V, E)$ is a bijection $\sigma : N \mapsto V$ that shows the two graphs to be identical except for the node labels. That is, $(i, j) \in A \Leftrightarrow (\sigma(i), \sigma(j)) \in E$. Construct an IP to test whether there is an G and H are isomorphic.
Solution: Preprocess to verify $|N| = |V|$ and $|A| = |E|$. Let i, j index nodes in G and u, v index nodes in H. Let binary variable $s_{i,u} = 1$ iff i is mapped to u (i.e., $\sigma(i) = u$). To ensure σ is a bijection, put constraints that force each $i \in N$ to be mapped to a member of V, and each $v \in V$ to be mapped to from some member of N.

$$\sum_{u \in V} s_{i,u} = 1 \ \ \forall i \in A; \sum_{j \in N} s_{j,v} = 1 \ \ \forall v \in V.$$

Forcing each edge in A to map to an edge in E is equivalent to forcing each non-edge of G to map to a non-edge of H. If the graph is very dense, $|N| > \binom{|V|}{2}$, enforce the latter with constraints

$$\forall (i, j) \notin A, \ \forall u \in V, v \in V : \quad s_{i,u} + s_{j,v} \leq 2 - F_{u,v}$$

where $F_{u,v}$ are binary data defined by $F_{u,v} = 1$ iff $(u, v) \in E$. If the graph is not that dense, enforce the former with constraints

$$\forall (i, j) \in A, \ \forall u \in V, \forall v \in V : \quad s_{i,u} + s_{j,v} \leq 1 + F_{u,v}.$$

It is not necessary to enforce both sets of constraints because $|A| = |E|$.

There is an implicit "and" between constraints in a linear program. When we write $A\mathbf{x} \geq \mathbf{b}$ we mean $A_{1,:}\mathbf{x} \geq b_1$ AND $A_{2,:}\mathbf{x} \geq b_2$ AND .... We would need integer variables to incorporate "OR"s into the formulation. Geometrically, "and" preserves convexity of the feasible region because intersections of convex sets are convex; "or" does not usually preserve convexity because unions of convex sets can be non-convex.

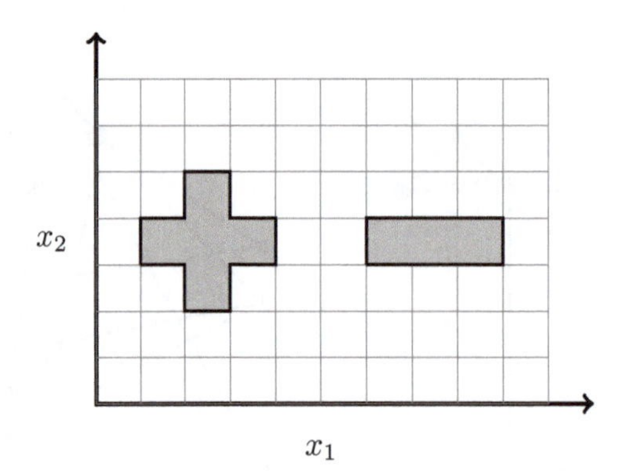

FIGURE 15.2: Auxiliary binary variables are needed to require (x_1, x_2) to be in the non-convex shaded region.

Non-convex Unions of Convex Regions

Suppose we want to require two continuous variables x_1, x_2 to represent a point in a "plus and minus sign" shaped region, shown in Figure 15.2. This is equivalent to requiring that x_1, x_2 be a point in at least one of the rectangular polyhedra defined by constraints $1 \leq x_1 \leq 4; 3 \leq x_2 \leq 4$ and $2 \leq x_1 \leq 3; 2 \leq x_2 \leq 5$ and $6 \leq x_1 \leq 9; 3 \leq x_2 \leq 4$. Introduce three binary variables y_1, y_2, y_3 corresponding to the three rectangles. Think of M as a very large number whose exact value we will specify later. Write:

$$\begin{array}{lll} 1 - My_1 & \leq x_1 & \leq 4 + My_1 \\ 3 - My_1 & \leq x_2 & \leq 4 + My_1 \\ 2 - My_2 & \leq x_1 & \leq 3 + My_2 \\ 2 - My_2 & \leq x_2 & \leq 5 + My_2 \\ 6 - My_3 & \leq x_1 & \leq 9 + My_3 \\ 3 - My_3 & \leq x_2 & \leq 4 + My_3 \\ & y_1 + y_2 + y_3 & \leq 2 \end{array}$$

The last constraint forces at least one of the y_i binary variables to equal zero. When $y_i = 0$, the x_1, x_2 values are constrained to be within the ith rectangle. Therefore, this set of constraints forces x_1, x_2 to be within at least one of the three rectangles. Conversely, if x_1, x_2 is within one of the rectangles, then the y_i for the other two rectangles can be set to 1. For large enough M the constraints associated with these two rectangles will effectively vanish. We will return to this example to discuss exactly what value to use for M.

Satisfying k out of m Sets of Constraints

In the plus-minus example, we wanted at least one set out of three sets of constraints to be satisfied. More generally, if we have m sets of constraints, and we want at least k out of the m sets to be satisfied, we employ m binary variables $y_1 \dots y_m$, with large M coefficients as above, and add the constraint $\sum_{i=1}^{m} y_i \leq m - k$. This constraint forces at least k of the y_i variables to equal zero, which in turn forces at least k of the constraint sets to be satisfied.

A Nongeometric Either-Or Example

Example 15.2 [Single Machine Total Tardiness] A set of jobs indexed by $j = 1, \ldots n$ must be run on a machine, one at a time but not in any pre-specified order. Job j takes time T_j to run, and has deadline D_j. Let c_j be the time at which job j is completed, which will depend on the order of jobs chosen. The tardiness of job j is $\max\{0, c_j - D_j\}$, i.e., zero if it finishes at or before the deadline, and the amount of time by which it is late otherwise. The problem is to sequence the jobs to minimize the total tardiness incurred.

Solution: We have several choices of decision variables. They could be integer variables $1 \leq x_j \leq n$, denoting job j's place in the sequence. They could be binary variables $x_{j,m}$, defined equal to 1 iff job j is the mth in the sequence. They could be binary variables $y_{i,j}$ defined to equal 1 iff job i immediately precedes j in the sequence. Any of these choices would be a complete set of decision variables, in the sense that an assignment of values to the variables determines the processing schedule. However, these choices are complete only under the assumption that the machine will never be idle until all jobs have run. We've assumed that as soon as a job is complete, the next job starts. As the problem is given, that assumption is OK. But if the problem is complicated by a scheduled maintenance downtime, or because some jobs are not available to run at time zero, these decision variables become less attractive. For this reason, and for computational advantage, I choose the following *continuous* decision variables:

$$c_j = \text{ the completion time of job } j, \ \forall j = 1, \ldots, n.$$

Define auxiliary variables τ_j to be the tardiness of job j. Then we have constraints

$$\tau_j \geq 0; \tau_j \geq c_j - D_j; c_j \geq T_j; \ \forall j$$

and objective function

$$\min \sum_{j=1}^{n} \tau_j$$

Rather than specifying a sequence explicitly, we simply prohibit the machine from running more than one job at a time. For each pair of distinct jobs i and j, either i completes before j starts, or vice versa. In math, either $c_i \leq c_j - T_j$ or $c_j \leq c_i - T_i$. Binary variables $w_{i,j} : 1 \leq i < j \leq n$ enforce these constraints.

$$\begin{aligned} M(1 - w_{i,j}) & \quad + c_j - T_j \geq c_i \\ M w_{i,j} & \quad + c_i - T_i \geq c_j \end{aligned}$$

This model readily accommodates several complications such as scheduled downtime or a high priority job treated as a job with fixed completion time. $M = \sum_{j=1}^{n} T_j$ is sufficiently large.

Fixed Charge Problems

A *fixed charge* occurs when there is a continuous variable x_i and a positive cost c_i that is incurred (i.e., charged) if and only if $x_i > 0$. For example, in a network design problem, x_e could represent the amount of flow on edge e, and c_e could be the cost of installing a physical link corresponding to e. In a facility location and goods distribution problem,

x_i could represent the amount of goods delivered from a particular facility, and c_i could represent the cost of building the facility. We need an upper bound on x_i, say $x_i \leq M_i$, to model a fixed charge as an IP. Introduce a new binary variable y_i which we want to equal 1 iff $x_i > 0$. Add the constraint $M_i y_i \geq x_i$ and add the term $c_i y_i$ to the objective function. *Question: Is the objective maximization or minimization?*[3]

Non-convex Costs

Chapter 3 shows how to model piecewise-linear convex costs with LP, despite their nonlinearity. However, costs are often concave increasing functions of a quantity such as the amount of a resource to purchase, or the number of units of an item to be produced. Market power and economies of scale are both apt to create concave cost structures. Purchases by players in the U.S.A. retail market provide many examples, as do electric power plant generation and most other industrial processes. The service industries are admittedly less likely to benefit economies of scale, yet may still exhibit non-convex costs, or, equivalently, non-concave returns for quantities of services rendered. Whenever costs are not convex, LP does not suffice and IP is needed.

Let $t \geq 0$ denote a continuous variable. To model a non-convex cost function $f(t)$ within an IP, approximate it with a piecewise linear function with values c_j at the breakpoints u_j for $j = 1, \ldots, m$, respectively. See Figure 15.3. Some IP software makes special provision for piecewise linear functions. If you use such an option, your only task is to select the breakpoints of your approximation. Here I'll model the piecewise linear function two different ways.

Method 1. Allocate t among the intervals $[u_j, u_{j+1}]$ with continuous variables $t_j : j = 1, \ldots, m-1$ subject to

$$0 \leq t_j \leq u_{j+1} - u_j \tag{15.4}$$

$$t = \sum_{j=1}^{m-1} t_j \tag{15.5}$$

$$t_j \leq (u_{j+1} - u_j) y_j \tag{15.6}$$

$$t_j \geq (u_{j+1} - u_j) y_{j+1} \tag{15.7}$$

$$y_j \quad \text{binary} \tag{15.8}$$

Equation (15.5) forces all of t to be allocated among the t_j. Constraint (15.6) forces $t_j = 0$ if binary variable $y_j = 0$, and by the contrapositive forces $y_j = 1$ if $t_j > 0$. Thus y_j acts as an indicator variable as to whether t_j is nonzero. However, (15.6) permits $t_j = 0, y_j = 1$. Constraint (15.7) forces $y_{j+1} = 0 \Rightarrow t_{j+1} = 0$ if t_j is strictly less than its upper bound $u_{j+1} - u_j$. Thus none of t can be allocated to t_j unless, for each $i < j$, t_i has been allocated its maximum amount. The cost $f(t)$ is then approximated by

$$c_1 + \sum_{j=1}^{m-1} \frac{c_{j+1} - c_j}{u_{j+1} - u_j} t_j. \tag{15.9}$$

Method 2. When $u_k \leq t \leq u_{k+1}$, $f(t)$ is approximated as the convex combination of c_k and c_{k+1},

$$f(t) \approx \frac{t - u_k}{u_{k+1} - u_k} c_k + \frac{u_{k+1} - t}{u_{k+1} - u_k} c_{k+1}. \tag{15.10}$$

[3] Minimization. If $x_i = 0$ the constraint permits $y_i = 0$ or $y_i = 1$.

Think of (15.10) as a convex combination of $c_1, \ldots, c_m$ with at most two nonzero weights, those on c_k and c_{k+1}. Then for additional continuous variables λ_j, (15.10) becomes

$$\begin{aligned}
t &= \sum_{j=1}^{m} \lambda_j u_j && (15.11)\\
f(t) &\approx \sum_{j=1}^{m} \lambda_j c_j && (15.12)\\
\lambda_j &\geq 0,\ j = 1, \ldots, m && (15.13)\\
\sum_{j=1}^{m} \lambda_j &= 1 && (15.14)\\
& && (15.15)
\end{aligned}$$

There are various ways to force the nonzero weights to be consecutive. Introduce binary variables $w_j : j = 1, \ldots, m$

$$\begin{aligned}
\sum_{j=1}^{m} w_j &= 2 && (15.16)\\
\lambda_j &\leq w_j \text{ binary for all } j && (15.17)\\
t &\geq u_k w_{k+1} \text{ for all } k && (15.18)\\
t &\leq u_k + (1 - w_{k-1})(u_m - u_k) && (15.19)\\
u_{k-1} + u_{k+1} &\leq 1 \text{ for all } 2 \leq k \leq m-1 && (15.20)\\
w_j & \quad \text{binary for all } j && (15.21)
\end{aligned}$$

Inequality (15.18) forces $w_{k+1} = 0 = \lambda_{k+1}$ whenever $t < u_k$; inequality (15.19) forces $w_{k-1} = 0 = \lambda_{k-1}$ whenever $t > u_k$. Inequality (15.20) rules out $\lambda_k = 0$ in the case $t = u_k$. See Problems 22, 23, and 24 for some other formulations.

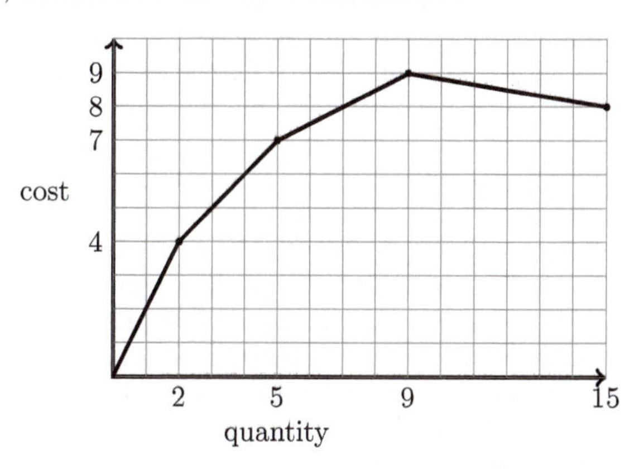

FIGURE 15.3: Binary variables require $t = \lambda_1 u_1 + \lambda_2 u_2$ for $u_1 = 0 \leq t < u_2 = 2$ and require $t = \lambda_3 u_3 + \lambda_4 u_4$ for $u_3 = 5 < t < u_4 = 9$.

Set Partitioning and Covering

Deliveries must be made by a fleet of trucks to locations $1, 2, \ldots, n$. Each truck will take a route that delivers to a subset of the locations. A large set $\mathcal{J}$ of possible routes has been

generated. The problem is to assign routes to trucks so that all deliveries are made, using the smallest possible number of trucks.
Define the data as a binary matrix A such that $A_{ij} = 1$ iff route $j \in \mathcal{J}$ delivers to location i. Define binary variables $x_j = 1$ if route $j \in \mathcal{J}$ is used, and $x_j = 0$ if not. The IP is a set partitioning model because it partitions the set of locations into subsets, each subset corresponding to a route.

$$\min \mathbf{1} \cdot \mathbf{x} \quad \text{subject to} \tag{15.22}$$

$$A\mathbf{x} = \mathbf{1} \tag{15.23}$$

$$\mathbf{x}_j \quad \text{binary for all } j \tag{15.24}$$

If different routes j have different costs c_j, change the objective (15.22) from minimizing the number of routes to minimizing $\mathbf{c} \cdot \mathbf{x}$, the sum of the route costs.

This model, though correct, can be challenging to solve. The requirement of an exact partition is unforgiving. Delivery problems can often be made computationally easier if removing a location from a feasible route cannot cause infeasibility. This is often the case because a vehicle could traverse the same physical route while skipping some deliveries. In such cases, constraint (15.23) may be replaced by the weaker condition

$$A\mathbf{x} \geq \mathbf{1} \tag{15.25}$$

The resulting IP is a set covering model because it finds a collection of subsets that covers (i.e., whose union contains) the set of locations. If deliveries may be skipped, any set covering can be converted to a partition by removing extra deliveries to the same location. Ordinarily, set covering IPs solve faster than their corresponding set partitioning IPs. However, if the objective (15.22) has been replaced by the more general minimization of $\mathbf{c} \cdot \mathbf{x}$, an optimal set covering solution does not necessarily lead to an optimal set partitioning solution.

For medium-sized fleet routing instances, not to mention the large instances faced by major airlines, it is computationally impractical to generate the entire set of possible routes $\mathcal{J}$. Column generation must be used just to solve the LP relaxation. There are two main choices once that LP is solved. The first choice is to "freeze" the set $\mathcal{J}$ and perform conventional branching on binary variables with fractional values. This choice is simpler computationally, but it has little chance of finding a true integer optimum, because it has only the subset $\mathcal{J}$ of the columns.

The second choice, called "branch-and-price", is to continue column generation at the nodes of the branch-and-bound tree. However, conventional branching becomes problematic. Suppose you branch on binary x_{88} with fractional value 0.5 in the LP relaxation. The branch $x_{88} = 1$ works fine, but the other does not. You remove the column $A_{:,88}$ from $\mathcal{J}$ along the branch $x_{88} = 0$. But from a column generation point of view, the column values $A_{:,88}$ are perfect to solve the LP. You must employ a more sophisticated type of branch. For example, for two locations t and u, one branch requires that t and u be serviced on the same route; the other branch requires that t and u be serviced on different routes [16]. (See also [284, 285] for relations to Dantzig-Wolfe decomposition.)

Logical Relationships: The Liars' Club I

In a school for liars, 4 students got different grades on an exam. The possible grades were A, B, C, D. They made the following statements:
Elgar: I got an A or a B. Grieg got a B or a C.
Franck: I got a better grade than Handel. Grieg did worse than Elgar.
Grieg: Franck or I got the A. Handel got a C.

Handel: Franck got a D. Grieg got a worse grade than Franck.

Unfortunately, their training was imperfect and only 7 of the 8 statements were false. Determine the grade of each student.

Solution: Define index sets $I = E, F, G, H$ for the people and $J = 1, 2, 3, 4$ for the grades. At the outset, we have to decide how we will represent the grades. We could use 4 general integer variables, x_E, x_F, x_G, x_H where the values $1, 2, 3, 4$ represent grades of A, B, C, D, respectively. We could use 16 binary variables, x_{ij} where $x_{ij} = 1$ iff person i receives grade j. As a general rule, IP software handles binary variables more easily than general integer variables, so we will use the latter representation for now. The constraints that everyone receives a different grade are represented as $\forall j \in J \sum_{i \in I} x_{ij} \leq 1$ ("= 1" is correct, too); everyone receives a grade is the constraint $\forall i \in I \sum_{j \in J} x_{ij} = 1$. It will be helpful to have a representation for each student's grade.

Introduce auxiliary variables $w_i, i \in I$:

$$w_i = x_{i1} + 2x_{i2} + 3x_{i3} + 4x_{i4}.$$

The w_i variables must have integer values in any feasible solution because the x_{ij} are binary. Yet it is better not to declare them as integer variables, but to leave them as continuous variables instead. In general, the fewer declared integer variables, the better.

Next, we need to require that at least one of the statements is true. Model this requirement by

$$\begin{aligned}
x_{E1} + x_{E2} &\geq 1 - My_1 \\
x_{G2} + x_{G3} &\geq 1 - My_2 \\
w_F + 1 &\leq w_H + My_3 \\
w_G &\geq 1 + w_E - My_4 \\
x_{F1} + x_{G1} &\geq 1 - My_5 \\
x_{H3} &\geq 1 - My_6 \\
x_{F4} &\geq 1 - My_7 \\
w_G &\geq 1 + w_F - My_8 \\
\sum_{k=1}^{8} y_k &\leq 7
\end{aligned}$$

Finally, we need to require that at least 7 of the 8 statements are false. Model this as follows:

$$\begin{aligned}
x_{E1} + x_{E2} &\leq 0 + Mz_1 \\
x_{G2} + x_{G3} &\leq 0 + Mz_2 \\
w_F - 1 &\geq w_H - Mz_3 \\
w_G + 1 &\leq w_E + Mz_4 \\
x_{F1} + x_{G1} &\leq 0 + Mz_5 \\
x_{H3} &\leq 0 + Mz_6 \\
x_{F4} &\leq 0 + Mz_7 \\
w_G + 1 &\leq w_F + Mz_8 \\
\sum_{k=1}^{8} z_k &\leq 1
\end{aligned}$$

You may have noticed that the y and z variables could be linked. I'll address this in Section 15.3.1.

Logical Relationships: The Liars' Club II

Now let's model the liar's club problem using general integer variables. On the surface these seem to be more natural decision variables. However, the model turns out to be much more awkward. This time the index set I equals $\{E, F, G, H\}$ as before. The decision variables are w_i = grade of person i, where a value of 1 represents A, 2 represents B, 3 represents C, and 4 represents D. For additional decision variables, let $y_1 \ldots y_8$ be binary variables that equal 1 if the corresponding statement is false, and 0 if the statement is true. Most of the constraints that force exactly seven of the eight statements to be false are very similar to those of the first formulation:

$$\begin{aligned}
w_E &\leq 2 + My_1 \\
2 - My_2 \leq w_G &\leq 3 + My_2 \\
w_F + 1 &\leq w_H + My_3 \\
w_G &\geq 1 + w_E - My_4 \\
&\ldots \\
w_E &\geq 2 - My_5; w_H \geq 2 - My_5 \\
w_H &\geq 2 - My_6; w_H \leq 2 + My_6 \\
w_F &\geq 4 - My_7 \\
w_G &\geq 1 + w_F - My_8 \\
\sum_{k=1}^{8} y_k &\leq 7
\end{aligned}$$

We will deal with the 5th statement later.

$$\begin{aligned}
w_E &\geq 3 - Mz_1 \\
w_F &\geq w_H + 1 - Mz_3 \\
w_G &\leq w_E - 1 + Mz_4 \\
w_F &\geq 2 - Mz_5; w_G \geq 2 - Mz_5 \\
w_F &\leq 3 + Mz_7 \\
w_G &\leq w_F - 1 + Mz_8 \\
\sum_{k=1}^{8} z_k &\leq 1
\end{aligned}$$

We will deal with the 6th statement later.

The biggest awkwardness comes when we have to require that no two students receive the same grade. To force E and F to have different grades, we must enforce one of the two constraints $E \leq F - 1$, $E \geq F + 1$. Introduce a new binary variable v_{EF} and write

$$E \leq F - 1 + Mv_{EF}; E \geq F + 1 - M(1 - v_{EF}).$$

We must do this for each pair of different students for a total of six new binary variables and 12 new constraints.

Now let's tackle the fifth statement of the first group of eight constraints. At first it looks awkward because it requires either $w_F = 1$ or $w_G = 1$ if $y_5 = 0$. The "or" makes things difficult. We can get around this awkwardness by considering the variables w_E and w_H.

Reading Tip: *Figure out how to model this constraint before reading further.*

The idea is that { F or G get an A} if and only if { both E and H don't get an A}. This replaces an "or" with an "and". So we write

$$w_E \geq 2 - My_5; w_H \geq 2 - My_5.$$

Finally, the sixth constraint of the second group of eight constraints must prohibit $w_H = 3$ when $z_6 = 0$. I don't know any way to avoid the "or" here. Introduce yet another additional binary variable v.

$$\begin{aligned} w_H &\leq 2 + Mv + Mz_6 \\ w_H &\geq 4 - M(1-v) - Mz_6 \end{aligned}$$

The logic is that when $z_6 = 1$ the variable w_H is not constrained. When $z_6 = 0$ the system reduces to the familiar way of forcing either $w_H \leq 2$ or $w_H \geq 4$.

The Liar's Club problem illustrates a common phenomenon. General integer variables permit the IP model to have fewer *decision* variables, by an order of magnitude, than a binary IP model. Thus, initially, it seems advantageous to model with non-binary integer variables. However, the general integer model then requires so many binary *auxiliary* variables that it ends up more cumbersome to write and to solve. This is especially true when underlying decisions are assignments or matchings between two sets (e.g., assigning rankings to people).

TSP: Traveling Salesman Problem, Part I

This famous problem asks for the shortest route that visits a set of points with given distances between them, and returns to its starting point. Distances may be physical distances, travel times, costs, etc. Regardless, in a TSP the objective value of a route must be the sum of the "distances" between pairs of points that are traversed consecutively. Many formulations of the TSP assume that the "distances," like physical distances, are symmetric. That is, the "distance" $c_{i,j}$ from i to j equals the "distance" $c_{j,i}$ from j to i. Asymmetric TSPs are usually considerably more difficult to solve.

I'll write two formulations. The first is simple but less effective computationally than the second. On the other hand, the first can handle directed arcs.

Reading Tip: *On a first reading, try to write your own formulation. If you can't, where do you get stuck? On a second reading, write at least one formulation.*

Let i index the points $\{1, 2, \ldots, n\}$. Let $c_{i,j}$ be the cost of moving from i to j. Define binary variables $x_{i,m}$ for all i and all $m = 1, 2, \ldots, n$ by the condition $x_{i,m} = 1$ iff point i is the mth point traversed by the route. Each point must be traversed once: $\sum_{m=1}^{n} x_{i,m} = 1 \; \forall i$. I need auxiliary variables to describe the objective. Let binary variable $y_{ij} : 1 \leq i \leq n; 1 \leq j \leq n; i \neq j$ equal 1 iff the route traverses j immediately after i. In graph theoretic terms, we want $y_{ij} = 1$ iff the (directed) arc (i, j) is traversed. These auxiliary y variables must be linked to the $x_{i,m}$ variables. Logically, $y_{ij} = \sum_{m=1}^{n} x_{i,m} x_{j,m+1}$, where the index value $m = n + 1$ is interpreted as $m = 1$. *Question: What is wrong with this equation as an IP constraint?*[4]

[4]It contains products of variables.

Reading Tip: *Figure out for yourself how to replace the nonlinear constraints with linear ones.*

As before, by convention interpret $m = n + 1$ as $m = 1$. Write linear constraints equivalent to the nonlinear equations:

$$y_{ij} \geq x_{i,m} + x_{j,m+1} - 1 \text{ for all } i \neq j; m = 1, \ldots n.$$

This (along with upper bound $y_{ij} \leq 1$) forces $y_{ij} = 1$ if, for some m, both $x_{i,m} = 1$ and $x_{j,m+1} = 1$. The objective is then to minimize $\sum_{i \neq j} c_{ij} y_{ij}$. *Question: What is the potential flaw in this formulation?*[5]

If we are not sure that the $c_{i,j}$ data will always be nonnegative, we should force $y_{i,j}$ variables that need not equal 1 to equal zero. Since all n traversed arcs are forced to have their corresponding $y_{i,j} \geq 1$, the constraint $\sum_{i \neq j} y_{ij} = n$ forces the other y_{ij} to equal zero. *Question: Why would the constraints $2y_{i,j} \leq x_{i,m} + x_{j,m+1}$ be an incorrect substitute for this constraint?*[6]

The model is then as follows:

- i and j index the points $\{1, 2, \ldots, n\}$ to be traversed.
- Data $c_{ij} = c_{ji} \geq 0$ is the cost or distance of a traversal from i to j.
- Binary decision variables $x_{i,m} = 1$ iff point i is the mth point visited in the route, $m = 1, \ldots, n$.
- Binary auxiliary variables $y_{i,j} = 1$ iff the route proceeds from i immediately to j.
- Objective: minimize $\sum_{i \neq j} c_{ij} y_{ij}$.
- Each point is visited once: $\sum_{m=1}^{n} x_{i,m} = 1$ for all i.
- Traversal from i to j incurs cost c_{ij}:
  $y_{ij} \geq x_{i,m} + x_{j,m+1} - 1$ for all $i \neq j; m = 1, \ldots n$.
- Start at point 1: $x_{1,1} = 1$.
- $x_{i,m}$ binary; $y_{i,j} \geq 0$; $y_{i,j} \leq 1$
- Traverse exactly n arcs: $\sum_{i \neq j} y_{ij} = n$.

Just as with LP, a correct statement of an IP model includes unambiguous definitions of variables and data, and any restrictions the data are assumed to satisfy. This model has a few subtleties that illustrate good principles. First, it restricts only the $x_{i,m}$ variables to be binary. The $y_{i,j}$ may be continuous nonnegative variables. This is valid because the costs are $c_{ij} > 0$. At an optimum solution, $y_{i,j} = 0$ unless traversal from i to j forces $y_{i,j} \geq 1$ in which case it will equal 1. Second, it is not logically necessary to set $x_{1,1} = 1$. It is permissible because any point may be thought of as the start and end of the cycle; it is desirable because it reduces the size of the search space by breaking a symmetry. Third, the last constraint is not necessary if all $c_{ij} \geq 0$. Similarly, the upper bounds $y_{ij} \leq 1$ are not necessary if all $c_{ij} \geq 0$. Moreover, from a problem-solving perspective there is no loss of generality in the requirement that all $c_{ij} > 0$. One could manipulate the true data by adding a constant M to every c_{ij}, which would not affect the relative values of different solutions.

[5]It does not force $y_{ij} = 0$ if j is not traversed immediately after i. It depends on $c_{ij} \geq 0$ and minimization to enforce that.

[6]Suppose the first move is from i to j. Then $x_{i,1} = x_{j,2} = 1$, forcing $y_{i,j} = 1$ as it should. But that is violated by the constraint $2y_{i,j} \leq x_{i,2} + x_{j,3}$ since $x_{i,2} = x_{j,3} = 0$.

However, I don't like to count on the user to manipulate the data if there are legitimate $c_ij < 0$ values, nor do I like to count on the data files to have all nonnegative values even if there are no legitimate negative values. The downside I'm avoiding is that without these constraints the model is apt to be invalid if some $c_{ij} < 0$.

TSP: Traveling Salesman Problem, Part II

Let $1 \leq i \leq n$ and $1 \leq j \leq n$ index the points. Define the binary variables $x_{ij} : 1 \leq i < j \leq n$ to equal 1 iff the route goes directly between points i and j (in either direction) and let $c_{ij} : 1 \leq i < j \leq n$ be the distance between points i and j.

The objective function is to minimize $\sum_{i<j} c_{ij}x_{ij}$. For constraints, we first have the simple fact that the route must employ exactly two edges incident on each point, because it must enter and exit each point once. These constraints are

$$\sum_{i=1}^{j-1} x_{ij} + \sum_{i=j+1}^{n} x_{ji} = 2 \ \forall j. \tag{15.26}$$

However, as Figure 15.4 shows, these constraints are not sufficient. The trouble is that the route might consist of several cycles that together visit all the points, rather than a single cycle that visits all the points. Such cycles are called *subtours*. To eliminate them, we add an exponential number of constraints, called *subtour elimination constraints.*

$$\sum_{i\in S, j\in S, i<j} x_{ij} \leq |S| - 1 \quad \forall S \subset \{1 \ldots n\} : \ \ 2 \leq |S| \leq \frac{n}{2} \tag{15.27}$$

The subtour elimination constraints forbid each set S of points from having a subtour, for all sets up to size $\frac{n}{2}$. *Question: Why don't we have to forbid subtours on larger sets?*[7]

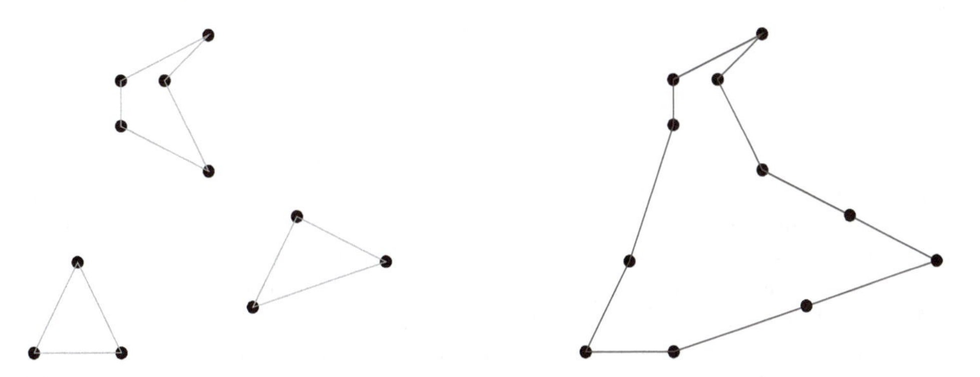

FIGURE 15.4: Without subtour elimination constraints, the minimum cost solution consists of several cycles (left) instead of one cycle (right).

Despite the awkwardness brought on by the large number of subtour elimination constraints, on large instances this formulation performs better in practice than other formulations. That will be explained in Section 15.3. First we have to see how IPs are solved, the topic of the next section.

---

[7]If there is a subtour larger than $\frac{n}{2}$ there must also be a subtour smaller than $\frac{n}{2}$.

Summary: Master the examples in this section. Then you will be ready to model most NP-complete problems as integer programs.

15.2 Solving IPs

Preview: This section describes, in general, how IPs are solved in practice, and the many ways LP helps.

An IP solver's most important tool is its LP solver. Associated with every IP is its *LP relaxation*, which is the same problem with the integrality constraints removed. The first thing (except scaling and pre-solving) an IP solver will do is to solve the LP relaxation. This relaxation's solution is the starting point for almost everything the IP solver does.

Let $A, \mathbf{b}, \mathbf{c}$ be an integer matrix and integer vectors, respectively. Let $\mathbf{x}^*$ be an optimal primal solution to the LP

$$\min \mathbf{c} \cdot \mathbf{x} \quad \text{such that} \qquad (15.28)$$

$$\mathbf{x} \in P \quad \equiv \quad \{\mathbf{x} : A\mathbf{x} \leq \mathbf{b}\}, \qquad (15.29)$$

which is the LP relaxation of the IP

$$\min \mathbf{c} \cdot \mathbf{x} \quad \text{such that} \qquad (15.30)$$

$$\mathbf{x} \quad \in \quad P; \qquad (15.31)$$

$$x_i \quad \text{integer for all } i \in I. \qquad (15.32)$$

The high level structure of an IP solver maintains a set of subproblems, arranged in a tree, the leaves of which compose the whole problem being solved. Figure 15.5 shows that the tree root represents the entire IP. The root problem *branches* into two subproblems, each represented by a child of the root. In general, the subproblem represented by a node branches into two smaller problems, represented by the node's children.

The IP solver always maintains the best feasible solution it has found so far. The solver iteratively chooses a subproblem at a leaf to focus on. It solves the LP relaxation of the subproblem. If the LP relaxation has an integer solution, that subproblem is solved. If the LP relaxation does not have a better value than that of the best feasible solution found so far, then the subproblem has been "fathomed" and is discarded. *Question: Why?*[8] Otherwise, the solver either *cuts* or *branches*. At this stage, the LP relaxation solution has at least one integer variable whose value is fractional *Question: Why?*[9] To cut, the solver finds one or more inequalities that are valid for the IP but are violated by the LP relaxation solution. It adds these inequalities to the model and re-solves the LP relaxation. To branch, the solver picks one such variable, say y_1 with fractional value f, and divides the subproblem into two further subproblems, one with constraint $y_1 \leq \lfloor f \rfloor$ and the other with constraint $y_1 \geq \lceil f \rceil$.

[8] No feasible solution to the subproblem can improve on the best solution found so far.
[9] Otherwise, it would be solved.

In the frequently occurring case that y_1 is a binary variable, the constraints are $y_1 = 0$ and $y_1 = 1$. This fixes the variable to 0 or 1, which in effect reduces the number of variables. In general, all or most cutting is done at the root of the tree of subproblems.

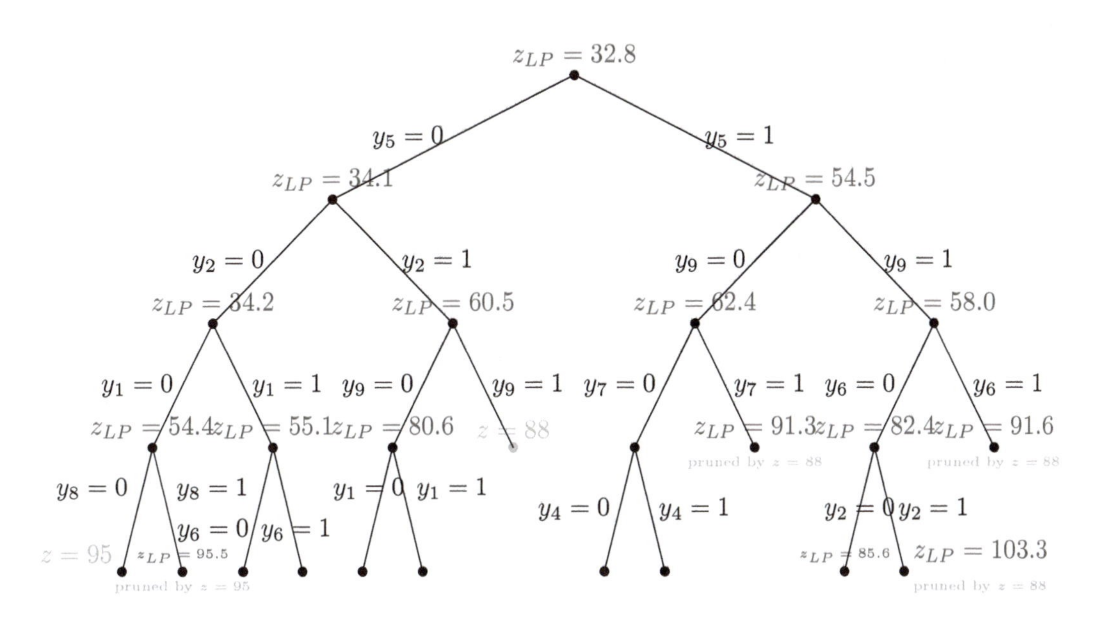

FIGURE 15.5: An IP solver maintains a tree of subproblems.

Whether branching or cutting, employ the dual simplex algorithm to re-solve LPs. *Question: Why can dual simplex be used?*[10] Solving from the warm start $\boldsymbol{\pi}^*$ is much, much faster than solving from scratch. In practice, if m is the number of constraints in the LP, and q is the average number of nonzeroes per column, the warm start will typically run in $O(q)$ time rather than $O(m)$ time. This speedup is essential to the practicality of solving IPs.

Branching on a binary variable is nice because it fixes the variable to 0 or 1, which in effect reduces the number of variables, and often allows the solver to fix other binary variables at zero or one.

The entire procedure is called *branch-and-bound.* Solving the LP relaxation is the bounding part; dividing the subproblem into two further subproblems is the branching part.

Many things are added to this main structure. A pre-processor or pre-solver simplifies the problem initially. Heuristics are run to find good feasible solutions or to improve on feasible solutions that have been found. Especially at the root of the tree, the solver looks for valid inequalities it can add to the problem formulation. Computations are made to decide which subproblem to focus on next and which fractional-valued variable to pick.

When the main emphasis of the procedure is to generate valid inequalities that cut off fractional subproblem LP relaxation solutions when they occur, and to branch as a last resort, the procedure may be called a cutting plane method instead. When the problem has a huge number of columns, as is common in set covering formulations, a different procedure called *branch-and-price* can be used, but it is beyond the scope of this book to describe.

[10] The dual LP solution $\boldsymbol{\pi}^*$ is still feasible because the only change to the LP is an added constraint.

15.3 Tighter Formulations

Now that you understand the overall process of IP solving, you should believe the following guideline:

The closer is the LP relaxation to the IP, the better.

For example, consider the following two formulations of the relationship $y = x_1$ OR x_2 for binary variables y, x_1, and x_2.

$$y \geq \frac{x_1 + x_2}{2}; \tag{15.33}$$

$$y \leq x_1 + x_2 \tag{15.34}$$

and

$$y \geq x_1 \tag{15.35}$$

$$y \geq x_2 \tag{15.36}$$

$$y \leq x_1 + x_2 \tag{15.37}$$

The two formulations are logically equivalent in that their integer feasible solution sets are identical. But their LP relaxations are not equivalent. Adding inequalities (15.35) and (15.36) of the second formulation yields inequality (15.33) of the first formulation. Inequalities (15.34) and (15.37) are identical. Therefore, the LP relaxation constraints of the second formulation imply the LP relaxation constraints of the first formulation. The converse is false, because the fractional solution $x_1 = 1, x_2 = 0, y = \frac{1}{2}$ satisfies (15.33) but not (15.35). We say that the second formulation is *tighter* than the former, because the feasible region of its LP relaxation is strictly contained in the LP relaxation of the other.

Here is a classic example from facility location. Let $i \in I$ index a set of possible locations for facilities (e.g., warehouses or factories). Let $j \in J$ index a set of delivery locations. The given cost data are transportation costs c_{ij} for sending a unit of goods from facility location i to delivery location j, and construction costs f_i for building a facility at location i. The other given data are demand amounts d_j at location j, and capacity amounts s_i for facility location i. The problem is to decide which facilities to build, and how much demand to satisfy from each built facility to each delivery location, so as to minimize total cost. Qualitatively speaking, the problem is to trade off construction costs against transportation costs. If you build too many facilities, the f_i costs will kill you; if you don't build enough facilities, the transportation costs will. Besides the data just defined, the model is

- Indices $i \in I$ of possible facility locations; $j \in J$ of delivery locations.
- $y_i, i \in I$ Binary variable, equals 1 iff facility i is built.
- $x_{ij}, i \in I, j \in J$ Continuous variable, equals the number of units of goods sent from facility i to delivery location j.

$$\min \sum_{i \in I} f_i y_i + \sum_{i \in I, j \in J} c_{ij} x_{ij} \quad \text{subject to}$$

$$\sum_i x_{ij} = d_j \;\; \forall j \in J$$

$$\sum_j x_{ij} \leq s_i y_i \;\; \forall i \in I$$

$$x_{ij} \geq 0; y_i \; \textit{binary}$$

This is a parsimonious model. By LP standards, it is the best model. But by IP standards, it would be better to *add* the following constraints:

$$x_{ij} \leq \min\{d_j, s_i\} y_i \;\; \forall i, j.$$

These additional constraints are valid. They mean that you can't send anything from i to j if you don't build facility i, and you can't send more than either the demand of j or the supply at i if you do build facility i. However, these constraints are logically unnecessary. Why should they be included? To see why, let's make the data concrete. Set $s_i = 100, i = 1, 2$ and $d_j = 25, j = 1, \ldots 4$. The additional constraints are $x_{ij} \leq 25y_i$. Consider the feasible solution to the LP relaxation of (15.38), $y_1 = y_2 = 0.5; x_{11} = x_{12} = x_{23} = x_{24} = 25$. This LP solution "cheats" by only charging half the cost of building the facilities. If the transportation costs are as shown in Figure 15.6, it would be the optimal LP solution. This solution violates four of the additional constraints including $x_{11} \leq 25y_1$. In particular, the fractional values $y_1 = y_2 = 0.5$ are too small to satisfy the constraints. The y_i would have to equal 1. Adding valid but logically unnecessary constraints can make the LP relaxation much closer to the IP. In general it is worth doing so, even though the LP becomes bigger and thus harder to solve.

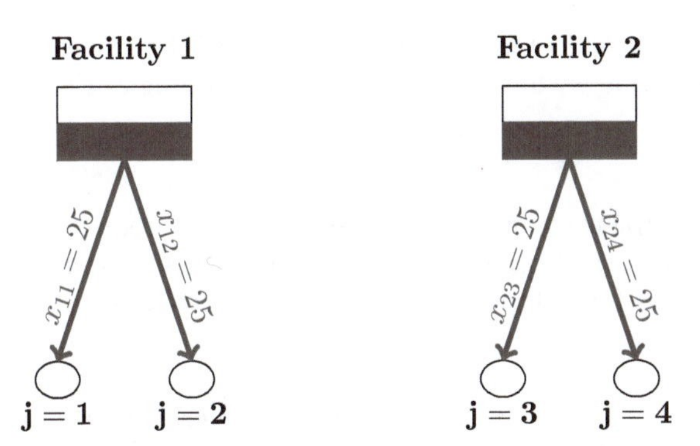

FIGURE 15.6: When delivery locations $j = 1$ and $j = 2$ are close to facility 1, and locations 3 and 4 are close to facility 2, the LP cheats by building half of each facility.

Let's continue by considering the use of "big M" values. In general, when using "big M" penalty values, you should select the smallest valid value for M. Don't use the same value of M for different constraints! Instead, for each constraint, use the minimum valid value. For example, in the "plus-minus" formulation, the first two constraints were $1 - My_1 \leq x_1 \leq 4 + My_1$. How small can we make the M values? No matter where x_1, x_2 is in the feasible region, it satisfies

$$1 \leq x_1 \leq 9 \text{ and } 2 \leq x_2 \leq 5. \tag{15.38}$$

Therefore, we can write $1 \leq x_1 \leq 4 + 5y_1$ for the first two constraints. Similarly we can write $3 - y_1 \leq x_2 \leq 4 + y_1$ for the next two constraints. *Question: How would we write the next four constraints?* [11] It is easy to prove that a smaller M value yields a tighter formulation. For example, sum the inequalities $x_1 \leq 4+5y_1$ and $0 \leq y_1$ to get the as loose or looser constraint $x_1 \leq 4+6y_1$, and verify that $y_1 = \frac{1}{6}, x_1 = 5$ satisfies the latter but not the former. *Question: Why is $0 \leq y_1$ part of the formulation?*[12]

Balas [14] invented a clever formulation for bounded non-convex regions which is tighter yet. I illustrate the formulation on the non-convex region of Figure 15.2. Make three "copies" of the x_i variables, x_i^1, x_i^2, x_i^3 corresponding to the three rectangles which together compose the non-convex region. Each rectangle is controlled by a binary variable y_1, y_2 or y_3. The constraint

$$y_1 + y_2 + y_3 = 1$$

ensures that exactly one of these binary variables will equal 1. Next, force membership in a rectangle if the associated binary variable $y_j = 1$. For rectangle 1, which is defined by $1 \leq x_1 \leq 4; 3 \leq x_2 \leq 4$, write

$$\begin{aligned} 1y_1 &\leq x_1^1 \leq 4y_1 \\ 3y_1 &\leq x_2^1 \leq 4y_1. \end{aligned}$$

When $y_1 = 1$ these constraints force the x_i^1 copy to be in the first rectangle. Moreover, when $y_1 = 0$ they force the x_i^1 variables to zero. Similarly, for the second and third rectangles, write

$$\begin{aligned} 2y_2 &\leq x_1^2 \leq 3y_2 \\ 2y_2 &\leq x_2^2 \leq 5y_2 \\ 6y_3 &\leq x_1^3 \leq 9y_3 \\ 3y_3 &\leq x_2^3 \leq 4y_3. \end{aligned}$$

These constraints, taken together, force all but one copy of the x_i variables to equal zero, and force the other copy to be within the corresponding rectangle. Finally, set

$$x_i = x_i^1 + x_i^2 + x_i^3 \,:\, i = 1,2$$

and the formulation is complete. In general, if the non-convex region consists of the union of m polytopes in $\Re^n$, make m "copies" of $x_1 \ldots x_n$, each associated with its own binary variable. Unfortunately, this formulation method does not extend to handle k out of m sets of constraints when $k > 1$.

Coefficient reduction: Let x_1, x_2, x_3, x_4 be nonnegative general integer variables. Consider the constraint $25x_1 + 37x_2 + 12x_3 + 11x_4 \leq 43$. It is logically equivalent to the constraint $2x_1 + 3x_2 + x_3 + x_4 \leq 3$. Which constraint is tighter? The latter is. This can be seen by dividing the former by 12.5. If we then round up the coefficients on $x_1 \ldots x_4$, we make the constraint tighter; if we round down the right-hand side, we make the constraint even tighter. In this way we arrive at the latter constraint.

As a final example of tighter versus looser constraints, in the liar's club problem, it would have been logically correct to express a constraint, "Elgar did better than Handel" as $w_E \leq w_H$. This inequality only says that Elgar did at least as well as Handel. It is logically sufficient because there are other constraints that force Elgar's grade to be different from Handel's. However, it is a looser formulation and it should not be used. This is because $w_E + 1 \leq w_H$ implies $w_E \leq w_H$ but not the reverse. The former inequality cuts off more

[11] $2 - y_2 \leq x_1 \leq 3 + 6y_2$ and $2 \leq x_2 \leq 5$.

[12] The LP relaxation of y_1 binary is $0 \leq y_1 \leq 1$.

from the feasible region of the LP relaxation. *Question: If there were no constraint that forced Elgar's grade to be different from Handel's, how could you force Elgar's grade to be better with a looser constraint than* $w_E + 1 \leq w_H$*?*[13]

15.3.1 Valid Inequalities

Sometimes through logical reasoning about the integer variables, or logical reasoning based on your knowledge of the problem being formulated, you can think of additional constraints that any feasible, or the optimal solution must satisfy. For example, in the liar's club problem in this chapter, one set of constraints forces at least one of the statements to be true, while another set of constraints forces at least seven of the eight statements to be false. Together these constraints force exactly seven of the eight statements to be false. Therefore, the statement that is forced to be true in the first set of constraints need not be forced to be false in the second set of constraints. Therefore, any feasible solution to the problem could satisfy the additional constraints that $z_k = 1 - y_k, k = 1 \ldots 8$. Going a step further, we could simply replace z_k by $1 - y_k$, thus reducing the number of integer variables by eight.

As another example, in the liar's club problem with w_i as decision variables, you know that $\sum_i w_i = 10$ since each person receives a different grade. Adding this constraint to the formulation might help it solve faster.

As another example, if you are pairing up teams for a tournament, you might have index sets I for the teams and K for the days of the tournament, and binary variables x_{ijk} where $i \in I; j \in I; i < j; k \in K$ to equal 1 if team i plays team j on day k. Besides the necessary constraints that a team must play exactly one game per day, $\sum_{j<i} x_{ji} + \sum_{j>i} x_{ij} = 1$, you could notice that for any three teams $h < i < j$ and any day k, $x_{hik} + x_{ijk} + x_{hjk} \leq 1$. (This is a kind of constraint called an "odd-set constraint"). This is a valid inequality that cuts away from the feasible region of the LP relaxation, and so it tightens the formulation. The only disadvantage to the inequality is that there are so many they may make the LP too large to solve quickly. This could be handled by including the constraints as "lazy" constraints, constraints that are only added when they cut off the LP solution, or by separation routines which are unfortunately beyond the scope of this text.

15.3.2 Facets

Every bounded IP can be reduced to an almost exactly equivalent LP. Do you see how to do it? Keep in mind that the reduction isn't a valid polynomial-time transformation as defined in Figure 14.3, since IP is NP-hard and LP is polynomial-time solvable,

For simplicity, suppose all variables in a given IP are required to be integers. Let $P \in \Re^n$ be the polytope of the LP relaxation of the IP. Let $Z(P) \equiv P \cap \mathcal{Z}^n$, the feasible solutions to the IP. Let $Q = conv(Z(P))$, the convex hull of the IP's feasible solutions. Then Q is a polytope and all of Q's extreme points are in $Z(P)$. By Theorem 4.4.8, for every $\mathbf{c} \in \Re^n$, the LP to minimize $\mathbf{c} \cdot \mathbf{x}$ subject to $\mathbf{x} \in Q$, possesses an optimal solution that has all integer values. Moreover, every basic feasible optimal solution to that LP has all integer values. Therefore, solving the LP is essentially equivalent to solving the given IP.

▶ *The construction of Q can be thought of as a transformation that takes P as input and produces Q as output. There are at least two ways it could fail to be a valid polynomial-time transformation.*

Reading Tip: *On a first reading, try to think of at least one possible failure. On a second reading, state both before continuing.*

---

[13] $w_E + 0.1 \leq w_H$ – but do not do this!

First, the length of the transformation's output might be more than an any polynomial function of the length of its input. If so, the transformation cannot possibly run in polynomial time in the length of the input, even if it runs in linear time in the length of the output. Second, even if the output is not long, it might be hard to compute it. We need some polyhedral theory from Chapter 4 and definitions from Chapter 14 to make these potential failures precise.

When we say that P is an input to the transformation, we cannot mean that the transformation is given an n-dimensional object. We mean that the transformation is given a description of P. The standard descriptive form, in accord with inequality (15.2), comprises a matrix A and a vector $\mathbf{b}$ that define the constraints, so that $P = \{\mathbf{x} : A\mathbf{x} \geq \mathbf{b}\}$. This form, together with the objective vector (15.1) and set I, adequately defines the IP that is the input to the transformation. However, it would be ambiguous to define the output form of Q merely as a matrix F and vector $\mathbf{g}$ such that $Q = \{\mathbf{x} : F\mathbf{x} \geq \mathbf{g}\}$. Question: Why? There will be infinitely many choices of F and $\mathbf{g}$, some of them much lengthier than other choices. Which should the transformation produce as output? By Theorem 8.2.7, the facet-defining inequalities of Q, together with $n - dim(Q)$ linearly independent equations, are both necessary and sufficient to specify the polytope Q. Equivalently, we may require the description of Q to be minimal. This requirement resolves the ambiguity in the specification of Q, up to scaling. Question: How would you resolve the scaling ambiguity?[14] *We can now describe the first possible failure clearly: Let A have m rows. The number of facets of polytope Q might exceed any polynomial function of m. This does occur both in theory and practice. It has long been known that the TSP polytope has more than polynomially many facets. In fact, the subtour elimination constraints (15.27) define facets and there are exponentially many of them.*

Theorem 4.2.2 offers a possible loophole. There might be a higher dimensional polyhedron $\hat{Q}$ that has polynomially many facets, and projects down to Q. In the case of the TSP polytope, Yannakakis [303] proved that no symmetric such $\hat{Q}$ exists. This was a remarkable result because it does not rely on any assumption such as $P \neq NP$. Later, Fiorini et al. [95] proved the stronger result, that any $\hat{Q}$ which projects to the TSP polytope, whether symmetric or not, must have exponentially many facets. Even more recently, Rothvoß [255] proved that the same is true for the matching polytope, despite the fact that optimization over the matching polytope can be done in polynomial time.

In principle, the length of Q's description could be large because the coefficients of F were large. This should seem unlikely to you, because coefficients are encoded in binary. Numbers of size $k_0^{k_1 n^{k_2}}$ for constants k_0, k_1, k_2 have polynomial length.

For the second potential failure, it is simple to prove that obtaining the most elementary information about Q from P is NP-hard.

Theorem 15.3.1 *Given integer matrix $A \in \mathcal{Z}^{m \times n}$ and vector $\mathbf{b} \in \mathcal{Z}^m$, let $P = \{\mathbf{x} : A\mathbf{x} \geq \mathbf{b}\}$. It is NP-hard to determine whether or not $Q \equiv conv(P \cap \mathcal{Z}^n)$ is empty.*

Proof: Given an instance of $3 - SAT$ with v variables and c clauses, model it as an IP with v variables and $c + 2v$ constraints. (The $2v$ constraints are $I\mathbf{x} \geq \mathbf{0}$ and $-I\mathbf{x} \geq -\mathbf{1}$.) Then Q is empty iff the 3-SAT instance cannot be satisfied. ■

The precise complexity class of determining whether a given inequality is facet-defining for Q was invented in a lovely paper by Papadimitriou and Yannakakis [243], and applied to facets in [242]). ◄

15.4 Tips

Multicore CPUs are now standard in PCs. Your IP software will be running multiple threads, each thread choosing a different subproblem from the tree of subproblems. A thread

[14] Since all data are rational, require F and $\mathbf{g}$ to be integer and such that in each constraint, the coefficients have no common factor other than 1.

that is trying to bound a subproblem needs to know the value of the best feasible solution found so far by any thread. Its decision as to whether to branch or not depends on that value. A thread that is choosing a variable to branch on wants to know how branching on different variables has performed elsewhere in the tree. Therefore, the future computation path of a thread depends on information received from other threads. However, communication between threads is not automatically synchronized with in-thread computations. Hence you can't expect the same solution performance on same instance when run twice (unless you tell the software to do so, for which you pay a performance penalty). This can be downright embarrassing if your formulation fails to replicate its terrific performance in front of your client. The moral of the story is: run your test example several times before showing it off. Better yet, save the log of the run of your test example, and show the log.

As I've explained already, use the smallest valid values for penalties. Your IP solver will automatically try to reduce penalty values, but often your domain knowledge will permit you to obtain a smaller penalty value than can the IP solver. In no case should you use a huge penalty of order, say, 10^8, because that can interfere with the software tolerances. The software might not distinguish between 10^{-8} and zero.

IP is NP-hard, so it will exhibit bizarre behavior at times. A tiny change in the data can result in a huge change in the solution, and/or in the time required for solution. Rearranging constraints or other trivial changes in the model can drastically change the solution time. I remember a sports scheduling model that took more than an hour to solve. I rearranged the variables and it solved in less than two minutes. The reverse could have happened just as easily.

It is very common for an IP solver to find an optimal solution fairly quickly, and then spend most of its CPU time proving that its solution is optimal. See Section 14.10.2 for some insight as to why that might be so.

If the IP solver is not finding good feasible solutions quickly, perhaps you, with your specialized knowledge of the problem, can provide a good heuristic solution at the outset. This will allow the solver to discard many subproblems and can dramatically speed up the solution process.

Valid inequalities: Cut generation is a tradeoff between cost of solving the LP subproblems and having a tighter LP relaxation. On top of that, you will see the occasional bizarre behavior of IPs. It is not very predictable whether using more valid inequalities will help or hurt you.

Tuning: Tuning means setting the parameters of the solver software to run more quickly on your model. Some software has automatic self-tuning. Or you can do the tuning yourself. Tuning is useful if you will be running the same model (with different data) multiple times. If there is a real user, you or the user will almost certainly run it multiple times. Things to try are, generally speaking:

- How much to use cuts. They trade off between slowing the LPs and being tighter. Tightness makes fathoming more likely, and often increases the chance of finding a good feasible integer solution.
- How much and when to use heuristics to find or improve good quality solutions.
- Variable selection (branching strategy). Especially if the objective has one or few terms, try changing the branching strategy.
- Reformulate the model, usually building a larger but tighter model.
- Settle for a larger optimality gap than the software default, which is typically 10^{-4}. If your data are accurate only to within 1%, why should you demand a gap smaller than that?

IP solvers often have a difficult time with models that have few terms in the objective function. The worst kind of IP objective, besides none at all, is to minimize a single variable, as one does in a minimax problem. If your solver is working very slowly, consider reformulating your problem to put more terms into the objective. You could relax some of the constraints and put penalty terms into the objective, for example.

Modelers are rarely better than commercial software in selecting which fractional variables to branch on. However, if you feel that you must, try branching up first on variables that appear in constraints like $\sum_i x_i \leq 1$ because fixing one variable to 1 forces a lot of other variables to 0. Branching down doesn't force any other variable to a value. Sometimes you know that a variable is really important. You can tell the software to give higher priority to branching on that variable. For example, give high priority to decisions made in the first time period in a multi-period model.

Symmetry is your enemy. Sometimes you can lessen it by perturbing the objective function coefficients, a built-in software option. Often you can lessen symmetry by adding constraints. Don't settle for breaking one symmetry if you can break others, too. But be careful that your symmetry breaks do not interfere with each other.

Example 15.3 [Vehicle Fleet Symmetry Breaking] An IP model assigns a route to each of a set of K vehicles indexed by k. If vehicles are interchangeable, the model has $K!$-fold symmetry. Suppose part of the objective function is to minimize the total travel distance. Let d_k be an auxiliary variable equal to the distance traveled by vehicle k. Two natural methods to break symmetry are:

- Perturb the $\sum_{k=1}^{K} d_k$ portion of the objective to

$$\sum_{k=1}^{K}(1 - k\delta)d_k,$$

  where δ is a small value such as $0.01/K$. This will assign shorter routes to lower indexed vehicles.

- Directly impose the constraints $d_k \leq d_{k+1}$ for $k = 1, 2, \ldots, K-1$. The logic of this method is cleaner, though it adds $K-1$ constraints to the model.

However, don't use both methods together, because they conflict. Likewise, you could break one symmetry by constraining vehicle 1 to serve delivery point 1. That would be valid on its own, but it conflicts with both of the above methods.

Summary: Some of the main guidelines for IP modeling are as follows. Integer programs are harder to solve than linear programs. Don't make a variable integer if it doesn't have to be. Sometimes it is enough to make a key subset of the variables integer, and other variables may then be forced to integer values. Binary integer variables are usually easier computationally than general integer variables. Try to break symmetries, which can slow down your software. A tighter formulation that uses more variables and/or more constraints may be better than a looser, smaller formulation.

15.5 Notes and References

This brief chapter provides just a glimpse into the rich world of IP. Two books that describe the history of IP research are [167] and [55], the latter through the lens of the TSP, whose history is inextricably linked with the history of IP in general, both in theory and computation. Standard references that cover both modeling and theory include [235] and [301]. For focus on modeling (including reformulation) and optimization, see the texts [299], [245], and [46]. Also see the expositions in [290] and [168]. For focus on theory that underlies modeling and algorithms, see [54] and[262].

15.6 Problems

E Exercises

1. Write constraints that force at least one of the binary variables x_1, x_2, x_3 to be false and at least one of them to be true.

2. Write constraints that force the binary variables x_1, x_2, x_3, x_4 to not all be equal to each other.

3. Write constraints that force binary variable y to equal 1 if and only if at least one of the binary variables x_1, x_2, x_3 is false and at least one of them is true.

4. Given that x_{ij} are binary and given constraint (15.26), show that constraint (15.27) is mathematically equivalent to requiring that at least two edges traverse between S and its complement,
$$\sum_{i \in S, j \notin S, i<j} x_{ij} + \sum_{j \in S, i \notin S} x_{ij} \geq 2.$$

5. Suppose distances in a TSP instance are not symmetric. That is, the cost c_{ij} of traveling from i to j may differ from c_{ji}, the cost of traveling from j to i. Alter the IP of Section 15.1 to model this problem. Begin by defining your variables precisely. Show how to modify the objective function and constraints (15.26) and (15.27).

6. Suppose x, y, and z are binary variables. Write linear inequalities that force z to equal x XOR y? ("XOR" means "exclusive or": x XOR y is true iff exactly one of x and y is true.)

7. Suppose x, y, and z are binary variables. Write linear inequalities that force z to equal "true" if $x = y$ and "false" if $x \neq y$.

8. Suppose x and y are continuous variables and z is a binary variable. Write linear inequalities that force $z = 0$ if $x > y$ and force $z = 1$ if $x < y$.

9. Model each of the following problems, all defined in Chapter 14, as IPs:
   (a) 3-SAT
   (b) Partition
   (c) X3C
   (d) Set Covering
   (e) Set Packing
   (f) Clique
   (g) Bin Packing
   (h) Graph 3-Coloring

10. Alter the IP model of graph isomorphism in Example 15.1 to apply to undirected graphs G and H.

11. Write explicit constraints that define the cost function in Figure 15.3. Use the method in the text.

12. Define variables and write constraints to model the cost function $f(t) = 0 : t = 0$; $f(t) = 5 + 18t : 0 < t \leq 100$.

13. Suppose your IP has no objective function, but it has at least one inequality constraint. How can you alter it to help software solve it faster?

14. Suppose your IP has no objective function and no inequality constraints. How can you alter it to help software solve it faster?

M Problems

15. Let x_1, x_2, x_3, and y be binary variables. Write constraints that force $y = 1$ iff at least one of the x_i equals zero and at least one of the x_i equals 1.

16. Let x_1, x_2, x_3, and y be binary variables. Write constraints that force $y = 1$ iff $x_1 \leq x_2 \leq x_3$.

17. Alter the IP model of graph isomorphism in Example 15.1 to apply to mixed graphs G and H. (A mixed graph has both directed and undirected edges.)

18. The graph automorphism problem is to determine whether there exists an isomorphism of a graph $G = (V, E)$ to itself, other than the trivial identity mapping $\sigma(v) = v \; \forall v \in V$. Model this problem as an IP. (Caution: The automorphism may map some vertices to themselves, as long as it does not map all vertices to themselves.)

19. Let x_1, x_2, x_3, y be binary variables. An IP has the constraint $9x_1 + 16x_2 + 5x_3 \leq 20$.
   - Prove that the constraint $10x_1 + 16x_2 + 5x_3 \leq 18$ is both valid and better than the given constraint.
   - Find a best possible constraint or constraints to replace the given constraint. Prove both validity and optimality of your answer.

20. Model each of the following problems, all defined in Chapter 14, as IPs. Observe that modeling a permutation tends to take more effort than modeling a partition.
   (a) MAX-CUT

(b) MIN BISECTION CUT

(c) Minimum Feedback Arc Set

(d) s,t-Hamiltonian Path

(e) Hamiltonian Path

(f) Longest Path

(g) 2-Machine Weighted Flowtime Minimization

21. Define variables and write constraints that describe the following non-convex cost function with a fixed charge:

$$f(t) = \begin{cases} 0 & t = 0 \\ 5 + 18t & 0 < t \leq 10 \\ 185 - 3(t-10) & 10 \leq t \leq 35 \end{cases}$$

22. Show that the following constraints are equivalent in their effect to constraints (15.17)–(15.20).
$\lambda_k \leq w_k + w_{k-1}$;
$\sum_{k=1}^{m} w_k = 1$;
$t \geq u_k w_{k+1}\ \forall k = 2, \ldots, m$ (same as (15.18)).

23. Construct constraints similar to those of Problem 22, equivalent to (15.17)–(15.20), that employ (15.19) rather than (15.18).

24. Complete the following model of piecewise linear costs. Describe the cost function as having value c_j at point u_j where $u_1 < u_2 < \ldots < u_m$. Assume $c_j > 0 \forall j$. Constrain binary variable z_j with inequalities $t \geq u_j + (u_m - u_j)z_j$; $t \leq u_j - (1 - z_j)(u_m - u_j)$.

25. For the two sets of constraints (15.33),(15.34) and (15.35),(15.36),(15.37), physically construct the polyhedra. Use cardboard or paper. Do not use a 3D printer. Compare them, and verify that one strictly contains the other.

26. Construct an instance of the set partitioning problem in which the objective is to minimize $\mathbf{c} \cdot \mathbf{x}$ rather than (15.22), such that there is an optimal set covering solution that cannot be converted to an optimal solution by removing multiple deliveries.

27. Model the following logic puzzle as an integer program. There are five houses in a row, each painted a different color. Their residents are of different nationalities, own different pets, drink different beverages and smoke different brands of cigarettes.

(a) The Englishman lives in the red house.

(b) The Spaniard owns the dog.

(c) Coffee is drunk in the green house.

(d) The Ukrainian drinks tea.

(e) The green house is immediately to the right of the ivory house.

(f) The Old Gold smoker owns snails.

(g) Kools are smoked in the yellow house.

(h) Milk is drunk in the middle house.

(i) The Norwegian lives in the first house.

(j) The man who smokes Chesterfields lives in the house next to the man with the fox.

(k) Kools are smoked in a house next to the house where the horse is kept.

(l) The Lucky Strike smoker drinks orange juice.

(m) The Japanese smokes Parliaments.

(n) The Norwegian lives next to the blue house.

Who drinks water? Who owns the zebra? (Use the objective function to identify the water drinker and the zebra owner).

28. A clique in a graph is a subset of vertices all pairs of which are connected by an edge. For example, a clique of 5 vertices contains 10 edges. For graph $G = (V, E)$, you wish to partition V into as few cliques as possible. Model this problem as an IP. Break symmetry as best you can.

29. Job j takes t_j time units to complete for $j = 1, \dots, n$. There are two identical machines which can be run simultaneously. Each machine can process one job at a time. Each job must be run on one of the two machines. The completion time c_j of job j is the time at which it is completed.

    - Model the problem of finding a schedule that minimizes $\max_{1 \le j \le n} c_j$ as an IP.
    - Model the problem of finding a schedule that minimizes $\sum_{1 \le j \le n} c_j$ as an IP.

30. Determine the least valid value for each M in the first liars' club formulation. Do the same for the second formulation.

31. Run the two liars' club formulations with presolve turned off and compare the branch and bound tree. Run the formulations with default settings (presolve turned on) and compare the results.

32. Write two different IP models to solve *Sudoku* puzzles (defined in Chapter 1), one using only binary variables and the other using general integer variables. Define indices, variables, and constraints. Which model is better?

33. Let $\mathcal{U}$ be an enlargement of the TSP polytope $\mathcal{T}$ defined by $\mathcal{U} = \{\mathbf{u} \ge \mathbf{0}, \exists \mathbf{t} \in \mathcal{T} \text{ subject to } \mathbf{u} \le \mathbf{t}\}$.

    - Prove that the constraint $u_{1,2} \ge 0$ defines a facet of $\mathcal{U}$.
    - Prove that the constraint $u_{1,2} \le 1$ defines a facet of $\mathcal{U}$.
    - Prove that the constraint $\sum_{j=2}^{n} u_{1,j} \le 2$ defines a facet of $\mathcal{U}$.

D Problems

34. Revisit Problem 15. Determine for each of your constraints whether or not it defines a facet. If not all are facets, find a different solution with only facets. *Hint: Enumerate the extreme points of the convex hull of the feasible integer solutions. At how many affinely independent extreme points must facet-defining constraint have to be binding?*

35. Revisit Problem 16.

    (a) Prove that the problem can be solved without introducing additional variables.

    (b) Find a solution without additional variables that has minimum possible number of constraints, and prove minimality. (Hint: Facets).

(c) Can you find a solution with more variables but fewer constraints (excluding lower and upper bounds on the additional variables)?

36. Revisit Problem 27. Let h, n, d, p, s, c index house locations, nationalities, drinks, smokes, pets, and house colors, respectively. Index $h = 1$ denotes the left-most house, $h = 2$ denotes the house adjacent to the left-most, etc. The other indices take values $1, \ldots, 5$ corresponding to the alphabetical ordering. Thus $p = 5$ denotes the zebra.

    Write *three* formulations of the problem and *compare their computational advantages and disadvantages, both with and without presolve.* The first formulation has binary decision variables $x_{h,n,d,p,s,c} = 1$ iff the house in location h has color c, houses pet of type p, and is occupied by a person of nationality n who likes drink d and smokes s. The second formulation has binary decision variables $u_{h,n}, v_{h,d}, w_{h,p}, x_{h,s}$, and $z_{h,c}$ which equal 1 iff the house in location h has person with nationality n, serves drink d, houses pet p, has smoke from cigarette s, and has color c, respectively. The third formulation has integer decision variables $u_h, v_h, w_h, x_h, y_h, z_h$ denoting, respectively, the index number of the nationality, drink, pet, smoke, and color of the house in location h.

37. A chain of approximately 3000 stores in North America wants to split the stores into groups of three. There will be a manager for each group. Managers must visit all stores in their group every day to monitor quality. To keep the manager tasks tolerable, no group may contain a pair of stores more than 50 miles apart. This constraint may make it impossible for every group to consist of exactly three stores. The chain's objectives are two-fold. They want to minimize the total roundtrip travel distance, because doing so reduces vehicle costs and improves manager retention. They also want to minimize the number of managers. The travel distance of a group of three is the sum of the three pairwise distances between the stores; the travel distance of a group of two is twice the distance between the stores. (Assume all pairwise distances are known.) Model the chain's problem as an IP that can be used to find different tradeoffs between the two objectives. Do the best you can to break the symmetries in your model.

38. For $\mathcal{U}$ as defined in Problem 33, prove that the subtour elimination constraint $u_{1,2} + u_{1,3} + u_{2,3} \leq 2$ is facet-defining.

39. Let $\{1, 2, \ldots, n\}$ be the vertices of a complete graph $G = (V, E)$. Denote its edges as (i, j) where $1 \leq i < j \leq n$. For any subset $T \subset E$ of the edges, define $x^T_{i,j} = 1$ if $(i, j) \in T$ and $x^T_{i,j} = 0$ if $(i, j) \notin T$. We say T is a *matching* iff for all $k \in V$, $\sum_{i<k} x_{ik} + \sum_{j>k} x_{kj} \leq 1$. Define the matching polytope $\mathcal{M}$ to be the convex hull of vectors representing matchings, $\mathcal{M} = conv(\{x^T : T \text{ is a matching}\})$.

    (a) Prove that for integer variables $x_{i,j} : 1 \leq i < j \leq n$ the constraints

    $$\begin{aligned} x_{i,j} + x_{j,k} &\leq 1, \\ x_{i,j} + x_{i,k} &\leq 1, \\ x_{i,k} + x_{j,k} &\leq 1, \\ &\text{for all } 1 \leq i < j < k \leq n; \\ \text{and} \qquad &\mathbf{0} \leq \mathbf{x} \leq \mathbf{1} \end{aligned}$$

    suffice to define $\mathcal{M}$.

    (b) Prove that the LP relaxation of the constraints in 39a strictly contains the LP relaxation defined by $\mathbf{0} \leq \mathbf{x} \leq \mathbf{1}$ and $\sum_{i<k} x_{ik} + \sum_{j>k} x_{kj} \leq 1$ for all k.

(c) Derive the constraint $x_{i,j} + x_{i,k} + x_{j,k} \leq 1$ from a positive linear combination of the constraints 39a and the integrality of $\mathbf{x}$.

(d) Generalize your solution to 39c to the set of all edges in a subgraph induced by an odd number of vertices.

(e) Derive the constraint $\sum_{i<k} x_{ik} + \sum_{j>k} x_{kj} \leq 1$ from the constraints 39a by repeatedly using the method in 39c.

(f) Prove for $n \geq 3$ that the constraint $x_{12} + x_{13} \leq 1$ does not induce a facet of $\mathcal{M}$.

(g) Prove that the constraint in 39c defines a facet of $\mathcal{M}$.

(h) Generalize your solution to 39g to the set of all edges in a subgraph induced by an odd number of vertices.

I Problems

40.
- The LP relaxation of the TSP Part II formulation is exponentially large. Why might you hope to solve it in polynomial time?
- Explain why the LP relaxation cannot be solved in polynomial time if the following problem cannot be solved in polynomial time: Given $\mathbf{x}$ satisfying the 2-matching constraints, find a violated subtour constraint if one or more exist.
- Solve the problem just stated in polynomial time. *Hint: Treat the* $\mathbf{x}$ *values as arc capacities in a network.*

41. Search for prices of different quantities of Valrhona cocoa or another item you like. Determine the delivery costs, too. Try to find at least one vendor with a minimum quantity requirement, and at least one with a fixed charge or charges for delivery. Model your options with IP.

42. For the two sets of constraints (15.33),(15.34) and (15.35),(15.36),(15.37), physically construct the polyhedra. Use cardboard or paper. Do not use a 3D printer. Compare them, and verify that one strictly contains the other.

43. Recall that a language $L \in NP$ iff there exists a polynomial-time verification algorithm A_L such that for all $s \in L$ there exists a string c (the certificate) such that A_L accepts the input s, c, and for all $s \notin L$ and all strings c A_L does not accept the input s, c. Clarification: On input s, c A_L must run in time polynomial in the length of s. Recall further that $\hat{L} \in co(NP)$ iff $\hat{L} = L^C$, the complement of some language $L \in NP$.

Define $DP \equiv \{H : H = L \cap M, L \in NP, M \in co(NP)\}$.

(a) For the alphabet $\{0, 1, \&\}$ define the language SAT-UNSAT to consist of all strings of form $s1\&s2$ where $s1$ is a satisfiable instance of SAT encoded in the alphabet $\{0, 1\}$ and $s2$ is an unsatisfiable instance of SAT encoded in the alphabet $\{0, 1\}$. Prove SAT-UNSAT is complete for DP. Be sure to correctly handle binary strings that do not encode valid instances of SAT.

(b) Prove that the following language is in DP. Integer matrix A and vector $\mathbf{b}$, integer vector $\mathbf{c}$ and scalar c_0, such that $\mathbf{c} \cdot \mathbf{x} \leq c_0$ is a facet-inducing inequality of the polyhedron $\{\mathbf{x} : A\mathbf{x} \leq \mathbf{b}\}$.

(c) Prove that the following language is DP-complete. Integer matrix A and vector $\mathbf{b}$, integer vector $\mathbf{c}$ and scalar c_0, such that c_0 is the optimum objective value of

$$
\begin{aligned}
\min \mathbf{c} \cdot \mathbf{x} \quad & \text{subject to} \\
A\mathbf{x} \leq \quad & \mathbf{b} \\
\mathbf{x} \quad & \text{integer}
\end{aligned}
$$

Chapter 16

Elementary Nonlinear Programming Theory

Preview: Nonlinear optimization over continuous variables is much harder than linear programming. This chapter presents some portions of the theory that derive from linear approximations of differentiable functions in the neighborhood of a point. These include the first order Karush-Kuhn-Tucker (KKT) conditions, which are very closely related to LP duality.

Definition 16.1 (NLP) *The general nonlinear programming problem NLP is to minimize* $f(\mathbf{x})$ *subject to* $g_j(\mathbf{x}) \geq 0 : j = 1, \ldots, m$*, where* f *and the* g_j *are arbitrary real-valued functions on* $\Re^n$.

It is pointless to study NLP as just defined because it is excessively general. Unsolved problems such as the Riemann Hypothesis and unsolvable problems such as the halting problem are all special cases of general NLP. We must restrict NLP with conditions on the functions f and g_j to have a chance at solving it. The trick is to find a good tradeoff between solvability and generality. You have already seen one good tradeoff in this book, namely ILP. It permits integrality constraints, which are inherently nonlinear, but otherwise requires linearity of functions. By convention and in practice, however, integer and nonlinear optimization are thought of as different fields that overlap only when there is some nonlinearity in addition to discreteness of variables, e.g., minimizing a polynomial function of integer variables. Therefore, we exclude ILP from this discussion. *Question: Identify the pun in this paragraph and the pun in the next.*[1]

At minimum, we require all functions to be continuous. Classical NLP requires existence, continuity, or differentiability of the first-order derivatives of the functions. At the time of this writing, convex optimization – which requires all functions to be convex but not necessarily differentiable – is the most successful tradeoff.

[1] 3rd word; 2nd word

16.1 Multivariate Calculus Prerequisites

Preview: This section explains the bits of elementary multivariate calculus and linear algebra that underly basic NLP theory. If you are already familiar with gradients, Hessians, and positive definite matrices, skip it.

Basic theory and algorithms for NLP follow naturally from the first few terms of the Taylor series for multiple variables. The Taylor series for a single variable function $f : \Re \mapsto \Re$ at the point x is:

$$f(x+t) = f(x) + f^{'}(x)t + f^{''}(x)\frac{t^2}{2} + f^{'''}(x)\frac{t^3}{6} + \ldots + f^{(k)}(x)\frac{t^k}{k!} + O(t^{k+1}) + \ldots \quad (16.1)$$

The sum of the first two terms is precisely the linearization of f at x. It defines the line tangent to f at x. Geometrically that is why $f(x) + f^{'}(x)t$ is a good approximation of $f(x+t)$ near x, that is, when $|t|$ is small (see Figure 16.1).

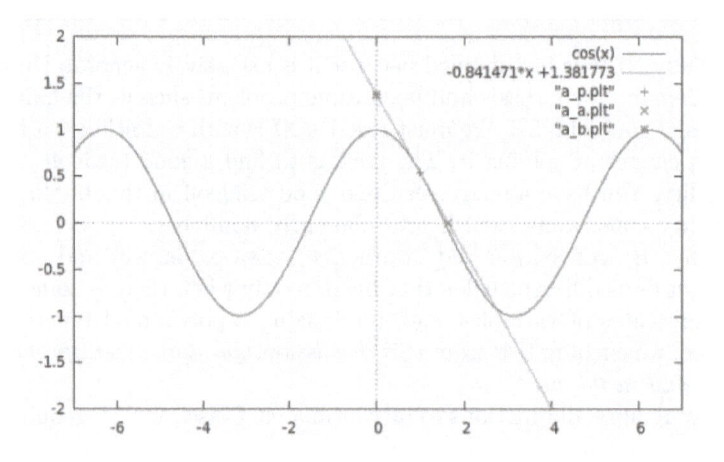

FIGURE 16.1: The line tangent to $f(x) = \cos x$ at $x = 1$ is a good approximation to $\cos x$ in a neighborhood of $x = 1$.

The sum of the first three terms is, in a sense, the best quadratic approximation to f near x. The analogous first three terms for the multivariate function $f : \Re^n \mapsto \Re$ at the point $\mathbf{x} \in \Re^n$ are in the equation below:

$$f(\mathbf{x}+\mathbf{t}) \approx f(\mathbf{x}) + \nabla f(\mathbf{x}) \cdot \mathbf{t} + \frac{1}{2}\mathbf{t}^T\nabla^2 f(\mathbf{x})\mathbf{t} + O((||\mathbf{t}||)^3). \quad (16.2)$$

The gradient $\nabla f(\mathbf{x})$ is the vector of partial derivatives of f at $\mathbf{x}$, $(\frac{\partial f}{\partial \mathbf{x}_1}, \frac{\partial f}{\partial \mathbf{x}_2}, \ldots, \frac{\partial f}{\partial \mathbf{x}_n})$. The sum of the first two terms is precisely the linearization of f at $\mathbf{x}$. It defines the hyperplane tangent to f at $\mathbf{x}$. As in the case $n = 1$, you can visualize the tangent hyperplane as a good approximation of $f()$ near $\mathbf{x}$, that is, when $||\mathbf{t}||$ is small (see Figure 16.2).

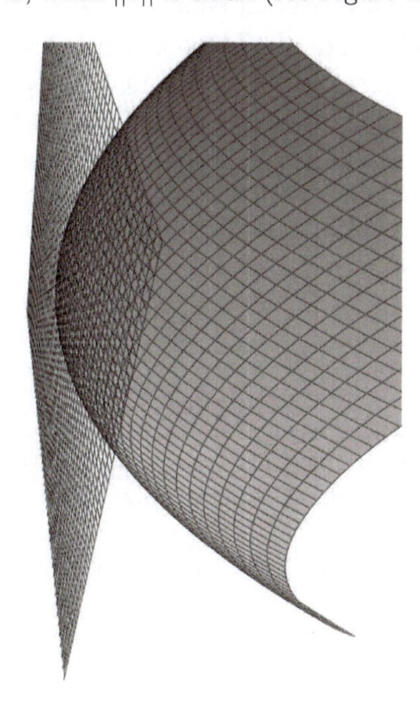

FIGURE 16.2: The hyperplane tangent to f at $\mathbf{x}$ is a good approximation to f in a neighborhood of $\mathbf{x}$.

In a minimization LP with objective function $\mathbf{c} \cdot \mathbf{x}$, moving in the direction $-\mathbf{c}$ decreases the objective value at the quickest rate. This follows from $\cos 0 = 1$ and $\mathbf{c} \cdot \mathbf{y} = ||c|| ||y|| \cos \theta$ where θ is the angle between the vectors $\mathbf{c}$ and $\mathbf{y}$. As illustrated before, in NLP we often approximate the objective function f in the neighborhood of a point $\mathbf{x}$ by the constant $f(\mathbf{x})$ plus the second Taylor series term, a linear function. Therefore, the same logic applies. From $\mathbf{x}$, moving in the direction $-\nabla f(\mathbf{x})$ decreases our approximation of f at the quickest rate. At any local minimum $\mathbf{y}$ of f, $\nabla f(\mathbf{y}) = 0$. Otherwise, for some δ, $\nabla f(\mathbf{y}) \cdot \nabla f(\mathbf{y}) > \delta > 0$ and for all $\epsilon > 0$, $f(\mathbf{y} - \epsilon \nabla f(\mathbf{y})) < f(\mathbf{y}) - \delta\epsilon$, prohibiting $\mathbf{y}$ from being a local minimum. Also, $\nabla f(\mathbf{y}) = 0$ at any local maximum of f as well. It would not be good to maximize f if you or your client wants to minimize f. Hence, to solve an NLP, we usually want additional information about f at values close to $\mathbf{y}$ when $\nabla f(\mathbf{y}) = 0$.

The *Hessian* $\nabla^2 f(\mathbf{x})$ is the matrix of partial derivatives of the gradient. It is the generalization of the second derivative to multivariate functions. The (i, j) element of the Hessian is $\frac{\partial^2 f}{\partial \mathbf{x}_i \partial \mathbf{x}_j}$. In the Taylor series, the Hessian is multiplied on the left by $\mathbf{t}$ as a row vector and on the right by $\mathbf{t}$ as a column vector, yielding a scalar. Let's unpack matrix multiplication to understand this term. For most functions $f(x, y) : \Re^2 \mapsto \Re$ that you will encounter, the duality property $\frac{\partial^2 f}{\partial x \partial y} = \frac{\partial^2 f}{\partial y \partial x}$ holds (see Section 1.2). Therefore, the Hessian is usually assumed to be symmetric. Then by definition of matrix multiplication,

$$\mathbf{t}^T \nabla^2 f(\mathbf{x}) \mathbf{t} = \sum_{i=1}^{n} \left(\frac{\partial^2 f}{\partial^2 x_i^2} \mathbf{t}_i^2 + 2 \sum_{j=i+1}^{n} \frac{\partial^2 f}{\partial x_i x_j} \mathbf{t}_i \mathbf{t}_j \right) \tag{16.3}$$

You should first make sure you understand the qualitative meaning of (16.3). Without loss of generality, examine the term $\frac{\partial^2 f}{\partial x_1 \partial x_2}\mathbf{t}_1\mathbf{t}_2$. As you move by $\mathbf{t}$ from $\mathbf{x}$, x_1 and x_2 become x_1+t_1 and x_2+t_2, respectively. Their first-order impacts on f, based on the gradient ∇f, are $t_1\frac{\partial f}{\partial x_1}$ and $t_2\frac{\partial f}{\partial x_2}$, respectively. (These are simply the first two terms in the dot product $\nabla f(\mathbf{x})\cdot\mathbf{t}$.) Their individual second-order impacts on f are $\frac{1}{2}$ times $t_i^2\frac{\partial^2 f}{\partial^2 x_i}$ for $i=1, i=2$. Thus the diagonal terms of the Hessian are like second derivatives of single-variable functions.

Now let's understand the off-diagonal terms of the Hessian. For separable functions of form $f(\mathbf{x})=\sum_{i=1}^n f_i(x_i)$ in which there are no interactions between pairs of variables, the Hessian's off-diagonal terms are zero. But for most functions, as you move x_1 to x_1+t_1 and x_2 to x_2+t_2, there is a second-order interaction effect on f of $\frac{\partial^2 f}{\partial x_1 \partial x_2}t_1t_2$. *Question: Why isn't this multiplied by $\frac{1}{2}$ like the impacts of the diagonal terms?*[2] These interactions must be included to get the second-order Taylor series approximation of $f(\mathbf{x}+\mathbf{t})$.

Example 16.1 Figure 16.3 represents the value of the function $f(x_1,x_2)=(x_1+2)(x_2)$ by a rectangular area with dimensions x_1+2 and x_2. As $(10,5)$ changes to $(11,7)$, the change in f is due to three nonzero components. They are

$$\frac{\partial f}{\partial x_1}(11-10)=5(11-10)=5 \text{ (1st-order effect)} \tag{16.4}$$

$$\frac{\partial f}{\partial x_2}(7-5)=12(7-5)=24 \text{ (1st-order effect)} \tag{16.5}$$

$$\frac{\partial^2 f}{\partial x_1 \partial x_2}(11-10)(7-5)=1(1)(2) \quad =2 \text{ (2nd-order effect)} \tag{16.6}$$

As you can see in Figure 16.3, the lower right rectangle corresponds to (16.4) and has area 5. The upper left rectangle (16.5) has area 24 and the upper right triangle corresponding to (16.6) has area 2.

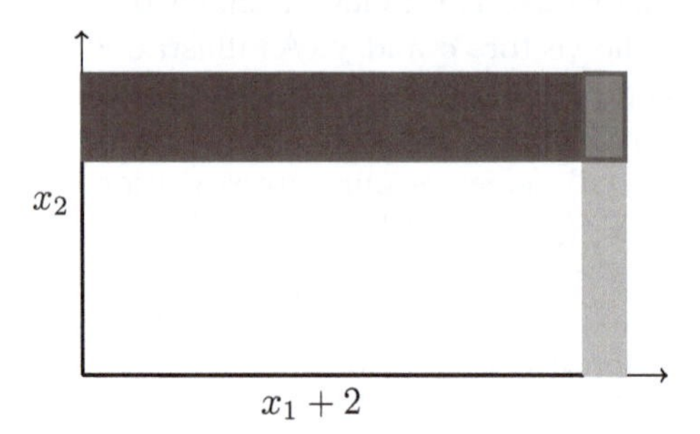

FIGURE 16.3: The small rectangle in the upper right corner is the addition to the value $f(x_1,x_2)=(x_1+2)(x_2)$ from $\frac{\partial^2 f}{\partial x_1 \partial x_2}t_1t_2$ as $\mathbf{x}$ moves from $(10,5)$ to $\mathbf{x}+\mathbf{y}=(10,5)+(1,2)=(11,7)$.

Once you understand Example 16.1 and Figure 16.3, I want you to acquire the geometric intuition that large positive off-diagonal terms in the Hessian may cause non-convexity. Fold a square piece of paper in half lengthwise, unfold it, and fold it in half widthwise.

[2] It actually is $\frac{1}{2}$ times $\frac{\partial^2 f}{\partial x_1 \partial x_2}t_1t_2$ plus the interaction effect $\frac{1}{2}$ times $\frac{\partial^2 f}{\partial x_2 \partial x_1}t_2t_1$. They have equal value which cancels out the $\frac{1}{2}$.

Grasping along the second fold with both hands, turn your hands slightly to bring the ends of the fold closer together, which will activate the first fold. (This is like the first step in some origami.) Each fold will present a "V" shape, consistent with a convex function. But convexity is violated along the diagonals, where the paper slopes upward too steeply. Thinking of the two folds as the x_1 and x_2 axes, and the paper center as the origin, the interaction $\frac{\partial^2 f}{\partial x_1 \partial x_2} x_1 x_2$ adds too much along the diagonals to achieve convexity. It should now be geometrically plausible that a 2×2 matrix $\begin{bmatrix} a & b \\ c & d \end{bmatrix}$ with $ad < bc$ will not yield convexity.

Let's return to the algebra. In single-variable calculus, let x be a point where $f^{'}(x) = 0$ and $f^{''}(x) > 0$ is a strict local minimum. The first three terms of the Taylor series at x are

$$f(x+t) \approx f(x) + f^{'}(x)t + \frac{1}{2}f^{''}(x)t^2 = f(x) + \frac{1}{2}f^{''}(x)t^2 > f(x) \text{ for all } t \neq 0,$$

whether or not $t > 0$. For small enough $|t|$, these terms dominate the rest of the Taylor series. That's what makes x a strict local minimum. We seek a generalization to functions $f : \Re^n \mapsto \Re$ at a point $\mathbf{x}$ where $\nabla f(\mathbf{x}) = \mathbf{0}$. The first three terms of the multi-dimensional Taylor series (16.2) give

$$f(\mathbf{x}+\mathbf{t}) \approx f(\mathbf{x}) + \nabla f(x) \cdot \mathbf{t} + \frac{1}{2}\mathbf{t}^T \nabla^2 \mathbf{t} = f(\mathbf{x}) + \frac{1}{2}\mathbf{t}^T \nabla^2 \mathbf{t}.$$

As in the single variable case, the third term dominates the sum of the higher order terms for sufficiently small $||\mathbf{t}|| > 0$. The generalization of the condition $f^{''}(x) > 0$ must be that $\mathbf{t}^T \nabla^2 f(\mathbf{x}) \mathbf{t} > 0 \ \forall \mathbf{t} \neq \mathbf{0}$. We call such a Hessian "positive definite."

Definition 16.2 *A square matrix M is positive definite (pd) iff $\mathbf{x}^T M \mathbf{x} > 0 \ \forall \mathbf{x} \neq \mathbf{0}$. M is positive semidefinite (psd) iff $\mathbf{x}^T M \mathbf{x} \geq \mathbf{0} \ \forall \mathbf{x}$.*

How can we tell whether or not M is positive definite? It is necessary that M be nonsingular, for otherwise $M\mathbf{x} = \mathbf{0}$ for some $\mathbf{x} \neq \mathbf{0}$. But the 1 by 1 matrix $[-1]$ shows nonsingularity to be insufficient. If M is a diagonal matrix (all off-diagonal terms are zero), all its diagonal terms must be strictly positive. Aside from that very special case, the luckiest situation is if we know M meets the conditions of the following proposition.

Proposition 16.1.1 *If square matrix L is nonsingular and $M = L^T L$, then M is positive definite. If L is singular, then M is positive semidefinite but not positive definite.*

Proof: Let $\mathbf{x} \neq \mathbf{0}$. Then $\mathbf{x}^T M \mathbf{x} = \mathbf{x}^T L^T L \mathbf{x} = (L\mathbf{x})^T L\mathbf{x} = L\mathbf{x} \cdot L\mathbf{x} = ||L\mathbf{x}||^2 \geq 0$. Hence M is psd. If moreover L is nonsingular, $L\mathbf{x} \neq 0$ whence $||L\mathbf{x}|| > 0$ and M is pd. ■

Geometrically, $\mathbf{x}^T M \mathbf{x}$ is the squared length of the vector that $\mathbf{x}$ gets mapped to by the linear transformation $\mathbf{x} \mapsto L\mathbf{x}$. If L is non-singular, then its image is full-dimensional and only $\mathbf{x} = \mathbf{0}$ gets mapped to $\mathbf{0}$.

The matrix $M = L^T L$ is symmetric because it equals its transpose $(L^T L)^T = L^T (L^T)^T = L^T L$. This symmetry suggests a converse to Proposition 16.1.1 which happily is true.

Theorem 16.1.2 *If M is symmetric and positive semi-definite, there exists a lower triangular matrix L with nonnegative diagonal elements such that $M = L^T L$. Moreover, if M is symmetric positive definite, such an L has strictly positive diagonal elements and is unique.*

The second part of Theorem 16.1.2 is so important it calls for a definition.

Definition 16.3 *The matrix product $L^T L$ in Theorem 16.1.2 is a ("the" if M is pd) Cholesky factorization of M.*

Since the Hessian matrix is extremely likely to be symmetric, these results give us a nice characterization of positive definite Hessians. If M is not symmetric, we sometimes can replace it with $\frac{1}{2}(M + M^T)$ to get a symmetric matrix that can be substituted for M.

There is a well-known characterization of symmetric pd and psd matrices in terms of their eigenvalues.

Theorem 16.1.3 *A symmetric matrix M is* positive definite *(positive semi-definite) iff all its eigenvalues are* strictly positive *(nonnegative).*

Question: Why is Theorem 16.1.3 restricted to symmetric matrices?[3] However, this characterization is not useful computationally. Determinants of submatrices of M also indicate whether or not M is pd. We need another definition first.

Definition 16.4 *Let $J \subset \{1, 2, \dots n\}$ be any nonempty ordered subset of the integer indices $1 \dots n$. The principal submatrix of M induced by P is denoted $M_{J,J}$. It is the submatrix of M obtained by extracting the rows of M indexed by J and then extracting the columns indexed by J. If $J = \{1, 2, \dots m\}$ for some $1 \leq m \leq n$, then $M_{J,J}$ is a leading principal submatrix of M.*

The standard test is given by the following theorem.

Theorem 16.1.4 *Let M be a symmetric matrix. Then M is pd iff all its leading principal submatrices have strictly positive determinants.*

Theorem 16.1.5 *Let $f : \Re^n \mapsto \Re$ be a continuously twice-differentiable function on a domain $\mathcal{D}$ that is convex and open. Then f is convex on $\mathcal{D}$ iff its Hessian $\nabla^2 f(\mathbf{x})$ is positive semi-definite for all $\mathbf{x} \in \mathcal{D}$. If moreover $\nabla^2 f(\mathbf{x})$ is positive definite everywhere in the domain, then f is strictly convex.*

The converse to the second part of Theorem 16.1.5 is false.

Here is another geometric interpretation of Definition 16.2. Treat M as representing the linear transformation $\mathbf{x} \mapsto M\mathbf{x}$. If the angle between $\mathbf{x}$ and its image $M\mathbf{x}$ is less than (less than or equal to) $\pi/2$ for all nonzero $\mathbf{x}$, then M is positive (semi-) definite.

Tips:

- People always assume the Hessian is symmetric.
- Some types of matrices such as variance-covariance matrices are mathematically proven to be symmetric and/or psd and/or pd. However, I've encountered data from users that failed to have these properties. Always check your data to be sure it has the properties your methods assume.

[3] Eigenvalues of non-symmetric matrices can be complex numbers.

- The Cholesky factorization is important for computation because solving a system of simultaneous equations for an upper or lower triangular matrix is quicker than for matrices in general. You've seen this in the LP material earlier in the text. It is faster and more stable numerically to compute $\boldsymbol{\pi} = B^{-1}\mathbf{y}$ from B and $\mathbf{y}$ by solving the equations $B\boldsymbol{\pi} = \mathbf{y}$ than by computing B^{-1} explicitly. If $B = L^T L$, first solve $L^T\mathbf{w} = \mathbf{y}$ and then solve $L\boldsymbol{\pi} = w$.

- If the Cholesky factorization of a pd matrix has a diagonal term very close to zero, or equivalently it has an eigenvalue close to zero, you can count on it as psd but may get numerical trouble treating it as pd. This is similar to an ill-conditioned matrix that has a determinant close to zero, but it is not as serious an issue.

- It is immediate from the definitions that M and Q pd (psd) imply M^T, $M + Q$ and βM are pd (psd) for all $\beta > 0$ ($\beta \geq 0$). That is why substituting $M' \equiv \frac{1}{2}(M + M^T)$ for M does no harm. Also, $\mathbf{x}^T M\mathbf{x} = \mathbf{x}^T M'\mathbf{x}$.

16.2 First-Order Karush-Kuhn-Tucker Conditions for NLP

For differentiable functions $f : \Re^n \mapsto \Re$, a zero gradient value $\nabla f(\mathbf{x}) = 0$ is a necessary condition for optimality of $\min f(\mathbf{x})$ if there are no constraints. This the generalization of the standard single variable necessary condition for maxima and minima, $f^{'}(x) = 0$, to n dimensions (Section 16.1).

Suppose differentiable functions g_j define constraints $g_j(\mathbf{x}) \geq 0$. The bad news is that no analogous conditions are guaranteed to be necessary. The good news is that analogous conditions known as the first-order KKT conditions are usually necessary in practice.

The idea of the KKT conditions is simple: *for a point to be optimal, one should not be able to move from it while improving the objective function and remaining feasible.* The conditions follow from this idea in two steps: Step 1 applies differential calculus to linearly approximate the functions; Step 2 applies LP duality to convert a "does not exist" statement into an equivalent "exists" statement.

Step 1: The point $\mathbf{x}^*$ should not be optimal if, for some vector $\mathbf{d}$, travel in direction $\mathbf{d}$ from $\mathbf{x}^*$ decreases $f(\mathbf{x})$ and does not decrease $g_j(\mathbf{x})$ for any constraint $g_j(\mathbf{x}) \geq 0$ that is binding at $\mathbf{x}^*$. We may ignore all g_j not binding at $\mathbf{x}^*$ because for sufficiently small $\epsilon > 0$, $g_j(\mathbf{x}^* + \epsilon\mathbf{d}) \geq 0$. Apply the principle that differentiable calculus approximates functions with affine functions to write this idea as $\nexists\mathbf{d}$ such that $\nabla f(\mathbf{x}^*) \cdot \mathbf{d} < 0$, and such that for all j, $\{g_j(\mathbf{x}^*) > 0 \text{ OR } \nabla g(\mathbf{x}^*) \cdot \mathbf{d} \geq 0\}$. *Question: Verify the translation of the idea into mathematics, and explain why $\mathbf{d} \neq \mathbf{0}$ is not included?*[4].

Fix $\mathbf{x}^*$. Let $\mathcal{B} = \{j : g_j(\mathbf{x}^*) = 0\}$ and let G be the $n \times |\mathcal{B}|$ matrix whose columns are the gradients $\nabla g_j(\mathbf{x}^*) : j \in \mathcal{B}$. Step 1 yields the following statement:

$$\nexists\ \mathbf{d} \quad \text{such that} \tag{16.7}$$

$$\nabla f(\mathbf{x}^*) \cdot \mathbf{d} < 0 \tag{16.8}$$

$$\mathbf{d}^T G \geq \mathbf{0} \tag{16.9}$$

At any $\mathbf{x}^*$ this statement asserts that a linear system of inequalities has no solution. For

[4]For small $\epsilon > 0$, $f(\mathbf{x} + \epsilon\mathbf{d}) \approx f(\mathbf{x}) + \epsilon\mathbf{d} \cdot \nabla f(\mathbf{x})$; $\mathbf{d} = 0 \Rightarrow \nabla f(\mathbf{x}^*) \cdot \mathbf{d} = 0$.

Step 2 apply part 2 of Theorem 8.1.1 to get the logically equivalent statement below:

$$\exists\, \boldsymbol{\lambda} \text{ such that} \tag{16.10}$$
$$G\boldsymbol{\lambda} = \nabla f(\mathbf{x}^*) \tag{16.11}$$
$$\boldsymbol{\lambda} \geq \mathbf{0} \tag{16.12}$$

Append the obviously necessary condition that $\mathbf{x}^*$ be feasible and rearrange terms to get the conditions in standard form.

First-Order Karush-Kuhn-Tucker (KKT) Conditions for NLP

$$\text{There exist } \boldsymbol{\lambda} \geq \mathbf{0}, \text{ and } \mathbf{x}^* \text{ such that} \tag{16.13}$$
$$g_j(\mathbf{x}^*) \geq 0 \text{ for all } j = 1, \ldots, m \tag{16.14}$$
$$\sum_{j=1}^{m} \lambda_j \nabla g_j(\mathbf{x}^*) = \nabla f(\mathbf{x}^*) \tag{16.15}$$
$$\text{For all } j = 1, \ldots, m: \ \lambda_j = 0 \text{ if } g_j(\mathbf{x}^*) > 0 \tag{16.16}$$

Kuhn and Tucker published their conditions in [196]. Years later Kuhn discovered that Karush had derived the same conditions in his 1939 M.S. thesis [175]. Thereafter Kuhn insisted on calling them the Karush-Kuhn-Tucker conditions. The optimization community has respected his wishes.

Question: What should this book's first example of the KKT conditions be like?[5]

Example 16.2 Let's find the minimum of $f(x) = \sin x$ for $\pi/6 \leq x \leq 5\pi/4$. In this example, depicted in Figure 16.4, π is the mathematical constant rather than a variable, and the variable x is a scalar. The constraints are defined by the functions $g_1(x) = x - \pi/6$; $g_2(x) = 5\pi/4 - x$. For consistency with the notation for the KKT conditions, I'll write $\nabla f(x)$ for $f'(x)$ and $\nabla g_j(x)$ for $g_j'(x)$. In this notation, $\nabla f(x) = \cos x$, $\nabla g_1(x) = 1$ and $\nabla g_2(x) = -1$ for all x. Complementary slackness (16.16) requires us to examine several cases.

$g_1(x^*) > 0$ **and** $g_2(x^*) > 0$ This case is the same as minimizing $f(x)$ without constraints. If a minimum is in the interior of the feasible region, the derivative of f must be zero there. By (16.16), $\lambda_1 = \lambda_2 = 0$. By (16.15) $\nabla f(x^*) = \cos x^* = 0$. By (16.14) $x^* = \pi/2$ is the only solution. Its objective value is $f(x^*) = \sin \pi/2 = 1$.

$g_1(x^*) = 0$ **and** $g_2(x^*) > 0$ In this case $g_1(x^*) = 0 \Rightarrow x^* = \pi/6$ and permits $\lambda_1 > 0$. The other dual variable $\lambda_2 = 0$ by (16.16) because $g_2(\pi/6) \neq 0$. We also verify primal feasibility by $g_2(\pi/6) \geq 0$. Next, we determine whether or not (16.15) and (16.13) can be satisfied. First we enforce (16.15). At $x^* = \pi/6$, $\nabla f(x^*) = \cos x^* = \cos \pi/6 = \sqrt{3}/2 = \lambda_1 \nabla g_1(\pi/6) = (\lambda_1)(1) \Rightarrow \lambda_1^* = \sqrt{3}/2$. Second, we verify (16.13) by $\lambda_1 = \sqrt{3}/2 \geq 0$. Hence this solution $x^* = \pi/6, \lambda_1^* = \sqrt{3}/2; \lambda_2^* = 0$ satisfies the KKT conditions with objective value $f(x^*) = \sin \pi/6 = 0.5$. In Figure 16.4 you can see geometrically that x^* is a local minimum even though $\nabla f(x^*) \neq 0$, because the direction that decreases $f(x)$ there – to the left – makes x infeasible.

[5]Simple, but not too simple.

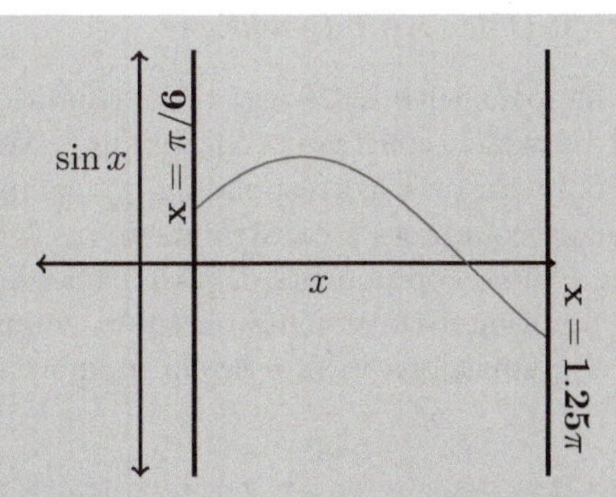

FIGURE 16.4: Minimizing $f(x) = \sin x$ subject to $\pi/6 \le x \le 5\pi/4$ has local optima at both the lower and upper bounds despite the derivative being nonzero there.

$g_1(x^*) > 0$ **and** $g_2(x^*) = 0$ In this case $g_2(x^*) = 0 \Rightarrow x^* = 5\pi/4$ and permits $\lambda_2 > 0$. We observe that $g_1(x^*) > 0$, giving us primal feasibility and forcing $\lambda_1^* = 0$. Enforcing (16.15) yields $\nabla f(x^*) = \cos x^* = \cos 5\pi/4 = -1/\sqrt{2} = \lambda_2^* \nabla g_2(\mathbf{x}^*) = -\lambda_2 \Rightarrow \lambda_2 = 1/\sqrt{2}$. Since $\lambda_2^* \ge 0$ we have dual feasibility. The solution is $x^* = 5\pi/4, \lambda_1^* = 0, \lambda_2^* = 1/\sqrt{2}$. The objective value is $f(x^*) = \sin 5\pi/4 = -1/\sqrt{2}$. As in the previous case, x^* is a local minimum despite $f'(x^*) \ne 0$, because x must increase to decrease $f(x)$ there, but increasing x from x^* renders x infeasible.

$g_1(x^*) = 0 = g_2(x^*)$ This case is primal infeasible.

The true minimum has the best of the values $1, 0.5, -1/\sqrt{2}$, namely, $x^* = 5\pi/4$.

Question: In what ways is the preceding a good first example?[6]

Example 16.3 Let's find the maximum of $\sin x$ subject to the same constraints as in the previous example. For the maximum, minimize $f(x) = -\sin x$. The constraint functions g_1, g_2 are the same as before. The four cases are defined as before, too.

$g_1(x^*) > 0$ **and** $g_2(x^*) > 0$ As before the solution is $x^* = \pi/2; \lambda_1 = \lambda_2 = 0$. The objective function value is $-\sin \pi/2 = -1$.

$g_1(x^*) = 0$ **and** $g_2(x^*) > 0$ As before, $x^* = \pi/6$ and $\lambda_2 = 0$. This time, however (16.15) yields $-\cos x^* = -\sqrt{3}/2 = (\lambda_1)(1) \Rightarrow \lambda_1 < 0$ which violates (16.13). You can see geometrically that $x^* = \pi/6$ cannot be a local maximum of $\sin x$ because increasing x from $\pi/6$ retains feasibility and increases $\sin x$.

$g_1(x^*) > 0$ **and** $g_2(x^*) = 0$ As before, $x^* = 5\pi/4$ and $\lambda_1 = 0$. This time, (16.15) yields $-\cos 5\pi/4 = -\lambda_2$, whence $\lambda_2 = -1/\sqrt{2} < 0$, which violates (16.13). The KKT conditions are not satisfied in this case.

Last, as before, $g_1(x^*) = g_2(x^*) = 0$ is infeasible. The only feasible KKT solution is $(x^*, \lambda_1, \lambda_2) = (\pi/2, 0, 0)$ with objective value -1.

---

[6]Simplicity: It has one variable. You can sketch the problem and see the optimal and sub-optimal solutions. The algebra is easy. Sufficient complexity: The constraints matter. The objective function is not linear. There are multiple local optima.

Tips and Remarks for First-Order KKT Conditions

- This is what typically happens with KKT first-order conditions: You consider a bunch of cases, each defined by which constraints are binding. When k out of the m constraints are binding, only those k corresponding λ_j components may be nonzero. Equation (16.15) therefore comprises n constraints on $n+k$ real variables, namely the n components of $\mathbf{x}^*$ and the k components of λ that may be nonzero. The k binding constraints provide k additional constraints on the n components of $\mathbf{x}^*$. Therefore, you typically solve $n+k$ equations in $n+k$ real unknowns for a case defined by k binding constraints.

Example 16.4

$$\begin{aligned} \text{Minimize } f(\mathbf{x}) = x_1^2 - 3x_2^2 &\text{ subject to} \\ g_1(\mathbf{x}) = x_1 &\geq 0 \\ g_2(\mathbf{x}) = x_2 &\geq 0 \\ g_3(\mathbf{x}) = -x_1 - 4x_2 + 10 &\geq 0 \end{aligned}$$

Equation (16.15) is $\nabla f(\mathbf{x}) = (2x_1, -6x_2) = \lambda_1(1,0) + \lambda_2(0,1) + \lambda_3(1,-4)$. Split this into the two equations:

$$2x_1 = \lambda_1 + \lambda_3; \qquad 6x_2 = -\lambda_2 + 4\lambda_3.$$

Complementary slackness (16.16) seems to requires us to examine 8 cases, defined by the 8 possible subsets of the constraints $g_i(\mathbf{x}) \geq 0 : i = 1,2,3$ that are binding.

Case 1: $g_3(\mathbf{x}) > 0$, which implies $\lambda_3 = 0$ by complementary slackness. The equations for the ∇f reduce to

$$2x_1 = \lambda_1 \tag{16.17}$$
$$6x_2 = -\lambda_2 \tag{16.18}$$

By complementary slackness, $x_1 = 0$ or $\lambda_1 = 0$. Combined with Equation (16.17), this implies $x_1 = \lambda_1 = 0$. Similarly, $\{x_2 = 0 \text{ or } \lambda_2 = 0\}$ and Equation (16.18) imply $x_2 = \lambda_2 = 0$. The only feasible solution consistent with $g_3(\mathbf{x}) > 0$ is $\mathbf{x} = (0,0)$ and $\lambda = (0,0,0)$ with objective value 0. This dispenses with 4 of the 8 possible cases.

Case 2: $g_3(\mathbf{x}) = 0 \Rightarrow x_1 = 4x_2 - 10$. Then $x_1 \geq 0 \Rightarrow x_2 \geq 2.5 > 0 \Rightarrow \lambda_2 = 0$. The equations for the gradient reduce to

$$2x_1 = \lambda_1 + \lambda_3 \tag{16.19}$$
$$6x_2 = 4\lambda_3 \tag{16.20}$$

Case 2a: $x_1 = 0$. Then $\lambda_1 \geq 0$ and $\lambda_3 \geq 0$ and Equation (16.19) imply $\lambda_3 = 0$. Then $x_2 = 0$ by Equation (16.20), which contradicts $x_2 \geq 2.5$. Hence Case 2a cannot occur.

Case 2b: $x_1 > 0 \Rightarrow \lambda_1 = 0 \Rightarrow$ (by Equation (16.20)) $2x_1 = \lambda_3 = \frac{6}{4}x_2$. Combining this with $x_1 = 4x_2 - 10$ gives $\frac{3}{2}x_2 = 8x_2 - 20 \Rightarrow x_2 = \frac{40}{13}$. Plugging in x_2's value gives $\mathbf{x} = (\frac{30}{13}, \frac{40}{13}) \geq \mathbf{0}$ and $\lambda = (0, 0, \frac{15}{13}) \geq \mathbf{0}$. The objective value is $\frac{-300}{13}$.

Of the two feasible solutions to the KKT conditions, the latter has the better objective value. Our solution is $x_1 = \frac{30}{13}$, $x_2 = \frac{40}{13}$ with objective value $\frac{-300}{13}$. The other solution $x_1 = x_2 = 0$ is actually a saddlepoint of $f(\mathbf{x})$ rather than a local minimum. For small $\epsilon > 0$, $\epsilon^2 = f(\epsilon, 0) > f(0,0) > f(0, \epsilon) = -3\epsilon^2$.

- In words, the KKT conditions say that the solution $\mathbf{x}$ is feasible and the gradient of the objective function is in the cone of the gradients of the binding constraints at $\mathbf{x}$. This should sound extremely familiar. It is the same as the LP optimality conditions of primal feasibility, complementary slackness and dual feasibility. It is the same because we have approximated the NLP at $\mathbf{x}$ as an LP. However, in NLP the "dual" you infer from the KKT conditions is different for different points $\mathbf{x}$, unlike in LP where one dual serves for all points. Generally speaking, NLPs duals have no duality gap only when they are convex [276]. See [94] and [253] for convex functions; see [57] and [81] for convex quadratic; see [292] for LP generalized to infinite dimensional function space; see [219] for an equivalence between Fenchel and Lagrange duality.

- The relationships between different primal and dual LP forms work for KKT, too. If the kth constraint is $g_k(\mathbf{x}) = 0$ (respectively ≤ 0) rather than $g_k(\mathbf{x}) \geq 0$, the kth dual variable λ_k is unrestricted (respectively ≤ 0) rather than nonnegative. If a primal variable x_i is constrained to be ≥ 0, the ith equation in (16.15) becomes an inequality ($\geq$) constraint.

- Some prefer to write (16.16) as

$$\lambda_j g_j(\mathbf{x}^*) = 0 \ \forall j = 1, \ldots, m, \tag{16.21}$$

which has the advantage of being purely algebraic. I prefer the form (16.16), which emphasizes its being a complementary slackness condition.

- If the feasible region is nonempty, f and the g_j are differentiable convex functions, and if moreover f is strictly convex and bounded below in the feasible region, the first-order KKT conditions will determine the unique optimal solution to NLP. For most other NLPs, the KKT conditions determine a set of candidate solutions that contains an optimal solution. The candidate solutions are often called "KKT points." Sometimes, however, there exists an optimal solution but no KKT point is optimal. See Example 16.5.

Example 16.5 [2-variable example where the first-order KKT conditions fail]

Minimize $f(\mathbf{x}) = 7x_1 + 10x_2$ subject to $g_1(\mathbf{x}) = 3x_1 - x^2 \geq 0$; $g_2(\mathbf{x}) = -5x_1 - x^2 \geq 0$. The only feasible point with respect to (16.14) is $x^* = (0,0)$. Then $\nabla f(\mathbf{x}^*) = (7,10)$, $\nabla g_1(\mathbf{x}^*) = (3,0)$ and $\nabla g_2(\mathbf{x}^*) = (-5,0)$ rendering (16.15) infeasible.

- By definition, problems with unbounded optima have no global optima. Such problems might lack local optima altogether. If so the first-order KKT conditions will not yield any local optima. Here is a modification of Example 16.4 to illustrate.

Example 16.6

$$\begin{aligned}\text{Minimize } f(\mathbf{x}) = x_1^2 - 16x_2^2 \quad &\text{subject to}\\ g_1(\mathbf{x}) = x_1 &\geq 0\\ g_2(\mathbf{x}) = x_2 &\geq 0\\ g_3(\mathbf{x}) = -x_1 - 4x_2 + 10 &\geq 0\end{aligned}$$

Equation (16.15) yields the two equations:

$$2x_1 = \lambda_1 + \lambda_3; \qquad 32x_2 = -\lambda_2 + 4\lambda_3.$$

Case 1: $g_3(\mathbf{x}) > 0 \Rightarrow \lambda_3 = 0$ is the same as Case 1 of Example 16.4, yielding the saddlepoint $\mathbf{x} = (0,0)$, $\lambda = (0,0,0)$, $f(\mathbf{x}) = 0$.

Case 2: As in Case 2 of Example 16.4, $g_3(\mathbf{x}) = 0 \Rightarrow x_2 = \frac{x_1+10}{4} \geq 2.5 > 0 \Rightarrow \lambda_2 = 0$.

Again, $x_1 = 0$ leads to a contradiction. Therefore, $\lambda_1 = 0$. The equations that place the gradient in the cone of binding constraint gradients reduce to

$$2x_1 = \lambda_3; \qquad 32x_2 = 4\lambda_3.$$

Plugging these into $g_3(\mathbf{x}) = 0$ implies $\frac{1}{8}\lambda_3 = x_2 = \frac{x_1+10}{4} = \frac{\lambda_3+20}{8}$, whence $0 = \frac{20}{8}$, a contradiction. The KKT conditions correctly yield no local minimum. Set $x_1 = 4M - 10$ and $x_2 = M$ to show that the problem has unbounded optimum while keeping $g_3(\mathbf{x}) = 0$.

If instead $f(\mathbf{x}) = x_1^2 - 15x_2^2$, the KKT conditions would yield the local minimum $\mathbf{x} = (0, \frac{5}{2})$ with $\boldsymbol{\lambda} = (15, 0, 15)$.

Example 16.7 [Linear Regression] Linear regression finds a "best" linear fit of a "dependent" real variable y to a vector of *independent* variables $\mathbf{x}$, given a set of observed data. The data consist of m sets of observed values $y^j, \mathbf{x}^j : j = 1, \ldots, m$. To predict y as an affine function of $\mathbf{x}$, $\alpha_0 + \sum_{i=1}^n \alpha_i x_i$, we choose the $n+1$ coefficients $\alpha_0, \ldots, \alpha_n$ to fit the m observed y^j values as closely as possible. Define the *residual* of the jth prediction to be $r^j \equiv y^j - \alpha_0 - \sum_{i=1}^n \alpha_i \mathbf{x}_i^j$. We saw in Chapter 3 how to choose coefficients $\alpha_0, \ldots, \alpha_n$ to minimize either $\max_{1 \leq j \leq m} |r_j|$ or $\sum_{j=1}^m |r_j|$. However, the standard measure of inaccuracy is the sum of the squared residuals, $\sum_{j=1}^m r_j^2$. One of the standard measure's advantages is computational: it only requires the solution of $n+1$ equations in $n+1$ unknowns. We now derive those equations.

To simplify notation, define $\mathbf{x}_0^j \equiv 1 \; \forall j$, let $\mathbf{x}^j$ denote the augmented vector $(\mathbf{x}_0^j, \mathbf{x}_1^j, \ldots, \mathbf{x}_n^j)$. We assume $m > n$, since otherwise we'd be overfitting the data. In practice m is usually larger than n by at least an order of magnitude. We also assume the data matrix $X \equiv [\mathbf{x}^1|\mathbf{x}^2|\ldots|\mathbf{x}^m]$ has full rank $n+1$. This assumption loses no generality, for if X has rank $p < n+1$, the data have redundancy. Remove $n+1-p$ redundant rows from X to get a new data matrix of full rank.

Let $\alpha = \alpha_0, \alpha_1, \ldots, \alpha_n$. Then the jth residual is $r^j = y^j - \alpha \cdot \mathbf{x}^j$. Our nonlinear optimization problem is to select α to minimize $f(\alpha) = \sum_{j=1}^n r_j^2$. The first-order KKT condition is

$$\nabla f(\alpha) = \sum_{j=1}^{m}(2\mathbf{x}_1^j(\alpha \cdot \mathbf{x}^j - y_j), \ldots, 2\mathbf{x}_n^j(\alpha \cdot \mathbf{x}^j - y_j)) = 0. \tag{16.22}$$

Hence the first-order condition (16.22) comprises a set of linear equations in α.
The Hessian, the matrix of second-order partial derivatives, turns out to be concisely expressible as the sum of matrix products of a column vector and its transpose, treated as an $(n+1) \times 1$ matrix times a $1 \times (n+1)$ matrix:

$$\nabla^2 f = 2\sum_{j=1}^{m} \mathbf{x}^j (\mathbf{x}^j)^T.$$

If the Hessian is positive definite, f is strictly convex and the first-order condition (16.22) is necessary and sufficient. Let $\mathbf{v} \in \Re^{n+1}$ be arbitrary. Then

$$\mathbf{v}^T\nabla^2(f)\mathbf{v} = 2\sum_{j=1}^{m} \mathbf{v}^T(\mathbf{x}^j(\mathbf{x}^j)^T)\mathbf{v} = 2\sum_{j=1}^{m}(\mathbf{v}^T(\mathbf{x}^j)(\mathbf{x}^j)^T\mathbf{v}) = 2\sum_{j=1}^{m}(\mathbf{v} \cdot \mathbf{x}^j)^2 \geq 0 \text{ for all } \mathbf{v}. \tag{16.23}$$

This proves that the Hessian is positive semi-definite. To prove it is positive definite we show $\mathbf{v}^T\nabla^2 f\mathbf{v} = 0$ iff $\mathbf{v} = 0$. Since each summand in (16.23) is nonnegative, $\mathbf{v}^T\nabla^2 f\mathbf{v} = 0$ iff $\mathbf{v}$ is orthogonal to $\mathbf{x}^j$ for all j. This is impossible unless $\mathbf{v} = 0$ because X has full rank. Hence the Hessian is positive definite. This completes our derivation.

We've used elementary NLP to derive and prove the validity of a standard statistical method. This is but one example of the large overlap between optimization and statistics.

16.2.1 Global Optimization: A Cursory Look

NLPs for which finding a local optimum is not adequate are called *global optimization* problems. With a few exceptions, global optimization comprises all of nonlinear optimization except convex optimization. Thus "non-convex optimization" and "global optimization" are synonymous.

Global optimization problems do not need constraints to make them very tough to solve. Many algorithms search the domain for local minima and select the best they find. It is a good idea to have a clear definition.

Definition 16.5 *Let f be a function $\Re^n \mapsto \Re$. The point $\mathbf{y} \in \Re^n$ is a local minimum (respectively strict local minimum) if for some $\epsilon > 0$, $||\mathbf{x} - \mathbf{y}|| < \epsilon \Rightarrow f(\mathbf{x}) \geq f(\mathbf{y})$ (respectively $0 < ||\mathbf{x} - \mathbf{y}|| < \epsilon \Rightarrow f(\mathbf{x}) > f(\mathbf{y})$).*

Here is a sufficient condition for local minimality.

Proposition 16.2.1 *Let f be continuously twice differentiable. If $\nabla f(\mathbf{y}) = \mathbf{0}$, and $\nabla^2 f(\mathbf{y})$ is positive definite, then $\mathbf{y}$ is a local minimum of f.*

Proof: This follows from (16.2). As happens so often in the messy world of NLP, the sufficient conditions just given are not necessary. In other words, the converse to Proposition 16.2.1 is false. *Question: There are simple one-dimensional counterexamples to Proposition 16.2.1. Can you think of one?*[7]

Proposition 16.2.2 *Let f be continuously twice differentiable. If $\mathbf{y}$ is a local minimum of f, and the domain of f contains an open set containing $\mathbf{y}$, then $\nabla f(\mathbf{y}) = \mathbf{0}$.*

Not much is known for carving out a useful subset of the global optimization problems. Minimizing a concave (or equivalently, maximizing a convex) function on a convex domain is one of few.

Theorem 16.2.3 *Let $f(\mathbf{x}) : \Re^n \mapsto \Re$ be a concave function. Let $C \subset \Re^n$ be compact and convex. Then there exists an extreme point of C that minimizes $f(\mathbf{x})$ over C.*

Proof: Since $-f$ is a convex function and hence continuous, f is continuous. By Theorem 19.2.3 f attains its minimum on C. Let the minimum be attained at $\mathbf{w} \in C$. If $\mathbf{w}$ is an extreme point of C it is the point we seek. Suppose then $\mathbf{w}$ is not an extreme point of C. By Theorem 4.5.2 C is the convex hull of its extreme points. By Theorem 8.1.8 $\mathbf{w} \in C$ implies that $\mathbf{w}$ is the convex combination of finitely many extreme points of C. Let $\mathbf{w} = \sum_{k=1}^{K} \alpha_k \mathbf{v}^k$ where $\sum_{k=1}^{K} \alpha_k = 1$ and for all $1 \leq k \leq K$, $\alpha_k \geq 0$ and $\mathbf{v}^k$ is an extreme point of C. Then by concavity,

$$f(\mathbf{w}) \geq \sum_{k=1}^{K} \alpha_k f(\mathbf{v}^k) \geq \min_{1 \leq k \leq K} f(\mathbf{v}^k). \tag{16.24}$$

Therefore, there exists an extreme point as good as $\mathbf{w}$, hence optimal. (Moreover, if f were strictly concave, the inequality (16.24) would be strict, which would contradict the supposition that $\mathbf{w}$ is not an extreme point. Then f could attain its minimum over C only at extreme points of C.) ∎

Local search from a large number of starting points has long been a popular class of heuristics for global optimization. Suppose $f(\mathbf{x})$ is Lipschitz continuous, meaning that there exists a constant K such that $||f(\mathbf{x}) - f(\mathbf{y})|| \leq K||\mathbf{x} - \mathbf{y}||$ for all $\mathbf{x}, \mathbf{y}$). Geometrically, f can't be arbitrarily steep. Perform local search from independent randomly chosen starting points in a bounded domain. Then for all $\epsilon > 0$, the probability converges to 1 that a solution whose value is within ϵ of optimal will be found, as the number of searches goes to infinity. That's not a fully satisfying performance guarantee. Moreover, it may be challenging to sample randomly within domains other than balls or hyper-rectangles. For example, sampling uniformly in a polytope can be done in polynomial time by a randomized algorithm, but the algorithm is complicateed, and as of this writing has not been de-randomized.

Finding a global optimum with certainty in finite time is generallly considered to be a strong performance guarantee. Many such global optimization algorithms perform some type of branch and bound. Conceptually, branch and bound in NLP is the same as in IP: repeatedly subdivide the solution space into regions (branching), searching for high quality solutions and eliminating regions that provably do not contain a solution better than the best found so far (bounding). The specifics are quite different. Branching typically entails subdividing a hyper-rectangle rather than fixing the value of a binary variable or splitting the range of a general integer variable. Bounding typically entails finding a relaxation tailored to the form of the objective $f(\mathbf{x})$ rather than simply using the LP relaxation of an IP. Solving small- to medium-sized NLPs routinely has only recently become practical, thanks to improvements in both algorithms and hardware.

[7] Minimize $f(x) = -x^4$.

Differences of convex functions, i.e., sums of a convex function and a concave function, called *d.c.* functions, provide a possible general framework for global optimization. The d.c. functions have generality because of the following fact: *The problem of minimizing a function f on a compact convex set can be approximated to arbitrary accuracy as the minimization of a d.c. function, as long as f is continuous [283].* The d.c. function structure helps with bounding because of the well-known nice properties of convex functions, and because concave functions, too, have some useful properties, including a duality [152] and that given by Theorem 16.2.3.

Summary: Even when functions are restricted to be differentiable, nonlinear optimization is much more difficult than linear optimization. Most of the basic theory has to do with characterizing local optima by the linear first-order local approximations of the objective and the constraints, and categorizing them with second-order information.

16.3 Problems

E Exercises

1. Write the KKT conditions for each of the following:

$$\begin{aligned} \max x \quad & \text{subject to } x^2 \leq 2 \\ \max x_1 + x_2 \quad & \text{subject to } x_1^2 + x_2^2 \leq 1 \\ \max \mathbf{x}^T\mathbf{x} \quad & \text{subject to } \sum_{i=1}^{n} x_i \leq 1; \mathbf{x} \geq 0 \\ \min \mathbf{x}^T\mathbf{x} \quad & \text{subject to } \sum_{i=1}^{n} x_i \geq 1; \mathbf{x} \geq 0 \end{aligned}$$

2. Solve each problem in Exercise 1 by inspection, and verify that your solutions satisfy the KKT conditions.

3. Write the KKT conditions for

$$\begin{aligned} \max \ & x^2yz + 2xy^2z + 3xyz^2 - x^2y^2z^2 \\ & \text{subject to} \\ & x + y + z \leq 1 \\ & x^2 + y^2 \leq 1 \\ & x^2 + z^2 \leq 1 \\ & y^2 + z^2 \leq 1 \end{aligned}$$

4. Suppose you have a fast algorithm to determine whether or not a symmetric matrix is pd. Explain how to use it to determine whether or not an asymmetric (square) matrix is pd.

5. Determine whether the following functions are strictly convex, convex but not strictly convex, or not convex.

   (a) $f(x) = 12 - 5x$

   (b) $f(x) = (12 - 5x)^2$

   (c) $f(x) = |x^3|$

   (d) $f(\mathbf{x}) = \mathbf{x}^T\mathbf{x}$

   (e) $f(x, y) = x + y - xy$

   (f) $f(\mathbf{x}) = \max_{1 \leq k \leq n} k^3 x_k$

   (g) $f(x, y, z) = 10x^4 + 10y^2 + 10z - xyz$

6. Formulate an NLP with $n = 4$ and $m = 5$ that if solved would prove Fermat's last theorem (proved in 1995 by Wiles [294]).

7. Show how to formulate a binary integer linear program as an NLP using only linear and quadratic functions.

8. Show how to formulate a general integer linear program as an NLP using only linear and floor functions ($\lfloor z \rfloor \equiv$ the largest integer $\leq z$.)

9. Verify that if $M' \equiv \frac{1}{2}(M + M^T)$, then $\mathbf{x}^T M x = x^T M' x \; \forall x$.

M Problems

10. Can a matrix containing a non-real (complex) element be pd? Can it be psd?

11. Apply the first-order KKT conditions to the problem

$$\min f(\mathbf{x}) \quad \text{subject to } g_j(x) \geq 0; g_j(x) \leq 0 \; \forall j = 1, \ldots, m.$$

Verify that your conditions are equivalent to Lagrange's criterion $\nabla L(\mathbf{x}, \lambda) = \mathbf{0}$.

12. Show that changing $g_1(\mathbf{x}) \geq 0$ to $g_1(\mathbf{x}) = 0$ changes $\lambda_1 \geq 0$ in (16.13) to λ_1 unrestricted. Compare with the relationship between primal constraints and dual variables in linear programming.

13. Show that changing x_1 unrestricted to $x_1 \geq 0$ changes the first equation of (16.15) to a $\geq$ constraint. (Hint: Add the constraint $g_{m+1}(\mathbf{x}) = x_1 \geq 0$.)

14. Find a function that is twice differentiable and strictly convex on an open set, but whose Hessian is not positive definite everywhere.

15. How many Cholesky factorizations does a matrix of all zeros have? A matrix of all 1's? The matrix $\begin{pmatrix} 1 & 0 \\ 0 & 0 \end{pmatrix}$?

16. Show how to formulate a binary integer linear program with a countably infinite number of variables as an NLP with finitely many variables.

17. Modify Example 16.5 so that there exists a feasible KKT point, but it is not optimal. Hint: Alter the feasible region to consist of two distinct points.

18. Show how to formulate a binary integer linear program as an NLP with all functions analytic (for all n the nth derivative exists).

19. Show how to formulate a general integer linear program as an NLP with all functions analytic.

20. State versions of the main theorems about convex functions for *concave* functions and the corresponding *negative (semi-)definite* matrices. Hints: f is convex iff $-f$ is concave; be careful with determinants – the determinant of $-I$ is 1, not -1, when I is a 2×2 matrix.

D Problems

21. Derive an algorithm akin to Newton's for finding a zero of $f : \Re \mapsto \Re$ that approximates f by a quadratic function. Assuming the third derivative is bounded, prove that your algorithm has superquadratic convergence, if you do not count the cost of finding square roots.

I Problems

22. Formulate an NLP that if solved would determine the truth of the Riemann Hypothesis.

23. Prove that every variance-covariance matrix is psd.

24. Prove that every symmetric psd matrix is a variance-covariance matrix.

25. Show how to formulate the halting problem as an NLP.

26. Devise an algorithm based on Gaussian elimination to determine whether or not a symmetric matrix is pd. Hint: Be careful not to introduce a factor of -1.

Chapter 17

Introduction to Nonlinear Programming Algorithms

Preview: The standard elementary nonlinear unconstrained optimization algorithms search for local optima, by either setting the derivative to zero, or iteratively improving their solution. The former depend on a method such as bisection or Newton's that numerically solves an equation $g(\mathbf{x}) = 0$. The latter, in each iteration, compute a direction $\mathbf{d}$ to move from the current solution $\mathbf{y}$, choose a step size α, and update the solution to $\mathbf{y} + \alpha\mathbf{d}$. For optimization problems subject to linear equality constraints, a promising direction $\mathbf{d}$ that violates feasibility can be altered by projection to retain feasibility. For problems with nonlinear or inequality constraints, the KKT conditions provide a finite number of cases to check for local optima. This chapter desecribes bisection, Newton (or Newton-Raphson), quasi-Newton, and gradient algorithms for single and multi-variable problems.

17.1 Bisection Search

Bisection is a basic search method used in both continuous and discrete settings. In discrete settings it is usually called *binary search.* Suppose we seek a zero of a continuous function $f : \Re \mapsto \Re$, that is, an x such that $f(x) = 0$. Suppose further we have numbers a, b such that $f(a) < 0$ and $f(b) > 0$. Bolzano's theorem [30] states the seemingly obvious fact that $f(x) = 0$ for at least one x between a and b. Bisection defines $c = \frac{a+b}{2}$ and computes $f(c)$. In the unlikely event $f(c) = 0$, terminate. Otherwise, if $f(c) < 0$ set a to c; if $f(c) > 0$ set b to c. In either case the inequalities $f(a) < 0 < f(b)$ are preserved. Bisection halves $|b-a|$ at each iteration. Hence the value c must be within $2^{-n}|b-a|$ of a zero of f during the nth iteration. Hence, bisection guarantees one more place of binary accuracy per iteration.

A rigorous proof is not trivial. From a modern perspective, Bolzano's theorem and its generalization, the Intermediate Value theorem (for every $\tau : f(a) < \tau < f(b)$ there exists x between a and b such that $f(x) = \tau$) follow from properties of connected sets and continuous functions.

The standard binary search scenario is to find a specific key value v in a sorted list of keys $L[1] \ldots L[N]$ from which one can quickly retrieve $L[i]$ for any i. It is assumed that it is easy to compare two key values to determine whether they are equal, and if not, which should precede the other. For example, i would have been a page number when your grandparents searched for a name in a phone book, or for a word in a dictionary. The name or word would have been the key. For another example, L could be an ordered list of transaction record codes, each linked to a file documenting that particular transaction. The index i would have no intrinsic meaning; it would merely denote the place of the code on the list. You would have to know in advance the code of the transaction whose documentation you seek. *Question: Without reading further, guess the algorithm.*

Algorithm 17.1.1 *I. If $v < L[1]$ or $v > L[N]$ terminate: v is not in the list.*
II. Set $low = 1; high = N$.
III. Set $mid = \lfloor \frac{low+high}{2} \rfloor$.
IV. If $L[mid] < v$ set $low = mid$;
Else if $L[mid] > v$ set $high = mid$;
Else terminate with success $L[mid] = v$.
IV. If $low >= high - 1$ terminate with failure, $L[low] < v < L[high]$
Else Go To III.

In some scenarios, Step IV would not represent a failure. For example, L could comprise a list of messages from a motion sensor, sorted by time. You might want to find the first message that was sent on or after a specific time, or the last message sent prior to a specific time. *Question: What would be the result of your search?*[1]

17.2 Newton's Method

Centuries after Newton invented this method for solving $f(x) = 0$, it remains among the core nonlinear programming algorithms. Newton's method follows naturally from this principle: *Differential calculus is all about approximating functions by affine functions.* The derivative $f'(x)$ of $f(x) : \Re \mapsto \Re$ at y yields the affine function $h_y(x) \equiv f(y) + f'(y)(x - y)$ which equals f at y and approximates $f(x)$ in a neighborhood of y. Newton's method iteratively replaces its current estimate y of a solution to $f(x) = 0$ by the solution to $h_y(x) = 0$. Thus, if $f(x)$ were an affine function, the method would solve $f(x) = 0$ in one step. Algorithm 17.2.1 specifies the algorithm. *Question: Suppose y^* solves $f(x) = 0$. Why does Step 4 watch out for increases in $|f(y)|$ rather than in $|y - y^*|$?*[2]

[1] $L[high]$ and $L[low]$, respectively.
[2] We don't know y^*.

Algorithm 17.2.1 Newton's Method to Solve $f(x) = 0$, $f : \Re \mapsto \Re$

1. *Initialize: Set $k = 0$ and $y^{(k)}$ to an initial estimate of the solution. Set $\epsilon_1, \epsilon_2 > 0$ to the desired tolerances.*
2. *Derivative: Compute $d^{(k)} = f^{'}(y^{(k)})$. If $|d^{(k)}| < \epsilon_2$ terminate with failure.*
3. *Update: Set $y^{(k+1)} = y^{(k)} - \frac{f(y^{(k)})}{d^{(k)}}$. Compute $f(y^{(k+1)})$.*
4. *Compare: If $k \geq 2$ and $|f(y^{(k+1)})| \geq \max\{|f(y^{(k)})|, f(y^{(k-1)})|$ terminate (the algorithm is not converging);*
   Else
   If $|f(y^{(k+1)})| \leq \epsilon_1$ terminate with success at solution $y^{(k+1)}$;
   Else
   set $k = k + 1$ and return to Step 2.

Example 17.1 Solve $g(x) = x^3 + x^2 = 10$. This is equivalent to finding a zero of $f(x) = x^3 + x^2 - 10$, as shown in Figure 17.1. Let's start with $x^{(0)} = 1$. $f(x^{(0)}) = -8$; $f^{'}(x^{(0)}) = 3 + 2 = 5$; $x^{(1)} = 1 - \frac{-8}{5} = 2.6$. $f(x^{(1)}) = 14.336$.
$f^{'}(x^{(1)}) = 25.48$. $x^{(2)} = 2.6 - \frac{14.336}{25.48} = 2.037$. $f(x^{(2)}) = 2.608$.
$f^{'}(x^{(2)}) = 16.52$. $x^{(3)} = 2.037 - \frac{2.608}{16.52} = 1.879$. $f(x^{(3)}) = 0.1663$.
$f^{'}(x^{(3)}) = 14.350$ $x^{(4)} = 1.879 - \frac{0.1663}{14.350} = 1.8674$. $f(x^{(4)}) = -0.00074$
The convergence rate is very rapid. In just four iterations the absolute value of the function drops from 8 to less than 10^{-3}, and the absolute gap between x and the exact value drops from approximately 0.867 to less than 10^{-4}. Although $|f(x^{(1)})| > |f(x^{(0)})|$, the algorithm does not diverge. In fact, the second value of x is closer to the zero than the first value. That is, $|x^{(0)} - 1.879| = 0.879 > 0.721 = |x^{(1)} - 1.879|$. On the other hand, Newton's algorithm would fail if it were started at $x = 0$. Bisection would converge from starting values 0 and 100, albeit slowly.

Example 17.2 Solve $f(x) = 0$ for $f(x) = 1/x$. Starting from $x_0 \neq 0$,
$f^{'}(x_0) = -1/x_0^2$;
$x_1 = x_0 - \frac{1/x_0}{-1/x_0^2} = 2x_0$. Hence $x_k = 2^k x_0$. In this example the step size increases each iteration. This occurs because $|f^{'}(x)|$ becomes arbitrarily close to 0, which violates part of the sufficient conditions we will derive for Newton's method to converge.

How to Use Newton's Method

- As we will see, Newton's method converges much faster than bisection. However, unlike bisection, Newton's method can fail to converge. In general, it will converge if its starting point is sufficiently close to a zero of $f(x)$. If you observe divergence, use bisection to get moderately close and then switch to Newton's method. This is an example of a *hybrid* algorithm.
- *Quasi-Newton Methods:* The other significant limitation of Newton's method is its reliance on the derivative of f. Bisection only assumes that f is continuous. This

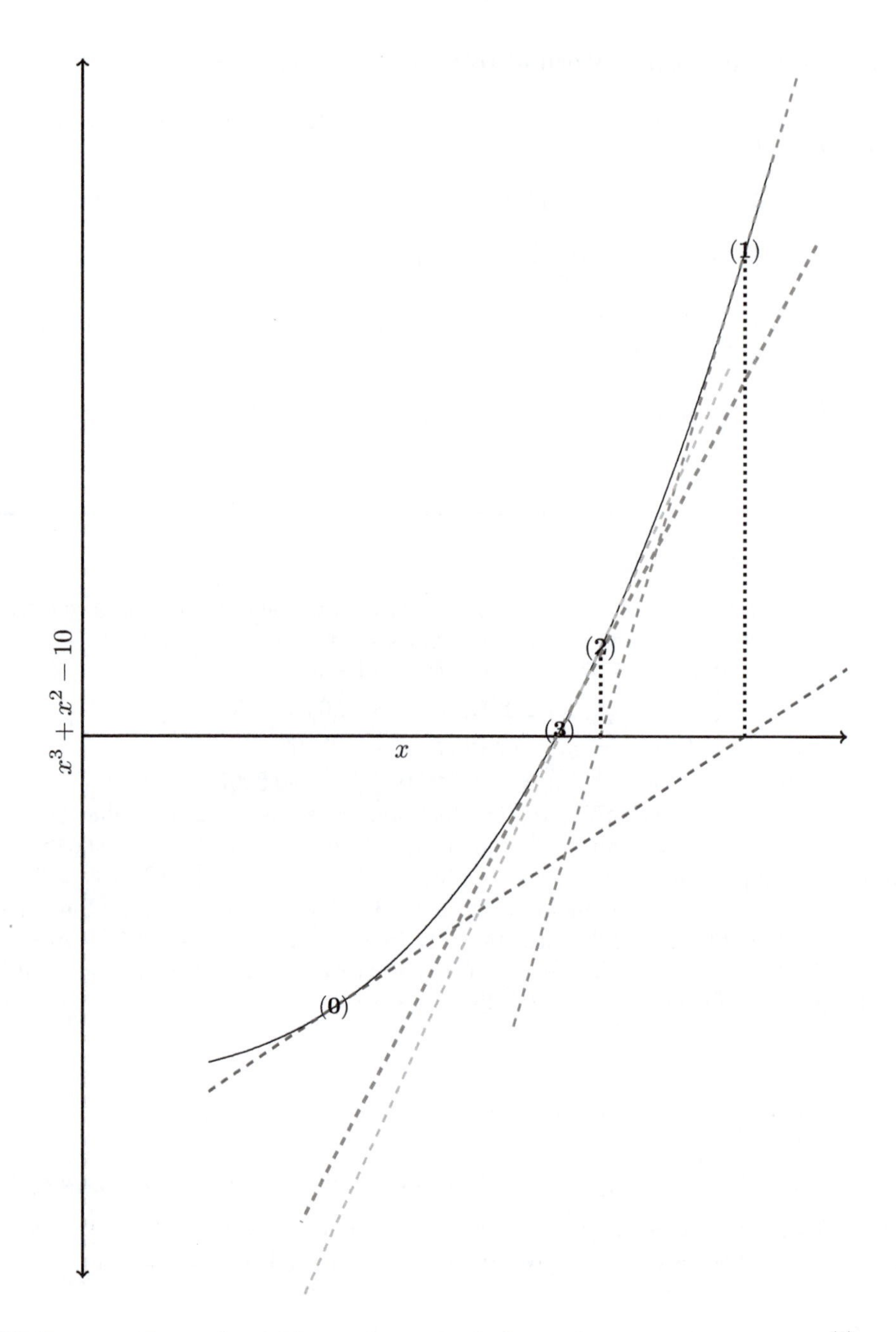

FIGURE 17.1: Example of Newton's method: Start from initial solution $x^{(0)} = 1$ to find a zero (a root) of $f(x) = x^3 + x^2 - 10$. The dashed line tangent to f at $(x^{(0)} = 1, f(x^{(0)}) = -8)$ has slope $f^{'}(1) = 5$. That tangent line intersects the x-axis at the next solution $x^{(1)} = 1 - \frac{-8}{5} = 2.6$ with function value $f(x^{(1)}) = 14.34$. There, the derivative $f^{'}(2.6) = 25.48$ implies that the tangent line is $y = 14.34 + 25.48(x - 2.6)$, which intersects the x-axis at $x^{(2)} = 2.6 - \frac{14.34}{25.48} = 2.04$. In two more iterations $|f(x^{(4)})| < 10^{-3}$.

limitation may be computational as well as theoretical. In some problems, $f(x)$ is differentiable but $f^{'}(x)$ is computationally expensive to evaluate compared with $f(x)$. For these cases, the *secant* method, Algorithm 17.2.2, is a popular compromise. It estimates the derivative by the slope between the two most recent solution points. Its convergence rate is slower than Newton's but faster than bisection. Figure 17.2 shows two iterations of the secant method for the function $f(x) = x^3 + x^2 - 6$ starting from the estimates $y^{(0)} = 0$ and $y^{(1)} = 2$. The iterations yield $f^{'}(y^{(1)}) \approx 6$; $y^{(2)} = 1$, $f(y^{(2)}) = -4$; $f^{'}(y^{(2)}) \approx 10$; $y^{(3)} = 1 - \frac{-4}{10} = 1.4$. Versions of Newton's method that estimate the second derivative are called *quasi-Newton* methods.

Algorithm 17.2.2 *Altering Newton's Method to the Secant Method*

- *Initialize with $k = 1$ and two distinct solution estimates $y^{(0)}$ and $y^{(1)}$.*
- *Compute an estimate of the derivative as $d = (f(y^{(k)}) - f(y^{(k-1)}))/(y^{(k)} - y^{(k-1)})$.*

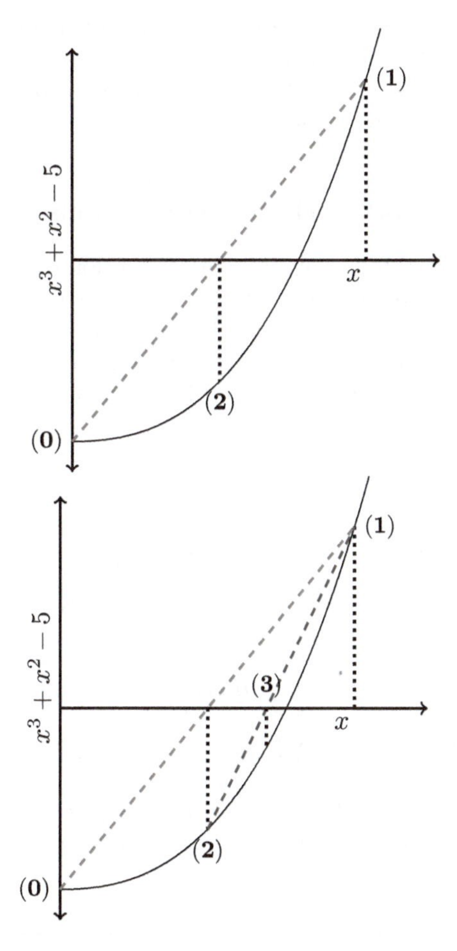

FIGURE 17.2: The quasi-Newton secant method estimates $f^{'}(y^{(1)}) \approx 12$ by the slope of the dashed line through $(y^{(0)} = 0, f(y^{(0)}) = -6)$ and $(y^{(1)} = 2, f(y^{(1)}) = 6)$. That line intersects the x-axis at the next solution estimate $y^{(2)} = 2 - \frac{6}{6} = 1$ with function value -4. The next estimate is $f^{'}(1) \approx \frac{6-(-4)}{2-1} = 10$, which results in $y^{(3)} = 1 + \frac{4}{10} = 1.4$.

- The divergence criterion in Step 4 would be too harsh if it were $|f(y^{(k+1)})| \geq |f(y^{(k)})|$. As Example 17.1 demonstrates, the shape of f may be more favorable on one side of the solution x^* than on the other side. Some implementations run a small fixed number of iterations to start, and thereafter terminate with failure if $|f^{(k)}|$ increases.

- Another way to increase the chance of convergence, at the cost of speed, is to change the update step to the more conservative $y^{(k+1)} = y^{(k)} - \frac{\beta f(y^{(k)})}{d^{(k)}}$ for some $\beta < 1$, usually $\frac{1}{2} \leq \beta < 1$.

- To find a minimum or maximum of $f(x)$, solve $f^{'}(x) = 0$. Newton's method then requires evaluation of $f^{"}(x)$. Hence, Newton's method for finding extrema of functions requires theoretical existence and practicable computation of both the first and second derivatives.

- I explicitly included numerical tolerance tests in the algorithm specification 17.2.1. I did so to emphasize their importance in nonlinear optimization. To exaggerate only slightly, you cannot do any computation in continuous nonlinear optimization without taking numerical issues into account.

17.2.1 Newton's Method in Higher Dimensions

Let's generalize Newton's Method to compute a solution to $f(\mathbf{x}) = 0$ where $f : \Re^n \mapsto \Re$. The affine function that approximates f at $\mathbf{y}$ is $h_{\mathbf{y}}(\mathbf{x}) \equiv f(\mathbf{y}) + \nabla f(\mathbf{y}) \cdot (\mathbf{x} - \mathbf{y})$. Unlike the case $n = 1$, for all $n \geq 2$ there are infinitely many solutions $\mathbf{x}$ to $h_{\mathbf{y}}(\mathbf{x}) = 0$. *Question: Can you describe the set of solutions?*[3] Which solution do we pick? The natural choice is the projection of $\mathbf{y}$ onto the hyperplane of solutions, which is the point on the hyperplane closest to $\mathbf{y}$. The statement of the resulting Algorithm 17.2.3 follows.

Algorithm 17.2.3 Newton's Method to Solve $f(\mathbf{x}) = 0, \quad f : \Re^n \mapsto \Re$

1. *Initialize: Set $k = 0$ and $\mathbf{y}^{(k)}$ to an initial solution. Set $\epsilon_1, \epsilon_2, \epsilon_3 > 0$ to the desired tolerances.*

2. *Gradient: Compute $\mathbf{d} = \nabla f(\mathbf{y}^{(k)})$. If $||\mathbf{d}|| < \epsilon_2$ terminate with failure.*

3. *Update: Set $\mathbf{y}^{(k+1)} = \mathbf{y}^{(k)} - \frac{f(\mathbf{y}^{(k)})}{\mathbf{d} \cdot \mathbf{d}}\mathbf{d}$. Compute $f(\mathbf{y}^{(k+1)})$.)*

4. *Compare: If $|f(\mathbf{y}^{(k)})| < \epsilon_3 |f(\mathbf{y}^{(k+1)})|$ terminate (the algorithm is diverging). Else if $|f(\mathbf{y}^{(k+1)})| \leq \epsilon_1$ terminate with success at solution $\mathbf{y}^{(k+1)}$; Else set $k = k + 1$ and return to Step 2.*

Now let's generalize Newton's method further to compute a solution to $g(\mathbf{x}) = \mathbf{0}$ where $g : \Re^n \mapsto \Re^n$. *Question: Why would we want to solve such a problem?*[4] De-

[3]The hyperplane defined by the linear equation in $\mathbf{x}$, $\nabla f(\mathbf{y}) \cdot \mathbf{x} = \nabla f(\mathbf{y}) \cdot \mathbf{y} - f(\mathbf{y})$.

note the n constituent functions of g as $g_1, \ldots, g_n$. That is, $g(\mathbf{x}) \equiv (g_1(\mathbf{x}), \ldots, g_n(\mathbf{x}))$. Our current solution is $\mathbf{y}$. The affine approximation to the kth constituent function $g_k(\mathbf{x})$ at $\mathbf{y}$ is $h_{k,\mathbf{y}}(\mathbf{x}) \equiv g_k(\mathbf{y}) + \nabla g_k(\mathbf{y}) \cdot (\mathbf{x} - \mathbf{y})$. We seek a single point $\mathbf{x}$ such that $h_{k,\mathbf{y}}(\mathbf{x}) = 0 \ \forall\ k = 1, \ldots, n$. For each k, $h_{k,\mathbf{y}}(\mathbf{x}) = 0$ defines a hyperplane. Hence we seek the intersection of n hyperplanes in $\Re^n$.

Let H be the Jacobian of g, the $n \times n$ matrix whose rows are the gradients at $\mathbf{y}$ of the constituent functions g_k. That is, $H_{k,:} = \nabla g_k(\mathbf{y})$. When we are optimizing a function $f : \Re^n \mapsto \Re$, $g = \nabla f$ and H is the **Hessian** matrix of f at $\mathbf{y}$. Let $\mathbf{b}_k = \nabla g_k(\mathbf{y}) \cdot \mathbf{y} - g_k(\mathbf{y})$. Our next iterate solution estimate $\mathbf{x}$ satisfies $H\mathbf{x} = \mathbf{b}$. Thus each iteration of Newton's method only requires differentiation and solving n simultaneous linear equations in n variables.

Algorithm 17.2.4 One Iteration of Newton's Method to Solve $g(\mathbf{x}) = \mathbf{0}$ for $g : \Re^n \mapsto \Re^n$

1. *Input: Solution* $\mathbf{y}$ *with function value* $g(\mathbf{y}) = (g_1(\mathbf{y}), g_2(\mathbf{y}), \ldots, g_n(\mathbf{y}))$.
2. *Jacobian: Compute matrix* H *defined by* $H[k,:] = \nabla g_k(\mathbf{y})$ *for* $k = 1, 2, \ldots, n$.
3. *Step: Compute* Δ *to solve the equations* $H\Delta = g(\mathbf{y})$.
4. *Output: New solution* $\mathbf{y} - \Delta$

How to Use Newton's Method in Higher Dimensions

- State-of-the-art numerical solution techniques are very complicated, especially for large n, sparse data, or ill-conditioned or badly scaled H. Nowadays MATLAB™ and other packages (such as R) have built-in sophisticated techniques. You can write the higher levels of your algorithm, but let your software handle matrix computations.

- Solving a sequence of linear systems is even more complicated. If H changes only slightly, it may be more efficient to change its factorization instead of recomputing from scratch. Sometimes it is faster to use the same H for several consecutive iterations and update only $\mathbf{y}$ each iteration.

- Newton's method's property of not necessarily converging if you don't start close to a solution is all-the-more-so a limitation for large n. For functions $g : \Re^n \mapsto \Re$, bisection is guaranteed to bring you close to a solution, as long as you have initial solutions $\mathbf{u}, \mathbf{v}$ with $g(\mathbf{u}) < 0$ and $g(\mathbf{v}) > 0$. However, bisection does not apply to functions that map to $\Re^n, n \geq 2$.

- I define the iterative solution Δ by the equation $H\Delta = g(\mathbf{y})$, where H is the Jacobian. Some textbooks define it by $\Delta = H^{-1}g(\mathbf{y})$, which is mathematically correct but computationally misleading. Inverting H is slower and less numerically stable. Don't ask MATLAB or your programming language's linear algebra library to invert H when all you need is to solve equations.

- Quasi-Newton methods require function values at $n + 1$ affinely independent points to estimate the gradient. It is both inaccurate and impractical to use the most recent

[4]To minimize or maximize a real-valued function $f : \Re^n \mapsto \Re$ we set its gradient $g(\mathbf{x}) \equiv \nabla f(\mathbf{x})$ to $\mathbf{0}$.

$n+1$ solutions *Question: Why*[5]*?*. If evaluating $\nabla g(\mathbf{y})$ is considerably more expensive than n evaluations of $g(\mathbf{y})$, use the points $\mathbf{y} + \epsilon e_i$ for $i = 1, \ldots, n$ with $\mathbf{y}$.

17.2.2 Convergence Rate of Newton's Algorithm

In general, if Newton's method converges, its accuracy is squared each iteration. That is, the number of decimal places of accuracy doubles from each estimated solution to the next. This is called a *quadratic* convergence rate. Our next theorem establishes quadratic convergence if $|f^{'}(x)|$ is bounded away from zero and the higher order derivatives are well-behaved.

Theorem 17.2.5 *Suppose f is analytic, $f(x^*) = 0$ and there exist constants $\epsilon > 0, M, \delta > 0$ such that $|f^{'}(x)| > \epsilon$ and $|f^{"}(x)| < M \quad \forall x : |x - x^*| < \delta$. Then there exists $\tau > 0$ such that for all $x : |x - x^*| < \tau$, Algorithm 17.2.1 converges to x^* from x. Moreover, the step sizes $\Delta_{(k)}$ satisfy $\Delta_{(k+1)} = O(\Delta^2_{(k)})$ and the estimates $x_{(k)}$ satisfy $|x_{(k)} - x^*| = O(\Delta_{(k)})$.*

Proof: *Concisely describe Algorithm 17.2.1 as $\Delta_{(k)} = \frac{f(x_{(k)})}{f^{'}(x_{(k)})}; x_{(k+1)} = x_{(k)} - \Delta_{(k)}$. We will show $\frac{\Delta_{(k+1)}}{\Delta_{(k)}} = O(\Delta_{(k)})$. It will follow by induction that $\sum_{j=k+2}^{\infty} |\Delta_{(j)}| = o(\Delta_{(k+1)})$. We may approximate f and its derivatives by Taylor series because f is analytic. From the Taylor series about $x_{(k)}$, with $\Delta_{(k)} = \frac{f(x_{(k)})}{f^{'}(x_{(k)})}$,*

$$\begin{aligned} f(x_{(k+1)}) &= f(x_{(k)} - \Delta_{(k)}) = f(x_{(k)}) - \Delta_{(k)} f^{'}(x_{(k)}) + \frac{\Delta^2_{(k)}}{2} f^{"}(x_{(k)}) \pm O(\Delta^3_{(k)}) \\ &= \frac{\Delta^2_{(k)}}{2} f^{"}(x_{(k)}) \pm O(|\Delta^3_{(k)}|); \\ f^{'}(x_{(k+1)}) &= f^{'}(x_{(k)} - \Delta_{(k)}) = f^{'}(x_{(k)}) - \Delta_{(k)} f^{"}(x_{(k)}) \pm O(\Delta^2_{(k)}). \end{aligned}$$

Since $|f^{"}(x_{(k)})|$ is bounded from above and $|f^{'}(x_{(k)})|$ is bounded from below,

$$\Delta_{(k+1)} = \frac{f(x_{(k+1)})}{f^{'}(x_{(k+1)})} \approx \frac{\Delta^2_{(k)}}{2} \frac{f^{"}(x_{(k)})}{f^{'}(x_{(k)}) - \Delta_{(k)} f^{"}(x_{(k)})} = O(\Delta^2_{(k)}).$$

The functions $|f(x)|$ and $|f^{'}(x)|$ are bounded in the range $|x - x^*| < \delta$ because f is analytic. Also, $|f^{'}(x)| > \epsilon \; \forall |x - x^*| < \delta$. Hence there exists $\tau > 0; \tau \le \delta$ such that $\frac{f(x)}{f^{'}(x)} < 0.5 \; \forall |x - x^*| < \tau$.
■

17.3 Unconstrained and Linearly Constrained Gradient Search

Many optimization algorithms search for a local optimum of a real-valued function $f()$ by repeating the following steps:

- Start at the current solution $\mathbf{x} \in \Re^n$;

[5]Inaccurate because solutions from the distant past are apt to be distant from the current solution; impractical because you need many solutions to start the algorithm, more than the typical number of iterations.

- Terminate if a stopping criterion is met;
- Otherwise, choose a direction $\mathbf{d}$ to move from $\mathbf{x}$;
- Choose a step length α;
- Update $\mathbf{x}$ to $\mathbf{x} + \alpha\mathbf{d}$.

We can iteratively seek to maximize $f(\mathbf{x}) : \Re^n \mapsto \Re$ in this way, guided by the first two terms in the Taylor expansion $f(\mathbf{x}+\mathbf{d}) \approx f(\mathbf{x}) + \nabla f(\mathbf{x}) \cdot \mathbf{d}$. In this first-order approximation, $\nabla f(\mathbf{x})$ points in the direction of greatest rate of increase in $f()$ from $\mathbf{x}$. Let $\mathbf{x}$ be the solution at the beginning of an iteration. Set

$$\mathbf{d} = \nabla f(\mathbf{x}) \tag{17.1}$$

a vector in the direction of maximum rate of decrease. Compute the step length α by solving, either exactly or approximately, the one-dimensional optimization problem

$$\alpha^* = \arg\max_{\alpha \geq 0} f(\mathbf{x} + \alpha\mathbf{d}). \tag{17.2}$$

Move α^* in direction $\mathbf{d}$ to get the next value of $\mathbf{x}$,

$$\mathbf{x}' = \mathbf{x} + \alpha^*\mathbf{d}. \tag{17.3}$$

The typical termination criterion is that $|f(\mathbf{x}') - f(\mathbf{x})|$ be sufficiently small. When minimizing, use the negative gradient $-\nabla f(\mathbf{x})$.

Example 17.3 [Unconstrained gradient search]
Minimize $f(\mathbf{x}) = (x_1 - x_2)^2 + 2(x_1 - x_3 - 10)^2 + 3(x_2 - x_3 + 4)^2$ starting from $(0, 0, 0)$. At each iteration $f(\mathbf{x}) = 248 + x_1^2 + x_2^2 - 2x_1x_2 + 2x_1^2 + 2x_3^2 - 4x_1x_3 - 40x_1 + 40x_3 + 3x_2^2 + 3x_3^2 - 6x_2x_3 + 24x_2 - 24x_3 = 248 + 3x_1^2 + 4x_2^2 + 5x_3^2 - 2x_1x_2 - 4x_1x_3 - 6x_2x_3 - 40x_1 + 24x_2 + 16x_3$. The negative gradient is

$$-\nabla f(\mathbf{x}) = (-6x_1 + 2x_2 + 4x_3 + 40, -8x_2 + 2x_1 + 6x_3 - 24, -10x_3 + 4x_1 + 6x_2 - 16)$$

First iteration: At the initial solution $(0, 0, 0)$ the objective equals 248. $\mathbf{d} = -\nabla f(0, 0, 0) = (40, -24, -16)$. Scaling $\mathbf{d}$ for convenience, $\mathbf{d} = (5, -3, -2)$. $f(\mathbf{x} + \alpha\mathbf{d}) = f(5\alpha, -3\alpha, -2\alpha) = 248 + \alpha^2(75 + 36 + 20 + 30 + 40 - 36) + \alpha(-200 - 72 - 32) = 248 + 165\alpha^2 - 304\alpha$. This is minimized at $2(165)\alpha - 304 = 0 \Rightarrow \alpha^* = \frac{304}{330} \approx 0.921$ The new solution is $0.921(5, -3, -2) = (4.61, -2.76, -1.84)$ with objective value 107.98.
Second iteration: $\mathbf{d} = -\nabla f(4.61, -2.76, -1.84) \approx (-0.53, -3.73, 4.27)$. $f(\mathbf{x} + \alpha\mathbf{d}) = f(4.61 - 0.53\alpha, -2.76 - 3.73\alpha, -1.84 + 4.27\alpha) = 248.32\alpha^2 + 32.43\alpha + c$, where c is a constant. This is minimized at $\alpha^* = \frac{32.43}{496.64} \approx 0.065$. The new solution is $(4.61 - 0.53(0.065), -2.76 - 3.73(0.065), -1.84 + 4.27(0.065)) = (4.57, -3.01, -1.56)$ with objective value 106.91716.
Third iteration: $\mathbf{d} = (0.302, -0.18, -0.121)$; $f(\mathbf{x} + \alpha\mathbf{d}) = 0.604\alpha^2 + 0.139\alpha + c$; $\alpha^* = 0.11515$. The new solution is $(4.60606, -3.02829, -1.57777)$ with objective value 106.90915.
Fourth iteration: $\mathbf{d} = (-0.00403, -0.02822, 0.03226)$; $\alpha^* = 0.06529$. The new solution is $(4.6058, -3.03013, -1.57567)$ with objective value 106.90909.

Example 17.4 In Example 17.3, convergence is quite rapid. If we start from $\mathbf{x} = (-10, 0, 10)$ instead of $(0, 0, 0)$, it takes twice as many iterations to reach the same degree of precision. The sequence of iterations is as shown below.

Iteration	x_1	x_2	x_3	objective
1	−10.0000	0.0000	10.0000	2008.00000
2	1.0733	1.26552	−2.3388	260.31523
3	3.4274	−2.7858	−0.6417	119.28801
4	4.3210	−2.6837	−1.6373	107.90799
5	4.5109	−3.0106	−1.5004	106.98970
6	4.5831	−3.0023	−1.5807	106.91560
7	4.5984	−3.0287	−1.5697	106.90962
8	4.6042	−3.0280	−1.5762	106.90913
9	4.60544	−3.0302	−1.5753	106.09090

17.3.1 Linear Equality Constraints

Now let's adapt gradient search to the linearly constrained optimization problem

$$\max f(\mathbf{x}) : \Re^n \mapsto \Re$$
$$\text{subject to}$$
$$A\mathbf{x} = \mathbf{b}$$

Inductively assume $A\mathbf{x} = \mathbf{b}$. To retain feasibility, the direction of movement $\mathbf{d}$ must satisfy $A\mathbf{d} = \mathbf{0}$. Therefore, instead of $\mathbf{d} = \nabla f(\mathbf{x})$, choose direction $\mathbf{d}$ to be the projection of $\nabla f(\mathbf{x})$ onto the subspace $\mathbf{v} : A\mathbf{v} = \mathbf{0}$.

Deriving the formula for the projection is reminiscent of Theorem 4.3.1, our first theorem of the alternative. Let's project vector $\mathbf{x}$ onto the subspace $\{\mathbf{v} : A\mathbf{v} = \mathbf{0}\}$. A look at the geometry shows what projection does. Figure 17.3 depicts the projections of two points, $(4, 7)$ and $(11, -2)$ onto the subspace $\{(v_1, v_2) : -v_1 + 2v_2 = 0\}$.

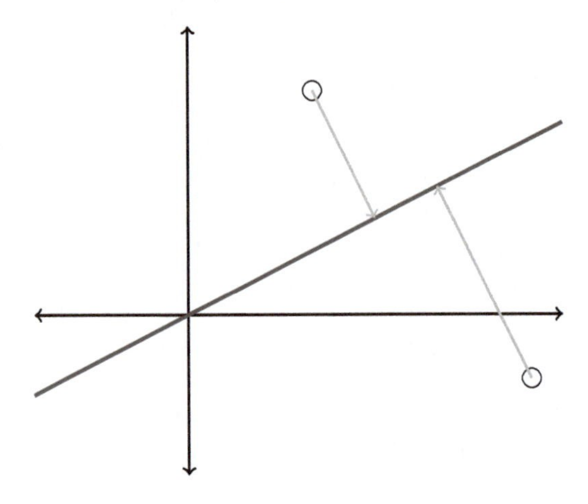

FIGURE 17.3: Move by a multiple of $(-1, 2)$ to project a point onto the subspace $\{(v_1, v_2) : (-1, 2) \cdot (v_1, v_2) = 0\}$.

As you can see in Figure 17.3, projection subtracts a (possibly negative) multiple of the constraint vector $(-1, 2)$ to the point to move it to the subspace. *Question: Why use a multiple of the constraint vector?*[6] In general, if a subspace is defined by more than one constraint, projection subtracts an appropriate multiple of each constraint vector. Therefore, we subtract from $\mathbf{x}$ a linear combination of the rows of A. Let $\boldsymbol{\pi}$ be the vector of multipliers of the linear combination $\boldsymbol{\pi}^T A$. We want

$$A(\mathbf{x} - A^T\boldsymbol{\pi}) = \mathbf{0} \Rightarrow A\mathbf{x} = AA^T\boldsymbol{\pi} \Rightarrow \boldsymbol{\pi} = (AA^T)^{-1}A\mathbf{x}.$$

Hence, the projection of $\mathbf{x}$ onto the subspace $\{\mathbf{v} : A\mathbf{v} = \mathbf{0}\}$ equals

$$\mathbf{x} - A^T\boldsymbol{\pi} = \mathbf{x} - A^T(AA^T)^{-1}A\mathbf{x} = (I - A^T(AA^T)^{-1}A)\mathbf{x}. \tag{17.4}$$

The steps of an iteration of gradient search subject to equality constraints are summarized as Algorithm 17.3.1.

Algorithm 17.3.1 (Gradient search step to maximize $f(\mathbf{x})$ subject to $A\mathbf{x} = \mathbf{b}$)
Input: solution $\mathbf{x}$ *such that* $A\mathbf{x} = \mathbf{b}$.
Step 1: gradient Set $\mathbf{d} = \nabla f(\mathbf{x})$.
Step 2: project Compute $\mathbf{d} = (I - A^T(AA^T)^{-1}A)\mathbf{d}$.
Step 3: stepsize Find $\alpha^* = \arg\max_{\alpha \geq 0} f(\mathbf{x} + \alpha\mathbf{d})$.
Output: $\mathbf{x} + \alpha^*\mathbf{d}$.

Summary: Bisection search finds a zero of a continuous function $f : \Re^n \mapsto \Re$, if given $\mathbf{x}$ and $\mathbf{y}$ such that $f(\mathbf{x}) > 0$ and $f(\mathbf{y}) < 0$. Its linear convergence rate ensures an additional bit of accuracy per iteration. The other elementary algorithms in this chapter approximate a nonlinear function at a point by the affine tangent function. Newton's method finds a zero of a differentiable function. Its quadratic convergence rate doubles the number of bits of accuracy per iteration. However, it may fail to converge from starting points too far from a zero. Gradient search finds a local minimum or maximum of a function $f : \Re^n \mapsto \Re$ by repeatedly moving in the direction $\pm\nabla f(\mathbf{x})$ of maximum rate of change of $f(\mathbf{x})$. Gradient search is adapted to optimize $f(\mathbf{x})$ subject to equality constraints $A\mathbf{x} = \mathbf{b}$ by projecting the gradient $\nabla f(\mathbf{x})$ onto the subspace $\{vv : A\mathbf{v} = \mathbf{0}\}$.

17.4 Notes and References

It would be the subject of an entire book to describe how to implement the algorithms in this chapter to be numerically well-behaved. The standard reference is by Philip Gill, Walter Murray, and Margaret Wright [121]. It may surprise you that it is not simple even

[6] Because the constraint vector is the defining orthogonal vector of the subspace.

to write numerically stable code for the quadratic formula [102, 103]. Moreover, it is subtle and complicated to precisely define the term "numerically stable." Roughly speaking, in the case of computing $\sqrt{x}$, the computed value y should be such that for small $\epsilon > 0$, there exist $\hat{x}, \hat{y}$ such that $|\hat{x} - x|/|x| < \epsilon$, $|\hat{y} - y|/|y| < \epsilon$, and $\hat{y} = \sqrt{\hat{x}}$. That is, there exists a small perturbation of x whose exact square root is a small perturbation of y [147].

17.5 Problems

E Exercises

1. Project $(3, 5)$ onto the following subspaces. Do each projection both geometrically and algebraically.

   (a) $\{(x_1, x_2) \in \Re^2 : x_1 = 0\}$
   (b) $\{(x_1, x_2) \in \Re^2 : x_1 = 8\}$
   (c) $\{(x_1, x_2) \in \Re^2 : x_2 = 5\}$
   (d) $\{(x_1, x_2) \in \Re^2 : x_1 - x_2 = 0\}$
   (e) $\{(x_1, x_2) \in \Re^2 : x_1 - x_2 = 10\}$
   (f) $\{(x_1, x_2) \in \Re^2 : x_1 + x_2 = 0\}$
   (g) $\{(x_1, x_2) \in \Re^2 : x_1 + x_2 = 6\}$
   (h) $\{(x_1, x_2) \in \Re^2 : x_1 + 3x_2 = -6\}$

2. Project $(3, 5, 7)$ onto the following subspaces. After you have computed the solution, try to visualize the projection geometrically.

   (a) $\{vx \in \Re^3 : x_1 = 0\}$
   (b) $\{vx \in \Re^3 : x_1 = 6\}$
   (c) $\{vx \in \Re^3 : x_2 = 0\}$
   (d) $\{vx \in \Re^3 : x_1 = x_2 = 0\}$
   (e) $\{vx \in \Re^3 : x_2 = x_3 = 0\}$
   (f) $\{vx \in \Re^3 : x_1 = x_3\}$
   (g) $\{vx \in \Re^3 : x_1 = x_2 = x_3\}$
   (h) $\{vx \in \Re^3 : x_1 = 10\}$
   (i) $\{vx \in \Re^3 : x_3 = 30\}$
   (j) $\{vx \in \Re^3 : x_1 + x_2 + x_3 = 6\}$
   (k) $\{vx \in \Re^3 : x_1 - 2x_2 + 4x_3 = 12\}$

3. Use bisection to search for the solution to $x - 7 = 0$ from initial lower and upper values 5 and 10. How small is the range after 2, 4, and 8 iterations? How many iterations would it take to reduce the gap to 10^{-5}?

4. Use bisection to find the cube root of 100 to three significant digits of accuracy, starting with lower and upper values 4 and 5.

5. Repeat Problem 3 using Newton's method starting from 10.

6. Repeat Problem 4 using Newton's method starting from 4.

7. Repeat Problem 4 using the secant method starting from 4 and estimating the derivative from the values at 4 and 5.

8. Repeat Problem 4 using the secant method starting from 5 and estimating the derivative from the values at 4 and 5. Why does the first iteration find the same point as if starting from 4, but a different point in the second iteration?

9. Compute two more iterations of the secant method for the example of Figure 17.2.

M Problems

10. Verify the following secant update equation: $x^{(n+1)} - x^{(n)} = (x^{(n)} - x^{(n-1)}) \frac{f(x^{(n)})}{f(x^{(n-1)})-f(x^{(n)})}$.

11. Perform two iterations of gradient search for the function in Example 17.3 starting from the point $\mathbf{x} = (10, 0, -10)$.

12. Perform two iterations of gradient search to minimize the function $2(x_1 - x_2 + 30)^2 + 4(x_1 - x_3 - 20)^2 + 11(x_2 - x_3 + 5)^2$

13. Prove that the projection formula (17.4) yields the point closest to the subspace as measured by the Euclidean norm.

D Problems

14. Suppose $f : [0, 1] \mapsto \Re$ is a nondifferentiable continuous function. A *local minimum* of f is an $x \in [0, 1]$ such that for some $\epsilon > 0$ $|y - x| < \epsilon$ and $y \in [0, 1]$ imply $f(y) \geq f(x)$. Invent a variant of bisection search that finds a local minimum of f with precision δ in $O(\log 1/\delta)$ function evaluations.

15. Prove that the bisection algorithm in Problem 3 cannot reach the exact value 7 (in a finite number of iterations) if computations are performed exactly. Hint: There is a short proof using symmetry.

16. Prove that the bisection algorithm in Problem 3 will reach the exact value 7 (in a finite number of iterations) if computations are performed in fixed precision.

17. What is the geometric meaning of the matrix product AA^T when projecting onto the subspace $\{\mathbf{v} : A\mathbf{v} = \mathbf{0}\}$? Hint: Normalize.

18. Verify that Algorithm 17.2.1 updates correctly in Step 3 in all four cases, $f(y^{(k)}) > 0, d > 0; \ldots; f(y^{(k)}) < 0, d < 0$. Generalize the secant method to find a zero of a function $f : \Re^n \mapsto \Re$. Write simple pseudo-code for the quadratic formula without taking any numerical issues into account. Identify as many places in your code as you can where there could be numerical trouble. Specialize Newton's method to an algorithm that computes the square root of a positive real number. Verify (16.23) by taking the partial derivatives of (16.22).

    (a) Give an intuitive explanation of your algorithm that you could explain to a child who only knows elementary arithmetic and the concept of a square root.

    (b) You seek $\sqrt{1001}$. Your initial estimate is $y^{(0)} = 32 = \sqrt{1024}$. Estimate the number of correct significant digits after 10 iterations.

(c) You seek $\sqrt{1001}$. For what range of values for $y^{(0)}$ will your algorithm converge?

(d) Prove that your answer to 18c is correct. *Hint: Convexity.*

(e) Modify your algorithm to find mth roots for positive integer m. Generalize your answers to 18a and 18c.

19. Think of a function f for which $f''(x)$ is computationally more expensive to evaluate than is $f'(x)$.

20. Speed up the algorithm of Problem 14 by maintaining a ratio other than 1 between the two line segment lengths. Find the optimum ratio and prove its optimality [213].

21. Prove order 1.5 convergence of the secant method. Assume a finite bound on the absolute value of the second derivative.

22. Under the assumption of Problem 21, analyze the performance of the quasi-Newton method that is the same as the secant method except that it estimates $f'(x^{(n)}) \approx \frac{f(x^{(n)})-f(y^{(n-1)})}{x^{(n)}-y^{(n-1)}}$ where $y^{(n-1)} \equiv \frac{1}{2}(x^{(n)}+x^{(n-1)})$. Does the doubled number of function evaluations per iteration yield a convergence rate better than order 1.5?

23. Under the assumption of Problem 21, analyze the performance of the quasi-Newton method that is the same as the secant method except that it estimates $f'(x^{(n)}) \approx \frac{f(x^{(n)})-f(w^{(n-1)})}{x^{(n)}-w^{(n-1)}}$ where $w^{(n-1)} \equiv \frac{n}{n+1}x^{(n)} + \frac{1}{n+1}x^{(n-1)}$. Does the doubled number of function evaluations per iteration yield a convergence rate better than order 1.5?

I Problems

24. Prove that determining whether $\mathbf{x}$ is a local maximum of a function $f : \Re^n \mapsto \Re$ is NP-hard.

25. Inspired by the bisection algorithm, prove that every odd-degree polynomial has at least one real root. (The largest exponent of a polynomial is its degree.)

26. Write a computer program to minimize $\sum_{\{i,j\}:1\leq i<j\leq 3} c_{ij}(x_i+a_{ij}-x_j)^2$ with gradient search. Run it starting from $(0,0,0)$ for $c_{ij} = i+j$; $a_{ij} = (i+1)(j+1)$.

Chapter 18

Affine Scaling and Logarithmic Barrier Interior-Point Methods

---

Preview: Interior point algorithms for linear programming employ nonlinear programming methods to move in the interior of the feasible region. This chapter presents affine scaling, the first such algorithm to challenge the dominance of the simplex method, and the logarithmic barrier method, which is widely used and is usually faster than simplex on large sparse LPs.

Why do we solve LPs by going all the way around the polyhedron instead of going through the middle? In the 1940s, John von Neumann and others proposed algorithms that move through the (relative) interior of the feasible polyhedron, but these proved to be computationally ineffective, at least as implemented on the hardware available at that time. The study of interior point algorithms was dormant until the 1980s, when Narendra Karmarkar [172]invented an algorithm with a lower order polynomial-time bound than the ellipsoid algorithm's. Using affine scaling, which from his perspective was a variant of his algorithm, he announced he solved LPs 50 to 100 times faster than the simplex method [173]. His claims shocked the math programming community, especially after computational tests by many researchers did exhibit a factor of 50 to 100 speed difference, but in favor of simplex. Karmarkar persisted in his claims. Many people were skeptical. The atmosphere was contentious. I remember asking George Dantzig, at the 18th joint ORSA/TIMS conference in Dallas, Texas in November 1984, what would come of the controversy. He said that there would be a competition. Each side would keep improving their algorithm, and after a while they would begin to use ideas from each other. "And when it's all settled, simplex will be faster on some problems, and Karmarkar's will be faster on some." Then he added, "I think Karmarkar's will be faster on staircase problems. And if it is, I sure as heck am going to use it." I was surprised and abashed – I was feeling more defensive about the simplex algorithm than its inventor. It was a great reminder to maintain a problem-solving attitude.

Dantzig's prediction proved to be essentially correct. Simplex implementations run much faster now than they did then, exclusive of hardware speedups [26]. And one of their important improvements is the presolve step (Section 2.3), which Karmarkar had used to great effect in his first implementations. Interior point algorithms became a major research area in optimization, and they are better at solving large staircase LPs than are simplex algorithms.

18.1 Affine Scaling

Affine scaling was invented by Robert Vanderbei, Marc Meketon, and Barry Freedman [286], Earl Barnes [15], and several others as a simple variation of Karmarkar's theoretically fast algorithm. The two just cited also proved convergence under different conditions. Affine scaling was the method of choice when researchers initially tried to replicate Karmarkar's reported computational results. Later, it was discovered that affine scaling had been invented years before by Dikin [78]. Some refer to the algorithm as "Dikin's". Here I use the more descriptive term.

We solve the problem

$$\max \mathbf{c} \cdot \mathbf{x}$$

subject to

$$A\mathbf{x} = \mathbf{b}$$
$$\mathbf{x} \geq \mathbf{0}$$

Affine scaling always generates points in the relative interior defined by $A\mathbf{x} = \mathbf{b}; \mathbf{x} > \mathbf{0}$. Let $\mathbf{y} > \mathbf{0}$ be the current solution, so $A\mathbf{y} = \mathbf{b}$. We'd like to move in the direction $\mathbf{c}$, but we must deal with two difficulties.

The first difficulty is that $\mathbf{c}$ may push us into one or more of the $\mathbf{x} \geq \mathbf{0}$ constraints instead of towards an optimum solution. Figure 18.1 illustrates this for the objective

$$\max 30x_1 + 40x_2$$

subject to

$$2x_1 + x_2 + x_3 = 6$$
$$x_1, x_2, x_3 \geq \mathbf{0}$$

I've drawn the feasible region, which lies positioned in $\Re^3$, projected to the two dimensions x_1, x_2. At $x_1 = 1.4, x_2 = 2.8$, variable $x_3 = 0.4$ functions as a slack variable for a constraint $2x_1 + x_2 \leq 6$. If we move in direction $(30, 40)$ we can only take a small step without hitting the $x_3 \geq 0$ constraint. Worse, that direction has little in common with the direction towards the optimum point $(0, 6)$. Intuitively, we ought to scale movements in different directions differently, so that a move to the constraint $x_3 \geq 0$, which has Euclidean length $\sqrt{.16^2 + .08^2} \approx 0.18$ counts as being the same order of magnitude as a move to the constraints $x_1 \geq 0$ and $x_2 \geq 0$, which have lengths 1.4 and 2.8, respectively. The simplest scaling, which is not based on the Euclidean metric, turns out to work: map (x_1, x_2, x_3) to $\frac{x_1}{1.4}, \frac{x_2}{2.8}, \frac{x_3}{0.4}$. In general, when the current solution is $\mathbf{y}$, map each point $\mathbf{x}$ to $(\frac{x_1}{y_1}, \frac{x_2}{y_2}, \ldots, \frac{x_n}{y_n})$. One reason this mapping works is that $\mathbf{x} \geq \mathbf{0}$ is the entire set of inequality constraints. Another reason is that the algorithm always maintains $\mathbf{x} > \mathbf{0}$.

Thus the current solution $\mathbf{y}$ is mapped to $\mathbf{1}$. Define the mapping as $p(\mathbf{x}) = Y^{-1}\mathbf{x}$, where Y is the $n \times n$ diagonal matrix with entries $\mathbf{y}$. Then

$$\mathbf{x} = Yp(\mathbf{x}),$$

and so

$$A\mathbf{x} = AYp(\mathbf{x}) = \mathbf{b}.$$

The objective value is

$$\mathbf{c}^T\mathbf{x} = \mathbf{c}^T Yp(\mathbf{x}) = (Y\mathbf{c})^T p(\mathbf{x}).$$

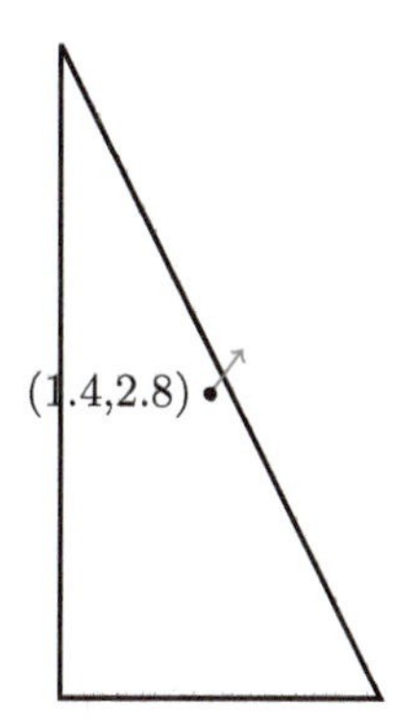

FIGURE 18.1: Movement in the gradient direction $(30, 40)$ from $(1.4, 2.8)$ is severely limited by the constraint $x_3 = 6 - 2x_1 - x_2 \geq 0$, and does little to move towards the optimum point $(0, 6)$.

Since $\mathbf{y} > \mathbf{0}$,

$$p(\mathbf{x}) > \mathbf{0} \Leftrightarrow Yp(\mathbf{x}) > \mathbf{0} \Leftrightarrow \mathbf{x} > \mathbf{0}.$$

In the scaled space, therefore, the problem is

$$\max (Y\mathbf{c})^T p(\mathbf{x}) \tag{18.1}$$

$$\text{subject to} \tag{18.2}$$

$$(AY)p(\mathbf{x}) = \mathbf{b} \tag{18.3}$$

$$p(\mathbf{x}) \geq \mathbf{0} \tag{18.4}$$

starting at $p(\mathbf{x}) = \mathbf{1}$.

Figure 18.2 continues the example of Figure 18.1 to illustrate the mapping. The 3D coordinates of the vertices of the feasible region $\{2x_1 + x_2 + x_3 = 6; x_i \geq 0\}$, whose projection onto the x_1, x_2 plane is shown in Figure 18.1, are $(3, 0, 0)$, $(0, 6, 0)$, and $(0, 0, 6)$. Figure 18.2 (left) shows the feasible region as it would look were we looking "down" at it from "above" its plane. To be precise, $(2, 1, 1)$ is orthogonal to the feasible region *Question: How do you verify it is orthogonal?*[1] Figure 18.2 (left) shows the feasible region as seen from $(1, 2, 2) + \theta(2, 1, 1)$ for large θ. Figure 18.2 (right) shows the region it is mapped to.

According to gradient search, we should move in the direction of the gradient $\mathbf{c}^T Y$. Doing so runs into the second difficulty: we must maintain the equality constraints (18.3). We overcame this difficulty in Section 17.3.1 by projecting the objective onto the subspace of vectors that do not affect the LHS of the constraints. The projection formula, Equation (17.4), was with respect to direction $\mathbf{c}$ and subspace $A\mathbf{x} = \mathbf{0}$. It produced a direction to move from the point $\mathbf{x}$.

$$\mathbf{x} + A^T \pi = \mathbf{x} - A^T (AA^T)^{-1} A\mathbf{x} = (I - A^T (AA^T)^{-1} A)\mathbf{x} \tag{17.4}$$

In the scaled space, the projection is with respect to direction Yc and constraints $(AY)p(\mathbf{x}) = \mathbf{0}$. It produces a direction to move from the point $p(\mathbf{x}) = \mathbf{1}$. Therefore, replace A by AY, $\mathbf{c}$ by $Y\mathbf{c}$, and $\mathbf{x}$ by $p(\mathbf{x})$ in Equation (17.4) to get direction

$$\mathbf{d} = (I - (AY)^T (AY(AY)^T)^{-1} AY) Y\mathbf{c} = Y(I - A^T (AY^2 A^T)^{-1} AY^2)\mathbf{c}$$

Let α be a scalar that controls the step size. Then the new solution in the scaled space is

$$\mathbf{1} + \alpha \mathbf{d}$$

[1] $(2, 1, 1) \cdot ((0, 6, 0) - (0, 0, 6)) = 0$ and $(2, 1, 1) \cdot ((0, 6, 0) - (3, 0, 0)) = 0$.

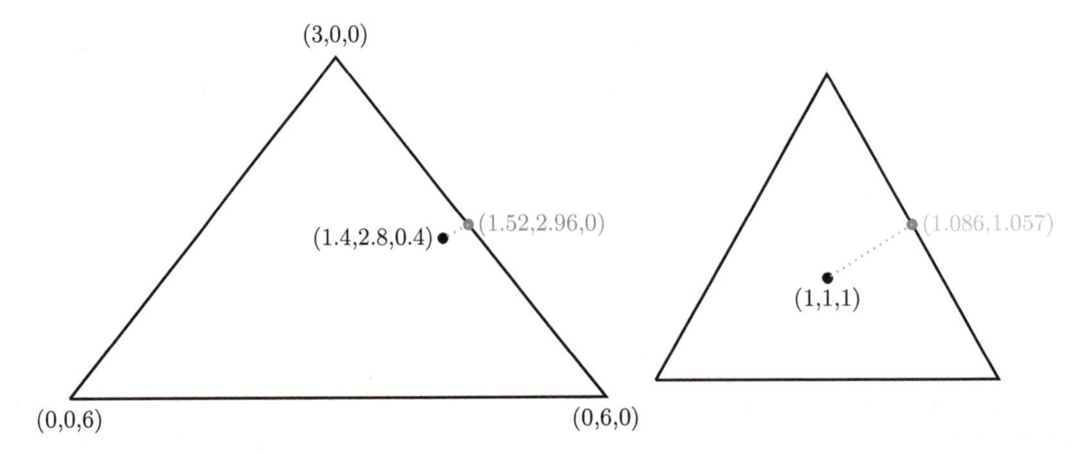

FIGURE 18.2: The current solution $(1.4, 2.8, 0.4)$ maps from the unscaled feasible region (left) to the triangle center (right). The solution $(1.52, 2.96, 0)$ on the perimeter, which the gradient points to from the current solution, maps to the point on the perimeter (right). The two mapped points are much farther apart than the original points. Dotted lines depict the line segments between these pairs of points.

which, unscaled back to the original space, equals

$$Y(\mathbf{1} + \alpha\mathbf{d}) = \mathbf{y} + \alpha Y^2(I - A^T(AY^2A^T)^{-1}AY^2)\mathbf{c}. \tag{18.5}$$

In the example above, using $\alpha = 0.01$, the vector $\mathbf{d}$ is $10(-4.9, 9.8, -.44)$. The new solution is $(0.91, 3.82, 0.36)$.

If a feasible $\mathbf{x} > \mathbf{0}$ is not available, one can set up a Phase 1 problem first. Another approach is to set up variables and constraints for both primal and dual feasibility, and minimize the difference between primal and dual objective values.

18.2 Logarithmic Barrier and the Central Path

If we try to solve LP as a constrained NLP, by applying the KKT conditions, we get exponentially many cases, each corresponding to a different set of binding constraints. As Dantzig [64] points out, this is computationally impractical. Instead, in this section we solve LP as an *unconstrained* NLP. We work with the following form of LP.

$$\min \mathbf{c} \cdot \mathbf{x} \quad \text{subject to} \tag{18.6}$$

$$A\mathbf{x} \geq \mathbf{b} \tag{18.7}$$

$$A \in \Re^{m \times n};\ \mathbf{c} \in \Re^n;\ \mathbf{b} \in \Re^m;\ m \geq n$$

We assume the LP has the following properties:

P1 Full Rank A has full rank.

P2 Full Dimension The polyhedron $\mathcal{P} = \mathbf{x} : A\mathbf{x} \geq \mathbf{b}$ has nonempty interior $int(\mathcal{P})$. Equivalently $\mathcal{P}$ contains a ball $B(\hat{\mathbf{x}}, \epsilon)$ for some $\hat{\mathbf{x}}$ and some $\epsilon > 0$.

P3 Bounded Optima The set of optimal solutions to the LP (18.6),(18.7) is nonempty and bounded.

As usual, artificial variables and/or perturbations enforce these assumptions entail no loss of generality. Define $\mathbf{y}$ to be the vector of slack variables $A\mathbf{x}-\mathbf{b}$. We employ $\mathbf{y}$ to make notation concise, not as variables *per se* in the LP. Replace the jth constraint of (18.7) with the weighted penalty term $-\alpha\log(A_{j:}\mathbf{x}-\mathbf{b}_j) = -\alpha\log y_j$ in the objective function. The logarithmic function $\log y_j$ adds a penalty that is beneficial or innocuous for $\mathbf{x}$ far from the hyperplane $y_j = 0$, but increases without bound as y_j approaches 0. That is why it is called a *barrier function.* Instead of solving (18.6),(18.7) directly, we solve

$$\min f_\alpha(\mathbf{x}) \quad = \quad \mathbf{c}\cdot\mathbf{x} - \alpha\sum_{j=1}^{m}\log(A_{j:}\mathbf{x}-\mathbf{b}_j) = \mathbf{c}\cdot\mathbf{x} - \alpha\sum_{j=1}^{m}\log y_j. \tag{18.8}$$

The justification for the barrier function is that for small enough α, optimal solutions to (18.8) are nearly optimal for the LP (18.6),(18.7).

Theorem 18.2.1 *Suppose properties P2 and P3 hold. Let $\mathbf{x}^*$ be an optimal solution to* (18.6),(18.7)*. For each $\alpha > 0$ let $\mathbf{x}^*_\alpha$ be an optimal solution to Equation* (18.8)*. Then*

$$\lim_{\alpha\to 0}\mathbf{c}\cdot\mathbf{x}^*_\alpha = \mathbf{c}\cdot\mathbf{x}^*.$$

Proof: *Let $\mathbf{y}^*, \mathbf{y}, \hat{\mathbf{y}}, \mathbf{y}^*_\alpha$ be the vector of slack values corresponding to $\mathbf{x}^*, \mathbf{x}, \hat{\mathbf{x}}, \mathbf{x}^*_\alpha$, respectively. Suppose $A\mathbf{y} > \mathbf{0}$, $A\hat{\mathbf{y}} > \mathbf{0}$ and $\mathbf{c}\cdot\mathbf{x} < \mathbf{c}\cdot\hat{\mathbf{x}}$. If $\sum_{j=1}^{m}\log\mathbf{y}_j \geq \sum_{j=1}^{m}\log\hat{\mathbf{y}}_j$, then $f_\alpha(\mathbf{x}) < f_\alpha(\hat{\mathbf{x}})\forall\alpha > 0$. If not, set*

$$\epsilon = \frac{c\cdot\hat{\mathbf{x}} - c\cdot\mathbf{x}}{\sum_{j=1}^{m}\log\hat{\mathbf{y}}_j - \sum_{j=1}^{m}\log\mathbf{y}_j}.$$

Then $\forall\alpha, 0 < \alpha < \epsilon$ we have $f_\alpha(\mathbf{x}) < f_\alpha(\hat{\mathbf{x}})$. In either case, for all sufficiently small positive α, $\mathbf{x}$ has better objective value than $\hat{\mathbf{x}}$ for (18.8).

By property P3, the infimum of $\mathbf{c}\cdot\mathbf{x}$ over $int(\mathcal{P})$, the interior of $\mathcal{P}$, is bounded. Therefore,

$$\lim_{\alpha\to 0}\mathbf{c}\cdot\mathbf{x}^*_\alpha = \inf_{\mathbf{x}\in int(\mathcal{P})}\mathbf{c}\cdot\mathbf{x}.$$

Next, we claim $\inf_{\mathbf{x}\in int(\mathcal{P})}\mathbf{c}\cdot\mathbf{x} = \mathbf{c}\cdot\mathbf{x}^*$. By property P2 there exists $\mathbf{x}\in int(\mathcal{P})$. For all $0\leq\lambda < 1$, the point $\lambda\mathbf{x}^* + (1-\lambda)\mathbf{x}\in\mathcal{P}$. Then $\lim_{\lambda\to 1}\mathbf{c}\cdot(\lambda\mathbf{x}^* + (1-\lambda)\mathbf{x}) = \mathbf{c}\cdot\mathbf{x}^*$, proving the claim.

We now have $\lim_{\alpha\to 0}\mathbf{c}\cdot\mathbf{x}^*_\alpha = \mathbf{c}\cdot\mathbf{x}^*$ as desired. ■

We can say more. Suppose $A\mathbf{x}\geq\mathbf{b}$ is a minimal characterization of $\mathcal{P}$ (see Chapter 8). Then the limit as $\alpha\to 0$ of $\mathbf{x}^*_\alpha$ equals the centroid of the face of $\mathcal{P}$ consisting of optimal solutions to the LP. This is because the centroid of a polyhedron is defined as the point that minimizes the product of the distances to the facets, which is the same as minimizing the sum of the logarithms of the slack variables corresponding to the facets.

Example 18.1 The feasible region of the LP in Figure 18.1 is $\{x_1, x_2, x_3 : 2x_1 + x_2 + x_3 = 6; x_i \geq 0\ \forall i\}$. Figure 18.3 depicts isoquants of the penalty function $\sum_{i=1}^{3}\log x_i$ as dashed contour lines.

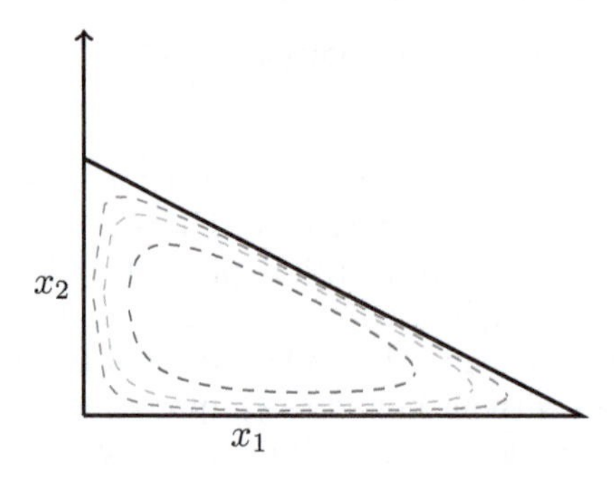

FIGURE 18.3: Isoquants of the barrier penalty function $\sum_{i=1}^{3} \log x_i$ keep the product of the distances to the three sides equal. This figure depicts the projection onto (x_1, x_2) of the triangle defined by constraints $2x_1 + x_2 + x_3 = 6$; $x_i \geq 0 : i = 1, 2, 3$.

Now let's see how to solve (18.8). For any fixed $\alpha > 0$

$$\nabla f_\alpha(\mathbf{x}) = \mathbf{c} - \alpha \sum_{j=1}^{m} \left\{ \frac{A_{j1}}{A_{j,:}\mathbf{x} - \mathbf{b}_j}, \ldots, \frac{A_{ji}}{A_{j,:}\mathbf{x} - \mathbf{b}_j}, \ldots, \frac{A_{jn}}{A_{j,:}\mathbf{x} - \mathbf{b}_j} \right\}. \tag{18.9}$$

Write this more compactly as

$$\nabla f_\alpha(\mathbf{x}) = \mathbf{c} - \alpha \sum_{j=1}^{m} \frac{1}{\mathbf{y}_j} A_{j,:} \tag{18.10}$$

being careful to remember that $\mathbf{y}_j$ is not a variable in its own right.

Let Y denote the diagonal matrix with diagonal terms $\mathbf{y}_1, \ldots, \mathbf{y}_m$. In matrix notation, the gradient is

$$\nabla f_\alpha(\mathbf{x}) = c - \alpha A^T Y^{-1} \mathbf{1}. \tag{18.11}$$

From Equation (18.10), the term in row i, column k of the Hessian is $\alpha \sum_{j=1}^{m} \frac{A_{ji}A_{jk}}{\mathbf{y}_j^2}$. The Hessian matrix is then conveniently described as

$$\begin{aligned}
\nabla^2 f_\alpha(\mathbf{x}) = &\ \alpha \textstyle\sum_{j=1}^{m} \frac{1}{\mathbf{y}_j^2} \left[\begin{pmatrix} A_{j1} \\ A_{j2} \\ \vdots \\ A_{jn} \end{pmatrix} (A_{j1}, A_{j2}, \ldots, A_{jn}) \right] && (18.12) \\
= &\ \alpha \textstyle\sum_{j=1}^{m} \frac{1}{\mathbf{y}_j^2} A_{j,:}^T A_{j,:} && (18.13) \\
= &\ \alpha \textstyle\sum_{j=1}^{m} \left[((Y^{-1}A)_{j,:})^T (Y^{-1}A)_{j,:} \right] && (18.14) \\
= &\ \alpha (Y^{-1}A)^T (Y^{-1}A) = \alpha A^T Y^{-1} Y^{-1} A. && (18.15)
\end{aligned}$$

By Proposition 16.1.1, the Hessian is symmetric and psd. Moreover, since A is full rank, the Hessian is positive definite. By Theorem 16.1.5 $f_\alpha(\mathbf{x})$ is strictly convex. Hence by Corollary 1.3.2 it has a unique minimum. We summarize these properties as a theorem.

Theorem 18.2.2 (Central Path Theorem) *Suppose Properties P1, P2, and P3 hold. Then for every $\alpha > 0$, $f_\alpha(\mathbf{x})$ is strictly convex and there is a unique solution $\mathbf{x}_\alpha^*$ to* (18.8). *These solutions compose the central path of the LP* (18.6),(18.7), *which by Theorem 18.2.1 converges to the centroid of the face of optimal solutions as $\alpha \to 0$.*

Corollary 18.2.3 *For every $\alpha > 0$ there exists a vector $\boldsymbol{\pi}(\alpha)$ of valid Lagrange multipliers for (18.8) such that $\boldsymbol{\pi}\alpha$ is a feasible solution to the LP dual of (18.6),(18.7), and $\boldsymbol{\pi}\alpha$ converges to an optimal dual solution as $\alpha \to 0$.*

Proof: *The Lagrangian is*

$$L(\mathbf{x}, \boldsymbol{\pi}) = \mathbf{c}^T \mathbf{x} - \alpha \sum_{i=1}^{n} \log x_i - \boldsymbol{\pi}^T (A\mathbf{x} - \mathbf{b}).$$

Its gradient is

$$\nabla L(\mathbf{x}, \boldsymbol{\pi}) = \mathbf{c} + \alpha(1/x_1, \ldots, 1/x_n) - A^T \boldsymbol{\pi}, A\mathbf{x} - \mathbf{b}).$$

Let X be the diagonal matrix with values $X_{i,i} = x_i$ as usual. When the gradient is zero,

$$\nabla L(\mathbf{x}, \boldsymbol{\pi}) = \mathbf{0} \Rightarrow A\mathbf{x} = \mathbf{b} \text{ and } A^T \boldsymbol{\pi} + \alpha X \mathbf{1} = \mathbf{c}.$$

Then $\boldsymbol{\pi}$ satisfies dual feasibility $A^T \boldsymbol{\pi} \leq \mathbf{c}$ because $\alpha \geq 0$ and $X\mathbf{1} \geq \mathbf{0}$.
To prove optimality in the limit as $\alpha \to 0$, let $\mathbf{x}^*$ be the optimal primal solution that the central path converges to, as ensured by Theorem 18.2.1. In order that it converges, for each i such that $\mathbf{x}_i^* > 0$, $\mathbf{x}(\alpha)_i$ must be bounded away from 0 eventually as $\alpha \to 0$. Hence there exist $\epsilon > 0$ and $\delta > 0$ such that $x_i^* > 0 \Rightarrow \mathbf{x}(\alpha)_i > \epsilon$ for all $\alpha \leq \delta$. ■

▶ *Corollary 18.2.3 does not state that every sequence of Lagrange multipliers converges as $\alpha \to 0$. Its statement is weaker in that it only asserts the existence of a convergent sequence; its statement is stronger in that it asserts that the convergence is to an optimal dual solution. A slightly stronger statement than the Corollary is also true. Every sequence of valid Lagrange multipliers contains a subsequence that converges to an optimal dual solution, as $\alpha \to 0$. This is because, by hypothesis, the dual feasible region is bounded and so every sequence must contain a convergent subsequence (by the Bolzano-Weierstrass Theorem, for which see any standard text in analysis).* ◀

The term "central path" suggests that the mapping from α to the unique minimum of (18.8) is continuous, but that property requires proof.

Theorem 18.2.4 *The function $g(\alpha) = \arg\min f_\alpha(vx) \equiv \arg\min \mathbf{c} \cdot \mathbf{x} - \alpha \sum_{j=1}^{m} \log(A_{j:}\mathbf{x} - \mathbf{b}_j)$ is continuous.*

Proof: *Let $\epsilon > 0$. Reserve the right to require ϵ to be sufficiently small. The structure of the proof is, first, to find $\mu > 0$ such that $||\mathbf{x} - g(\alpha)|| = \epsilon \Rightarrow f_\alpha(\mathbf{x}) \geq f_\alpha(g(\alpha)) + \mu$. In words, points at distance ϵ from $g(\alpha)$ have worse (larger) f_α value by at least μ. This will be true because $f_\alpha(\mathbf{x})$ is strictly convex and we can bound the effect of its Hessian away from zero (Lemma 18.2.5). Second, find a sufficiently small $\delta_0 > 0$ such that $|f_{\alpha+\delta}(\mathbf{x}) - f_\alpha(\mathbf{x})| < \mu/2$ for all $0 < \delta < \delta_0$ and for all $\mathbf{x}$ in the closed ball $\bar{B}(g(\alpha), \epsilon)$. This will be true because for each fixed $\mathbf{x}$, f viewed as a function of α is linear, and we can bound its slope away from zero (Lemma 18.2.6). Together, these inequalities will imply $|f_{\alpha+\delta}(\mathbf{x}) > f_{\alpha+\delta}(g(\alpha))$ for all $\mathbf{x}$ at distance ϵ from $g(\alpha)$. Third, the convexity of $f_{\alpha+\delta}(\mathbf{x})$ implies $|f_{\alpha+\delta}(\mathbf{x}) > f_{\alpha+\delta}(g(\alpha))$ for all $\mathbf{x}$ at distance greater than ϵ from $g(\alpha)$. By definition of $g()$, $f_{\alpha+\delta}(g(\alpha + \delta)) \leq f_{\alpha+\delta}(g(\alpha))$. Therefore, $g(\alpha + \delta)$ cannot be at distance ϵ or more from $g(\alpha)$. That's the definition of continuity of $g()$ at α. The precise argument follows.*

Lemma 18.2.5 *Given $\alpha > 0$ and given a sufficiently small value $\epsilon > 0$, there exists $\mu > 0$ such that $f_\alpha(\mathbf{x}) > f_\alpha(g(\alpha)) + \alpha\mu\epsilon^2$ for all $||\mathbf{x} - g(\alpha)|| = \epsilon$.*

Proof of Lemma 18.2.5: *The gradient $\nabla f_\alpha = 0$ at its minimum, which is defined to be $g(\alpha)$. Then the Taylor series expansion of f_α about $g(\alpha)$ is $f_\alpha(g(\alpha) + \mathbf{d}) = f_\alpha(g(\alpha)) + 0 + \frac{1}{2}\mathbf{d}^T\nabla^2 f_\alpha \mathbf{d} + O(\alpha\epsilon^3)$. Let ϵ be sufficiently small that the $O(\alpha\epsilon^3)$ portion is negligible compared with the $\theta(\alpha\epsilon^2)$ second derivative term. Let $H = Y^{-1}A$. As Y is nonsingular, H has full rank. By Equation (18.14), $\nabla^2 f_\alpha = \alpha H^T H$. Employing the non-negligible portion of the Taylor series gives is*

$$f_\alpha(g(\alpha) + \mathbf{d}) - f_\alpha(g(\alpha)) \approx \frac{\alpha}{2}\mathbf{d}^T H^T H\mathbf{d} = \frac{\alpha}{2}H\mathbf{d} \cdot H\mathbf{d} = \frac{\alpha}{2}||H\mathbf{d}||^2.$$

Let $\mathbf{d}^*$ minimize $||H\mathbf{d}||$ subject to $||\mathbf{d}|| = 1$. Since H has full rank, $||H\mathbf{d}^*|| > 0$. By scaling, $\epsilon\mathbf{d}^*$ minimizes $||H\mathbf{d}|$ subject to $||\mathbf{d}|| = \epsilon$. Set $\mu = \frac{1}{4}||H\mathbf{d}^*||$. Then for all $||\mathbf{x} - g(\alpha)|| = ||\mathbf{d}|| = \epsilon$,

$$f_\alpha(\mathbf{x}) - f_\alpha(g(\alpha)) = \frac{1}{2}\mathbf{d}^T\nabla^2 f_\alpha \mathbf{d} + O(\alpha\epsilon^3) > \frac{\alpha}{4}||H\mathbf{d}|| \geq \frac{\alpha\epsilon^2}{4}||H\mathbf{d}^*|| = \mu\alpha\epsilon^2$$

as desired for the lemma. ■

Lemma 18.2.6 *Let $\eta > 0, \alpha > 0$, and $\epsilon > 0$, and let the closed ball $\bar{B}(g(\alpha), \epsilon)$ of radius ϵ about $g(\alpha)$ be in the strict interior of polyhedron $\mathcal{P}$. Then there exists $\delta_0 > 0$ such that for all $0 < \delta < \delta_0$, $|f_\alpha(\mathbf{x}) - f_{\alpha+\delta}(\mathbf{x})| < \eta/2$ for all $\mathbf{x} \in \bar{B}(g(\alpha), \epsilon)$.*

Proof of Lemma 18.2.6: *For any fixed $\mathbf{x} \in \bar{B}(g(\alpha), \epsilon)$, $f_\alpha(\mathbf{x})$ is linear in α with slope $-\sum_{j=1}^m \log y_j$. The slope is a continuous function of $\mathbf{x}$ in the compact set $\bar{B}(g(\alpha), \epsilon)$ and hence is uniformly continuous there and attains its maximum absolute value, say, $\beta > 0$. Set $\delta_0 = \eta/2\beta$. Then for all $\mathbf{x} \in \bar{B}(g(\alpha), \epsilon)$ and $0 < \delta < \delta_0$,*

$$|f_{\alpha+\delta}(\mathbf{x}) - f_\alpha(\mathbf{x})| \leq \beta\delta < \beta\delta_0 = \eta/2$$

as desired for the lemma. ■

Now let $\mathbf{x}$ satisfy $||\mathbf{x} - g(\alpha)|| = \epsilon$ and let $0 < \delta < \delta_0$. Let μ be as guaranteed by Lemma 18.2.5. Apply Lemma 18.2.6 with $\eta = \alpha\mu\epsilon^2$ to both $f_\alpha(\mathbf{x})$ and $f_\alpha(g(\alpha))$ to get

$$|f_{\alpha+\delta}(\mathbf{x}) - f_\alpha(\mathbf{x})| < \alpha\mu\epsilon^2/2; \tag{18.16}$$

$$|f_{\alpha+\delta}(g(\alpha)) - f_\alpha(g(\alpha))| < \alpha\mu\epsilon^2/2. \tag{18.17}$$

By these inequalities and Lemma 18.2.5,

$$f_{\alpha+\delta}(\mathbf{x}) > f_\alpha(\mathbf{x}) - \alpha\mu\epsilon^2/2 > f_\alpha(g(\alpha)) + \alpha\mu\epsilon^2/2 > f_{\alpha+\delta}(g(\alpha)) \geq f_{\alpha+\delta}(g(\alpha+\delta)) \tag{18.18}$$

where the last inequality comes from the definition of $g()$.

This immediately tells us that $\mathbf{x} \neq g(\alpha+\delta)$. For any $\mathbf{w}$ farther than ϵ from $g(\alpha)$, employ the convexity of $f_{\alpha+\delta}(\mathbf{w})$ assured by Theorem 18.2.2. Suppose $||\mathbf{w} - g(\alpha)|| > \epsilon$. Then there exist $\lambda > 0$ and $\mathbf{x}$ with $||\mathbf{x} - g(\alpha)|| = \epsilon$ such that $\mathbf{x} = \lambda\mathbf{w} + (1 - \lambda)g(\alpha)$. By inequality (18.18) and convexity,

$$f_{\alpha+\delta}(g(\alpha)) < f_{\alpha+\delta}(\mathbf{x}) \leq \lambda f_{\alpha+\delta}(\mathbf{w}) + (1 - \lambda)f_{\alpha+\delta}(g(\alpha)).$$

As $\lambda > 0$, this implies $f_{\alpha+\delta}(\mathbf{w}) > f_{\alpha+\delta}(g\alpha) \geq f_{\alpha+\delta}(g(\alpha + \delta))$, whence $\mathbf{w} \neq g(\alpha + \delta)$. Therefore, $||g(\alpha + \delta) - g(\alpha)|| < \epsilon$. ■

If we knew how to solve the log-barrier formulation of LP (18.8) for arbitrary $\alpha > 0$ in a single step we would not need the central path. We would solve once for a tiny positive α. However, Newton's method requires a starting point close to the solution. The idea of a central-path-following LP algorithm is to start at a value of $\alpha^{(1)}$ for which an approximate value $\mathbf{x}^{(1)} \approx \mathbf{x}^*_{\alpha^{(1)}}$ is known, and generate a sequence of points close to the central path corresponding to a sequence $\alpha^1, \alpha^2, \ldots$ converging to 0. Algorithm 17.2.4 starts from $\mathbf{x}^{(k)}$ to generate the next point $\mathbf{x}^{(k+1)}$. Applying that algorithm with the Hessian (18.15) and gradient (18.11) gives the update step, which is the heart of the following log-barrier central path following algorithm. This is a phase 2 algorithm which assumes a starting feasible solution is known. Its parameter $\rho < 1$ controls the rate at which α decreases.

Primal Log-Barrier Central Path Algorithm

1. Input: feasible interior solution $\mathbf{x}$ for the LP (18.6),(18.7) and corresponding penalty value α with respect to the objective (18.8).
2. Store $\tilde{\mathbf{x}} = \mathbf{x}$.
3. Centering: Set $\mathbf{y} = A\mathbf{x} - \mathbf{b}$; set diagonal matrix entries $Y_{i,i} = y_i$.
   Calculate Δ to satisfy $\alpha A^T Y^{-1} Y^{-1} A \Delta = \mathbf{c} - \alpha A^T Y^{-1} \mathbf{1}$.
   Compute new solution $\mathbf{x}' = \mathbf{x} - \theta\Delta$.
4. If $||\mathbf{x}' - \mathbf{x}|| > \epsilon$, set $\mathbf{x} = \mathbf{x}'$ and go to 3.
5. Termination check: If $||\mathbf{x} - \tilde{\mathbf{x}} < \epsilon$ terminate with solution $\mathbf{x}$.
6. Path Following: Decrease $\alpha = \rho\alpha$.
7. Go to 2.

- The parameter θ modifies the Newton step size. The smaller is θ, the more conservative is the step.
- For this introductory, purely primal version I employed the primal termination criteria to be that $||\mathbf{x}' - \mathbf{x}||$ and $||\tilde{\mathbf{x}} - \mathbf{x}||$ be sufficiently small. Primal-dual versions monitor three measures: primal feasibility, dual feasibility, and complementary slackness.
- Ordinarily, barrier algorithm termination is followed by "crossover", a simplex-type crashing procedure to produce an optimal basic feasible solution. Crossover is absolutely necessary if the LP is being solved as the relaxation of an IP.
- The algorithm depends heavily on a good factorization of $A^T(Y^{-1})^2 A$ to calculate Δ efficiently. The Y matrix has no impact on density. However, a single dense column in A^T renders the entire matrix $(A^T A)$ dense. *Question: Why?*[2]
- Degeneracy does not slow this or other interior-point algorithms. However, dual degeneracy at the optimum increases the number of iterations required by crossover.
- In practice, it takes $O(\log n)$ iterations on average for the algorithm to converge.

[2]Every term in the matrix $\mathbf{1}\mathbf{1}^T$ equals 1. For every i, j such that $a_i \neq 0, a_j \neq 0$, $(\mathbf{a}\mathbf{a}^T)_{i,j} \neq 0$.

I've introduced the logarithmic barrier, central path, Lagrange multiplier dual path, and Newton step in a simple Phase 2 primal algorithm as a prelude to a primal-dual one-phase version. The latter version is more complicated but is computationally superior, and also more appealing mathematically because of its use of duality. To begin its exposition, I want you to see how it greatly differs from the simplex method.

Conceptual Overview of Primal-Dual Single-Phase Interior Point Algorithm

Add slack variables to a standard minimization LP to get the primal and dual problems below.

$$\min \mathbf{c}^T\mathbf{x} \qquad \max \mathbf{b}^T\boldsymbol{\pi} \tag{18.19}$$

subject to

$$A\mathbf{x} - \bar{\mathbf{x}} = \mathbf{b}; \qquad A^T\boldsymbol{\pi} + \bar{\boldsymbol{\pi}} = \mathbf{c}; \tag{18.20}$$

$$\mathbf{x}, \bar{\mathbf{x}} \geq \mathbf{0}. \qquad \boldsymbol{\pi}, \bar{\boldsymbol{\pi}} \geq \mathbf{0}. \tag{18.21}$$

Simplex algorithms need a starting solution that satisfies the equations (18.20), and enforce those equations throughout their execution. They also begin with, and maintain throughout, the complementary slackness conditions:

$$\mathbf{x}_i\bar{\boldsymbol{\pi}}_i = 0 \quad \text{for all } i; \tag{18.22}$$

$$\boldsymbol{\pi}_i\bar{\mathbf{x}}_i = 0 \quad \text{for all } j. \tag{18.23}$$

They work towards satisfying the nonnegativity constraints (18.21). The primal (dual) simplex method needs to be given a primal (dual) feasible solution to start, and keeps the primal (dual) variables nonnegative. At each iteration, it chooses a direction to move from its current solution, and moves as far as possible in that direction without violating feasibility.

Conceptually, the primal-dual interior point algorithm is the complete opposite. It enforces the nonnegativity constraints (18.21) strictly, throughout its execution. It does not need to be given a feasible solution to start. It does not enforce the equations (18.20), nor does it enforce the complementary slackness conditions (18.22, 18.23). At each iteration, it chooses a direction to move from its current solution, but it never moves as far as possible in that direction without violating feasibility.

Primal-dual interior algorithm

Replace the primal nonnegativity constraints $\mathbf{x} \geq \mathbf{0}, \bar{\mathbf{x}} \geq \mathbf{0}$ with logarithmic barrier weight α to create the problem

$$\min \mathbf{c}^T\mathbf{x} - \alpha\sum_{i=1}^{n}\log \mathbf{x}_i - \alpha\sum_{i=1}^{m}\log \bar{\mathbf{x}}_i$$

$$\text{subject to } A\mathbf{x} - \bar{\mathbf{x}} - \mathbf{b} = \mathbf{0}.$$

Form the Lagrangian

$$L(\mathbf{x}, \bar{\mathbf{x}}, \boldsymbol{\pi}) = \mathbf{c}^T\mathbf{x} - \alpha\sum_{i=1}^{n}\log \mathbf{x}_i - \alpha\sum_{i=1}^{m}\log \bar{\mathbf{x}}_i - \sum_{j=1}^{m}\boldsymbol{\pi}_j(A_{j,:}\mathbf{x} - \bar{\mathbf{x}}_j - \mathbf{b}_j)$$

and the components of its gradient ∇L

$$\frac{\partial L}{\partial \mathbf{x}_i} = c_i - \frac{\alpha}{x_i} - \sum_{j=1}^{m} \boldsymbol{\pi}_j A_{j,i}$$
$$\frac{\partial L}{\partial \bar{\mathbf{x}}_j} = \frac{\alpha}{\bar{\mathbf{x}}_j} - \boldsymbol{\pi}_j$$
$$\frac{\partial L}{\partial \boldsymbol{\pi}_j} = A_{j,:}\mathbf{x} - \bar{\mathbf{x}}_j - \mathbf{b}_j$$

Set $\nabla L = \mathbf{0}$ to get equations that, if solved, would yield the unique point on the central path at α.

$$c_i = \frac{\alpha}{x_i} - \boldsymbol{\pi}^T A_{:,i} \forall i \Rightarrow \mathbf{c} = A^T \boldsymbol{\pi} + \alpha \begin{pmatrix} 1/x_1 \\ \vdots \\ 1/x_n \end{pmatrix}$$
$$\boldsymbol{\pi} = \alpha \begin{pmatrix} 1/\bar{\mathbf{x}}_1 \\ \vdots \\ 1/\bar{\mathbf{x}}_n \end{pmatrix}$$
$$A\mathbf{x} - \bar{\mathbf{x}} - \mathbf{b} = \mathbf{0} \Rightarrow \bar{\mathbf{x}} = A\mathbf{x} - \mathbf{b}$$

Define the dual slack variables by

$$\bar{\boldsymbol{\pi}} \equiv \mathbf{c} - A^T \boldsymbol{\pi}$$

Rewrite $\nabla L = \mathbf{0}$ with the four equations

$$\bar{\mathbf{x}} = \mathbf{b} - A\mathbf{x} \tag{18.24}$$
$$\bar{\boldsymbol{\pi}} = \mathbf{c} - A^T \boldsymbol{\pi} \tag{18.25}$$
$$\mathbf{x}_i \bar{\boldsymbol{\pi}}_i = \alpha \tag{18.26}$$
$$\boldsymbol{\pi}_j \bar{\mathbf{x}}_j = \alpha \tag{18.27}$$

Equations (18.24)–(18.27) are prettily symmetric between primal and dual. Solving them will give us the point on the central path corresponding to α. *Question: What is the meaning of Equations* (18.26) *and* (18.27) *as* $\alpha \to 0$*?*[3] Corollary 2.1.4 tells us that complementary slackness ensures optimality, given primal and dual feasibility. This observation should make it intuitively plausible that the central path leads to an optimal solution as $\alpha \to 0$.

Apply the multi-dimensional Newton method, Algorithm 17.2.4, to solve Equations (18.24)–(18.27). As usual, let $\mathbf{X}, \bar{\mathbf{X}}, \boldsymbol{\Pi}, \bar{\boldsymbol{\Pi}}$ be matrices with entries $\mathbf{x}, \bar{\mathbf{x}}, \boldsymbol{\pi}, \bar{\boldsymbol{\pi}}$, respectively, on the diagonal and zero entries elsewhere. The components of the Jacobian $\nabla^2 L$ of ∇L are readily seen from the equations' LHS minus RHS. Taking the partial derivatives with respect to the variables in the order $\mathbf{x}, \bar{\mathbf{x}}, \boldsymbol{\pi}, \bar{\boldsymbol{\pi}}$:

$$\nabla^2 L = \begin{bmatrix} A & I & 0 & 0 \\ 0 & 0 & A^T & I \\ \bar{\boldsymbol{\Pi}} & 0 & 0 & \mathbf{X} \\ 0 & \boldsymbol{\Pi} & \bar{\mathbf{X}} & 0 \end{bmatrix} \tag{18.28}$$

By Algorithm 17.2.4, the step direction $\Delta = (\delta\mathbf{x}, \delta\bar{\mathbf{x}}, \delta\boldsymbol{\pi}, \delta\bar{\boldsymbol{\pi}})$ is defined by the $2(n+m)$ by $2(n+m)$ linear system

$$\nabla^2 L \Delta = -\nabla L. \tag{18.29}$$

[3]For small positive values of α they enforce an "almost-complementary" slackness. They enforce complementary slackness in the limit as $\alpha \to 0$.

The new solution is

$$(\mathbf{x}, \bar{\mathbf{x}}, \boldsymbol{\pi}, \bar{\boldsymbol{\pi}}) + \theta\Delta \tag{18.30}$$

where $0 < \theta \leq 1$ controls the step size. It has been observed that $\theta = 1$ sometimes violates nonnegativity, which forces us to take a more conservative step.

The major limitation of Newton's method is that it needs a starting point that is not far from the point it is seeking. Newton steps serve two purposes in the primal-dual interior point algorithm. The first is to get (very close to) to the point on the central path for the current value of α. These are called *centering* steps. The second is to move (approximately) along the central path in the direction of decreasing α. Such steps are called *path following.*

Algorithm 18.2.7 (Primal-Dual Log-Barrier Algorithm)

1. *Choose initial strictly positive solution values* $\mathbf{x}, \boldsymbol{\pi}$ *and parameter* α.
2. *Centering: Perform Newton steps according to Equations (18.29,18.30) until sufficiently close to central path as defined by Equations (18.24–18.27).*
3. *Terminate if criteria are met (sufficiently small complementarity value* α*, primal infeasibility* $A\mathbf{x} - \bar{\mathbf{x}} - \mathbf{b}$*, and dual infeasibility* $A^T\boldsymbol{\pi} + \bar{\boldsymbol{\pi}} - \mathbf{c}$*.)*
4. *Reduce* α *to new value* αt.
5. *Path following: Update solution* $\mathbf{x}, \boldsymbol{\pi}$ *(e.g., with a Newton step) to be close to central path for new value of* α.
6. *Adjust value of* α *to be more accurate for* $\mathbf{x}, \boldsymbol{\pi}$.
7. *Go to Centering.*

- A typical value for t is $t = 0.1$.
- Some versions of the algorithm perform only path following steps. Software implementations generally use predictor-corrector methods, which adjust α by quadratically approximating the central path. These methods were introduced to optimization by Sanjay Mehrotra [226, 227].
- Computing Δ involves a matrix product AD^2A^T where D is a diagonal matrix. Even one dense column in A makes this dense and can greatly increase the computation time.
- Some software either detects dense columns or lets you identify dense columns to be handled specially. However, removing several from the main algorithm can hurt the numerical stability.
- If your model has a very dense column, try to eliminate it by reformulating your LP, even if doing so requires additional variables and constraints.
- What is a dense column? In practice, about 10 or fewer nonzeros is sparse, and 100 nonzeros is dense. In between can be relative. Also, 50 nonzeros when most other columns have 10 isn't very dense, but 50 nonzeros when most other columns have 4 is dense enough that you might want the column treated specially.

- Interior point algorithms can have trouble if the face of optimal solutions is not bounded, because they will try to move to the centroid of an unbounded set. Simplex algorithms aren't troubled by this.

- On the other hand, simplex algorithms can get terribly bogged down by highly degenerate LPs, but interior point algorithms aren't slowed by degeneracy.

- Log barrier usually starts to be faster than simplex at somewhere between 10,000 and 100,000 variables plus constraints.

The following example mimics the derivation of the primal-dual barrier algorithm on a small numerical case, and it applies the algorithm from start to optimal solution.

Example 18.2 The feasible region $\{\mathbf{x} \in \Re^2 : \mathbf{x} \geq \mathbf{0}; -2x_1 - x_2 \leq -6\}$ is the same as in Figure 18.1. The objective is to minimize $10x_1 - 30x_2$. The dual is to maximize $-6\pi_1$ subject to constraints $\pi_1 \geq 0; -2\pi_1 \leq 10; -\pi_1 \leq -30$. After adding slack variable y_1 to the primal and slack variables δ_1, δ_2 to the dual, we get the following primal and dual.

(P)	**(D)**
$\min 10x_1 - 30x_2$	$\max -6\pi_1$
subject to	subject to
$-2x_1 - x_2 - y_1 = -6$	$-2\pi_1 + \delta_1 = 10$
	$-\pi_1 + \delta_2 = -30$
$x_1, x_2, y_1 \geq 0$	$\pi_1, \delta_1, \delta_2 \geq 0$

The one-dimensional dual is easy to solve, with $\pi^* = 30$ and $\delta_1^* = 70, \delta_2^* = 0$. By complementary slackness, $x_1^*\delta_1^* = 0 \Rightarrow x_1^* = 0$, and $\pi^* y_1^* = 0 \Rightarrow y_1^* = 0$. Hence the optimal solutions are $x_1^* = 0; x_2^* = 6; y_1^* = 0$ and $\pi_1^* = 30, \delta_1^* = 70; \delta_2^* = 0$, respectively, both with objective value -180.
Replace the primal nonnegativity constraints with logarithmic barrier functions:

$$\min 10x_1 - 30x_2 - \alpha \log x_1 - \alpha \log x_2 - \alpha \log y_1 \text{ subject to} \tag{18.31}$$
$$y_1 + 2x_1 + x_2 - 6 = 0 \tag{18.32}$$

The Lagrangian of (18.31,18.32) is

$$L(x_1, x_2, y_1, \pi_1) = 10x_1 - 30x_2 - \alpha \log x_1 - \alpha \log x_2 - \alpha \log y_1 + \pi_1(y_1 + 2x_1 + x_2 - 6).$$

The gradient of the Lagrangian is

$$\nabla L(x_1, x_2, y_1, \pi_1) = \begin{pmatrix} 10 - \frac{\alpha}{x_1} + 2\pi_1 \\ -30 - \frac{\alpha}{x_2} + \pi_1 \\ -\frac{\alpha}{y_1} + \pi_1 \\ y_1 + 2x_1 + x_2 - 6 \end{pmatrix}.$$

Introduce variables $\delta_i \equiv \frac{\alpha}{x_i}$ for $i \in \{1, 2\}$. In terms of the six variables $x_1, x_2, y_1, \pi_1, \delta_1, \delta_2$, the optimality condition $\nabla L(x_1, x_2, y_1, \pi_1) = \mathbf{0}$ is equivalent to the following six equations. *Question: The gradient ∇L has only four terms. How can there* *be six equations?*[4]

$$f_1(x_1,x_2,y_1,\pi_1,\delta_1,\delta_2) \equiv x_1\delta_1 - \alpha = 0 \tag{18.33}$$
$$f_2(x_1,x_2,y_1,\pi_1,\delta_1,\delta_2) \equiv x_2\delta_2 - \alpha = 0 \tag{18.34}$$
$$f_3(x_1,x_2,y_1,\pi_1,\delta_1,\delta_2) \equiv y_1\pi_1 - \alpha = 0 \tag{18.35}$$
$$f_4(x_1,x_2,y_1,\pi_1,\delta_1,\delta_2) \equiv -2x_1 - x_2 - y_1 + 6 = 0 \tag{18.36}$$
$$f_5(x_1,x_2,y_1,\pi_1,\delta_1,\delta_2) \equiv -2\pi_1 + \delta_1 - 10 = 0 \tag{18.37}$$
$$f_6(x_1,x_2,y_1,\pi_1,\delta_1,\delta_2) \equiv -\pi_1 \quad + \delta_2 + 30 = 0 \tag{18.38}$$

I've written the optimality conditions in the form $f(x_1,x_2,y_1,\pi_1,\delta_1,\delta_2) = \mathbf{0}$ because we will use Newton's method for finding a zero of a function $f : \Re^n \mapsto \Re^n$ to values that satisfy the optimality conditions. To use Newton's method, Algorithm 17.2.4, we will need the gradient of f, which is

$$\nabla f = \begin{bmatrix} \frac{\partial f_1}{\partial x_1} & \frac{\partial f_1}{\partial x_2} & \frac{\partial f_1}{\partial y_1} & \frac{\partial f_1}{\partial \pi_1} & \frac{\partial f_1}{\partial \delta_1} & \frac{\partial f_1}{\partial \delta_2} \\ \frac{\partial f_2}{\partial x_1} & \frac{\partial f_2}{\partial x_2} & \frac{\partial f_2}{\partial y_1} & \frac{\partial f_2}{\partial \pi_1} & \frac{\partial f_2}{\partial \delta_1} & \frac{\partial f_2}{\partial \delta_2} \\ \vdots & \vdots & \vdots & \vdots & \vdots & \vdots \\ \vdots & \vdots & \vdots & \vdots & \vdots & \vdots \\ \frac{\partial f_6}{\partial x_1} & \frac{\partial f_6}{\partial x_2} & \frac{\partial f_6}{\partial y_1} & \frac{\partial f_6}{\partial \pi_1} & \frac{\partial f_6}{\partial \delta_1} & \frac{\partial f_6}{\partial \delta_2} \end{bmatrix} = \begin{bmatrix} \delta_1 & 0 & 0 & 0 & x_1 & 0 \\ 0 & \delta_1 & 0 & 0 & 0 & x_2 \\ 0 & 0 & \pi_1 & y_1 & 0 & 0 \\ -2 & -1 & -1 & 0 & 0 & 0 \\ 0 & 0 & 0 & -2 & 1 & 0 \\ 0 & 0 & 0 & -1 & 0 & 1 \end{bmatrix} \tag{18.39}$$

For initial values I choose $x_1 = 3, x_2 = 4, y_1 = 2, \pi_1 = 40, \delta_1 = 100, \delta_2 = 5$. These values must be strictly positive. I purposely choose an infeasible solution to illustrate that the algorithm does not require two phases. *Question: How is it infeasible?*[5]
To choose an initial value for α, start with the three complementarity products $x_1\delta_1 = 300, x_2\delta_2 = 20, y_1\pi_1 = 80$. Their geometric mean is approximately 78.3. The rule of thumb $t = 0.1$ which would reduce α by a factor of 10 suggests $\alpha = 8$. I choose $\alpha = 10$ to make the first computations easier to follow. At $\alpha = 10$,

$$f(3,4,2,40,100,5) = \begin{pmatrix} 290 \\ 10 \\ 70 \\ -6 \\ 10 \\ -5 \end{pmatrix}.$$

and, from (18.39)

$$\nabla f(3,4,2,40,100,5) = \begin{bmatrix} 100 & 0 & 0 & 0 & 3 & 0 \\ 0 & 5 & 0 & 0 & 0 & 4 \\ 0 & 0 & 40 & 2 & 0 & 0 \\ -2 & -1 & -1 & 0 & 0 & 0 \\ 0 & 0 & 0 & -2 & 1 & 0 \\ 0 & 0 & 0 & -1 & 0 & 1 \end{bmatrix} \tag{18.40}$$

Solve $f(3,4,2,40,100,5)+\nabla f(3,4,2,40,100,5)\mathbf{d} = \mathbf{0}$ for the direction to move. The solution (calculated by my computer) is $\mathbf{d} = -(2.17, 0.268, 1.392, 7.165, 24.33, 2.165)$. Our new solution is $(3,4,2,40,100,5) + \theta\mathbf{d}$. At this iteration, choosing step size parameter

$\theta = 1$ works well, because it keeps all variables strictly positive. *Question: What would be the optimum value of θ if f were linear?*[6] Thus our new solution vector is

$$\begin{pmatrix} x_1 \\ x_2 \\ y_1 \\ \pi_1 \\ \delta_1 \\ \delta_2 \end{pmatrix} = \begin{pmatrix} 3 \\ 4 \\ 2 \\ 40 \\ 100 \\ 5 \end{pmatrix} - \begin{pmatrix} 2.17 \\ 0.268 \\ 1.392 \\ 7.165 \\ 24.33 \\ 2.165 \end{pmatrix} = \begin{pmatrix} 0.83 \\ 3.73 \\ 0.61 \\ 32.8 \\ 75.7 \\ 2.84 \end{pmatrix}.$$

The leftmost part of Figure 18.4 shows the movement of the first step.
For the second iteration, reduce α by a factor of 10 to $\alpha = 1$. Do not apply Equation (18.33) to alter the value of δ_1. We are relying on Newton's method to change the values of the variables. At $\alpha = 1$,

$$f\begin{pmatrix} x_1 \\ x_2 \\ y_1 \\ \pi_1 \\ \delta_1 \\ \delta_2 \end{pmatrix} = f\begin{pmatrix} 0.83 \\ 3.73 \\ 0.61 \\ 32.8 \\ 75.7 \\ 2.84 \end{pmatrix} = \begin{pmatrix} 61.8 \\ 9.59 \\ 19.0 \\ 0 \\ 0.1 \\ 0.04 \end{pmatrix}.$$

Observe that after merely one iteration, $f_4 \approx f_5 \approx f_6 \approx 0$. The value of f_4 is an indication of primal infeasibility; the values of f_5 and f_6 indicate dual infeasibility. These infeasibilities have all but disappeared. On the other hand, the values of f_1, f_2, and f_3, which indicate violation of complementarity slackness, are far from 0. It is typical for this algorithm to converge in primal and dual feasibility more rapidly than in complementary slackness.
The gradient of f is now

$$\nabla f(0.83, 3.73, 0.61, 32.8, 75.7, 2.84) = \begin{bmatrix} 75.7 & 0 & 0 & 0 & 0.83 & 0 \\ 0 & 2.84 & 0 & 0 & 0 & 3.73 \\ 0 & 0 & 32.8 & 0.61 & 0 & 0 \\ -2 & -1 & -1 & 0 & 0 & 0 \\ 0 & 0 & 0 & -2 & 1 & 0 \\ 0 & 0 & 0 & -1 & 0 & 1 \end{bmatrix} \tag{18.41}$$

Solving $f(0.83, 3.73, 0.61, 32.8, 75.7, 2.84) + \nabla f(0.83, 3.73, 0.61, 32.8, 75.7, 2.8)\mathbf{d} = \mathbf{0}$ by computer gives step $\mathbf{d} = (-0.73, 1.96, -.505, -4.0, -8.11, -4.0)$. With $\theta = 1$ the next solution is $(0.10, 5.69, 0.10, 28.8, 67.6, -1.2)$. We could have chosen $\theta < 1$ to keep $\delta_2 > 0$, but our derivation does not require that.
For the third iteration, reduce α again by a factor of 10 to $\alpha = 0.1$. The next step yields $\mathbf{d} = (-0.10, 0.316, -0.10, 1.284, 2.57, 1.28)$. This time $\theta = 1$ would make $x_1 = 0$ and $y_1 = 0$. We require $x_1 > 0$ and $y_1 > 0$. Therefore, our step must be scaled by slightly less than 1. Reduce 1 by 5% (5% and 10% are typical values) to get $\theta = 0.95$. The new solution is $(0.005, 5.99, 0.005, 29.93, 69.87, 5.99)$. In just a few iterations, the algorithm's primal solution $x_1 = 0.005, x_2 = 5.99$ has come very close to the optimal primal solution $x_1^* = 0, x_2^* = 6$. Likewise, in the dual, $\pi_1 = 29.93 \approx 30 = \pi_1^*$.
Figure 18.4 depicts the algorithm's progress in the primal.

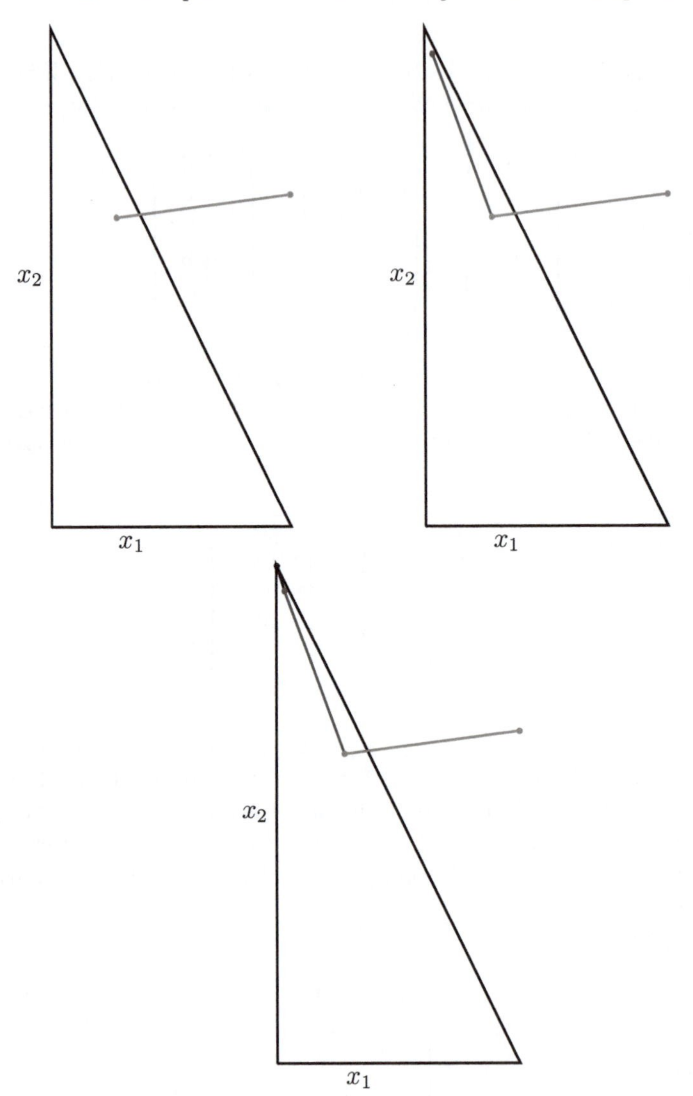

FIGURE 18.4: Trajectory of the log barrier algorithm in the primal.

Let's see why the algorithm finds an optimal solution despite performing computations only to polynomial length accuracy. First, it must be true that for sufficiently small $\alpha > 0$ the solution must be so close to an optimal solution that values close to zero can be rounded down to zero to obtain an optimal solution. Second, there must exist a point at which the algorithm can terminate. The point should be strictly positive in both primal and dual. Yet it also must be such that the almost-complementary-slackness constraints $x_i\bar{\pi}_i = \alpha$ and $\pi_j\bar{x}_j = \alpha$ are satisfied at the extremely small value of α that mathematically guarantees the rounding property just stated. To be precise, we have the following theorem.

Theorem 18.2.8 *Suppose both primal and dual (18.19–18.21 are full-dimensional and have bounded optimal solutions (Properties* **P2** *and* **P3**). **Let matrix** A **have dimensions**

[4] The first two equations enforce the definitions of the δ_i; the other four set the four terms of ∇L to zero.
[5] $-2x_1 - x_2 - y_1 = -12$, rather than equaling -6.
[6] $\theta = 1$ would make $f() = 0$.

$m \times n$. Let $\mathcal{L}$ be the length of a binary encoding of A, b, and c. Then there exist $\epsilon^* > 0$ and $\alpha^* > 0$ such that for all $0 < \alpha \le \alpha^*$, there exists an optimal solution to the primal-dual pair, $\mathbf{s}^* \equiv (\mathbf{x}^*, \bar{\mathbf{x}}^*, \boldsymbol{\pi}^*, \bar{\boldsymbol{\pi}}^*)$ such that the solution $\mathbf{s}^\diamond \equiv (\mathbf{x}^\diamond, \bar{\mathbf{x}}^\diamond, \boldsymbol{\pi}^\diamond, \bar{\boldsymbol{\pi}}^\diamond)$ to Equations (18.24)–(18.27) satisfies:

$$s_k^* = 0 \Leftrightarrow s_k^\diamond < \epsilon \qquad \textbf{for all } k. \tag{18.42}$$

Moreover, α^* can be taken to be $\alpha = \theta(2^{-Lmn})$.

In words, Theorem 18.2.8 says that there exists $\epsilon > 0$ such that for all sufficiently small $\alpha > 0$, rounding down to zero the components of the corresponding central path point that are less than ϵ determines the components that equal zero in an optimal solution to the LP.
Proof: *The existence of $\epsilon > 0$ was proved in the analysis of the ellipsoid algorithm in Section 10.2. See Corollary 10.2.2. By hypothesis there exists M such that $||s||_\infty < M$ for all optimal s. Let $\alpha = \epsilon/M$. Then $x_i\bar{\pi}_i = \alpha$ and $x_i < M$ imply $x_i\bar{\pi}_i <$.*
■

Sparseness is important to all LP algorithms. But it has a special relationship with interior point algorithms. Like the simplex algorithms, interior point algorithms repeatedly solve systems of simultaneous linear equations such as $M\mathbf{d} = \mathbf{f}$. In simplex algorithms, the matrix M changes significantly during a sequence of iterations. In interior point algorithms, the numerical values of M change from iteration to iteration, but the locations of the nonzeros stay the same. For instance, the matrix of partial derivatives of the Lagrangian in Equation (18.39) is the same in every iteration of Example 18.2. Only the values of the primal and dual variables within the matrix change. Therefore, a systematic way to solve the simultaneous linear equations during the first iteration will work equally well in all future iterations. Thus, interior point algorithms perform a factorization of $M = AA^T$ prior to the first iteration, to be used throughout. The factorization is not purely numeric; it is called a symbolic factorization. Problem 13 illustrates how different orderings of operations can require significantly different amounts of computation, independent of the specific numeric values except for which are nonzero.

Many theoretical analyses of interior point algorithms prove $O(\sqrt{m})$ or $O(\sqrt{m}\log m)$ upper bounds on the number of iterations. See, for example, Monteiro, Adler, and Resende [233], and Renegar [248]. In practice, the number of iterations required seems usually to be $O(\log m)$. There are substantial differences between the algorithms that are theoretically analyzed and the implementations that are most successful computationally. One difference is that the latter take more aggressive, (i.e., longer) steps. Another difference is that the latter generally use predictor-corrector methods as introduced by Sanjay Mehrotra [227], which complicate the theoretical analysis and are often not considered. Less obvious is that handling sparseness well is essential to good computational performance. On sparse instances, commercial LP codes usually run in time close to linear in the number of nonzeros. Such performance is not captured by present-day worst-case analysis.

Summary: As we saw in the previous chapter, the principal algorithms for nonlinear optimization work by linearizing. Ironically, the best known method for solving very large linear programs is a nonlinear programming algorithm! Both the log-barrier central path and the affine scaling algorithms employ the original constraint matrix A in every iteration, in a matrix product ADA^T, where D is a diagonal matrix that depends on the current solution. The persistence of this term opens up the possibility of a preliminary factorization step that exploits sparsity and speeds up every iteration. On the other hand, a single dense column in A can make all of AA^T dense. Since the central path always leads to the centroid of the face of optimal solutions, the log-barrier algorithm finishes with simplex-type steps to reach a basic optimal solution.

18.3 Notes and References

There always seems to be a tradeoff between what is theoretically required for guaranteed convergence and what is computationally efficient. The affine scaling algorithm furnishes a rare example in which the most generous stepsize that guarantees convergence has been determined. Small step sizes slow the algorithm. However, too large a step size such as 0.999 can be proved to fail to converge correctly [221]. Tsuchiya [281] proved convergence for $\frac{2}{3}$ even in the presence of primal degeneracy (later extended to both primal and dual degeneracy [282]). Hall and Vanderbei [141] proved that the lower bound value $\frac{2}{3}$ is tight.

The log-barrier approach to LP dates back at least to Frisch [109]. Huard [154] identified points in the central path. Karmarkar [172] ignited the explosion of interest in interior point algorithms, not so much with his theoretical upper bound which improved slightly on Khachian's, as with his subsequent reports of computational results (see for example [273]). Within a few years, regardless of unverified computational results, it became clear that Karmarkar's algorithm was a profound contribution to linear programming algorithms. Its number of iterations as a function of problem size grew much more slowly than the simplex method's. It took advantage of sparseness via a single albeit expensive (symbolic) factorization rather than by a factorization that has to be continually updated. Another impact was not apparent immediately. Karmarkar had cleverly used a potential function, the log of $\prod_{i=1}^{n} \mathbf{c}^{\mathbf{x}}/x_i$ to prove polynomial time convergence. This potential function was not part of the algorithm itself, but it turned out to be related to the central path. Key properties of the central path were proved by Dave Bayer and Jeff Lagarias [18, 19]. Nimrod Megiddo [224] discovered the connection between Karmarkar's algorithm, which by then was recognized as being computationally effective, and the central path, which until then had not been associated with practical computation. Gill *et al.* [120] showed an equivalence between projected Newton methods and Karmarkar's algorithm . Polynomial-time algorithms based on Newton's method and central paths were soon devised by Renegar [248] and others. Kojima, Mizuno, and Yoshise's [191] primal-dual algorithm with proof of convergence, together with Monteiro and Adler's [232] improved bound and Lustig's [218] single phase technique greatly influenced the overall algorithmic structure of log-barrier implementations.

There are now many polynomial-time interior point algorithms for LP, and new ones continue to be invented. One recent algorithm [52] essentially reduces LP to matrix multiplication, so that its run time is the same order as that of the fastest available matrix multiplication algorithm.

18.4 Problems

E Exercises

1. Let M be an $m \times n$ matrix, and let vector $\mathbf{d} \in \Re^n$. Define D to be the diagonal matrix with diagonal entries $\mathbf{d}$. ($D_{i,i} = d_i$ and $D_{i,j} = 0$ for all $i \neq j$.) Describe the matrix product MD in words.

2. Let M be an $m \times n$ matrix, and let vector $\mathbf{y} \in \Re^m$. Define Y to be the diagonal matrix with diagonal entries $\mathbf{y}$. Describe the matrix product YM in words.

3. Continue the affine scaling algorithm example of Figures 18.1 and 18.2 with one more iteration, using $\alpha = 0.1$ and $\alpha = 0.5$.

4. Show that the centroid of a square is the intersection of its diagonals.

5. Show that the centroid of an equilateral triangle is the intersection of its angle bisectors.

6. Suppose A is a nonnegative 2000 by 1000 matrix, mostly sparse, except for column $A_{:,666}$ which has density 10%. That is, $A_{:,666}$ has 200 nonzero entries. How many entries of the matrix product AA^T does this column, alone, cause to be nonzero?

M Problems

7. Verify that the KKT conditions applied to LP yield exponentially many cases.

8. Given Exercise 4, use a scaling argument to prove that the centroid of a rectangle is the intersection of its diagonals.

9. Find the centroid of a 3-4-5 right triangle.

10. The greatest reduced cost rule for the entering (exiting) variable renders the primal (dual) simplex method susceptible to scaling of the columns (rows). Determine whether or not the affine scaling algorithm is susceptible to column scaling and/or row scaling.

11. Determine whether or not the primal log-barrier algorithm is susceptible to column and/or row scaling.

12. Determine whether or not the primal-dual log-barrier algorithm is susceptible to column and/or row scaling.

13. This problem will help you see why a symbolic factorization, computed once, can be used repeatedly. Suppose the nonzero elements of square matrix E are $E_{i,i} : i = 1, \ldots, n$ and $E_{i,i+1} : i = 1, \ldots, n-1$. Compare the number of arithmetic operations needed to solve $E\mathbf{x} = \mathbf{b}$ in the following two ways:

    - $x_n = b_n / E_{n,n}$; $x_{n-1} = (b_{n-1} - E_{n-1,n}x_n)/E_{n-1,n-1}$, etc.
    - $x_2 = (b_1 - E_{1,1}x_1)/E_{12}$; $x_3 = (b_2 - E_{2,2}x_2)/E_{2,3} = (b_2 - \frac{E_{2,2}}{E_{1,2}}(b_1 - E_{1,1}x_1))/E_{2,3}$, etc.

Observe that your numbers do not depend on the numerical values of $\mathbf{b}$, nor even on the specific values of the nonzero elements of E.

14. Suppose Properties P1 and P2 hold. Prove the converse to Theorem 18.2.2, that if there is a unique solution to (18.8), then Property P3 must hold. Therefore, the logarithmic barrier problem has an optimal solution iff the LP has a nonempty bounded set of optimal solutions.

15. Suppose the primal (18.6),(18.7) and its dual are both feasible, and suppose further that the dual contains an interior point. Prove that Property P3 holds.

16. Assume properties P1, P2, and P3. Prove the analog of Theorem 18.2.1 for the dual solution in the primal-dual log-barrier model. That is, prove that in the limit as $\alpha \to 0$, π converges to an optimal dual solution.

D Problems

17. Prove that for all $\alpha > 0$ the central path at α is *orthogonal* to the set of isoquants of the barrier penalty function.

18. Let $\mathbf{x}^n > \mathbf{0} : n = 1, 2, 3, \ldots$ be a sequence of feasible solutions to an LP such that $\lim_{n\to\infty} x_1^n = 0$ and $\lim_{n\to\infty} x_j^n > 0$ for all $j \neq 1$. Let the objective vector $\mathbf{c} > \mathbf{0}$. Prove that (in the unscaled space) the step vectors $\mathbf{d}^n$ are such that $\lim_{n\to\infty} d_1^n = 0$.

19. Consider the problem

$$\min \mathbf{x}^T M \mathbf{x} + \mathbf{c}^T \mathbf{x}$$

$$\text{subject to}$$

$$A\mathbf{x} \geq \mathbf{b}$$

where M is a positive definite matrix.

    (a) Assume the feasible region is a polytope with nonempty interior. Introduce the usual log-barrier function with parameter α. Prove that for any $\epsilon > 0$, an optimal solution to the unconstrained problem has objective function value within ϵ of the true optimum for all sufficiently small $\alpha > 0$.

    (b) Introduce slack variables to the inequalities $A\mathbf{x} \geq \mathbf{b}$ and write the Lagrangian.

    (c) Write the conditions for the gradient of the Lagrangian to be $\mathbf{0}$.

    (d) Explain why your equations could be solved numerically in reasonable time despite the objective being quadratic rather than linear.

20. Prove that affine scaling with infinitesimally small steps can take superpolynomial time if started close enough to a vertex of the Klee-Minty polytope [225].

21. Prove that the central path may visit exponentially many vertices of the Klee-Minty cube if a large number of redundant constraints are added to the model [77].

I Problems

22. Let A be an $m \times n$ matrix, where m and n are both large. Populate it with 4 random positive values in each column, in random rows independent of other columns. Set all other entries of A to zero. Find a very close approximation of the expected number of nonzeroes of the matrix product AA^T. Calculate the consequent expected density.

Chapter 19

Appendices

19.1 Linear Algebra and Vector Spaces

19.1.1 Linear Combinations, Dimension

Let $\mathbf{x}^j \in \Re^n$ for $j = 1 \dots m$. As usual, x_i^j denotes the ith component of the vector $\mathbf{x}^j$. That is, $\mathbf{x}^j = (x_1^j, x_2^j, \dots, x_n^j)$. A *linear combination* of $\mathbf{x}^1 \dots \mathbf{x}^m$ is a weighted sum of these vectors, $\sum_{j=1}^m \alpha_j \mathbf{x}^j$ where $\alpha_j \in \Re\ \forall j$. Geometrically, the linear combinations are all the points you can get to from the origin by taking steps of any lengths (including negative lengths) in any of the directions given by the vectors $\mathbf{x}^1 \dots \mathbf{x}^m$. In matrix notation, define the matrix X by

$$X_{:,j} \equiv \mathbf{x}^j,$$

the $n \times m$ matrix whose columns are $\mathbf{x}^1 \dots \mathbf{x}^m$. For every $\alpha = (\alpha^1, \alpha^2, \dots \alpha^m) \in \Re^m$, the product $X\alpha$ is a linear combination of $\mathbf{x}^1 \dots \mathbf{x}^m$.

The columns of matrix X are *linearly independent* if $X\alpha = 0 \Rightarrow \alpha = 0$; otherwise they are *linearly dependent.* Equivalently, the columns of X are linearly dependent if at least one of them can be expressed as a linear combination of the others. *Question: Must this be true of all columns?*[1]

If $\mathbf{y} = X\alpha$ we say that $\mathbf{y}$ is in the *span* of the columns of X, or it is in the *subspace* generated by the columns of X. The subspace S generated by the columns of X is closed under taking linear combinations. That is, any linear combination of elements of S (linear combinations of the columns of X) is itself an element of S (a linear combination of the columns of X). Conversely, for any subset $S \subseteq \Re^n$ that is closed under taking linear combinations, there exists a matrix X whose columns generate S, and so S is a subspace. We think of the origin as a special subspace that is generated by the empty set of columns.

Let S be a subspace. The *dimension* of S, denoted by $dim(S)$, is the smallest number of vectors whose span is S. A subspace S must contain a subset of at least $dim(S)$ linearly independent vectors. *Question: Why?*[2] In fact, $dim(S)$ equals the maximum size of any set of linearly independent vectors in S. If the columns of X are linearly independent and span S, they form a *basis* for S. It follows from the preceding statements that every basis of S has size $dim(S)$. If the columns of the $m \times n$ matrix X are linearly independent but do not span S, then there exists a vector $\mathbf{x}^{m+1} \in S$ that, together with the columns of X, forms a linearly independent set. In matrix terms, the columns of the matrix $[X|\mathbf{x}^{m+1}]$ are linearly independent. Therefore, any set of linearly independent vectors in S that do not form a basis for S can be augmented to create a basis.

[1] No, for example $(1,0), (0,1), (0,2)$.

[2] A minimum size set of vectors that spans S must be linearly independent, for otherwise a vector could be removed from that set without affecting the span.

19.1.2 The Dot Product, Cosines, Projections, Cauchy-Schwartz Inequality

Let $\mathbf{x} = \{x_1, x_2, \ldots, x_n\} \in \Re^n$ and let $\mathbf{y} = \{y_1, y_2, \ldots, y_n\} \in \Re^n$. The dot product of $\mathbf{x}$ and $\mathbf{y}$, denoted $\mathbf{x} \cdot \mathbf{y}$, is

$$\mathbf{x} \cdot \mathbf{y} = \mathbf{x}^T \mathbf{y} = \sum_{i=1}^{n} x_i y_i.$$

The dot product is symmetric, i.e., $\mathbf{x} \cdot \mathbf{y} = \mathbf{y} \cdot \mathbf{x}$. The dot product of a vector with itself is the square of the vector's Euclidean length, $\mathbf{x} \cdot vx = ||\mathbf{x}||^2$. Some textbooks define the inner product of two vectors $\mathbf{x}, \mathbf{y} \in \Re^n$ as $||\mathbf{x}||||\mathbf{y}|| \cos\theta$ where θ is the angle between $\mathbf{x}$ and $\mathbf{y}$. For $\mathbf{x}$ and $\mathbf{y}$ in $\Re^n$, apply the law of cosines to the triangle defined by $0, \mathbf{x}$, and $\mathbf{y}$. We get $||\mathbf{x}-\mathbf{y}||^2 = ||\mathbf{x}||^2 + ||\mathbf{y}||^2 - 2||\mathbf{x}||||\mathbf{y}|| \cos\theta$. The left-hand side gives $||\mathbf{x}-\mathbf{y}|| = (\mathbf{x}-\mathbf{y}) \cdot (\mathbf{x}-\mathbf{y})$ $= \mathbf{x} \cdot \mathbf{x} + \mathbf{y} \cdot \mathbf{y} - 2\mathbf{x} \cdot \mathbf{y} = ||\mathbf{x}||^2 + ||\mathbf{y}||^2 - 2\mathbf{x} \cdot \mathbf{y}$. Hence $\mathbf{x} \cdot \mathbf{y} = ||\mathbf{x}||||\mathbf{y}|| \cos\theta$. Observe the method of proof. The points $\mathbf{x}, \mathbf{y}$ and $\mathbf{0}$ live in n-dimensional space, but these three points define a triangle in a two-dimensional subspace which is isomorphic to the Euclidean plane. Therefore, we can use tools from standard Euclidean geometry, if we are careful, even when we are in a high-dimensional space.

From $|\cos\theta| \leq 1$ we have the Cauchy-Schwartz inequality,

$$|\mathbf{x} \cdot \mathbf{y}| \leq ||\mathbf{x}||||\mathbf{y}||.$$

Memorize it; it is possibly the most frequently used inequality in mathematics.

Two vectors $\mathbf{x}$ and $\mathbf{y}$ are orthogonal if $\mathbf{x} \cdot \mathbf{y} = 0$. A basis is orthogonal if all pairs of its members are orthogonal. An orthogonal basis is orthonormal if $\mathbf{x} \cdot \mathbf{x} = 1$ for each member $\mathbf{x}$.

The projection of a vector $\mathbf{x}$ on a vector $\mathbf{y}$ (both in $\Re^n$) is $(\mathbf{x} \cdot \mathbf{y})\mathbf{y}/||\mathbf{y}||^2 = \frac{\mathbf{x} \cdot \mathbf{y}}{\mathbf{y} \cdot \mathbf{y}}\mathbf{y}$. Let $\mathbf{b}^1, \ldots \mathbf{b}^k$ be an orthonormal basis for a subspace S. Then the projection of $\mathbf{x}$ onto S is the sum

$$\sum_{j=1}^{k} (\mathbf{x} \cdot \mathbf{b}^j)\mathbf{b}^j.$$

Application: Suppose we wish to maximize $f(\mathbf{x})$ on the set $\mathbf{x} : A\mathbf{x} = \mathbf{b}$. We have a candidate point $\mathbf{x}^0$ such that $A\mathbf{x}^0 = \mathbf{b}$. Let $\mathbf{p} = \nabla f(\mathbf{x}^0)$, the gradient of $f()$ at $\mathbf{x}^0$. Then $f()$ will increase if we move in the direction $\mathbf{p}$ from $\mathbf{x}^0$. However, unless we are incredibly lucky, it will not be true that $A\mathbf{p} = 0$. Therefore, moving in the direction $\mathbf{p}$ will render us infeasible. So we project $\mathbf{p}$ onto the subspace $\mathbf{x} : A\mathbf{x} = \mathbf{0}$.

The most useful geometric property of the dot product in optimization is that $\mathbf{c} \cdot \mathbf{y}$ is a measure of how far $\mathbf{y}$ is in the direction $\mathbf{c}$. If we are at a point $\mathbf{x}$, and we move in any direction $\mathbf{y}$ such that $\mathbf{c} \cdot \mathbf{y} > 0$, then $\mathbf{c} \cdot (\mathbf{x} + \mathbf{y}) > \mathbf{c} \cdot \mathbf{x}$. If $\mathbf{y}$ is orthogonal to $\mathbf{c}$, then $\mathbf{c} \cdot (\mathbf{x} + \mathbf{y}) = \mathbf{c} \cdot \mathbf{x}$. When we minimize a linear function $\mathbf{c} \cdot \mathbf{x}$, moving in the direction $\mathbf{y}$ is beneficial when $\mathbf{c} \cdot \mathbf{y} < 0$, neutral when $\mathbf{c}$ and $\mathbf{y}$ are orthogonal, and harmful when $\mathbf{c} \cdot \mathbf{y} > 0$. When we minimize a nonlinear real-valued function $f(\mathbf{x})$ whose gradient at $\mathbf{x}$ is $\nabla f(\mathbf{x})$, moving from $\mathbf{x}$ in the direction $\mathbf{y}$ is beneficial when $\nabla f(\mathbf{x}) \cdot \mathbf{y} < 0$ and harmful when $\nabla f(\mathbf{x}) \cdot \mathbf{y} > 0$. This guideline holds for sufficiently small movements, but becomes less accurate the larger the stepsize since the gradient of $f()$ changes as we move from $\mathbf{x}$ unless f is an affine function, i.e., of form $c_0 + \mathbf{c} \cdot \mathbf{x}$.

19.1.3 Determinants, Inverses

Let A be an $n \times n$ matrix. Theorem: the following are logically equivalent.

1. There exists a matrix A^{-1} such that $AA^{-1} = I = A^{-1}A$.

2. The determinant of A is nonzero.

3. $A\mathbf{x} = 0$ has unique solution $\mathbf{x} = 0$.

4. The column rank of A is n.

5. The row rank of A is n.

An elementary row operation on a matrix consists of adding a multiple of one row to another row, or exchanging two rows. One way you may have learned to calculate the inverse of a matrix A is to form the block matrix $[AI]$ and perform elementary row operations until A has been transformed to I. Then whatever is in the location that I used to be is now A^{-1}.

Exercises: Let $A \in R^{4\times 6}$. Express, as a product of two matrices, the matrix resulting from A by:

- multiplying the 4th row by 8.

- adding the 2nd row to the 4th row.

- subtracting $\frac{1}{5}$ times the 4th row from the 1st row.

- adding i times the 2nd row to row i for all $i = 1, 2, 3, 4$ (resulting in one matrix, not four).

- subtracting j times *column* 4 from column j for all $j = 1, 2, 3, 4, 5, 6$.

Explain why the method for computing inverses, described above, works correctly.

19.1.4 Gaussian Elimination

To solve a system of n linear equations in n unknowns we can use the Gaussian elimination procedure. This procedure has an interesting history. According to Wilkinson, who worked with Turing, the procedure was known to be numerically unstable, in the sense that computational errors due to rounding could propagate exponentially badly. Wilkinson [298] discovered that it works well in practice, in a desperate effort to solve an 18-by-18 system. See also [297].

In Gaussian elimination to solve $A\mathbf{x} = \mathbf{b}$, we perform elementary row operations on the matrix $[A\mathbf{b}]$ until A is triangular (step 1). Then we back-substitute, finding variable values one at a time with $O(n)$ work per variable (step 2). The first step of Gaussian elimination suffices to calculate the determinant of A, since the determinant of a triangular matrix is the product of its diagonal terms.

19.1.5 Vector Spaces

By a scalar we usually mean a real number, though often it is OK to use complex numbers. A real vector space, or a vector space over the reals, or a linear space over the reals, is a set W of elements ("vectors") together with operations addition $+ : W \times W \mapsto W$ and multiplication: $\Re \times W \mapsto W$ such that:

1. $x + y = y + x \forall x \in W, y \in W$

2. $(x + y) + z = x + (y + z) \forall x \in W, y \in W, z \in W$.

3. $\exists 0 \in W$ such that $x + 0 = x \; \forall x \in W$, and 0 is unique.

4. $\forall x \in W \exists -x \in W$ such that $x + -x = 0$.

5. The result of multiplying 1 by any vector x is x, $1x = x \; \forall x \in W$.

6. $\alpha(\beta x) = (\alpha\beta)x \forall \; x \in W \forall \alpha, \beta \in \Re$.

7. $\alpha x + \beta x = (\alpha + \beta)x \forall \; x \in W \forall \alpha, \beta \in \Re$.

8. $\alpha(x + y) = \alpha x + \alpha y \; \forall x, y \in W \forall \alpha \in \Re$.

It follows from these definitions that $-0 = 0$; $\alpha 0 = 0 \forall \alpha \in \Re$; $-x = (-1)x \; \forall x \in W$; $0x = 0 \; \forall x \in W$.

If we changed our definition of scalar to a rational number, a complex number, an integer, or a positive integer, for which cases would the definition of vector space still work?

The vector space is a generalization of n-dimensional space where vectors consist of ordered $n - tuples$ of real numbers and the addition and multiplication operations are defined as follows:

$(x_1, x_2, \ldots x_n) + (y_1, y_2, \ldots y_n) = (x_1 + y_1, x_2 + y_2, \ldots x_n + y_n)$.

$\alpha(x_1, x_2, \ldots x_n) = (\alpha x_1, \alpha x_2, \ldots \alpha x_n)$.

Proving that under these definitions we get a vector space is trivial. If instead of $n-tuples$ of real numbers we had used $n - tuples$ of rational numbers, complex numbers, integers, real numbers that sum to 0, real numbers that sum to 1, for which cases would we still have a vector space?

Let $C_{[0,1]}$ denote the set of continuous functions $[0, 1] \mapsto \Re$. Define $f + g$ as $(f + g)(x) = f(x) + g(x)$ and define $(\alpha f)(x) = \alpha f(x)$. This forms a vector space over the reals.

For a finite set of elements $x^1 \ldots x^m$ of a real vector space, a linear combination of these elements is a sum

$$\sum_{i=1}^{m} \alpha^i x^i$$

where $\alpha^i \in \Re \forall i$. If for a set of elements there exist values $\alpha^i, i = 1 \ldots m$, not all zero (the zero in $\Re$), such that the associated linear combination equals 0 (the zero in W), the vectors are linearly dependent. If on the other hand

$$\sum_{i=1}^{m} \alpha^i x^i = 0 \Rightarrow \alpha^i = 0 \forall i$$

then the vectors are linearly independent. For an infinite set of vectors, we say that the set is linearly independent if every finite subset is linearly independent. (Notice that infinite sums of vectors are not defined.) We follow the convention that an empty sum equals the zero vector, $\sum_{i \in \phi} x^i = 0$.

Exercise: Let W be the elements of $C_{[0,1]}$. Prove that the set of functions $f(x) = x^0, x^1, x^2, x^3, \ldots$ is a linearly independent set.

Optimization problem to think about: Given a set of vectors $X = x^1 \ldots x^t$ and corresponding nonnegative real weights $w^1 \ldots w^t$, find a subset $S \subset X$ such that $\sum_{i: x^i \in S} w^i$ is maximized subject to the constraint that S is linearly independent.

19.1.6 Dimension, Basis

The dimension of a vector space W is n if W contains a linearly independent subset of cardinality n, but not one of cardinality $n+1$. If for all n W contains n linearly independent vectors, then W is infinite-dimensional. The span of a set X of vectors is the set of all linear combinations of (finite subsets of) X. By the convention above, the span of an empty set is the vector 0. A basis of W is a linearly independent set of vectors whose span is W. For example, the functions $f^i(x) = x^i$ for $i = 0, 1, 2, 3, \ldots$ form a basis of the vector space of (single variable) polynomial functions. Exercise: Find a basis for the space of two-variable polynomial functions.

Suppose $X = x^1 \ldots x^m$ is a maximal linearly independent set of vectors of a finite dimensional vector space W. Maximal means that no vector can be added to the set while retaining the property of linear independence. Then the span of X is W, i.e., X is a basis.

Proof: If y is not in the span of X we claim $X \bigcup y$ is linearly independent. For if not, there exist $\alpha_0, \alpha_1, \ldots \alpha_n$ where $n = |X|$, not all zero, such that $\alpha_0 y + \sum_{j=1}^{m} \alpha_j x^j = 0$. Now if $\alpha_0 = 0$ the set X would not be l.i., a contradiction. Therefore, $(-\alpha_0)y = \sum_{j=1}^{m} \alpha_j x^j$ whence (since $\alpha_0 \neq 0$) $y = \sum_{j=1}^{m} (\alpha_j / - \alpha_0) x^j$, i.e., y is in the span of X, a contradiction.

Suppose $x^1 \ldots x^m$ and $y^1 \ldots y^n$ are both maximal linearly independent sets of vectors of a vector space W. Then $n = m$ and we call n the dimension of W.

Proof: Without loss of generality, let $m < n$. Then each of the vectors y^i is in the span of $x^1, \ldots, x^m$ for $i = 1 \ldots n$. Hence for each $i \exists \beta_i^j$ not all zero such that

$$y^i = \sum_{j=1}^{m} \beta_i^j x^i.$$

Now we have each of n l.i. vectors expressed as a linear combination of $m < n$ vectors. The matrix whose i, j term is β_i^j has m columns and n rows so its rows cannot be linearly independent, which contradicts the linear independence of $y^1 \ldots y^n$.

The point of this proof is that the idea of dimension generalizes from standard linear algebra on vectors of real numbers to real vector spaces. We don't need the geometric ideas of length and position for this theorem to be true. Next we introduce the idea of an inner product, which generalizes some of the geometric nature of $\Re^m$.

19.1.7 Inner Products and Norms

An inner product in a real vector space W is a function from $W \times W \mapsto \Re$, denoted $< v, w >$ such that

1. $< v, w >=< w, v > \ \forall v, w \in W$
2. $< w, w >\geq 0 \ \forall w \in W$
3. $< w, w >= 0 \Leftrightarrow w = 0$
4. $< \alpha v + \beta w, x >= \alpha < v, x > + \beta < w, x >$.

An inner product always defines a norm, a function from $W \mapsto \Re^+$, which equals $||w|| \equiv \sqrt{< w, w >}$. Example: The dot product defines an inner product in the vector space $\Re^n$. Example: If f and g are continuous real-valued functions on $[0, 1]$, then $< f, g >\equiv \int_0^1 f(t)g(t)dt$ defines an inner product.

19.1.8 Linear Functions and Matrices

A linear function on a real vector space W to a real vector space V, $f : W \mapsto V$ is a function satisfying the following property. $f(\alpha\mathbf{x}+\beta\mathbf{y}) = \alpha f(\mathbf{x})+\beta f(\mathbf{y}) \ \forall \mathbf{x}, \mathbf{y} \in W, \forall \alpha, \beta \in \Re$. Suppose X is a basis for W. Then if we know $f(\mathbf{x})$ for each $\mathbf{x} \in X$, we know $f(vy)$ for all

$\mathbf{y} \in W$. Moreover, the range of f in V is closed under multiplication by scalars and under vector addition, i.e., the range of f is a subspace of V. (Prove these statements).

Linear functions on the traditional vector space $\Re^m$ are represented by matrices. The ith column of the matrix equals $f(e_i)$ where e_i is the ith unit vector. For function f represented by matrix M we write $M\mathbf{x} = f(\mathbf{x})$. The term $M\mathbf{x}$ is an $n - vector$, where M is an $n \times m$ matrix and $f(\mathbf{x}) = f(\sum_i x_i e_i) = \sum_i x_i f(e_i) = \sum_i x_i M_{:,i}$ by the linearity property of linear functions. Therefore, the definition of the product of a matrix times a vector is not arbitrary. It is derived logically as the representation of the linear function. It is trivial but very useful to remember that $M\mathbf{x}$ is a weighted sum of the columns of M, where the coefficients of $\mathbf{x}$ are the weights. It is equally trivial and equally useful to remember that $\mathbf{y}^T M$ is a weighted sum of the rows of M, where the coefficients of $\mathbf{y}$ are the weights.

More generally, a linear function f on a vector space W is described by a list $f(\mathbf{x}^1), f(\mathbf{x}^2)\ldots$ where $\mathbf{x}^1, \mathbf{x}^2, \ldots$ are a basis for W.

The definition of multiplication of matrices is not arbitrary either. Suppose U, V, W are vector spaces and $f : W \mapsto V$ and $g : V \mapsto U$ are linear functions. Then $g \circ f$, the composition of g with f, is defined as $(g \circ f)(x) \equiv g(f(x))$. (Prove that $g \circ f$ is a linear function). So $g \circ f$ should be represented by a matrix. Let M represent f and let N represent g. Matrix multiplication is defined so that $NM\mathbf{x} = g(f(x))$. (Prove that this is so. Hint: Consider $g(f(e_i))$ where e_i is the ith basis vector).

19.2 Point Set Topology

Definition 19.1 *A metric space is a set $\mathcal{S}$ and a distance function or metric $d : \mathcal{S} \times \mathcal{S} \mapsto \Re$ with the properties:*

1. $d(t, s) = d(s, t) \geq 0 \ \forall s, t \in \mathcal{S}$.
2. $d(t, s) = 0 \Leftrightarrow t = s \ \forall s, t \in \mathcal{S}$.
3. $d(r, s) + d(s, t) \geq d(r, t) \forall r, s, t \in \mathcal{S}$.

Elements of $\mathcal{S}$ are called points. The value $d(t, s)$ is called the distance from t to s, or the distance between t and s. Property 3 is called the *triangle inequality.*

When $\mathcal{S} = \Re^n$, the most commonly used metrics are

- The L_2 or Euclidean metric, $d(\mathbf{x}, \mathbf{y}) = \sqrt{\mathbf{x} \cdot \mathbf{y}} = \sqrt{\sum_{j=1}^n (x_j - y_j)^2}$. This is the default choice of metric.
- The L_1 or taxicab metric, $d(\mathbf{x}, \mathbf{y}) = \sum_{j=1}^n |x_j - y_j|$.
- The L_∞ or sup metric, $d(\mathbf{x}, \mathbf{y}) = \max_{j=1}^n |x_j - y_j|$.

These are all specific cases of the metric induced by the L_p norm $||x||_p = (\sum_{j=1}^n x_j^p)^{1/p}$.

Definition 19.2 *A set of points $S \subset \mathcal{S}$ is open iff for all $t \in S$, there exists $\epsilon > 0$ (which may depend on t) such that $d(t, s) < \epsilon \Rightarrow s \in S$. That is, at each $t \in S$, for sufficiently small $\epsilon > 0$ all points closer than ϵ to t are also in S.*

Let $\epsilon > 0$. The open ball $B(t, \epsilon)$ around $t \in \mathcal{S}$ is $\{s \in \mathcal{S} : d(t, s) < \epsilon\}$, the set of points in $\mathcal{S}$, including t, that have distance strictly less than ϵ from t. The parameter ϵ must be strictly positive.

A set $S \subset \mathcal{S}$ is *open* if for every point $s \in S$ there exists an open ball around s contained in S.

Example 19.1 Let $\mathcal{S} = \Re$. Let $d(t, s) = |t - s|$. The open ball $B(t, \epsilon)$ is the interval $(t - \epsilon, t + \epsilon)$. The subset S consisting of the interval $(0, 10)$ is open because for any $t \in (0, 10)$ there exists a small enough $\epsilon > 0$ such that $0 < t - \epsilon$ and $t + \epsilon < 10$ (i.e., $B(t, \epsilon) \subset (0, 10)$).

Question: Specify a valid value for ϵ.[3]

Example 19.2 Let $\mathcal{S} = \Re^2$. Let $d(t, s)$ be the Euclidean metric. Let $S = \{(t, 0) : 0 < t < 10\}$. S seems to be the same set as in the preceding example. But S is not open. The point $t = (4, 0)$ is in S but for any $\epsilon > 0$ the ball $B(t, \epsilon)$ contains point $(4, \epsilon/2)$ which is not in S.

Example 19.3 Let $\mathcal{S} = \Re^2$. Let $S = \{(t, 0) : 0 < t < 10\}$ as in the previous example. Let $d(t, s)$ be the *discrete metric* defined as $d(t, t) = 0$ for all t and $d(t, s) = 1$ for all $s \neq t$. Let's verify that $d()$ is a valid distance function. The first two properties are immediate. For the triangle inequality, if $r = t$, then $d(r, t) = 0 \leq d(r, s) + d(s, t)$; if $r \neq t$ at least one of $d(r, s)$ and $d(s, t)$ equals 1 which is greater than or equal to $d(r, t) = 1$.
Now, let's prove S is open with respect to the discrete metric. For any point $t \in S$ let $\epsilon = 0.1234$. The ball $B(t, \epsilon)$ is the singleton set $\{t\}$, which is in S.

Thus the property that a set S in a metric space $\mathcal{S}, d()$ is open is a relationship among S, the set $\mathcal{S}$, and the metric $d()$. *Question: Why did I choose the value* 0.1234*?*[4]

The standard definition of a closed set is that its complement is open. Fortunately, that definition is equivalent to a more direct definition.

Definition 19.3 *Let $\mathcal{S}, d()$ be a metric space. A subset $S \subset \mathcal{S}$ is closed if its complement $\bar{S}$ is open. Equivalently, S is closed if*

$$\{\textit{for all } \epsilon > 0, B(t, \epsilon) \cap S \neq \phi\} \Rightarrow \{t \in S\}.$$

In other words, if S contains points arbitrarily close to t, then it contains t as well.

Example 19.4 The sets $\{\mathbf{x} \in \Re^n : A\mathbf{x} \leq \mathbf{b}\}$ and $\{\mathbf{x} \in \Re^n : ||\mathbf{x}|| \leq \alpha\}$ are both closed. The set $\{x \in \Re : 0 < x \leq 1\}$ is neither closed nor open.

---

[3] $\epsilon = \min\{t/2, (10 - t)/2\}$.

[4] I chose it to emphasize that in this example the exact value doesn't matter much. Any $0 < \epsilon < 1$ would do.

An infinite sequence of (not necessarily distinct) points $s_1, s_2, \ldots$ converges to t if for all $\epsilon > 0$, eventually the sequence stays within ϵ of t.

Example 19.5 The sequences
$2, -2, \frac{3}{2}, -\frac{3}{2}, \frac{4}{3}, -\frac{4}{3}, \frac{5}{4}, -\frac{5}{4}, \ldots$
and
$\frac{1}{2}, \frac{3}{4}, \frac{7}{8}, \frac{15}{16}, \ldots$
both converge to 1. The sequence
$1, 0, 1, 1, 0, 1, 1, 1, 0, 1, 1, 1, 1, 0, \ldots$
does not converge to any value.

A set S is closed if it has the following property: if a sequence of points in S converges to t, then t must also be in S.

The language of infinitesimals captures the intuition of point-set topology well. Without venturing formally into nonstandard analysis, you may find the following helpful:

- S is open if for every $t \in S$, all points infinitesimally close to t are also in S.
- S is closed if it has the following property: if S contains a point infinitesimally close to t, then S must contain t.
- A function $f : \mathcal{S} \mapsto \mathcal{V}$ is continuous if it has the following property: if s and t are infinitesimally close, then $f(s)$ and $f(t)$ are infinitesimally close.

The formal definition of a continuous function is less intuitive.

Definition 19.4 *Let $\mathcal{S}, \mathcal{V}$ be metric spaces. The function $f : \mathcal{S} \mapsto \mathcal{V}$ is continuous if $\{V \subset \mathcal{V}$ is open$\}$ implies $\{f^{-1}(V)$ is open $\}$. Equivalently, f maps closed sets to closed sets, i.e., if S is closed, then $f(S)$ is closed.*

Imagine a sequence $s_1, s_2, s_3, \ldots$ of points in $S \subset \mathcal{S}$ that converges to $t \in \mathcal{S}$. As the sequence progresses, its points get arbitrarily close to each other. Next, imagine the images of those points, $f(s_1), f(s_2), f(s_3), \ldots$ in the space $\mathcal{V}$. If f is continuous, the image points get arbitrarily close to each other, and converge to $f(t)$. If S is closed, $t \in S$, and so $f(t) \in f(S)$.

The "epsilon-delta" definition of continuity combines Definitions 19.4 and 19.2. The function $f : \mathcal{S} \mapsto \mathcal{V}$ is continuous at $s \in \mathcal{S}$ if for all $\epsilon > 0$ there exists $\delta > 0$ such that $t \in B(s, \delta) \Rightarrow f(t) \in B(f(s), \epsilon)$. If f is continuous at all $s \in S \subset \mathcal{S}$ (respectively $s \in \mathcal{S}$), we say f is continuous on S (respectively continuous).

Uniform continuity is a slightly stronger property than continuity. The difference is that δ may not depend on s.

Definition 19.5 *The function $f : \mathcal{S} \mapsto \mathcal{V}$ is uniformly continuous if for all $\epsilon > 0$ there exists $\delta > 0$ such that for all $s \in \mathcal{S}$ and $t \in \mathcal{S}$, $t \in B(s, \delta) \Rightarrow f(t) \in B(f(s), \epsilon)$.*

Example: The function $f : \Re \mapsto \Re$, $f(s) = 9s$ is uniformly continuous. For given $\epsilon > 0$, set $\delta = \epsilon/9$. Then $|s - t| < \delta \Rightarrow |f(s) - f(t)| = 9|s - t| < 9\delta = \epsilon$. The function $g : \Re \mapsto \Re$, $g(s) = s^2$ is not uniformly continuous. For $\epsilon = 1$, consider any $\delta > 0$. The points $s = 1/\delta$ and $t = s + \delta/2$ differ by less than δ. Yet, $g(t) - g(s) = s\delta + \delta^2/4 = 1 + \delta^2/4 > 1 = \epsilon$.

Compactness is an important property whose formal definition is, unfortunately, inscrutable to the novice.

Definition 19.6 *$S \subset \mathcal{S}$ is compact iff for every set $\mathcal{O}$ of open sets whose union $\cup_{O \in \mathcal{O}}$ contains S, there exists a finite subset of $\mathcal{O}$, $O_1, O_2, \ldots, O_k$, whose union contains S.*

Compactness is often phrased as "every open cover contains a finite subcover." Fortunately, we can use a much simpler definition in $\Re^n$ with the Euclidean (or sup or L_1) metric.

Theorem 19.2.1 *A set $S \subset \Re^n$ is compact iff it is closed and bounded.*

Compactness enhances continuity.

Theorem 19.2.2 *If f is continuous on S and S is compact, then f is uniformly continuous.*

The key theorem of this section is about minimizing or maximizing a continuous real-valued function on a compact set.

Theorem 19.2.3 *Let $\mathcal{S}, d()$ be a metric space. Let $f : \mathcal{S} \mapsto \Re$ be a continuous function. Let S be a compact subset of $\mathcal{S}$. Then $f()$ attains its minimum on S. That is, there exists $t^* \in S$ such that $f(t^*) \leq f(t)$ for all $t \in S$.*

Proofs of these theorems can be found in any analysis text.

Corollary 19.2.4 *Let $S \subset \Re^n$ be closed and bounded. Let $f : \Re^n \mapsto \Re$ be continuous. Then f attains its minimum (and its maximum) on S.*

19.3 Summary

For $m \times n$ matrix A, m-vector $\mathbf{b}$, and n-vector $\mathbf{c}$, the primal seeks a nonnegative linear combination $\mathbf{y} = A\mathbf{x}$ of the columns of A whose multipliers $\mathbf{x}$ have minimum $\mathbf{c} \cdot \mathbf{x}$ subject to $\mathbf{y} \geq \mathbf{b}$. The dual seeks a nonnegative linear combination $\rho = A\pi$ of the rows of A whose multipliers π have maximum $\mathbf{b} \cdot \pi$ subject to $\rho \leq \mathbf{c}$. The value of any feasible solution to the primal (resp. dual) is an upper (resp. lower) bound on the value of any feasible solution to the dual (resp. primal).

Exactly one of the following holds for these two LPs:
I. Neither primal nor dual has a feasible solution.
II. One of the two is infeasible, and the other has unbounded optimum.
III. Both primal and dual are feasible. Every pair $\mathbf{x}^*, \pi^*$ of optimal solutions to the primal and dual respectively satisfies complementary slackness and $\mathbf{c} \cdot \mathbf{x}^* = \mathbf{b} \cdot \pi^*$. There exists a pair that satisfies strict complementary slackness. At optimality, the slack variables of an LP equal the marginal cost of enforcing the constraints of the other LP. Predictions based on these values are either accurate or optimistically inaccurate.

A set in $\Re^n$ is convex iff it contains all convex combinations of its points. Every (closed) convex set in $\Re^n$ can be (strictly) separated from any point not in the set by a hyperplane. Extreme points of a subset of $\Re^n$ are not convex combinations of other subset members. All polyhedra are convex and have finitely many extreme points. Only the empty polyhedron and polyhedra containing a line have no extreme points.

The feasible region of an LP is a polyhedron, and vice versa. It can be characterized in two fundamentally different ways: first, by the constraints of the LP. If it is d-dimensional, its minimal such characterization is a set of $n - d$ linearly independent equality constraints, and a finite set of linear inequalities that is unique up to scaling. These linear inequalities define the facets of the polyhedron. They are the valid inequalities whose intersection with the polyhedron is $d-1$-dimensional. The intersection is called a facet. The extreme points are the set of points in the polyhedron that lie on at least d facets. Equivalently, extreme points are those in the polyhedron for which n linearly independent constraints hold with equality.

Second, a polyhedron is the set sum of a polytope and a polyhedral cone. A polytope is a bounded polyhedron. It is equal to the set of convex combinations of its finitely many extreme points. The polyhedral cone can be characterized as the set of nonnegative linear combinations of a finite set of vectors, which are extreme directions of the polyhedron.

The set of optimal solutions to an LP is always a face of the feasible region. Conversely, for every face there exists an objective vector $\mathbf{c}$ for which that face is the set of optimal solutions. In particular, every extreme point is the unique optimizer of some objective.

The simplex method traverses a sequence of adjacent extreme points. It uses the dual variables to choose which facet to leave to reach the next extreme point. If a tie-breaking rule such as Bland's or perturbation/lexicography is used, it terminates in a finite number of steps at an optimal solution. It can take exponentially many steps in the theoretical worst-case but is fast in practice.

In practice, the dual simplex method is more widely used than the original primal method. It is especially suited to re-solving a solved LP to which a constraint that cuts off the optimal solution has been added. That is why it is a fundamental component of algorithms for integer programs.

Network models are a special case of LP models. They are easy to understand visually, and can be solved faster than LPs in general. The reduced cost at a node equals the cost of disposing an additional unit of supply from there. The extreme points of network models have all integer values if the bounds on flows are integers. This integrality property combines with LP strong duality to prove Hall's, Dilworth's, and some other combinatorial duality theorems. The dual to maximum flow from s to t is the minimum cut between s and t, which is the minimum possible sum of arc capacities from S to T of a partition S, T of the nodes such that $s \in S$ and $t \in T$.

There are many specialized algorithms for shortest path and maximum flow problems. They depend in part on sophisticated data structures. They are faster than general LP algorithms in both theory and practice.

The primary classification of an algorithm's speed is whether it is "fast," meaning that its runtime is bounded by a polynomial in the size of its input, or whether it is "slow." The primary theoretical classification of a problem is whether it is "easy," meaning that it can be solved by a fast algorithm, or whether it is "hard." This classification guides you as to what kind of algorithm to use, and how strong a guarantee of optimality or solution quality you can reasonably expect to obtain. In practice, the NP-complete problems are treated as "hard."

The most useful NP-complete problem in practice is integer programming. It is natural to model problems as IPs, and the software for solving IPs is very advanced. IP algorithms rely heavily on the speed of the dual simplex algorithm in re-solving the LP relaxations after constraints have been added, and on the generation of facet-defining and other valid inequalities for the convex hull of the feasible IP solutions.

The ellipsoid method is a theoretically fast algorithm for LP. Moreover, it establishes the polynomial equivalence between separation of a point from the feasible region, and optimization over that region. In each iteration it projects the current ellipsoid to a unit ball, encloses half the ball in a new ellipsoid, and projects back. In practice it is rarely competitive with other algorithms except possibly subgradient search.

Elementary nonlinear optimization is all about finding local optima by approximating nonlinear functions with affine functions (linear functions plus a constant) computed from their derivatives. Newton's method for finding zeros of a differentiable function $f : \Re \mapsto \Re$ is the first such algorithm. It finds local optima of g by setting $g^{'} = f$. Its quadratic convergence rate doubles the solution precision each iteration. The secant method, a variation, approximates $f^{'}$ from the evaluation of $f()$ at two points. Its convergence is superlinear but subquadratic.

These methods generalize to functions $f : \Re^n \mapsto \Re$ via the Taylor series $f(\mathbf{v} + \mathbf{x}) = f(\mathbf{v}) + \nabla f(\mathbf{v}) \cdot \mathbf{x} + \frac{1}{2}\mathbf{x}^T \nabla^2 f(\mathbf{v})\mathbf{x} + \ldots$. The second term in the series tells us that the $\nabla f(\mathbf{v})$ $(-\nabla f(\mathbf{v}))$ is the direction of maximum instantaneous increase (decrease) in f at $\mathbf{v}$. Gradient search algorithms seek local optima by choosing a search direction based on ∇f. Once the direction is chosen, computing a step size is a one-dimensional optimization subproblem.

To perform gradient search subject to equations $A\mathbf{x} = \mathbf{b}$, project the objective gradient ∇f onto the subspace $\{\mathbf{y} : A\mathbf{y} = \mathbf{0}\}$. Feasibility constraints $g_j(\mathbf{x}) \geq 0$ make $\nabla f(\mathbf{x}) = \mathbf{0}$ not necessary for $\mathbf{x}$ to be a local optimum. A feasible point $\mathbf{x}$ is a local minimum if there is no direction to move that decreases f and retains feasibility. Approximating f and g_j with their derivatives, no $\mathbf{d}$ exists such that $\nabla f(\mathbf{x}) \cdot \mathbf{d} < 0$ and, for all j, either $g_j(\mathbf{x}) > 0$ or $\nabla g(\mathbf{x}) \cdot \mathbf{d} \geq 0$. The KKT conditions contain the logically equivalent dual statement that the gradient $\nabla f(\mathbf{x})$ is a nonnegative linear combination of the gradients of the constraints that are binding at $\mathbf{x}$.

Interior point algorithms can be theoretically fast, and their implemented versions are now faster on average than simplex algorithms on large LPs. They also parallelize more readily than simplex algorithms, an important advantage with multicore processors. Projective scaling algorithms perform gradient search in the strict relative interior $\{\mathbf{x} : A\mathbf{x} = \mathbf{b}; \mathbf{x} > \mathbf{0}\}$ of the feasible region. Their key innovation is to scale the current solution $\mathbf{x}$ away from the $\mathbf{x} > \mathbf{0}$ constraints to $\mathbf{x}diag(\mathbf{x})^{-1} = \mathbf{1}$, calculate the projection $\mathbf{d}$ of the gradient $\mathbf{c}$ onto $\{\mathbf{y} : A\mathbf{y} = \mathbf{0}\}$, and reverse the scaling to get direction $\mathbf{d}diag(\mathbf{x})$. They are guaranteed to converge if the step size is not too large. Logarithmic barrier algorithms are currently the dominant interior point techniques for LP. They replace inequality constraints $\mathbf{x} \geq \mathbf{0}$ with objective penalty term $-\theta \sum_i \log x_i$. The set of optimal solutions parameterized by θ, called the central path, converges to an optimal LP solution as $\theta \to 0$. The algorithms take Newton's method steps to stay close to the central path as θ decreases. In practice, barrier algorithms converge in $O(\log n)$ steps. However, if there are multiple optima they converge to the centroid of the face of optimal solutions, whereas an optimal basic feasible solution is often desired. Barrier algorithms therefore append a simplex-type procedure to "cross over" to a basic solution.

Bibliography

[1] Tobias Achterberg, Robert E Bixby, Zonghao Gu, Edward Rothberg, and Dieter Weninger. Presolve reductions in mixed integer programming. *INFORMS Journal on Computing*, 2019.

[2] Tobias Achterberg and Roland Wunderling. Mixed integer programming: Analyzing 12 years of progress. In *Facets of Combinatorial Optimization*, pages 449–481. Springer, 2013.

[3] Ilan Adler. Personal communication, 2004.

[4] Ravindra K Ahuja, Kurt Mehlhorn, James Orlin, and Robert E Tarjan. Faster algorithms for the shortest path problem. *Journal of the ACM (JACM)*, 37(2):213–223, 1990.

[5] Ravindra K Ahuja and James B Orlin. A fast and simple algorithm for the maximum flow problem. *Operations Research*, 37(5):748–759, 1989. Manuscript dated 1986.

[6] Ravindra K Ahuja, James B Orlin, and Robert E Tarjan. Improved time bounds for the maximum flow problem. *SIAM Journal on Computing*, 18(5):939–954, 1989.

[7] Erling D Andersen and Knud D Andersen. Presolving in linear programming. *Mathematical Programming*, 71(2):221–245, 1995.

[8] Esther M Arkin and Christos H Papadimitriou. On negative cycles in mixed graphs. *Operations Research Letters*, 4(3):113–116, 1985.

[9] Sanjeev Arora. Polynomial time approximation schemes for Euclidean TSP and other geometric problems. In *Proceedings of 37th Conference on Foundations of Computer Science*, pages 2–11. IEEE, 1996.

[10] Sanjeev Arora and Boaz Barak. *Computational Complexity: A Modern Approach.* Cambridge University Press, 2009.

[11] Sanjeev Arora, Carsten Lund, Rajeev Motwani, Madhu Sudan, and Mario Szegedy. Proof verification and the hardness of approximation problems. *Journal of the ACM (JACM)*, 45(3):501–555, 1998.

[12] Sanjeev Arora and Shmuel Safra. Probabilistic checking of proofs: A new characterization of NP. *Journal of the ACM (JACM)*, 45(1):70–122, 1998.

[13] Bengt Aspvall and Richard E Stone. Khachiyan's linear programming algorithm. *Journal of Algorithms*, 1(1):1–13, 1980.

[14] Egon Balas. Disjunctive programming. In *Annals of Discrete Mathematics*, volume 5, pages 3–51. Elsevier, 1979.

[15] Earl R Barnes. A variation on Karmarkar's algorithm for solving linear programming problems. *Mathematical Programming*, 36(2):174–182, 1986.

[16] Cynthia Barnhart, Ellis L Johnson, George L Nemhauser, Martin WP Savelsbergh, and Pamela H Vance. Branch-and-price: Column generation for solving huge integer programs. *Operations Research*, 46(3):316–329, 1998.

[17] John Bartholdi, Lakshmi Narasimhan, and Craig Tovey. Recognizing equilibria in spatial voting games. *Social Choice and Welfare*, 8:183–197, 1991.

[18] Dave A Bayer and Jeffrey C Lagarias. The nonlinear geometry of linear programming. i. affine and projective scaling trajectories. *Transactions of the American Mathematical Society*, 314(2):499–526, 1989.

[19] Dave A Bayer and Jeffrey C Lagarias. The nonlinear geometry of linear programming. ii. Legendre transform coordinates and central trajectories. *Transactions of the American Mathematical Society*, 314(2):527–581, 1989.

[20] Mokhtar S Bazaraa and John J Jarvis. *Linear Programming and Network Flows.* John Wiley & Sons, 1977.

[21] E. M. L. Beale. Cycling in the dual simplex algorithm. *Naval Research Logistics Quarterly*, 2(4):269–276, 1955.

[22] J.F. Benders. Partitioning procedures for solving mixed-variables programming problems. *Numerische Mathematik*, 4:238–252, 1962.

[23] Dimitris Bertsimas and James B. Orlin. A technique for speeding up the solution of the Lagrangean dual. *Mathematical Programming*, 63(1-3):23–45, 1994.

[24] Daniel Bienstock, Fan Chung, Michael L. Fredman, Alejandro A. Schäffer, Peter W. Shor, and Subhash Suri. A note on finding a strict saddlepoint. *Am. Math. Monthly*, 98(5):418–419, April 1991.

[25] Robert Bixby and Edward Rothberg. Progress in computational mixed integer programming—a look back from the other side of the tipping point. *Annals of Operations Research*, 149(1):37–41, 2007.

[26] Robert E Bixby. Solving real-world linear programs: A decade and more of progress. *Operations Research*, 50(1):3–15, 2002.

[27] Robert E Bixby, Mary Fenelon, Zonghao Gu, Ed Rothberg, and Roland Wunderling. Mixed-integer programming: A progress report. In *The Sharpest Cut: The Impact of Manfred Padberg and his Work*, pages 309–325. SIAM, 2004.

[28] Robert G. Bland, Donald Goldfarb, and Michael J. Todd. The ellipsoid method: A survey. *Operations Research*, 29(6):1039 – 1091, 1981.

[29] Manuel Blum, Robert W. Floyd, Vaughan R. Pratt, Ronald L. Rivest, and Robert Endre Tarjan. Time bounds for selection. *J. Comput. Syst. Sci.*, 7(4):448–461, 1973.

[30] B. Bolzano. Purely analytic proof of the theorem that between any two values which give results of opposite sign, there lies at least one real root of the equation. In William Bragg Ewald, editor, *From Kant to Hilbert Volume 1: A Source Book in the Foundations of Mathematics*, pages 225–248. Oxford University Press, 1996. Originally published as Bolzano, B. 1817, Rein analytischer Beweis des Lehrsatzes, dasszwischen je zwey Werthen, die ein entgegen gesetzes Resultat gewähren, wenigstens eine reele Wurzel der Gleichung liege. Prag: Gottlieb Haase, 1817.

[31] Oleg V Borodin, Alexei N Glebov, Mickaël Montassier, and André Raspaud. Planar graphs without 5-and 7-cycles and without adjacent triangles are 3-colorable. *Journal of Combinatorial Theory, Series B*, 99(4):668–673, 2009.

[32] AL Brearley, Gautam Mitra, and H Paul Williams. Analysis of mathematical programming problems prior to applying the simplex algorithm. *Mathematical Programming*, 8(1):54–83, 1975.

[33] Ronald L Breiger. The duality of persons and groups. *Social Forces*, 53(2):181–190, 1974.

[34] R.W. Brockett. Dynamical systems that sort lists, diagonalize matrices and solve linear programming problems. *Proceedings of the 27th Conference on Decision and Control*, pages 799–803, December 1988.

[35] Asad A Butt and Robert T Collins. Multi-target tracking by Lagrangian relaxation to min-cost network flow. In *Proceedings of the IEEE Conference on Computer Vision and Pattern Recognition*, pages 1846–1853, 2013.

[36] Michael Carter and Craig Tovey. When is the classroom assignment problem hard? *Operations Research*, 40:S28–S39, 1992.

[37] Barun Chandra, Howard Karloff, and Craig Tovey. New results on the old k-opt algorithm for the tsp. *SIAM Journal on Computation*, 28:1998–2029, 1999.

[38] Vijay Chandru and Michael A. Trick. The search for optimal Lagrange multipliers. *Journal of the Indian Institute of Science*, 78(2):131–151, 1998. (First circulated as unpublished manuscript in 1987.).

[39] Abe Charnes. Optimality and degeneracy in linear programming. *Econometrica*, 20:160–170, April 1952.

[40] Abraham Charnes and William W Cooper. Chance-constrained programming. *Management Science*, 6(1):73–79, 1959.

[41] Abraham Charnes, William W Cooper, and Robert O Ferguson. Optimal estimation of executive compensation by linear programming. *Management Science*, 1(2):138–151, 1955.

[42] Abraham Charnes, William W Cooper, and Bob Mellon. Blending aviation gasolines–a study in programming interdependent activities in an integrated oil company. *Econometrica: Journal of the Econometric Society*, pages 135–159, 1952.

[43] Abraham Charnes and William Wager Cooper. Management models and industrial applications of linear programming. *Management Science*, 4(1):38–91, 1957.

[44] Abraham Charnes, Alexander Henderson, and William Wager Cooper. *An Introduction to Linear Programming*. John Wiley & Sons, 1955.

[45] Gary Chartrand, Albert Polimeni, and Ping Zhang. *Mathematical Proofs: A Transition to Advanced Mathematics*. Boston, 2008.

[46] Der-San Chen, Robert G Batson, and Yu Dang. *Applied Integer Programming*. Wiley Online Library, Hoboken, NJ, 2010.

[47] Joseph Cheriyan and SN Maheshwari. Analysis of preflow push algorithms for maximum network flow. *SIAM Journal on Computing*, 18(6):1057–1086, 1989.

[48] Boris Cherkassky and Andrew Goldberg. Negative-cycle detection algorithms. *Mathematical Programming*, 85:277–311, 1999.

[49] Nicos Christofides. Worst-case analysis of a new heuristic for the travelling salesman problem. Technical report, Carnegie-Mellon Univ Pittsburgh Pa Management Sciences Research Group, 1976.

[50] Vasek Chvátal. Edmonds polytopes and a hierarchy of combinatorial problems. *Discrete Mathematics*, 4:305–337, 1973.

[51] Vasek Chvátal. *Linear Programming.* W. H. Freeman, New York, NY, 1983.

[52] Michael B Cohen, Yin Tat Lee, and Zhao Song. Solving linear programs in the current matrix multiplication time. In *Proceedings of the 51st Annual ACM SIGACT Symposium on Theory of Computing*, pages 938–942, 2019.

[53] Wikimedia Commons. File:tokyo subway map.png — wikimedia commons, the free media repository, 2015. [Online; accessed 29-November-2018].

[54] Michele Conforti, Gérard Cornuéjols, Giacomo Zambelli, et al. *Integer Programming*, volume 271. Springer, 2014.

[55] William J Cook. *In Pursuit of the Traveling Salesman: Mathematics at the Limits of Computation.* Princeton University Press, Princeton, New Jersey, 2011.

[56] Thomas Cormen, Charles Leiserson, and Ronald Rivest. *Introduction to Algorithms.* The MIT Press and McGraw-Hill Co., New York, 1990.

[57] Richard W Cottle. Symmetric dual quadratic programs. *Quarterly of Applied Mathematics*, 21(3):237–243, 1963.

[58] Richard W Cottle. Minimal triangulation of the 4-cube. *Discrete Mathematics*, 40(1):25–29, 1982.

[59] Richard W Cottle. Quartic barriers. *Computational Optimization and Applications*, 12(1-3):81–105, 1999.

[60] Gregory E Coxson. The p-matrix problem is co-np-complete. *Mathematical Programming*, 64(1-3):173–178, 1994.

[61] Antonella Cupillari. *The Nuts and Bolts of Proofs: An Introduction to Mathematical Proofs.* Academic Press, 2011.

[62] Daniel Dadush and Sophie Huiberts. A friendly smoothed analysis of the simplex method. In *Proceedings of the 50th Annual ACM SIGACT Symposium on Theory of Computing*, pages 390–403, 2018.

[63] John P. D'Angelo and Douglas B. West. *Mathematical Thinking: Problem-Solving and Proofs.* Prentice-Hall Upper Saddle River, 2nd edition, 2000.

[64] George Dantzig. *Linear Programming and Extensions.* Princeton University Press, 1963.

[65] George B. Dantzig. A theorem on linear inequalities. (Statement and proof of von Neuman's October 1947 invention of LP duality.), January 1948.

[66] George B Dantzig. Programming of interdependent activities: II mathematical model. *Econometrica, Journal of the Econometric Society*, pages 200–211, 1949.

[67] George B Dantzig. A proof of the equivalence of the programming problem and the game problem. In T.C. Koopmans, editor, *Cowles Commission Monograph 13: Activity Analysis of Production and Allocation*, chapter 13, pages 330–338. John Wiley & Sons, Inc, New York, 1951.

[68] George B Dantzig. Reminiscences about the origins of linear programming. Technical report, Stanford Univ Ca Systems Optimization Lab, 1981.

[69] George B Dantzig. Impact of linear programming on computer development. Technical report, Stanford Univ Ca Systems Optimization Lab, 1985.

[70] George B Dantzig. Linear programming. *Operations Research*, 50(1):42–47, 2002.

[71] George B. Dantzig, Alex Orden, and Philip Wolfe. The generalized simplex method for minimizing a linear form under linear inequality restraints. *Pacific Journal of Mathematics*, 5(2):183–195, 1955.

[72] George B. Dantzig and Philip Wolfe. The decomposition principle for linear programs. *Operations Research*, 8:101–111, 1960.

[73] George Bernard Dantzig. *Linear Programming and Extensions.* Princeton University Press, 1963.

[74] George Bernard Dantzig. Reminiscences about the origins of linear programming. In *Mathematical Programming: The State of the Art*, pages 78–86. Springer, 1983.

[75] George Bernard Dantzig, Lester Randolph Ford Jr, and Delbert Ray Fulkerson. A primal–dual algorithm. Technical report, Rand Corp Santa Monica Ca, 1956.

[76] Constantinos Daskalakis, Paul W Goldberg, and Christos H Papadimitriou. The complexity of computing a Nash equilibrium. *SIAM Journal on Computing*, 39(1):195–259, 2009.

[77] Antoine Deza, Tamás Terlaky, and Yuriy Zinchenko. Central path curvature and iteration-complexity for redundant klee—minty cubes. In *Advances in Applied Mathematics and Global Optimization*, pages 223–256. Springer, 2009.

[78] II Dikin. Iterative solution of problems of linear and quadratic programming. In *Doklady Akademii Nauk*, volume 174, pages 747–748. Russian Academy of Sciences, 1967.

[79] RP Dilworth. A decomposition theorem for partially ordered sets. *Annals of Math.*, 51:161–166, 1950.

[80] E.A. Dinits. Algorithm of solution to problem of maximum flow in network with power estimates. *Doklady Akademii Nauk SSSR*, 194(4):754, 1970.

[81] WS Dorn. Self-dual quadratic programs. *Journal of the Society for Industrial and Applied Mathematics*, 9(1):51–54, 1961.

[82] Richard James Duffin, George Bernard Dantzig, and Ky Fan. *Linear Inequalities and Related Systems.* Princeton University Press, 1956.

[83] B. Curtis Eaves. Computing kakutani fixed points. *SIAM Journal Applied Mathematics*, 21:236–244, 1971.

[84] B. Curtis Eaves. Homotopies for computation of fixed points. *Mathematical Programming*, 3:1–22, 1972.

[85] B. Curtis Eaves and Herbert Scarf. The solution of systems of piecewise linear equations. *Mathematics of Operations Research*, 1:1–27, 1976.

[86] Dorit S Hochbaum (editor). *Approximation Algorithms for NP-Hard Problems.* PWS Publishing Co., 1997.

[87] Jack Edmonds. Paths, trees and flowers. *Canadian Journal of Mathematics*, 17:449–467, 1965.

[88] Jack Edmonds and Richard Richard Karp. Theoretical improvements in algorithmic efficiency for network flow problems. *Journal of the Association of Computing Machinery*, 19(2):248–264, April 1972.

[89] Eugene (Jenő) Egerváry. On combinatorial properties of matrices. *Matematikai és Fizikai Lapok*, 38:16–28, 1931. (In Hungarian.).

[90] Paul Erdös and George Szekeres. A combinatorial problem in geometry. *Compositio Mathematica*, 2:463–470, 1935.

[91] Julius Farkas. Uber die theorie der einfachen ungeichungen. *J. Reine Angew. Math.*, 124:1–24, 1902.

[92] Yahya Fathi and Craig Tovey. Affirmative action algorithms. *Mathematical Programming*, 34:292–301, 1986.

[93] Werner Fenchel. On conjugate convex functions. *Canad. J. Math*, 1(73-77), 1949.

[94] Werner Fenchel and Donald W. Blackett. *Convex Cones, Sets, and Functions.* Princeton University, Department of Mathematics, Logistics Research Project, 1953.

[95] Samuel Fiorini, Serge Massar, Sebastian Pokutta, Hans Raj Tiwary, and Ronald De Wolf. Linear vs. semidefinite extended formulations: exponential separation and strong lower bounds. In *Proceedings of the 44th Annual ACM Symposium on Theory of Computing*, pages 95–106. ACM, 2012.

[96] Marshall L Fisher. An applications oriented guide to Lagrangian relaxation. *Interfaces*, 15(2):10–21, 1985.

[97] Lester R Ford Jr and Delbert R Fulkerson. A simple algorithm for finding maximal network flows and an application to the Hitchcock problem. Technical Report P-743, The RAND Corporation, Santa Monica, California, 1955.

[98] Lester R Ford Jr and Delbert Ray Fulkerson. Maximal flow through a network. Technical Report P-605, The RAND Corporation, 1954.

[99] Lester R Ford Jr and Delbert Ray Fulkerson. Constructing maximal dynamic flows from static flows. *Operations Research*, 6(3):419–433, 1958.

[100] Lester R Ford Jr and Delbert Ray Fulkerson. *Flows in Networks.* Princeton University Press, 2015. (First published in 1962.).

[101] John Forrest and Donald Goldfarb. Steepest-edge simplex algorithms for linear programming. *Mathematical Programming*, 57(1-3):341–374, 1992.

[102] George E. Forsythe. What is a satisfactory quadratic equation solver? Technical report, Stanford University, Dept. of computer science, 1967.

[103] George E. Forsythe. Pitfalls in computation, or why a math book isn't enough. *American Mathematical Monthly*, pages 931–956, 1970.

[104] András Frank. On Kuhn's Hungarian method—a tribute from Hungary. *Naval Research Logistics Quarterly*, 52(1):2–5, 2005.

[105] Michael L. Fredman and Robert E. Tarjan. Fibonacci heaps and their uses in improved network optimization algorithms. *Journal of the Association of Computing Machinery*, 34:596–615, 1987.

[106] Oliver Friedmann. *Exponential lower bounds for solving infinitary payoff games and linear programs.* PhD thesis, Ludwig-Maximilians-Universität München, 2011.

[107] Oliver Friedmann. A subexponential lower bound for Zadeh's pivoting rule for solving linear programs and games. In *International Conference on Integer Programming and Combinatorial Optimization*, pages 192–206. Springer, 2011.

[108] Oliver Friedmann, Thomas Dueholm Hansen, and Uri Zwick. Subexponential lower bounds for randomized pivoting rules for the simplex algorithm. In *Proceedings of the Forty-Third Annual ACM Symposium on Theory of Computing*, pages 283–292. ACM, 2011.

[109] Ragnar Frisch. La résolution des problèmes de programme linéaire par la méthode du potentiel logarithmique. *Cahiers du Seminaire D'Econometrie*, pages 7–23, 1956.

[110] Harold N. Gabow. Scaling algorithms for network problems. *Journal of Computer and Systems Sciences*, 31:148–168, 1985.

[111] David Gale, Harold W Kuhn, and Albert W Tucker. Linear programming and the theory of games. In T. C. Koopmans, editor, *Cowles Commission Monograph 13: Activity Analysis of Production and Allocation.* John Wiley & Sons, Inc, New York, 1951.

[112] Michael Garey and David Johnson. *Computers and Intractability: A Guide to the Theory of NP-completeness.* W.H. Freeman & Co., San Francisco, 1979.

[113] Michael R Garey, David S Johnson, and Larry Stockmeyer. Some simplified np-complete problems. In *Proceedings of the Sixth Annual ACM Symposium on Theory of Computing*, pages 47–63. ACM, 1974.

[114] Bernd Gärtner and Jiri Matousek. *Approximation Algorithms and Semidefinite Programming.* Springer Science & Business Media, 2012.

[115] Saul Gass. Comments on the possibility of cycling with the simplex method. *Operations Research*, 27(4):848–852, 1979.

[116] Arthur Geoffrion. An introduction to structured modeling. *Management Science*, 33:547–588, 1987.

[117] Arthur M Geoffrion. Primal resource-directive approaches for optimizing nonlinear decomposable systems. *Operations Research*, 18(3):375–403, 1970.

[118] Arthur M Geoffrion. Lagrangean relaxation for integer programming. In *Approaches to Integer Programming*, pages 82–114. Springer, 1974.

[119] John R Gilbert and Tim Peierls. Sparse partial pivoting in time proportional to arithmetic operations. *SIAM Journal on Scientific and Statistical Computing*, 9(5):862–874, 1988.

[120] Philip E Gill, Walter Murray, Michael A Saunders, John A Tomlin, and Margaret H Wright. On projected Newton barrier methods for linear programming and an equivalence to Karmarkar's projective method. *Mathematical Programming*, 36(2):183–209, 1986.

[121] Philip E Gill, Walter Murray, and Margaret H Wright. *Practical Optimization.* SIAM, 2019.

[122] P.C. Gilmore and R.E. Gomory. A linear programming approach to the cutting stock problem. *Operations Research*, 9:849–859, 1961.

[123] P.C. Gilmore and R.E. Gomory. A linear programming approach to the cutting stock problem: part ii. *Operations Research*, 11:863–888, 1963.

[124] Michel X Goemans and David P Williamson. A general approximation technique for constrained forest problems. *SIAM Journal on Computing*, 24(2):296–317, 1995.

[125] Michel X Goemans and David P Williamson. The primal-dual method for approximation algorithms and its application to network design problems. In Dorit Hochbaum, editor, *Approximation Algorithms for NP-hard Problems*, pages 144–191. PWS Publishing Co., 1997.

[126] Israel Gohberg, Thomas Kailath, and Vadim Olshevsky. Fast Gaussian elimination with partial pivoting for matrices with displacement structure. *Mathematics of Computation*, 64(212):1557–1576, 1995.

[127] Andrew V. Goldberg and Satish Rao. Beyond the flow decomposition barrier. *Journal of the Association of Computing Machinery*, 45(5):783–797, 1998.

[128] Andrew V. Goldberg and Robert E. Tarjan. A new approach to the maximum-flow problem. *Journal of the Association of Computing Machinery*, 35(4):921–940, 1988.

[129] Andrew V. Goldberg and Robert E. Tarjan. Efficient maximum flow algorithms. *Communications of the ACM*, 57(8):82–89, 2014.

[130] Donald Goldfarb and John Reid. A practicable steepest-edge simplex algorithm. *Mathematical Programming*, 12(1):361–371, 1977.

[131] Ralph E. Gomory, Ellis L. Johnson, and Lisa Evans. Corner polyhedra and their connection with cutting planes. *Mathematical Programming*, Series B 96:321–339, 2003.

[132] Jacek Gondzio. Presolve analysis of linear programs prior to applying an interior point method. *INFORMS Journal on Computing*, 9(1):73–91, 1997.

[133] Paul Gordan. Ueber die auflösung linearer gleichungen mit reellen coefficienten. *Mathematische Annalen*, 6(1):23–28, 1873.

[134] Curtis Greene and Daniel J Kleitman. The structure of Sperner k-families. *Journal of Combinatorial Theory, Series A*, 20(1):41 – 68, 1976.

[135] Martin Grötschel, Lászlo Lovász, and Alexander Schrijver. The ellipsoid method and its consequences in combinatorial optimization. *Combinatorica*, 1(2):169–197, 1981.

[136] Martin Grötschel, Lászlo Lovász, and Alexander Schrijver. Corrigendum to our paper "The ellipsoid method and its consequences in combinatorial optimization". *Combinatorica*, 4(4):291–295, 1984.

[137] Martin Grötschel, László Lovász, and Alexander Schrijver. *Geometric Algorithms and Combinatorial Optimization*, volume 2. Springer Science & Business Media, 2012.

[138] Herbert Grötzsch. Ein dreifarbensatz für dreikreisfreie netze auf der kugel. *Wiss. Z. Martin-Luther-Univ. Halle-Wittenberg Math.-Natur. Reihe*, 8:109–120, 1959.

[139] Pablo Guerrero-Garcia and Ángel Santos-Palomo. On Hoffman's celebrated cycling LP example. *Computers & Operations Research*, 34:2709–2717, 2007.

[140] Steve Hackman and Robert Leachman. A general framework for modeling production. *Management Science*, 35:478–495, 1989.

[141] Leslie A Hall and Robert J Vanderbei. Two-thirds is sharp for affine scaling. *Operations Research Letters*, 13(4):197–201, 1993.

[142] Phillip Hall. On representatives of subsets. *Journal London Mathematics Society*, 10:26–30, 1935.

[143] Paula Harris. Pivot selection methods of the Devex LP code. *Mathematical Programming*, 5(1):1–28, 1973.

[144] J. Håstad. Some optimal inapproximability results. *Journal of the Association of Computing Machinery*, 48:798–859, 2001.

[145] Michael Held and Richard M Karp. The traveling-salesman problem and minimum spanning trees. *Operations Research*, 18(6):1138–1162, 1970.

[146] Michael Held and Richard M Karp. The traveling-salesman problem and minimum spanning trees: Part ii. *Mathematical Programming*, 1(1):6–25, 1971.

[147] Nicholas J. Higham. *Accuracy and Stability of Numerical Algorithms*. Society for Industrial and Applied Mathematics, Philadelphia, PA, USA, second edition, 2002.

[148] Claus Hillermeier. *Nonlinear Multiobjective Optimization: A Generalized Homotopy Approach*. Birkhäuser Verlag, Basel-Boston-Berlin, 2001.

[149] Frank L Hitchcock. The distribution of a product from several sources to numerous localities. *Journal of Mathematics and Physics*, 20(1-4):224–230, 1941.

[150] A.J Hoffman and D.E Schwartz. On partitions of a partially ordered set. *Journal of Combinatorial Theory, Series B*, 23(1):3 – 13, 1977.

[151] Alan J. Hoffman. Cycling in the simplex algorithm. Technical Report 2974, National Bureau of Standards, 1953.

[152] Reiner Horst and Nguyen V Thoai. DC programming: overview. *Journal of Optimization Theory and Applications*, 103(1):1–43, 1999.

[153] Kevin Houston. *How to Think Like a Mathematician: A Companion to Undergraduate Mathematics*. Cambridge University Press, 2009.

[154] Pierre Huard. Resolution of mathematical programming with nonlinear constraints by the method of centers. *Nonlinear Programming*, pages 207–219, 1967.

[155] Robert B Hughes. Minimum-cardinality triangulations of the d-cube for d= 5 and d= 6. *Discrete Mathematics*, 118(1-3):75–118, 1993.

[156] Robert B Hughes and Michael R Anderson. Simplexity of the cube. *Discrete Mathematics*, 158(1-3):99–150, 1996.

[157] Braden Hunsaker, Ellis Johnson, and Craig Tovey. Polarity and the complexity of the shooting experiment. *Discrete Optimization*, 5:541–549, 2008.

[158] Kiyoshi Ishihata. *Algorithms and Data Structures.* Iwanamishoten, 1989.

[159] Alon Itai, Christos H Papadimitriou, and Jayme Luiz Szwarcfiter. Hamilton paths in grid graphs. *SIAM Journal on Computing*, 11(4):676–686, 1982.

[160] William Jacobs. The caterer problem. *Naval Research Logistics, Quarterly*, 1:154–165, 1954.

[161] David Johnson. The NP-completeness column: an ongoing guide. *Journal of Algorithms*, 1982–1992.

[162] David S Johnson. *Near-optimal bin packing algorithms.* PhD thesis, Massachusetts Institute of Technology, 1973.

[163] David S Johnson. The NP-completeness column: an ongoing guide. *Journal of Algorithms*, 2(4):393–405, 1981. This is the first of 23 that appeared in *Journal of Algorithms.*

[164] David S Johnson. The NP-completeness column: Finding needles in haystacks. *ACM Transactions on Algorithms (TALG)*, 3(2):24–es, 2007. This is the 26th and last of David Johnson's columns.

[165] David S Johnson, Christos H Papadimitriou, and Mihalis Yannakakis. How easy is local search? *Journal of Computer and System Sciences*, 37(1):79–100, 1988.

[166] David Judin and Arkadi Nemirovski. Informational complexity and effective methods of solution for convex extremal problems. *Ekonomika i Matematicheskie Metody*, 12:357–369, 1976.

[167] Michael Jünger, Thomas M Liebling, Denis Naddef, George L Nemhauser, William R Pulleyblank, Gerhard Reinelt, Giovanni Rinaldi, and Laurence A Wolsey. *50 Years of Integer Programming 1958-2008: From the Early Years to the State-of-the-art.* Springer Science & Business Media, 2009.

[168] Volker Kaibel. Extended formulations in combinatorial optimization. *arXiv preprint arXiv:1104.1023*, 2011.

[169] LV Kantorovich. Mathematical methods for production organization and planning. *Leningr. Gos. Univ., Leningrad*, 1939.

[170] Howard Karloff and Uri Zwick. A 7/8-approximation algorithm for max 3sat. In *Proceedings 38th Annual Symposium on Foundations of Computer Science*, pages 406–415. IEEE, 1997.

[171] Howard J. Karloff and Prabhakar Raghavan. Randomized algorithms and pseudorandom numbers. *Journal of the Association of Computing Machinery*, 40(3):454–476, July 1993.

[172] Narendra Karmarkar. A new polynomial-time algorithm for linear programming. In *Proceedings of the Sixteenth Annual ACM Symposium on Theory of Computing*, pages 302–311, 1984.

[173] Narendra Karmarkar. Some comments on the significance of the new polynomial time algorithm for linear programming. Technical report, AT&T Bell Laboratories, Murray Hill, New Jersey, 1984.

[174] Richard M Karp. A characterization of the minimum cycle mean in a digraph. *Discrete Mathematics*, 23(3):309–311, 1978.

[175] William Karush. *Minima of functions of several variables with inequalities as side constraints.* PhD thesis, University of Chicago, 1939. M. Sc. dissertation.

[176] A.V. Karzanov. Determining the maximal flow in a network by the method of preflows. *Soviet Math. Doklady*, 15:434–437, 1974.

[177] DG Kelly. Some results on random linear programs. *Methods of Operations Research*, 40:351–355, 1981.

[178] Leonid G Khachian. A polynomial algorithm in linear programming. *English Translation, Soviet Math. Dokl.*, 20:191–194, 1979.

[179] Leonid G. Khachian. Polynomial algorithms in linear programming. *USSR Computational Mathematics and Mathematical Physics*, 20(1):53–72, 1980.

[180] Valerie King, Satish Rao, and Rorbert Tarjan. A faster deterministic maximum flow algorithm. *Journal of Algorithms*, 17(3):447–474, 1994.

[181] V. Klee and G. Minty. How good is the simplex algorithm? In O. Shisha, editor, *Inequalities — III*, pages 159–175. Academic Press, New York, 1972.

[182] Victor Klee and Peter Kleinschmidt. The d-step conjecture and its relatives. *Mathematics of Operations Research*, 12(4):718–755, 1987.

[183] Victor Klee and David W Walkup. The d-step conjecture for polyhedra of dimension d¡ 6. *Acta Mathematica*, 117(1):53–78, 1967.

[184] Donald E. Knuth. *The Art of Computer Programming: Fundamental Algorithms.* Addison-Wesley Publishing Company, Reading, Massachusetts, 1973.

[185] Donald E. Knuth. class notes, January 1978.

[186] Donald Ervin Knuth. *The Art of Computer Programming: Fundamental Algorithms*, volume 1. Addison-Wesley, Reading, Mass., 3rd edition, 1997.

[187] Donald Ervin Knuth. *The Art of Computer Programming: Seminumerical Algorithms*, volume 2. Addison-Wesley, Reading, Mass., 3rd edition, 1997.

[188] Donald Ervin Knuth. *The Art of Computer Programming: Sorting and Searching*, volume 3. Addison-Wesley, Reading, Mass., 2nd edition, 1998.

[189] Donald Ervin Knuth. *The Art of Computer Programming: Combinatorial Algorithms, Part 1*, volume 4A. Addison-Wesley, Upper Saddle River, New Jersey, 2011. See https://www-cs-faculty.stanford.edu/ knuth/taocp.html for current publication status of books and fascicles.

[190] Masakazu Kojima and Shinji Mizuno. Computation of all solutions to a system of polynomial equations. *Mathematical Programming*, 25:131–157, 1983.

[191] Masakazu Kojima, Shinji Mizuno, and Akiko Yoshise. A primal-dual interior point algorithm for linear programming. In *Progress in Mathematical Programming*, pages 29–47. Springer, 1989.

[192] Dénes König. Graphs and matrices. *Matematikai és Fizikai Lapok*, 38:116–119, 1931. (In Hungarian.).

[193] Tjalling C Koopmans. An analysis of production as an efficient combination of activities. In Tjallng C. Koopmans, editor, *Cowles Commission Monograph 13: Activity Analysis of Production and Allocation*, pages 33–97. John Wiley & Sons, Inc, New York, 1951.

[194] Thoddi C. Kotiah and David I. Steinberg. Occurrences of cycling and other phenomena arising in a class of linear programming models. *Communications of the ACM*, 20(2):107–112, February 1977.

[195] Mark Krein and David Milman. On extreme points of regular convex sets. *Studia Mathematica*, 9:133–138, 1940.

[196] Harold Kuhn and Albert Tucker. Nonlinear programming. In *Proceedings of the Second Berkeley Symposium on Probability and Statistics*, pages 481–492. Berkeley: University of California Press, 1951. (Symposium held August 1950.).

[197] Harold W Kuhn. The Hungarian method for the assignment problem. *Naval Research Logistics Quarterly*, 2(1-2):83–97, 1955.

[198] Harold W Kuhn. Variants of the Hungarian method for assignment problems. *Naval Research Logistics Quarterly*, 3(4):253–258, 1956.

[199] Harold W. Kuhn. Discussion. In *Proceedings of the IBM Scientific Symposium on Combinatorial Problems: March 16–18, 1964*, pages 118–121. IBM Data Processing Division, White Plains, New York, 1966. Follows the article *The Traveling Salesman Problem* by Ralph E. Gomory.

[200] Harold W Kuhn. On the origin of the Hungarian method. *History of Mathematical Programming*, pages 77–81, 1991.

[201] Harold W. Kuhn. On the origin of the Hungarian method. In Jan Karel Lenstra, Alexander H. G. Rinnooy Kan, and Alexander Schrijver, editors, *History of Mathematical Programming: A Collection of Personal Reminiscences*, pages 77–81. Elsevier Science Publishers B.V., 1991. (Postscript of the article.).

[202] Harold W Kuhn. A tale of three eras: The discovery and rediscovery of the Hungarian method. *European Journal of Operational Research*, 219(3):641–651, 2012.

[203] Harold W. Kuhn and R. E. Quandt. An experimental study of the simplex method. In N. C. Metropolis, editor, *Experimental Arithmetic, High-Speed Computing and Mathematics*, pages 107–124. American Math. Society, 1963.

[204] Jon Lee. Classroom note: Hoffman's circle untangled. *SIAM Review*, 39:98–105, 1997.

[205] Carleton E Lemke. The dual method of solving the linear programming problem. *Naval Research Logistics Quarterly*, 1(1):36–47, 1954.

[206] Carlton E Lemke. Bimatrix equilibrium points and mathematical programming. *Management Science*, 11(7):681–689, 1965.

[207] Carlton E Lemke and Joseph T Howson, Jr. Equilibrium points of bimatrix games. *Journal of the Society for Industrial and Applied Mathematics*, 12(2):413–423, 1964.

[208] Hendrik W Lenstra Jr. Integer programming with a fixed number of variables. *Mathematics of Operations Research*, 8(4):538–548, 1983.

[209] Wassily W Leontief. Quantitative input and output relations in the economic systems of the United States. *The Review of Economic Statistics*, pages 105–125, 1936.

[210] Wassily W Leontief. *The Structure of American Economy, 1919-1939: An Empirical Application of Equilibrium Analysis.* Oxford University Press, New York, 1951.

[211] Beatrice List, Markus Maucher, Uwe Schöning, and Rainer Schuler. Randomized quicksort and the entropy of the random source. In *Proceedings of the 11th Annual International Conference on Computing and Combinatorics*, COCOON'05, pages 450–460, Berlin, Heidelberg, 2005. Springer-Verlag.

[212] Donna C. Llewellyn, Craig Tovey, and Michael Trick. Finding saddlepoints of two-person, zero sum games. *Am. Math. Monthly*, 95(10):912–918, 1988.

[213] Donna C Llewellyn, Craig A Tovey, and Michael A Trick. Local optimization on graphs. *Discrete Applied Mathematics*, 23, 1989. Erratum, volume 27, 1993.

[214] Lynn Loomis and Shlomo Sternberg. *Advanced Calculus.* Jones and Bartlett Publishers International, 1968. (Available online at Sternberg's website.).

[215] László Lovász. *Combinatorial Problems and Exercises*, volume 361. American Mathematical Soc., 1993.

[216] David Lubell. A short proof of Sperner's lemma. *Journal of Combinatorial Theory*, 1(2):299, 1966.

[217] David Luenberger. *Linear and Nonlinear Programming.* Kluwer, 1973. Later editions published by Springer.

[218] Irvin J Lustig. Feasibility issues in a primal-dual interior-point method for linear programming. *Mathematical Programming*, 49(1-3):145–162, 1990.

[219] Thomas L. Magnanti. Fenchel and Lagrange duality are equivalent. *Mathematical Programming*, 7(1):253–258, 1974.

[220] Patrick Scott Mara. Triangulations for the cube. *Journal of Combinatorial Theory, Series A*, 20(2):170–177, 1976.

[221] Walter F Mascarenhas. The affine scaling algorithm fails for stepsize 0.999. *SIAM Journal on Optimization*, 7(1):34–46, 1997.

[222] James V. McConnell. *Understanding Human Behavior: An Introduction to Psychology.* Holt, Rinehart, and Winston, 1983.

[223] Richard D. McKelvey. Covering, dominance, and institution free properties of social choice. *American Journal of Political Science*, 30:283–314, 1986.

[224] Nimrod Megiddo. Pathways to the optimal set in linear programming. In *Progress in Mathematical Programming*, pages 131–158. Springer, 1989.

[225] Nimrod Megiddo and Michael Shub. Boundary behavior of interior point algorithms in linear programming. *Mathematics of Operations Research*, 14(1):97–146, 1989.

[226] Sanjay Mehrotra. On finding a vertex solution using interior point methods. *Linear Algebra and its Applications*, 152:233–253, 1991.

[227] Sanjay Mehrotra. On the implementation of a primal-dual interior point method. *SIAM Journal on Optimization*, 2(4):575–601, 1992.

[228] Robert Meusel, Sebastiano Vigna, Oliver Lehmberg, and Christian Bizer. Graph structure in the web—revisited: a trick of the heavy tail. In *Proceedings of the 23rd International Conference on World Wide Web*, pages 427–432. ACM, 2014.

[229] Hermann Minkowski. *Geometrie der Zahlen.* B.G.Teubner, Leipzig and Berlin, 1896. (Reprinted 1953 by Chelsea Publishing Co. NY.).

[230] Hermann Minkowski. Geometrie der zahlen. *Bulletin of American Mathematical Society*, 21(3):131–132, 1914. Review by L.E. Dickson.

[231] Joseph SB Mitchell. Guillotine subdivisions approximate polygonal subdivisions: A simple polynomial-time approximation scheme for geometric tsp, k-mst, and related problems. *SIAM Journal on Computing*, 28(4):1298–1309, 1999.

[232] Renato DC Monteiro and Ilan Adler. Interior path following primal-dual algorithms. part i: Linear programming. *Mathematical Programming*, 44(1-3):27–41, 1989.

[233] Renato DC Monteiro, Ilan Adler, and Mauricio GC Resende. A polynomial-time primal-dual affine scaling algorithm for linear and convex quadratic programming and its power series extension. *Mathematics of Operations Research*, 15(2):191–214, 1990.

[234] Oskar Morgenstern and John Von Neumann. *Theory of Games and Economic Behavior.* Princeton University Press, 1944.

[235] George Nemhauser and Lawrence Wolsey. *Integer and Combinatorial Optimization.* John Wiley and Sons, 1989.

[236] Arkadi Nemirovski. Personal communication, 2015.

[237] James B. Orlin. Max flows in O (nm) time, or better. In *Proceedings of the Forty-Fifth Annual ACM Symposium on Theory of Computing*, pages 765–774. ACM, 2013.

[238] Ping-Qi Pan. Efficient nested pricing in the simplex algorithm. *Operations Research Letters*, 36(3):309–313, 2008.

[239] Christos H Papadimitriou. On the complexity of the parity argument and other inefficient proofs of existence. *Journal of Computer and System Sciences*, 48(3):498–532, 1994.

[240] Christos H Papadimitriou. *Computational Complexity.* John Wiley & Sons Ltd., 2003.

[241] Christos H. Papadimitriou and Kenneth Steiglitz. *Combinatorial Optimization: Algorithms and Complexity.* Prentice-Hall, Englewood Cliffs, NJ, 1982.

[242] Christos H Papadimitriou and David Wolfe. The complexity of facets resolved. Technical report, Cornell University, 1985.

[243] Christos H Papadimitriou and Mihalis Yannakakis. The complexity of facets (and some facets of complexity). *Journal of Computer and System Sciences*, 28(2):244–259, 1984.

[244] Loren Platzman. Personal communication, 2014.

[245] Yves Pochet and Laurence A Wolsey. *Production Planning by Mixed Integer Programming.* Springer Science & Business Media, 2006.

[246] George Polya. *How to Solve It: A New Aspect of Mathematical Method.* Princeton University Press, 2014.

[247] William Prager. On the caterer problem. *Management Science*, 3(1):15–23, 1956.

[248] James Renegar. A polynomial-time algorithm, based on Newton's method, for linear programming. *Mathematical Programming*, 40(1-3):59–93, 1988.

[249] Ron Rivest, Adi Shamir, and Lawrence Adleman. A method for obtaining digital signatures and public-key cryptosystems. *Communications of the ACM*, 21:120–126, 1978.

[250] Neil Robertson and PD Seymour. Graph Minors. XIII. The Disjoint Paths Problem. *J. Combin. Theory Ser. B*, 63:65–110, 1995.

[251] Ralph Tyrrell Rockafellar. *Conjugate Duality and Optimization*, volume 16. SIAM, 1974.

[252] Ralph Tyrrell Rockafellar. *Convex Analysis.* Princeton University Press, 2015. (First published in 1970.).

[253] Ralph Tyrrell Rockafellar et al. Extension of Fenchel's duality theorem for convex functions. *Duke Mathematical Journal*, 33(1):81–89, 1966.

[254] Adam N. Rosenberg. Numerical solution of systems of simultaneous polynomial equations using continuous homotopy methods. Ph.D. dissertation, Stanford University, 1983.

[255] Thomas Rothvoß. The matching polytope has exponential extension complexity. *Journal of the ACM (JACM)*, 64(6):1–19, 2017.

[256] Alexander M Rush and MJ Collins. A tutorial on dual decomposition and Lagrangian relaxation for inference in natural language processing. *Journal of Artificial Intelligence Research*, 45:305–362, 2012.

[257] John F Sallee. A triangulation of the n-cube. *Discrete Mathematics*, 40(1):81–86, 1982.

[258] Francisco Santos. A counterexample to the Hirsch conjecture. *Annals of Mathematics*, pages 383–412, 2012.

[259] Herbert Scarf. The approximation of fixed points of a continuous mapping. *SIAM Journal Applied Mathematics*, 15:1328–1343, 1967.

[260] Michael J. Schramm. *Introduction to Real Analysis.* Dover Publications, 2008.

[261] Alexander Schrijver. Finding k disjoint paths in a directed planar graph. *SIAM Journal on Computing*, 23(4):780–788, 1994.

[262] Alexander Schrijver. *Theory of Linear and Integer Programming.* John Wiley & Sons, 1998.

[263] Alexander Schrijver. On the history of the transportation and maximum flow problems. *Mathematical Programming*, 91(3):437–445, 2002.

[264] Alexander Schrijver. *Combinatorial Optimization: Polyhedra and Efficiency*, volume 24. Springer Science & Business Media, 2003.

[265] Alexander Schrijver. On the history of combinatorial optimization (till 1960). *Handbooks in Operations Research and Management Science*, 12:1–68, 2005.

[266] Naum Z Shor. Cut-off method with space extension in convex programming problems. *Cybernetics and Systems Analysis*, 13(1):94–96, 1977.

[267] Daniel D. Sleator and Robert Endre Tarjan. A data structure for dynamic trees. *Journal of Computer and System Sciences*, 26(3):362–391, 1983.

[268] Daniel Solow. *How to Read and Do Proofs: An Introduction to Mathematical Thought Processes.* John Wiley & Sons, New York City, USA, 2nd edition, 2002.

[269] Emanuel Sperner. Ein satz über untermengen einer endlichen menge. *Mathematische Zeitschrift*, 27(1):544–548, 1928.

[270] Richard Steinberg. The state of the three color problem. In *Annals of Discrete Mathematics*, volume 55, pages 211–248. Elsevier, 1993.

[271] E. Steinitz. Bedingt konvergente reihen und konvexe systeme. (schluß.). *Journal für die reine und angewandte Mathematik*, 146:1–52, 1916.

[272] Richard E. Stone and Craig A. Tovey. The simplex and projective scaling algorithms as iteratively reweighted least squares methods. *SIAM Review*, 33:220–237, 1991. Printer error correction in volume 33, no. 3.

[273] Gilbert Strang. Karmarkar's algorithm and its place in applied mathematics. *The Mathematical Intelligencer*, 9(2):4–10, 1987.

[274] Eva Tardös, Michael Trick, and Craig Tovey. Layered augmenting path algorithms. *Mathematics of Operations Research*, 11:362–370, 1986.

[275] Antonis Thomas. Exponential lower bounds for history-based simplex pivot rules on abstract cubes. *arXiv preprint arXiv:1706.09380*, 2017.

[276] John F Toland. Duality in nonconvex optimization. *Journal of Mathematical Analysis and Applications*, 66(2):399–415, 1978.

[277] Eric Tollefson, David Goldsman, Anton Kleywegt, and Craig Tovey. Optimal selection of the most probable multinomial alternative. *Sequential Analysis*, 33(4):491–508, 2014.

[278] John A Tomlin. On scaling linear programming problems. In *Computational Practice in Mathematical Programming*, pages 146–166. Springer, 1975.

[279] Craig Tovey. A simplified NP-complete satisfiability problem. *Discrete Applied Mathematics*, 8:85–89, 1984.

[280] Craig A Tovey. Nature-inspired heuristics: Overview and critique. In *Recent Advances in Optimization and Modeling of Contemporary Problems*, pages 158–192. INFORMS, 2018.

[281] Takashi Tsuchiya. Global convergence property of the affine scaling methods for primal degenerate linear programming problems. *Mathematics of Operations Research*, 17(3):527–557, 1992.

[282] Takashi Tsuchiya and Masakazu Muramatsu. Global convergence of a long-step affine scaling algorithm for degenerate linear programming problems. *SIAM Journal on Optimization*, 5(3):525–551, 1995.

[283] Hoang Tuy. DC optimization: theory, methods and algorithms. In *Handbook of Global Optimization*, pages 149–216. Springer, 1995.

[284] François Vanderbeck. On Dantzig-Wolfe decomposition in integer programming and ways to perform branching in a branch-and-price algorithm. *Operations Research*, 48(1):111–128, 2000.

[285] François Vanderbeck and Martin WP Savelsbergh. A generic view of Dantzig-Wolfe decomposition in mixed integer programming. *Operations Research Letters*, 34(3):296–306, 2006.

[286] Robert J Vanderbei, Marc S Meketon, and Barry A Freedman. A modification of Karmarkar's linear programming algorithm. *Algorithmica*, 1(1-4):395–407, 1986.

[287] Vijay Vazirani. *Approximation Algorithms*. Springer, 2003.

[288] Daniel J Velleman. *How to Prove it: A Structured Approach*. Cambridge University Press, 2006.

[289] Roman Vershynin. Beyond Hirsch conjecture: Walks on random polytopes and smoothed complexity of the simplex method. *SIAM Journal on Computing*, 39(2):646–678, 2009.

[290] Juan Pablo Vielma. Mixed integer linear programming formulation techniques. *Siam Review*, 57(1):3–57, 2015.

[291] Lev Vygotsky. Zone of proximal development. *Mind in Society: The Development of Higher Psychological Processes*, 5291:157, 1987.

[292] Gideon Weiss. A simplex based algorithm to solve separated continuous linear programs. *Mathematical Programming*, 115(1):151–198, 2008.

[293] Hermann Weyl. Elementare theorie der konvexen polyeder. *Commentarii Mathematici Helvetici*, 7(1):290–306, 1934.

[294] Andrew Wiles. Modular elliptic curves and Fermat's last theorem. *Annals of Mathematics*, 141(3):443–551, 1995.

[295] James Hardy Wilkinson. *The Algebraic Eigenvalue Problem*, volume 662. Oxford Clarendon, 1965.

[296] James Hardy Wilkinson. Linear algebra. *Handbook for Automatic Computation Volume II*, 1971.

[297] James Hardy Wilkinson. *Rounding Errors in Algebraic Processes.* Courier Corporation, 1994.

[298] James Hardy Wilkinson. Some comments from a numerical analyst. In *ACM Turing Award Lectures.* ACM, 2007.

[299] H Paul Williams. *Model Building in Mathematical Programming.* John Wiley & Sons, 2013.

[300] David P Williamson and David B Shmoys. *The Design of Approximation Algorithms.* Cambridge University Press, 2011.

[301] Laurence A. Wolsey. *Integer Programming.* John Wiley & Sons, 1998.

[302] Gene Woolsey. The incredible shrinking bug separation model. *ACM SIGMAP Bulletin*, 18:20–23, 1975.

[303] Mihalis Yannakakis. Expressing combinatorial optimization problems by linear programs. In *Proceedings of the Twentieth Annual ACM Symposium on Theory of Computing*, STOC '88, pages 223–228, New York, NY, USA, 1988. ACM.

[304] Norman Zadeh. Theoretical efficiency of the Edmonds-Karp algorithm for computing maximal flows. *Journal of the Association of Computing Machinery*, 19(1):184–192, January 1972.

Index

A

Absolute values, 62, 73
Acyclic graphs, 343–344
Affine combination, 112
Affine function, 12, 482
Affine hull, 112
Affine independence, 244–245
Affine scaling, 495–498, 512
Aircraft maintenance problem, 312–313
Algorithm running time, 262–266, *See also* Polynomial time algorithms
Alternative, theorems of the, 121–123, 231–234, 238
Alternative pairs, 231–232
"AND," 436, 437–438
Approximation algorithms, 416–417, 418
Arc-node incidence matrix, 315–317
Assignment problem, 192, 193–194, 202
Augmenting path (AP) algorithms, 366–374
Auxiliary variables, 74–76

B

Balanced tree, 269
Barrier algorithms, 94, 130, 498–504
Basic feasible solutions (BFS), 125–126, 147–148
- degeneracy and, 154–158
- dual simplex method, 181
- finding initial BFS, 162–167
- network models, 323–326, 331–332
- network problems, 308
- representing polyhedra, 132
- *See also* Simplex method

Basic solutions, 25
Basis, spanning trees in network models, 319–323
Basis of a vector space, 518–519
Basis reinverting, 201
Bellman-Ford algorithm, 351–354
Benders decomposition, 199
Big-M method, 167
Binary heaps, 276–281, 359
Binary search algorithm, 268–269, 481–482
Binary trees, 276–278
Binary variables, 436
Binding constraints, 43, 125
Bin packing, 402, 417, 423
Bipartite graph, 314, 377
Bipartite matching problem, 377–378
Bland's rule, 157–158
Blending problem, 85–86
Bolzano's theorem, 481
Boolean operations, 436
Bounded set, definition, 111
Branch-and-bound trees, 26, 449
Branch-and-price procedure, 449

C

Calculus, *See* Differential calculus; Multivariate calculus
Cargo shipping example, 81–84
Cauchy-Schwartz inequality, 516
Central path and interior point algorithm, 500–503, 512
Central path following methods, 130
Changes in constraints, 211–212
Charnes, Abraham, 102
Cholesky factorization, 468, 469
Classification model, 76–78
Classroom scheduling problems, 422–423
Clique problem, 399, 402, 436
Closed sets, 522
Coloring problems, 401, 411–416, 421, 423, 425, 436
Column density, 506
Column generation, 94–95, 221–225
Column geometry, 250–253
Column space, 250
Combinatorial optimization, primal-dual algorithm, 191–194
Compactness, 522–523
Complementary slackness, 35–38, 53, 100
- degeneracy and, 130, 234–238
- shadow prices and, 213

Complexity, *See* Computational complexity; NP-hard problems
Computational complexity, 387–388
  approximation algorithms, 416–417, 418
  complexity proofs, 394, *See also* NP-hardness proofs
  Dantzig's comment, 417
  dynamic programming, 418–421
  easy and hard problems, 388–390
  example hard and easy problems, 398–402
  fast and slow algorithms, 388
  formal definitions, 424–429
  integer programming, 387–388, 398–399
  nonlinear programming algorithms, 491–492
  other kinds of difficulty, 429
  practical considerations, 429–430
  P-space completeness, 428
  recognizing complexity, 397–402
  search problems, 428
  terminology, 387, 390
  Yes/No form, 391–393, 426
  See also NP-complete problems; NP-hardness proofs; NP-hard problems; Polynomial time algorithms
Computational issues, simplex method, 200–202
Concave functions, 18
  piecewise linear function, 82–83, 219
Cone, 44–46, 50, 134
Connected graph, 319
Co-NP-complete problems, 425–427
Constraint generation, 221, 223
Constraints, 1
  binding or tight, 43, 125
  complementary slackness, 35–38, *See also* Complementary slackness
  convexity, 251
  costs of enforcing, 23
  dealing with infeasibility, 91
  dual constraints and reduced cost calculation, 327–329
  integer programming sets of, 438
  LP problem formulation, 61–62, 67–71
  LP relaxation, 450–452
  nonnegativity, 71
  predicting changes in, 211–212
  "soft" and"hard," 212
  theorems of the alternative, 121–123
  unwritten unwritable (nonlinear), 80–81
  See also Costs; Linear programming
Continuity and compactness, 523
Continuous function
  definition, 522
  nonlinear programming requirement, 463
  optimal solution existence and, 7
  self-dual parametric algorithm, 199
Converse representation of pointed polyhedra, 133
Converse representation of polytopes, 132
Convex combination, 112–114
Convex costs, integer programming non-convex costs, 440–441
Convex functions, 17–20
  piecewise linear function, 82–83, 219
Convex hull, 112
Convexity and hard problems, 429
Convexity constraint, 251
Convex optimization and nonlinear programming, 463
Convex regions, non-convex unions of, 438
Convex sets, 89, 112, 523
  Helly's theorem, 239–240
  separation, 241–242
Cooper, William, 102
Costs, 21, 23
  dual variables and shadow prices, 95–96
  full tableau simplex method, 159–162, 181–183
  integer programming, 440–441
  non-convexity in, 89, 440–441
  reduced costs in network models, 326–330
  shadow prices, 213
  See also Constraints; Shortest path models
CPU time, 94
Crashing algorithm, 127–128
Cube, 254
Cutting plane methods, 26, 388, 425, 449
Cutting stock problem, 223–225
Cycle completion for reduced costs, 326–327
Cycle paths, 318–319
  Hamiltonian cycle model, 394–396, 400–401
  shortest paths and negative cycles, 350–356

Cycling, simplex method degeneracy and, 155–158

D

Dantzig, George, 55–56, 147–154, 167–168, 250, 253, 417, 428, 495
Dantzig-Wolfe decomposition, 199–200
Debugging models, 91
Decision variables, 67–69
Degeneracy, 95, 130–131, 154–158
  Bland's rule, 157–158
  column generation and, 222
  complementary slackness and, 234–238
  perturbation/lexicographic method, 155–157
  removing degenerate temporary variables, 165–166
  shadow prices and, 216–218
Dense column, 506
Dense data, 92
Dependent variables, 84
Determinants, 517
  matrix operations and, 270
  polynomial time algorithms, 272
Devex pricing, 202
Diet problem, 62–63, 85–86, 100, 150
Differential calculus, 12, 482
Dijkstra's algorithm, 356–360
Dikin's algorithm, 496
Dilworth's theorem, 379–383
Dimensions, 51
  affine independence, 244–245
  definition for sets, 245
  matrix, 270
  of a subspace, 515
  of vector spaces, 518–519
Directed arcs or edges, 306
Directed path, 318
Direction of a polyhedron, 126
Dot product, 12, 15, 15, 42–43, 114–115, 150, 516, 519, *See also* Inner product
Dual constraints, computing reduced costs, 327–329
Dual degeneracy, 95, 130–131
Dual feasibility, 43*f*
Duality, 232
  area-maximization example, 8–9
  classic primal/dual pairs, 62–66
  facet and vertices exchange (polarity), 253–255
  general definition, 9, 26
  LP presolving, 53
  in matrices, 15
  maximum flow and minimum cut, 368
  network problem integer solutions, 315–317
  partial derivatives, 10–11
  separation property of polyhedra, 114–116
  shortest path models, 347–350
  in Sudoku, 13–14
  See also Strong duality; Weak duality
Dual LP models, 33–35, 523
  converting between primal and dual, 39–41
Dual simplex method, 26, 92, 155, 181–191, 202, 216, 449, 524
  example demonstrating geometry, 183–191
  primal-dual algorithm, 191–194, 202
  re-solving LPs and solving IPs, 449
Dual variables, 95–96
  shadow prices, 95–101, 211
  See also Shadow prices
Dynamic programming (DP), 418–421

E

Economies of scale, 440
Edmonds-Karp modification, 369–372
Eigenvalues, 468
Ellipsoid algorithm, 26, 293–300, 299, 524
Enforcers, 411
Epigraph, 18
Equipartition problem, 411, 418
Euclidean algorithm, 267
Euclidean metric, 520
Exact-3-cover (X3C), 401, 409–410, 419–420
Exact-3-SAT problem, 399
Exponential time algorithms, 265, 387–388, *See also* NP-hard problems
Extreme directions, 124, 133
Extreme points, 123–131, 134

F

Faces, 246–249
Facets, 243–249, 453–454
  faces, 246–249
  polarity, 253–255
Facility location problem, 450–452
Factor-? approximation algorithms, 417
Factoring large numbers, 428

Feasible region, 1, 3, 33
cone, 44–46, 50
polyhedra, 50, 111, 523–524, *See also* Polyhedra
visualizing, 42
Feasible solutions, 2–3
non-convexity, 89
theorems of the alternative, 121–123
See also Basic feasible solutions; Polyhedra
Fibonacci heaps, 358–359
Finitely generated cone, 134
First-order Karush-Kuhn-Tucker (KKT) conditions, 469–475
Fixed charge problem, 439–440
Fleet planning problems, 410
Flow decomposition theorem, 321
Floyd-Warshall algorithm, 354–356
Forcing, 411
Ford and Fulkerson's augmenting path algorithm, 366–372
4-Coloring problem, 401
Fourier-Motzkin elimination, 116–121
Full tableau method, 159–162, 181–183

G

Gadgets, 411–416
Gaussian elimination, 271–273, 517
General integer programs, 436, 445
Generated cone, 44
Global optimization, 475–477
Gordan's theorem, 232
Gradients, 12, 464–465
Gradient search algorithm, 488–491, 525
Graph coloring, 401, 411–416, 421, 423, 425, 436
Graph concepts and properties, 318–321
acyclic graphs, 343–344
bipartite graph, 314, 377
definition, 84, 306
isomorphism, 437
negative cycles, 350–356
See also Shortest path models
Greatest divisor algorithm, 267

H

Halfspace, 33
Hall's theorem, 378
Halting problem, 424
Hamiltonian cycle, 394–396, 400–401, 402–408, 423
Hamiltonian path problems, 400–408, 423
Hamiltonian s—t path, 351
Hard constraints, 212
Hard problem solving, *See* Computational complexity; NP-complete problems; NP-hard problems
Hash tables, 269–270
Heap property, 278
Heaps, 276–281, 358–359
Heapsort, 268
Helly's theorem, 238–241
Hessians, 465–469, 500
Homotopy methods, 26, 195, 199
Hyperplane, 50*f*, 51
classification model, 77
column geometry, 252
separation, 241–242
supporting, 246
tangent, 464–469

I

Ill-conditioned matrices, 93
Incident, 318
Indegree, 318
Independent sets problems, 392
Independent variables, 84
Infeasibility, dealing with, 91
Infeasible solutions, 2–3
Infinitesimals, 522
Inner product, 297, 516, 519, *See also* Dot product
Instance of a problem, 261, 262, 390
Integer programming (IP), 435–436
bipartite matching, 377–378
either-or-example, 439
fixed charge problem, 439–440
general integer programs, 436, 445
hard and easy problems, 398–399
hard problem solving and, 387–388
Lagrangian relaxation method, 199
logical relationships (Liars' Club examples), 442–445, 453
mixed integer program (MIP), 435
network problems, 308
NP-complete problems and feasibility, 425–426
NP-hard problems, 388, 398–399, 435, 436–438
presolving, 56
set partitioning and covering, 441–442
sets of constraints, 438

traveling salesman problem (TSP), 445–447
unions of convex regions, 438
See also Integer programming
Integer programs (IPs), solving, 448–449
branch-and-bound trees, 26
LP relaxation, 448–451
reduction to LP, 453–454
tighter formulations, 450–455
tips and software considerations, 454–456
valid inequalities, 453, 455
Integer solutions to network problems, 315–317, 326
Interior point algorithms, 495, 525
affine scaling, 495–498, 512
conceptual overview, 504
degeneracy and, 236
logarithmic barrier, 498–504, 525
logarithmic barrier and primal-dual, 504–511
log-barrier approach history, 512
number of iterations, 511
sparseness and, 511
Intermediate Value theorem, 481
Intersecting line segments, 238–239
Invariant properties, 343
Inventory selection to minimize waste, 341–342
Inventory shrink, 74–75
Inverse, product form of, 201
Inverting a matrix, 271, 272, 517
Isomorphism, 437
Isoquants, 43*f*, 47
Iterative improvement algorithms, 25

K

Kakutani fixed point theorem, 428
Kantorovich, Leonid, 55
Karmarkar, Narendra, 495, 496, 512
Karush-Kuhn-Tucker (KKT) conditions, 469–475
k-coloring, 261, 436
Klee-Minty example, 287–292
Knapsack problem, 225, 400, 409, 417, 418–421, 436
Köenig-Egerváry theorem, 379
Koopmans, T. C., 56
Krein-Milman theories, 124
Kuhn, Harold, 202

L

Lagrange multipliers, 21
Lagrange's method, 20–26
Lagrangian, 21–22, 37, 121–122, 501
Lagrangian relaxation method, 199
Law of conservation of difficulty, 231
Lemke, Carlton, 202
Lemke-Howson pivoting algorithm, 202
Lexicography, 157, 168
Liars' Club problems, 442–445, 453
Linear algebra, 515–520
Linear combinations, 515
Linear equations
polynomial time algorithms, 270–273
systems, alternative pairs, 231–232
Linear functions and matrices, 519–520
Linearization, 79–80
Linearly independent columns, 515
Linearly independent points, 244–245
Linear programming (LP), 41, 523
alternative forms, 39–40
basic solutions, 25
converting between primal and dual, 39–41
definition, 33
ellipsoid algorithm and polynomial time behavior, 293–300
fundamental theorem of, 129
higher dimensions, 51
historical development, 55–56, 102, 428
integer program reduction and solution, 453–454
lower and upper bounds, 175–181
LP relaxation, 223–224, 448–451
maximum flow and minimum cut, 374–383
network flow problem, 307–308, *See also* Network models
polyhedra, *See* Polyhedra
presolving, 52–55
primal and dual models, 33–35, 523
primal-dual algorithm, 191–194
software, 52
solving integer programs, 448–449
symmetry, 38
threshold, 391–392
visualizing, 41–52, 66–67
See also Simplex method
Linear programming, modeling methods, 72–81
auxiliary variables, 74–76

classification model, 76–78
debugging, 91
linearization, 79–80
problem visualization, 66–67
unwritten unwritable constraints, 80–81
Linear programming, problem formulation and solving, 61–62, 90
blending example, 85–86
cargo shipping example, 81–84
classic primal/dual pairs, 62–66
constraints, 61–62, 67–71
defining decision variables, 67–69
linear regression example, 84
modeling methods, 72–81
multicommodity flow example, 84–85
production scheduling example, 86–87
using software, 90–95
work scheduling example, 87–89
Linear programming software, dealing with problems with, 90–95
Linear regression, 84, 474–475
Line balancing problem, 392, 408, 423, 429
Line segment intersections, 238–239
Local optimum, 2, 20
Logarithmic barrier algorithm, 498–504, 525
historical development, 512
primal-dual algorithm, 504–511
Logical relationships, Liars' Club examples, 442–445, 453
Lower and upper bounds, 175–181
LP relaxation, 223–224, 448–451

M
Machine line balancing problem, 392, 408, 423, 429
Machine scheduling problem, 392, 409
Maintenance problem, 312–313
Manufacturing problems
combined blending problem, 79
cutting stock, 223
hard problem solving and, 411, 429
LP problem formulation, 63–64
primal/dual formulation, 63–64
shadow prices, 100
Matchmaking problem, 400
Mathematical programming-based NP-hard problem solving, 418
Matrices
duality in, 14–15
polynomial time algorithms, 270–273
representation of linear functions, 519–520
submatrices, 13
Matrix addition, 270
Matrix inversion, 271, 272, 517
Matrix multiplication, 270, 520
Max 2-SAT problems, 393, 399
MAX-CUT problem, 399
Maximum clique problem, 399, 402
Maximum flow problem, 365–366, 383, 524
bipartite matching, 377–378
Dilworth's theorem, 379–383
Edmonds-Karp modification, 369–372
Ford and Fulkerson's augmenting path algorithm, 366–369
improved algorithms, 372–374
Köenig-Egerváry theorem, 379
linear program formulation, 374–383
residual capacity graph, 367
"Max min," 62
Memory issues, 92
Mergesort, 268
Metric space, 520
Minimum cost circulation problem, 313
Minimum cost flow problem, 309–315, 347
Minimum cut problem, 365, 368, 383, 399
bipartite matching, 377–378
linear program formulation, 374–383
Minimum feedback arc set, 401
Minimum spanning tree problem, 400
Min-max problem, 62, 72–74
Mixed integer program (MIP), 435
Modeling languages, 94
Modeling methods, 72–81
Mountain climbing analogy, 231
Multivariate calculus
nonlinear programming prerequisites, 464–469
partial derivatives, 10–11

N
NAE-3-SAT problem, 399, 436
Nash equilibria, 202, 428
Negative cycles, 350–356
Negative weight, 44, 46
Network models, 89, 305–310, 423, 524
advantages, 307–308
basic solutions, 323–326
computing reduced costs, 326–330
integer solutions, 315–317, 326
minimum cost flow problem, 309–315

shortest path models, 339–342
visualization, 308
See also Network simplex algorithm
Network simplex algorithm, 318–326, 330–332
computing basic solutions, 323–326
computing reduced costs, 326–330
graph concepts, 318–321
spanning trees, 319–323
Newton's method for square root computation, 275–276
Newton's nonlinear programming algorithm, 482–488, 525
Node capacities, 313–314
Nodes of binary trees, 360
Non-convexity in feasibility or costs, 89
Non-convex optimization, 475–477
Non-deterministic polynomial (NP), 427
Nonlinear programming (NLP) algorithms, 481, 524–525
bisection search, 481–482
computational complexity, 491–492
gradient search, 488–491, 525
interior point algorithms, affine scaling, 495–498
interior point algorithms, logarithmic barrier, 498–504, 525
Newton's method, 482–488, 525
quasi-Newton methods, 485–488
Nonlinear programming (NLP) theory, 463, 475–477
definition, 463
global optimization, 475–477
gradients and Hessians, 464–469
Karush-Kuhn-Tucker conditions, 469–475
linear regression example, 474–475
multivariate calculus prerequisites, 464–469
Nonlinear unwritten unwritable constraints, 80–81
Nonnegativity constraints, 71
Norms, 519
"NOT," 436
NP-complete problems, 388–389, 524
co-NP-complete problems, 425–427
definition, 427
integer programs, 425–427, 435
practically simple problems, 429
NP-hardness proofs, 393–397
common pitfalls examples, 402–408
examples, 408–416
gadgets and transforming 3-SAT to 3-coloring, 411–416
general form, 396–397
Hamiltonian cycle model, 394–396
NP-hard problems, 389
approximation algorithms, 416–417
definitions, 426–427
example hard and easy problems, 398–402
general NP-hardness proof procedure, 396–397, *See also* NP-hardness proofs
integer programming, 388, 398–399, 435, 436–438, *See also* Integer programming
recognizing complexity, 397–402
solving, 417–424
traveling salesman problem (TSP), 394
unary or strongly, 423
unsolvable problems, 424–429
See also Computational complexity; NP-complete problems
NP-hard problem solving
dynamic programming, 418–421
math programming, 418
special structure, 421–424
Numerical instability issues, 93

O

Objective function, 1, 42, 166–167, *See also* Linear programming
Objective value, 3
Offset cone, 44
Optimal dual solution, 44
Optimality principle for shortest paths, 345–346
Optimal solution, 3, 7
checking against reality, 94
extreme points and, 123–131
Optimal value, definition, 3
Optimization problems, 1–8
Optimization problems, approaches to solving, 16–20
ellipsoid algorithm, 26
iterative improvement, 25–26
Lagrange's method, 20–26
See also Integer programming; Linear programming; Simplex method
"OR," 436, 437–438

Orientable path and cycle, 319
Outdegree, 318

P

Padding and forcing transformations, 410–411
Parameter specialization technique, 408–409
Parametric programming, 218–220
Partial derivatives, 10–11, 465
Partially ordered set (poset), 379–383
Partition problem, 400, 409, 410, 418, 422, 436, 441–442
Paths, 318, 339
 augmenting path models, 366–372
 Hamiltonian problems, 351, 394–396, 400–408, 423
 See also Shortest path models
PCP theorem, 417, 427
Penalties, 25, 91
 auxiliary variables and, 75–76
 solving integer programs, 451
Permuting, 437
Perturbation/lexicographic method, 155–157, 168
Phase I, simplex method, 162–166
Phase II, simplex method, 162, 166–167
Piecewise linear concave/convex function, 82–83, 219
Pointed polyhedron, 124, 127, 133
Point set topology, 520–523
Polarity, 253–255, 299
Polyhedra, 111–114, 523–524
 definition, 111, 131
 degeneracy, 130–131
 direction of, 126
 extreme points and optimal solutions, 123–131, 134
 faces, 246–249
 facets, *See* Facets
 Fourier-Motzkin elimination and linear transformations, 116–121
 Helly's theorem, 238–241
 pointed, 124, 127, 133
 polarity, 253–255
 projection of, 120, 242–243
 representation of, 131–134, 242–243
 separation, 114–116, 241–242
 shooting experiments, 249
 theorems of the alternative, 121–123
"Polynomial Hirsch Conjecture," 288
Polynomial time algorithms, 261, 267, 387
 "affirmative action" selection rules, 288
 definition, 265
 ellipsoid algorithm, 293–300
 Euclid's algorithm, 267
 fast and slow algorithms, 388
 heaps, 276–281
 interior point algorithms, 512
 pseudo-polynomial dynamic programming, 419
 searching, 268–270
 simplex method worst-case behavior, 287–293
 sorting, 267–268
 square roots, 273–276
 superpolynomial time algorithms, 387–388
 See also Computational complexity
Polytopes, 111–114
 polarity, 254–255
 representation of, 132
 simplexes, 243–244
 See also Polyhedra
Presolving integer programming, 56
Presolving linear programming, 52–55
Primal-dual algorithm, 191–194, 202
Primal/dual formulation pairs, 62–66
Primal-dual interior algorithm, 504–511
Primal LP models, 33–35, 523
 converting between primal and dual, 39–41
 LP software considerations, 92
 See also Linear programming
Primal simplex algorithm, 181
 dual and primal algorithm geometries, 183–191
 primal-dual algorithm, 191–194, 202
 self-dual parametric algorithm, 199
 See also Simplex method
Principle of optimality, 345–346
Problem, definition, 261, 390
Product form of the inverse, 201
Production minimum cost flow model, 311–312
Production scheduling with inventory problem, 86–87
Projection of a polyhedron, 120–121, 242–243
Project planning, 428
Pseudo-polynomial algorithms, 418–421
P-space completeness, 428

Q
Quadratic convergence rate, 488
Quasi-Newton methods, 485–488
Quicksort, 268

R
Random number generator, 268
Realistic cases, 390, 430
Recession cone, 134
Reinverting the basis, 201
Representation theorem, 242–243, 254
Residual capacity graph, 367
Riemann Hypothesis, 463
Rock-paper-scissors game, 64–65
Roshambo game, 64–65
Rounding, 63–64
Running time of algorithms, 262–266, *See also* Polynomial time algorithms

S
Saddlepoint, 21–22, 37, 64
Scalars, 517–518
Search algorithms, 268–270, 428, 481–482, 488–491, 525
Secant method, 485
Self-dual parametric algorithm, 194–199
Sensitivity analysis, 215–216, 218–220
Separation, 114–116, 241–242, 299
Sequencing, 437
Set covering, 401, 417
Set packing, 401
Shadow prices, 95–101, 211
    degeneracy and, 216–218
    mathematical justification, 212–215
Shooting experiments, 249
Shortest path models, 339–342, 524
    acyclic graphs, 343–344
    Bellman-Ford algorithm, 351–354
    Dijkstra's algorithm, 356–360
    Floyd-Warshall algorithm, 354–356
    graphs with negative cycles, 350–356
    minimum cost flow problem, 347
    primal and dual linear programs, 347–350
    principle of optimality, 345–346
    See also Traveling salesman problem
Shortest path with obstacles problem, 401
Shrink rate, 74–75
Simple path, 318
Simplexes, 243–244
Simplex method, 25
    column geometry, 250–253
    computational issues, 200–202
    cutting plane algorithm, 26, 388
    degeneracy and, 154–158, 234–235, *See also* Degeneracy
    dual, *See* Dual simplex method
    example demonstrating geometry, 183–191
    finding initial basic feasible solution, 162–167
    full tableau method, 159–162, 181–183
    historical development, 167–168, 250
    LP software considerations, 92
    primal simplex algorithm, 181
    time bounds, 265
    time run behavior, 287–293
    visualizing, 250–253
    See also Linear programming
Simplex method, variants, 175
    Dantzig-Wolfe decomposition, 199–200
    dual and primal algorithm geometries, 183–191
    dual simplex method, 25, 92, 155, 181–191, 202, 216, 449, 524
    lower and upper bounds, 175–181
    primal-dual algorithm, 191–194
    self-dual parametric algorithm, 194–199
Simpson's formula, 271
Single machine total tardiness, 439
Slack constraints, 35–36, *See also* Complementary slackness
Slack variables, 39–40, 213
Soft constraints, 212
Software
    bug detection, 424
    dealing with problems using, 90–95
    IP solving considerations, 454–456
    LP presolving, 52
Solutions, 2–3, 8
    basic, 25
    optimal, 3
    See also Basic feasible solutions; Feasible solutions
Sorting algorithms, 267–268
Spanning tree, 319–323, 400
Sparse data, 92
Sparseness and interior point algorithms, 511
Specialization, 409–410, 411
Square root computation, 273–276
Staff scheduling model, 87–89

Staircase problem, 495
Steiner tree, 400, 417
Stochasticity, 428
Strassen, Volker, 270
Strict complementary slackness, 38
Strictly convex function, 18–19
String, 426
Strong duality, 36–37, 157, 167, 231–234, 238
Strongly connected path, 319
Submatrix, 13
Suboptimal solution, 3
Subpath, 345–346
Subset Sum problem, 411, 418
Sudoku, 13–14
Superoptimal solution, 3
Superpolynomial time, 387–388
Sup metric, 520
Supporting hyperplane, 246
Symmetry, 38, 224, 456
Systems of equalities or inequalities, alternative pairs, 231–232
Systems of linear equations, Gaussian elimination, 517
Systems of linear inequalities, Fourier-Motzkin elimination, 116–121

T

Taxicab metric, 520
Taylor series, 12, 464
Temporary variables
  need for, 201–202
  removing degenerate variables, 165–166
Theorems of the alternative, 121–123, 231–234, 238
3-Coloring problem, 401, 411–416, 425
3-Dimensional matching (3DM) problem, 399–400, 409, 422
3-Partition problem, 400, 423
3-SAT problem, 399, 411–416, 417, 423, 436, 454
Threshold linear programming, 391–392
Tight constraints, 43, 125
Time bounds of algorithms, 262–266, *See also* Polynomial time algorithms
Topological sort, 343–344
Total unimodular (TU) property, 315
Transformation for NP-hardness proofs, 396–397
Transport problems
  cargo shipping LP example, 81–84
  fleet planning, 410
  minimum cost flow model, 310–311, 314
  multicommodity flow LP example, 84–85
  set partitioning and covering, 441–442
Traveling salesman problem (TSP), 394, 401, 409, 410, 417, 423, 424, 445–447
Treap, 269
Trees
  balanced tree search algorithm, 269
  binary, 276–278
  branch-and-bound, 449
  definitions, 319–321
  network simplex algorithm, 319–323
  Steiner tree, 417
  Steiner tree/minimum spanning tree problems, 400
Triangle inequality, 520
Tripartite matching problem, 399–400
2-Coloring problem, 401
2-Machine line balancing, 392, 408, 423
2-Machine weighted flowtime minimization, 400
Two person zero sum game, 64–66
2-SAT problem, 399

U

Unary NP-hard problems, 423
Unbounded problems, 7, 91
Undirected arcs or edges, 306
Unwritten unwritable constraints, 80–81

V

Valid inequalities, 453, 455
Value of a solution, 3
Variables, 1
  auxiliary, 74–76
  binary, 436, *See also* Integer programming
  defining decision variables, 67–69
  independent and dependent, 84
  LP software considerations, 94
  slack, 39–40
Vector spaces, 517–519
Vertex-disjoint paths, 402

W

Warm start, 215, 449
Waste minimization, 341–342

Weak duality, 35, 53
Word problems, 61
Work scheduling LP example, 87–89

X
x*, 43

Y
Yes/no decisions, 222
Yes/No form, 391–393, 426
Yin-Yang duality, 27

Z
Zadeh, Norman, 288